Chemistry

CLAUDE H. YODER

FRED H. SUYDAM

FRED A. SNAVELY

Franklin and Marshall College

Chemistry

 HARCOURT BRACE JOVANOVICH, INC.

New York   Chicago   San Francisco   Atlanta

ISBN: 0-15-506465-7
Library of Congress Catalog Card Number: 74-24771

Printed in the United States of America

Illustrations by Eric G. Hieber Associates, Inc.

Photo Credits

Part I Photomicrograph of polyethylene glycol polymers (Carbo-wax).

Manfred Kage from Peter Arnold

Part II Surface of a liquid in which a sequence of chemical reactions is proceeding in an array of spiral patterns.

Photo by Fritz Goro

Part III Spiral Nebula (Messier 101) in Ursa Major. Photographed with the 200-inch Hale Telescope at Mt. Palomar, California.

Courtesy of Mount Wilson and Palomar Observatories

To three teacher chemists:

W. CONARD FERNELIUS

L. CARROLL KING

JEROLD J. ZUCKERMAN

Preface

In writing this book, our aim has been to present the basic principles of modern chemistry in sufficient depth to serve students majoring in the various areas of science and at the same time to satisfy the needs of serious students whose major fields of interest lie elsewhere. Consequently, this is not a book on inorganic chemistry or on simplified physical chemistry, nor is it a book on chemical mathematics. It is a book on chemical principles, presented on a level that is understandable, yet challenging, to today's college student.

The various "branches" of chemistry have been integrated into the subject matter as a logical and normal part of the study of chemistry. For example, the nomenclature, structure, and behavior of "organic" compounds are not gathered together in a separate chapter but are dealt with at appropriate places throughout the book. Similarly, there is no chapter on thermodynamics as such; rather, the concepts of thermodynamics are presented and used wherever their inclusion is of value in making a phenomenon more understandable. Ideas and techniques basic to analytical chemistry find their place, too, as integral parts of the study of matter. An effort has also been made to provide examples that relate chemical principles to students' everyday experiences and to some of the problems of the world in which they live.

The particular sequential organization of material presents the fundamentals of chemistry as a single, continuous, and logical story. New concepts are introduced as they become useful in telling the story, and frequent recall and application of earlier concepts are employed throughout.

This book is divided into three parts. Part One, The Structure and Composition of Matter, deals with the nature of matter, beginning with the simplest particles and progressing toward the more complex. Description of the fundamental particles and their discovery leads to atoms and elementary substances and then to compounds, with an emphasis on chemical bonding. This is followed by treatment of the macroscopic aspects of matter, with chapters on the states of matter and on solutions. The relationship between structure and properties is stressed throughout.

Part Two, Reactions of Matter, is a study of the transformations that matter undergoes. After an introduction to the nature of chemical reactions, rates of reactions and the extent of reactions are investigated, making use of both kinetic and thermodynamic considerations. Chemical reactions are then discussed on the basis of classification into four types—ion combination, proton transfer, electron sharing, and electron transfer.

Part Three, The Chemistry of the Elements, stresses chemical behavior and its rationalization in terms of theoretical concepts. In investigating various facets of the chemistry of the elements, the student reviews most of the concepts presented in Parts One and Two and has an opportunity to apply them in rationalizing and predicting trends in the chemistry of the elements.

Because secondary school chemistry courses vary widely in content and in depth of coverage, we have assumed no single body of knowledge as a common background of students entering the first-year college chemistry course. A student with no previous course in chemistry will find here all the ideas and facts necessary for mastery of the material of the text. The only mathematical background required is algebra.

In writing *Chemistry*, we tried to avoid creating the impression that scientists know everything about the nature of matter. We hope we have succeeded in demonstrating that many questions remain unanswered, and that the answers we do have are based on intellectual models. These models are useful in helping us to visualize and understand complex phenomena, but they are always subject to change.

This book was developed out of our own experience in teaching general chemistry at Franklin & Marshall College. We are grateful to those people who reviewed the manuscript in detail during various stages of its development: Larry Berliner, Ohio State University; Carroll DeKock, Oregon State University; Robert Fay, Cornell University; Steve Hadley, University of Utah; William Jolly, University of California, Berkely; Bob Pinnell, Claremont College; Leonard Spicer, University of Utah. In addition, we would like to thank Carl Hansen, Joint Institute for Laboratory Astrophysics, and Clifford Matthews, University of Illinois at Chicago Circle, for their in-depth analyses and critiques of Chapter 18. Finally, we are indebted to Franklin & Marshall College for sabbatical leaves during which much of this book was written and to Carol Strausser who expertly typed the manuscript in its several drafts.

Claude H. Yoder
Fred H. Suydam
Fred A. Snavely

Contents

ix

7

COMPOUNDS III: THE COVALENT MODEL   102

8

COMPOUNDS IV: MOLECULAR GEOMETRY   152

9

MATTER IN BULK I: SOLIDS AND LIQUIDS   183

10

MATTER IN BULK II:
GASES AND TRANSITIONS BETWEEN STATES   218

11

MATTER IN BULK III: SOLUTIONS   252

PART TWO   The Reactions of Matter

12

RATES OF REACTION   294

13

EXTENT OF REACTION   314

14

TYPES OF CHEMICAL REACTIONS I: ION-COMBINATION REACTIONS   359

15

TYPES OF CHEMICAL REACTIONS II: PROTON-TRANSFER REACTIONS   381

16

TYPES OF CHEMICAL REACTIONS III: ELECTRON-SHARING REACTIONS   431

17

TYPES OF CHEMICAL REACTIONS IV: ELECTRON-TRANSFER REACTIONS   450

PART THREE   The Chemistry of the Elements

18

THE ORIGIN OF THE ELEMENTS   494

19

HYDROGEN   522

20

THE REPRESENTATIVE ELEMENTS I: THE *s*-FILLERS   541

21

THE REPRESENTATIVE ELEMENTS II: THE *p*-FILLERS   558

22

TRANSITION ELEMENTS: THE *d*-FILLERS   592

APPENDIXES

Chemistry

1 *Introduction*

Although often associated with "magic"—the sleight-of-hand change of a colorless solution to a colored one or the explosive violence of a chemical reaction—the art and science of chemistry appears to have evolved from a desire to understand nature and to convert natural materials into more useful substances. Efforts to transform these desires into reality have been remarkably successful. Consider, for example, the effect of synthetic fibers such as nylon, dacron, and polyesters on the clothing needs of the world; the impact of alloys, plastics, rubber, and paints on the quality and availability of housing and transportation; the effect of fertilizer, pesticides, and food preservatives on the supply and storage of food; and the value of countless pharmaceuticals such as antibiotics and analgesics, which have eased human suffering and increased life expectancy.

While chemistry has had a tremendous influence in altering our environment to make it more comfortable, it has rapidly become apparent that every alteration affects the overall ecology in some way. Hence, every change, every new product must be introduced carefully, with ample forethought to possible adverse ecological effects.

Many of these advances are a direct result of our expanded knowledge of the laws, concepts, and theories of chemistry. These concepts allow not only the development of the technology needed for the production of numerous consumer products but also the rationalization, if not actual understanding, of many chemical processes. Indeed, our knowledge of the principles of chemistry is sufficiently sophisticated to permit investigation of the most complex and also the most fascinating chemical system—the human body. Important advances have been made in the understanding of many biological processes on the molecular level: the structure of nucleic acids, the mechanism involved in the transmission of the genetic code, the molecular processes that permit vision, to name just a few.

THE SCOPE OF CHEMISTRY

Exactly what is encompassed by the term *chemistry* is not easy to define, for the boundaries change with time, and the range of problems that attract the attention of chemists widens as knowledge increases. Broadly defined, *chemistry is that branch of science which deals with the composition and structure of the matter of which the universe is made and with the changes in composition and structure that matter undergoes.*

When one considers that *matter* is defined as *anything that has mass and occupies space* (and therefore such diverse items as a tree, petroleum, a living cell, and the gases of the atmosphere are all samples of matter), this definition of chemistry seems to be almost all-inclusive. And yet even this definition requires elaboration, for when matter undergoes changes in structure and composition, these changes are accompanied by a transfer of energy, and thus energy and its changes also become an integral part of the concerns of chemistry.

Any attempt to differentiate the area of chemistry from other branches of science soon leads to the realization that the boundaries are not clear-cut. The change in matter that occurs when a nerve impulse is transferred from one neuron to another is of interest to the chemist, but obviously it is also the concern of the biologist, and perhaps the experimental psychologist as well; the structure of the matter of which the earth's crust is composed falls in the domain of geology as well as chemistry; and the energetics of nuclear processes capture the attention of both physicists and chemists. Indeed, such areas of overlap are fast becoming the rule, rather than the exception, and most new knowledge about the universe is being gained through cooperative efforts of all the branches of natural science.

Imprecise, then, as our definition may be, we shall nonetheless adopt it as the working definition upon which this introductory study of chemistry is based. In the chapters that follow we shall examine the current concepts of the structure and composition of matter, beginning with the fundamental, submicroscopic particles and progressing to matter on the macroscopic level. Then, with this understanding as a background, we shall investigate the changes that matter undergoes.

BASIC TERMS

Chemistry, like any other branch of learning, has its own special vocabulary, and some degree of mastery of that vocabulary is essential to an understanding of the subject. Included in the vocabulary of chemistry are a number of basic terms that are useful in describing matter.

Mass and Weight

In our definition of matter, we used the term *mass*. *Mass* and *weight* are often used synonymously, although strictly speaking this usage is not correct. The *mass* of a body is a measure of the quantity of matter in that body; it accounts for the tendency of the body to remain in its state of rest or motion (its *inertia*). *Weight*, on the other hand, is the force of the attraction of gravity exerted on a body. Therefore, the *weight* of a body varies with the force of gravity; the *mass* of that body is invariable. The *weight* of a prize bass measured on a simple spring scale will depend on the altitude and latitude at which it is weighed; the *mass* of the fish is independent of location. While the mass of an object cannot be determined directly from its weight unless the earth's gravitational attraction at the location at which the weight was measured is known, it can be determined easily by comparison with another object of known mass on an analytical balance, since then the attraction of gravity is the same on both objects. Since *mass* is normally determined by a process of *weighing,* the interchangeable usage of the terms *mass* and *weight* has become firmly entrenched; this usage will cause us no serious problem.

Substances and Mixtures

All specimens of matter can be categorized as either *substances* or *mixtures*. A *substance* is a form of matter that has a definite, invariable composition and a specific set of characteristics by which it can be identified. If a specimen of matter contains two or more substances, it is called a *mixture.* For example, calcium carbonate is a *substance*; all samples of calcium carbonate have the same composition, no matter what their origin. Furthermore, calcium carbonate has specific characteristics that distinguish it from all other substances. Limestone, on the other hand, is a mixture, since it contains, in addition to calcium carbonate, a number of other substances, including magnesium carbonate, silicon dioxide, and aluminum oxide. Furthermore, the relative amounts of these various substances differ among different limestone samples; that is, the composition of limestone is variable.

Elementary and Compound Substances. Substances may be classified as either *elementary substances* or *compound substances.* (Sometimes these terms are shortened simply to *elements* and *compounds,* although this may lead to some confusion in the case of elements, since, as we shall see in a subsequent chapter, the term *element* also has another usage, in which it means a kind of atom.) During the early development of chemistry (in the seventeenth and eighteenth centuries), when these terms first came into use, the two types of substances were distinguished from each other on a purely empirical basis. If the substance in question could be decomposed or broken down into two or more simpler substances, it was called a *compound substance.* If a substance could not be so decomposed it was classified as an *elementary substance.* To illustrate, the substance mercuric oxide can rather easily be converted into two simpler substances, mercury and oxygen, but neither of these can be decomposed by any ordinary means to any simpler substances. Therefore, mercuric oxide is classified as a *compound substance,* and both mercury and oxygen are *elementary substances.* As we shall see in the following chapter, more precise definitions can be given for these terms.

Homogeneous and Heterogeneous Mixtures. If a mixture is uniform throughout in composition and characteristics, it is said to be *homogeneous.* If this is not the case, then the mixture is *heterogeneous. Heterogeneous* mixtures usually have distinct boundaries between the various substances of which they are composed. Suppose one mixes some small pieces of iron with water (two substances). The result is a mixture, because its composition is variable. Furthermore, it is a *heterogeneous mixture,* because it is not uniform throughout. On the other hand, if one mixes salt in water, the salt dissolves, and the result is a *homogeneous mixture*—still a mixture because its composition can be varied, but homogeneous because it is uniform throughout. Homogeneous mixtures are commonly called solutions.

These basic categories of matter are illustrated schematically in Figure 1-1.

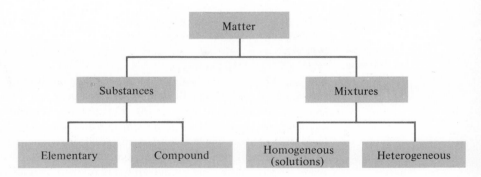

FIGURE 1-1

Classification of Matter

Properties   The specific characteristics by which a given substance is distinguished from all others are called the *properties* of that substance. Properties such as melting point, color, density, and electrical conductivity, which can be observed without changing the composition of the substance, are called *physical properties. Chemical properties,* on the other hand, are those that describe the changes in composition that the substance may undergo. Thus, for the elementary substance sulfur, the observations that it is a pale yellow solid with a specific gravity of 2.07, a melting point of 112.8°C, and a boiling point of 444.6°C, and that it does not conduct an electric current are all observations of its *physical properties.* However, the fact that it combines with oxygen to form a different substance with its own specific set of properties (sulfur dioxide) is a statement of a *chemical property*.

States of Matter

Any sample of matter at a given pressure and temperature must exist in one of three physical forms: solid, liquid, or gas. These forms are referred to as the *states of matter*.

Physical and Chemical Changes

The changes which substances may undergo are classified as either physical or chemical. *Physical changes* are those that do not involve any change in the composition of the substance. They include changes of state (melting, evaporation, and so on) and changes in shape or size (such as when a bar of iron is ground to a fine powder). A *chemical change* in matter is one in which its composition undergoes change; substances are converted into other substances. The combination of sulfur and oxygen to form sulfur dioxide is an example of a chemical change. Chemical changes are commonly referred to as *chemical reactions*.

Systems and Phases

The term *system* is frequently used in science simply to designate the particular body of matter under immediate consideration; it may be the contents of a test tube, the contents of a landlocked sea, or the contents of a living cell. A *phase* is a part of any system that is physically and chemically homogeneous. Thus, we may speak of one-phase systems, two-phase systems, and so forth. A homogeneous mixture is a one-phase system, but a heterogeneous mixture is a multiphase system. Consider, by way of illustration, a sealed bottle, half full of water and containing a few pieces of iron. The contents of that bottle are a *three-phase system,* consisting of a solid phase (iron), a liquid phase (water), and a gaseous phase (the air above the water).

New

Energy

Like mass, *energy* is a familiar concept but one that is not easily defined in simple terms. It is usually defined as *the capacity to do work*. Since *work* is the product of a force and the distance through which it moves ($w = f \times d$), and *force* is defined as the product of a mass and its acceleration ($f = m \times a$), then *energy* is the capacity to move matter from one place to another. *Kinetic energy* is the energy a body possesses by virtue of its motion, and the amount it possesses is equal to one-half its mass times its velocity squared ($KE = \frac{1}{2}mv^2$). *Potential energy,* on the other hand, is energy a body possesses because of its position relative to other bodies. Thus, a flowerpot hurtling down from a fifth-floor windowsill has *kinetic* energy as well as *potential* energy, but the same flowerpot poised motionless on the edge of the windowsill has *potential* energy alone.

New

The energy (either kinetic or potential) possessed by the flowerpot is an example of *mechanical* energy, but energy exists in other forms as well, including heat, light, sound, electrical energy, and chemical energy. Each of these forms of energy may be converted into some other, and it is through these transformations that energy may be observed. All other forms of energy may be transformed completely into heat, and for this reason energy changes are usually described in terms of heat. The basic unit for expressing a quantity of heat is the *calorie,* defined as *the amount of heat required to raise the temperature of exactly one gram of water from 14.5 to 15.5 degrees Celsius.*

When chemical reactions occur, the change in composition of the matter is usually accompanied by a change in its *chemical energy,* manifested by the liberation or absorption of energy, most commonly in the form of heat. If a reaction liberates energy from the system, it is said to be *exothermic.* In the course of such a reaction, chemical energy is transformed into thermal energy (heat), and the products of the reaction must therefore possess less chemical energy than the starting materials (the *reactants*). Conversely, in an *endothermic* reaction, energy is absorbed from the surroundings, which means that heat is transformed into chemical energy and the products of the reaction possess more energy than the reactants. The quantity of energy liberated or absorbed in a given reaction depends, of course, on the quantities of the substances involved.

Temperature

There is sometimes a tendency to confuse the terms *heat* and *temperature.* Heat is a form of energy and is measured in calories, units of *quantity. Temperature,* on the other hand, *is a measure of the intensity of heat, or degree of hotness.* Whereas the amount of *heat* possessed by a body is dependent on the mass of the body, its *temperature* is not. Temperature is measured in arbitrary units based on some convenient reference point. The temperature scale most frequently used in science is the Celsius (or centigrade) scale, on which the normal freezing point and the normal boiling point of water are assigned values of 0 and 100, respectively. The difference between these temperatures is divided into 100 equal divisions, and each division is referred to as a degree Celsius ($^\circ$C). On the Fahrenheit scale, the normal freezing and boiling points of water are 32° and 212°, respectively. Since 100 Celsius degrees are equivalent in temperature difference to 180 Fahrenheit degrees (a ratio of 5:9), temperatures may be converted from one system to the other by the relationship:

$$^\circ C = \tfrac{5}{9}(^\circ F - 32) \quad \text{or} \quad ^\circ F = \tfrac{9}{5}(^\circ C) + 32$$

EXAMPLE: Residential water heaters are commonly set to provide water at a temperature of 140°F. What is this temperature on the Celsius scale?

Solution:

$$^\circ C = \tfrac{5}{9}(140 - 32) = \tfrac{5}{9} \times 108 = 60^\circ C$$

EXAMPLE: The melting point of lead is 328°C. Express this temperature on the Fahrenheit scale.

Solution:

$$°F = \tfrac{9}{5}(°C) + 32 = (\tfrac{9}{5} \times 328) + 32$$

$$= 590 + 32 = 622°F$$

Measurement

The system used for the measurement of matter by scientists the world over is the *metric system.* This is a decimal system of measurement in which the basic unit of length is the *meter,* the basic unit of weight is the *gram,* and the basic unit of volume is the *liter.* Units of the metric system and conversion factors between this system and the English system (foot–pound–quart) are presented in Appendix 1.

p. 621

EXAMPLE: If a man is 5 ft 9 in. tall, what is his height in centimeters (cm)?

Solution:

$$5 \text{ ft } 9 \text{ in.} = (5 \times 12) + 9 = 69 \text{ in.}$$

Since 1 in. = 2.54 cm, the man's height is

$$69 \text{ in.} \times 2.54 \text{ cm/in.} = 175 \text{ cm}$$

Get familiar with

EXAMPLE: 1.0 oz of lead weighs how many milligrams (mg)?

Solution:

$$1.0 \text{ oz} = \frac{1.0}{16} \text{ lb}$$

$$1 \text{ lb} = 453.5 \text{ g}$$

Therefore,

$$1.0 \text{ oz} = \frac{1.0}{16} \text{ lb} \times 453.5 \text{ g/lb}$$

$$= 28.34 \text{ g}$$

But, since 1 g = 1000 mg,

$$1.0 \text{ oz} = \frac{1.0}{16} \times 453.5 \times 1000 = 28,340 \text{ mg}$$

EXAMPLE: How many milliliters (ml) are there in exactly one gallon?

Solution:

$$1 \text{ liter} = 1000 \text{ ml} = 1.057 \text{ qt}$$

$$1 \text{ gal} = 4 \text{ qt}$$

$$1 \text{ gal} = \frac{4 \text{ qt}}{1.057 \text{ qt/liter}} \times 1000 \text{ ml/liter} = 3784 \text{ ml}$$

The length, weight, and volume of a sample of a substance are not properties of that substance, for these measurements do not distinguish it from other substances. However, *the mass of a given volume of a substance at a fixed set of conditions* (temperature and pressure) is a property called *density*. Density may be expressed in any combination of units of mass per unit of volume, but most commonly it is expressed as grams per milliliter (g/ml) or grams per cubic centimeter (g/cm^3) for solids and liquids, and as grams per liter (g/liter) for gases. The density of a substance varies with temperature and pressure, because changes in these conditions cause variation in the volume of a given mass.

Density, mass, and volume, then, are related by the expression $d = m/V$, and if we know any two of these for a given sample of matter, we can easily calculate the value of the third for that sample.

EXAMPLE: What is the density of a liquid, 15.0 ml of which weighs 13.2 g?

Solution:

$$d = \frac{13.2 \text{ g}}{15.0 \text{ ml}} = 0.880 \text{ g/ml}$$

EXAMPLE: A certain liquid has a density of 1.20 g/ml. What is the volume occupied by exactly 1 lb of the liquid?

Solution:

$$V = \frac{m}{d} = \frac{453.5}{1.20 \text{ g/ml}} = 378 \text{ ml}$$

EXAMPLE: If a gas has a density of 2.34 g/liter, what is the weight of 25 liters of the gas?

Solution:

$$m = Vd = 25 \text{ liters} \times 2.34 \text{ g/liter} = 58.5 \text{ g}$$

METHODS OF SCIENCE

While the types of problems that interest chemists are as numerous as the complexities of the matter that confront them, the methods used to solve the problems have much in common. The biochemist concerned with the composition and shape of proteins frequently solves his problems by making use of the same intellectual processes used by the nuclear chemist in his attempt to understand the nucleus of the atom. Although the *scientific method* is a common thread woven through most of the history of science, some important advances in our knowledge of the universe have also been made by sheer chance. For example, the discovery of radioactivity by H. Becquerel (Chapter 2) is generally believed to have been purely accidental and not a result of any systematic intellectual method.

As an illustration of the scientific method, let us follow the thoughts and actions of a scientist who happens upon the box pictured in Figure 1-2. The box is apparently constructed from some heavy metal and has three rather stiff ropes protruding from holes labeled A, B, and C.

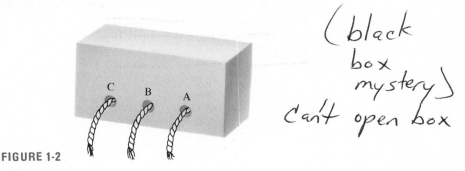

(black box mystery) can't open box

FIGURE 1-2

His curiosity aroused, our scientist attempts to open the box to discover its purpose and inner workings. He soon discovers, however, that he lacks the necessary tools and must content himself with observations made from outside the box.

An initial tug on rope A produces a movement of rope A but no apparent movement of ropes B and C. This observation triggers the thought that the box may contain three unconnected, independent ropes. This mental visualization is our scientist's first *model* (also called a *theory* or *hypothesis*) of the nature of the interior of the box.

Now, he reasons, if this model is correct, each rope should move to its limit without affecting (moving) any of the other ropes. Thus, in order to test his model, he pulls rope A a distance of 6 feet and is surprised to find that rope B has moved out a short distance. This distance he measures as 4 inches. He now pulls rope B a distance of 1 foot and observes an advance of 18 feet for rope A. He repeats these experiments a number of times using various distances and finds that rope A always moves 18 times as far as rope B. Throughout these experiments rope C has not moved.

Since the relationship between the distances moved by ropes A and B does not depend upon how far either rope is pulled—that is, the relationship is a general one—this relationship can be expressed as a mathematical *law*. If *x* represents the distance traversed by rope A, and *y* the distance traversed by rope B, the law becomes

$$x = 18y$$

Apparently the movement of rope B during the initial pull of rope A went unnoticed because of the high 1:18 distance ratio. In any case, the first model must now be modified. Ropes A and B are *not* unrelated, although rope C still appears to be independent of the other two.

Because of the specific interdependency of ropes A and B, the scientist now proposes the model pictured in Figure 1-3. In this model, rope A is wound around a drum attached to an axle; rope B is wound around the axle itself. The circumference of the drum is greater than the circumference of the axle (in fact, 18 times as great), and rope A is therefore played out at a greater rate than rope B. Rope C remains unattached and independent.

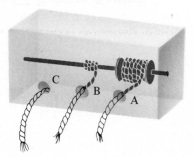

FIGURE 1-3

While this model of the interior certainly accounts for all of the observations, our scientist is not quite satisfied. He tests his new model by pulling rope C and observing ropes A and B (there is no effect). The experiments possible with these ropes are now almost exhausted, and the model appears to be satisfactory.

Let us now review the investigation of the box. Two models were proposed: The first, based on only one observation, did not explain later experiments and was therefore discarded; the second, a result of a greater number of observations, provides an explanation for all the experimental data. This model also permits the formulation of new questions and predictions. For example: Can rope C be withdrawn from the box completely, or is it held inside by a knot at the end? Are the ends of ropes A and B attached to the drum and the axle? If the model is correct, then when rope A is withdrawn to its limit, rope B may not have reached its limit, and (assuming that the ropes are attached) a further pull on rope B may wind rope A back into the box. These questions suggest additional experiments which might never have been conceived without the help of the model.

While our scientist has spent a considerable amount of energy investigating an almost trivial problem, his approach to the problem contains many of the same features found in the methods of scientists the world over. These methods generally begin with experimental observation and lead to the formulation of a law, which is then rationalized by a model or theory. The model is subsequently used to generate new experiments designed to test its validity. The outcome of the new experiments may enhance the validity of the model, or may cause it to be modified or discarded altogether.

Because of the central role of models, it is important to be cognizant of a number of their characteristics. First, the scientist usually draws on his own experiences in fashioning a theory. In our example, it might be suggested that the ropes are controlled by elves residing in the box, but the existence of such creatures has no scientific basis. As we shall see later in this book, many models designed to account for the behavior of matter so small that it has never been seen are based on the behavior of macroscopic bodies, such as billiard balls, which lie within the realm of everyone's experience.

On the other hand, some models are mathematical and abstract in nature. For example, the mathematical nature of the contemporary model of the electron makes many of its features difficult to visualize. Some scientists feel that the most significant scientific discoveries occur within the realm of mathematics.

It is also important to realize that a given set of experiments and observations can usually be explained by more than one model. Our scientist could have developed a model based on gears rather than drums, and in fact there are a number of alternate models that will satisfactorily account for the behavior of the ropes. As data and observations accumulate, one of a set of equally good models may become more satisfactory than the rest, or the choice of model may be based on considerations of simplicity or symmetry.

Finally, the fact that models may not, and very likely do not, correspond to reality cannot be overemphasized. Since, in our example, the box cannot be opened, the scientist will probably never know if the box really does contain a drum and axle. When the model is intended as a picture or visualization of matter at the submicroscopic, molecular level, the problem is even more acute. Atoms cannot possibly be either billiard balls or mathematical abstractions, nor is it likely that atoms behave *like* billiard balls. And yet, the billiard-ball model of atoms is at the heart of the determination of the structure of the nucleic acids DNA and RNA, the revelation of the genetic code, and all of its biological implications. Thus, while the correspondence between the model and reality may not be very high, the benefits of the model—the development of new experiments, the discovery of new laws of nature, and so forth—may be very great indeed.

SUGGESTED READINGS   Feynman, R. *The Character of Physical Law.* Cambridge, Massachusetts: MIT Press, 1965.

Garrett, A. B. "The Discovery Process and the Creative Mind." *Journal of Chemical Education,* Vol. 41, No. 9 (September 1964), pp. 479–482.

Mandleberg, C. J. *Topics in Modern Chemistry.* London: Cleaver-Hume Press, 1963. Chapter 1.

Pinkerton, R. C., and C. E. Gleit. "The Significance of Significant Figures." *Journal of Chemical Education,* Vol. 44, No. 4 (April 1967), pp. 232–234.
Wolfenden, J. H. "The Role of Chance in Chemical Investigations." *Journal of Chemical Education,* Vol. 44, No. 5 (May 1967), pp. 299–303.

PROBLEMS   A listing of the conversion factors relevant to these problems is provided in Appendix 1.

1. Indicate clearly the difference between the terms in each of the following pairs:
    (a) weight and mass
    (b) substance and mixture
    (c) homogeneous and heterogeneous
    (d) physical properties and chemical properties
    (e) potential energy and kinetic energy
    (f) heat and temperature
    (g) endothermic and exothermic

2. Using the values given in Appendix 1, calculate the mass, in pounds, of an electron, a proton, and a neutron.

3. A U.S. dime is approximately 1 mm thick and weighs about 2.5 g.
    (a) How many dimes would make a stack one foot high?
    (b) How many pounds would the stack weigh?

4. If a man has a body temperature of $38.9°C$, does he have a fever? (Normal body temperature is $98.6°F$.)

5. At what temperature do the Fahrenheit and Celsius scales coincide?

6. If the density of water is 1 g/ml, how much does a gallon of water weigh?

7. The density of lead is 11.4 g/cm^3, and the density of tin is 7.17 g/cm^3. Which weighs more, a cube of lead measuring 2.0 cm on a side, or a cube of tin measuring 1.0 in. on a side?

8. If 2.54 ml of a certain liquid weighs 1.789 g, what is the density of the liquid?

9. Mercury has a density of 13.55 g/ml. What volume is occupied by $\frac{1}{4}$ lb of mercury?

10. Make the following conversions:
    (a) One mile (1760 yd) into kilometers (km)
    (b) Six feet into meters (m)
    (c) One quart into milliliters (ml)
    (d) 1000 Å into nanometers (nm)
    (e) One decimeter (dm) into inches
    (f) Five pints into liters
    (g) One ounce into milligrams (mg)

11. A 100-g portion of skim milk (about $\frac{3}{8}$ cup) contains approximately 34 cal, and the same portion of whole milk contains 60 cal. If you drink 4

cups of milk per day, how much is your daily intake of calories reduced by using skim milk rather than whole milk?

12. Calculate the cost per gram of gold if its price is $160 per ounce.

13. In the metric system, energy is expressed in kilojoules (kJ) rather than kilocalories (kcal). The H—H bond energy is 103 kcal/mole. What is its value in kJ?

14. In the rope box shown in Figure 1-2, suppose rope C moved out a distance of 4 in. when rope A was pulled out 1 ft. Modify the model in Figure 1-3 to account for this relationship.

PART ONE

The Structure
and Composition of Matter

2 *Elements I:*
The Atomic Model

Our present-day understanding of the nature of matter—and therefore the birth of modern chemistry—may be dated from the development of John Dalton's *atomic theory* in the first decade of the nineteenth century. The basic concept of this theory, however, was not original with Dalton and in fact preceded him by many centuries.

ATOMISM

In the fourth and fifth centuries B.C. the Greek philosophers Democritus, Empedocles, Leucippus, and others expressed the belief that matter was composed of extremely small particles, and that if one were able to break up a sample of matter into increasingly smaller pieces he would eventually come to these particles which could not be further subdivided. These indivisible "building blocks" of matter came to be called *atoms,* from the Greek *atomos,* meaning uncuttable. Aristotle and his followers, on the other hand, maintained that matter was continuous and therefore could be divided endlessly into ever smaller pieces. Thus two schools of thought regarding the basic nature of matter came into being—the Aristotelian or continuous school and the Democritean or atomistic school—and philosophical debates over the merits of each continued through the centuries that followed.

The writings of the Arab philosophers of the tenth through the twelfth centuries A.D. indicate a rather wide acceptance of the idea of atomism, extending the concept even into religion. On the other hand, the alchemists Avicenna (tenth century) and Thomas Aquinas (thirteenth century) strongly rejected atomism in favor of the Aristotelian point of view. The atomic theory was used by a number of scientists in their work in the seventeenth and eighteenth centuries, including Robert Boyle, Isaac Newton, and William Higgins.

For more than 2000 years, then, the concept of the atom was largely a matter of speculation. An attempt to explain the riddle of the universe, it was based on philosophical thought—often metaphysical and religious in nature—not on the data of experience. It was not until scientific experimentation had come into its own—until careful observation of facts gave rise to statements of natural laws, and these laws required explanation—that atomism could progress from philosophical speculation to an accepted, useful chemical theory.

SOME LAWS OF MATTER

By the beginning of the nineteenth century, scientific investigation had led to the formulation of a number of laws dealing with the composition of matter. Three of the most fundamental are the Law of Conservation of Matter, the Law of Constant Proportions, and the Law of Multiple Proportions.

The Law of Conservation of Matter (first published by Antoine Lavoisier in 1789) states that *in any chemical reaction the mass of the system remains constant.* Thus, if substances A and B undergo a chemical reaction producing substances C and D, then the total mass of A and B consumed is equal to the mass of C and D produced, within the precision of our ability to weigh. Another way in which this law is sometimes expressed is *matter is neither created nor destroyed in a chemical reaction.*

The Law of Constant Proportions (also called the Law of Constant Composition) is generally attributed to Joseph Proust (1799). It states that *different samples of the same pure substance contain elements in the same proportions by weight.* If, for example, we analyzed a number of different samples of pure calcium carbonate taken from a variety of sources, all the samples would be found to contain 40.04 percent calcium, 12.01 percent carbon, and 47.96 percent oxygen.

A third law dealing with the composition of matter was published by John Dalton in 1805. It is the *Law of Multiple Proportions: When two elements combine to form more than one compound, the weights of one element that combine with the same weight of the other are in the ratio of small whole numbers.* By way of illustration, consider the two different compound substances water and hydrogen peroxide. Each of these substances is composed only of the elements hydrogen and oxygen, but in water the ratio by weight of oxygen to hydrogen is 8 to 1, whereas in hydrogen peroxide it is 16 to 1. Thus, the weights of oxygen that are combined in the two compounds with one part by weight of hydrogen are in the ratio of 8 to 16, or 1 to 2; that is, small whole numbers.

DALTON'S ATOMIC THEORY

John Dalton's atomic theory, which first appeared in print in 1808, consisted essentially of the following postulates:

1. Matter is composed of extremely small, indivisible particles called atoms. (It appears that Dalton visualized atoms as tiny, hard spheres.)
2. There are a number of different kinds of atoms, each kind having its own size, mass, and specific set of properties.
3. Each different kind of atom corresponds to a different element; i.e., all atoms of a given element are of the same kind, but atoms of different elements are of different kinds.
4. Compounds are composed of atoms of different elements combined with each other in simple ratios.
5. Chemical reactions involve only the combination, separation, or rearrangement of atoms. Atoms are not created, destroyed, or changed in any way.

(✱ ✱ ✱ see bottom ✱ ✱ ✱)

It is readily apparent that Dalton's model explained the fundamental laws of composition. If atoms have specific masses, and if chemical reactions involve only rearrangements of atoms, then matter must adhere to the Law of Conservation of Matter. Furthermore, if compounds are simply combinations of atoms of different elements in fixed ratios, then in every sample of the same compounds the elements must always be present in the same proportions by mass, and the Law of Constant Proportions is explained. Finally, the Law of Multiple Proportions is explained by the combination of one atom of one element with one atom of a second element to form a compound, or of one atom of the first with two atoms of the second to form a different compound, and so forth. The laws of composition of matter do not constitute *proof* of the atomic theory, but the atomic theory provides one satisfactory explanation of the laws (which are statements based on observation).

Unsophisticated as it may seem today, John Dalton's atomic theory represented a giant step forward in the development of chemistry. It provided acceptable explanations of observed facts and enabled chemists to make predictions that could then be tested experimentally; it laid the foundation for a system of relative masses of the elements so that chemical reactions could be studied on a quantitative basis; and it paved the way for the multitude of scientific investigations that have led to our present concept of matter and that are still continuing.

However, any theory is subject to refinement and modification as new observations are made, and Dalton's theory was no exception. While its basic tenets—that matter is made up of atoms, and chemical reactions are simply rearrangements of these atoms—are almost universally accepted today, evidence accumulated since Dalton's day has shown some of the details of his theory to be inadequate; in particular, his belief in the indivisibility of atoms.

(✱✱✱ ✱✱✱)

EVIDENCE OF SUBATOMIC PARTICLES

The observations that led to a gradual refinement and extension of the atomic theory are numerous indeed, and a detailed description of them all is beyond the purpose of this book. However, a brief account of some of the more significant breakthroughs is helpful in acquiring an understanding of modern atomic theory.

The Electron   Near the middle of the nineteenth century, a new field of investigation into the structure of matter was opened by the invention of the gas discharge tube, forerunner of the modern television picture tube. Basically, this device consists of a closed glass tube, into the ends of which are sealed two metal plates to serve as electrodes. A high voltage (5,000–10,000 volts) is imposed across the electrodes. If the air (or other gas) in the tube is at atmospheric pressure, nothing is observed. However, if air is pumped from the tube, the gas that remains begins to glow and conducts the electric current. At very low pressures the glow fades, although current continues to flow, and the end of the glass tube at the anode (positive electrode) fluoresces (emits light). Rays are emitted at the cathode (the negative electrode) and are directed toward the anode; the fluorescence at the anode end of the tube is caused by the rays passing through holes in the anode and striking the glass (Figure 2-1). When the rays are allowed to impinge on certain substances (for instance, zinc sulfide), the fluorescence is intensified.

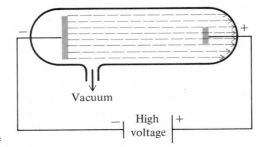

Vacuum

FIGURE 2-1

Gas Discharge Tube

High voltage

These rays, first reported by Julius Plücker in 1859, were shown by Plücker and Johann Hittorf to be similar to light, in that they cast a shadow, but also to be deflected by a magnet, which suggested that they were electrically charged. They were given the name *cathode rays* by Eugen Goldstein in 1876. In 1879 Sir William Crookes reported the results of a series of extensive and ingenious experiments with the cathode ray tube. Not only did he confirm all the observations of earlier workers, he also showed that cathode rays are bent toward the positive pole of an electric field, indicating that they are negatively charged. In another experiment he placed a paddle wheel in the path of the rays and discovered that the paddle wheel turned, from which he concluded that the rays are really streams of particles with definite mass. (Some twenty years later, J. J. Thomson showed that the masses are too small to move a paddle wheel and suggested that the motion was caused by the heating effect of the collisions, an explanation which has since been confirmed.)

The accumulated evidence indicated that cathode rays consist of very small, negatively charged particles that are emitted from the substance of which the cathode is made. To these particles the name *electron* was applied. (The word had been introduced in 1874 by G. J. Stoney as a designation for the fundamental particle of electricity.) This view became firmly entrenched when, in 1897, J. J. Thomson succeeded in measuring the charge-to-mass ratio of electrons by observing the deflection of cathode rays in electrical and magnetic fields applied at right angles to one another. Thomson's findings showed that this ratio remained constant regardless of the particular gas in the tube or the specific metal of which the cathode was made. This suggested that the electron was a universal particle found in all matter. In a series of experiments extending from 1908 to 1917, for which he won the Nobel Prize in physics, R. A. Millikan successfully measured the absolute charge of the electron (4.80×10^{-10} electrostatic units) and made it possible to calculate its mass. The electron's mass is 9.1×10^{-28} gram, indicating a particle so light that it weighs approximately 1/1840 as much as the lightest atom, the hydrogen atom.

Here, then, was solid evidence—pieced together from the work of many investigators over a period of a half century—of the incorrectness of Dalton's theory of the indivisibility of the atom. The cathodes are elementary substances (metals). Electrons are emitted from the cathodes. If substances are composed only of atoms, then electrons must be emitted from atoms. Since electrons weigh only 1/1840 as much as even the lightest atom, the atom can no longer stand as the ultimate, indivisible particle.

The Proton

Since matter is electrically neutral and therefore, presumably, so are atoms, it was logical to suggest that the emission of electrons from atoms should leave a positively charged residue. That is, if atoms contain these negative particles, might we not also expect atoms to contain positive particles?

Partial evidence for this view was obtained by Goldstein in 1886, using a gas discharge tube in which the cathode was constructed with a number of holes through it. Goldstein observed rays in the end of the tube beyond the cathode, and he called these *canal rays,* since they passed through the holes (or canals) in the cathode (Figure 2-2). Unlike the cathode rays (which are not apparent unless they strike an object that fluoresces, such as glass), canal rays possess a luminosity of their own. Furthermore, the color of the luminosity varies with the particular gas in the tube.

Investigating Goldstein's observations more fully, Wilhelm Wien succeeded in showing in 1898 that canal rays are deflected toward the negative pole in an electric

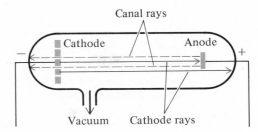

FIGURE 2-2

Production of Canal Rays

field and must therefore be positively charged. Proceeding on the assumption that canal rays were streams of positive particles (and using a technique similar to Thomson's in his studies of the electron), Wien determined the charge-to-mass ratio of these particles. His results, published in 1902, show that the charge-to-mass ratio is not the same for all particles but varies with the particular gas in the tube.

The phenomena observed by Goldstein, Wien, and others are now interpreted as follows: As electrons leaving the cathode (that is, the cathode rays) travel toward the anode, they collide with atoms of the gas in the tube. These collisions cause electrons to be dislodged from the gas atoms, and the dislodged electrons are attracted toward the anode. The atoms from which electrons have been removed (*ions*) are now positively charged and therefore move toward the negatively charged cathode. It is these positive ions that form the canal rays.

Obviously the charge-to-mass ratio of any particular ion depends on the mass of the neutral atom as well as the number of electrons removed from it to form the ion. However, Wien's results (as later refined by Thomson and by F. W. Aston) showed that when the gas in the tube is hydrogen—the element whose atoms have the lowest mass of all atoms—the charge-to-mass ratio of the ions formed is approximately 1/1840 of the value for an electron. If one assumes that the magnitude of the charge on this positive particle is equal to that of an electron's negative charge, then it can easily be calculated that the mass of the positive particle obtained when one electron is removed from a hydrogen atom is 1.67×10^{-24} gram. This positive particle has been named the *proton*.

The discovery of the proton provided further evidence against the indivisibility of the atom. But it did more than that; it provided scientists with a simple model of atomic structure. A hydrogen atom could now be considered to consist of two parts: a positively charged proton, accounting for essentially all the mass of the atom, and a negatively charged, extremely light electron.

Around the turn of the century, several attempts were made to extend this model of the hydrogen atom to a generalized model that would apply to all atoms. One such model was created by Phillip Lenard in 1903. Lenard had observed that cathode rays pass through a foil of aluminum and was impressed with the apparent "emptiness" of atoms. Accordingly, he proposed that the atom is a shell, largely empty except for a number of what he called *dynamids* at the center (Figure 2-3). Each dynamid consisted of a single positive charge and a single negative charge, the number of dynamids being proportional to the mass of the atom. A second model,

FIGURE 2-3

The *Dynamid* Atomic Model (Lenard)

elaborated by Thomson in 1904, regarded the atom as consisting of a uniformly positive sphere of matter in which electrons were imbedded (Figure 2-4). This has been frequently referred to as the "raisin muffin" or "plum pudding" model. Neither of these models explained all the facts satisfactorily, but scientific thought had taken a step away from Dalton's structureless atom toward an atom with a describable composition.

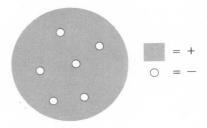

FIGURE 2-4

The "Raisin Muffin" Atomic Model (Thomson)

Radioactivity
During the same period in which the electron and proton were being characterized, another area of investigation was developing which led to further evidence for the existence of subatomic particles.

The French physicist Henri Becquerel reported in 1896 his discovery that compounds containing the element uranium spontaneously emit radiation that darkens photographic plates, induces conductivity in gases, and can pass through objects such as heavy paper or thin metal foils.. This phenomenon, which became known as *radioactivity,* was later shown by Marie and Pierre Curie and others to be a property of certain other elements as well.

Immediately following its discovery, radioactivity was studied by a number of investigators, foremost of whom was Ernest Rutherford. One of the most revealing of these early investigations may be described as follows. A narrow beam of radiation was obtained by confining a radioactive substance in a thick lead block with a small hole drilled into the cavity. Since the radiation cannot penetrate through lead, this arrangement permitted a single directed beam of radiation which could be detected by means of a photographic plate. When the beam was made to pass through an electric field, it was found that some of the radiation was deflected slightly toward the negative electrode, some was deflected through a larger angle toward the positive electrode, and some continued on its course unaffected by the electric field. It was concluded that the radiation emitted by a radioactive element is not homogeneous, but of three kinds: one positively charged, one negatively charged, and the third without charge (Figure 2-5). These three types of radiation were designated simply by the first three letters of the Greek alphabet: alpha (α), beta (β), and gamma (γ) rays.

Subsequent study of these rays showed that *beta rays* are streams of electrons, identical in all properties to the particles of cathode rays. *Alpha rays* consist of particles with a positive charge of a magnitude equal to twice that of a proton and with a mass four times that of a proton. We now recognize the alpha particles

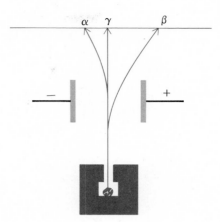

FIGURE 2-5

Separation of α, β, and γ Rays

to be helium ions—that is, helium atoms from which two electrons have been removed. The electrically neutral *gamma rays* are electromagnetic vibrations, similar to light but with a very short wavelength.

Here, then, is still further evidence of the existence of particles smaller than atoms, for in radioactivity we see a spontaneous, natural process—not controlled by man—whereby atoms undergo a self-disintegration to emit particles having masses smaller than the atoms themselves.

The Nucleus   As was mentioned earlier, there were a number of attempts in the opening years of the twentieth century to devise a model that would represent the arrangement of the positivity and negativity within the atom. The discovery of radioactivity, and particularly the α particle, with its ability to penetrate thin sheets of metal, provided a new line of investigation toward the solution of this problem. Although a number of scientists of the period became involved in these "alpha-scattering" experiments, major credit is usually given to Rutherford, since it was he who joined his own observations to those of others to provide a structural model of the atom that replaced all earlier ones. Figure 2-6, while it is not intended to represent any specific experiment, illustrates in a generalized way the nature of these α-particle investigations.

A source of α particles is arranged so that the alpha radiation is directed in a narrow beam onto a screen coated with some material which fluoresces when struck by α particles; for instance, zinc sulfide. When nothing is placed in the path of the beam, the positively charged α particles continue in a straight line and are observed as a glowing spot on the screen. When a very thin foil of a heavy metal (such as gold or copper) is placed in the path of the beam, any changes in the paths of α particles caused by the atoms of the metal foil can be detected on the screen. Most of the α particles continue on their course, unaffected by the metal atoms. Some, however, are deflected from their paths through a variety of angles, even as much as $180°$.

In his interpretation of these results, which he published in 1911, Rutherford

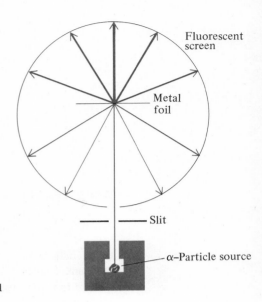

FIGURE 2-6

Scattering of Alpha Particles by Metal Foil

reasoned somewhat as follows: Since α particles bear a positive charge, they will be deflected from their paths by repulsion of other positive charges. However, because the kinetic energy of an α particle is very large (the mass, charge, and velocity of α particles were known), only a particle of considerable mass and high charge within the metal atom could cause deflection through such large angles. If the positive and negative particles (protons and electrons) were distributed uniformly throughout an atom, they would exert little or no influence on the α particles, which would therefore pass directly through the atom without deflection (Figure 2-7).

Rutherford postulated that all the positivity of the atom resides within a particle in the center of the atom. This particle, called the *nucleus,* accounts for

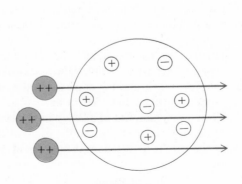

FIGURE 2-7

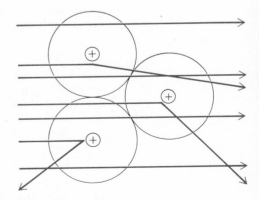

FIGURE 2-8

Paths of Alpha Particles through Nuclear Atoms

essentially all the mass of the atom but has a very small volume. The electrons are distributed throughout the extranuclear part of the atom (*extra-*, outside).

That this *nuclear model* of the atom explains the observed α-particle scattering is illustrated in Figure 2-8. In order to be deflected, an α particle must come relatively close to a massive, highly charged nucleus. The electrons with their extremely low masses will have no noticeable effect on α particles. The nucleus occupies such a small fraction of the volume of an atom that few α particles are deflected. The closer the path of an α particle to a nucleus, the greater is the angle of deflection, a deflection of 180° representing a "head-on" collision.

The Neutron The nuclear atomic model that emerged from the foregoing evidence—a positive nucleus, which could be thought of as an aggregate of protons, and an equal number of extranuclear electrons—had one major shortcoming: It did not account for the *masses* of atoms. A comparison of the hydrogen and helium atoms provides a simple illustration. The hydrogen atom consists of a proton (the hydrogen nucleus) and one electron; its mass, therefore, is essentially the mass of a proton. The second lightest atom, helium, has a nuclear charge twice that of a proton, and therefore the helium nucleus may be thought of as consisting of two protons. In order for the helium atom to be neutral it must have two extranuclear electrons. Then it would have a mass twice that of a hydrogen atom, but in fact the helium atom's mass is approximately four times that of hydrogen.

Some scientists theorized that the nucleus contains more protons than its total charge would indicate but that some of these protons are neutralized by electrons in the nucleus. Following this view, the mass of the helium atom is explained by assuming the nucleus to consist of four protons and two electrons (in addition to the two extranuclear electrons). Others postulated the existence of yet another kind of subatomic particle to account for mass.

Finally, in 1932, James Chadwick (a former student of Rutherford) bombarded the element beryllium with α particles and discovered that beryllium atoms were converted into carbon atoms, and that this transformation was accompanied by the emission of a previously unknown particle. This new particle had a mass almost exactly that of a proton but no charge, so it was called a *neutron*.

Thus, the last essential piece of the puzzle had been put into place and the nuclear model of the atom became firmly entrenched. In the years since, a number of additional subatomic particles have been discovered (positron, neutrino, meson, and so forth), but these are not essential for an explanation of most of the physical and chemical properties of matter.

MODERN ATOMIC THEORY

Let us now examine the present concept of the atom—a concept that may indeed be changed drastically in the future but that in our present state of knowledge is the most satisfactory, workable model available to us.

Atoms, the basic building blocks of all matter, are exceedingly small, having diameters of the order of 1 to 3 Å (1 angstrom unit, 1 Å, = 10^{-8} cm). Their size may perhaps be appreciated by the realization that a single grain of sand contains about 10^{18} atoms. (If 300,000,000 planets each had a population equal to that of the earth, their combined population would be approximately 10^{18}.)

Each atom consists of one positively charged nucleus and one or more negatively charged electrons. Since an atom is neutral, the nuclear charge is equal to the total charge of the electrons.

The nucleus accounts for only a very small fraction of the volume of the atom. Nuclei have diameters of about 0.0001 Å. If an atom could be enlarged until its nucleus were the size of a tennis ball, the atom would be a sphere nearly a mile in diameter. The size of an electron is estimated to be in the same range as that of a nucleus, but the electron is extremely light, weighing only about 1/1840 as much as the lightest atomic nucleus. The nucleus, therefore, accounts for essentially all of the mass of the atom.

Electrons may be thought of as particles in rapid motion, but with wavelike properties. In effect they fill the extranuclear part of the atom with a diffuse cloud, sometimes described as a "mushy sphere." This may be visualized by comparing it with a multiple-exposure photograph of a moving object, where the density is greatest in the area traversed most frequently by the object. The density of the electron cloud decreases as the distance from the nucleus increases, so that the boundaries of the atom are not sharply defined (Figure 2-9). The electronic structure of atoms is dealt with in detail in Chapter 5.

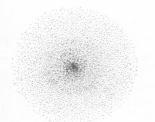

FIGURE 2-9

Atomic Number

An atom of the lightest element, hydrogen, has a proton as its nucleus and one electron forming the electron cloud. Since the magnitudes of the nuclear charges of all atoms are whole-number multiples of that of the hydrogen nucleus, it is convenient to consider each nucleus as containing some definite number of protons. For example, the sodium atom, with a nuclear charge eleven times as great as that of the hydrogen atom, is considered to possess 11 protons in its nucleus and therefore to have 11 electrons forming the electron cloud.

The nuclear charge, equal to the number of protons in the atom, is called the *atomic number.* It is this number—not the atomic mass, as Dalton believed—that distinguishes one element from another. Atoms with the same atomic number are all atoms of the same element; atoms with different atomic numbers are atoms of different elements.

The terms *elementary substance* and *compound substance* introduced in Chapter 1 may now be given more theoretical definitions based on our atomic model: An *elementary substance* is a substance whose atoms all have the same atomic number; a *compound substance* contains atoms with different atomic numbers.

Mass Number

All atoms except the hydrogen atom have nuclear masses greater than can be accounted for by the protons alone. Since these mass differences are essentially whole-number multiples of the mass of a neutron, we may consider atomic nuclei to contain—in addition to their protons—some number of neutrons. These nuclear particles—protons and neutrons—taken together are referred to as *nucleons,* and the number of nucleons in an atomic nucleus is called the *mass number.* Thus,

Mass number − Atomic number = Number of neutrons

A particular atom with an atomic number of 17 and a mass number of 35, therefore, contains 17 protons and 18 neutrons in its nucleus, and has 17 extra-nuclear electrons.

To designate a specific atom and its nuclear contents, the symbol for the element is written with the atomic number as a subscript and the mass number as a superscript, both to the left of the symbol. Table 2-1 illustrates this notation with a few examples.

TABLE 2-1		*Protons*	*Neutrons*	*Electrons*
	$^{4}_{2}\text{He}$	2	2	2
	$^{12}_{6}\text{C}$	6	6	6
	$^{80}_{35}\text{Br}$	35	45	35
	$^{127}_{53}\text{I}$	53	74	53
	$^{227}_{89}\text{Ac}$	89	138	89

Isotopes

While all atoms of the same element have the same atomic number, they do not necessarily have the same mass number. For example, the element chlorine occurs in nature with atoms of two different masses, one represented by $^{35}_{17}\text{Cl}$, the other by $^{37}_{17}\text{Cl}$. Note that both atoms have 17 protons and 17 electrons, but one has only 18 neutrons, whereas the other has 20. Both are atoms of the same element, and they have the same chemical properties, but they differ slightly in those physical properties which depend on mass. Atoms that have the same atomic number but different mass numbers are called *isotopes.*

As a further example of the occurrence of isotopes, consider the element hydrogen. Most hydrogen atoms have the composition we have indicated above; that is, the nucleus is simply a proton and we can designate them as $^{1}_{1}\text{H}$. However, about one out of every 5,000 hydrogen atoms has a mass twice that of the others. The nucleus of this heavier atom must be considered to have a neutron as well as a proton and is therefore designated as $^{2}_{1}\text{H}$. This heavier isotope is called *deuterium* and its nucleus is sometimes referred to as a *deuteron.*

For a few of the elements only one isotopic form is known, but most elements have a number of isotopes—as many as 10 in the case of the element tin.

Atoms may also have the same mass number but different atomic numbers; for example, $^{214}_{82}\text{Pb}$ and $^{214}_{83}\text{Bi}$. These atoms, which do not have the same properties because they are not atoms of the same element, are called *isobars.*

ELEMENTS: NAMES AND SYMBOLS

There are 105 different elements known today; that is, atoms with 105 different atomic numbers. Not all of these may be found in nature. Some have been made synthetically in recent years, including a few which may once have occurred in the earth's crust but have disappeared through radioactive disintegration. Some of the elements were discovered and characterized many years before John Dalton; two (elements 104 and 105) have been synthesized so recently that official names have not yet been established for them.

Although the elements may be distinguished from each other by reference to their atomic numbers, each (except for the last two) has been given a name. The origins of the names range widely, from mythological characters (thorium, tantalum, neptunium) to the names of notable scientists (curium, einsteinium, mendelevium). Some elements have been named for an outstanding property: bromine (bad odor), phosphorus (light bearer), argon (lazy one). Nations have been honored (polonium, germanium, americium), as have small villages (magnesium for the ancient town of Magnesia in Asia Minor; strontium for the Scottish village of Strontian).

As a matter of convenience in describing chemical reactions, a system of chemical shorthand has been developed in which each element has been assigned a symbol. This system of symbols was first introduced by the Swedish chemist Jöns Berzelius in 1813, has been extended as new elements were discovered, and is now in use by international agreement. The symbols are the initial letters of the names of the elements, and in cases of duplication, an additional letter taken from the name; for example, C for carbon, Cl for chlorine, Ca for calcium, Cd for cadmium, etc. However, not all the symbols are based on the English name. A number are derived from the Latin name for the element—for example, Hg for mercury (*hydrogyrum*), K for potassium (*kalium*), Fe for iron (*ferrum*)—and one comes from the German name, W for tungsten (*wolfram*). The elements are listed alphabetically, together with their symbols and atomic numbers, on the inside front cover of this book.

For most elements the symbol has two meanings: it represents an atom of that element (that is, an atom with a specific atomic number), but it also represents the *elementary substance* composed of those atoms. For example, the symbol Cu is used to designate an atom with an atomic number of 29, but it is also used to represent a *piece of copper*. There are some exceptions to this dual use of the symbol, however. Some elementary substances do not normally exist as collections of separate atoms, but as molecules—discrete groupings of two or more atoms linked together. In these cases, a numerical subscript is added to the symbol for the element to indicate the number of atoms in each molecule. For example, N is the symbol for the element nitrogen and represents atoms with atomic number 7. But the elementary substance called nitrogen—a colorless, odorless gas—exists in the form of diatomic (two-atom) molecules. To designate gaseous nitrogen, therefore, we write N_2. Some other common elementary substances in the same category are oxygen (O_2), hydrogen (H_2), fluorine (F_2), chlorine (Cl_2), bromine (Br_2), iodine (I_2), and phosphorus (P_4).

ATOMIC WEIGHTS

With the acceptance of Dalton's postulates that the masses of atoms are different for each element and that chemical reactions are simply rearrangements of atoms, it became apparent that knowledge of atomic masses was of great importance in the investigation of chemical processes. The actual masses of individual atoms are extremely small, however. As indicated earlier in this chapter, the lightest atom weighs only 1.67×10^{-24} g, and even the heaviest atom known weighs only about 260 times that. Furthermore, chemists are not able to carry out reactions with only a few atoms but must use amounts of substances that contain very large numbers of atoms. (Recall that a grain of sand contains approximately 10^{18} atoms.)

Efforts were made, therefore, to determine the *relative* masses of atoms. If some arbitrary value is assigned to the weight of an atom of element Q, then by experimentally determining what weight of element X combines with a given weight of Q, and knowing the ratio in which atoms of Q and X combine, one can assign a value to the weight of an atom of X *relative* to the arbitrary weight assigned to Q. These *relative* weights are called *atomic weights*. Since oxygen combines directly with nearly all the other elements, it is not surprising that it was chosen as the reference standard early in the development of atomic weight systems. After a number of attempts to find a suitable reference value, 16 was chosen as the arbitrary weight of an oxygen atom, because this particular value gave atomic weights for many other elements which were very near whole numbers, and gave hydrogen, the lightest element, an atomic weight of approximately 1. Thus, the atomic weight of an element was established as the weight of an atom of that element relative to the weight of an oxygen atom taken as exactly 16 units. This atomic weight scale served chemists well for more than a century and has only recently been replaced.

The discovery of the existence of isotopes altered the concept of atomic weights somewhat. It was found that all atoms of the same element are *not* identical in mass. Most natural samples of an element are mixtures of two or more isotopes. In working with substances in the laboratory one is dealing with enormous numbers of atoms, with the various isotopes present in fixed ratios. An experimentally determined atomic weight, therefore, is really an average of the masses of the various isotopes of the element, weighted according to their relative abundance. Even in the reference standard, oxygen, the assigned value of exactly 16 units did not apply to any single atom, but to a weighted average of the naturally occurring isotopes, $^{16}_{8}O$, $^{17}_{8}O$, and $^{18}_{8}O$.

The realization that atomic weights are really *average* weights of atoms did not reduce the usefulness of the oxygen-based atomic weight scale, for chemists usually work with elements in their naturally occurring mixtures. However, as interest in nuclear reactions increased, and scientists began to deal with pure isotopes, it was found convenient to develop a different atomic weight system based on the assignment of exactly 16 units to the oxygen-16 isotope ($^{16}_{8}O$). Atomic weights on this physical scale have slightly different values from atomic weights on the older chemical scale, and this difference frequently led to confusion.

In 1961, by international agreement, both standards were abandoned in favor of a scale in which the carbon-12 isotope ($^{12}_{6}C$) is assigned a weight of exactly 12 units. The *atomic mass unit* (abbreviated amu) is now defined as a mass equal to 1/12 the mass of the $^{12}_{6}C$ atom. The atomic weights of the elements based on this reference standard are listed inside the front cover of this book.

Based on the carbon-12 scale, the masses of both the proton and neutron are very nearly 1 amu. More precisely, the mass of the proton is 1.0073 amu, the neutron is 1.0087 amu, and the electron is 0.00055 amu.

THE GRAM-ATOM AND AVOGADRO'S NUMBER

Chemists cannot work with individual atoms; they must work with visible amounts of matter containing huge numbers of atoms. They must weigh substances in grams, not in atomic mass units. Obviously, it is necessary to be able to relate the weights of single atoms to the weights of bulk samples.

Suppose we have two hypothetical elements, A and B, whose atomic weights are, respectively, 2 and 5; that is, an average atom of B weighs 2.5 times as much as an average atom of A. Now, suppose we make a pile containing 1000 atoms of A and another pile containing 1000 atoms of B. It is obvious that the weights of the two piles will be in the same ratio as the weights of the individual atoms: 2 to 5. The pile of 1000 B atoms weighs 2.5 times as much as the 1000 A atoms. This relationship will hold no matter how many atoms are in the piles, as long as the two piles contain equal numbers of atoms. Looking at this from the other direction, 2 pounds of A atoms must contain the same number of atoms as 5 pounds of B atoms, or 2 grams of A atoms must contain the same number of atoms as 5 grams of B atoms. No matter what units they are measured in, if the weights of the samples of the two elements are in the same ratio as the weights of single atoms of the elements, then the two samples must contain the same number of atoms. Because in science weight is usually expressed in grams, we attach special significance to *that amount of an element whose weight in grams is numerically equal to its atomic weight.* This amount is referred to as a *gram-atomic weight,* or simply a *gram-atom.* Thus, since the atomic weight of calcium is 40.08, one gram-atom of calcium weighs 40.08 g, and since the atomic weight of aluminum is 26.9815, one gram-atom of aluminum weighs 26.9815 g.

It follows that one gram-atom of any element contains the same number of atoms as one gram-atom of any other element. This number has been determined by several different methods and has been established as 6.023×10^{23} atoms/g-atom. It is known as *Avogadro's number,* in honor of the nineteenth-century Italian chemist, Amadeo Avogadro.

Avogadro's number can be used to calculate the number of atoms in any given weight of an element.

EXAMPLE: How many atoms are there in 50.0 g of copper?

Solution: The atomic weight of copper is approximately 63.5. Therefore, 1 g-atom of copper weighs 63.5 g.

The number of gram-atoms in 50.0 g of copper is

$$\frac{50.0 \text{ g}}{63.5 \text{ g/g-atom}}$$

Since each gram-atom of copper contains Avogadro's number of atoms, to obtain the number of atoms we multiply the number of gram-atoms by 6.023×10^{23} atoms/g-atom.

$$\frac{50.0}{63.5}\text{g-atoms} \times 6.023 \times 10^{23} \ \frac{\text{atoms}}{\text{g-atom}} = 4.74 \times 10^{23} \ \text{atoms}$$

Avogadro's number also enables one to calculate the weight of a single atom of any element (and also, therefore, the weight of any given number of atoms).

EXAMPLE: What is the weight of a single phosphorus atom?

Solution: One gram-atom of phosphorus weighs approximately 30.97 g. Since this is the weight of Avogadro's number of phosphorus atoms, the weight of one phosphorus atom must be

$$\frac{30.97 \text{ g/g-atom}}{6.023 \times 10^{23} \ \text{atoms/g-atom}} = 5.14 \times 10^{-23} \ \text{g/atom}$$

SUGGESTED READINGS

Ihde, A. J. *The Development of Modern Chemistry.* New York: Harper & Row, 1964. Chapters 4, 18.

Partington, J. R. *A History of Chemistry.* London: Macmillan, 1964. 4 vols. Vol. 4, Chapter 27.

Weeks, Mary E. *Discovery of the Elements.* 6th ed. Easton, Pennsylvania: Journal of Chemical Education, 1956.

PROBLEMS

1. Indicate the difference between the terms in each of the following pairs:
   (a) nucleus and nucleon
   (b) atomic number and mass number

(c) isotopes and isobars

(d) atomic weight and mass number

(e) alpha particles and beta particles

(f) atomic number and atomic weight

2. Indicate the number of protons and the number of neutrons in each of the following nuclei:

$$^{3}_{1}H \qquad ^{31}_{15}P \qquad ^{34}_{16}S \qquad ^{206}_{82}Pb \qquad ^{99}_{43}Tc \qquad ^{235}_{92}U$$

3. How many atoms are there in
   (a) 1.0 g of gold?
   (b) 1 lb of iron?
   (c) 5.0 liters of helium with a density of 0.18 g/liter?
   (d) 2.5 g-atoms of sodium?

4. Calculate the weight of a single atom of each of the following:
   (a) silicon   (b) bromine
   (c) barium   (d) lead

5. What is the weight of one million silver atoms?

6. Which contains the larger number of atoms—1 g-atom of aluminum or 1 g-atom of lead?

7. Which weighs more—0.5 g-atom of calcium or 0.2 g-atom of barium?

8. Which weighs more—one billion tin atoms or one billion lead atoms?

9. If 1 g-atom of sodium weighs 23 g, what is the weight of 1 lb-atom of sodium, 1 kg-atom of sodium, and 1 ton-atom of sodium?

10. Calculate the number of gram-atoms of carbon in a 1.5-carat diamond. (Diamond is pure carbon, and one carat weighs 200 mg.)

11. How many gram-atoms of zinc are there in 15.5 g of the pure metal?

12. Commercial bronze contains 90 percent copper and 10 percent zinc (by weight). What is the atomic ratio of copper to zinc?

13. From the average diameter of a nucleus and that of an atom, calculate the volume of an atom and compare it with that of the nucleus.

3 *Compounds I: Formulas, Equations, and Stoichiometry*

As we have seen in Chapter 2, compounds are substances formed by the combination of atoms of two or more elements, usually in the ratio of small whole numbers. The properties of compounds, their internal structures, and the nature of the binding between their constituent atoms will be discussed in subsequent chapters. This chapter deals mainly with the composition of compounds and the weight relationships that exist within them.

FORMULAS

The composition of a compound is conveyed by its *formula*, a combination of the chemical symbols for the elements present in the compound. There are a number of different types of formulas, the simplest of which is the *empirical formula*. The empirical formula reveals two items of information: (a) what elements are present in the compound and (b) the combining ratio of the atoms of those elements. For example, the subscript 2 in the formula $MgCl_2$ indicates that this compound contains twice as many chlorine atoms as magnesium atoms. Similarly, the formula H_2SO_4 represents a compound consisting of atoms of hydrogen, sulfur, and oxygen in a ratio of 2 to 1 to 4, respectively.

The *formula weight* of a compound is analogous to the atomic weight for an element. It is simply the sum of the atomic weights of the elements shown in the empirical formula. The atomic weight of each element in the compound must be multiplied by the number of atoms of that element indicated by the empirical formula (that is, by the subscript to the symbol). Thus, for magnesium chloride, $MgCl_2$, the formula weight is $24.3 + 2(35.5)$, or 95.3; for K_2CO_3, the formula weight is $2(39) + 12 + 3(16)$, or 138.

The *gram-formula weight* has the same relationship to the formula weight of a compound as the gram-atomic weight has to the atomic weight of an element. That is, a gram-formula weight of a compound is that amount of the compound whose weight in grams is numerically equal to its formula weight. Thus, one gram-formula weight of K_2CO_3 is 138 g.

It follows, then, that a formula reveals not only the ratio of *atoms* in a compound, but also the ratio of *gram-atoms*. The formula $MgCl_2$ tells us that each gram-formula weight of the compound (95.3 g) contains 1 g-atom of magnesium (24.3 g) and 2 g-atoms of chlorine (2 $\times$ 35.5 g). Furthermore, recalling the significance of Avogadro's number, note that one gram-formula weight of $MgCl_2$ contains 6.023×10^{23} magnesium atoms and $2 \times 6.023 \times 10^{23}$ chlorine atoms.

These weight relationships enable us to calculate the weight of any element in any given weight of any compound of that element, as long as we know the empirical formula of the compound.

EXAMPLE: What weight of potassium is contained in 50.0 g of K_2SO_4?

Solution: The formula weight of K_2SO_4 is $2(39.1) + 32.1 + 4(16.0) = 174.3$. Therefore, in every 174.3 g of the compound there are 78.2 g of potassium, and in 50.0 g of the compound there must be

$$\frac{78.2}{174.3} \times 50.0 \text{ g} = 22.4 \text{ g K}$$

A common application of this relationship is the calculation of the *percentage composition* of a compound; that is, the percentage by weight of each element in the compound.

EXAMPLE: What is the percentage composition of $CHCl_3$?

Solution: One gram-formula weight of the compound is $12.0 + 1.0 + 3(35.5)$ $= 119.5$ g. There are 12.0 g of carbon in every 119.5 g of the compound, and therefore the percentage of carbon in the compound (the number of grams of carbon in 100 g of the compound) is

$$\frac{12.0}{119.5} \times 100 = 10.0\% \text{ C}$$

By the same reasoning, the percentages of H and of Cl in $CHCl_3$ are

$$\frac{1.0}{119.5} \times 100 = 0.84\% \text{ H}$$

$$\frac{3 \times 35.5}{119.5} \times 100 = 89.1\% \text{ Cl}$$

By means of a calculation which is essentially the reverse of that in the preceding example, the empirical formula of a compound can be derived when its percentage composition is known. Suppose, for example, a new compound is synthesized or is isolated from some natural source. Its percentage composition can be determined experimentally in the laboratory, and from that information its empirical formula can be calculated.

Consider a compound containing only carbon and hydrogen with a percentage composition of 92.31 percent C and 7.69 percent H. These are percentages, or parts per hundred, by weight, and so 100 g (chosen to make the calculations easier) of this compound must contain 92.31 g carbon and 7.69 g hydrogen. Now the empirical formula is just an expression of the relative number of gram-atoms of the elements present in the compound, and the number of gram-atoms of carbon and hydrogen in this sample can be found by dividing these weights by the weight of one gram-atom of the elements. Thus, in 100.0 g of this compound, there are

$$\frac{92.31 \text{ g}}{12.0 \text{ g/g-atom}} = 7.69 \text{ g-atoms of carbon}$$

and

$$\frac{7.69 \text{ g}}{1.00 \text{ g/g-atom}} = 7.69 \text{ g-atoms of hydrogen}$$

The simplest ratio of the number of gram-atoms of carbon to hydrogen is then 1:1, and the empirical formula can be expressed as CH.

A slightly more complex calculation is necessary for a compound whose percentage composition is 50.04 percent C, 5.59 percent H, 44.37 percent O. Using the procedure above, for a 100-g sample we get 4.167 g-atoms C, 5.546 g-atoms H, and 2.773 g-atoms O. In order to obtain the simplest gram-atom ratio, divide the number of gram-atoms of each element by the number of gram-atoms of the least abundant element:

$$\frac{4.167}{2.773} = 1.503 \frac{\text{g-atoms C}}{\text{g-atoms O}}$$

$$\frac{5.546}{2.773} = 2.000 \frac{\text{g-atoms H}}{\text{g-atoms O}}$$

The formula is, then, $C_{1.503}H_{2.000}O_{1.000}$, which can be expressed in whole numbers by multiplying each subscript by 2. The resulting empirical formula is $C_3H_4O_2$.

The rounding off (3.006 to 3.000) for the number of carbons is justified because the percentage composition of a compound is experimentally determined and therefore subject to varying degrees of error. This sort of experimental error is usually more or less random, and if the empirical formulas of most compounds were calculated from the average of a large number of percentage-composition determinations, the numbers would come quite close to exact whole numbers.

In some compounds the atoms are bound together in discrete units called *molecules*. For compounds of this type it is important to know not only the ratio of the different atoms present but also the actual number of atoms in each molecule. This information is given by the *molecular formula*. For example, the molecular formula for carbon tetrachloride, CCl_4, tells us not only that the compound consists of carbon atoms and chlorine atoms in a ratio of 1:4, but also that each molecule of the compound is an aggregate of one carbon atom and four chlorine atoms. For carbon tetrachloride the empirical formula and the molecular formula are identical, but for many other compounds that is not the case. Both acetylene and benzene have the empirical formula CH, indicating both compounds consist of carbon and hydrogen in an atomic ratio of 1:1. But the molecular formula for acetylene is C_2H_2 and that for benzene is C_6H_6, and these are two quite different compounds, each with its own specific set of properties. The molecular formula is always the same as, or some whole-number multiple of, the empirical formula; for instance, $C_2H_2 = 2(CH)$ and $C_6H_6 = 6(CH)$. The calculation of a formula from percentage composition, as just described, can lead only to the empirical formula; further information—the molecular weight—is required for calculation of the molecular formula.

The *molecular weight* is the sum of the atomic weights of the elements appearing in the molecular formula. Thus, the molecular weight of acetylene (C_2H_2) is $2(12) + 2(1) = 26$, and the molecular weight of benzene (C_6H_6) is $6(12) + 6(1) = 78$. The *gram-molecular weight* is equal to the sum of the gram-atomic weights of the elements in the molecular formula. Hence, 26 g of acetylene is one gram-molecular weight of the compound. Since one gram-atom of an element contains Avogadro's number of atoms, one gram-molecular weight of a compound must contain Avogadro's number of molecules.

One gram-molecular weight of a compound is also referred to as a *mole*. Thus, one mole of acetylene is 26 g of acetylene and contains 6.023×10^{23} molecules. The term *mole* was at one time restricted to molecular substances, but in current usage it can refer to any substance. For example, 58.5 g is 1 g-formula wt of NaCl and is therefore also 1 mole of NaCl; 12 g is 1 g-at. wt. of carbon and is therefore also 1 mole of carbon. This broader use of the term *mole* may lead to confusion in connection with elementary substances that exist in the form of molecules. For example, "one mole of oxygen" may mean one gram-atomic weight of oxygen (Avogadro's number of O atoms, 16 g) or it may mean one gram-formula weight (Avogadro's number of O_2 molecules, 32 g). Care must be taken to distinguish between these two meanings.

The relationships between these various terms are demonstrated in the following examples.

EXAMPLE: How many molecules are there in 50.0 g of CO_2?

Solution: The molecular weight of CO_2 is $12 + 2(16) = 44$. Therefore, 1 mole of CO_2 weighs 44 g, and 50.0 g is equal to

$$\frac{50.0}{44} \text{ moles}$$

Since each mole contains 6.02×10^{23} molecules, the number of molecules in 50.0 g of CO_2 is

$$\frac{50.0 \text{ g}}{44 \text{ g/mole}} \times 6.02 \times 10^{23} \frac{\text{molecules}}{\text{mole}} = 6.84 \times 10^{23} \text{ molecules}$$

EXAMPLE: How many grams of sodium is contained in 2.5 moles of Na_2CO_3?

Solution: Each mole of Na_2CO_3 contains 2 g-atoms of Na, and each gram-atom of Na weighs approximately 23 g. Therefore,

$$2.5 \text{ moles} \times 2 \frac{\text{g-atom}}{\text{mole}} \times 23 \frac{\text{g}}{\text{g-atom}} = 115 \text{ g}$$

CHEMICAL EQUATIONS

Symbols and formulas are also used to describe chemical reactions by statements called *chemical equations.* Consider the simple reaction in which carbon and oxygen combine to form a third substance, carbon dioxide. This process is described in words as: carbon *plus* oxygen *yields* carbon dioxide. The equation for this reaction is derived as follows. Since carbon is an elementary substance, it is designated by its symbol, C. The elementary substance oxygen, however, does not exist in its normal state as a collection of individual atoms, but rather as diatomic molecules, and we must indicate this fact by writing O_2. The compound carbon dioxide is designated by its formula, CO_2. Using an arrow to replace the word "yields" in our word description, we may write the equation:

$$C + O_2 \rightarrow CO_2$$

This chemical equation is more informative than the word description, because it tells us not only what substances are involved in the reaction but also the composition of these substances. Furthermore, the equation tells us the ratio (by atoms or molecules) in which the various substances react; that is, one atom of carbon reacts with one molecule of oxygen to form one molecule of carbon dioxide.

In any chemical equation, all substances consumed in the process, called the *reactants,* are placed to the left of the arrow, and all substances produced in the process, called the *products* of the reaction, are placed to the right of the arrow. An equals sign is sometimes used in place of the arrow.

Chemical reactions are simply rearrangements of atoms (as Dalton observed, atoms are neither created nor destroyed), and this fact must be observed in writing chemical equations. The same number of atoms of each element must appear in the products as in the reactants. (Thus, in the equation for the reaction of carbon and oxygen, one carbon atom and two oxygen atoms appear on each side of the arrow.) If this condition is not met by simply writing the symbols and formulas for the reactants and products, then the statement is not truly an equation, and further steps must be taken.

As an illustration, consider the combustion of methane, in which methane and oxygen react to form carbon dioxide and water. Writing the proper formulas for the substances involved results in the statement

$$CH_4 + O_2 \rightarrow CO_2 + H_2O$$

This statement is not really an equation—an expression of equality—for, of the three elements involved, only carbon has the same number of atoms on both sides of the arrow. (An expression of this kind is sometimes referred to as a *skeletal* equation or an *unbalanced* equation.) In order to remedy this situation we must supply the proper numerical coefficients. To begin with, since four hydrogen atoms appear on the left side of our statement and only two on the right, we may assume that one molecule of methane will give rise to two molecules of water, and we write:

$$CH_4 + O_2 \rightarrow CO_2 + 2H_2O$$

Now both the carbon and hydrogen are balanced, but the oxygen is not; there are four oxygen atoms on the right-hand side but only two on the left. If we place a 2 in front of the formula for oxygen, the statement will then read

$$CH_4 + 2O_2 \rightarrow CO_2 + 2H_2O$$

and we have achieved a *balanced* equation, which states that one molecule of methane reacts with two molecules of oxygen to form one molecule of carbon dioxide and two molecules of water.

It should perhaps be noted that in obtaining a balanced equation one may not change any numerical subscripts, which indicate the composition of a substance. For example, we may not balance the hydrogen atoms in the above illustration by changing CH_4 to CH_2. Nature has decreed that the substance we call methane has a composition represented by the formula CH_4, and a substance with the formula CH_2, if it did exist, would not be methane but a different compound entirely.

The following examples further illustrate the balancing procedure.

EXAMPLE: Balance the equation

$$PCl_3 + H_2O \rightarrow HCl + H_3PO_3$$

Solution: Probably the simplest starting point is to balance the chlorine atoms:

$$PCl_3 + H_2O \rightarrow 3HCl + H_3PO_3$$

Phosphorus and chlorine are now balanced; hydrogen and oxygen are not. By giving H_2O a coefficient of 3, the balancing is completed:

$$PCl_3 + 3H_2O \rightarrow 3HCl + H_3PO_3$$

EXAMPLE: Balance the equation

$$Ba(OH)_2 + HNO_3 \rightarrow Ba(NO_3)_2 + H_2O$$

Solution: The nitrogen atoms may be balanced by placing a 2 in front of HNO_3:

$$Ba(OH)_2 + 2HNO_3 \rightarrow Ba(NO_3)_2 + H_2O$$

Giving H_2O a coefficient of 2 balances both the oxygen and hydrogen and results in a balanced equation:

$$Ba(OH)_2 + 2HNO_3 \rightarrow Ba(NO_3)_2 + 2H_2O$$

EXAMPLE: Balance the equation

$$C_6H_6 + O_2 \rightarrow CO_2 + H_2O$$

Solution: The most obvious starting point is perhaps with the carbon atoms. Since there are six carbon atoms on the left, there must be six on the right. Place a 6 before CO_2:

$$C_6H_6 + O_2 \rightarrow 6CO_2 + H_2O$$

Giving H_2O a coefficient of 3 will balance the hydrogens:

$$C_6H_6 + O_2 \rightarrow 6CO_2 + 3H_2O$$

The only element remaining to be balanced is oxygen. Since we have 15 oxygen atoms on the right and only 2 on the left, we need a coefficient of $7\frac{1}{2}$ for O_2:

$$C_6H_6 + 7\tfrac{1}{2}O_2 \rightarrow 6CO_2 + 3H_2O$$

The equation is now balanced. However, if we wish to think of an equation as representing numbers of molecules involved in the reaction, then fractional coefficients should be avoided. This can be accomplished by multiplying each coefficient by 2, resulting in the following balanced equation:

$$2C_6H_6 + 15O_2 \rightarrow 12CO_2 + 6H_2O$$

It is sometimes desirable to have a chemical equation convey additional information about the reaction. In particular, it is often important to indicate the physical state of a reactant or product, and certain symbols are used for this purpose.

The formation of a gaseous substance is indicated by placing an upward-pointing arrow immediately after the formula of the substance. For example, when calcium carbonate and hydrochloric acid react, carbon dioxide is given off as a gas. To indicate this fact, the equation may be written as

$$CaCO_3 + 2HCl \rightarrow CaCl_2 + H_2O + CO_2 \uparrow$$

In some solution reactions, a product is formed that is insoluble in the reaction medium and therefore *precipitates* in the solid state. This *precipitate* may be designated by a downward-pointing arrow following the formula or, alternatively, by underlining the formula:

$$AgNO_3 + NaCl \rightarrow NaNO_3 + AgCl \downarrow$$

or

$$\underline{AgCl}$$

The physical states of all substances appearing in an equation may be shown by writing in parentheses after the formula the letters g, l, or s, indicating gas, liquid, or solid, respectively. For example,

$$NH_3(g) + HCl(g) \rightarrow NH_4Cl(s)$$

Similarly, a reactant or product in aqueous solution may be so designated by use of the abbreviation aq.

$$Zn(s) + 2HCl(aq) \rightarrow ZnCl_2(aq) + H_2(g)$$

Whether heat is absorbed or evolved may be shown by simply writing the word "heat" as if it were a reactant or a product. The alchemical symbol for fire (Δ) is often used as a substitute for "heat."

$$2C_2H_2 + 5O_2 \rightarrow 4CO_2 + 2H_2O + heat \quad (or \; \Delta)$$

STOICHIOMETRY

As we have seen, the coefficients in a balanced chemical equation tell us the relative *numbers of molecules* (or atoms or ions) of each substance consumed or produced in the reaction. Because a mole of any substance contains the same number of molecules as a mole of any other substance, the coefficients also tell us the relative *numbers of moles* of each substance involved in the reaction. Furthermore, since a mole of a substance is equal to the formula weight of that substance in grams, chemical equations are expressions of the weight relationships between substances in a chemical reaction. This area of calculations dealing with weight relationships in matter and its reactions is called *stoichiometry,* from the Greek meaning "measurement of elements."

Chemical equations enable us to calculate the number of moles, number of molecules, and weight of the various substances taking part in chemical reactions.

Mole–Mole Relationships

Since the coefficients in a chemical equation are in the same ratio as the number of moles of substances reacting or produced in the reaction, then if the number of moles of any one reactant or product is given, one can use the equation to determine the number of moles of each of the other substances involved in the reaction.

EXAMPLE: In the combustion of benzene according to the equation

$$2C_6H_6 + 15O_2 \rightarrow 12CO_2 + 6H_2O$$

how many moles of CO_2 will be produced in the complete combustion of 3.2 moles of C_6H_6?

Solution: The balanced equation indicates that C_6H_6 and CO_2 are related by the molar ratio $2:12$. One mole of C_6H_6 gives 6 moles of CO_2. Therefore, 3.2 moles of C_6H_6 will yield

$$3.2 \times 6 = 19.2 \text{ moles of } CO_2$$

Through the use of Avogadro's number, the relationship between moles can be extended to molecules.

EXAMPLE: Using the equation for the combustion of benzene given in the previous example, how many molecules of water are produced from the reaction of 0.20 mole of C_6H_6?

Solution: Every 1 mole of C_6H_6 yields 3 moles of H_2O; therefore, 0.20 mole of C_6H_6 will react to produce

$$(0.20 \times 3) \text{ mole of } H_2O$$

Since 1 mole contains 6.02×10^{23} molecules, the number of water molecules produced is

$$0.20 \times 3 \times 6.02 \times 10^{23} = 3.6 \times 10^{23} \text{ molecules}$$

Weight–Weight Relationships

By combining the mole ratios given in the equation with the relationship between the mole and the formula weight, one can calculate the weight of substances reacting with, or produced from, a given weight of any substance involved in the reaction.

EXAMPLE 1: In the hydrolysis of phosphorus trichloride, what weight of HCl can be produced from 15.0 g of PCl_3? The equation for the reaction is

$$PCl_3 + 3H_2O \rightarrow 3HCl + H_3PO_3$$

Solution: The amount of PCl_3 reacting may be expressed in terms of moles. The formula weight of PCl_3 is approximately 137. Therefore,

$$15.0 \text{ g of } PCl_3 = \frac{15.0}{137} \text{ moles } PCl_3$$

The equation for the reaction shows that for every 1 mole of PCl_3 consumed, 3 moles of HCl are formed. Thus,

$$\text{Number of moles of HCl produced} = 3 \times \frac{15.0 \text{ g}}{137 \text{ g/mole}}$$

To convert the number of moles of HCl to the number of grams of HCl, we need only multiply by the number of grams in one mole.

$$\frac{15.0}{137} \times 3 \text{ moles} \times 36.5 \frac{\text{g}}{\text{mole}} = 12.0 \text{ g HCl}$$

EXAMPLE 2: What weight of aluminum is required to produce 3.0 g of hydrogen according to the following equation?

$$2Al + 3H_2SO_4 \rightarrow Al_2(SO_4)_3 + 3H_2$$

Solution:

$$3.0 \text{ g } H_2 = \frac{3.0}{2.0} \text{ moles } H_2$$

For each mole of H_2 produced, two-thirds of a mole (gram-atom) of Al is required. Therefore, for 3.0 g of hydrogen, the number of moles of Al required is:

$$\left(\frac{3.0}{2.0} \times \frac{2}{3}\right) \text{ moles of Al}$$

Since each mole of aluminum weighs 27.0 g, the weight of aluminum required is:

$$\left(\frac{3.0}{2.0} \times \frac{2}{3}\right) \text{ moles} \times 27.0 \frac{\text{g}}{\text{mole}} = 27 \text{ g Al}$$

Note that weights in units other than grams need not be converted into grams in order to carry out these weight–weight calculations. We usually deal with weights in grams and therefore with moles, or *gram*-formula weights. We can just as easily deal with *pound*-formula weights, *ton*-formula weights, and so on. The weights derived from chemical equations are simple *weight ratios,* and they can be used with any units of weight as long as the units are used consistently.

EXAMPLE: What weight of CO is required to produce 100 pounds of iron by the reaction described by the following equation?

$$Fe_2O_3 + 3CO \rightarrow 2Fe + 3CO_2$$

Solution: Since the atomic weight of Fe is 55.8,

$$100 \text{ lb Fe} = \frac{100}{55.8} \text{ lb-at. wt. Fe}$$

Since the molar ratio of Fe to CO in the reaction is 2:3, each pound-formula weight of Fe requires 1.5 lb-formula wt of CO. Therefore, the number of pound-formula weights of CO required to produce 100 lb of Fe is

$$\left(\frac{100}{55.8} \times 1.5 \right) \text{ lb-formula wt CO}$$

Each pound-formula weight of CO weighs approximately 28.0 lb. Therefore,

$$\left(\frac{100}{55.8} \times 1.5 \right) \text{ lb-formula wt} \times 28.0 \frac{\text{lb}}{\text{lb-formula wt}} = 75.3 \text{ lb CO}$$

CALCULATION OF YIELD

When a compound is prepared by means of a chemical reaction (either in the laboratory or on a larger, commercial scale), the amount of product obtained (the *yield*) is, of course, of major importance. From the equation for the reaction in question and the stoichiometric relationships considered above, one can calculate the yield that *ought* to be obtained on the basis of the quantity of reactants used. In practice, however, the amount of product obtained (the *actual yield*) is nearly always less than the calculated amount (the *theoretical yield*). This difference may be due to a number of reasons. Many reactions do not go to completion; that is, equilibrium is established with a large amount of the reactants still present. In some cases, side reactions may occur—reactions that are not described by the equation and that lead to unexpected products. Furthermore, a considerable fraction of the product may be lost during the process of separating it in pure form from the reaction mixture.

The "efficiency" of a reaction as a preparative method may be indicated by the *percentage yield*. This is the ratio of the actual yield to the theoretical yield, expressed as percentage.

$$\text{Percentage yield} = \frac{\text{Actual yield}}{\text{Theoretical yield}} \times 100$$

There is another matter to be considered in the practical application of stoichiometric calculations. In actually carrying out a reaction, the relative amounts of reactants used may be different from those required by the balanced equation. If more of one reactant is present than is required by the other reactant(s), then the substance in excess simply does not react. The theoretical yield must be calculated on the basis of the reactant that is present in limiting quantity.

EXAMPLE: Copper(II) phosphide (Cu_3P_2) may be prepared by reaction of phosphine with an aqueous solution of copper sulfate, according to the equation:

$$3CuSO_4 + 2PH_3 \rightarrow Cu_3P_2\downarrow + 3H_2SO_4$$

When 10.0 g of PH_3 was added to a solution containing 50.0 g of $CuSO_4$, 20.0 g of Cu_3P_2 was obtained. Calculate (a) the theoretical yield and (b) the percentage yield.

Solution Part (*a*): We must first determine which is the limiting quantity— 50.0 g of $CuSO_4$ or 10.0 g of PH_3.

$$50.0 \text{ g } CuSO_4 = \frac{50.0}{159.5} \frac{g}{g/mole} = 0.31 \text{ mole } CuSO_4$$

$$10.0 \text{ g } PH_3 = \frac{10.0}{34.0} \frac{g}{g/mole} = 0.29 \text{ mole } PH_3$$

The equation indicates that $CuSO_4$ and PH_3 react in the ratio 3:2. Thus, the number of moles of PH_3 required to react with the 0.31 mole of $CuSO_4$ is approximately 0.2 mole. Hence, the 0.29 mole of PH_3 is more than enough to react with all the $CuSO_4$ present. Phosphine is in excess, and the 50.0 g of $CuSO_4$ is the limiting quantity.

Each mole of $CuSO_4$ consumed results in one-third mole of Cu_3P_2 produced. Therefore, the number of moles of Cu_3P_2 formed is:

$$\left(\frac{50.0}{159.5} \times \frac{1}{3}\right) \text{ mole } Cu_3P_2$$

Since the formula weight of Cu_3P_2 is 252.5, the number of grams of Cu_3P_2 formed (that is, the theoretical yield) is:

$$\left(\frac{50.0}{159.5} \times \frac{1}{3}\right) \text{ mole} \times 252.5 \frac{g}{mole} = 26.4 \text{ g}$$

Solution Part (*b*): The actual yield is 20.0 g and the theoretical yield, as calculated, is 26.4 g. The percentage yield, therefore, is

$$\frac{20.0}{26.4} \times 100 = 75.8\%$$

SUGGESTED READING   Kieffer, William F. *The Mole Concept in Chemistry*. New York: Reinhold, 1963. Chapters 3 and 4.

PROBLEMS   1. What is the weight of aluminum in 1 kg of each of the following compounds?
   (a) $AlCl_3$
   (b) Al_2O_3
   (c) $KAl(SO_4)_2 \cdot 12H_2O$

2. Determine the percentage composition of the following compounds:
   (a) LiBr   (b) CH_4
   (c) FeF_3 (d) K_2CO_3
   (e) $CoCl_2$

3. Calculate the empirical formulas of compounds with the following percent compositions:
   (a) 46.68% N, 53.32% O
   (b) 41.81% Na, 58.19% O
   (c) 72.36% Fe, 27.64% O
   (d) 85.62% C, 14.38% H
   (e) 75.92% C, 6.37% H, 17.70% N
   (f) 58.53% C, 4.09% H, 11.38% N, 25.99% O

4. A sample of chloroform ($CHCl_3$) weighing 150 g contains:
   (a) How many moles of $CHCl_3$?
   (b) How many chloroform molecules?
   (c) What weight of carbon?
   (d) How many chlorine atoms?
   (e) How many gram-atoms of hydrogen?

5. A certain compound has an empirical formula of CH_2O and a molecular weight of approximately 90. How many moles and how many molecules are there in 50.0 g of the compound? How many carbon atoms are there in 50.0 g of the compound?

6. Determine the molecular formulas of the following compounds:
   (a) 78.14% B, 21.86% H, molecular weight = 28
   (b) 40.00% C, 6.72% H, 53.29% O, molecular weight = 180

7. Balance the following equations:
   (a) $SO_2 + O_2 \rightarrow SO_3$
   (b) $PCl_5 \rightarrow PCl_3 + Cl_2$
   (c) $Na + H_2O \rightarrow NaOH + H_2$
   (d) $CaH_2 + H_2O \rightarrow Ca(OH)_2 + H_2$
   (e) $Fe_3O_4 + H_2 \rightarrow FeO + H_2O$
   (f) $Al + H_2SO_4 \rightarrow Al_2(SO_4)_3 + H_2$
   (g) $HClO \rightarrow HClO_3 + HCl$
   (h) $SbCl_3 + H_2S \rightarrow Sb_2S_3 + HCl$
   (i) $C_8H_{18} + O_2 \rightarrow CO_2 + H_2O$
   (j) $MgNH_4PO_4 \rightarrow Mg_2P_2O_7 + NH_3 + H_2O$
   (k) $NH_3 + NaClO \rightarrow NaCl + H_2O + N_2$
   (l) $H_3PO_4 + NaOH \rightarrow Na_2HPO_4 + H_2O$
   (m) $AgNO_3 + Cl_2 \rightarrow AgCl + N_2O_5 + O_2$
   (n) $Ca_3(PO_4)_2 + SiO_2 + C \rightarrow CaSiO_3 + P + CO$
   (o) $Ca_3P_2 + H_2O \rightarrow PH_3 + Ca(OH)_2$
   (p) $Mg_3As_2 + HCl \rightarrow AsH_3 + MgCl_2$
   (q) $CO_2 + K \rightarrow K_2CO_3 + C$
   (r) $Na_2CO_3 + C + N_2 \rightarrow NaCN + CO$
   (s) $KNO_2 + K \rightarrow K_2O + N_2$
   (t) $Ag_2S + KCN \rightarrow KAg(CN)_2 + K_2S$
   (u) $Ca_3N_2 + H_2O \rightarrow Ca(OH)_2 + NH_3$

(v) $ZnS + O_2 \rightarrow ZnO + SO_2$

(w) $H_2C_2O_4 \rightarrow H_2O + CO_2 + CO$

(x) $Pb + Na + C_2H_5Cl \rightarrow Pb(C_2H_5)_4 + NaCl$

(y) $V_2O_5 + HCl \rightarrow VOCl_3 + H_2O$

(z) $(NH_4)_2Cr_2O_7 \rightarrow Cr_2O_3 + N_2 + H_2O$

8. In the complete combustion of 10.0 g of acetylene, according to the equation

$$2C_2H_2 + 5O_2 \rightarrow 4CO_2 + 2H_2O$$

(a) How many moles of CO_2 are produced?

(b) What weight of H_2O is produced?

(c) How many molecules of O_2 are consumed?

9. The fermentation of sucrose (cane sugar) to produce ethyl alcohol and carbon dioxide can be represented by the equation

$$C_{12}H_{22}O_{11} + H_2O \rightarrow 4C_2H_5OH + 4CO_2$$

(a) What is the theoretical weight of ethyl alcohol produced from the fermentation of exactly one pound of sucrose?

(b) What weight of CO_2 gas will be liberated during the fermentation of one pound of sucrose?

10. What weight of $BaCl_2 \cdot 2H_2O$ can be prepared from 50.0 g $BaCO_3$ by the following reaction?

$$BaCO_3 + 2HCl + H_2O \rightarrow BaCl_2 \cdot 2H_2O + CO_2$$

11. According to the equation

$$2NH_3 + 3Mg \rightarrow Mg_3N_2 + 3H_2$$

what weight of magnesium metal is required to prepare 100 g of magnesium nitride?

12. The following reaction, carried out at high temperature, is an important step in the commercial preparation of nitric acid:

$$4NH_3(g) + 5O_2(g) \rightarrow 4NO(g) + 6H_2O(g)$$

(a) What weight of NO can be produced theoretically from 50 kg of NH_3?

(b) What weight of O_2 will be consumed in reacting with 10 moles of NH_3?

(c) What weight of water is produced from the reaction of 10 moles of NH_3?

(d) How many moles of O_2 will react with 1.0 kg of NH_3?

13. On the basis of the equation

$$2Al(s) + 3H_2SO_4(aq) \rightarrow Al_2(SO_4)_3(aq) + 3H_2(g)$$

(a) What weight of Al is required to produce 1.0 lb of H_2?

(b) What weight of pure sulfuric acid will be required to react completely with 15.0 g of Al?

(c) What weight of $Al_2(SO_4)_3$ is produced from 0.50 g-atom of Al?

14. What weight of $FeSO_4$ will be obtained from 50 kg of FeS_2 according to the following equation, if the process gives a 78 percent yield?

$$2FeS_2 + 7O_2 + 2H_2O \rightarrow 2FeSO_4 + 2H_2SO_4$$

15. What weight of acetylene can be obtained from 0.5 lb of calcium carbide, if the CaC_2 is only 92 percent pure?

$$CaC_2 + 2H_2O \rightarrow C_2H_2 + Ca(OH)_2$$

16. When 25.0 g of HI is treated with an excess of H_2SO_4, 20.5 g of I_2 is obtained. The reaction is

$$8HI + H_2SO_4 \rightarrow H_2S + 4H_2O + 4I_2$$

Calculate the theoretical yield and the percentage yield.

17. If 20.0 g of PCl_5 react with an excess of water to give an 89 percent yield of H_3PO_4, what weight of H_3PO_4 is obtained? The reaction is

$$PCl_5 + 4H_2O \rightarrow H_3PO_4 + 5HCl$$

18. If 100 g of Ag and 100 g of S are heated together, what is the theoretical yield of Ag_2S, according to the equation

$$2Ag + S \rightarrow Ag_2S$$

19. What is the theoretical yield of Ag_2CrO_4 obtained from mixing 0.050 mole of $AgNO_3$ with 0.050 mole of K_2CrO_4? The equation is

$$2AgNO_3 + K_2CrO_4 \rightarrow Ag_2CrO_4 + 2KNO_3$$

20. If 42 g of $COCl_2$ was obtained from 25 g of CO and 35 g of Cl_2, what was the percentage yield?

$$CO + Cl_2 \rightarrow COCl_2$$

4 Elements II: Classification of the Elements

If each element had its own specific and unique properties, it would be almost impossible to learn enough about the behavior of matter to develop a predictive science. Fortunately, this is not the case, and chemistry has become a major science within the past century. Certain similarities and regularities do exist among elemental substances, and they serve as the basis for a detailed, very useful classification of the elements.

Probably the first notable step in the evolution of this classification was provided by Johann Döbereiner in 1829. He pointed out that certain elements can be listed in groups of three with similar chemical properties. He further observed that when each group of three is listed in order of increasing atomic weight, the atomic weight of the middle member is approximately the arithmetic mean of the atomic weights of the other two. Some of these groups of three, which became known as *Döbereiner's triads,* are listed in Table 4-1. Perhaps the major significance of Döbereiner's triads was that they established the idea of groups of elements having similar properties that are in some way a function of atomic weight.

Before any significant extension of Döbereiner's classification could be made,

TABLE 4-1 Döbereiner's Triads with Approximate Atomic Weights			
Cl 35	Ca 40	Li 7	S 32
Br 80	Sr 88	Na 23	Se 79
I 127	Ba 137	K 39	Te 128

additional developments had to occur: more elements had to be discovered so that relationships would be apparent, and atomic weights had to be known with greater accuracy. By the 1860s these conditions had been met.

In 1864 John Newlands noted that when the known elements were listed according to increasing atomic weights, similar properties recurred in every seventh element. Seeing the similarity between this phenomenon and the musical scale, he named this relationship the *Law of Octaves*. The relationship is illustrated in Figure 4-1 for the first 17 elements of Newlands' table. Elements in the same vertical column have similar properties and are seven elements apart. Although his work was given little credence by the scientific community of his day, Newlands deserves credit for recognizing the periodic recurrence of properties based on increasing atomic weight—an idea that formed the basis for later classification of the elements.

FIGURE 4-1

Illustration of Newlands' Law of Octaves

H	Li	Be	B	C	N	O
F	Na	Mg	Al	Si	P	S
Cl	K	Ca				

In 1869 Dmitrii Mendeleev in Russia and Lothar Meyer in Germany, working independently of each other, published classification tables that were based on the same principles and were very similar. Because of the greater detail in Mendeleev's work, his bold predictions for elements not yet known at that time, and his publication of an improved table in 1871, history accords him—rather than Meyer— the major credit for this important breakthrough.

Mendeleev's classification was based on his *Periodic Law: The chemical properties of the elements are not arbitrary but vary in a systematic way according to atomic weight.* When the elements are listed according to increasing atomic weights, similar properties recur periodically, so that the elements fall logically into groups, all the elements within any group having similar properties.

A periodic table—not identical to Mendeleev's, but based on his 1871 table—is shown in Figure 4-2. The shaded blocks indicate elements that had not yet been discovered in 1871. Mendeleev predicted the existence of some of the missing elements by leaving blank spaces in his chart. The vertical column headed Group VIIIB did not appear at all, since none of the elements in that group were known until more than two decades after Mendeleev's publication. The elements cerium, terbium, erbium, thorium, and uranium also created a problem that Mendeleev was unable to solve.

The power of Mendeleev's classification has been demonstrated dramatically since 1871 by the accuracy of the predictions it enabled him to make. Scandium, gallium, and germanium are three of the elements that had not yet been discovered. Mendeleev's strict adherence to the law of periodicity that he had formulated enabled him not only to predict the existence of these elements, but even to predict a number of their physical and chemical properties. All three of these elements were discovered during the next fifteen years, and their properties were found to be

	I		II		III		IV		V		VI		VII		VIII	
	A	B	A	B	A	B	A	B	A	B	A	B	A	B	A	B
H 1.0																He 4.0
Li 6.9			Be 9.0			B 10.8		C 12.0		N 14.0		O 16.0		F 19.0		Ne 20.2
Na 23.0			Mg 24.3			Al 27.0		Si 28.1		P 31.0		S 32.1		Cl 35.5		Ar 39.9
K 39.1			Ca 40.1		Sc 45.0		Ti 47.9		V 50.9		Cr 52.0		Mn 54.9		Fe 55.8 Co 58.9 Ni 58.7	
	Cu 63.5		Zn 65.4		Ga 69.7		Ge 72.6		As 74.9		Se 79.0		Br 79.9			Kr 83.8
Rb 85.5			Sr 87.6		Y 88.9		Zr 91.2		Nb 92.9		Mo 95.9		Tc 98.9		Ru 101.1 Rh 102.9 Pd 106.4	
	Ag 107.9		Cd 112.4		In 114.8		Sn 118.7		Sb 121.8		Te 127.6		I 126.9			Xe 131.3
Cs 132.9			Ba 137.3		La* 138.9		Hf 178.5		Ta 180.9		W 183.9		Re 186.2		Os 190.2 Ir 192.9 Pt 195.1	
	Au 197.0		Hg 200.6		Tl 204.4		Pb 207.2		Bi 209.0		Po (210)		At (210)			Rn (222)
Fr (223)			Ra 226.0		Ac** (227)											

*

Ce 140.1	Pr 142.9	Nd 144.5	Pm (147)	Sm 150.4	Eu 152.0	Gd 157.3	Tb 158.9	Dy 162.5	Ho 164.9	Er 167.3	Tm 168.9	Yb 173.0	Lu 175.0

**

Th 232.0	Pa 231.0	U 238.0	Np 237.0	Pu (242)	Am (243)	Cm (247)	Bk (247)	Cf (251)	Es (254)	Fm (253)	Md (256)	No (254)	Lr (257)

FIGURE 4-2

Periodic Table Based on Mendeleev's Table of 1871. Colored squares are elements not known in 1871. (Atomic weights are currently accepted values rounded to nearest 0.1 unit. Numbers in parentheses are mass numbers of most stable isotope.)

remarkably close to those predicted by Mendeleev. Table 4-2 compares some properties predicted for germanium with those currently accepted as correct.

Another illustration of the great value of the periodic table—which could not be foreseen by Mendeleev—may be found in the discovery of the inert gases (helium, neon, argon, krypton, xenon, and radon). When argon was discovered in 1894 by Lord Raleigh and Sir William Ramsay, it was apparent that it did not belong in any of the vacant positions in the periodic table. This fact suggested the existence of a whole additional group of elements and spurred investigators to search for them. Within a half-dozen years the entire group had been discovered.

TABLE 4-2
Comparison of Properties
Predicted by Mendeleev (1871)
with Those Currently
Accepted for Germanium

Property	Predicted	Current
Color	dark gray	grayish-white
Atomic Weight	72	72.59
Density (g/cm^3)	5.5	5.35
Atomic volume (cm^3/g-atom)	13	13.5
Specific heat (cal/g/$^\circ$C)	0.073	0.074
Formula of oxide	XO_2	GeO_2
Density of oxide (g/cm^3)	4.7	4.703
Formula of chloride	XCl_4	$GeCl_4$
Boiling point of chloride	below 100°C	86°C
Density of chloride (g/cm^3)	1.9	1.844

Naturally, the periodic table has been modified since Mendeleev first formulated it. Additional elements have been added as they were discovered, the problem of the position of the actinides and lanthanides has been solved, and some of Mendeleev's misinterpretations have been corrected. The most significant alteration involves the very basis upon which the periodic law was founded. Mendeleev and his contemporaries believed that periodicity was a function of atomic weight. We now realize that periodicity is based on *atomic number,* not atomic weight. Mendeleev could not possibly have reached this conclusion, since the proton was not discovered until 1902, and atomic numbers were not determined until 1913 (by Henry Moseley). Mendeleev had found it necessary in several instances to deviate from strict adherence to a listing according to increasing atomic weight in order to make certain elements fall into the proper groups; for instance, the inversion of iodine and tellurium. These deviations he explained as being due to inaccurate atomic weights. In this regard he was incorrect; the determination of atomic numbers explained the inversions. In light of this later development, then, the *Periodic Law* must be restated: *The properties of the elements are periodic functions of their atomic numbers.*

Throughout the years the periodic table has taken a variety of forms. There has been the short form (following closely Mendeleev's 1871 table), the long form (in both vertical and horizontal versions), spiral tables, three-dimensional helical tables, and a number of others. The periodic table most commonly used today is a version of the long form, shown in Figure 4-3 and inside the back cover of this book. (Numbers above the symbols are atomic numbers and numbers below the symbols are atomic weights.)

THE MODERN PERIODIC TABLE

The version of the periodic table used throughout the remainder of this book (Figure 4-3) consists of seven horizontal rows called *periods* and sixteen vertical

IA	IIA	IIIB	IVB	VB	VIB	VIIB	VIII	VIII	VIII	IB	IIB	IIIA	IVA	VA	VIA	VIIA	0
1 H 1.0079																1 H 1.0079	2 He 4.00260
3 Li 6.941	4 Be 9.01218											5 B 10.81	6 C 12.011	7 N 14.0067	8 O 15.9994	9 F 18.99840	10 Ne 20.179
11 Na 22.98977	12 Mg 24.305											13 Al 26.98154	14 Si 28.086	15 P 30.97376	16 S 32.06	17 Cl 35.453	18 Ar 39.948
19 K 39.098	20 Ca 40.08	21 Sc 44.9559	22 Ti 47.90	23 V 50.9414	24 Cr 51.996	25 Mn 54.9380	26 Fe 55.847	27 Co 58.9332	28 Ni 58.70	29 Cu 63.546	30 Zn 65.38	31 Ga 69.72	32 Ge 72.59	33 As 74.9216	34 Se 78.96	35 Br 79.904	36 Kr 83.80
37 Rb 85.4678	38 Sr 87.62	39 Y 88.9059	40 Zr 91.22	41 Nb 92.9064	42 Mo 95.94	43 Tc 98.9062	44 Ru 101.07	45 Rh 102.9055	46 Pd 106.4	47 Ag 107.868	48 Cd 112.40	49 In 114.82	50 Sn 118.69	51 Sb 121.75	52 Te 127.60	53 I 126.9045	54 Xe 131.30
55 Cs 132.9054	56 Ba 137.34	57 La* 138.9055	72 Hf 178.49	73 Ta 180.9479	74 W 183.85	75 Re 186.207	76 Os 190.2	77 Ir 192.22	78 Pt 195.09	79 Au 196.9665	80 Hg 200.59	81 Tl 204.37	82 Pb 207.2	83 Bi 208.9804	84 Po (210)	85 At (210)	86 Rn (222)
87 Fr (223)	88 Ra 226.0254	89 Ac** (227)	104 (260)	105 (260)													

*Lanthanum Series

58 Ce 140.12	59 Pr 140.9077	60 Nd 144.24	61 Pm (147)	62 Sm 150.4	63 Eu 151.96	64 Gd 157.25	65 Tb 158.9254	66 Dy 162.50	67 Ho 164.9304	68 Er 167.26	69 Tm 168.9342	70 Yb 173.04	71 Lu 174.97

**Actinium Series

90 Th 232.0381	91 Pa 231.0359	92 U 238.029	93 Np 237.0482	94 Pu (244)	95 Am (243)	96 Cm (247)	97 Bk (247)	98 Cf (251)	99 Es (254)	100 Fm (257)	101 Md (258)	102 No (255)	103 Lr (256)

FIGURE 4-3

Modern Periodic Table (Long Form). (Numbers in parentheses are mass numbers of most stable or most common isotope.)

groups or *families.* The elements within a group are similar in chemical properties. It should be noted that hydrogen does not show a close resemblance to any other element and might be placed in a group by itself. In our chart it is located at the top of both Group IA and Group VIIA, since in some of its properties it is similar to the elements of both these groups. With the exception of Group VIII, each group contains only one element within a period. The Group VIII triads, as they are sometimes called, were combined in one group by Mendeleev because they were thought to be very similar in properties, and that arrangement has been continued in modern periodic tables.

The *periods* are generally referred to by number, beginning at the top; that is, period 1 consists of H and He, period 2 consists of Li through Ne, and so forth. The *groups* may be referred to by number and letter; for instance, Group IIA, Group VIB. It is important to note that the A and B designations in Figure 4-3 will be used throughout this book. This is largely a matter of choice, however, and this usage is not uniform in all textbooks.

Some groups are also designated by names: Group IA elements (excluding H) are called the *alkali metals;* Group IIA elements are called the *alkaline earth metals* (historically this name was applied only to Ca, Sr, and Ba, but in current usage it applies to all the members of the group); the elements of Group VIIA (excluding H) are called *halogens;* and those of Group 0 are called *inert gases* or *noble gases* (names meant to indicate their lack of reactivity toward other elements). Groups may also be referred to in terms of the first element in the column: the nitrogen group, the oxygen group, and so on.

The elements of all the A groups are often called the *representative elements,* because their properties adhere closely to the regular progression on which the periodic table is based. The B-group elements and Group VIII, which interrupt the representative elements in periods 4, 5, and 6, are called *transition elements.* These elements show considerably more deviation from the regular progression than do the representative elements.

The elements with atomic numbers 58 through 71 actually belong in the sixth period between lanthanum (atomic number 57) and hafnium (atomic number 72). These 14 elements, all of which have very similar properties, are called the *lanthanum series* or simply *lanthanides.* They are placed in a separate row, rather than being included in the main body of the table, simply to avoid making the table inconveniently wide. The elements with atomic numbers 90 through 103 are called the *actinium series* or *actinides,* and they occupy the same position in the seventh period as the lanthanides do in the sixth. Taken together, the lanthanides and actinides are sometimes called the *inner transition elements,* since they interrupt transition series.

It is convenient for some purposes to classify elements as *metals* and *non-metals.* The division between the two is indicated roughly by the stepped heavy line from the left of B to the right of Po in our table. The elements to the left of that line are called metals; those to the right are called nonmetals. As is described further on in this chapter, the metallic character (*metallicity*) of elements is a relative property, and the division between metals and nonmetals is not sharply defined. The elements bordering the stepped line are sometimes called *metalloids* because they show metallic behavior in some of their properties and nonmetallic in others.

PERIODIC RELATIONSHIPS

It was mentioned above that elements within a given group have very similar chemical properties, and that this is particularly true of the representative elements. This means that members of the same group behave alike in the way they combine with other elements to form compounds, and that the compounds behave similarly as well. For example, given the fact that calcium combines with oxygen in an atomic ratio of 1:1 to form the compound calcium oxide (CaO), one can correctly predict that strontium will also form a compound with oxygen in which the atomic ratio is 1:1 (SrO). Furthermore, calcium oxide and strontium oxide can be expected to have similar chemical properties. Likewise, knowing that carbon and chlorine form a compound in which one carbon atom is combined with four chlorine atoms (CCl_4) enables one to predict that silicon and the other Group IVA elements also form compounds with chlorine in the atomic ratio of 1:4 ($SiCl_4$, $GeCl_4$, and so on.) In addition to these group similarities there are a number of property trends that can be related directly to the periodic table.

Atomic Radius

The exact size of an atom is very difficult to determine. Recall that the atom's radius is really the distance the electron cloud extends from the nucleus, and that the density of the cloud diminishes gradually as it gets farther from the nucleus, so that the boundary of the cloud is not clearly defined. Furthermore, because atoms are so small, it is not possible to measure the radius of a single atom directly. Rather, we must work with matter in bulk—large collections of atoms—where the size of an atom is influenced by electrical interactions with neighboring atoms. Thus, the size of an atom in a metallic crystal (Cu, Au) or in a molecule (Cl_2, N_2) is

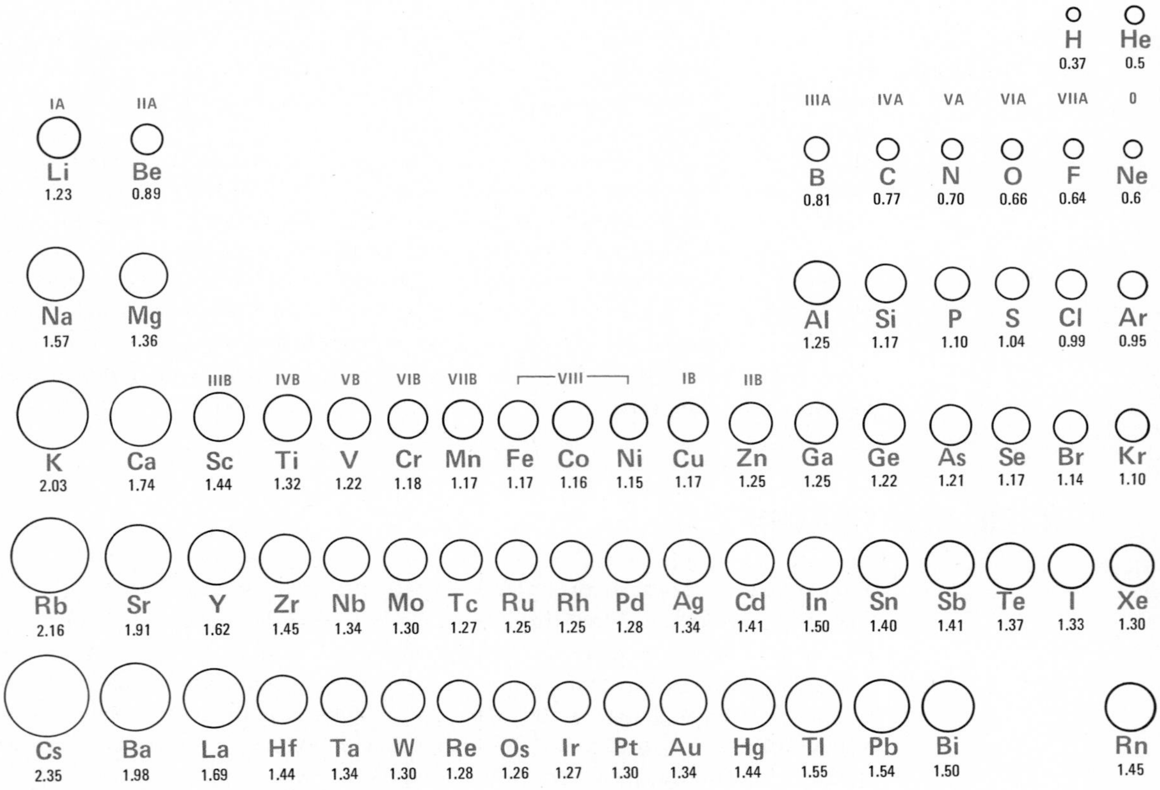

FIGURE 4-4

Atomic Radii (Angstrom Units)

not the same as it would be if the atom were isolated. In a crystal or molecule the atomic radius must be considered to be one-half the distance between the centers of two adjacent atoms.

Nevertheless, the atomic radii of most elements have been measured, and, although the reported values vary with the method of measurement and should be considered only approximate, these measurements are useful as a relative comparison of sizes. Figure 4-4 shows the relative sizes of atoms for 70 elements, arranged as in the periodic table. Radii are given in angstrom units.

Certain periodic trends in the atomic radius are obvious. *Within a group the atomic radius increases with increasing atomic number;* that is, it increases from top to bottom in the periodic table. *Within a period the atomic radius tends to decrease with increasing atomic number;* that is, in general it decreases from left to right in the periodic table. It can be seen in Figure 4-4 that the representative elements follow this generalization quite closely, whereas the transition elements show less regularity.

Ionization Energy   When atoms combine to form compounds they sometimes lose one or more electrons, becoming positively charged *ions.* This process of electron removal requires the input of energy (it is endothermic); that is, work must be done in removing the negative electron from the attractive influence of the positive nucleus. The amount of energy required for electron removal varies with the particular element, but it also depends on the physical state of the element, because neighboring atoms have a greater influence in the solid state, for example, than in the liquid or gaseous states. Therefore, in order that the energies required to ionize the various elements can be compared, they are measured in reference to the isolated gaseous atoms.

The *ionization energy* of an element is defined as *the energy required to remove an electron from an atom in the gaseous state to produce an ion, also in the gaseous state.* The process may be represented as follows, using sodium as an example:

$$Na(g) \rightarrow Na^+(g) + e^-$$

The ionization energy (also called the ionization potential) may be expressed in a number of different units, but the most convenient is calories (or kilocalories) per gram-atom.

The periodicity of ionization energies can be seen in Figure 4-5, in which the first 36 elements and their approximate ionization energies are listed. The trends revealed in these elements continue throughout the entire periodic table. *Within a group the ionization energy decreases from top to bottom.* In general, *ionization energy increases within a period from left to right.* A notable exception to the latter statement is the consistently lower ionization energies of the elements of Groups IIIA and VIA.

In the foregoing discussion we have dealt only with the energy required to remove *one* electron from a neutral atom; this is known as the *first* ionization

H
314

He
567

Li   Be
124   215

B   C   N   O   F   Ne
191   260   335   314   402   497

Na   Mg
119   176

Al   Si   P   S   Cl   Ar
138   188   254   239   300   363

K   Ca   Sc   Ti   V   Cr   Mn   Fe   Co   Ni   Cu   Zn   Ga   Ge   As   Se   Br   Kr
100   141   151   158   155   156   171   182   181   176   178   216   138   183   226   225   273   323

FIGURE 4-5

First Ionization Energies of the First 36 Elements (kcal/g-atom)

energy. It is also possible to determine the energy necessary to remove a second electron, a third electron, and so on. These amounts of energy are referred to as the *second ionization energy,* the *third ionization energy,* and so forth. Again using sodium as an illustration, the second ionization energy is the energy that must be supplied to accomplish the process:

$$Na^+(g) \rightarrow Na^{2+}(g) + e^-$$

The more electrons removed from an atom (that is, the greater the positive charge on the ion), the more energy is required to remove yet another electron. In other words, in every case the *second* ionization energy is greater than the *first,* and the *third* is greater than the *second.* The relationship is illustrated in Table 4-3 with the first three ionization energies of the second-period elements. It will be noted that, with two exceptions, the second and third ionization energies follow the same general trend within the period as the first ionization energy; they increase from left to right. The exceptions lie in the Group IA element Li, which has the *highest* second ionization energy, and the Group IIA element Be, which has the *highest* third ionization energy.

TABLE 4-3
Ionization Energies
of the Second-Period Elements

	Ionization Energies (kcal/g-atom)		
Element	*1st*	*2nd*	*3rd*
Li	124	1744	2823
Be	215	420	3548
B	191	580	875
C	260	562	1104
N	335	683	1094
O	314	811	1267
F	402	807	1445
Ne	497	947	1500

Electron Affinity Just as atoms may *lose* electrons to form *positive* ions during a chemical reaction, others may *gain* electrons to form *negative* ions. With few exceptions, when an atom gains an electron, energy is given off (the process is exothermic). Following the same conventions used in describing ionization potential, *electron affinity* is defined as *the amount of energy evolved when a gaseous atom gains an electron to become a gaseous negative ion.* The process may be illustrated with chlorine:

$$Cl(g) + e^- \rightarrow Cl^-(g)$$

Electron affinity may be expressed conveniently in terms of kilocalories per gram-atom.

Values of electron affinity are difficult to determine. Since relatively few elements actually form stable negative ions, accurate values of electron affinities are available for only a few elements. On the basis of available information, however, electron affinities appear in general to *increase from left to right within a period.* This parallel relationship between ionization energy and electron affinity is to be expected. The more difficult it is to remove an electron from a given atom (the higher the ionization energy), the stronger is the attraction of that atom for electrons, and therefore the greater is the amount of energy given off when the atom gains an electron (the higher the electron affinity).

Electronegativity   The electron-attracting tendencies of the elements are more generally described in terms of the property called *electronegativity.* Electronegativity, which is related to both ionization energy and electron affinity, is defined simply as *an element's attraction for electrons within a molecule* (see p. 104).

While it is not possible to assign absolute values of electronegativities to the various elements, a scale of *relative* values can be established on the basis of an arbitrary value assigned to one element. Several different electrogenativity scales have been proposed, but the one most commonly used was created by Linus Pauling. This scale, which is based on experimentally determined *bond energies*

H													C	N	O	F
2.1																
Li	Be	B											C	N	O	F
1.0	1.5	2.0											2.5	3.0	3.5	4.0
Na	Mg	Al											Si	P	S	Cl
0.9	1.2	1.5											1.8	2.1	2.5	3.0
K	Ca	Sc	Ti	V	Cr	Mn	Fe	Co	Ni	Cu	Zn	Ga	Ge	As	Se	Br
0.8	1.0	1.3	1.5	1.6	1.6	1.5	1.8	1.8	1.8	1.9	1.6	1.6	1.8	2.0	2.4	2.8
Rb	Sr	Y	Zr	Nb	Mo	Tc	Ru	Rh	Pd	Ag	Cd	In	Sn	Sb	Te	I
0.8	1.0	1.2	1.4	1.6	1.8	1.9	2.2	2.2	2.2	1.9	1.7	1.7	1.8	1.9	2.1	2.5
Cs	Ba	La–Lu	Hf	Ta	W	Re	Os	Ir	Pt	Au	Hg	Tl	Pb	Bi	Po	At
0.7	0.9	1.1–1.2	1.3	1.5	1.7	1.9	2.2	2.2	2.2	2.4	1.9	1.8	1.8	1.9	2.0	2.2
Fr	Ra	Ac	Th	Pa	U	Np–Lr										
0.7	0.9	1.1	1.3	1.5	1.7	1.3										

FIGURE 4-6

Electronegativities of the Elements. Reprinted from Linus Pauling, *The Nature of the Chemical Bond.* Third Edition. © 1960 by Cornell University. Used by permission of Cornell University Press.

(Chapter 7), assigns the value of 4.0 to the most electronegative element, fluorine. The other elements are given values that describe their electronegativities relative to that of fluorine.

Figure 4-6 is a listing of Pauling electronegativity values. The periodicity of electronegativities is readily apparent. In general, *electronegativity increases from left to right within a period and from bottom to top within a group.* There are exceptions to this generalization, of course, particularly among the transition elements.

Metallicity and Nonmetallicity

We have already referred to the division of the elements into metals and nonmetals. Historically, the distinction between the two categories was based largely on physical properties. (Metals exhibit the properties of luster, malleability, ductility, conductivity, and so forth, while nonmetals do not.) In modern usage, the definitions are based more on chemical properties: Metals tend to form positive ions; nonmetals tend to form negative ions. From this point of view, *metallicity* and *nonmetallicity* are entirely relative and are related to ionization energy, electron affinity, and electronegativity. The lower an element's ionization energy, electron affinity, and electronegativity, then the more metallic that element is. Conversely, nonmetallicity goes with high values of these three properties. In other words, *metallicity increases from top to bottom within a group and from right to left within a period,* while *nonmetallicity increases in the opposite directions.* Figure 4-7 summarizes the progression of some of these periodic properties.

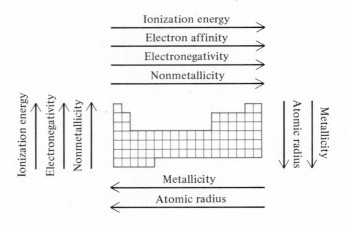

FIGURE 4-7

Some Periodic Trends, Increasing in the Direction of the Arrows

Ionic Radius

The sizes of the stable ions formed when atoms lose or gain electrons also show periodic regularities. Positive ions, which are considerably smaller than the neutral atoms from which they are formed, show an increase in radius as a group is descended but a decrease in radius within a period from left to right. Negative ions are larger than their parent neutral atoms, but show the same trends within group and period as positive ions do. In Table 4-4, the elements of groups IA, IIA, VIA, and VIIA will serve to illustrate these size relationships.

TABLE 4-4
Atomic and Ionic Radii of Some Representative Elements (Angstrom Units)

IA		IIA		VIA		VIIA	
Li	Li^+	Be	Be^{2+}	O	O^{2-}	F	F^-
1.23	0.74	0.89	0.3	0.66	1.40	0.64	1.33
Na	Na^+	Mg	Mg^{2+}	S	S^{2-}	Cl	Cl^-
1.57	1.02	1.36	0.72	1.04	1.84	0.99	1.81
K	K^+	Ca	Ca^{2+}	Se	Se^{2-}	Br	Br^-
2.03	1.38	1.74	1.00	1.17	1.98	1.14	1.96
Rb	Rb^+	Sr	Sr^{2+}	Te	Te^{2-}	I	I^-
2.16	1.49	1.91	1.16	1.27	2.21	1.33	2.20
Cs	Cs^+	Ba	Ba^{2+}				
2.35	1.70	1.98	1.36				

SUGGESTED READINGS

Partington, J. R. *A History of Chemistry.* London: Macmillan, 1964. 4 vols., Vol. 4., Chapter 26.

Mazurs, E. G. *Types of Graphic Representation of the Periodic System of Chemical Elements.* LaGrange, Illinois: E. Mazurs, 1957.

Ihde, A. J. *The Development of Modern Chemistry.* New York: Harper & Row, 1964. Chapter 9.

Kauffman, G. B. "American Forerunners of the Periodic Law." *Journal of Chemical Education,* Vol. 46, No. 3 (March 1969), pp. 128–135.

PROBLEMS

1. Indicate the meaning of each of the following terms as it applies to the periodic table in Figure 4-3.
   (a) The fourth period
   (b) The oxygen group
   (c) Group VIIB
   (d) The iron triad
   (e) The inert gases
   (f) The inner transition elements
   (g) Metals versus nonmetals
   (h) Representative elements
   (i) The transition elements

2. Describe the periodic trends of
   (a) atomic radii
   (b) first ionization energies
   (c) electronegativities
   (d) metallic properties

3. If in the future an additional *representative* element is synthesized, what is the lowest atomic number it can have?

4. Show how the following groups illustrate Döbereiner's triads:
   (a) K, Rb, Cs   (b) P, As, Sb
   (c) Ar, Kr, Xe

5. The space in the periodic chart for atomic number 87 was not filled in until 1939. In that year the very unstable element francium was discovered among the natural decay products of actinium. Using the following information, predict the density, melting point, and boiling point of Fr.

Element	Density	mp (°C)	bp (°C)
Li	0.53	179	1331
Na	0.97	98	892
K	0.86	64	766
Rb	1.53	39	701
Cs	1.90	28	690

5 Elements III: The Electronic Structure of Atoms

The realization that many properties of the elements are a periodic function of atomic number provided considerable impetus for the development of new models of the electronic structure of atoms. While Rutherford's model remains the basis for our contemporary belief that the atom consists of a nucleus and one or more extranuclear electrons, it leaves unanswered many questions about the electronic structure of the atom. Are the electrons fixed in space about the nucleus? If they are stationary, what prevents them from being attracted into the positively charged nucleus? Or are they moving about the nucleus like a miniature solar system?

This last suggestion seemed most likely in the early part of this century, because, it was reasoned, the rotation of the electron around the nucleus could provide sufficient energy to keep the electron away from the nucleus. There were, however, two serious objections to this proposal. According to the laws of physics, such a system of moving charges should continuously radiate light of all frequencies. Moreover, this radiation would be accompanied by a loss in energy of the charged bodies, and the electron should therefore spiral inward and eventually collide with the nucleus. Now, under certain conditions atoms do emit light, but, as we shall see later, this light contains only certain frequencies—by no means *all* the frequencies. Moreover, at room temperature most elements emit no light whatsoever and are quite stable.

The facts, therefore, could not be reconciled with the solar system model, and scientists at the beginning of the twentieth century were in a quandary. It remained to Niels Bohr, a Danish physicist, to provide a model that is in at least partial agreement with experiment.

THE NATURE OF LIGHT AND EMISSION SPECTRA

Many ideas were vital to the construction of his model, but Bohr probably drew most heavily upon the work of another physicist, Max Planck, and the fact that elements emit light of only certain frequencies. Up to the middle of the seventeenth century, light was believed to be a stream of particles that entered the eye and caused the sensation of "seeing." In the eighteenth and nineteenth centuries a number of experiments pointed to a wave nature of light. For example, if a single light source, say a light bulb, is directed at an opaque partition containing two parallel slits, the light emerging from the two slits will form a band pattern that can be viewed on a screen placed behind the partition (Figure 5-1). This behavior is almost impossible to rationalize with the particle model, but if light is assumed to be a wave, very much like the waves produced when a pebble is dropped into a still pond, then the band pattern can be rationalized as follows. The source sends out concentric waves (each black arc in Figure 5-1 represents the crest of a wave) which strike the partition and form a set of identical concentric waves at each slit. At

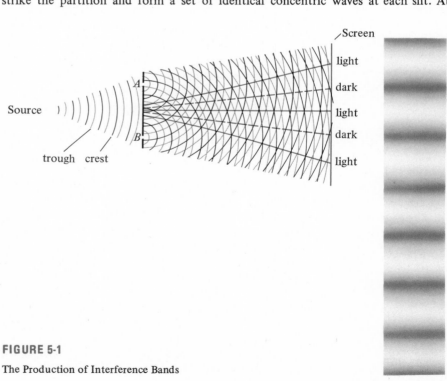

FIGURE 5-1

The Production of Interference Bands

some points on the screen, the crests of waves from slit *A* meet crests of waves from slit *B*, or troughs meet troughs, and at these points the waves *reinforce* one another and form a bright band. At other points crests meet troughs and *cancel* the effect of one another, and no light is observed at these points.

The wavelength (denoted by Greek *lambda*, λ) of such waves can be defined as the distance from crest to crest (Figure 5-2), while their frequency (denoted by Greek *nu*, *ν*) is the number of crests per second that pass a given point.

FIGURE 5-2

The Wavelength of Wave Motion

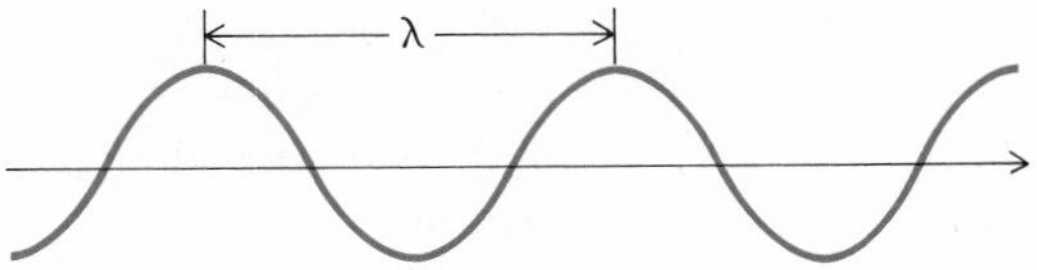

Light was later found to travel at a constant speed of 3×10^{10} cm/sec, which means that one wave (using the pebble-in-the-water analogy) could travel 186,000 miles in one second. The wavelength, frequency, and velocity of light (*c*), are related by the equation

$$\lambda \nu = c$$

After agreeing that certain properties of light could be explained by the wave theory, scientists began to wonder about what it was that was waving. It is certainly clear that when a pebble is dropped into water it is the water itself that oscillates up and down, but what about light? Some physicists speculated that a mysterious material, a "luminiferous ether," was the medium for the oscillations. In 1888, Heinrich Hertz verified a prediction made by James Maxwell in 1873 that light could be produced from electromagnetic oscillations in a loop of wire. This and other experiments have led scientists to adopt the concept that light can be thought of as perpendicularly oscillating electric and magnetic fields.

The term *light* is often used to mean ordinary visible light, the electromagnetic radiation that strikes our retinas and allows us to "see." The same term can also be used to refer to any electromagnetic radiation, such as high-energy X-rays, low-energy radar waves, and so forth. The various types of radiation that make up the total electromagnetic spectrum will be discussed in Chapter 8.

The wave theory of light reigned supreme for a number of years, but in the first decade of the twentieth century, the physicists Max Planck and Albert Einstein found the wave theory unsatisfactory for explaining certain experimental observations. One of these concerned the electrons released from the surface of certain metals upon exposure to ultraviolet light. That electrons were released was not surprising, since light is an oscillating electric field and must surely interact with the negatively charged electron. The surprising feature of the experiment lay in the relationship between the intensity of light and the velocity of the liberated electrons. The wave theory predicted that when the intensity or amount of light falling on the metal was increased, the electrons would leave the surface with a greater velocity. Experiments showed that an increase in intensity affected only the number of electrons leaving the metal; it was necessary to increase the frequency of the light to increase the velocity of the electrons.

According to Planck and Einstein, this could be explained by assuming that light consists of tiny bundles of electromagnetic energy. Each bundle of energy, called a *photon,* has an energy given by an equation derived by Planck:

$$E = h\nu$$

where ν is the frequency of light and h is a constant (since called *Planck's constant*) with a value of 6.6×10^{-27} erg-sec. The greater the number of photons hitting the metal, the greater the number of electrons knocked out. The higher the frequency of the photons, the greater their energy and the greater the jolt imparted to the electron.

While the new photon theory was eminently successful in explaining this photoelectric effect and several other phenomena, it was incapable of providing insights to phenomena such as interference bands. The wave and photon models, then, are complementary. Neither adequately explains all phenomena, and each one works when the other does not. Therefore, both models remain in use and both will be necessary until a better model is devised.

Another phenomenon of considerable importance to Bohr was the observation that when energy is imparted to an element, not only will light be emitted, but, more important, this emitted light contains only certain frequencies. For example, when hydrogen gas is exposed to an electrical arc, light is emitted, and when this light has been separated into its constituent frequencies (see Chapter 8) by passage through a prism, the line spectrum shown in Figure 5-3 is recorded. Each

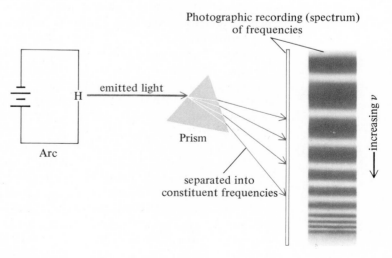

FIGURE 5-3

The Line Spectrum of Hydrogen

line on this spectrum corresponds to light of a different frequency; as the frequency increases, the lines get closer and closer together.

Every element has its own unique line spectrum, and it would seem likely, therefore, that these spectra are in some way characteristic of an atom's electronic structure. That is to say, any successful model of electronic structure must certainly provide an explanation for the line spectra of the elements.

THE BOHR MODEL

In 1913, armed with Planck's photon model and a knowledge of line spectra, Niels Bohr proposed the following model of the hydrogen atom. Around the single proton, which is the nucleus of the hydrogen atom, revolves one electron in a circular orbit. The major difference between the Bohr model and the earlier Rutherford "solar system" model is that only certain orbits (designated in Figure 5-4 as $n = 1$, $n = 2$, and so on) are available to the electron. Bohr's refinement of the model, though based on Planck's notion of bundles of energy, was dramatically revolutionary and resulted in his receiving the Nobel Prize for physics in 1922. The Bohr model of the atom is analogous to a small boy twirling overhead a stone attached to a string, who finds that the stone will revolve about him when the string has a length of 1 foot, 4 feet, 9 feet, and so forth, but not when the string is 18 inches long, 47 inches long, or 50 inches long.

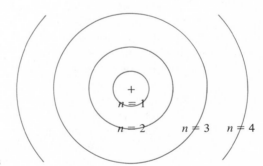

FIGURE 5-4

The Bohr Orbits

Another crucial point in Bohr's model is that the electron is not restricted to one particular orbit but can jump from one to another. Since energy is required to move the electron away from the nucleus (to overcome the electrostatic attraction), a jump from an orbit near the nucleus to one further from the nucleus requires energy, while a jump from a far to a near orbit emits energy. In order to determine the energies required for these orbit jumps, or transitions, Bohr calculated the energy of the electron in each orbit to be

$$E = -\frac{2\pi^2 m e^4}{n^2 h^2}$$

(1)

where m is the mass of the electron (9.1×10^{-28} g); e is the charge on the electron (4.8×10^{-10} eu); h is Planck's constant; and n identifies the orbit, the orbit nearest the nucleus being $n = 1$, the next one further from the nucleus $n = 2$, and so on. If the values for all the constants are inserted into the equation, the result is

$$E = -\frac{21.7 \times 10^{-12}}{n^2} \text{ erg}$$

or, in calories,

$$E = -\frac{51.8 \times 10^{-20}}{n^2} \text{ cal}$$

The energy of the electron in the orbit closest to the nucleus, where $n = 1$, is -51.8×10^{-20} cal. The negative sign indicates that the electron is attracted, rather than repelled, by the nucleus.

When $n = 2$, the electron is in the second orbit and its energy is

$$-\frac{51.8}{(2)^2} \times 10^{-20} = -12.95 \times 10^{-20} \text{ cal}$$

Energies for a number of n values are plotted in Figure 5-5, which shows that the spacings between the energy levels get smaller as the energy increases. As n approaches infinity, the distance between the electron and the nucleus increases, until finally at $n = \infty$ the electron is removed from the atom. At this point the energy of the system is zero. Thus, the energy of the electron and its distance from the nucleus are *quantized;* that is, both the energy and the distance may have certain values, but not others. This situation is somewhat analogous to our child who finds that he can swing the stone overhead only when the length of string is 1 foot, 4 feet, and so on, but that even when the string is, say, 4 feet long he can swing the stone at only one speed.

The similarity between the spacings of the energy levels shown in Figure 5-5 and the spacings in the line spectrum of hydrogen suggests that an explanation of the line spectrum may lie in the energy levels permitted the Bohr atom. And indeed it does. At room temperature the electron resides in the lowest energy orbit, $n = 1$, called the *ground* energy level. When the hydrogen atom is exposed to a source of energy, the electron may obtain just the right amount of energy to produce a jump to a higher energy level. This transition to a higher level is referred to as *excitation.* After the electron has made the transition to higher energy level, it can then fall back down to a lower level. In doing so, it emits energy in the form of a photon.

Suppose, for example, the electron has been excited to the $n = 4$ level. It may now fall back, or *relax*, to the $n = 3$, the $n = 2$, or the $n = 1$ levels. If it falls back to $n = 3$, its energy decreases from -3.24×10^{-20} cal to -5.76×10^{-20} cal, a change of 2.52×10^{-20} cal or (since 1 cal = 4.19×10^7 ergs) 10.6×10^{-13} erg. According to the law of conservation of energy, this energy must now appear in another form—an emitted photon with an energy of 10.6×10^{-13} erg. The frequency of this photon can be calculated from the Planck equation

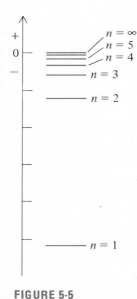

FIGURE 5-5

Energy Levels
for the Hydrogen Atom

$$E = h\nu$$

$$\nu = \frac{E}{h} = \frac{10.6 \times 10^{-13} \text{ erg}}{6.6 \times 10^{-27} \text{ erg-sec}} = 1.6 \times 10^{14} \text{ sec}^{-1}$$

This, according to the Bohr model, is the origin of the line spectrum for hydrogen: The electron is excited by an energy source into some higher orbit, and when it relaxes it emits a photon, which registers as a line in the spectrum. When a collection of many hydrogen atoms is excited, some atoms experience excitation to very high energy orbits, others to lower energy orbits. When relaxation occurs, the electrons of some atoms will fall back to the next lower level, others to the third lower level, and so on. The many lines in the spectrum, therefore, are a result of numerous electronic transitions.

Bohr was able to account for most of the lines in the spectrum of hydrogen in this way. Indeed, the energy of the ground state of hydrogen as calculated by equation (1) on page 65 is in excellent agreement with the experimentally determined ionization energy of hydrogen (which is, after all, the energy required to remove an electron from the ground state to the $n = \infty$, or zero, energy state). That it explained much of the hydrogen spectrum was clearly a major triumph for the Bohr model. The victory was short-lived, however, because it soon became apparent that the model could not explain some of the lines in the hydrogen spectrum, and it was also inadequate for atoms containing more than one electron.

THE WAVE MODEL

The wave model, the currently accepted theory of electronic structure, began with a suggestion by the French physicist Louis-Victor de Broglie. If light, which was traditionally conceived of as a wave motion, could also be interpreted as having the qualities of a particle (the photon), then, suggested de Broglie, it is possible that the electron, traditionally treated as a particle, is also endowed with wave properties. This proposal was later verified by the interference experiment discussed above. De Broglie related the wave properties of the electron to its particle properties with the relationship

$$\lambda = \frac{h}{mv}$$

where λ is the wavelength of the wave, m is the mass of the electron, v is its velocity, and h is Planck's constant. An electron moving at a speed of 10^7 cm/sec would have an associated wavelength of 70 Å, a very small but detectable wavelength.

This, then, was the seed that fell upon the fertile minds of physicists like Werner Heisenberg, Erwin Schrödinger, and P. A. M. Dirac and blossomed into the vastly important new subject of wave mechanics.

The development of the contemporary wave model of the atom has been quite mathematical and abstract in nature, and we will attempt to provide here only a qualitative understanding of its principal features.

Let us visualize an electron, then, as something which, because of the peculiarities and inadequacies of our model, we must endow with both particle and wave properties, these properties being related by the de Broglie equation. Since an atom is a rather complex three-dimensional object, let us first examine a much simpler one-dimensional problem: an electron confined to move along a line, say from $x = 0$ to $x = d$ on the x axis of Figure 5-6a. The electron has some of the characteristics of a wave, and we shall assume, therefore, that a wave exists in this region. We will represent the amplitude of the wave, measured along the y axis in such a way that the vertical center of each oscillation falls on the x axis ($y = 0$). Every trough then lies below the x axis (y is negative) and every crest lies above the x axis (y is positive).

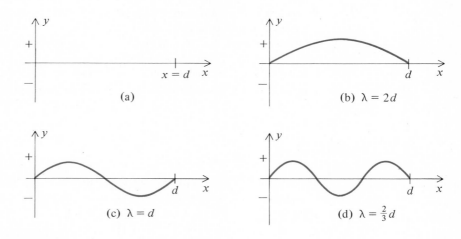

FIGURE 5-6

Waves Associated with Electron Confined to a Line

Now we must impose a condition on our waves, which we will justify later: At the boundaries of the region, that is, at $x = 0$ and $x = d$, the amplitude of the wave shall be zero ($y = 0$). Even with this boundary condition, there are a number of waves that can occur in the region. Figure 5-6 shows several of these waves. The first, Figure 5-6b, is one half of a complete oscillation and (recall that wavelength is the distance between two crests) has a wavelength of twice the distance from 0 to d, or $2d$. The second wave, Figure 5-6c, is a complete oscillation and therefore has a wavelength of d. The third wave (Figure 5-6d) is one and one-half oscillations and has a wavelength of $\frac{2}{3}d$. For *any* wave occurring in this region, then, the wavelength can be expressed as

$$\lambda = \frac{2d}{n}$$

where n is an integer from 1 to infinity. For the first wave, $n = 1$; for the second $n = 2$, and so on.

The energy of the electron, as we have seen in the Bohr model, is of great importance. Can we now, by treating the electron as a wave, and from our knowledge of how the waves must fit into this region of space, obtain an equation giving the energy of the electron? We have previously found that the energy of electromagnetic radiation is directly proportional to its frequency ($E = h\nu$) and that frequency is inversely proportional to wavelength ($\nu = c/\lambda$). The energy of electromagnetic radiation is therefore also inversely proportional to its wavelength. Thus, the greater the n value for the wave, the smaller the wavelength and the higher the energy.

The energy can be determined more quantitatively by first combining the de Broglie equation and our equation for wavelength.

$$\lambda = \frac{2d}{n} \quad \text{(for waves confined to a line)}$$

$$\lambda = \frac{h}{mv} \quad \text{(de Broglie relation)}$$

Thus,

$$\frac{2d}{n} = \frac{h}{mv}$$

and, solving for v,

$$v = \frac{nh}{2md}$$

Finally, if we treat the electron as a particle with a kinetic energy (in the region from $x = 0$ to $x = d$ it has *only* kinetic energy) of $\frac{1}{2}mv^2$, we find

$$E = \tfrac{1}{2}mv^2$$

$$v = \frac{nh}{2md}$$

so

$$E = \frac{m}{2}\left(\frac{nh}{2md}\right)^2 \quad \text{or} \quad E = \frac{n^2h^2}{8md^2}$$

The energy of the electron is therefore a function of a number of constants and the integer n. Figure 5-7 shows the relative energies and the wave associated with each energy level for $n = 1, 2, 3,$ and 4.

It is now apparent that our electron can have only certain energies, these energies being prescribed by the number n. Hence, the energies are quantized, and n

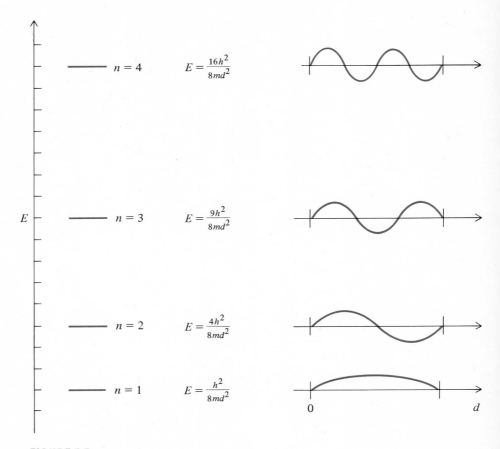

FIGURE 5-7

Energy Levels for Electron Confined to a Line

is referred to as a *quantum number*. This quantization of energy is a natural consequence of the wave character of the electron *and* the boundary conditions imposed upon these waves.

Now we return to the same question that we encountered in our discussion of light. What is waving? What do these electron waves mean? The best explanation of the physical significance of the wave was provided by Max Born: The square of the amplitude of the wave at a given point is proportional to the probability of finding the electron at that point. Figure 5-8 shows the $n = 1$ and $n = 2$ waves and their squared values. Since the square of the amplitude for the $n = 1$ wave is greatest in the middle of the region, this is the place where the electron is most likely to be found; in other words, the electron spends most of its time in the center of the region. For the $n = 2$ electron, the points of greatest probability lie on either side of

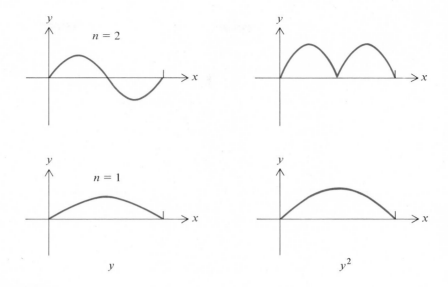

FIGURE 5-8

Relative Location Probabilities for Electron Confined to a Line

the center of the region, and the probability of finding the electron at the very center of the region is zero.

We can now rationalize our decision to restrict the waves to those with zero amplitudes at the boundaries of the region. If the electron is to be confined between $x = 0$ and $x = d$, it must have zero probability of being on or beyond the boundaries: hence, the amplitude at $x = 0$ and $x = d$ must be zero.

Another problem that arises when the position of an electron is examined is known as the *Heisenberg uncertainty principle.* According to Heisenberg, the position and the velocity of the electron cannot both be known with certainty. If the position of the electron is pinpointed, the velocity becomes very uncertain. If the velocity is accurately known, the position is virtually unknown. In order to observe the position of the moving electron we must be able to "see" it, which means that we must shine light on it. One of the laws of optics tells us that the smaller the object being observed, the shorter the wavelength of the light needed to observe it. Since an electron is extremely small, we need extremely short wave-length light to "see" it in some hypothetical microscope. According to the Planck relationship, short-wavelength electromagnetic radiation has very high energy, and thus when the high-energy photon hits the moving electron so that we can observe it, the photon imparts a considerable jolt to the electron.

Let us assume that we note the position of the electron before it is moved by the photon. Now we must determine the velocity of our electron, and this can be accomplished by observing its position after a certain interval of time. We would then know how far it has traveled during that interval; that is, we would know its velocity. But we have already changed the energy, and therefore the velocity of the electron is not the same as the first time we observed it. Thus, we can never know the electron's original velocity, because in pinpointing its position we change its velocity. On the other hand, we might decrease the energy of the photon, and consequently increase its wavelength, so that the velocity of the electron would be altered only slightly. However, our "picture" of the electron would then become much less distinct—the uncertainty in its *position* would increase. This perversity seems to be inherent in the world of subatomic particles and can be summarized by the old cliche, "you can't have your cake and eat it too."

Clearly, the wave-mechanical vision of the electron is a strange one: sometimes the electron is treated as a particle, at other times as a wave. Its position can be dealt with only in terms of probabilities, and its energy is restricted to certain values. The strangeness of the vision, particularly the apparent wave-particle duality, is probably due in part to the inadequacies of the models, as indicated in our discussion of the nature of light. On the other hand, it is not unreasonable to suppose that the behavior of submicroscopic particles does not parallel the behavior of the macroscopic world.

The application of the wave model to the simplest of all atoms, the hydrogen atom, is vastly more complex than the description of an electron oscillating along a line. The results, however, have the same form. Since an atom is a three-dimensional object, three quantum numbers, rather than one, are required to specify the energy and probability of finding the electron at any given point in space. The three quantum numbers are *n, l,* and *m.*

1. *n,* the *principal quantum number,* can have any integral value from $n = 1$ to $n = \infty$. For the hydrogen atom, *n* determines the total energy of the electron according to an equation identical to that derived by Bohr.

$$E = -\frac{2\pi^2 m e^4}{n^2 h^2}$$

It also specifies the most probable separation between the electron and the nucleus. The greater the value of *n,* the more likely the electron will be to spend most of its time farther from the nucleus. The letters *K, L, M,* and so on are sometimes used to denote principal quantum number values of 1, 2, 3, etc.

2. *l* is the *angular momentum quantum number.* Its value depends upon *n* and can be any integer from 0 to $n - 1$. If *n* for a given electron is 1, *l* can only be zero; if $n = 2$, *l* can be either 0 or 1. This quantum number describes the angular momentum of the electron and also helps specify how far from the nucleus the electron is likely to be. Most important, however, the value of *l* determines the relative probability of finding an electron at any point on a hypothetical sphere surrounding the nucleus. (This we will refer to as the *angular* location probability pattern.) Suppose, for example, an electron is designated by $l = 0$. The probability of

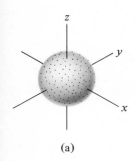

(a)

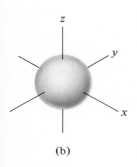

(b)

FIGURE 5-9

Location Probability Pattern
for an $l = 0$ Electron

finding this electron at equally spaced points on a sphere surrounding the nucleus is shown in Figure 5-9, where relative probabilities are indicated by the size of the dots. Since all dots are of equal size, the probabilities are equal. This same information is often conveyed by the more picturesque electron-cloud picture (Figure 5-9b). Note that neither picture offers any information about how far from the nucleus the electron is most likely to be.

3. m is the *magnetic quantum number.* Its value depends upon l and can be any integer from $-l$ to $+l$, including zero. If an electron has an l quantum number of 2, its m value could be -2, -1, 0, $+1$, or $+2$. It is known that a moving electrical charge produces a magnetic field. The value of m is related to the magnetic field produced by the electron. When the atom is located in an external magnetic field, m helps to determine the energy of the atom. Probably a more important characteristic, however, is that it determines the spatial orientation of the electron cloud, as we shall see below.

Since the values for the three quantum numbers are interrelated, only certain sets of the three are possible. For example, if $n = 1$, l must be zero and m must be zero; only one set containing a principal quantum number of 1 is possible. Table 5-1 shows that there are four sets of numbers for $n = 2$: (2, 0, 0), (2, 1, -1), (2, 1, 0), and (2, 1, 1). Each set is a unique description of the energy and location probabilities of an electron. The electron in a hydrogen atom might be characterized by the set $n = 1$, $l = 0$, $m = 0$, or by the set $n = 2$, $l = 1$, $m = -1$, or by any other set, depending on its state of excitation. Each unique description—that is, each set of n, l, m numbers—is called an *orbital.* To simplify reference to these orbitals, the letters s, p, d, f are used to designate l values of 0, 1, 2, and 3, respectively, while m values are designated by the letters shown in Table 5-1. Thus, the $n = 1$, $l = 0$, $m = 0$

TABLE 5-1 Possible Sets of n, l, m Quantum Numbers for $n = 1$, $n = 2$, $n = 3$	n	l	m	Orbital designation
	1	0	0	$1s$
	2	0	0	$2s$
		1	-1	
			0	$2p_x, 2p_y, 2p_z$
			$+1$	
	3	0	0	$3s$
		1	-1	
			0	$3p_x, 3p_y, 3p_z$
			$+1$	
		2	-2	
			-1	
			0	$3d_{xy}, 3d_{yz}, 3d_{xz}, 3d_{x^2-y^2}, 3d_{z^2}$
			$+1$	
			$+2$	

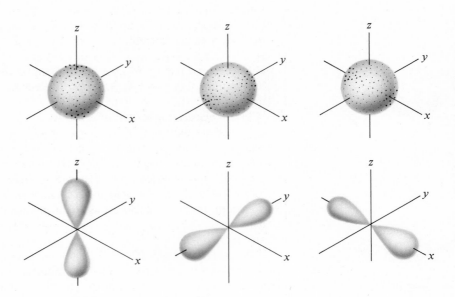

FIGURE 5-10

Location Probability Patterns for the *p* Orbitals

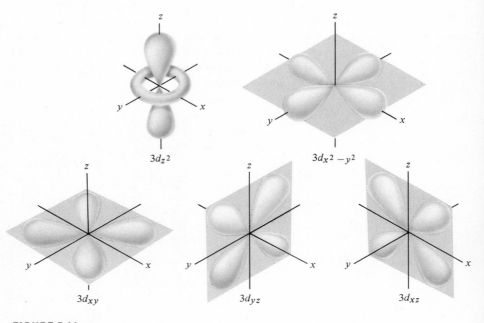

FIGURE 5-11

Location Probability Patterns for the *d* Orbitals

orbital is referred to as a 1*s* orbital; the $n = 2$, $l = 0$, $m = 0$ orbital is a 2*s* orbital; when $n = 3$, $l = 1$, $m = 0$, the orbital is a 3*p* orbital; and so forth.

We can now discuss more specifically how the orbital designation specifies the energy and location of an electron. The location probabilities for the *s* orbitals have already been given (Figure 5-9). Every *s* orbital, whether 1*s*, 2*s*, 3*s*, and so forth, has the same angular location probability pattern. However, since the principal quantum number determines how far from the nucleus the electron is likely to be, a 2*s* electron will have a greater probability of being farther from the nucleus than a 1*s* electron. The angular location probabilities for *p* orbitals are given as dot diagrams and as electron cloud pictures in Figure 5-10. A change in the *m* quantum number simply shifts the orientation of the orbital from one axis to another. All *p* orbitals have the same probability patterns, but, as noted above, *n* determines distance from the nucleus.

Only the electron-cloud pictures are given for the *d* orbitals in Figure 5-11. The *f* orbitals are even more complex and are not shown.

In addition to characterizing the electron's location probability, the orbital also specifies the total energy of the electron. The energy levels for the hydrogen atom are shown in Figure 5-12. This atom is quite unique in that all the orbitals within a given principal quantum level (*n*) have the same energies. When two or more orbitals have the same energy, they are said to be *degenerate*. For all other atoms, only the orbitals within a given *l* level (2*p*, 3*p*, 3*d*, and so on) are degenerate. If these atoms are placed in a magnetic field, then all of the degeneracies are removed; that is, each orbital has a different energy.

The wave model of the hydrogen atom, then, also explains line spectra: The electron can reside in any of the energy levels. At room temperature the electrons of the vast majority of hydrogen atoms are in the lowest energy level, the 1*s* level. At higher temperatures the electron may be in an excited level, a 2*s*, 2*p*, 3*s*, and so

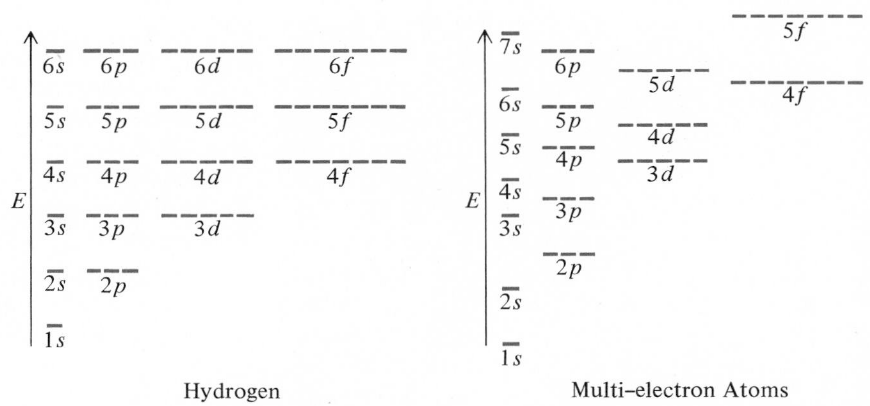

Hydrogen   Multi–electron Atoms

FIGURE 5-12

Schematic Representation of the Energy Levels for Hydrogen and More Complex Atoms on, and when it reverts to a lower energy level, energy is released in the form of a photon. The energy of the photon is equal to the difference in energy between the two energy levels involved in the transition.

When hydrogen is excited in a magnetic field, the resulting line spectrum contains several features that still remain unexplained. These features can be rationalized if a fourth quantum number is introduced. This quantum number, s, seems to represent the spin of the electron, analogous to the spin of the earth on its axis, and can have only two values, $+\frac{1}{2}$ and $-\frac{1}{2}$. The two values can be interpreted as designations of spin in either the clockwise or counterclockwise directions. Thus, an electron in any orbital will have a spin quantum number of either $+\frac{1}{2}$ or $-\frac{1}{2}$, and therefore every electron is fully characterized by four quantum numbers.

The wave model can also be applied to atoms more complex than hydrogen. The location probability patterns of the orbitals for these atoms are the same as those for hydrogen. Figure 5-12 shows the relative energies of the orbitals for most atoms. The exact energies for a given atom depend upon a number of factors: nuclear charge, number of electrons, and so forth.

It is possible to assign every electron in a given atom to an orbital according to the energy-level diagram of Figure 5-12. When the *ground-state electron configuration* is being determined, every electron must be given the lowest possible energy. This must be done, however, within the bounds of a very fundamental law of nature, the *Pauli exclusion principle.* The Pauli principle states that each electron within a given atom must have a unique set of the four quantum numbers. That is, two electrons can have the same set of $n, l,$ and m values only if their s values differ. Consequently, only two electrons may occupy the same orbital, and to do so they must have different spin quantum numbers.

Consider the second simplest atom, the helium atom, with an atomic number of 2, which signifies that it has two electrons. Since we are interested in the ground-state electron configuration, the two electrons should be placed in the lowest possible energy level. This is, of course, the $1s$ orbital, and we can place both electrons in it if their spins are opposed (different s values). This configuration is written as $1s^2$, where the superscript 2 is used to indicate the number of electrons in the orbital. The lithium atom, with three electrons, may have only two in the $1s$ level; the other goes to the next lowest level, the $2s$. The lithium configuration is then $1s^2 2s^1$ (the superscript 1 is often omitted).

Using the same procedure, the configuration of beryllium is $1s^2 2s^2$ and that of boron is $1s^2 2s^2 2p$. With carbon a problem arises: Since there are three $2p$ orbitals of equal energy, should the sixth electron be placed into the p orbital that already contains an electron, to give the configuration $1s^2 2s^2 2p_x^2$ (the use of p_x rather than p_y or p_z orbitals is perfectly arbitrary, since their energies are the same), or should the electron be placed in one of the other p orbitals to give the configuration $1s^2 2s^2 2p_x^1 2p_y^1$? If both $2p$ electrons occupied the same orbital they would also move within the same volume of space, and repulsions between them would be greater than if they were in two different $2p$ orbitals. The second configuration is therefore more favorable. This principle, based on experimental evidence, has been formalized as *Hund's rule: When filling a set of degenerate*

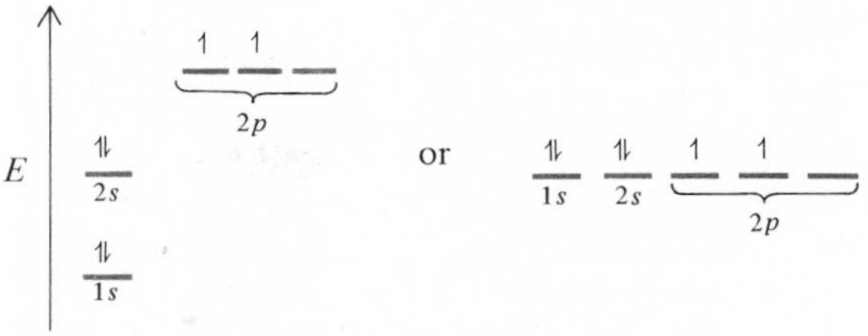

$$E \quad \begin{array}{cc} \overline{\uparrow} \;\; \overline{\uparrow} \\ 2p \end{array}$$

$$\begin{array}{l} \underline{\uparrow\downarrow} \\ 2s \end{array} \qquad \text{or} \qquad \underline{\uparrow\downarrow}\;\;\underline{\uparrow\downarrow}\;\;\overline{\uparrow}\;\;\overline{\uparrow} \\ \quad 1s \quad\; 2s \quad\;\; 2p$$

$$\begin{array}{l} \underline{\uparrow\downarrow} \\ 1s \end{array}$$

FIGURE 5-13

The Electron Configuration
of Carbon

energy levels, the electrons enter the orbitals singly, with spins in the same direction (same s number), until the set *is half-filled.* In order to emphasize the spin aspect of this rule, we write the electron configuration of carbon as shown in Figure 5-13. Here the direction, up or down, of the arrow signifies the sign of the *s* value, plus or minus. This same rule is followed to give nitrogen the configuration $1s^2 2s^2 2p_x{}^1 2p_y{}^1 2p_z{}^1$ or

$$\underline{\uparrow\downarrow}\;\;\underline{\uparrow\downarrow}\;\;\underline{\uparrow}\;\;\underline{\uparrow}\;\;\underline{\uparrow} \\ 1s \quad\; 2s \quad 2p_x \; 2p_y \; 2p_z$$

and the $2p$ orbitals are now half-filled.

The oxygen atom has eight electrons, and the last electron to be placed into the orbitals must now be paired with an electron in one of the $2p$ orbitals; in fluorine a fifth electron is added to the $2p$ level; and finally, with neon, whose configuration is $1s^2 2s^2 2p^6$, the level is completely filled.

If we now turn our attention back to the periodic chart on page 52, we find that we have just described the first two rows of elements. The first-period elements, hydrogen and helium, have their electrons in the first principal quantum level ($n = 1$), while the second-period elements, lithium through fluorine, have electrons in the first and second principal quantum levels. The electrons in the second principal quantum level of the second-period elements are the electrons farthest from the nucleus and therefore are the electrons involved in chemical bonding. For this reason, these electrons are often called *valence electrons,* and the $n = 2$ level is the *valence level* or *shell* for the second-period elements. (The valence level of the first-period elements is the $n = 1$ level.)

We can now continue this process of building the elements, a process sometimes referred to as *aufbau* (German, "building up"). The third-period elements fill the $3s$ and $3p$ orbitals; the electron configuration of phosphorus is, for example, $1s^2 2s^2 2p^6 3s^2 3p_x{}^1 3p_y{}^1 3p_z{}^1$. The energy-level diagram (Figure 5-12) indicates that the $4s$ orbital is slightly lower in energy than the $3d$ orbitals. Hence, rather than continuing to fill the third principal quantum level, the outermost electron of the first member of the fourth period, potassium, occupies the $4s$ orbital. The $4s$ orbital is filled at calcium, and at scandium the $3d$ orbitals begin to be occupied. Thus, the configuration for scandium is $1s^2 2s^2 2p^6 3s^2 3p^6 3d^1 4s^2$. (Note that in the electron configuration the orbitals are written in order of increasing n and l values, not necessarily in order of increasing energy.) Since there

TABLE 5-2
Electron Configurations of the Atoms

Atomic Number	Element	Configuration	Atomic Number	Element	Configuration
1	H	$1s$	53	I	$[Kr]4d^{10}\,5s^2\,5p^5$
2	He	$1s^2$	54	Xe	$[Kr]4d^{10}\,5s^2\,5p^6$
3	Li	$[He]2s$	55	Cs	$[Xe]6s$
4	Be	$[He]2s^2$	56	Ba	$[Xe]6s^2$
5	B	$[He]2s^2\,2p$	57	La	$[Xe]5d6s^2$
6	C	$[He]2s^2\,2p^2$	58	Ce	$[Xe]4f5d6s^2$
7	N	$[He]2s^2\,2p^3$	59	Pr	$[Xe]4f^3 6s^2$
8	O	$[He]2s^2\,2p^4$	60	Nd	$[Xe]4f^4 6s^2$
9	F	$[He]2s^2\,2p^5$	61	Pm	$[Xe]4f^5 6s^2$
10	Ne	$[He]2s^2\,2p^6$	62	Sm	$[Xe]4f^6 6s^2$
11	Na	$[Ne]3s$	63	Eu	$[Xe]4f^7 6s^2$
12	Mg	$[Ne]3s^2$	64	Gd	$[Xe]4f^7 5d6s^2$
13	Al	$[Ne]3s^2\,3p$	65	Tb	$[Xe]4f^9 6s^2$
14	Si	$[Ne]3s^2\,3p^2$	66	Dy	$[Xe]4f^{10}\,6s^2$
15	P	$[Ne]3s^2\,3p^3$	67	Ho	$[Xe]4f^{11}\,6s^2$
16	S	$[Ne]3s^2\,3p^4$	68	Er	$[Xe]4f^{12}\,6s^2$
17	Cl	$[Ne]3s^2\,3p^5$	69	Tm	$[Xe]4f^{13}\,6s^2$
18	Ar	$[Ne]3s^2\,3p^6$	70	Yb	$[Xe]4f^{14}\,6s^2$
19	K	$[Ar]4s$	71	Lu	$[Xe]4f^{14}\,5d6s^2$
20	Ca	$[Ar]4s^2$	72	Hf	$[Xe]4f^{14}\,5d^2\,6s^2$
21	Sc	$[Ar]3d4s^2$	73	Ta	$[Xe]4f^{14}\,5d^3\,6s^2$
22	Ti	$[Ar]3d^2\,4s^2$	74	W	$[Xe]4f^{14}\,5d^4\,6s^2$
23	V	$[Ar]3d^3\,4s^2$	75	Re	$[Xe]4f^{14}\,5d^5\,6s^2$
24	Cr	$[Ar]3d^5\,4s$	76	Os	$[Xe]4f^{14}\,5d^6\,6s^2$
25	Mn	$[Ar]3d^5\,4s^2$	77	Ir	$[Xe]4f^{14}\,5d^7\,6s^2$
26	Fe	$[Ar]3d^6\,4s^2$	78	Pt	$[Xe]4f^{14}\,5d^9\,6s$
27	Co	$[Ar]3d^7\,4s^2$	79	Au	$[Xe]4f^{14}\,5d^{10}\,6s$
28	Ni	$[Ar]3d^8\,4s^2$	80	Hg	$[Xe]4f^{14}\,5d^{10}\,6s^2$
29	Cu	$[Ar]3d^{10}\,4s$	81	Tl	$[Xe]4f^{14}\,5d^{10}\,6s^2\,6p$
30	Zn	$[Ar]3d^{10}\,4s^2$	82	Pb	$[Xe]4f^{14}\,5d^{10}\,6s^2\,6p^2$
31	Ga	$[Ar]3d^{10}\,4s^2\,4p$	83	Bi	$[Xe]4f^{14}\,5d^{10}\,6s^2\,6p^3$
32	Ge	$[Ar]3d^{10}\,4s^2\,4p^2$	84	Po	$[Xe]4f^{14}\,5d^{10}\,6s^2\,6p^4$
33	As	$[Ar]3d^{10}\,4s^2\,4p^3$	85	At	$[Xe]4f^{14}\,5d^{10}\,6s^2\,6p^5$
34	Se	$[Ar]3d^{10}\,4s^2\,4p^4$	86	Rn	$[Xe]4f^{14}\,5d^{10}\,6s^2\,6p^6$
35	Br	$[Ar]3d^{10}\,4s^2\,4p^5$	87	Fr	$[Rn]7s$
36	Kr	$[Ar]3d^{10}\,4s^2\,4p^6$	88	Ra	$[Rn]7s^2$
37	Rb	$[Kr]5s$	89	Ac	$[Rn]6d7s^2$
38	Sr	$[Kr]5s^2$	90	Th	$[Rn]6d^2\,7s^2$
39	Y	$[Kr]4d5s^2$	91	Pa	$[Rn]5f^2 6d7s^2$
40	Zr	$[Kr]4d^2\,5s^2$	92	U	$[Rn]5f^3 6d7s^2$
41	Nb	$[Kr]4d^4\,5s$	93	Np	$[Rn]5f^4 6d7s^2$
42	Mo	$[Kr]4d^5\,5s$	94	Pu	$[Rn]3f^6 7s^2$
43	Tc	$[Kr]4d^5\,5s^2$	95	Am	$[Rn]5f^7 7s^2$
44	Ru	$[Kr]4d^7\,5s$	96	Cm	$[Rn]5f^7 6d7s^2$
45	Rh	$[Kr]4d^8\,5s$	97	Bk	$[Rn]5f^9 7s^2$
46	Pd	$[Kr]4d^{10}$	98	Cf	$[Rn]5f^{10}\,7s^2$
47	Ag	$[Kr]4d^{10}\,5s$	99	Es	$[Rn]5f^{11}\,7s^2$
48	Cd	$[Kr]4d^{10}\,5s^2$	100	Fm	$[Rn]5f^{12}\,7s^2$
49	In	$[Kr]4d^{10}\,5s^2\,5p$	101	Md	$[Rn]5f^{13}\,7s^2$
50	Sn	$[Kr]4d^{10}\,5s^2\,5p^2$	102	No	$[Rn]5f^{14}\,7s^2$
51	Sb	$[Kr]4d^{10}\,5s^2\,5p^3$	103	Lr	$[Rn]5f^{14}\,6d7s^2$
52	Te	$[Kr]4d^{10}\,5s^2\,5p^4$			

are five $3d$ orbitals, there are a total of 10 elements whose valence electrons are $3d$ electrons—the transition elements.

Inspection of Table 5-2, which tabulates the electron configurations for the elements, shows two anomalies in electron configuration among the first row of transition elements: chromium has an outer configuration of $3d^5 4s^1$ rather than $3d^4 4s^2$, and copper is $3d^{10} 4s^1$ rather than the expected $3d^9 4s^2$. Both anomalies reflect an *extra stability associated with half-filled or filled degenerate energy levels.* In other words, for chromium the extra stability of a half-filled $3d$ level more than compensates for the energy used in promoting an electron from the $4s$ to the $3d$ level. The extra stability of the filled level is also partly responsible for the inert behavior of the inert gases.

When the first row of the transition elements is complete the five $3d$ orbitals have been filled, and at the element gallium the $4p$ orbitals begin to fill; they are completely filled at the inert gas krypton. The aufbau trend of the elements from potassium to krypton is now repeated for the elements rubidium to xenon: the $5s$ orbitals are filled in rubidium and strontium, the $4d$ orbitals from yttrium to cadmium, and the $5p$ orbitals from indium to xenon.

The $6s$ orbitals are next in energy according to the scheme in Figure 5-12, and the highest energy electrons in cesium and barium are therefore in $6s$ orbitals. Figure 5-12 shows the $4f$ orbitals immediately above the $6s$ in energy, but an exception to this generalization occurs at element 57, lanthanum. The $5d$ orbital of lanthanum is lower in energy than the $4f$ orbitals, as one might expect, since lanthanum appears in the periodic table as a transition metal. The scheme now holds again, however, for element 58, cerium, which has the configuration $1s^2 2s^2 2p^6 3s^2 3p^6 3d^{10} 4s^2 4p^6 4d^{10} 4f^1 5s^2 5p^6 5d^1 6s^2$. From praseodymium to lutetium the seven $4f$ orbitals are filled, and the lanthanide elements, therefore, are the f-fillers.

We now return in our aufbau of the elements to the $5d$ orbitals and complete the third row of transition metals. Thallium then begins the $6p$ orbitals, which are filled at the inert gas radon. The next element, francium, has the radon configuration, denoted [Rn], plus a single $7s$ electron. The trend in configurations of the elements cesium to lutetium is now repeated with the series francium to laurentium.

The general trends in orbital filling are summarized in Figure 5-14, which illustrates the relationship between the form of the periodic chart and electron configurations. Elements of the first period fill the $1s$ orbital, elements of the second period fill orbitals in the second principal quantum level, elements of the third period fill orbitals in the third principal quantum level, elements of the fourth period fill orbitals in the fourth and third ($3d$) quantum levels, and so on. Elements within the same group have the same valence electron configurations. For example, in Group IIIA, boron has the configuration $1s^2 2s^2 2p^1$, aluminum is $1s^2 2s^2 2p^6 3s^2 3p^1$, and gallium is $1s^2 2s^2 2p^6 3s^2 3p^6 3d^{10} 4s^2 4p^1$. The valence electron configuration for all three can be denoted as $ns^2 np^1$, where n is 2 for boron, 3 for aluminum and 4 for gallium.

The configurations and their relationship to the periodic chart can be summarized in still another way: Of the representative elements (A families), groups I

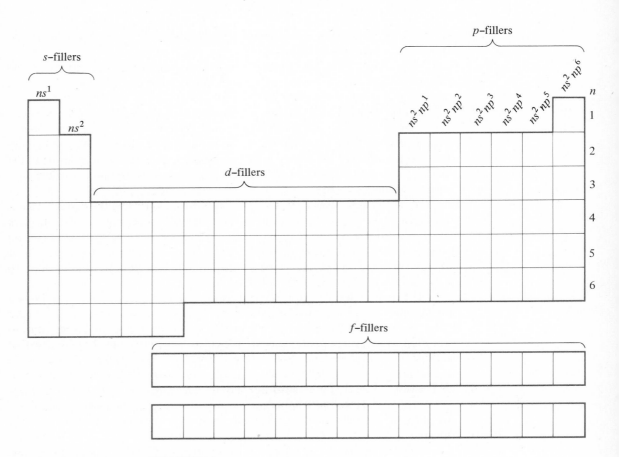

FIGURE 5-14

Relation of Electron Configurations to the Periodic Chart

and II are the *s*-fillers, while groups III, IV, V, VI, VII, and the inert gases are the *p*-fillers. The transition elements are the *d*-fillers, and the inner transition groups are the *f*-fillers.

Close scrutiny of the experimentally determined configurations presented in Table 5-2 will reveal a number of exceptions to the generalizations given above. We will not attempt to explain these exceptions. It must also be stressed that the configurations given are those for the neutral atoms. The energy level scheme for an ion, for example, is usually different from that given in Figure 5-12 for the atoms. In fact, the electrons lost when an ion is formed from an atom are always the electrons the greatest distance from the nucleus. In other words, the electrons lost are those at the end of the electron configuration. For example, the zinc atom has the configuration $1s^2 2s^2 2p^6 3s^2 3p^6 3d^{10} 4s^2$; the configuration of the Zn^{2+} ion is $1s^2 2s^2 2p^6 3s^2 3p^6 3d^{10}$. The configuration of tin is $1s^2 2s^2 2p^6 3s^2 3p^6 3d^{10} 4s^2 4p^6$

$4d^{10} 5s^2 5p^2$; the configuration of the Sn^{2+} ion is $1s^2 2s^2 2p^6 3s^2 3p^6 3d^{10} 4s^2 4p^6 4d^{10} 5s^2$.

RATIONALIZATION OF PERIODIC TRENDS

We can now use the wave model to rationalize the periodic trends discussed earlier. Since there is a small, but finite, probability of finding an electron at even large distances from the nucleus, the size of an atom is not clearly defined in the wave model. Nevertheless, it is possible to assume that atomic sizes as measured by X-ray diffraction and other techniques (see Chapter 9) reflect the major portion of the electron density of an atom. The variation in size (increase down a group, decrease across a period) can be explained as follows: The "size" of an atom is determined by the location probabilities for its electrons, especially the outermost electrons. Since the greater the value of the principal quantum number, the farther from the nucleus the electron is likely to be, the increase in size down a group is due to an increase in the principal quantum number of the outermost electrons down the group.

The valence electrons of all the elements within a given period have the same principal quantum number and therefore have their maximum probabilities at *roughly* the same distance from the nucleus. In moving from left to right in a given period, the nuclear charge increases, however. This increase in nuclear charge results in a greater attraction between the positively charged nucleus and the negatively charged valence electrons. This increased attraction in turn results in a shrinkage of the distance between the nucleus and the outermost electrons; that is, a decrease in the size of the atoms. Nuclear charge also increases down a given group, but the shrinkage due to this is outweighed by the increase in n. Moreover, a given outer electron does not experience the full charge of the nucleus. There are other electrons in the atom, and their electron density is interposed between the outer electron(s) and the nucleus. This shields the outer electron from the full charge of the nucleus, and in general the shielding due to a lower filled quantum level is far more effective than the shielding due to electrons within the same quantum level.

The decrease in ionization energy down a group has a similar explanation. The principal quantum number of the outermost electron (the one that is lost upon ionization) determines both its location probability and its energy. As a group is descended, n increases, the energy of the electron increases, and thus the energy required to remove the electron decreases. As a period is crossed from left to right, the distance between the outermost electron and the nucleus decreases, and the strength of the attraction increases; the energy required to remove an electron therefore increases.

While ionization energy generally increases across a period, Figure 4-5 shows two exceptions among the second-period elements. The first exception occurs at boron, which has a lower, rather than higher, first ionization energy than beryllium. This is because the outermost electron in boron is a $2p$ electron, while the outermost electron in beryllium is a 2s electron. Since the 2p orbitals have a higher energy than the 2s orbitals, and this difference in energy is not outweighed by the increased nuclear charge of boron, less energy is required to remove the 2p electron. The second exception occurs at oxygen, which has a lower ionization energy than nitrogen. This may be associated with the special stability of the half-filled 2p level of nitrogen (more energy is required to ionize nitrogen).

Variations in electronegativity can be explained in much the same way. Electronegativity decreases down a group because the atomic size increases; thus an external electron is maintained farther from the nucleus and is therefore less attracted to it. Electronegativity increases across a period because of the decrease in size and consequent increase in attraction for an external electron.

The metallicity of an element is related to its tendency to form positive ions. Since the ionization energy is a quantitative measure of this tendency, trends in metallicity can be rationalized from ionization energies: The greater the ionization energy, the smaller the tendency to form a positive ion and the smaller the metallicity of the element.

SUGGESTED READINGS

Hoffmann, B. *The Strange Story of the Quantum.* New York: Harper and Brothers, 1947.

Maybury, R. H. "The Language of Quantum Mechanics." *Journal of Chemical Education,* Vol. 39, No. 7 (July 1962), pp. 367–373.

Gamow, G. "The Principle of Uncertainty." *Scientific American,* Vol. 198, No. 1 (January 1958), pp. 51–57.

Garrett, A. B. "The Bohr Atomic Model: Niels Bohr." *Journal of Chemical Education,* Vol. 39, No. 10 (October 1962), pp. 534–535.

Garrett, A. B. "Quantum Theory: Max Planck." *Journal of Chemical Education,* Vol. 40, No. 5 (May 1963), pp. 262–263.

PROBLEMS

1. Calculate:
   (a) the frequency of light with a wavelength of 1000 Å.
   (b) the energy of light with a wavelength of 1000 Å.
   (c) the energy (in kilocalories) required for the transition from the ground state to the first excited state in one gram-atom of hydrogen atoms.

2. Distinguish carefully between:
   (a) an orbit and an orbital
   (b) frequency and wavelength
   (c) the Heisenberg uncertainty principle and the Pauli exclusion principle
   (d) a wave and a particle
   (e) quantum number and energy level
   (f) energy level and electron configuration.

3. Distinguish clearly between the Bohr model and the wave model in terms of:
   (a) nature of electron   (b) position of electron
   (c) energy of electron   (d) size of atom

4. Discuss the peculiarities of life in a world in which velocity is quantized.

5. Assume that the energy levels of a 100-g ball contained in a 1-meter, one-dimensional box are given by the equation for the electron confined to a line and calculate the spacing between the lowest energy levels. Would the behavior of the ball be affected in any observable way by this quantization?

6. Assume that most of the electron density of a hydrogen atom is contained within a sphere having a radius of 0.5 Å. Apply the formula developed for an electron confined to a line to the hydrogen atom and calculate the energies of the first three energy levels.

7. Calculate the energy of the first three levels of an electron in a hydrogen atom. Compare the energies and their spacings with those obtained in question 6.

8. Calculate the ionization energy of hydrogen in kcal/mole and compare with the experimental value.

9. If in a given collection of hydrogen atoms only the first four energy levels are populated, how many lines will be observed on the emission spectrum of this collection?

10. List all possible sets of n, l, m, and s for $n = 4$.

11. Why is there only one s orbital but three p orbitals in every major quantum level (except $n = 1$)?

12. For how many orbitals is the principal quantum number, n, equal to 5? Of those orbitals for which $n = 5$, how many have $l = 3$? how many have $m = +1$?

13. Distinguish clearly in terms of energy and electron probability distribution between a $2s$ and a $2p$ orbital for both hydrogen and carbon.

14. List the orbitals up to the $4p$ orbitals in order of increasing energy for the Ti^{2+} ion.

15. Make a plot in three-dimensional perspective of the electron probability pattern for a $1s$ and a $2s$ orbital. Use the size of dots to indicate relative probabilities.

16. List and explain the factors which, according to the wave model: (a) determine the size of an atom, and (b) determine the ionization energy of an atom.

17. Write electron configurations for:

(a) B	(b) N
(c) P	(d) K
(e) Ca^{2+}	(f) Fe
(g) Zn^{2+}	(h) Hg
(i) Br	(j) W
(k) U	(l) La
(m) Rn	(n) Te^{2-}
(o) Ag^+	(p) Lu

18. Carefully rationalize the following:
(a) K is larger than Na.

(b) Ca is smaller than K.

(c) The first ionization energy of F is greater than that of O.

(d) The first ionization energy of Br is less than that of Cl.

(e) O has a greater electron affinity than C.

(f) Lead is more metallic than carbon.

19. Arrange the elements in each of the following sets in order of increasing (a) ionization energy, (b) electronegativity, and (c) atomic radius.

(1) Li, Na, K, Rb   (2) Cl, Br, Se

(3) N, O, F   (4) He, Ne, Ar, Kr

20. Carefully account for each of the following:

(a) There is a change from metallic to nonmetallic character in going from sodium to chlorine in the third period.

(b) A positive ion is smaller than the atom from which it is derived, while a negative ion is larger than the atom from which it is derived.

(c) Much less energy is required to remove an electron from a gaseous fluoride ion than from a gaseous fluorine atom.

6 *Compounds II: The Ionic Model*

Our discussion of matter in the foregoing chapters has been concerned primarily with the internal structure and properties of *atoms*. Most of the matter on earth consists, however, not of single atoms but of combinations of atoms, that is, of *compounds*. In Chapters 6 through 8 we will use our knowledge of atoms together with the understanding of chemical composition and stoichiometry gained in Chapter 3 to explore the internal microscopic details of the nature of compounds.

Table 6-1 presents the melting points, boiling points, and relative electrical conductivities at the melting point of some of the compounds formed between chlorine and the third-period elements. It is obvious from the table that these compounds can be divided into two categories: one group's members have high melting and boiling points and conduct electricity in their molten state; the others have considerably lower melting and boiling points and do not conduct electricity

TABLE 6-1 Some Physical Properties of the Third-Period Chlorides	$NaCl$	$MgCl_2$	$AlCl_3$	$SiCl_4$	PCl_3	SCl_2	Cl_2
Melting point (°C)	801	708	ca. 190	−70	−91	−78	−101
Boiling point (°C)	1413	1412	ca. 190	58	76	59	−35

The molten compound conducts an electric current.

in their molten state. These two categories are useful descriptions for many types of compounds, not just the chlorides of the third-period elements.

The internal features of a compound that are responsible for the properties that place it in one category or the other have been discussed and studied by chemists for centuries. Our present understanding of these features is based upon two models: the *ionic* model and the *covalent* model.

Compounds of the first type, those with high melting points and so on are classified as *ionic*. According to the ionic model, these compounds consist of charged particles called *ions*. Thus the compound sodium chloride can be thought of as an orderly collection of positively charged, spherically symmetrical sodium ions and negatively charged, spherically symmetrical chloride ions.

Probably the two most compelling pieces of evidence for this model are (1) the electrical conductivity of "ionic" compounds in the liquid state, and (2) X-ray diffraction patterns of these compounds. When a battery is connected to molten sodium chloride as in Figure 6-1, an electric current flows through the wires. We will find in Chapter 17 that this flow of electrons produces a chemical reaction in the sodium chloride, but for now the salient feature of this experiment is that in order to have a flow of charged particles (electrons) in the wires, we must have a concomitant flow of charged particles in the molten sodium chloride. The charged particles in the melt are presumably sodium and chloride ions.

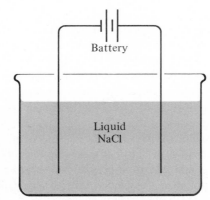

FIGURE 6-1

From the pattern that results when a stream of X-rays is diffracted by a crystal (see Chapter 9), information about the relative locations of the nuclei of compounds can be obtained. X-ray patterns of compounds like NaCl suggest that the electron density surrounding the nuclei in these compounds is spherically symmetrical. In addition, it has been demonstrated that in NaCl the sodium and chlorine nuclei are surrounded by 10 and 18 electrons, respectively. Since the neutral atoms have 11 and 17 electrons, respectively, the existence of ions in these compounds seems to be the best explanation of these data.

The ionic model also explains the high melting and boiling points of the compounds in this category. The temperature at which a compound boils, for example, is a measure of the amount of energy necessary to separate the "particles" of the compound to greater distances from one another, from their relative positions in the liquid state to their relative positions in the gaseous state. (Since gases are much less dense in general than liquids, the distances between particles in the gaseous state are much larger than those in the liquid state.) The amount of energy required for this change in state, then, depends on the nature of the forces between the "particles." In ionic compounds the "particles" being separated are ions, which are attracted to one another by very strong electrostatic forces. As we shall see below, the "particles" of a covalent compound are usually molecules. The forces between two uncharged molecules are considerably weaker than between ions.

According to the *covalent* model the atoms of compounds in the second category are held together by electron density shared between them. The binding in covalent compounds is quite directional; that is, the atoms are held at more or less fixed angles and distances relative to one another. In CCl_4, for example, four chlorine atoms are bonded to one carbon atom in such a way that the angle between any two chlorine atoms (the Cl–C–Cl angle) is $109°$. Atoms also tend to form only a certain number of covalent bonds: Carbon usually forms four bonds, but chlorine only one. This limitation results in aggregates of atoms that move and function as a unit. These aggregates are called *molecules*.

Evidence for the covalent model includes: (1) the nonconductivity of many liquids, which indicates the absence of charged particles; (2) the highly directional character of the "bonds" in molecules in solid, liquid, and gaseous states; and (3) the existence of molecules as evidenced by X-ray diffraction and electron microscopy (very large molecules have been "seen" in the electron microscope).

The rather low melting and boiling points for many covalent compounds have already been explained. However, not all covalent compounds have low melting and boiling points. Some consist of very long chains of atoms all connected by covalent bonds. As these compounds pass from one state to another, covalent bonds are severed, and this process requires a great deal of energy.

The two models—ionic and covalent—are represented pictorially in Figure 6-2, where the enclosed areas represent regions of high electron density. The ionic

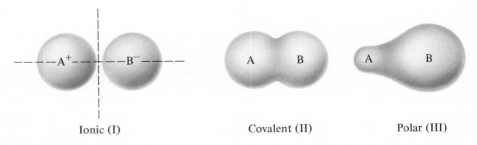

Ionic (I)   Covalent (II)   Polar (III)

FIGURE 6-2

model assumes that there is no electron density in the plane (vertical line) perpendicular to both the internuclear axis (dashed line) and the plane of the paper. The covalent model, on the other hand, shows electron density evenly shared between the A and B nuclei. For most compounds the disposition of electron density is somewhere between these two extremes. Such a case is shown in Figure 6-2c, where nucleus B is surrounded by *almost* one extra electron and A has lost *almost* one electron, and there is definitely a sharing of some electron density. This bond is usually referred to as a *polar bond,* or as having some ionic character.

We will now examine each of these models in more depth, attempting at each stage to rationalize the details of the internal, microscopic structure of compounds. The remainder of this chapter will be devoted to the ionic model; the covalent model will be discussed in Chapter 7.

TYPES OF IONS

Among the first questions to arise in any discussion of the ionic model are: What kinds of compounds can be classified as ionic, and what is the nature of the ions formed by the various elements?

The formation of a binary (two-element) ionic compound from its constituent elements must inevitably be accompanied by a transfer of electrons from the element forming the positive ion (the *cation*) to the element forming the negative ion (the *anion*). Thus, to a first approximation, ionic compounds are formed when elements that lose their electrons easily—elements that have low ionization energies—combine with elements that gain electrons easily—ones that have high electron affinities. Since metals have low ionization energies, it is not surprising that most of the compounds that have appreciable ionic character contain metallic elements. Similarly, the very electronegative nonmetals readily gain electrons, and consequently most ionic compounds have anions that contain these elements.

The second question—that is, what kinds of ions are formed by the various elements—can be answered by a combination of (a) the previously formulated rule that filled quantum levels have a special stability (inert gas rule, p. 79) and (b) the ionization energies of the elements. Table 6-2 presents the ionization energies for sodium, magnesium, chlorine, and iron. Sodium has a low first ionization energy and a very high second ionization energy. It also has an electron configuration of

TABLE 6-2 Selected Ionization Energies (kcal/mole)		*1st*	*2nd*	*3rd*	*4th*
	Na	119	1091		
	Mg	176	347	1848	
	Cl	300	549	920	1230
	Fe	182	373	707	

$1s^2 2s^2 2p^6 3s^1$, which means that it will have a filled second quantum level as its outer shell if it loses one electron. Thus, both ionization energies and the inert gas rule would predict that the most likely ion for sodium (and all of the alkali metals) to form is the plus one ion, designated Na^+. For magnesium and the alkaline earth metals, the same reasoning predicts the formation of plus two ions. Chlorine, on the other hand, certainly could lose an electron—the first ionization energy is not prohibitively high—but this would not lead to a filled quantum level, since the electron configuration of chlorine is $1s^2 2s^2 2p^6 3s^2 3p^5$. However, if it acquires an electron, chlorine will obtain the inert gas configuration. The halogens, then, are usually present in ionic compounds as the minus one anions. Iron has an electron configuration of $1s^2 2s^2 2p^6 3s^2 3p^6 3d^6 4s^2$, and thus an inert gas configuration cannot be achieved by loss or gain of any small number of electrons. The first, second, and third ionization energies are at most moderately high, and thus it would seem that iron could lose one, two, or three electrons. In fact, iron is found most commonly as the 2+ or 3+ ion.

Several qualifying remarks are necessary at this point. First, the feasibility of forming any particular ionic compound is a function of many factors, such as the stability of the particular arrangement adopted by the ions in the solid, as well as the ease of formation of the cation (and anion). These other factors will be discussed later in this chapter. Second, the existence of ions of high charge, greater than 2+ or 2−, becomes more improbable the higher the charge. Even compounds that appear to contain a 3+ ion, for example $AlCl_3$, may be rather covalent. A rationalization for this will also be presented later. Third, while the discussion above has been directed toward binary compounds, there are, of course, many ionic compounds that contain more than two elements; KNO_3 and $CuSO_4$ each contain three. In most of these compounds the anion is an oxyanion; that is, the anion

TABLE 6-3 Common Cations		
Group IA metals	1+
Group IIA metals	2+
Al	3+
Sn	2+
Pb	2+
Cr	2+, 3+
Mn	2+
Fe	2+, 3+
Co	2+, 3+
Ni	2+
Cu	1+, 2+
Zn	2+
Cd	2+
Ag	1+
Au	1+, 3+
Hg	1+[a], 2+

[a] Actually Hg_2^{2+}

consists of oxygen covalently bonded to another fairly electronegative element. In $CuSO_4$, the anion is $SO_4{}^{2-}$; in $Mg(ClO_4)_2$ the anion is $ClO_4{}^-$. The names and formulas of such ions will be discussed in the next section of this chapter. Since the bonding within these ions is covalent, their structure will be discussed in Chapters 7 and 8.

It should be obvious from our discussion of iron that the charges of the ions of some elements cannot easily be predicted and must therefore be committed to memory. Table 6-3 presents the most commonly encountered metals and their common ions. Remember, however, that compounds containing some of these "ions" are quite covalent.

THE NOMENCLATURE OF IONIC COMPOUNDS

Chemists frequently refer to compounds by name rather than by formula, and it is therefore necessary to know how to derive names from formulas and vice versa. The name of any ionic compound is a composite of the names of its cation and anion.

Cation Names   Traditionally, cations of an element that form only one ion—for example, sodium— are named exactly like the element; thus, Na^+ is the sodium ion. Cations of elements that can form more than one ion—for example, iron—are named by adding a suffix to the root of the name from which the chemical symbol was derived. These roots are given in Table 6-4.

TABLE 6-4
Name Roots
of Some Elements

Element	Symbol	Name Root
Gold	Au	Aur-
Chromium	Cr	Chrom-
Cobalt	Co	Cobalt-
Copper	Cu	Cupr-
Iron	Fe	Ferr-
Manganese	Mn	Mangan-
Mercury	Hg	Mercur-
Lead	Pb	Plumb-
Tin	Sn	Stann-

To distinguish between two possible ions of the same element, the suffix *-ous* is added to the root to indicate the ion of lower charge, while the suffix *-ic* indicates the ion of higher charge. The Fe^{2+} ion is therefore the ferrous ion and Fe^{3+} is the ferric ion. When there are more than two possible ions, this traditional system becomes rather unwieldy, and the International Union of Pure and Applied Chemistry (IUPAC) has proposed an alternate system. The IUPAC rule designates an ion by simply giving the name of the element followed in parentheses by the charge on the ion in Roman numerals. Thus, Fe^{3+} is iron(III).

A very common cation not included in the rules above is the ammonium ion, NH_4^+.

Anion Names

The common anions can be derived—on paper if not always in the laboratory—from acids. For this purpose, an acid can be defined as a substance that contains one or more hydrogen atoms that can be removed as H^+ ions by some chemical means. Generally, a hydrogen ion will not be easily removed unless it is attached to an element more electronegative than carbon. It is also generally true that the removable hydrogens are attached to the most electronegative atoms in the covalent acid. For example, the acid with molecular formula HNO_2 has the structural formula $H\cdots O\cdots N\cdots O$, in which the hydrogen is attached to the more electronegative atom, oxygen. Anions are produced when one or more hydrogen ions are removed from the acid; the name of the anion is based on the name of the parent acid.

The first step in learning to name the common anions, then, is to learn the names of the acids from which they are derived. Table 6-5 lists such acids under two categories: those that contain oxygen (oxy acids) and those that do not.

Note that the names of the first group of acids contain the prefix *hydro-* attached to the root of the name of the element (or group of elements) to which the hydrogen is attached, plus the suffix *-ic* plus *acid.* The names of most of the *oxy* acids are derived from the name of the central element, with a suffix that indicates the relative number of oxygen atoms. Thus, in the nitric acid molecule, three oxygen atoms are attached to a nitrogen according to the formula

$$O \\ \vdots \\ O\cdots N\cdots OH$$

The nitrogen is then the central atom, which accounts for the *nitr-* portion of the name. The suffixes *-ic* and *-ous* are used to designate more or less oxygen, respectively, relative to another oxy acid containing the same central element. When there are only two common acids containing the same central element, the nomenclature is straightforward. For example, HNO_3 is nitric acid; HNO_2, with fewer oxygens, is nitrous acid; H_2SO_4 is sulfuric acid; and H_2SO_3 is sulfurous acid.

When there are more than two oxy acids of the same element, the prefix *per-* is added to the name of the *-ic* acid to indicate more oxygen, while the prefix *hypo-* is added to the name of the *-ous* acid to designate less oxygen. The chlorine oxy acids are a good example of this nomenclature:

$HClO_4$ perchloric acid
$HClO_3$ chloric acid
$HClO_2$ chlorous acid
$HClO$ hypochlorous acid

TABLE 6-5	*Acids*		*Anions*	
Names				
of Some Acids	HF	hydrofluoric acid	F^-	fluoride ion
and Their Anions	HCl	hydrochloric acid	Cl^-	chloride ion
	HBr	hydrobromic acid	Br^-	bromide ion
	HI	hydroiodic acid	I^-	iodide ion
	H_2S	hydrosulfuric acid	HS^-	hydrogen sulfide ion
			S^{2-}	sulfide ion
	HCN	hydrocyanic acid	CN^-	cyanide ion
	Oxy Acids			
	HNO_3	nitric acid	NO_3^-	nitrate ion
	HNO_2	nitrous acid	NO_2^-	nitrite ion
	H_2SO_4	sulfuric acid	HSO_4^-	hydrogen sulfate ion
			SO_4^{2-}	sulfate ion
	H_2SO_3	sulfurous acid	HSO_3^-	hydrogen sulfite ion
			SO_3^{2-}	sulfite ion
	H_3PO_4	phosphoric acid	$H_2PO_4^-$	dihydrogen phosphate ion
			HPO_4^{2-}	hydrogen phosphate ion
			PO_4^{3-}	phosphate ion
	H_2CO_3	carbonic acid	HCO_3^-	hydrogen carbonate ion
			CO_3^{2-}	carbonate ion
	$H_2C_2O_4$	oxalic acid	$HC_2O_4^-$	hydrogen oxalate ion
			$C_2O_4^{2-}$	oxalate ion
	$HC_2H_3O_2$	acetic acid	$C_2H_3O_2^-$	acetate ion
	$HClO_4$	perchloric acid	ClO_4^-	perchlorate ion
	$HClO_3$	chloric acid	ClO_3^-	chlorate ion
	$HClO_2$	chlorous acid	ClO_2^-	chlorite ion
	$HClO$	hypochlorous acid	ClO^-	hypochlorite ion

If there is only one common oxy acid of an element, it is given the *-ic* designation; for example, H_2CO_3, carbonic acid. The names of some of the acids in Table 6-5 are not derived according to this system, and at this juncture it is best to simply memorize them.

Having examined the nomenclature of acids, we are now in a position to derive the names of anions. When hydrogen ions are removed from acids that do not contain oxygen, the resulting ions are of the type:

Cl^- (take H^+ away from HCl)
S^{2-} (take $2H^+$ from H_2S)

These ions are named by dropping the *hydro-* prefix of the parent acid and replacing the *-ic* suffix with *-ide*. Thus, from hydrochloric we get chloride as the name of the ion Cl^-. Table 6-6 lists four anions whose names have the *-ide* ending

TABLE 6-6	H^-	hydride
Some Additional Anions	O^{2-}	oxide
	N^{3-}	nitride
	OH^-	hydroxide

but are derived from compounds not usually considered as acids. Note that the names of the monatomic ions are based on the name root of the element.

When hydrogen ions are removed from oxy acids, the resulting anions are given the name of the parent acid with its suffix changed—from *-ic* to *-ate* or from *-ous* to *-ite*. The chlorine oxy anions are then:

$$ClO_4^- \quad \text{perchlorate}$$
$$ClO_3^- \quad \text{chlorate}$$
$$ClO_2^- \quad \text{chlorite}$$
$$ClO^- \quad \text{hypochlorite}$$

If less than all of the removable hydrogens of an acid are lost to form the anion, the number of hydrogens remaining are denoted by the word hydrogen for one, dihydrogen for two, trihydrogen for three, and so forth. The HCO_3^- ion is called the hydrogen carbonate ion; the $H_2PO_4^-$ ion is the dihydrogen phosphate ion. An older system adds the prefix *bi-* to name ions that have retained one of two hydrogens; HCO_3^- in this system is the bicarbonate ion.

We are now able to put the entire name of an ionic compound together by simply adding the cation name to the anion name. The compound with empirical formula $MgCl_2$ is called magnesium chloride; $RbNO_3$ is rubidium nitrate.

To name or write the formula for some compounds, it is obviously necessary to know the charge on the ions. In the compound $FeCl_3$, the charge on the iron ion is determined by recognizing first that the sum of the charges of all the ions must be zero because the compound is electrically neutral. Now, since the chloride ion always has a charge of 1− and since there are three chloride ions for every iron ion, the iron ion must have a charge of 3+. This compound is therefore ferric chloride, or iron(III) chloride.

Formulas are derived from names in a similar fashion. Cobaltous nitrite is $Co(NO_2)_2$—the cobaltous ion has a charge of 2+, whereas the nitrite ion is only 1−, so there must be twice as many nitrite ions as cobaltous ions in order to preserve electrical neutrality.

POTENTIAL ENERGY OF IONIC CONFIGURATIONS

Now let us examine the ionic model from a microscopic view. Imagine sitting on a sodium ion in the middle of a crystal of sodium chloride. This ion, according to data from X-ray diffraction studies, is surrounded by six chloride ions. The sodium, then, is situated at the center of the solid geometrical figure, the octahedron, with chloride ions at the apices. There are, of course, equal numbers of sodium ions and chloride ions in NaCl, and so each chloride ion is also surrounded by six sodium ions. This configuration is pictured in Figure 6-3. Now, our objective in sitting on the central sodium ion is to experience the forces that exist between the ions, the forces that hold the ions together and make NaCl a stable compound. Before analyzing these forces, however, we will digress to a consideration of one of the fundamental laws of electrostatics.

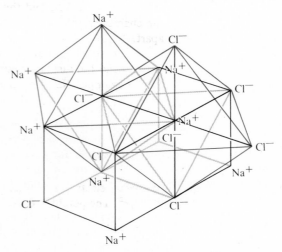

FIGURE 6-3

A Portion of the NaCl Structure

It is well known that oppositely charged objects—one with an excess of electrons and one with a deficiency of electrons—are attracted to one another, while objects with the same charge repel one another. In the latter part of the eighteenth century, the French physicist Auguste Coulomb put these observations on a quantitative basis with his experiments on the force between charged bodies. He found that the force between two charged bodies varies directly as the product of the charges on the two bodies and inversely as the square of the distance between the two bodies. Mathematically, this becomes

$$F = \frac{kq_1q_2}{r^2}$$

where F is the force between the two bodies, q_1 and q_2 are the charges on the first and second bodies, respectively, r is the distance between the bodies, and k is a proportionality constant.

The charge can be expressed in a number of different units. Probably the most fundamental is the *atomic unit* (au), which is the charge on a single proton; the charge on an electron has the same magnitude but a negative sign. Two other, more practical, units are (a) the *coulomb,* which is equal to 6.2×10^{18} atomic units, and (b) the *electrostatic unit* (esu), which is equal to 2.1×10^9 au. When q is expressed in electrostatic units, the proportionality constant in the Coulomb expression can be dropped.

Another way to express the attraction or repulsion between charges is in terms of their potential energy. As noted in Chapter 1, potential energy is the capacity of a system to do work by virtue of its position; in the centimeter-gram-second system of units this is expressed in ergs. A motionless boulder perched on the edge of a cliff has the potential to do work because as soon as it is released it possesses kinetic energy, which enables it to knock over obstacles in its path, for example. Two charges held a fixed distance apart also have potential energy. If these charges have opposite signs and if they were able to move, they would move toward one another, and this motion could produce work. If, on the other hand, the charges are of the same sign, work must be done to prevent them from flying apart.

Let us restrict our attention now to two ions separated by a distance r (where r is actually the distance between the centers of the ions)—one ion with a charge of 1+, for example, Na^+; the other with a charge of 1−, for example Cl^-.

If we use electrostatic units for the charge, the force between these ions is given by q_1q_2/r^2. The potential energy, V, of the ions is given by the expression

$$V = \frac{q_1q_2}{r}$$

Note carefully two aspects of this equation: First, if the ions are oppositely charged, as in our present example, the potential energy is negative (because one of the q's is negative). Secondly, as r gets larger, V approaches zero, and when r is infinity, V is zero. Consequently, the potential energy for any given r is the negative of the work necessary to separate the ions to infinity.

The energy of the two ions is shown in Figure 6-4 for two different values of

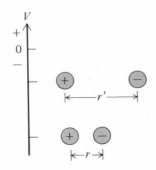

FIGURE 6-4

r. Obviously, the arrangement with the larger *r*, *r'*, has the higher energy (greater work is possible if the ions were to collapse toward one another). Thus, energy must be supplied to separate the ions, and the lower energy configuration is more stable.

Let us now calculate the potential energy of our two ions when *r* is 2.80 Å, the experimentally determined distance between the centers of the ions in solid sodium chloride. The charge on the sodium ion is one atomic unit, or, since 1 au = 4.8×10^{-10} esu, 4.8×10^{-10} esu. The charge on the chloride ion is -4.8×10^{-10} esu. The potential energy is then

$$V = \frac{(4.8 \times 10^{-10} \text{ esu}) (-4.8 \times 10^{-10} \text{ esu})}{2.8 \times 10^{-8} \text{ cm}} = -8.2 \times 10^{-12} \text{ esu}^2/\text{cm}$$

One esu^2/cm is equal to one erg, and the potential energy of this configuration of ions is -8.2×10^{-12} erg.

The potential energy of our sodium ion seat is, however, not the simple energy just evaluated. Our ion is attracted to its six equidistant chloride neighbors but repelled by the sodium ions diagonally removed from it at a greater distance, attracted to chloride ions even further removed, and so on. The total potential energy of our sodium ion, then, is a sum of an infinite number of terms all of the type $A(q_1 q_2/r)$, some with a negative sign, some with a positive sign. In the first term, representing attraction to six neighbors, A would be 6 and *r* the distance between the sodium and the chlorines. Fortunately, this infinite series converges— to the value $1.75(q_1 q_2/r)$, which in this case equals

$$-1.75 \frac{(4.8 \times 10^{-10})^2}{2.8 \times 10^{-8}}$$

The number 1.75 is called the *Madelung constant* and is the sum of the *A*-type coefficients of all the terms in the infinite series. The Madelung constant for any solid whose ions are arranged geometrically in this fashion will have the same numerical value.

It is convenient at this point to calculate the potential energy of a mole of sodium and chloride ions, and this is

$$V_{\text{mole}} = -(6.02 \times 10^{23}) (1.75) (8.2 \times 10^{-12}) \text{ ergs}$$

$$= -8.6 \times 10^{12} \text{ ergs}$$

Actually this is still not quite the potential energy of a mole of crystalline NaCl. Our calculation up to this point has considered the ions only as point charges. In fact, the electron clouds of the real ions repel one another somewhat, and this leads to an increase in the potential energy of about 10 percent.

Chemists generally prefer to express energy in terms of *calories*. One calorie is the amount of heat energy necessary to raise the temperature of one gram of water one degree centigrade. One erg is equivalent to 2.4×10^{-8} cal. Thus, the potential energy calculated for a mole of NaCl, increased by 10 percent, becomes -19×10^4 cal or -190 kcal. This is the negative of the work required to separate one mole of sodium and chloride ions from their positions in the crystal to infinite distances from one another.

In reality, it is impossible to separate the ions by infinite distances, but the separation they undergo as the compound passes from the solid phase to the gaseous phase is very nearly equivalent, since the distance between the ions in the gaseous state is much greater than the distance in the solid. The *energy evolved when the solid is formed from its constituent ions in the gaseous state* is called the *lattice energy* and can be determined from experimental data. For sodium chloride, the experimentally determined lattice energy is 184 kcal/mole. The good agreement between the calculated and experimental lattice energies is additional evidence for the existence of ions in these compounds.

Our treatment above suggests that lattice energy is a function primarily of three factors: (1) the charge on the ions, (2) the distance between the ions, and (3) the arrangement of ions in the solid (which determines the Madelung constant). In the alkali halides, the Madelung constant remains virtually constant (1.75 to 1.76, depending on the type of lattice), and the decrease in lattice energy in the series LiF > NaF > KF and NaF > NaCl > NaBr, as shown in Table 6-7, is due solely to an increase in ionic distance within these series (the greater the distance, the lower the absolute value of the potential energy). The Madelung constant for the alkaline earth oxides and sulfides is also 1.75; the much larger lattice energies for these compounds are obviously due, then, to the greater charge on the oxide and sulfide ions. Since q_1 and q_2 for the alkaline earth oxides and sulfides are both twice what they are for the alkali halides, the alkaline earth lattice energies should be roughly four times as large. Indeed, NaCl and CaS both have an interionic distance of 2.8 Å, and the lattice energy of CaS is very nearly four times that of NaCl.

TABLE 6-7 Selected Lattice Energies and Interionic Distances	Compound	Lattice Energy (kcal/mole)	Distance (Å)
	LiF	241	2.01
	NaF	216	2.31
	KF	192	2.66
	NaCl	184	2.81
	NaBr	176	2.98
	MgF_2	695	2.02
	MgO	938	2.10
	MgS	788	2.54
	CaO	841	2.40
	CaS	726	2.83

The lattice energy, then, is a measure of the stability of the crystalline solid with respect to its constituent gaseous ions. It would appear, therefore, that the greater the charge on the ions and the smaller the ions, the more favorable the formation of the solid. However, the stabilities (and ease of formation) of crystal-line ionic substances are usually measured relative to their constituent elements, not the gaseous ions. Hence, as indicated before, the ease of formation of an ionic compound from its elements is a function of the energy required to form the gaseous cation and anion (and this depends primarily on the ionization energy of the metal and the electron affinity of the nonmetal) as well as the energy released in the formation of the ionic lattice. Thus, while ions of high charge and small size result in large lattice energies, a large amount of energy is also required for their formation from the elements. The calculations necessary for the prediction of the stability of ionic compounds will be discussed in Part Three of this book.

IONIC LATTICES

The particular geometrical arrangement (the lattice) of ions adopted by an ionic compound will be the one for which the potential energy of the solid is least. The potential energy (lattice energy), in turn, is a function of three factors:

1. The cation-anion distance; the smaller this distance, the lower the potential energy.
2. The coordination number—that is, the number of equidistant ions of opposite charge; the greater the coordination number, the lower the energy.
3. The cation-cation and anion-anion distances; the smaller these distances, the greater the repulsions and the higher the energy of the crystal.

Thus, in general, the greater the number of anions that can be squeezed around a cation without causing significant repulsions between the anions, the more stable the crystal.

In order to see how these factors work in more detail, let us examine the most common lattice structures of simple ionic compounds of the AB type (1:1 ratio of cation to anion). These lattices are the cesium chloride lattice, the zinc blende (ZnS) lattice, and the sodium chloride lattice, portions of which are shown in Figures 6-5, 6-6, and 6-7.

In the cesium chloride lattice, each ion has a coordination number of eight: each cesium ion is surrounded by a cube of chloride ions—chlorides at the apices of the cube—and each chloride ion is surrounded by eight cesium ions. In actuality the ions are very nearly touching (as shown for NaCl in Figure 6-6), so that the large cesium ion fits snugly into the center of the chloride cube. Now, when the size of the cation is decreased, as when the potassium ion replaces the cesium ion, the anions move even closer, resulting in more repulsion, and the smaller cation "rattles around" in the hole. In order to minimize anion-anion repulsion and the cation-anion separation, the crystal lattice changes to the sodium chloride structure. In this lattice the coordination number is six—the cation is surrounded by six anions at

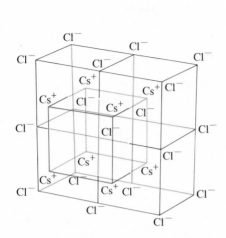

FIGURE 6-5

The Cesium Chloride Lattice

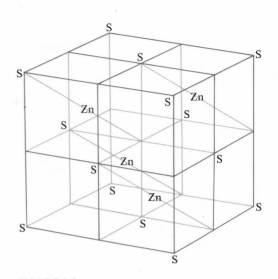

FIGURE 6-6

The Zinc Blende Lattice

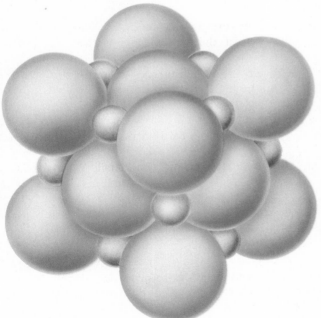

Ions represented as
spheres in close contact

FIGURE 6-7

The Sodium Chloride Lattice

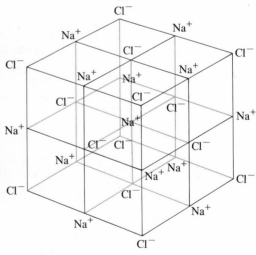

Positions of centers
of ions

the apices of an octahedron—and the coordination number, therefore, has decreased, but the potential energy has also decreased. Thus, the size of the cation relative to the size of the anion controls the arrangement of ions. In fact, by treating the ions of AB-type compounds as hard spheres, it can be shown with simple geometry that when the ratio of cation to anion radii, $r+/r-$, is between 1 and 0.73, the cesium chloride structure is favored, whereas a ratio between 0.73 and 0.41 is calculated for the sodium chloride lattice. When the ratio is even lower, between 0.41 and 0.24, the zinc blende structure, where the coordination number is four, is favored. Table 6-8 lists the crystal lattices of some AB-type compounds and their radius ratios as derived from the ionic radii of Appendix 2. The data here clearly illustrate that the radius ratio rules provide only a rough guide to crystal structure.

TABLE 6-8
Structures of Some Ionic Compounds

Compound	Radius Ratio, r^+/r^-
CsCl Structure	
CsCl	0.94
CsBr	0.87
TlCl	0.83
TlI	0.68
NaCl Structure	
LiI	0.34
NaI	0.46
RbI	0.68
CsF	1.28
MgO	0.51
MgS	0.39
Zinc Blende Structure	
AgI	0.52
BeS	0.16
ZnS	0.41

PROBLEMS

1. Consider a compound with empirical formula AB_2. What physical or chemical properties would help to classify this compound as "ionic" or "covalent"? How?

2. The most stable ions for three neighboring elements, Ni, Cu, and Zn, are the 2+ ions. Only for copper are compounds containing the 1+ ion somewhat stable. Explain this observation by making use of the ionization energies of these elements.

3. Name the following compounds:
   (a) $HClO_2$ (b) H_3PO_3 (c) HNO_2
   (d) K_2SO_3 (e) $NaHCO_3$ (f) $FeCl_2$
   (g) NiO (h) $Cu(OH)_2$ (i) $AgCN$
   (j) $HgSO_4$ (k) $Ca_3(PO_4)_2$ (l) CoF_2
   (m) $Al(NO_3)_3$ (n) CsI (o) $MgCO_3$
   (p) $LiClO_4$ (q) $Mn(IO_3)_2$

4. Write a formula for each of the following compounds:
   (a) potassium carbonate   (b) cesium chlorate
   (c) hypochlorous acid   (d) sodium bromide
   (e) hydrosulfuric acid   (f) strontium sulfite
   (g) copper(II) phosphate   (h) cuprous chloride
   (i) nickel(II) permanganate   (j) chromic sulfate
   (k) mercuric nitrate   (l) lithium bromate
   (m) indium(III) oxide   (n) aluminum hydroxide
   (o) bismuth(III) chloride   (p) ferric nitrite
   (q) silver sulfide   (r) gold(I) cyanide
   (s) hydrofluoric acid   (t) rubidium hydrogen sulfite
   (u) barium dihydrogen phosphate

5. Make rough plots on the same graph of the variation of the potential energy of two ions as a function of the distance between them for (a) a 1+ cation and a 1− anion and (b) a 2+ cation and a 2− anion.

6. Calculate the total potential energy in ergs of two protons and two electrons situated at the corners of a square as follows:

7. Account for the following trends in lattice energies:
   (a) $LiF > NaF > KF$ (b) $NaF > NaCl > NaBr$
   (c) $MgF_2 > NaF$ (d) $MgO > MgF_2$

8. Strontium oxide has the sodium chloride structure. Calculate its lattice energy in kilocalories per mole. Use the sum of the ionic radii as the interionic distance.

9. Magnesium oxide crystallizes in the same lattice arrangement as sodium chloride. Predict whether the melting point and lattice energy of MgO are higher or lower than those of NaCl. Give reasons for your predictions.

10. Use the radius ratio rules to predict the lattice structure of the following:
    (a) KBr (b) NaI
    (c) $BeSe$ (d) MgO

7 Compounds III: The Covalent Model

THE STABILITY OF COVALENT COMPOUNDS

The electrostatic forces that stabilize ionic compounds are readily understood, but the reasons for the stability of covalent compounds are less apparent. Consider the very simple molecule H_2, one mole of which is 103 kcal lower in energy than two moles of its constituent hydrogen atoms. The distance between the two protons in the molecule (remember that a hydrogen nucleus is a proton) is much smaller than the distance between the nuclei of two hydrogen atoms in the gaseous state. This is shown in Figure 7-1. The smaller internuclear distance in the molecule should result in a higher energy for H_2 because of the greater electrostatic repulsion.

The greater repulsions in the H_2 molecule are, however, only one factor in

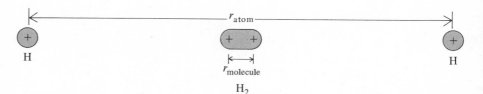

FIGURE 7-1

Relative Internuclear Distances for Gas-Phase H and H_2

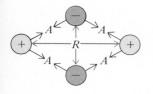

A = attraction

R = repulsion

FIGURE 7-2

A Representation of the
Electrostatics of the
H_2 Molecule

.the difference between the potential energies of the molecule and the separated atoms. Since a considerable amount of the electron density in H_2 resides between the nuclei, nearly two electrons are attracted to each nucleus. The electrostatic attractions in the molecule are also greater than the attractions within and between two free hydrogen atoms. The total attractions and repulsions are represented very artificially in Figure 7-2. (The electrons are not, of course, stationary; nor are there exactly two electrons located in exactly this way between the nuclei.) This representation illustrates in a qualitative way the predominance of the attractions over the repulsions. It is this predominance of nuclei-electron attractions, due to the buildup of electron density between the nuclei, that is responsible for the lower potential energy of the H_2 molecule. Of course, the total energy of any system is the sum of the potential and kinetic energies, but it is a result of wave mechanics that, although the kinetic energy of the H_2 molecule is greater than the kinetic energy of the atoms, the increase in kinetic energy is only half of the decrease in potential energy.

While we have now succeeded in rationalizing the stability of H_2 and covalent molecules in general, we are far from developing models of the covalent bond that will enable us to explain and/or predict the properties of these substances. Before turning our attention to several of these models, we will discuss some of these properties.

MOLECULAR PARAMETERS

Since most covalent compounds consist of molecules—aggregates of atoms, intimately and (usually) strongly bound, that function as units in the solid, liquid, and gaseous states—it will be useful at this point to examine some physical properties of molecules.

Bond Length   For our present purposes we will assume that the atoms constituting a molecule remain in fixed positions relative to one another. The distance between the centers of two adjacent atoms which is called the *bond length,* can be determined by X-ray diffraction and other means. Thus, in CH_4, in which four hydrogen atoms are attached to a central carbon atom, the distance between any hydrogen nucleus and the carbon nucleus has been found to be 1.094 Å. When the molecule contains more than two elements—for example, formaldehyde, H_2CO—there will naturally be more than one bond length.

Bond Angle   *Bond angle* can be defined as the angle between any two adjacent internuclear lines. In CH_4, the angle between any two adjacent C—H bonds—strictly speaking, the hypothetical lines connecting the carbon and hydrogen nuclei—has been determined to be 109.5°. (Bond angles can also be determined by X-ray diffraction.) In carbon dioxide, $O \cdots C \cdots O$, the bond angle is $180°$; in other words, the atoms lie on a single line and the molecule is said to be *linear*. In formaldehyde there are two different bond angles: the $H \cdots C \cdots O$ angle is $119°$, the $H \cdots C \cdots H$ angle is $122°$ (Figure 7-3). Note the use of dotted lines to portray internuclear lines.

FIGURE 7-3

Bond Angles in Formaldehyde

Bond Energy

For a diatomic molecule, the *bond energy* is the *amount of energy necessary to dissociate one mole of the gaseous compound into its gaseous atoms.* Thus, the dissociation of one mole of H_2 into hydrogen atoms requires 103 kcal.

$$H_2 \rightarrow 2H \qquad \text{bond energy} = 103 \text{ kcal/mole}$$

The energy required to dissociate H_2O into its neutral constituent atoms is 221 kcal/mole.

$$H_2O \rightarrow 2H + O$$

Since there are two O—H bonds, the *average* O—H bond energy in water is $221/2 =$ 110.5 kcal/mole. The bond energy for any given set of atoms, say OH, depends upon the nature of the remainder of the molecule, but generally the variation from one molecule to another is small enough that a set of empirical average bond energies can be tabulated (see Appendix 5).

Bond Polarity

The electron density between the atoms of most covalent compounds is not shared equally by the atoms. In a bond containing atoms A and B, the electron density will be shared equally only if A and B are the same or have the same electronegativities. Since, as Linus Pauling originally defined it, electronegativity is "the power of an atom within a molecule to attract electrons to itself," it seems reasonable to assume that the greater the difference in electronegativity between A and B, the more unequal the sharing of electrons will be (see Chapter 4 for the Pauling electronegativity scale). This is illustrated in Figure 7-4 for four situations: I, where A and B are the same; II, where the difference in the electronegativities of A and B is moderate; III, where the difference in the electronegativities of A and B is great; and IV, where the difference in the electronegativities is so large that B has completely captured an electron from A. The enclosed areas in Figure 7-4 represent the electron density of two electrons.

It is sometimes convenient to think of the bond pictured in the intermediate cases as a mixture, or hybrid, of the equal-sharing case (I) and the ionic case (IV). Thus, a 60–40 hybrid of I and IV might produce II, while 40–60 hybridization of I and IV might result in III. The bond in II, then, would be said to have 40 percent ionic character, and the bond in III would have 60 percent ionic character.

The inequality of the electron distribution, or polarity, of a diatomic molecule is often measured by its *dipole moment.* Mathematically, the dipole moment, μ, of a neutral molecule is given by the product of the amount of charge, q, which has been unequally distributed in the molecule and the distance, d, between the unequal charges:

$$\mu = qd$$

The dipole moment of case II can be calculated as follows: An ionic character of 40 percent means that essentially 0.4 electron has been transferred from A to B and q is therefore $0.4 \times (4.8 \times 10^{-10}$ esu) or 1.9×10^{-10} esu. If we assume, for computational simplicity, that the AB internuclear distance is 1.0 Å, or $d = 1.0 \times 10^{-8}$ cm, the dipole moment is

$$\mu = (1.9 \times 10^{-10} \text{ esu}) \times (1.0 \times 10^{-8} \text{cm}) = 1.9 \times 10^{-18} \text{ esu-cm}$$

The quantity 1.0×10^{-18} esu-cm is called a Debye (1.0 D), in honor of Peter Debye, who did much of the early work on dipole moments. Note that if one electron had been completely transferred from A to B (case IV), the dipole moment would be 4.8 D. The percent ionic character of case II could be calculated from the dipole moment as

$$\frac{1.9}{4.8} \times 100 = 40\%$$

Experimentally, dipole moments are obtained by observing the effect of the compound on the capacitance (the ability to store charge) of a capacitor containing the compound in solution. The plates of the capacitor are oppositely charged, as shown in Figure 7-5. Molecules with equal electron distribution, that is, no dipole moments, will orient themselves randomly in the electric field of the capacitor. On the other hand, molecules with uneven electron distributions have positive and negative ends (they do *not,* however, contain ions) and therefore align themselves with the electric field of the capacitor. The greater the separation of charge in the molecule, the greater the alignment in the capacitor and the greater the effect on the capacitance of the capacitor. The measured capacitance of the solution can be converted via a series of mathematical relationships to dipole moment.

I A = B

II B > A

III B >> A

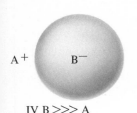

IV B >>> A

FIGURE 7-4

Electron Sharing
as a Function
of Electronegativity
Differences

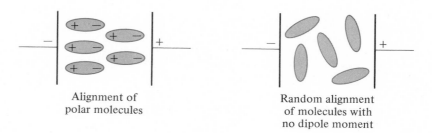

Alignment of
polar molecules

Random alignment
of molecules with
no dipole moment

FIGURE 7-5

Orientation of Polar Molecules in Capacitor

TABLE 7-1
Some Dipole Moments

Compound	Dipole Moment (D)
H_2	0
HI	0.38
HBr	0.79
HCl	1.03
HF	1.98
N_2	0
NO	0.16
CO_2	0
H_2O	1.87
BF_3	0
CF_4	0

Table 7-1 presents dipole moments of some molecules. In the hydrogen halide series, the increase in internuclear distance d from HF to HI due to the increase in size of the halogen is far outweighed by the decrease in ionic character (and thus q) in the same order. The table also contains dipole moments for several polyatomic molecules.

The reader should at this point be somewhat dismayed by the zero dipole moment for, say, CO_2. Surely the carbon-oxygen bonds in this covalent molecule are polar: The difference in electronegativity between C and O is rather large. However, bond polarity is not the only factor determining the dipole moment of a polyatomic molecule; the spatial disposition of the bonds is also critical. The charge distribution of the linear CO_2 molecule is shown in Figure 7-6, where the negative ends of the C–O bonds are indicated by the heads of arrows and the positive ends are located at the carbon. Thus, while each C–O bond is polar and has a *bond* dipole, the two bond dipoles cancel, and the *molecule* has no net dipole moment. The same type of reasoning can be used to explain the zero dipole moments of BF_3 and CF_4. However, these molecules have different structures and will be discussed later.

FIGURE 7-6

Charge Distribution in CO_2 $O{\leftarrow}C{\rightarrow}O$

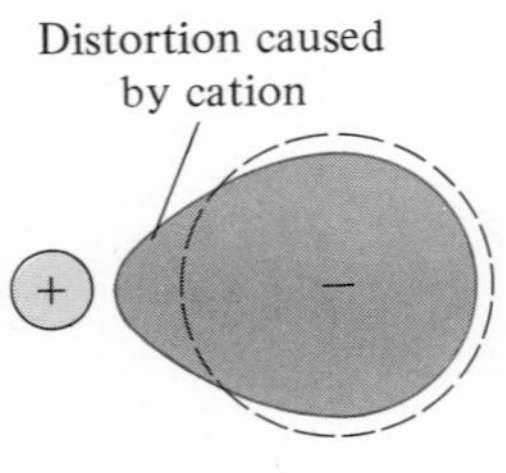

Distortion caused
by cation

FIGURE 7-7

Our treatment of bond polarity and the transition from covalent to ionic character has been based on electronegativity differences. An alternate approach of particular value in the prediction of relative ionic character begins by assuming the presence of ions in the compound in question. It then considers the effect of the cation on the electron density of the anion. If the cation is small and highly charged, it will attract some of the electron density of the anion into the region between the nuclei, resulting in shared electron density (Figure 7-7).

A set of rules based on these ideas has been formulated by K. Fajans. According to these rules, covalency is favored by:

1. A decrease in the size and an increase in the charge of the cation; or, equivalently, an increase in charge density (charge per volume) of the cation.
2. An increase in size and charge of the anion. The larger the anion and the greater its charge, the greater its susceptibility to distortion.
3. A cation without an inert gas electron configuration, because its nuclear charge is less adequately shielded by its *d*-electrons than the nuclear charge of an inert gas cation of similar size. Thus, transition metal cations distort anions more strongly than inert gas cations of the same size.

These rules are illustrated by the compounds in Table 7-2, where melting points and boiling points are used as a crude guide to covalent character (recall from Chapter 6 that most covalent compounds have melting and boiling points lower than those of ionic compounds).

TABLE 7-2
Illustration of Fajans' Rules

Compound	Series Illustrates Variation in	Melting Point[a] (°C)	Boiling Point (°C)
NaCl	Cation size and	801	1413
$MgCl_2$	charge	708	1412
$AlCl_3$		178 (s)	
$BeCl_2$	Cation size	405	∿550
$MgCl_2$		712	1412
$CaCl_2$		772	>1600
$SnCl_2$	Cation size and	246	652
$SnCl_4$	charge	−33	114
AlF_3	Anion size	1291 (s)	
$AlBr_3$		98	263
$CaCl_2$	Type cation	772	>1600
$CdCl_2$		568	960

[a] s = sublimes

LEWIS MODEL

One of the earliest theories of covalent bonding was developed by G. N. Lewis just before the advent of wave mechanics. It remains one of the simplest and most useful models of the electronic structure of covalent materials. Lewis believed that the inert gas rule, which was of such importance in the ionic model, could probably also be applied to covalent situations. Since the *valence* shell of a representative element is the outermost quantum level that is populated by electrons, these atoms achieve an inert gas configuration when eight electrons are present in their valence shell (the *octet rule*). This octet of electrons, according to Lewis, can be attained by the *sharing* of electrons. Since the valence shell of hydrogen is the first quantum shell, it cannot achieve an octet, but can attain the inert gas configuration of helium with two electrons.

The Lewis theory can be illustrated with the hydrogen fluoride molecule. Fluorine contains seven electrons in its valence shell and needs only one more to fill its octet. Hydrogen has one electron in its valence shell and also requires only one to fill its shell. Thus, if each atom shares one electron with the other, both will be satisfied. This is generally represented by an *electron dot formula,* which shows the disposition of electrons about each atom. Figure 7-8 illustrates several ways of writing the electron dot formula for HF. In Figure 7-8a the hydrogen electron is designated by an x and the fluorine electrons by dots. But, of course, electrons are indistinguishable, and in Figure 7-8b all are designated by dots. In Figure 7-8c the shared pair of electrons is represented by a dash, and this is the convention we will use in most cases.

FIGURE 7-8

Electron Dot Formulas for HF

	H $\times$ $\overset{\cdot\cdot}{\underset{\cdot\cdot}{F}}$:	H:$\overset{\cdot\cdot}{\underset{\cdot\cdot}{F}}$:	H—$\overset{\cdot\cdot}{\underset{\cdot\cdot}{F}}$:
	(a)	(b)	(c)

Let us now examine the electron dot formula of HF more closely. First, note that the two electrons shared by both atoms count as electrons of the valence shell of both atoms. That is, the shared pair is part of the hydrogen valence shell and also part of the fluorine valence shell. This shared *pair* of electrons is called a *bond* and provides the electron density responsible for the decrease in potential energy of the molecule relative to the individual constituent atoms.

Second, observe that the six electrons around the fluorine are arranged in groups of two. This is to indicate that the spins of these electrons are paired, and since the wave model predicts that paired electrons can approach more closely than unpaired electrons, it is reasonable to suggest spatial proximity by these groupings. Consequently, the octet of electrons can be thought of as four pairs of electrons. The three pairs of electrons around the fluorine are referred to as *nonbonded* electrons or *lone* pairs.

Third, the four pairs of electrons constituting any octet can be visualized as being positioned at the apices of a tetrahedron, since this arrangement of electrons minimizes the repulsions between them. This is obviously a blatantly artificial particle conception, but, as we shall see, it does allow some interesting predictions. Such a tetrahedron of electrons with the nucleus at the center is shown in Figure 7-9a. Figure 7-9b represents the tetrahedron of electrons about the fluorine nucleus

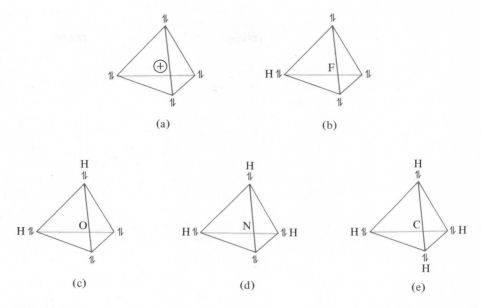

(a)   (b)

(c)   (d)   (e)

FIGURE 7-9

(the symbol F here denotes the fluorine nucleus and its $1s$ electrons) with the hydrogen nucleus sharing a pair.

In order to write the electron dot formula for any molecule, it is necessary to know its structural formula. One rule for establishing the structural formula can be illustrated with the water molecule. For H_2O there are two conceivable arrangements of the atoms:

$$H \cdots H \cdots O \quad \text{and} \quad H \cdots O \cdots H$$

According to the Lewis theory, each bond between atoms in a covalent molecule is formed by a pair of shared electrons, one from the valence shell of each of the bonded atoms. Since the valence shell of hydrogen can accommodate only two electrons, the first structural arrangement, in which hydrogen is attached to *two* atoms, is quite unlikely. (We shall see later that there are a few compounds with hydrogen "attached" to two elements and that other theories are necessary for an adequate description of the bonding in these compounds.) In general, the structural formula for the compound must be known, or must be established by rules similar to the one just stated, before an appropriate electron dot formula can be written.

The structural formula for water is, then, $H \cdots O \cdots H$, and we can proceed to write the electron dot formula. Oxygen has six valence electrons, each hydrogen has one, and there are therefore eight electrons to distribute in such a way as to place an octet of electrons about oxygen and two about each hydrogen. Since each H–O bond must contain two electrons, the proper formula can only be H–$\overset{..}{O}$–H. The most favorable distribution of the electrons about the oxygen is again at the apices of a tetrahedron, and this is shown in Figure 7-9c.

The principal value of any theory of bonding is its usefulness in the explanation and prediction of various properties of matter. The greater the number of properties it correctly predicts, the better the theory. It should be clear from Figure 7-9c that the Lewis theory predicts a certain bond angle for H_2O. The angle between lines drawn from the center of a tetrahedron to the apices is 109.5°. The electrons about the oxygen are almost certainly not located at the apices of a perfect tetrahedron, however; because of the greater repulsions between nonbonded electron pairs relative to bonded pairs, the tetrahedron in H_2O is probably distorted. For the present we will ignore this subtlety and assume that the bond angle predicted for H_2O by the Lewis theory is 109.5°. The experimental value for the bond angle in H_2O is 104.5°, and there is reasonable agreement between the predicted and experimental values.

By a similar process, the electron dot formulas for ammonia, NH_3, and methane, CH_4, can easily be written as

$$\begin{array}{cc} & \text{H} \\ & | \\ \text{H}-\overset{..}{\text{N}}-\text{H} \quad \text{and} \quad \text{H}-\overset{}{\underset{|}{\text{C}}}-\text{H} \\ | & | \\ \text{H} & \text{H} \end{array}$$

Their predicted three-dimensional structures are also presented in Figure 7-9. Note that the predicted bond angles for the isoelectronic molecules (molecules with the same number of electrons) in Figure 7-9 are all the same—109.5°.

The diatomic molecules F_2, O_2, and N_2 present a somewhat different case. The electron dot formula for F_2 is easily seen to be

$$:\overset{..}{\underset{..}{\text{F}}}-\overset{..}{\underset{..}{\text{F}}}:$$

Oxygen, however, has six valence electrons, so each oxygen must share two electrons with its partner, yielding the formula

$$:\overset{..}{\text{O}}::\overset{..}{\text{O}}: \quad \text{or} \quad :\overset{..}{\text{O}}=\overset{..}{\text{O}}:$$

And in N_2, since nitrogen has only five valence electrons, each atom must share *three* electrons with its partner, producing the formula

$$:\text{N}\overset{..}{\underset{..}{}}\text{N}: \quad \text{or} \quad :\text{N}\equiv\text{N}:$$

Thus, the oxygen molecule is held together by two bonds—a double bond—and, since this means a greater amount of electron density between the nuclei and consequently a lower potential energy, the molecule should be more stable than a similar molecule held together by only a single bond, such as F_2. The N_2 molecule contains a triple bond and should therefore be even more stable than O_2.

Here again, then, is a test of the Lewis theory. The theory predicts that the order of stability of the three molecules should be $N_2 > O_2 > F_2$. The experimental measure of the stability of a molecule with regard to dissociation is bond energy, and the bond energies of these molecules are: N_2, 225 kcal/mole; O_2, 118 kcal/mole; F_2, 37 kcal/mole. The theory's predictions have been borne out by experimental data.

The electron positions for F_2, O_2, and N_2 can again be visualized with sets of tetrahedra. The configurations of Figure 7-10 were constructed by surrounding each atom with a tetrahedron of electrons in such a way that the shared electrons become part of both tetrahedra. In F_2, the tetrahedra are joined at the apices; in O_2, on an edge; and in N_2, on a face.

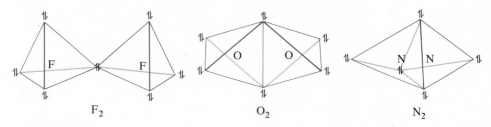

FIGURE 7-10

Lewis Electron Configurations for N_2, O_2, and F_2

Let us now examine some polyatomic molecules. There are two possible structural formulas for carbon dioxide:

$$O \cdots C \cdots O \quad \text{or} \quad C \cdots O \cdots O$$
$$\text{I} \qquad\qquad\qquad \text{II}$$

Reasonable electron dot formulas can be written for both structures, and so, as far as our present discussion extends, there is no way of deciding between them. It is usually true, however, that for compounds of general molecular formula AB_n— CO_2, SO_3, CCl_4, NH_3, and so on—the A atom is the central atom and the B atoms are attached only to it, not to each other. Thus, I is the structure of CO_2. For this structure there are two good electron dot formulas, shown here as III and IV. Formula V is not a good electron dot formula because the valence shell of carbon does not contain eight electrons.

$$:\ddot{O}=C=\ddot{O}: \qquad :O\equiv C-\ddot{O}: \qquad :\ddot{O}-C-\ddot{O}:$$
$$\text{III} \qquad\qquad \text{IV} \qquad\qquad \text{V}$$

An intriguing question at this point is: Which electron dot formula, III or IV, best describes the *true* electronic structure of CO_2? Formula III predicts two equivalent CO bonds, bonds that are the same in every regard—length, strength, etc. Formula IV predicts two unequal linkages; the CO linkage with the triple bond would be stronger and also shorter (as we shall see later) than the CO linkage with the single bond. In fact, the experimentally determined structure for CO_2 has two exactly equivalent linkages, and therefore electron dot formula III provides the best Lewis description of CO_2.

Does this electronic structure provide any information about the bond angle in CO_2? Remembering that atoms linked by double bonds have tetrahedra placed edge to edge, we can draw the spatial electronic structure for CO_2 presented in Figure 7-11. The joined edges of the tetrahedra are like hinges, giving each tetrahedron freedom of motion. The most stable arrangement of these tetrahedra, however, will be a linear one—the middle points of the outer and bonded edges fall on the same line—because this arrangement produces the least amount of repulsion between electrons of neighboring tetrahedra. As stated previously, the bond angle in CO_2 is 180°, and thus the Lewis theory again correctly predicts a molecular parameter.

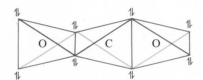

FIGURE 7-11

The structural formula of the nitrite ion can be deduced from the rule for compounds or ions of type AB_n, as $[O\cdots N\cdots O]^-$. This species has a 1− charge, and therefore a total of $6 + 5 + 6 + 1 = 18$ electrons must be distributed about the three nuclei in such a way as to give each atom an octet of valence electrons. A suitable electronic structure is

$$\left[\ddot{\underset{..}{O}} \diagdown \overset{\overset{\displaystyle ..}{N}}{} =\underset{..}{\ddot{O}} \right]^-$$

The minus sign is placed outside the brackets to indicate that the negative charge is associated with the entire grouping, not necessarily with any one atom. Again the electron dot formula predicts unequal N—O linkages, whereas in fact the linkages in NO_2^- are equivalent in every way. A means of modifying the theory to account for this experimental fact will be discussed later in this chapter.

**Exceptions
to the Octet Rule**

The majority of compounds obey, or appear to obey, the octet rule. Compounds whose electronic structures are not adequately represented by electron dot formulas that obey the octet rule fall into one of two categories: those with atoms whose valence shells contain fewer than eight electrons and those with atoms whose valence shells contain more than eight electrons.

A number of compounds of the elements of Group IIA and Group IIIA can be placed in the first category. Consider, for example, the species $BeCl_2$ and BCl_3.

For both compounds, there are at least two possible electron dot formulas.

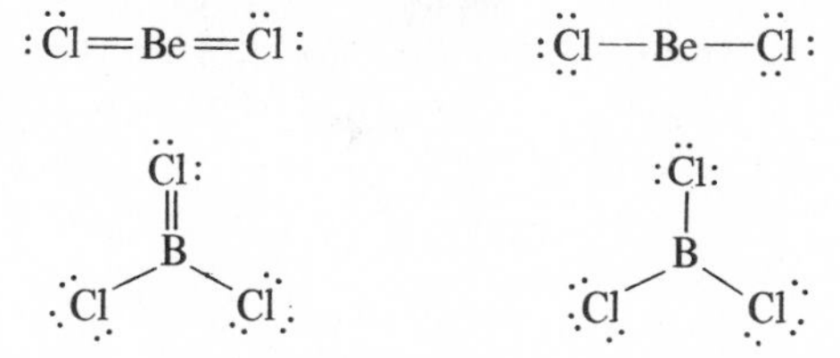

$$:\ddot{C}l = Be = \ddot{C}l : \qquad\qquad :\ddot{C}l - Be - \ddot{C}l :$$

The left-hand representations satisfy the octet rule for each atom, while those on the right portray electron-deficient central atoms. For BCl₃ there is evidence based on bond lengths and energies that suggests that the first formula is the better description. For BeCl₂, on the other hand, evidence indicates that the electron-deficient representation is the better one.

The valence shells of the atoms of the second period are restricted to a maximum of eight electrons. Elements of periods 3 through 6, however, have valence quantum levels that can hold more than eight electrons. For instance, phosphorus has its valence electrons in the third major quantum level, where the *s, p,* and *d* orbitals are available for bonding; phosphorus can therefore accept more than eight electrons in its valence shell.

In order to retain the sanctity of the electron-pair bond, the following electron dot formula must be written for PCl₅:

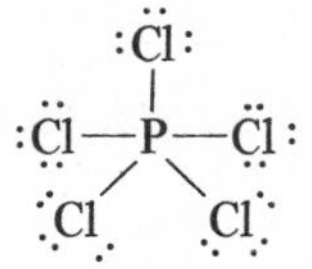

Here phosphorus has 10 electrons in its valence shell. The species $SiF_6{}^{2-}$ contains six silicon-fluorine bonds and has the following electron dot formula:

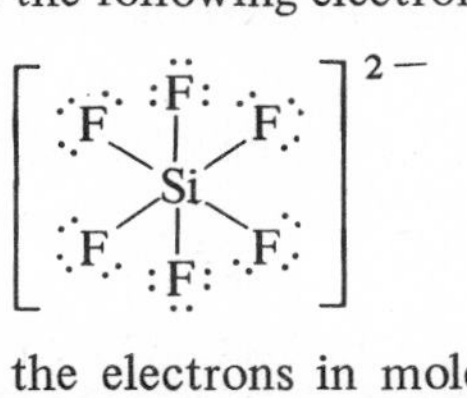

The spatial disposition of the electrons in molecules that do not obey the octet rule is governed by the same factor that determines the tetrahedral arrangement in the octet structures: minimization of the repulsions between electron pairs. The geometry of molecules will be treated in more detail in Chapter 8.

NOMENCLATURE

Our knowledge of the Lewis model now enables us to write structural formulas, and we will therefore take up the naming of covalent compounds before continuing with an analysis of the other models.

Binary Compounds   Simple covalent compounds containing two elements are easily named by giving first the name of the more electropositive element. This is followed by the *ide* form (root + *-ide,* see page 92) of the other element, with a prefix to indicate the number of atoms of this element in the molecular formula. The following serve as illustration:

CO	carbon monoxide
CS_2	carbon disulfide
BF_3	boron trifluoride
$SiBr_4$	silicon tetrabromide
PCl_5	phosphorus pentachloride
SF_6	sulfur hexafluoride
B_2Cl_4	diboron tetrachloride

In the last example a prefix is also used to indicate the number of atoms of the more electropositive element.

Some very common compounds are known exclusively by their common or trivial names; for example: H_2O, water, and NH_3, ammonia.

The nomenclature of covalent acids and anions is discussed in Chapter 6.

Carbon Compounds   Carbon forms a vast variety of covalent compounds, many of them naturally occurring in biological systems. Besides their obvious importance to plant and animal life, these carbon compounds offer examples of a wide variety of structural problems. We will therefore discuss the nomenclature of these compounds in some detail.

The vast majority of these compounds are composed of only a small number of elements: carbon, hydrogen, oxygen, nitrogen, and the halogens. In order to write their structural formulas, it is important to remember that in general carbon forms four bonds, nitrogen forms three bonds, oxygen forms two bonds, and hydrogen and the halogens form one bond. With these rules in mind, we find that for a compound of molecular formula CH_4O there is only one possible structural formula:

$$
\begin{array}{c}
\text{H} \\
| \\
\text{H--C--O--H} \\
| \\
\text{H}
\end{array}
$$

whereas for C_2H_6O there are two possible structures:

$$
\begin{array}{c}
\text{H} \quad\; \text{H} \\
| \qquad | \\
\text{H--C--O--C--H} \\
| \qquad | \\
\text{H} \quad\; \text{H}
\end{array}
\quad \text{and} \quad
\begin{array}{c}
\text{H} \; \text{H} \\
| \quad | \\
\text{H--C--C--O--H} \\
| \quad | \\
\text{H} \; \text{H}
\end{array}
$$

Since the number of possible structural formulas increases as the complexity of the molecule increases, it becomes quite important to be able to provide an unambiguous designation for a given molecule.

Structural formulas for many large carbon compounds become quite unwieldy. For example, 2,2-dimethylbutane, a relatively simple molecule, has the rather awkward formula

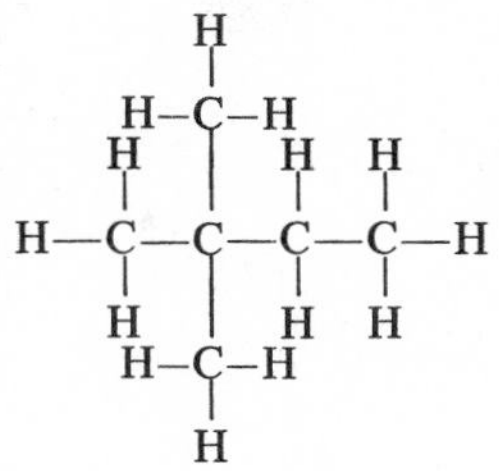

$$
\begin{array}{c}
\text{H} \\
| \\
\text{H}-\text{C}-\text{H} \\
\quad\;\; | \\
\text{H} \quad\; | \quad\;\; \text{H} \quad \text{H} \\
| \quad\;\; | \quad\;\; | \quad\; | \\
\text{H}-\text{C}-\text{C}-\text{C}-\text{C}-\text{H} \\
| \quad\;\; | \quad\;\; | \quad\; | \\
\text{H} \quad\; | \quad\;\; \text{H} \quad \text{H} \\
\quad\;\; \text{H}-\text{C}-\text{H} \\
\quad\quad\; | \\
\quad\quad\; \text{H}
\end{array}
$$

A condensed version of this formula is

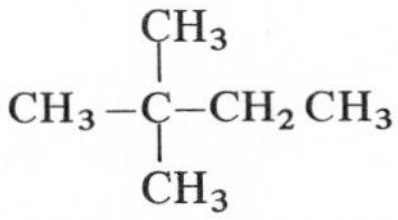

$$
\begin{array}{c}
\text{CH}_3 \\
| \\
\text{CH}_3-\text{C}-\text{CH}_2\,\text{CH}_3 \\
| \\
\text{CH}_3
\end{array}
$$

where the hydrogens are written after the carbon to which they are attached; for example, CH_3. It is also important to note that the above structure is identical to

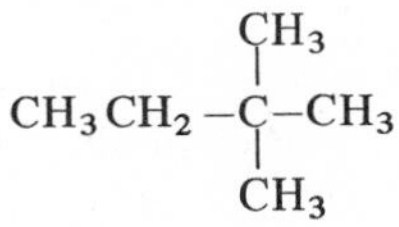

$$
\begin{array}{c}
\text{CH}_3 \\
| \\
\text{CH}_3\,\text{CH}_2-\text{C}-\text{CH}_3 \\
| \\
\text{CH}_3
\end{array}
$$

The one is simply the other rotated 180° in the plane of the paper.

There are two different systems for naming carbon compounds. The *IUPAC system* is based on rules drawn up by the International Union of Pure and Applied Chemistry and is a truly international system. The *common* system is older and less systematic but is still in common usage.

Hydrocarbons

The simplest type of carbon compounds, the hydrocarbons, contain carbon atoms linked to one another and also to hydrogen. There are three main kinds of hydrocarbons: (1) *alkanes,* in which all the carbon-carbon linkages are single bonds, (2) *alkenes,* in which one or more of the carbon-carbon linkages are double bonds, and (3) *alkynes,* in which one or more of the carbon-carbon linkages are triple bonds. Alkenes and alkynes are sometimes referred to as "unsaturated" because of their tendency to react with hydrogen to form the "saturated" alkane.

Alkanes. The formulas and names of a series of straight-chain alkanes (alkanes whose carbon atoms can be written on a straight line) are given in Table 7-3. The IUPAC and common names of these compounds are the same. All the names end in *ane,* and from pentane to decane the names are derived from the Greek word for the number of carbon atoms in one molecule of the compound.

Molecules often contain an alkane minus a hydrogen as part of their structure. These *groups* are named by replacing the ending -ane in the name of the

	Compound	Name	Name of Radical
TABLE 7-3 **Straight-Chain** **Alkanes**	CH_4	methane	methyl
	$CH_3 CH_3$	ethane	ethyl
	$CH_3 CH_2 CH_3$	propane	propyl
	$CH_3 CH_2 CH_2 CH_3$	butane	butyl
	$CH_3 (CH_2)_3 CH_3$	pentane	pentyl
	$CH_3 (CH_2)_4 CH_3$	hexane	hexyl
	$CH_3 (CH_2)_5 CH_3$	heptane	heptyl
	$CH_3 (CH_2)_6 CH_3$	octane	octyl
	$CH_3 (CH_2)_7 CH_3$	nonane	nonyl
	$CH_3 (CH_2)_8 CH_3$	decane	decyl

alkane by -yl. For example, CH_3CH_3 is ethane, and CH_3CH_2 is the ethyl group. The names of these *alkyl* groups or radicals are also given in Table 7-3. The following compound contains several such groups.

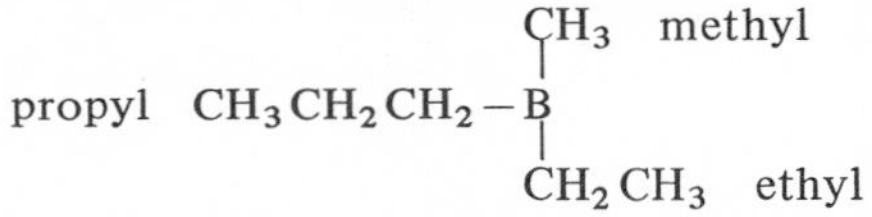

$$propyl \quad CH_3 CH_2 CH_2 - \overset{\overset{\textstyle CH_3 \quad methyl}{|}}{\underset{\underset{\textstyle CH_2 CH_3 \quad ethyl}{|}}{B}}$$

Alkanes can also be branched, as for example the compounds here designated A and B.

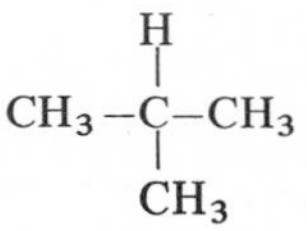

$$CH_3 - \overset{\overset{\textstyle H}{|}}{\underset{\underset{\textstyle CH_3}{|}}{C}} - CH_3 \qquad CH_3 - \overset{\overset{\textstyle H}{|}}{\underset{\underset{\underset{\textstyle CH_3}{|}}{\underset{\textstyle CH_2}{|}}}{C}} - CH_2 CH_3$$

$$A \qquad\qquad\qquad B$$

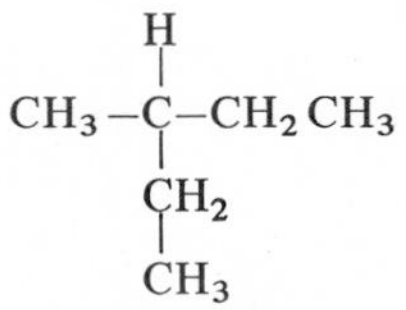

In the *IUPAC system*, branched hydrocarbons are named by first selecting the longest chain of carbon atoms that *could* be written in a straight line, and then numbering the carbon atoms of the longest chain starting from the end closest to the branch. For compound A, we will choose to number the chain as follows:

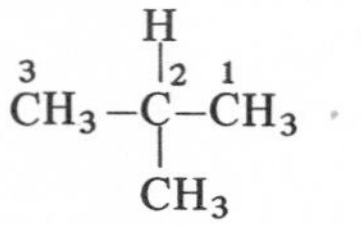

$$\overset{3}{C}H_3 - \overset{\overset{\textstyle H}{|}}{\underset{\underset{\textstyle CH_3}{|}}{C}}{\overset{2}{}} - \overset{1}{C}H_3$$

For compound B the longest chain is shown below, both as written above and as it could be written in a straight line.

$$CH_3 - \overset{\overset{\textstyle H}{|}}{\underset{\underset{\underset{\textstyle CH_3}{5|}}{\underset{\textstyle CH_2}{4|}}}{\overset{3}{C}}}\overset{2}{} - \overset{1}{C}H_2 CH_3 \qquad \overset{5}{C}H_3 - \overset{4}{C}H_2 - \overset{\overset{\textstyle H}{3|}}{\underset{\underset{\textstyle CH_3}{|}}{C}}{\overset{2}{}} - \overset{}{C}H_2 - \overset{1}{C}H_3$$

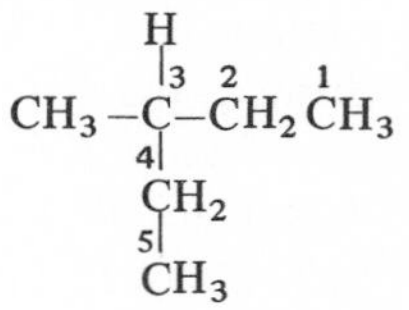

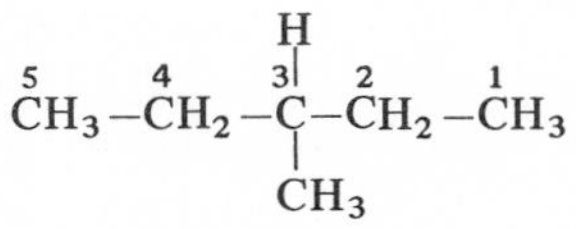

In both **A** and **B** the branch occurs midway in the longest chain, and so it is immaterial from which end we number.

The base name of the branched hydrocarbon depends on the number of carbons in the longest chain; for **A** the base name is propane, for **B** the base name is pentane. Finally, the group at the branch (also called the substituent because it substitutes for a hydrogen of the corresponding straight-chain alkane) is named, and its position on the longest chain is designated by the number of the carbon atom to which it is attached. The IUPAC name for compound **A** is 2-methylpropane; the IUPAC name for compound **B** is 3-methylpentane.

A more complicated molecule is represented by formula **C**, which shows the proper numbering of the longest chain. The numbering was begun at the end closest

$$
\begin{array}{c}
\overset{2}{\underset{|}{\text{CH}_3}} \\
\overset{1}{\text{CH}_3} - \overset{2}{\underset{|}{\text{C}}} \underline{\quad\quad} \overset{3}{\text{CH}} - \text{CH}_2\,\text{CH}_3 \\
\underset{|}{\text{CH}_3} \quad \overset{4}{\underset{|}{\text{CH}_2}} \\
\overset{5}{\underset{|}{\text{CH}_2}} \\
\overset{6}{\underset{|}{\text{CH}_3}}
\end{array}
$$

C

to branching. The substituent groups are two methyl groups attached to the number 2 carbon and one ethyl group attached to the number 3 carbon. When there are more than one of any given substituent, the number is designated by the prefixes di- (2), tri- (3), tetra- (4), penta- (5), hexa- (6). Thus, the IUPAC name for **C** is 3-ethyl-2,2-dimethylhexane. Notice several conventions in writing the name: (a) Numbers are separated from words by hyphens, (b) numbers are separated from each other by commas, (c) a number is given for the location of each group (the name 3-ethyl-2-dimethylhexane is not in accord with this convention), and (d) groups are listed in alphabetical order without regard for the prefixes di-, tri-, and so forth. The most important rule in naming any compound is, of course, that the name provide an unambiguous guide to the structure of the compound.

As a further example, consider one of the constituents of gasoline:

$$
\begin{array}{c}
\overset{2}{\underset{|}{\text{CH}_3}} \qquad \overset{4}{\underset{|}{\text{CH}_3}} \\
\overset{1}{\text{CH}_3} - \overset{2}{\underset{|}{\text{C}}} - \overset{3}{\text{CH}_2} - \overset{4}{\underset{|}{\text{C}}} - \overset{5}{\text{CH}_3} \\
\underset{}{\text{CH}_3} \qquad \underset{}{\text{H}}
\end{array}
$$

The IUPAC name is 2,2,4-trimethylpentane.

The carbon atoms of alkanes can also be arranged in rings; for example:

$$
\begin{array}{c}
\text{H}_2\,\text{C} - \text{CH}_2 \\
|\qquad\quad| \\
\text{H}_2\,\text{C} - \text{CH}_2
\end{array}
$$

Such compounds are termed *cyclic* alkanes and are named by adding the prefix *cyclo-* to the name of the parent alkane. The compound above is cyclobutane.

In the common system, straight-chain alkanes are often denoted by the prefix

Group	Name
TABLE 7-4 **Common Names of Some Alkyl Groups**	
$CH_3 CH_2 CH_2 -$	*n*-propyl
$CH_3 -\underset{\underset{CH_3}{\vert}}{CH}-$	isopropyl
$CH_3 CH_2 CH_2 CH_2 -$	*n*-butyl
$CH_3 -\underset{\underset{CH_3}{\vert}}{CH}-CH_2 -$	isobutyl
$CH_3 CH_2 \underset{\underset{CH_3}{\vert}}{CH}-$	*sec*-butyl
$CH_3 -\overset{\overset{CH_3}{\vert}}{\underset{\underset{CH_3}{\vert}}{C}}-$	*tert*-butyl (*t*-butyl)

normal, abbreviated as *n-,* while a $CH_3-\underset{\underset{CH_3}{\vert}}{CH}-$ group at the end of a chain is designated by the prefix *iso.* The base name of the *iso* branched hydrocarbon is derived from the *total* number of carbons in the molecule. Thus, the common name for 2-methylpropane is isobutane.

The common names for groups (radicals) containing one or two carbons are the same as the IUPAC names. The names of the two possible three-carbon groups and the four possible four-carbon groups are given in Table 7-4. Note the two new designations, *secondary* (*sec-*) and *tertiary* (*tert-*), used to denote branching at the point of attachment. A *secondary* group has two carbons bonded to the carbon that is the point of attachment, while a *tertiary* group has three carbons attached to this carbon.

Some Substituted Alkanes. Halogenated alkanes are alkanes in which one or more hydrogens have been replaced by halogens; for example, $CH_3 Cl$.

In the IUPAC system these compounds are named according to the rules established above, using the following names for the halogen substituents: F, fluoro; Cl, chloro; Br, bromo; I, iodo. The compound $CH_3 Cl$ is chloromethane; chloroform, which has the formula $CHCl_3$, is trichloromethane; the compound

$$CH_3 -\overset{\overset{CH_3}{\vert}}{\underset{\underset{I}{\vert}}{C}}-CH_3$$

is 2-iodo-2-methylpropane; the anesthetic "halothane," $CF_3 CHClBr$, is 2-bromo-2-chloro-1,1,1-trifluoroethane.

In the common system these compounds are named as alkyl halides; the name of the alkyl group is followed by the name of the halogen with the -ide ending (see page 92). Thus, CH_3Cl is methyl chloride, and

$$\begin{array}{c} CH_3 \\ | \\ CH_3-C-CH_3 \\ | \\ I \end{array}$$

is *t*-butyl iodide.

When the nitro group, NO_2, is a substituent, the compound is named by the IUPAC rules already stated. The compound CH_3NO_2 is nitromethane.

Compounds containing the hydroxy group, OH, and the amino group, NH_2, can be named in the same way, but are usually named by rules given in the sections on alcohols and amines.

Alkenes. Except for the simplest alkene, $CH_2{=}CH_2$, which has the common name ethylene, we will consider only the IUPAC name for hydrocarbons containing carbon-carbon double bonds. As with alkanes, the longest chain is located and numbered in such a way that the double bond is closest to the number one carbon. The -ane ending of the parent alkane is replaced with -ene to denote an alkene, and the position of the double bond (actually the number of the carbon of the double bond that is closest to the number one carbon) is denoted by a numeral at the beginning of the name.

Examples are given below.

$$\begin{array}{cc} \quad\ \ \overset{\textstyle H}{\underset{\textstyle |}{}} & \quad\ \ \ \ \overset{\textstyle H}{\underset{\textstyle |}{}} \\ CH_3\,C{=}CH_2 & CH_3\,CH_2\,C{=}CH_2 \\ \text{propene} & \text{1-butene} \end{array}$$

$$\begin{array}{cc} \ \ \ \overset{\textstyle H}{\underset{\textstyle |}{}}\ \overset{\textstyle H}{\underset{\textstyle |}{}} & \ \ \ \overset{\textstyle H}{\underset{\textstyle |}{}}\ \overset{\textstyle H}{\underset{\textstyle |}{}} \\ CH_3\,CHC{=}CCH_3 & CH_3\,C{=}CCH_3 \\ \ \ \ |\ \ \ \ \ \ \ \ \ \ \ \ \ \ & \text{2-butene} \\ \ \ \ CH_3 & \\ \text{4-methyl-2-pentene} & \end{array}$$

Note that (a) the name propene is unambiguous—that is, no number is needed to show the position of the double bond; and (b) the double bond is given preference in deciding which way to number the longest chain; thus,

$$\begin{array}{c} \ \ \ \overset{\textstyle H}{\underset{\textstyle |}{}}\ \overset{\textstyle H}{\underset{\textstyle |}{}} \\ CH_3\,CHC{=}CCH_3 \\ \ \ \ | \\ \ \ \ CH_3 \end{array}$$

is *not,* according to the IUPAC rules, 2-methyl-3-pentene.

A very special type of cyclic alkene is the compound benzene. The following formula is, as we shall see during our discussion of the valence bond model, a rather inadequate description of the bonding in this compound.

For this reason the ring is usually written as 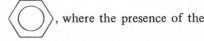, where the presence of the carbon and hydrogen atoms is understood. Compounds containing the benzene ring are termed *aromatic,* and the benzene ring is called the *aryl* portion of the compound.

Substituted benzenes can be named by the usual IUPAC rules; for example,

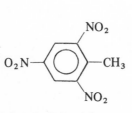

FIGURE 7-12

Orientation of Substituents on Benzene

is 1,3,5-trimethylbenzene. In the common system, adjacent substituents are designated by the prefix *ortho-* (*o*), substituents two atoms removed by *meta-* (*m*), and substituents directly opposite by *para-* (*p*) (Figure 7-12). Thus, the moth repellant

can be named as 1,4-dichlorobenzene or as *para*-dichlorobenzene.

Several methyl-substituted benzene compounds have trivial names. Methylbenzene is called toluene in both the IUPAC and common systems. Dimethylbenzene is almost always referred to as xylene. The two methyl groups can be *ortho, meta,* or *para* to one another, and thus there are three different xylenes. The common explosive TNT is 2,4,6-trinitrotoluene. Note that the carbon to which the methyl group is attached is the number one carbon.

When the benzene ring is a substituent, it is called the *phenyl* group. Thus, the compound $(C_6H_5)_3CH$ is named triphenylmethane.

Alkynes.   The IUPAC rules for hydrocarbons containing triple carbon-carbon bonds are identical to the alkene rules except that the -ane ending of the parent alkane is replaced by -yne to indicate the presence of the triple bond. Propyne is

2,4,6-trinitrotoluene

$CH_3C\equiv CH$; 4,4,4-trichloro-1-butyne is $Cl_3CCH_2C\equiv CH$. The simplest alkyne, $HC\equiv CH$, has the common name acetylene.

Alcohols

Certain substituents or groups change the chemical behavior of hydrocarbons quite drastically. These groups are called *functional groups*. Alcohols contain the OH functional group. The general formula for alcohols is ROH, where R represents a hydrocarbon group.

In the IUPAC system, alcohols are named by first determining the longest chain of carbon atoms that contains the OH group. This chain is then numbered in such a way as to give the OH group the lowest number. The name is produced from the name of the parent hydrocarbon by replacing the *e* ending by *ol*. Thus, CH_3OH is methanol. The position of the OH group on the chain is indicated by a number in

front of the name. The three-carbon alcohol $CH_3\overset{\displaystyle H}{\underset{\displaystyle OH}{C}}CH_3$ is 2-propanol. In the

$$\overset{5}{CH_3}\overset{4}{CH}=\overset{3}{CH}\overset{2}{C}\overset{1}{H_2}OH$$
$$|$$
$$CH_2$$
$$|$$
$$CH_3$$

2-ethyl-3-penten-1-ol

marginal example, note (a) that the name is based not on the longest chain, but on the longest chain containing the functional group, (b) the convention for identifying both unsaturation and functional group, and (c) that the functional group has preference over the unsaturation in determining how to number the chain.

Some alcohols have two or more OH groups. These are named by adding the prefix di-, tri-, and so on, to indicate the number of OH groups. Ethylene glycol, an ingredient in some types of antifreeze, is $HOCH_2CH_2OH$ and therefore has the IUPAC name 1,2-ethanediol.

The common names for alcohols are useful only for the less complex species. The names are the common name for the alkyl group that is attached to the OH group plus the word *alcohol*, as illustrated by the following examples:

$$CH_3CH_2OH \qquad CH_3CH_2\overset{\displaystyle CH_3}{\underset{\displaystyle H}{C}}OH \qquad CH_3\overset{\displaystyle H}{\underset{\displaystyle CH_3}{C}}OH$$

ethyl alcohol   *sec*-butyl alcohol   isopropyl alcohol

The aromatic alcohol ⟨O⟩—OH has the common name phenol.

Ketones

Ketones contain the grouping $C-\overset{\displaystyle O}{\overset{\|}{C}}-C$, as in the simplest ketone with the trivial name acetone, $CH_3-\overset{\displaystyle O}{\overset{\|}{C}}-CH_3$. The IUPAC rules for naming ketones are identical to those for naming alcohols except that the -e ending of the hydrocarbon parent is replaced by *-one*. The IUPAC name for acetone is therefore propanone. The compound

$$CH_3-\underset{\underset{CH_3}{|}}{\overset{\overset{H}{|}}{C}}-\overset{\overset{O}{\|}}{C}-CH_3$$

is named 3-methyl-2-butanone.

The common name for a ketone is derived from the common names of groups attached to the C=O (the *carbonyl* group) plus the word *ketone* as illustrated in the following:

$$\overset{\overset{O}{\|}}{CH_3CCH_3}$$
dimethyl ketone

$$CH_3\underset{\underset{CH_3}{|}}{\overset{\overset{O}{\|}}{CH}CCH_3}$$
methyl isopropyl ketone

Carboxylic Acids

The functional group of carboxylic acids is the carboxy group, $-\overset{\overset{O}{\|}}{C}-OH$. Their general formula is $R\overset{\overset{O}{\|}}{C}OH$.

The IUPAC name for acids is obtained in the usual way from the parent hydrocarbon (the parent must include the carboxy carbon) by replacing the final *e* with *oic* followed by the word *acid*. Thus, the simplest acid—where R equals hydrogen—is methanoic acid. The compound

$$Cl_3CC\overset{\diagup O}{\underset{\diagdown OH}{}}$$

is trichloroethanoic acid.

The common names of acids are generally derived from the names of their natural sources.

The common names of some carboxylic acids are given below.

HCOOH	formic acid
CH_3COOH	acetic acid
CH_3CH_2COOH	propionic acid
$CH_3CH_2CH_2COOH$	butyric acid
$CH_3CH_2CH_2CH_2COOH$	valeric acid
$CH_2=CHCOOH$	acrylic acid
⬡—COOH	benzoic acid

Salts and Esters

The OH hydrogen of carboxylic acids is rather easily removed to form carboxylate anions. The names of these anions follow the usual rules for anions derived from *-ic* oxy acids. The names of compounds containing these anions are also obtained from the ionic nomenclature rules of Chapter 6. Thus, the salt $CH_3\overset{\overset{\displaystyle O}{\|}}{C}ONa$ can be called either sodium ethanoate (IUPAC) or sodium acetate (common).

Carboxylic acids react with alcohols to form compounds of the type $R\overset{\overset{\displaystyle O}{\|}}{C}{-}OR$, called *esters.* Esters are named in both systems as alkyl salts; that is, exactly like the salts discussed above, except that an alkyl group replaces the metal cation (esters are, however, *not* ionic). Thus,

$$CH_3\overset{\overset{\displaystyle O}{\|}}{C}OCH_3$$

is methyl ethanoate (IUPAC) or methyl acetate (common), and

$$H\overset{\overset{\displaystyle O}{\|}}{C}O\underset{\underset{\displaystyle CH_3}{|}}{\overset{\overset{\displaystyle CH_3}{|}}{C}}H$$

is isopropyl formate.

Aldehydes

Aldehydes contain the group $-C\begin{smallmatrix}\diagup\!\!\diagup O\\ \diagdown H\end{smallmatrix}$. Since this group, like the carboxyl group, contains three bonds to the carbon, it must be a terminal group. The IUPAC names for aldehydes are derived in the usual fashion by replacing the final *e* of the parent hydrocarbon with *al.*

$$HC\begin{smallmatrix}\diagup\!\!\diagup O\\ \diagdown H\end{smallmatrix} \quad \text{is methanal}$$

The common name is formed from the common name of the parent acid by dropping the ending *-ic acid* and attaching the suffix *aldehyde.* Thus,

$$CH_3CH_2CH_2C\begin{smallmatrix}\diagup\!\!\diagup O\\ \diagdown H\end{smallmatrix} \quad \text{is butyraldehyde}$$

$$CH_3C\begin{smallmatrix}\diagup\!\!\diagup O\\ \diagdown H\end{smallmatrix} \quad \text{is acetaldehyde}$$

Ethers

Ethers contain the grouping C—O—C; their general formula is ROR'. Simple ethers can be named by the IUPAC system by treating the longest alkyl group as the parent hydrocarbon and then using the names

$$CH_3 O—$$ methoxy
$$CH_3 CH_2 O—$$ ethoxy
$$CH_3 CH_2 CH_2 O—$$ propoxy
etc.

for the RO group. The common anesthetic $CH_3 CH_2 OCH_2 CH_3$ is termed, then, ethoxyethane.

The common name consists of the names of the alkyl groups followed by the word *ether*.

$$CH_3 OCH_2 CH_2 CH_3 \quad \text{is methyl propyl ether}$$

$$(CH_3 CH_2)_2 O \quad \text{is ethyl ether}$$

Amines

Amines have the general formula R—N—R″, where no more than two R groups may be hydrogens. When two R groups are hydrogens, as in $CH_3 NH_2$, the compound is a *primary* amine; when one R group is hydrogen, as in $(CH_2 CH_2)_2 NH$, the amine is a *secondary* amine; when there is no hydrogen attached to the nitrogen, as in $CH_3 NCH_3$, the compound is a *tertiary* amine.

CH₃

The IUPAC name is derived by the same procedure used for ethers: $CH_3 CH_2 NH_2$ is aminoethane, $CH_3 CH_2 NH$ is methylaminoethane.

CH₃

The common name is formed by naming the alkyl groups attached to nitrogen, followed by the word *amine* as a suffix:

$$CH_3 CH_2 NH_2 \quad \text{is ethylamine}$$

$$CH_3 \underset{\underset{CH_3}{|}}{\overset{\overset{H}{|}}{C}}-N(CH_3)_2 \quad \text{is dimethylisopropylamine}$$

THE VALENCE BOND MODEL

The creation of wave mechanics in the 1920s led to the development of two covalent models—the valence bond model and the molecular orbital model. The valence bond theory was formulated primarily by W. Heitler and F. London in the early 1930s and extended and popularized by Linus Pauling, who wrote the classic treatise *The Nature of the Chemical Bond*.

In a real sense the valence bond theory is a post-wave mechanics modification of the Lewis theory. That is, the idea of the electron-pair bond is retained, but its origin is described in terms of atomic orbitals—a wave concept. Thus, valence bond theory utilizes as its basic assumption the idea that a bonding orbital (a bond) between two atoms is formed when an atomic orbital on one atom physically overlaps an atomic orbital on the other atom. The two electrons populating this bonding orbital must have their spins paired (Pauli principle).

There are, of course, different types of atomic orbitals, and their overlap in various combinations leads to a number of types of bonding orbitals. Some of the possible combinations of two orbitals are pictured in Figure 7-13, where the two nuclei involved are labeled X and Y. The overlaps pictured in Figure 7-13a, b, and c produce a buildup of electron density on the internuclear x axis; this kind of bond is called a sigma (σ) bond. (We have arbitrarily defined the internuclear axis as the x axis; this convention will be followed throughout.) The overlap between two p_z orbitals (Figure 7-13d) results in concentrations of electron density above and below the internuclear axis. This type of bond (remember, there can be only two electrons in this set of overlapped orbitals) is termed a pi (π) bond. The overlap of two p_y orbitals also produces a π bond. In this case the electron density is concentrated above and below the plane of the paper.

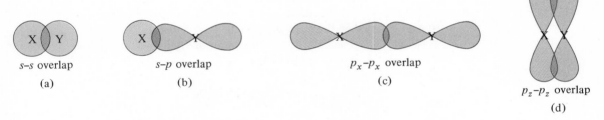

s–s overlap

(a)

s–p overlap

(b)

p_x–p_x overlap

(c)

p_z–p_z overlap

(d)

FIGURE 7-13

Overlap of Atomic Orbitals

The bonding orbital has no physical meaning unless populated by two electrons. In terms of electron population, orbitals can overlap to form bonds under the following circumstances: (a) when an orbital containing *one* electron on atom X can overlap with an orbital containing *one* electron on atom Y, or (b) when an orbital containing *two* electrons on atom X can overlap with an *empty* orbital on atom Y. These two situations are shown below using the convention of Chapter 5 to denote orbitals. Note that the electrons in both cases have their spins in opposite directions.

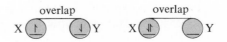

Therefore, to describe the bonding of any species in terms of valence bond theory, we begin with the valence electron configuration of each atom in the molecule and then decide which orbitals can overlap. The molecules F_2, O_2, and N_2 provide a convenient starting point for our discussion of concrete examples.

The fluorine atoms in F_2 each have one half-filled p-orbital that can be used to form a single bond.

The electron dot formula corresponding to this description is $:\ddot{F}-\ddot{F}:$ The wave-mechanical reason for writing the dots in pairs should now be evident: A pair of dots corresponds to a pair of electrons in an orbital. The overlap of bonding orbitals is of the type shown in Figure 7-13c. In addition to these orbitals, each fluorine has two electrons in a $2s$ orbital, two in a $2p_y$ orbital, and two in a $2p_z$ orbital. These are the nonbonding fluorine electrons.

The student might at this point ask why the p_x orbitals were overlapped rather than the p_z's or p_y's. Overlap of the p_z (or p_y) orbitals would result in a π bond. Experimentally, it has been shown that a π bond is rarely found without a σ bond. This can be explained by noting that the p_x orbitals extend *toward* one another in the molecule and therefore overlap to a greater extent (thereby forming a stronger bond) than the p_z or p_y orbitals.

We begin our valence bond description of O_2 by writing the valence electron configurations for both oxygen atoms.

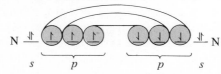

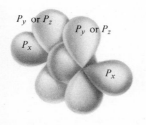

FIGURE 7-14

There are two half-filled orbitals on each atom that can be overlapped to form a double bond. Thus, in addition to the overlapped p_x orbitals, we now also have a π bond formed by overlap of either the p_y or p_z orbitals. This is shown in Figure 7-14, where the horizontal lines portray overlap of the p_y or p_z orbitals. The corresponding electron dot formula is $:\ddot{O}=\ddot{O}:$.

The description of N_2 follows the same procedure.

N $\frac{\text{⇅}}{s}$ $\overbrace{\frac{\text{↑}}{p}\frac{\text{↑}}{}\frac{\text{↑}}{}}$ $\overbrace{\frac{\text{↓}}{p}\frac{\text{↓}}{}\frac{\text{↓}}{}}$ $\frac{\text{⇅}}{s}$ N

There are now three half-filled p orbitals to overlap, resulting in a σ bond and two π bonds. The two π bonds are oriented at right angles to each other as shown in the following end-on view of N_2 (Figure 7-15a). Figure 7-15a is misleading, however,

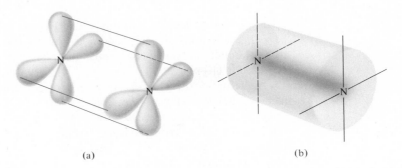

(a)   (b)

FIGURE 7-15

The Overlap of p_y and p_z Orbitals in N_2

because arguments based on wave mechanics indicate that the π-electron density is actually cylindrically symmetrical as shown in Figure 7-15b.

Bond Order

We have previously asserted that the greater the electron density between two atoms, the greater the electron-nuclei attractions and therefore the lower the potential energy of the molecule. If we define *bond order* as one-half the number of electrons between the nuclei (indeed, bond order is just a more sophisticated way of expressing the number of bonds between two atoms), we can now state an important generalization: *The greater the bond order, the higher the bond dissociation energy and the smaller the bond length.* Theoretically, this is reasonable since the greater the electron-nuclei attractions, the stronger the bond and the smaller the internuclear distance.

Table 7-5 presents the bond energies and bond lengths, along with our bond orders for N_2, O_2, and F_2. These parameters are also presented for ethyne, ethene, and ethane, which also have bond orders of 3, 2, and 1, respectively. These and similar data clearly substantiate our generalization. A comparison of the bond energies and bond lengths of F_2 and ethane, which have the same bond orders, also reveals, however, that the generalization must be used with care. It certainly cannot

TABLE 7-5

	Bond Order	Bond Energy (kcal/mole)	Bond Length (Å)
F_2	1	37	1.42
O_2	2	118	1.21
N_2	3	225	1.09
H_3C-CH_3	1	83	1.53
$H_2C=CH_2$	2	147	1.34
$HC\equiv CH$	3	194	1.20

be interpreted to mean, for example, that two bonds with the same bond order will have the same bond energies and bond lengths. This would only be approximately true for bonds containing very similar atoms (similar electronegativities, size, and so on).

The data of Table 7-5 can also be used to develop another generalization: namely, that *species with the same number of electrons (isoelectronic) have the same bond orders and consequently roughly similar molecular parameters.* Diatomic fluorine and ethane have the same number of electrons and the same bond orders; ethene is isoelectronic with O_2 and both have C—C and O—O bond orders of 2; ethyne is isoelectronic with N_2 and both have C—C and N—N bond orders of 3. As we have seen in our discussion of the Lewis model and thus far in our discussion of the valence bond model, the number of valence electrons determines the type of bonding adopted by the molecule, and this generalization, therefore, is not surprising.

Table 7-6 presents bond lengths and bond energies for other compounds isoelectronic with O_2 and N_2. Here, the similarities in bond lengths for isoelectronic species are obvious; the bond energies, however, are more widely divergent.

TABLE 7-6
Selected
Isoelectronic Species

Species	Bond Length (Å)		Bond Energy (kcal/mole)	
$H_2C=O$	CO	1.21	CO	149
$HN=O$	NO	1.21	NO	*ca.* 126
$HC≡N$	CN	1.16	CN	154
CN^-		1.15		—
CO		1.13		257
NO^+		1.06		—

Hybridization

Our treatment of valence bond theory to this point leads to a description of the bonding in water that is not in agreement with the structural parameters for this molecule. The valence configuration of oxygen indicates the possibility of overlap of the two 2*p* orbitals with the hydrogen 1*s* orbitals as shown in Figure 7-16a. These two *p* orbitals are oriented at 90° to one another and the angle between the two oxygen-hydrogen bonds should therefore also be 90°; but, as was noted earlier, the experimentally determined bond angle is 104.5°.

When this discrepancy was first noted, it was suggested that oxygen, because of its greater electronegativity, attracts a considerable amount of the electron density in the O—H bonds, leaving the hydrogen nuclei rather exposed. The partially exposed nuclei would then electrostatically repel each other, resulting in a widening of the bond angle. This proposed repulsion was later shown to be inadequate to account for a widening of 15°.

The explanation that is currently accepted as an integral part of valence bond theory is based on the concept of *hybridization* of atomic orbitals. During our discussion of the modern view of the atom (Chapter 5) we found that the orbital is a mathematical construct. It is a consequence of wave mechanics that a set of orbitals can be mathematically combined according to certain rules to produce a new, different set of orbitals. It can be shown that if, for example, an s orbital, a p_x orbital, a p_y orbital, and a p_z orbital are combined according to the rules of quantum mechanics, four different combinations are possible and each combination describes a new orbital. These four new orbitals are named according to the orbitals that were used in their creation and thus are termed sp^3 orbitals. There are, then, four sp^3 orbitals, or, since this combining process is usually referred to as hybridization, four sp^3 hybrid orbitals.

The spatial disposition of the four sp^3 orbitals is completely different from the spatial disposition of the individual s and p orbitals. As is shown in Figure 7-17, the sp^3 orbitals have their maximum electron density directed toward the apices of a tetrahedron. Each of the lobes shown there is one sp^3 orbital. Figure 7-17 also reveals that each of the four orbitals has the same energy. Indeed, the four sp^3 hybrid orbitals are identical in every way except their orientation in space.

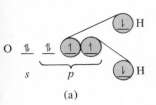

(a)

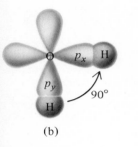

(b)

FIGURE 7-16

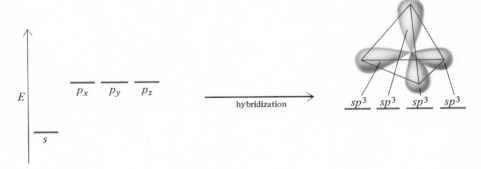

FIGURE 7-17

Other combinations of orbitals are also possible: Hybridization of one s and one p orbital produces two sp hybrids, which are directed toward the opposite ends of a straight line, as in Figure 7-18a. The angle between the orbitals is therefore $180°$. A combination of one s with two p orbitals produces three sp^2 orbitals whose regions of maximum electron density are directed toward the apices of an equilateral triangle (Figure 7-18b). The angle between any two adjacent sp^2 orbitals is therefore $120°$. A combination of one s orbital, three p orbitals, and one d orbital produces five sp^3d orbitals, which are directed toward the apices of a trigonal bipyramid, Figure 7-18c. The trigonal bipyramid is a unique geometrical figure, since it alone (of the figures discussed here) produces two different angles between the orbitals. The angle between the orbitals in the equatorial plane is $120°$, while the angle between an axial orbital and an equatorial orbital is $90°$. The energies of the five sp^3d orbitals are also unique in that there are two energy levels, with the

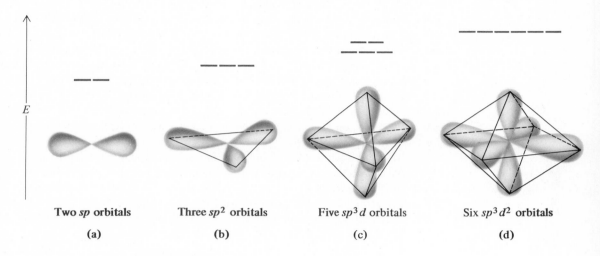

E

| Two *sp* orbitals | Three *sp*^{2} orbitals | Five *sp*^{3}*d* orbitals | Six *sp*^{3}*d*^{2} orbitals |
| (a) | (b) | (c) | (d) |

FIGURE 7-18

Some Common Types of Hybridization

two axial orbitals having the higher energy. The other common type of hybridization is *sp*^{3}*d*^{2}. These orbitals are directed toward the apices of an octahedron (Figure 7-18d). The angle between any two adjacent *sp*^{3}*d*^{2} orbitals is 90°.

Before we leave our general discussion of hybridization we must add one qualifying remark: Orbitals can be combined effectively only when they have similar energies. Thus, we would not expect a carbon atom to utilize *sp*^{3}*d* orbitals in bonding, because this would mean that the valence shell 2*s* and the three 2*p* orbitals would combine with a *3d* orbital, which is of much higher energy. Hybridization is restricted, therefore, to valence shell orbitals.

Let us now return to our dilemma with the bond angle in H$_2$O. If the oxygen uses *sp*^{3} hybrid orbitals to overlap with the 1*s* orbitals of the two hydrogen atoms (Figure 7-19) we would expect the tetrahedral angle of 109.5°. This prediction is certainly closer to the actual value of 104.5° than our previous prediction of 90°. Since the electron density of bonded electrons is more confined (because of electrostatic attraction to *two* nuclei) than the electron density of nonbonded

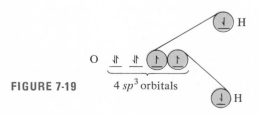

FIGURE 7-19   4 *sp*^{3} orbitals

electrons, the repulsion between nonbonded electrons (which are also in sp^3 orbitals) and bonded electrons is greater than the repulsion between bonded electrons (this will receive further elaboration in Chapter 8). Therefore, we would expect the angle to be somewhat less than the exact tetrahedral angle.

An orbital picture of H_2O is given in Figure 7-20a. Note the similarity between the valence bond and the Lewis description of H_2O.

The isoelectronic species NH_3 and CH_4 can be treated in the same way; that is, using sp^3 hybrid orbitals on the central atom. Figure 7-20 also shows the bonding in these species. The experimental bond angles in NH_3 and CH_4 are $107°$ and $109.5°$, respectively.

The derivation of the valence bond description of methane deserves elaboration. There are only two unpaired electrons in the valence configuration of carbon:

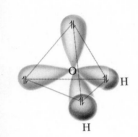

(a) H_2O

$$C\ \underline{1\!\downarrow}\ \underline{1}\ \underline{1}\ \underline{}$$
$$\quad\ s\ \ p\ \ p\ \ p$$

Methane has four CH bonds, and thus we need *four* unpaired electrons on the carbon to overlap with the unpaired electrons on the four hydrogens. The bond angle of $109°$ is an obvious manifestation of sp^3 hybridization, and thus we can hybridize the carbon orbitals and populate these orbitals according to Hund's first rule:

$$C\ \underbrace{\underline{1}\ \underline{1}\ \underline{1}\ \underline{1}}_{4\ sp^3\ \text{orbitals}}$$

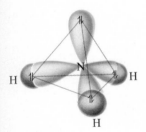

(b) NH_3

After hybridization, as a consequence of the equal energies of the sp^3 orbitals, there are four unpaired electrons on carbon. An alternative analysis of the change in electronic structure of carbon involves the promotion of an electron from an s orbital to a p orbital prior to hybridization.

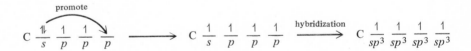

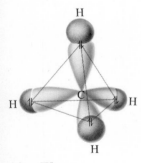

(c) CH_4

FIGURE 7-20

Let us now examine ethane, ethene, and ethyne. The $109°$ $C\cdots C\cdots H$ angle in ethane is certainly suggestive of sp^3 hybridization at the carbon atoms. Thus, after hybridization to sp^3, we have the orbital overlap shown in Figure 7-21.

The $C\cdots C\cdots H$ angle in ethene is $120°$, which is indicative of sp^2 hybridization. We will, therefore, create three sp^2 hybrids at each carbon, leaving one p orbital at each carbon. Because of Hund's first rule, the electron configuration at each carbon should be

$$C\ \underbrace{\underline{1\!\downarrow}\ \underline{1}\ \underline{1}}_{sp^2}\ \underbrace{\underline{}}_{p}$$

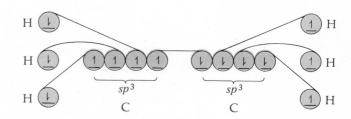

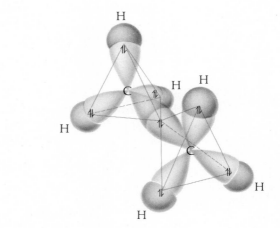

FIGURE 7-21

Valence Bond
Description of Ethane

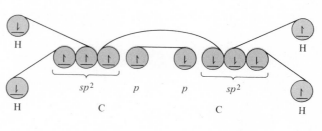

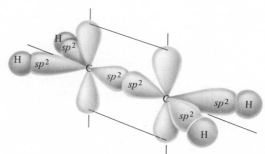

FIGURE 7-22

Valence Bond
Description of Ethene

but this provides only two unpaired electrons, whereas we need at least three for the two C—H bonds and one C—C linkage. Promotion of one electron to the p orbital provides the orbital schematics shown in Figure 7-22. The geometry of the molecule is fixed by the overlap of the two p orbitals, which must be parallel in order to overlap effectively.

The C⋯C⋯H angle in ethyne is 180°. The description illustrated in Figure 7-23 therefore employs sp hybridization at each carbon for the formation of the C—H and C—C σ bonds. The two π bonds are formed by overlap of the two p_y orbitals and the two p_z orbitals. (The p_x was used to form the sp hybrids.)

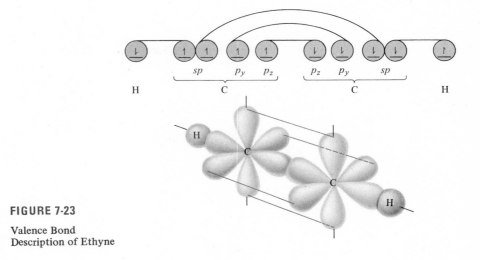

FIGURE 7-23

Valence Bond
Description of Ethyne

The use of hybridization in the valence bond model is further illustrated by the following molecules:

1. Beryllium chloride in the gaseous state has a bond angle of 180°. An appropriate description, therefore is

The corresponding electron dot formula is :C̈l–Be–C̈l:. The bonding in this molecule could also be illustrated as with the corresponding electron dot formula :C̈l=Be=C̈l:. In either case, *sp* hybridization is adopted by the beryllium atom.

2. Boron trichloride has a bond angle of 120°. The appropriate hybridization is therefore sp^2, as shown below.

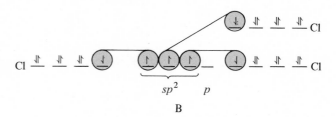

B

3. The bonding in phosphorus pentachloride can be analyzed as follows: The electron configuration of phosphorus shows only three unpaired electrons. Five unpaired electrons are necessary, however, to bond to the chlorine unpaired electrons. The valence shell of phosphorus is the third quantum shell, which contains *d* orbitals in addition to the valence *s* and *p* orbitals, and the desired five unpaired electrons can therefore be obtained by promoting an *s* electron to a *d* orbital. The energy required for this promotion is recovered from the energy released when the five P–Cl bonds are formed.

$$\text{P}\;\underset{s}{\uparrow\downarrow}\;\underset{p}{\uparrow}\;\underset{p}{\uparrow}\;\underset{p}{\uparrow}\quad\rightarrow\quad\text{P}\;\underset{s}{\uparrow}\;\underset{p}{\uparrow}\;\underset{p}{\uparrow}\;\underset{p}{\uparrow}\;\underset{d}{\uparrow}$$

If the *p* orbitals of the chlorines were now overlapped with these orbitals, at least three bonds would be separated by angles of 90° (because of overlap with the three perpendicular *p* orbitals). The actual structure of phosphorus pentachloride is trigonal bipyramidal, with two different bond angles, 90° and 120°. This, as we have seen previously, is the geometry of five sp^3d hybrid orbitals. The final step, then, is to hybridize the one *s*, three *p*, and one *d* orbital and allow the hybrid orbitals to interact with the single half-filled *p* orbital of each chlorine.

$$\text{P}\;\underbrace{\uparrow\;\;\uparrow\;\;\uparrow\;\;\uparrow\;\;\uparrow}_{5\;sp^3d\;\text{orbitals}}$$

This type of analysis does not in any way represent a mechanism for the formation of PCl_5 from monatomic phosphorus and chlorine; that is, promotion of electron, hybridization, overlap. It does however, provide a useful, though artificial, visualization of orbital interactions (the valence bond model) in molecules.

4. Our approach to the species SiF_6^{2-} begins by arbitrarily assigning the two extra electrons to the silicon. Thus, we write electron configurations for the central fictitious Si^{2-} ion and six fluorine atoms. Since this species has

octahedral geometry, two *d* orbitals are hybridized with the *s* and the three *p* orbitals to produce six sp^3d^2 hybrids. After hybridization the Si^{2-} ion has the configuration

$$Si^{2-} \quad \underbrace{\underline{\uparrow} \; \underline{\uparrow} \; \underline{\uparrow} \; \underline{\uparrow} \; \underline{\uparrow} \; \underline{\uparrow}}_{sp^3d^2}$$

which clearly shows six half-filled orbitals that can be overlapped with the single half-filled orbital on each fluorine. The same final disposition of electrons can be obtained by beginning the valence bond visualization with one silicon atom, four fluorine atoms, and two fluoride ions.

5. The description of SF_4 follows the analyses outlined in items 3 and 4, above. This molecule differs from PCl_5 and $SiF_6{}^{2-}$ in that the central atom has four bonding orbitals and one nonbonding orbital. It is generally true that the number of hybrid orbitals formed by the central element will be the same as the total number of σ-bonding and nonbonding orbitals (see Chapter 8, page 153). The hybridization in SF_4 is therefore sp^3d. The spatial orientation of the orbitals is shown in Figure 7-24. The lone pair is placed in an equatorial orbital because this arrangement minimizes the repulsion between bonded and nonbonded pairs. (This will be elaborated on in Chapter 8.)

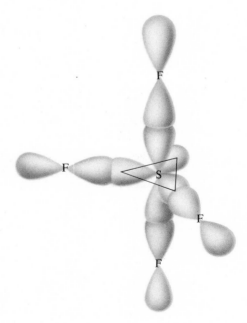

FIGURE 7-24

Valence Bond Description of SF_4

The ozone molecule, O_3, one of the constituents of smog, provides a convenient **Resonance** introduction to yet another feature of the valence bond model. This molecule contains two identical oxygen-oxygen linkages with bond lengths of 1.28 Å. The bond angle is 117°.

We begin, as usual, with the valence electron configuration for each oxygen.

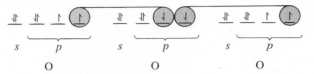

The central oxygen must bond to two oxygens, and the overlap pictured above is possible. The electron dot formula corresponding to this description is

$$:\ddot{O}-\ddot{O}-\ddot{O}:$$

from which it is immediately obvious that the octet rule is not obeyed for the outer atoms.

A description that does obey the octet rule can be obtained by transferring an electron from the central oxygen to one of the outer oxygens (naturally, this is just a kind of electron bookkeeping system and should in no way be interpreted as the formation of ions). If, at the same time, we recognize that the 117° bond angle is indicative of sp^2 hybridization, we can describe the molecule as

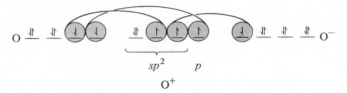

While this description certainly obeys the octet rule and is in agreement with the 117° bond angle, it does not account for the observed equal bond lengths. The formula above predicts two unequal linkages, the double bond being shorter than the single bond in accord with our bond-order generalization.

The description is rather arbitrary at any rate, for we could have decided to place the double bond between the central oxygen and the *other* oxygen. If we write both structures

and then combine or hybridize the two structures, we get a hybrid that has equal bonds. This process, especially for the purpose of "seeing" the equalization of bonds, is probably best visualized as a superimposition of the one structure on the other, as in superimposing two transparent images (Figure 7-25).

This fusion of two (or more) valence bond representations is completely analogous to the concept of hybridization, except that entire electronic descriptions rather than just atomic orbitals are mathematically combined. The process is

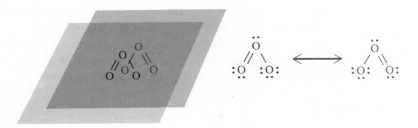

FIGURE 7-25

Superimposition of O_3 Resonance Forms

known as *resonance* hybridization; the individual structures that are hybridized are known as *resonance forms* or *resonance contributors*. The double-headed arrow shown in Figure 7-25 is used to indicate resonance hybridization.

The formation of resonance hybrids is sometimes likened to the creation of a rhinoceros by hybridization of a unicorn and a dragon. The "resonance contributors" in this analogy are mythical creatures and have no physical reality. The same is true for resonance forms, such as either of the O_3 contributors. This analogy is not quite perfect, however, because the *real* creature, the rhinoceros, is presumed to result from the hybridization. In the case of molecular resonance hybridization, the resulting electronic hybrid is very likely not a perfect description of the electronic structure of the *real* molecule. In fact, the most that can be said of the resonance hybrid is that it is a better representation of the electronic structure of the *real* molecule than either of the individual resonance contributors.

Let us now return to our ozone hybrid. Each of the resonance forms has one π bond. The resonance hybrid, however, has these two π electrons delocalized, or smeared out, over two linkages. Each O—O linkage then has a total of three electrons, or a bond order of 1.5. Our bond order–bond length generalization would now predict a bond length somewhere between the length observed for an oxygen-oxygen single bond and that of an oxygen-oxygen double bond. The bond length in O_2, to which we have previously ascribed a bond order of 2, is 1.21 Å, while the O—O bond length in hydrogen peroxide, which has an O—O linkage best described as a single bond, is 1.48 Å. Thus, the resonance structure provides a reasonable explanation for equivalent O—O linkages and the measured bond length of 1.28 Å.

The delocalization of electrons as represented by resonance has yet another consequence. It can be shown that the delocalization, or smearing of electrons between more than two atoms generally results in a decrease in energy. In a molecule that is best represented as a resonance hybrid—that is, one that has some electron delocalization—the difference between the energy expected for a given resonance contributor and the *true* energy of the molecule is called the *resonance energy*. This is presumably a measure of the degree to which the molecule has been stabilized by delocalization.

The energy expected for the dissociation of one mole of the resonance form O=O−O to three moles of monatomic oxygen is 237 kcal. The dissociation of one mole of *ozone* to three moles of monatomic oxygen actually requires 265 kcal. Ozone is therefore 27 kcal more stable than predicted (on the basis of one resonance form); that is, the resonance energy of ozone is 27 kcal/mole.

Dinitrogen oxide, N_2O (which is used as an anaesthetic and as the propellant gas in "whipped" cream containers), is a linear molecule with a N−N bond length of 1.13 Å and an N−O bond length of 1.19 Å. The linear structure is indicative of *sp* hybridization at the central nitrogen and, by reasoning similar to that used for ozone, we can write the following electron dot formulas:

$$\overset{\ominus}{:\!\!\overset{..}{N}}\!=\!\overset{\oplus}{N}\!=\!\overset{..}{O}: \qquad :N\!\equiv\!\overset{\oplus}{N}\!-\!\overset{..}{\underset{..}{O}}\!:^{\ominus} \qquad \overset{2\ominus}{\overset{..}{N}}\!-\!\overset{\oplus}{N}\!\equiv\!O\!:^{\oplus}$$

$$\text{(I)} \qquad\qquad \text{(II)} \qquad\qquad \text{(III)}$$

The charges in the formulas above are *formal charges* and are determined for any given atom by counting its nonbonded electrons plus one-half of its shared electron pairs. This sum is then subtracted from the number of valence electrons in the neutral atom. The difference is the formal charge. For example, in structure I, the valence shell of the end nitrogen contains four nonbonded electrons and four shared electrons. The valence shell of a neutral nitrogen atom contains five electrons. The formal charge of this nitrogen is then $5 - (4 + 4/2) = -1$. As the name implies, this number does not represent the actual distribution of electrons, because electronegativity differences have not been taken into account. In many cases, the number does represent an approximate idea of the electron distribution. Formal charge in no way implies the existence of ions, and is usually encircled to avoid confusion.

The question at this point is which structure, or combination of structures, is the best representation of the electronic structure of N_2O; i.e., which structure, or combination of structures, provides the best explanation of the physical and chemical properties of N_2O? To answer this question, we again make use of our bond-order generalization.

The nitrogen-nitrogen triple bond in N_2 has a bond length of 1.10 Å; the nitrogen-nitrogen double bond in $CH_3-N=N-CH_3$ has a length of 1.24 Å; the nitrogen-oxygen double bond in H−N=O has a length of 1.21 Å; the nitrogen-oxygen single bond in H_2N-OH has a length of 1.46 Å. Based on these data, the bond lengths of 1.13 Å and 1.19 Å for the N−N and N−O linkages in N_2O would suggest an N−N bond order of 3 or slightly less and an N−O bond order of 2 or slightly greater.

None of the electronic structures I, II, or III show N−N and N−O bond orders of about 3 and 2, respectively. In addition, structure III would appear to be quite unlikely because of the positive formal charges on adjacent atoms (the charge distribution implied by these formal charges would be electrostatically unfavorable). A resonance hybrid of forms I and II, however, will produce a reasonable representation of the electronic structure of N_2O. Assuming that the two forms contribute equally, the hybrid would have N−N and N−O bond orders of 2.5 and

1.5, respectively. Since compounds whose electron dot formulas indicate no formal charge are being compared to formulas with formal charges, the bond length-bond order rule can be expected to hold only approximately.

The formate ion, HCO_2^-, has a C–O bond length of 1.26 Å, and an O···C···O bond angle of about 125°. The central carbon apparently uses sp^2 hybrid orbitals and thus we can write

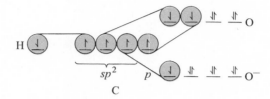

This description, although it explains the bond angle, does not account for the equal C–O linkages. The formation of a resonance hybrid does, however, produce

$$
\left[
\begin{array}{cc}
\overset{\displaystyle :\ddot{O}:}{\underset{\displaystyle H}{\overset{\displaystyle \|}{C}}}\!\!\!\!\diagdown \ddot{O}:^{\ominus} & \longleftrightarrow & \overset{\displaystyle :\ddot{O}:^{\ominus}}{\underset{\displaystyle H}{\overset{\displaystyle |}{C}}}\!\!\!\!\diagup^{\!\!\!\diagdown}_{\!\!\!\ddot{O}:}
\end{array}
\right]^-
$$

formate ion resonance

equal C–O bonds with bond orders of 1.5. The bond length in formaldehyde, which certainly has a C–O bond order of 2, is 1.23 Å. A bond order of 1.5 is then not unreasonable, since the corresponding bond length would be expected to be greater than 1.23 Å.

The amide linkage,

$$
-\!\!\overset{\displaystyle O}{\underset{\displaystyle N}{C}}
$$

is of particular interest because of its presence in amino acids and proteins. The dimensions of the simplest amide, formamide, are given below. The bond lengths and angles found in proteins are nearly identical to those of this molecule. The formamide molecule is completely planar, and apparently has sp^2 hybridization at both carbon and nitrogen. One of the interesting features of the molecule is its C–N bond length of 1.34 Å, which is smaller than that expected for a C–N single bond in this environment (1.47 Å) but larger than the C–N double bond length in the molecule $(CH_3)_2C=NOH$ (1.29 Å).

$$
\begin{array}{c}
\text{H} \qquad \overset{1.34\ \text{Å}}{\downarrow} \quad \text{N} \\
1.24\ \text{Å}\,\diagdown\; :\!C\!\cdot\!\cdot\!N: \;)\; 119° \\
\text{O} \quad \underset{124°}{} \quad \text{H}
\end{array}
$$

formamide

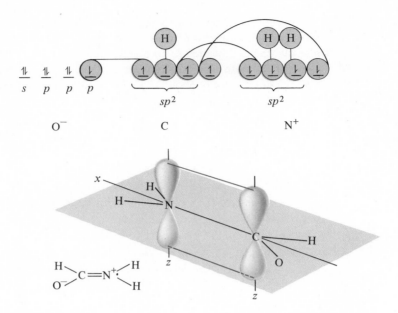

FIGURE 7-26

Valence Bond Description of One Resonance Contributor for Formamide

The resonance concept, as illustrated, again provides a satisfactory description of the molecule. The C—N bond order of the hybrid, assuming equal contributions of forms I and II, is 1.5. Note that overlap of the two p_z orbitals—one on carbon, the other on nitrogen—as shown in Figure 7-26 for contributor II, requires the two portions of the molecule, H···C···O and H···N···H, to be coplanar.

formamide resonance

The C···C···C bond angle in the unique molecule, benzene, is 120°, and the C—C bond lengths are all 1.40 Å. The carbon atoms are obviously sp^2 hybridized, and these hybrids are used to form the three σ bonds at each carbon. Each carbon also has an electron in a p_z orbital, and a suitable electron dot formula for benzene is

This formula would suggest, however, unequal C—C linkages: three bonds shorter than the other three. The equality of the bonds can be accounted for by resonance hybridization of the two structures below.

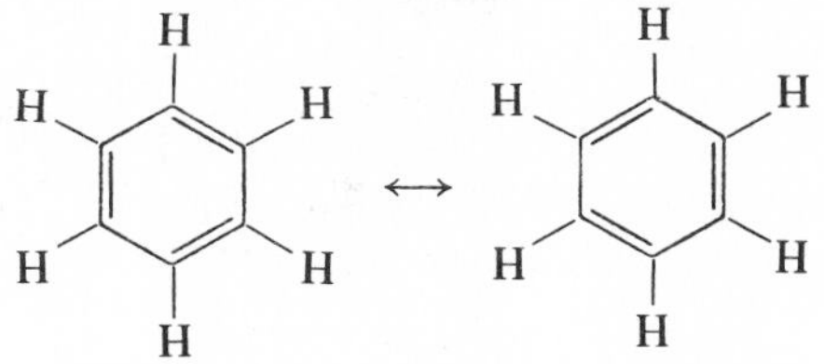

The superimposition of these structures produces a C—C bond order of 1.5. The length expected for a C—C single bond is 1.54 Å, whereas the C—C double bond length is 1.33 Å. The observed value of 1.40 Å lends credence to the 1.5 bond order and the description of benzene as a resonance hybrid.

The resonance structures above suggest that any π electron (there are six of them) in benzene is delocalized over all six C—C linkages, and this is implied by the

abbreviated formula for benzene, 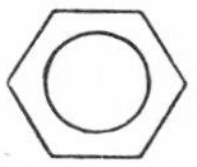 , where the presence of the carbons and hydrogens is understood.

The delocalization of π electrons leads, of course, to a lowering of energy relative to that expected for a single resonance contributor. This resonance energy is estimated to be 37 kcal/mole.

The compound naphthalene, sometimes used as a moth repellant, has the structure

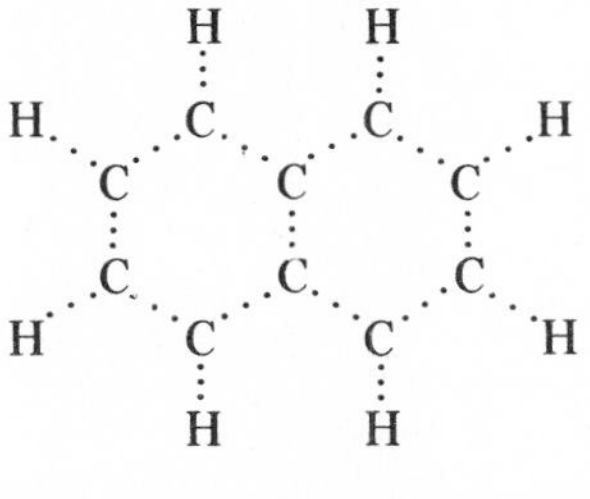

Three electron dot formulas can be written for this system, as illustrated, and the electronic structure is best described as a hybrid of all three. Here, then, delocaliza-

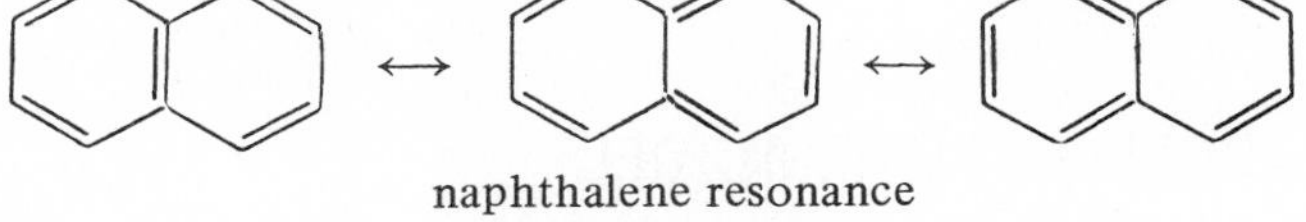

naphthalene resonance

tion of the ten π electrons occurs over eleven bonds, and the resonance energy should be even greater than that of benzene; it is estimated to be 75 kcal/mole.

Two benzene-like electron dot formulas can also be written for cycloocta-tetraene. However, this molecule does indeed have two different C—C bond lengths, and should not then be described as a resonance hybrid of the two forms. In addition, it is not planar, as is benzene, and apparently its nonplanarity prevents the overlap of the *p* orbital of a given carbon with the *p* orbitals of *both* neighboring carbons.

cyclooctatetraene

As a final example of resonance, we shall consider the aromatic alcohol phenol, for which the following electron dot formulas can be written.

A careful analysis of possible electron dot formulas shows that the structures III, IV, and V are also feasible. These structures show delocalization of electrons from the oxygen to the *ortho* and *para* positions of the benzene ring. The best representation of the molecule, then, is as a resonance hybrid of all five structures. Structures III, IV, and V very likely do not contribute as heavily as I and II and are believed to add only 7 kcal/mole to the resonance energy. The delocalization of electrons away from the oxygen does affect the chemical behavior of the compound, as we shall see in Chapter 15.

alternative structures for phenol

MOLECULAR ORBITAL MODEL

The basic assumption of the molecular orbital model is that for every electron in a molecule there exists an orbital that is associated to some extent with the entire molecule. Each molecular orbital, just like an atomic orbital, is designated by its energy and the spatial distribution of its electron density. Indeed, the concept of molecular orbitals is completely analogous to the concept of atomic orbitals.

Since the mathematical equations of the wave model cannot be solved exactly for species as complex as molecules, the energies and electron density distributions of molecular orbitals are obtained by an approximate mathematical method—the combination of atomic orbitals. This process is very similar to hybridization, except that the orbitals involved in the combination are on *different* atoms.

Consider the mathematical construction of the two lowest energy molecular orbitals for the H_2 molecule. These molecular orbitals are formed by addition and subtraction of the lowest energy atomic orbitals, the $1s$ orbitals. Figure 7-27a represents the electron density of two separated $1s$ atomic orbitals as a function of distance from the nucleus (the nuclei are represented by the letters A and B). When the orbitals are allowed to overlap and are added, the electron density between the nuclei increases, as in Figure 7-27b. When the orbitals are subtracted, the electron density is decreased in the region midway between the nuclei (Figure 7-27c).

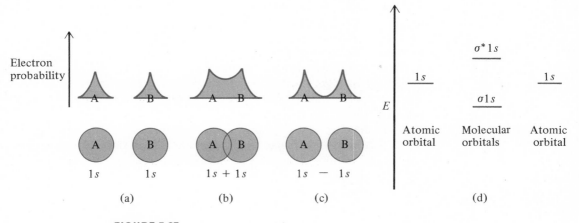

FIGURE 7-27

Electron Density Distributions and Energies of the $\sigma 1s$ and σ^*1s Molecular Orbitals

The molecular orbital that results from the addition of the $1s$ atomic orbitals obviously produces a buildup of electron density between the nuclei. We have previously found that this leads to a lowering of the potential energy relative to the isolated atoms. This orbital is therefore a *bonding* orbital and is usually called the $\sigma 1s$ orbital. The term σ refers to the concentration of electron density along the internuclear axis; the $1s$ denotes the atomic orbitals from which it was constructed.

The molecular orbital that results from the subtraction of the $1s$ atomic orbitals has no electron density in the plane midway between the nuclei. When electrons occupy this orbital, there is no decrease in potential energy; in fact, there is an increase in energy, and this orbital is therefore properly described as *antibonding*. The designation for this orbital is σ^*1s, where the asterisk denotes antibonding.

The energies of these two molecular orbitals relative to the energies of the atomic orbitals are also presented in Figure 7-27d. Note that the bonding orbital is lower in energy than the atomic orbitals, while the opposite is true for the antibonding orbital. The extent to which the energy of the molecular orbital is raised or lowered relative to its constituent atomic orbitals depends on the energies of the atomic orbitals and how effectively they overlap. It is also useful to note that the number of molecular orbitals obtained from a combination of atomic orbitals is the same as the number of atomic orbitals; thus, two molecular orbitals are obtained from two $1s$ atomic orbitals.

These two orbitals certainly suffice for a description of the ground state (lowest energy level) of H_2. The hydrogen molecule contains two electrons, and the aufbau principle and the Pauli exclusion principle dictate, just as for filling atomic orbitals, that the two electrons occupy the lowest energy $\sigma 1s$ orbital.

More orbitals are needed to describe the molecular configurations of diatomic molecules with a greater number of electrons, such as O_2. These orbitals are constructed from the other atomic orbitals—$2s$, $2p$, and so on.

Six molecular orbitals result when three p orbitals on one atom combine with three p orbitals on the other atom. Combination of the p_x orbitals (the internuclear axis is again the x axis) results in bonding and antibonding $\sigma 2p_x$ orbitals. The p_z orbitals produce two $\pi 2p_z$ orbitals; the electron density in both bonding and antibonding orbitals is above and below the internuclear line. The p_y orbitals also produce two $\pi 2p_y$ orbitals that have their maximum density areas rotated $90°$ relative to the $\pi 2p_z$ orbitals. The energies of the $\pi 2p_y$ and $\pi 2p_z$ orbitals are identical; the same is true for the $\pi^* 2p_y$ and $\pi^* 2p_z$ orbitals.

Figure 7-28 shows the relative energies and electron density patterns for the ten lowest energy molecular orbitals for homonuclear diatomic molecules and ions. Higher energy orbitals could be formed by combination of the $3s$, $3p$, etc., orbitals, but the molecular orbitals given here are sufficient for a description of diatomic species containing up to 20 electrons.

We are now in a position to write the electron configurations for some simple molecules. We have already seen that the H_2 molecule has two electrons in the $\sigma 1s$ orbital. In keeping with the convention for writing atomic electron configurations, we can denote this molecular configuration as $(\sigma 1s)^2$. This molecule, therefore, has two bonding electrons and consequently a bond order of 1.

The helium diatomic molecule, He_2, would be expected to have its four electrons in the lowest energy orbitals, and following the Pauli principle we can write the configuration $(\sigma 1s)^2 (\sigma^* 1s)^2$. This molecule would contain, then, two bonding electrons and two antibonding electrons. If we assume that the antibonding electrons exactly cancel the binding effect of the bonding electrons, the bond order in this molecule would be zero. This prediction of instability is borne out by the nonexistence of a stable species of He_2. We will, in general, assume that in molecular orbital theory the bond order for any given linkage is given by $1/2 \times$ (number of bonding electrons minus number of antibonding electrons).

The nitrogen molecule has a total of 14 electrons, and the configuration is therefore $(\sigma 1s)^2 (\sigma^* 1s)^2 (\sigma 2s)^2 (\sigma^* 2s)^2 (\pi 2p_y)^2 (\pi 2p_z)^2 (\sigma 2p_x)^2$. Since there are ten

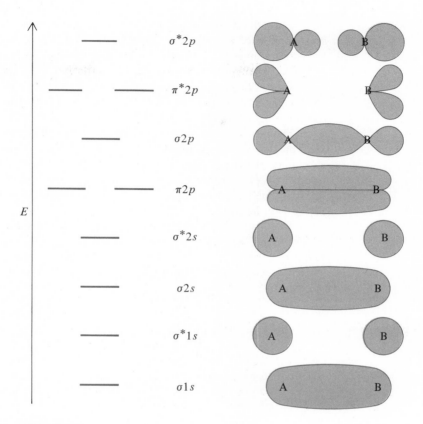

FIGURE 7-28

Electron Density Distribution and Energies of Simple Diatomic Molecular Orbitals

bonding electrons and four antibonding electrons, the N_2 bond order is $(10-4)/2$ = 3, which is the same as the bond order obtained by the valence bond model.

The configuration for O_2 is $(\sigma 1s)^2(\sigma *1s)^2(\sigma 2s)^2(\sigma *2s)^2(\pi 2p_y)^2(\pi 2p_z)^2$ $(\sigma 2p_x)^2(\pi *2p_y)^1(\pi *2p_z)^1$. The single occupancy of the $\pi *$ orbitals is necessitated by Hund's first rule and results in the prediction that O_2 should have unpaired electrons and thus exhibit paramagnetic behavior (see Chapter 22). The valence bond description of O_2, on the other hand, predicts no unpaired electrons. Since O_2 is indeed paramagnetic, this was considered one of the early triumphs of molecular orbital theory. Both theories predict a bond order of 2.

The species O_2^+ has one less electron than O_2 and thus has only one electron in a $\pi *$ orbital. The bond order is therefore $(10-5)/2 = 2.5$. The superoxide ion, O_2^-, has one electron more than O_2, and since this electron must also enter a $\pi *$ orbital, the bond order is $(10 - 7)/2 = 1.5$. The peroxide ion, O_2^{2-}, with 18 electrons, has a filled $\pi *$ level, and consequently a bond order of 1. Clearly, molecular orbital theory accounts very naturally for fractional bond orders. The bond lengths of O_2^+, O_2, O_2^-, and O_2^{2-} are 1.17 Å, 1.21 Å, 1.28 Å, and 1.49 Å, respectively, in agreement with our bond order–bond length generalization.

Diatomic fluorine is isoelectronic with O_2^{2-} and therefore also has a bond order of 1. The two additional electrons necessary for the Ne_2 molecule would occupy the σ^*2p orbital, thereby producing a bond order of zero. This species is, in fact, unknown.

The molecular orbital energy-level diagram (Figure 7-28) can also be used to describe the electron configuration of simple heteronuclear (different nuclei) diatomic molecules or ions. The odd-electron molecule of nitrogen oxide has a bond length of 1.15 Å, which lies between that of N_2, 1.10 Å, and that of O_2, 1.21 Å. The molecule is isoelectronic with O_2^+ and therefore has a bond order of 2.5, in accord with its bond length.

The major difference, then, between the molecular orbital model and the valence bond model is that each electron in a molecular orbital description is assumed to be associated to some extent with every atom of the molecule, whereas in the valence bond model some electrons are localized about specific nuclei.

SUGGESTED READINGS

Christian, J. D., "Strength of Chemical Bonds." *Journal of Chemical Education*, Vol. 50, No. 3 (March 1973), pp. 176–177.

Pauling, L. *The Nature of the Chemical Bond*. 3rd ed. New York: Cornell University Press, 1960.

Sebera, D. K. *Electronic Structure and Chemical Bonding*. New York: Blaisdell Publishing Co., 1964.

PROBLEMS

1. Provide both an IUPAC and common name where possible for each of the following compounds:

(a) $CH_3CH_2CH_2CH_2CH_3$

(b) $CH_3\overset{\overset{\displaystyle CH_3}{|}}{C}HCH_2CH_3$

(c) $CH_3-\overset{\overset{\displaystyle CH_3}{|}}{C}Br_2$

(d) CH_3I

(e)
$$\begin{array}{ccc} & CH_2 & \\ H_2C & & CH_2 \\ | & & | \\ H_2C & & CH_2 \\ & CH_2 & \end{array}$$

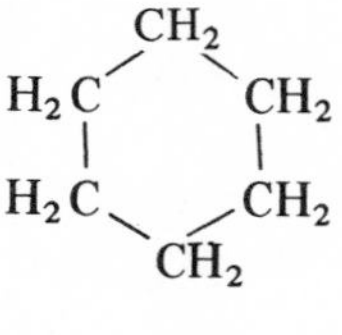

(f) $CH_3CH{=}CHCH_2CH_3$

(g) 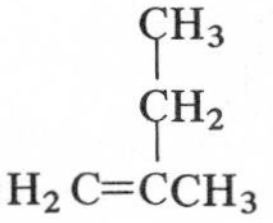 $H_2C=CCH_3$
(with CH_3 and CH_2 substituents shown above)

(h) $CH_3-C\equiv CH$

(i) 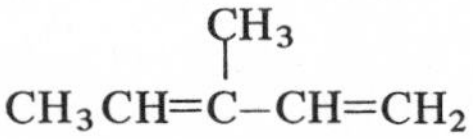 $CH_3CH=C-CH=CH_2$
(with CH_3 substituent)

(j) 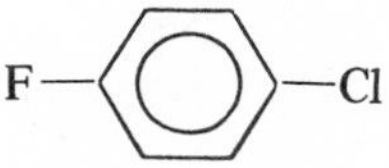 F—⟨O⟩—Cl

(k) $CH_3CH_2CH_2CH_2CH_2CH_2OH$

(l) F_3COH

(m) 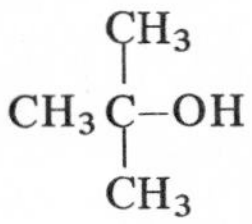 $CH_3\underset{CH_3}{\overset{CH_3}{C}}-OH$

(n) 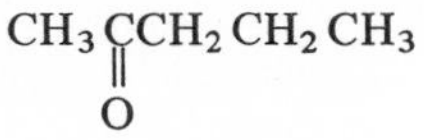 $CH_3\underset{O}{C}CH_2CH_2CH_3$

(o) $H_2\underset{Cl}{C}-\underset{Cl}{C}-\overset{O}{C}CH_3$ (with H on middle carbon)

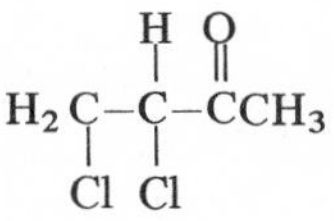

(p) 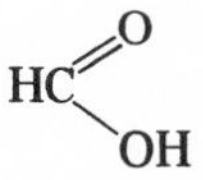 $HC\overset{O}{\underset{OH}{}}$

(q) 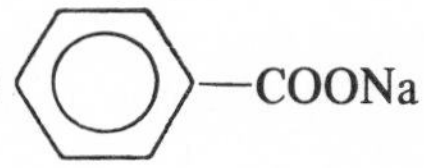 ⟨O⟩—COONa

(r) 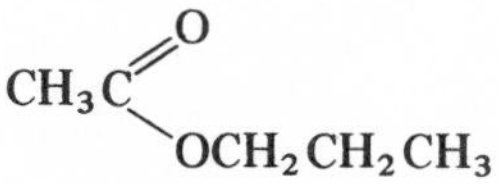 $CH_3C\overset{O}{\underset{OCH_2CH_2CH_3}{}}$

(s) $H_2C=O$

(t) $CH_3OCH_2CH_2CH_3$

(u) 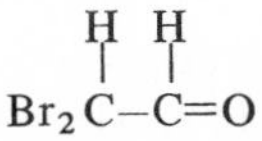 $Br_2C-C=O$ (with H H on carbons)

(v) $(CH_3CH_2CH_2CH_2)_3N$

(w) $(CH_3)_2CHNH_2$

2. Write a structural formula for each of the following:
   (a) 2,2-dimethylpropane
   (b) 1-chloro-2,2-dimethyl-pentane
   (c) 1-chloro-3,4-dinitro-butane
   (d) 1,1,2-trimethylcyclo-butane
   (e) 2,4-pentadiene
   (f) 1,1-dichloroethene
   (g) ethyne
   (h) 2,6-difluorotoluene
   (i) isopropyl alcohol
   (j) cyclopentanol
   (k) *t*-butyl alcohol
   (l) 3-methyl-1-penten-3-ol
   (m) 2-methyl-3-pentanone
   (n) butyric acid
   (o) 1,3-hexanediol
   (p) *o*-nitrobenzoic acid
   (q) sodium propionate
   (r) n-propyl formate
   (s) ethyl 2-methylpropanoate
   (t) 2,2-dimethylpropanal
   (u) ethylmethylamine
   (v) *t*-butyldimethylamine

3. Name the following compounds:
   (a) NO
   (b) N_2O
   (c) NO_2
   (d) N_2O_4
   (e) CS_2
   (f) PCl_5
   (g) SF_6
   (h) $SiBr_4$

4. Which compound in each of the following pairs has the greater amount of ionic character in its bonds?

(a) LiBr, BeBr$_2$ (b) BF$_3$, NF$_3$
(c) LiF, LiI   (d) CF$_4$, CI$_4$
(e) SiO$_2$, SiS$_2$ (f) CCl$_4$, SnCl$_4$
(g) SnCl$_2$, SnCl$_4$ (h) Tl$_2$O$_3$, Tl$_2$O
(i) F$_2$, IF   (j) FeCl$_2$, FeCl$_3$

5. Write suitable electron dot formulas for the following.

Ions
(a) carbonate   (b) borate (BO$_3^{3-}$)
(c) chlorate   (d) perchlorate
(e) sulfate   (f) hydrogen carbonate
(g) sulfite   (h) hydroxide
(i) acetate   (j) ammonium
(k) oxalate   (l) nitrate
(m) fluoride   (n) phosphate

Molecules
(o) ethyne   (p) propanone
(q) 1,3-butadiene   (r) methanol
(s) acetic acid   (t) carbon dioxide
(u) arsenic pentafluoride   (v) iodine pentafluoride
(w) beryllium chloride   (x) boron tribromide
(y) tin tetrachloride   (z) sulfur tetrafluoride

6. Which ions in question 5 are isoelectronic?

7. The following species (or their derivatives) are often used as bond length standards against which other species can be compared in order to determine approximate bond orders.

Species	Standard for Bond	Length of Bond (Å)
CH$_3$CH$_3$	C–C	1.54
H$_2$C=CH$_2$	C=C	1.34
HC≡CH	C≡C	1.20
CH$_3$OH	C–O	1.43
H$_2$CO	C=O	1.23
CO	C≡O	1.13
CH$_3$NH$_2$	C–N	1.47
H$_2$C=NH	C=N	1.29
CN$^-$	C≡N	1.15
H$_2$NOH	N–O	1.46
HNO	N=O	1.21
NO$^+$	N≡O	1.06
H$_2$NNH$_2$	N–N	1.45
HN=NH	N=N	1.25
N$_2$	N≡N	1.10
HOOH	O–O	1.48
O$_2$	O=O	1.21

(a) Show that only one electron dot formula that obeys the octet rule can be written for each species.

(b) Rationalize the bond length order C–C > C–N > C–O; C=C > C=N > C=O; C–C > C=C > C≡C.

8. For binary compounds it can be shown that dividing the total number of valence electrons in a molecule by eight gives the number of sigma bonds to the central atom. Any remainder is equal to the number of nonbonding electrons residing on the central atom. Why is it that this approach is not valid for binary hydrogen compounds?

9. The C≡O bond is stronger than the N≡N bond. Explain. However, carbon monoxide is more reactive than N_2. Explain.

10. Attempts have been made to prepare the species SO_4 and NH_4. Write electron dot formulas for each species and discuss the probable stability of each species.

11. Aqueous ammonia has sometimes been represented as molecular $NH_4 OH$. Can you draw a "good" electron dot structure for this species?

12. Write an electron dot formula for the cyanamide ion, CN_2^{2-}. What common molecule is isoelectronic with this ion? What would you predict as the shape of this ion?

13. Provide a valence bond description, complete with diagram showing overlap of orbitals, hybridization, and resonance where necessary for each of the following. Each description must be consistent with the data for the species as found in Appendix 6.

(a) NCO^-	(b) N_3^-	(c) SO_2
(d) $ClNO$	(e) CO_3^{2-}	(f) SO_3
(g) $O_2 NCl$	(h) HN_3	(i) $NCCN$
(j) $H_2 CCO$	(k) CO_2	(l) PCl_3
(m) $H_2 Se$	(n) NO_2	(o) I_3^-
(p) HCO_2^-		

14. Both ICN and SCN⁻ are linear species. Give valence bond descriptions for each that are consistent with their shape.

15. The salt $Na_2 N_2 O_3$ was first reported in 1896. Recently, X-ray diffraction studies have shown that the anion is planar and has the following bond lengths and angles:

$$a = 1.31 \text{ Å}, \quad b = 1.32 \text{ Å}, \quad c = 1.20 \text{ Å}, \quad d = 1.35 \text{ Å}$$

Provide a valence bond description of the bonding in this ion that is, as far as possible, in agreement with these parameters.

16. The nitro group in nitrobenzene has equal nitrogen-oxygen bond lengths. Account for this fact with suitable resonance structures.

17. The nitrogen-nitrogen bond length in N_2O_4 is 1.75 Å, which is considerably greater than the nitrogen-nitrogen single-bond length of 1.45 Å found for hydrazine (H_2NNH_2). Explain, using appropriate electron dot formulas.

18. The distance between the two middle carbons of 1,3-butadiene is 1.46 Å. The normal carbon-carbon single-bond length is 1.54 Å. Rationalize with appropriate resonance structures.

19. Contrast the Lewis and valence bond models in terms of (a) their theoretical bases and (b) the way in which they rationalize molecular geometry.

20. Describe the bonding in the following species with the molecular orbital model. Determine the bond order and predict the stability of each species.

    (a) He_2 (b) Li_2 (c) Be_2
    (d) N_2 (e) O_2 (f) O_2^+
    (g) O_2^- (h) O_2^{2-} (i) NO
    (j) NO^+ (k) F_2

21. Rationalize the following bond lengths based on your description in question 20.

O_2^-	1.28	NO	1.15
O_2	1.21	NO^+	1.06
O_2^+	1.17	N_2	1.10

22. Describe the bonding in NO with the Lewis model, the valence bond model, and the molecular orbital model. Make each description as consistent as possible with the fact that the bond length in NO is 1.15 Å. Carefully evaluate the success of each of the models in this case.

23. Carefully explain:
    (a) Why covalent compounds may be more stable than their constituent elements.
    (b) The basis for the generalization that the greater the bond order the shorter and stronger the bond.
    (c) The relationship between bond polarity, geometry, and dipole moment.
    (d) The basis and application of Fajans' rules.
    (e) The difference between a sigma bond and a pi bond.
    (f) The similarity between hybridization and resonance.
    (g) How and why isoelectronic species are similar.
    (h) The determination and meaning of formal charge.
    (i) The relationship between resonance and delocalization of electrons.
    (j) The difference between a bonding orbital and an antibonding orbital.

24. The molecular orbital energy levels for the *valence* electrons of CO_2 and approximate electron-density diagrams for each set of orbitals are given in Figure 7-29. The determination of bond order for polyatomic molecules

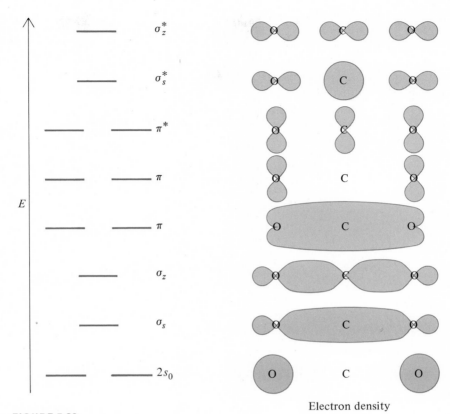

Electron density

FIGURE 7-29

(from molecular orbital theory) is not quite as simple as for diatomic molecules. The definition of bond order is the same:

$$\frac{\text{Number of bonding electrons} - \text{Number of antibonding electrons}}{2}$$

but there are frequently also nonbonding electrons—electrons that are essentially localized on one atom and do not contribute to, nor detract from, bonding between atoms. Another difficulty is that one must be careful to divide the total bond order (as defined above) between the proper number of bonds.

(a) Four *orbitals* in Figure 7-29 are nonbonding orbitals. By examining the electron-density diagrams, decide which orbitals are nonbonding and label them n.

(b) Populate the orbitals with the valence electrons of CO_2 and determine the bond order of each C—O bond.

(c) Describe the bonding in CO_2 with the valence bond model and then focus your attention on the pi bonds and decide how the molecular orbital "picture" differs from the valence bond "picture."

8 Compounds IV: Molecular Geometry

GEOMETRY OF MOLECULES

While the nature of the atoms and the type of bonding present in a compound are the major determinants of its chemical and physical properties, the three-dimensional geometry of molecules is a feature of prime importance in a covalent compound. The marked difference in physiological properties of the two drugs levorphan and dextrophan is just one example of the rather dramatic influence of structure on chemical behavior. The two compounds differ only in that their structures are mirror images of one another, and yet levorphan is more strongly analgesic and addictive than morphine, whereas dextrophan is neither addictive nor an analgesic.

levorphan   dextrophan

In the previous chapter the concept of hybridization was used to rationalize molecular geometry. Because of the importance of the structure of molecules, it is frequently desirable to have at hand a method to predict geometries rather than rationalize them. Over the past several decades, chemists have developed such a method, which is based upon the hypothesis that a molecule will adopt the shape that minimizes the electrostatic repulsions between electrons.

Since electrons are negatively charged and therefore repel one another, it is certainly reasonable to assume that their electron clouds will be as far apart as possible. Furthermore, wave mechanics reveals that the close approach of electrons of opposite spin is more likely than the close approach of electrons of the same spin. Thus, electrons of opposite spin will tend to group in pairs, and a molecule will adopt a geometry that minimizes repulsions between these pairs.

In order to predict this geometry for a given molecule, we need to know how many. pairs of electrons are in the valence shell of each atom. This can be determined most simply from the electron dot formula for the molecule. Not all of the electron pairs shown in the electron dot formula are considered, however. Pi bonds are directed along one or more sigma bonds, and therefore we assume that the *geometry about a given atom is determined by the number of sigma-bonded and nonbonded pairs of electrons at that atom.*

As a first example, consider beryllium chloride. Two plausible electron dot formulas are

$$:\ddot{C}l-Be-\ddot{C}l: \quad \text{and} \quad :\ddot{C}l=Be=\ddot{C}l:$$

Both show two sigma-bonded pairs but no nonbonded pairs about the central beryllium atom. Hence, two electron pairs must be positioned about beryllium in a way that will minimize the repulsions between them. Clearly, this can be accomplished by placing them on opposite sides of the beryllium on a straight line. Chlorine atoms are, of course, also attached to the lone pairs, and the molecule can be predicted to be *linear* with a bond angle of 180°. This is indeed the experimentally determined structure of gaseous $BeCl_2$.

The electron dot formula for boron trichloride can be written as

In either case, there are three sigma-bonded electron pairs, and by the application of simple geometry it can be shown that the separation of the three pairs is maximized if these pairs point toward the corners of an equilateral triangle. The bond angle should therefore be 120°, and the molecular shape can be characterized as *trigonal planar*.

The nitrate ion can be described as a resonance hybrid of three electron dot formulas (see diagram). Each formula shows only three *sigma*-bonded pairs of electrons, and the shape of this ion is also *trigonal planar*. A third example of a species with three electron pairs directed toward the corners of a triangle is the nitrite ion (see diagram). One of the pairs is a nonbonded pair, however, and since the shape of a molecule or ion is determined by the positions of the *atoms,* this ion is described simply as *angular* or *bent.*

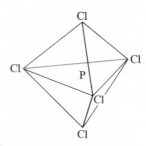

nitrate ion

nitrite ion

The repulsions between *four* electron pairs are minimized if they are directed toward the corners of a regular tetrahedron. Methane has four sigma-bonded pairs and is therefore a *tetrahedral* molecule. The central nitrogen atom of ammonia is surrounded by three sigma-bonded pairs and one nonbonded pair of electrons. These four electron pairs are tetrahedrally disposed about the central atom, but the arrangement of the *atoms* in ammonia is described as *pyramidal.*

Water is another example of a molecule whose central atom is tetrahedrally surrounded by electron pairs. Hydrogen atoms are attached to two of the pairs, and the shape of the molecule is denoted as *bent.*

Empirically, it has been found that most compounds of the type AB_5 that have five bonded and no nonbonded pairs of electrons surrounding the central atom A have a *trigonal bipyramidal* structure (with bond angles of 90° and 120°) as shown here for phosphorus pentachloride. It can be assumed, therefore, that this configuration minimizes the repulsions between five pairs of electrons.

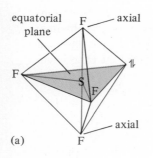

(a)

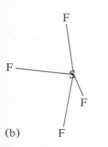

(b)

FIGURE 8-1

phosphorus pentachloride

It has also been found that when one or more of the five pairs of electrons are nonbonded pairs, these nonbonded pairs are positioned on the equatorial plane of the trigonal bipyramid. Sulfur tetrafluoride, therefore, has the shape given in Figure 8-1. Actually, the structure of SF_4 is not quite that shown in Figure 8-1a; instead, the angle between the "axial" fluorines is somewhat less than 180°; that is, they "lean" toward the other fluorines, as shown in Figure 8-1b. This distortion of the

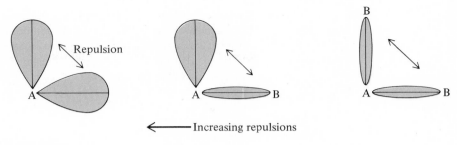

FIGURE 8-2

The Variation of Bonded, Nonbonded Repulsions

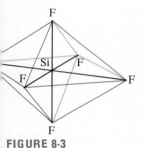

FIGURE 8-3

The Geometry of SiF_6^{2-}

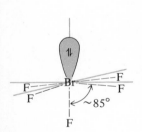

FIGURE 8-4

The Geometry of BrF_5

trigonal bipyramid can be explained in terms of the repulsions between the electron pairs. Because of their attraction to two nuclei rather than one, the electron density of bonded pairs is more closely confined to the internuclear axis than the electron density of nonbonded pairs (see Figure 8-2). Therefore, two bonded pairs do not repel one another as strongly as a nonbonded pair and a bonded pair. For the same reason, the repulsion between two nonbonded pairs is greater than that between a nonbonded pair and a bonded pair. (This is also illustrated in Figure 8-2.)

Chlorine trifluoride is an example of a molecule with five electron pairs, two of which are nonbonding, about the central atom. Both nonbonded pairs reside in the equatorial plane, giving the molecule a T shape. The actual bond angle of 87°, slightly less than 90°, can again be rationalized by nonbonded-bonded repulsions.

The repulsions between six pairs of electrons can be minimized by placing them at the apices of an octahedron. In the ion SiF_6^{2-}, for example, the silicon is surrounded by an octahedron of fluorines and the shape of the ion is designated as *octahedral* (bond angles of 90°; see Figure 8-3).

When one of the six pairs of electrons is nonbonding, the structure of the molecule can be described as a *square pyramid*. Bromine pentafluoride is an example of such a molecule, and its structure is shown in Figure 8-4. The deviation of the bond angle from 90° can be attributed to greater repulsions between the lone pair and the Br—F bonds than those between adjacent Br—F bonds.

Xenon tetrafluoride also has six pairs about the central xenon, but two of these are nonbonded pairs. The repulsions between these two pairs are minimized if they are positioned as far apart as possible. As shown in Figure 8-5, this results in a square *planar* geometry for XeF_4.

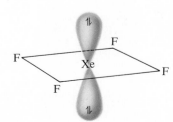

FIGURE 8-5

The Geometry of XeF_4

The principles of this predictive approach can be summarized as follows: The geometry of a molecule or ion is determined by repulsions between the sigma and nonbonded electron pairs about the central atom. The shape adopted is the one that minimizes these repulsions. A secondary determinant of structure is that the repulsions between bonded and nonbonded electron pairs vary: nonbonded-

TABLE 8-1
Summary of Geometry of Species of Type AB_n

Number of Sigma-Bonding and Nonbonding Electron Pairs about Central Atom	Number of Nonbonding Pairs	Examples	Shape	Hybridization	Approximate Bond Angle
2	0	$BeCl_2$ (g) CO_2, $HgCl_2$, N_3^-	linear	sp	$180°$
3	0	BCl_3, CO_3^{2-}, NO_3^-	trigonal planar	sp^2	$120°$
3	1	NO_2^-	bent	sp^2	$120°$
4	0	CH_4, BF_4^-, SO_4^{2-}	tetrahedral	sp^3	$109.5°$
4	1	NH_3, OH_3^+	pyramidal	sp^3	$109°$
4	2	H_2O, SCl_2	bent	sp^3	$109°$
5	0	PCl_5, $AsCl_5$	trigonal bipyramidal	sp^3d	$90°, 120°$
5	1	SF_4, $TeCl_4$	see-saw	sp^3d	$90°, 120°$
5	2	BrF_3	T-shaped	sp^3d	$90°$
5	3	I_3^-	linear	sp^3d	$180°$
6	0	SiF_6^{2-}	octahedral	sp^3d^2	$90°$
6	1	BrF_5	square pyramidal	sp^3d^2	$90°$
6	2	ICl_4^-, XeF_4	square planar	sp^3d^2	$90°$

nonbonded > nonbonded-bonded > bonded-bonded. Table 8-1 provides a tabular summary of molecular geometry and the corresponding hybridization.

EXAMPLE: Predict the shape of $OPCl_3$.

Solution: Two acceptable electron dot formulas for this compound are illustrated here as I and II. Both formulas show four sigma-bonded electron pairs and no nonbonded pairs about the central atom. In order to minimize repulsions, these pairs must be directed toward the corners of a tetrahedron. The molecule could be described as tetrahedral, although it cannot have the shape of a regular tetrahedron because there are two *different* kinds of atoms (and therefore two different bond lengths) surrounding the central atom. The bond angles would be predicted to be approximately 109°.

I   II

EXAMPLE: Explain the fact that BF_3 does not have a dipole moment, whereas both NF_3 and ClF_3 do.

Solution: The magnitude of the dipole moment of a molecule depends upon (a) the difference in electronegativities between the atoms within the molecule and (b) the geometry of the molecule. Since the electronegativity difference is greater between boron and fluorine than between nitrogen and fluorine or between chlorine and fluorine, the B—F bonds in BF_3 are almost certainly more polar than the N—F bonds in NF_3 or the Cl—F bonds in ClF_3.

The fact that BF_3 has no dipole moment whereas NF_3 and ClF_3 do can therefore be attributed to a difference in geometries. Examination of electron dot formulas for the three molecules shows that boron trifluoride has no nonbonded electron pairs, nitrogen trifluoride has one, and chlorine trifluoride has two. Boron trifluoride is therefore a trigonal planar molecule with 120° bond angles; NF_3 has its electron pairs directed toward the corners of a tetrahedron and is a pyramidal molecule; ClF_3 has its five electron pairs directed towards the corners of a trigonal bipyramid and is a T-shaped molecule.

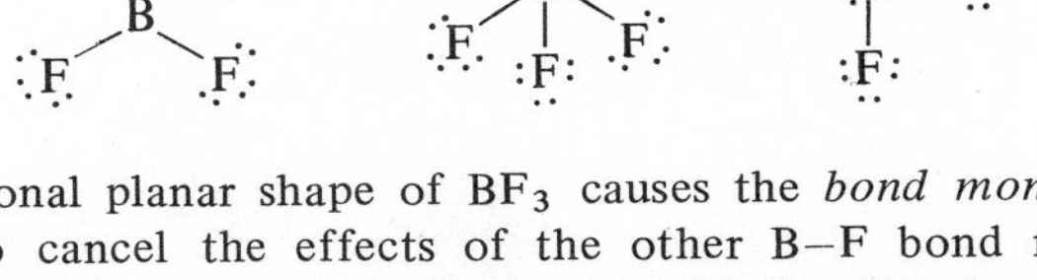

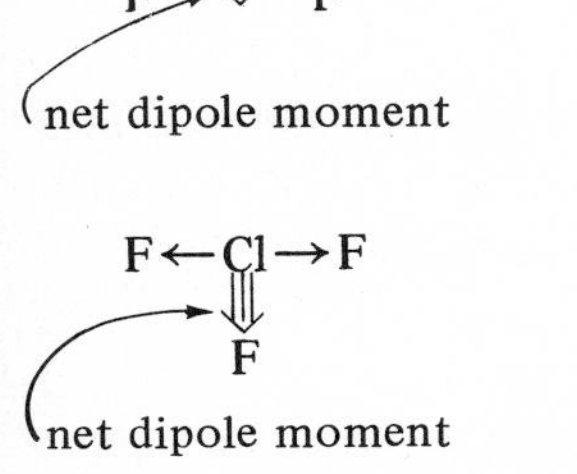

no net dipole moment

net dipole moment

net dipole moment

FIGURE 8-6

The Structures and Dipole Moments of BF_3, NF_3, and ClF_3

The trigonal planar shape of BF_3 causes the *bond moments* of each B—F bond to cancel the effects of the other B—F bond moments; the molecule therefore has no net dipole moment. On the other hand, in NF_3 and ClF_3 the bond moments reinforce one another to some extent and produce net dipole moments for these molecules (Figure 8-6).

DYNAMICS OF MOLECULAR GEOMETRY

Our previous discussion of structure and bonding may have produced the impression that molecules are rather rigid, static objects containing atoms held at fixed positions within the molecule. Experimental evidence suggests, however, that the atoms within molecules, and the molecules themselves, are in constant motion. These motions can be divided into the following categories: translational motion of the entire molecule, rotational motion of the molecule, vibrational motion of atoms about their molecular loci, and rotations of groups within the molecule.

When a molecular substance is in either the liquid or gaseous state, each molecule is free to move throughout the entire volume of the substance. This motion is termed *translation* and is pretty much a random straight-line motion. When the substance is in the solid state, the molecules are fixed in their lattice positions (they do vibrate about these positions, however) and thus there is no net translational movement.

While a molecule in the liquid or gaseous states is translating through space it also undergoes a tumbling motion. This rotation occurs about an imaginary line (axis) through the center of mass of the molecule, as illustrated for nitrogen oxide in Figure 8-7.

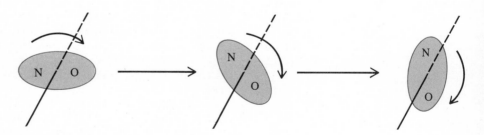

FIGURE 8-7

Rotation of NO

The atoms within a molecule are also in continual vibratory motions. In a polyatomic molecule these vibrations can be quite complex; in a simple diatomic molecule, the oscillations can be likened to the motion of two balls attached by a spring (Figure 8-8).

FIGURE 8-8

Vibration of Diatomic Molecule

It has also been shown experimentally that within polyatomic molecules there is rotation of groups that are attached by *single* bonds. In ethane, for example, the two methyl groups rotate relative to one another as shown in Figure 8-9. This rotation of groups is a consequence of the cylindrical symmetry of sigma bonds (Figure 8-9b). That is, the distribution of electron density of a sigma bond is not affected by the orientation of attached atoms, and therefore rotation of groups does not affect the energy of the bond.

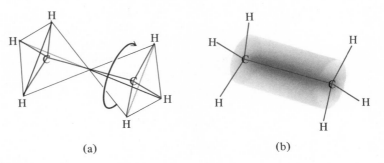

(a)   (b)

FIGURE 8-9

Rotation of Methyl Groups in Ethane

As an example of more complex rotations, consider 1-propanol. If the O—H bond of this molecule is taken as fixed in space, two of the many possible conformations of the molecule are as presented in Figure 8-10. Here each solid line indicates a bond in the plane of the paper; a heavy triangular line, a bond extending above the plane of the paper; and a dotted line, a bond extending below the plane of the paper. Note that the word *conformations* is used to denote spatial orienta-

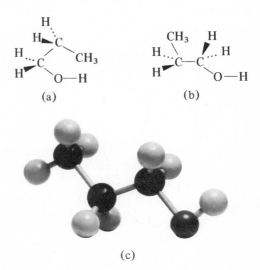

FIGURE 8-10

Some Conformations of 1-Propanol

tions resulting from rotation about single bonds. The student is advised to investigate the conformational possibilities for polyatomic molecules by using molecular models. The conformation of Figure 8-10b is illustrated in Figure 8-10c with a model of the ball-and-stick variety.

Now let us examine the possibility of rotation about double bonds. For example, do the CH_2 groups of ethene rotate relative to one another? The valence bond description of ethene depicts the formation of the pi bond via overlap of p_z orbitals on the carbon atoms (Figure 8-11a). If one CH_2 group were to rotate $90°$ relative to the other, the p_z orbitals would be perpendicular, rather than parallel, and their electron densities would no longer overlap (Figure 8-11b). Thus, rotation of the CH_2 groups necessarily destroys the pi bond and is therefore energetically unfavorable. At room temperature, then, groups do not rotate about double bonds. At elevated temperatures, where sufficient energy is available to make the breaking of the pi bond possible, even this type of rotation may occur.

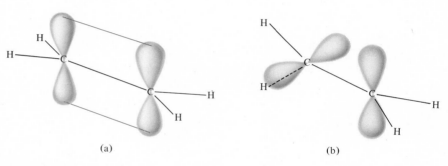

(a)   (b)

FIGURE 8-11

Rotation of CH_2 Groups in Ethene

One of the consequences of the lack of rotation about double bonds is the existence of *geometrical isomers* for some compounds.

ISOMERISM

The word isomerism is derived from the Greek word *isomeres*, meaning "of equal parts." In chemistry the word is used to designate two or more compounds that contain the same (equal) atoms but differ in the arrangement of these atoms. The arrangement of atoms can differ in (a) the way in which the atoms are linked together and (b) their spatial orientation. Correspondingly, there are two principal types of isomers: *Structural isomers* have the same molecular formula but different structural formulas, whereas *stereoisomers* have the same structural formula but different three-dimensional arrangements of the atoms.

TABLE 8-2 Structural Isomers with Molecular Formula $C_4H_{10}O$		
	$CH_3\,CH_2\,CH_2\,CH_2\,OH$	1-butanol
	$CH_3\,CHCH_2\,OH$ $\quad\;\; \mid$ $\quad\;\; CH_3$	2-methyl-1-propanol
	$\qquad CH_3$ $\qquad \mid$ $CH_3\,CH_2\,CHOH$	2-butanol
	$\quad\;\; CH_3$ $\quad\;\; \mid$ $CH_3\,C{-}OH$ $\quad\;\; \mid$ $\quad\;\; CH_3$	2-methyl-2-propanol
	$CH_3\,CH_2\,OCH_2\,CH_3$	ethyl ether
	$CH_3\,OCH_2\,CH_2\,CH_3$	methyl *n*-propyl ether
	$CH_3\,OCHCH_3$ $\qquad \mid$ $\qquad CH_3$	methyl isopropyl ether

As an illustration of structural isomerism, consider the seven compounds that have the molecular formula $C_4H_{10}O$. Their structural formulas are given in Table 8-2. Within this set of *structural* isomers, the alcohols are sometimes said to be *functional group* isomers of the ethers; that is, the alcohols have the same molecular formula as the ethers but a different functional group. The compounds 1-butanol, 2-methyl-1-propanol, and 2-methyl-2-propanol are also termed *chain* isomers because they have the same molecular formula and functional group but different branching on the alkyl chain. The two propyl ethers are also chain isomers.

The compounds with molecular formula $C_2H_4Cl_2$ provide another example of structural isomerism. The dichloroethanes differ in the position of the chlorine atoms and are therefore also sometimes referred to as *position* isomers.

$$H_2C{-}CH_2$$
$$\mid \quad\; \mid$$
$$Cl \quad Cl \qquad\qquad\qquad Cl_2\,CHCH_3$$
$$\text{1,2-dichloroethane} \qquad \text{1,1-dichloroethane}$$

There are also several types of *stereoisomerism.* The three-dimensional orientation of atoms is influenced by rotation about single bonds and lack of rotation about double bonds. While rotation about single bonds produces different conformations of atoms, and one conformation may be more stable than another, the rotation is so rapid at room temperature that conformational isomers cannot be isolated. Figure 8-12 shows two conformations for 1,2-dichloroethane. Conformation (b) is the more stable of the two, because it minimizes the repulsions between the chlorine atoms, but rapid rotation prevents its isolation at room temperature.

The absence, at room temperature, of rotation about double bonds results in a second type of stereoisomerism—*geometrical isomerism.* The structural formula of

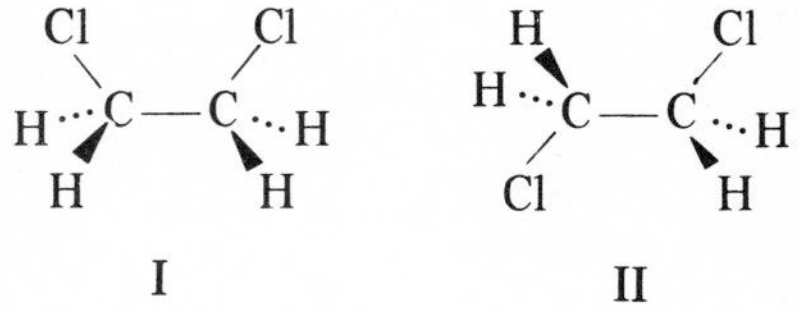

FIGURE 8-12

Two Conformations for 1,2-Dichloroethane

2-butene is $CH_3CH{=}CHCH_3$. There are, however, two compounds with this formula. One compound has both methyl groups on the same side of the double bond (remember that the carbon skeleton must be planar because of the sp^2 hybridization of the central carbons) and is termed *cis*-2-butene, while the other compound has the methyl groups on opposite sides of the double bond and is termed *trans*-2-butene. This *cis-trans* nomenclature is derived from the Latin: *cis*, on this side; *trans*, across.

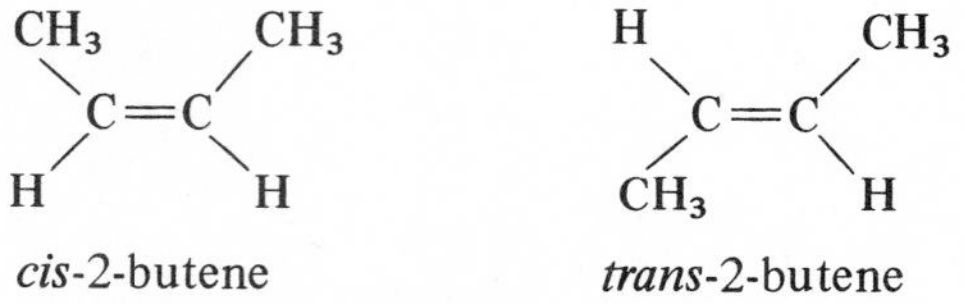

cis-2-butene   *trans*-2-butene

The presence of a double bond is not a sufficient condition for geometrical isomerism, however. While two compounds have the structural formula of 2-butene, there is only one compound with the structural formula of 1-butene, $CH_3CH_2CH{=}CH_2$. As shown in Figure 8-13, shifting the ethyl group from one side of the double bond to the other does not produce a different compound. Rotation of structure I through 180° about the C=C axis (dotted line) results in II, and thus the two structures are identical. Hence, a necessary condition for geometrical isomerism in alkenes, is the presence of two *different* groups attached to each of the double-bonded atoms. In Figure 8-14, where A, B, D, and E represent different groups such as H, Cl, CH_3, CH_2CH_3, and C_6H_5, only structures (e) and (f) exhibit geometrical isomerism. Geometrical isomers often have different physical and chemical properties. For example, the boiling point of *trans*-butene is 2.5°C and that of *cis*-butene is 1.0°C.

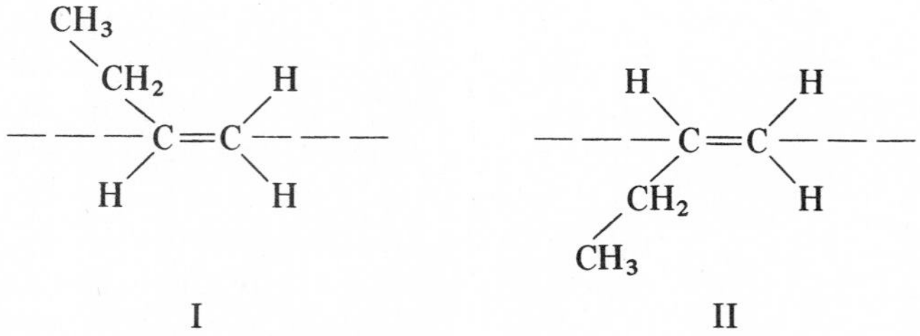

FIGURE 8-13   I   II

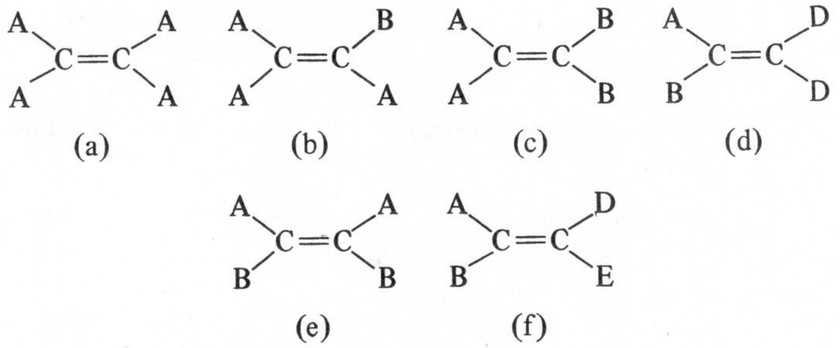

FIGURE 8-14

Geometrical Isomerism in Alkenes
Structures (e) and (f) exhibit geometrical isomerism

Geometrical isomerism is not restricted to alkenes. In the geometrical isomers

cis   *trans*

the lone pair of electrons on each nitrogen functions as the other "group" in the configuration. Many compounds containing a carbon-nitrogen double bond also exhibit geometrical isomerism. Oximes have the general formula $RON=CR'R''$ and can exist as isomers when R' and R'' are different groups.

A fascinating example of the presence of geometrical isomerism in biological processes is provided by the visual pigment, rhodopsin, found in the retina of the eye. Rhodopsin consists of an alkene attached to a protein. The alkene, called retinal, can exist as a number of geometrical isomers. When it is firmly attached to the protein it exists in the form shown in Figure 8-15. This form is referred to as 11-*cis*-retinal because all the hydrogen and methyl groups attached to the alkene

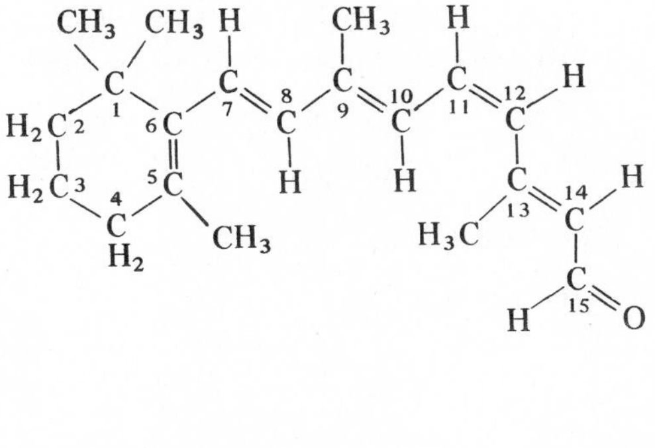

FIGURE 8-15

11-*cis*-Retinal

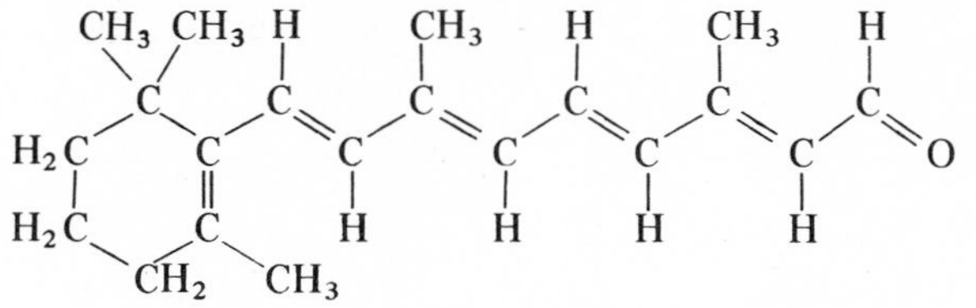

FIGURE 8-16

All-*trans* Retinal

chain are *trans* to one another except for those at the number 11 and 12 carbons. When light strikes the eye, sufficient energy is imparted to the 11-*cis* isomer to produce a rotation about the number 11–number 12 carbon-carbon double bond. This rotation is in effect a rotation of the entire group shown in the tinted portion of the figure and results in the all-*trans* isomer (Figure 8-16). It is this isomerization and accompanying changes in the attachment to the protein that allow the perception of light by the eye.

Geometrical isomerism can also occur in compounds, such as the square planar platinum compound $Pt(NH_3)_2Cl_2$, that do not contain double bonds. In this compound the chlorine atoms can be either adjacent to one another (*cis*) or across from one another (*trans*). Isomerism in transition metal compounds will be discussed in Chapter 22.

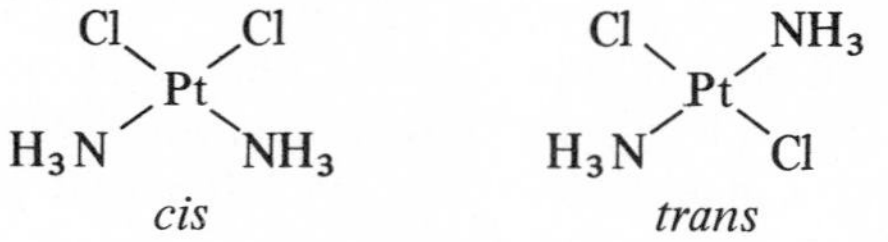

Cl   Cl   Cl   NH₃

Pt   Pt

H₃N   NH₃   H₃N   Cl

cis   *trans*

A third type of stereoisomerism occurs with compounds that are nonsuperimposable mirror images. Consider the compound bromochlorofluoromethane. As shown in Figure 8-17, there are two possible arrangements of the three halogens

FIGURE 8-17

Two Arrangements of the Atoms in CHBrClF

Cl   H   Br   Br   H   Cl

C   C

F   F

about a fixed C–H bond. If the image of the one is reflected in a mirror (Figure 8-18), the image seen is that of the other configuration. Since these two configura-

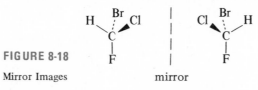

FIGURE 8-18

Mirror Images   mirror

tions cannot by any manner of rotation be made superimposable, they are not identical configurations.

The two analogous configurations of bromochloromethane, on the other hand, are superimposable and thus are identical. Thus, a sufficient condition for nonsuperimposable mirror images is the presence of *four different* groups attached to a carbon atom.

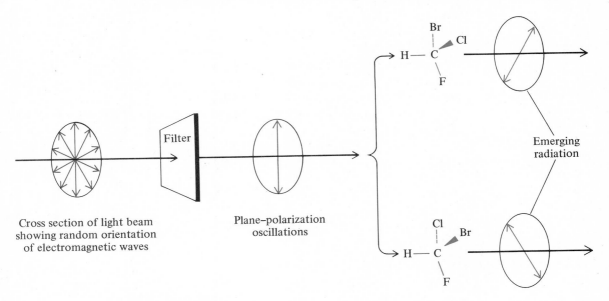

mirror

Experimentally, it has been observed that all except one of the physical properties of nonsuperimposable mirror images are identical. The one exception is the effect of the two configurations on plane-polarized light. As discussed in Chapter 5, light consists of electromagnetic waves. As shown in Figure 8-19, these waves vibrate in all possible orientations along the path of the light beam. If the beam of light is passed through appropriate filters, all of the oscillations except those confined to one plane are removed. The resulting electromagnetic radiation is plane-polarized light. When this light is passed through a substance whose molecules have one of the two configurations, it is found that the plane of the emerging radiation is rotated relative to the plane of the entering radiation. The same

FIGURE 8-19

Interaction of Nonsuperimposable Mirror Images with Plane-Polarized Light behavior is observed for the mirror image, except that the plane is rotated in the opposite direction (but by the same amount). For this reason, nonsuperimposable mirror images are termed *optical isomers,* or *enantiomers* (from the Greek: *enantios,* opposite; *mer,* form). The isomers are usually distinguished by the designations *levo* (Latin: *laevus,* left) and *dextro* (Latin: *dexter,* right), depending on the direction in which they rotate light.

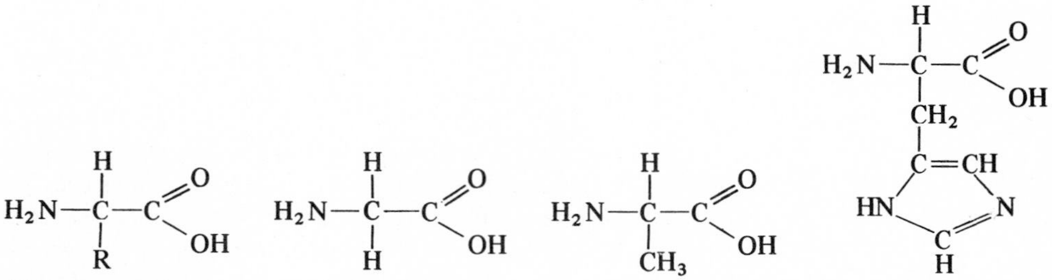

(a) general formula   (b) glycine   (c) alanine   (d) histidine

FIGURE 8-20

Some Amino Acids

A biologically important example of optical isomerism is provided by the constituents of proteins, the amino acids. The general formula for the amino acids is given in Figure 8-20a. All of the amino acids except the simplest one, glycine, where R = H, have four different groups attached to the central carbon and can therefore exist in one or both of the two optically active configurations. Because of some peculiarity of chemical evolution, all the naturally occurring amino acids have the same configuration. This configuration has been determined to be the one illustrated. The mirror image of this configuration has never been found in nature.

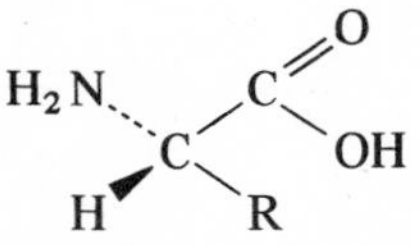

general configuration of amino acid

ABSORPTION OF RADIANT ENERGY

When energy in the form of electromagnetic vibrations impinges on matter, some of that energy—that is, certain of the vibrations—is absorbed. Which specific vibrations a given substance will absorb is determined by the particular structure of that substance. Thus, two different compounds absorb different frequencies of radiation, and each compound has its own unique *absorption spectrum*. The measurement of these spectra is called *absorption spectroscopy*.

In order to gain an understanding of this phenomenon it is necessary to have some comprehension of (a) the nature of the electromagnetic spectrum, (b) the various energies possessed by molecules, and (c) the relationship between these two.

The Electromagnetic Spectrum   Of the various kinds of electromagnetic vibrations (the nature and characteristics of which have been discussed in Chapter 5), the one with which everyone is most familiar is light, and it is therefore convenient to begin our discussion of the electromagnetic spectrum by considering the composition of visible light. Ordinary white light—whether sunlight or the light from an incandescent bulb—is composed of a conglomeration of vibrations of many different wavelengths. This can be demonstrated by use of a simple apparatus such as that shown in Figure 8-21.

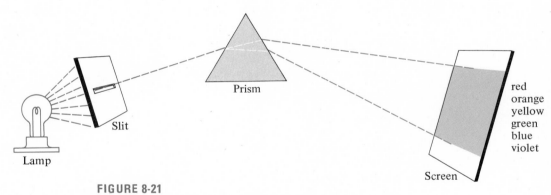

FIGURE 8-21

Dispersion of White Light: The Visible Spectrum

Light from an incandescent bulb is passed through a slit in order to produce a narrow beam. This beam is allowed to fall obliquely onto the side of a glass triangular prism. As the light beam passes from the air to the more dense glass, it is bent (*refracted*). But the component vibrations that make up the beam are not all refracted through the same angle, for they do not all have the same frequency; those vibrations with the greatest frequency (shortest wavelength) are refracted most. As the light emerges from the prism into the air, this time entering a less dense medium, more of this unequal bending occurs. As a result, if the light is allowed to fall upon a screen, one sees on the screen not a narrow band of white light but an enlarged, continuous band of light showing a "rainbow" of colors. Through refraction the white light has been separated into its component vibrations according to the wavelengths of those vibrations. The components of white light thus displayed on the screen together comprise what is called the *visible spectrum*.

If a portion of that spectrum were passed through a slit and refracted through another prism, further separation of the component waves could be accomplished.

Those electromagnetic vibrations which can be detected by the human eye, and are therefore part of the visible spectrum, have wavelengths ranging from approximately 4000 Å to 7500 Å (400 to 750 nm). Although the human visual apparatus is unable to distinguish between two vibrations that have nearly the same wavelength, it can detect differences between relatively large bands of wavelengths. Each of these bands elicits a somewhat different response from the visual nervous system, and each different response is equated with a different color. The wavelength bands of the visible spectrum corresponding to the six major colors are given in Table 8-3. The divisions between colors are not abrupt, of course, and the wavelengths given are only approximate.

TABLE 8-3
Colors of the
Visible Spectrum

Color	Wavelength Band (Å)
Violet	4000–4500
Blue	4500–5000
Green	5000–5700
Yellow	5700–5900
Orange	5900–6200
Red	6200–7500

In addition to the visible radiation, sunlight also contains waves that have wavelengths too large to be visible to the eye but that can be detected by heat-sensitive instruments. These are called *infrared* rays. Also present in sunlight are waves too short to be seen but detectable by means of photographic plates, and these constitute the part of the electromagnetic spectrum called the *ultraviolet.*

Electromagnetic vibrations also exist at either end of the radiant energy spectrum of sunlight. They comprise both the very short wavelengths (therefore, high frequency and high energy) of gamma rays and X-rays and the very long wavelengths (low frequency and low energy) of Hertzian waves, which include radio, television, and radar waves.

The visible spectrum, then, represents only a very small portion of the total electromagnetic spectrum, which is diagrammed in Figure 8-22. It will be recalled from an earlier discussion (p. 63) that wavelength and frequency are related inversely and that frequency is directly proportional to energy. Therefore, wavelengths of the vibrations increase but frequency and energy of the vibrations decrease from left to right in Figure 8-22.

Molecular Energies   Any molecule (or ion) possesses energy, and this energy may be divided into several distinct types. *Nuclear energy*—energy associated with the composition of the molecule's nuclei—accounts for nearly all of the total energy of the molecule.

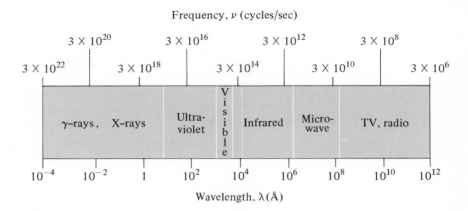

FIGURE 8-22

The Electromagnetic Spectrum

Changes in nuclear energy will be discussed in Chapter 18. The remainder of the energy may be divided into the following categories: *electronic energy,* energy possessed by the molecule by virtue of the distribution of its electrons; *vibrational energy,* resulting from the vibratory motions of the atoms within the molecule; *rotational energy,* associated with the rotation of the molecule about its center of mass; and *translational energy,* the energy possessed by the molecule because of its translational (linear) motion.

Each of these four types of energy is quantized (Chapter 5); that is, if the molecule loses or gains energy in any of these categories, the energy transfer must occur in the form of definite energy packets, called *quanta.* The quanta associated with changes in translational energy are so small that for practical purposes translational energy may be considered continuous, and translational energy changes are directly related to the absolute temperature. However, changes in the other three categories of energy—electronic, vibrational, and rotational—involve quanta of magnitudes that correspond to frequencies within the electromagnetic spectrum. Thus, the electronic, vibrational, and rotational energies of a molecule may be raised above the ground state by the absorption of electromagnetic vibrations. But a molecule can absorb only those frequencies of radiation that correspond to energy transitions possible for that particular molecule. Hence, each different substance has its own unique absorption spectrum.

The quanta involved in electronic energy transitions are quite large compared to those for vibrational or rotational transitions, with the rotational quanta being the smallest. This relationship is illustrated schematically in Figure 8-23, where ΔE_{rot}, ΔE_{vib}, and ΔE_{elect} represent transitions from the ground state to the next higher rotational, vibrational, and electronic states, respectively. Because of this difference in the magnitude of the quanta of the three types of energy transitions, each of the three absorbs energy from a different portion of the electromagnetic spectrum.

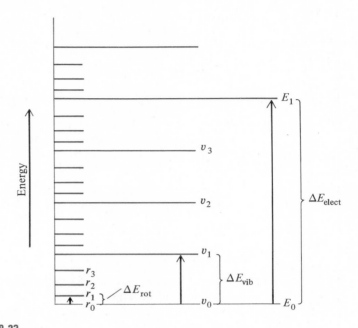

FIGURE 8-23

Schematic Comparison of Electronic (E), Vibrational (v), and Rotational (r) Energy Levels

Electronic Spectra   Transitions in electronic energy are usually of such magnitude that they correspond to frequencies of radiant energy in the ultraviolet and visible portions of the electromagnetic spectrum. Therefore, molecules absorb specific, characteristic wavelengths from this spectral region.

Suppose we had a source of ultraviolet radiation of a single wavelength—say, 2000 Å—and we allowed that radiation to pass through a thin layer of some substance. (Radiation of a single wavelength is referred to as *monochromatic*.) Suppose, further, that we had some means of measuring the intensity of the beam both before it entered the substance and after it had passed through it. We could then determine the percentage of the radiation transmitted by the substance or, alternatively, the percentage absorbed by it. Now, after making this measurement, suppose we repeated the process using a monochromatic beam of wavelength 2001 Å, then 2002 Å, and so on through a large number of beams of progressively increasing wavelengths. We could then make a plot of our measured data, plotting wavelength (λ) versus percent transmission (or percent absorption). The resulting plot, illustrated hypothetically in Figure 8-24, is an absorption spectrum of the substance under investigation. (Since the portion of the electromagnetic spectrum employed is in the ultraviolet region, this plot may also be referred to as an *ultraviolet spectrum*.) Note that our hypothetical compound transmits completely (shows no absorption) at wavelengths below approximately 2500 Å and above 3500 Å, and that it has an absorption maximum at about 2900 Å. Therefore, the energy

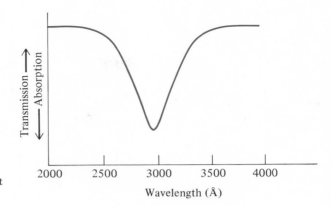

FIGURE 8-24

Hypothetical Ultraviolet
Absorption Spectrum

of radiation of wavelength 2900 Å corresponds to an electronic energy transition of the compound.

The instrument used in measuring absorption spectra is called a *spectro-photometer*. Most modern spectrophotometers provide automatic, gradual change of the wavelength, which is synchronized with a recorder chart, so that the spectrum can be plotted directly. Figure 8-25 shows the ultraviolet spectra of several compounds.

It will be noted in the spectra of both our hypothetical compound (Figure 8-24) and the real compounds (Figure 8-25) that absorptions occur not as sharp lines but as rather broad bands. This observation may raise the following question: Since an electronic transition is quantized, and therefore only radiation with a particular frequency can be absorbed, why does an absorption peak not appear as a single line at a specific wavelength? The answer is to be found in a consideration of the energy-level diagram of Figure 8-23. Within each electronic energy level there are many vibrational and rotational levels, and transitions can occur from a number of the vibrational and rotational levels of the electronic ground state to a number of the vibrational or rotational levels of the excited electronic state. Thus, an absorption peak really consists of many absorption lines so closely spaced that it is beyond the power of the spectrophotometer to resolve them and they appear as one continuous, smooth band.

The relationship between molecular structure and electronic spectra is perhaps best rationalized by application of molecular orbital theory (Chapter 7). Only the valence electrons of a molecule are excited by ultraviolet or visible radiant energy. Hence, we need not deal with the electrons of the filled inner levels but will consider only three categories of electrons: those involved in a sigma bond (σ), those involved in a pi bond (π), and nonbonding electrons (that is, unshared pairs of electrons), which we will designate n.

According to molecular orbital theory, a pi-bonding molecular orbital (π) is at a higher energy level than a corresponding sigma-bonding molecular orbital (σ), while the energy of a nonbonding molecular orbital (n) is higher than both. On the other hand, a sigma antibonding orbital (σ^*) is at higher energy than the corre-

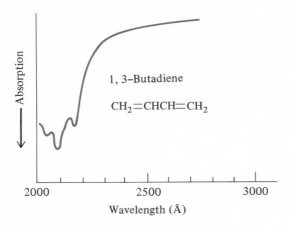

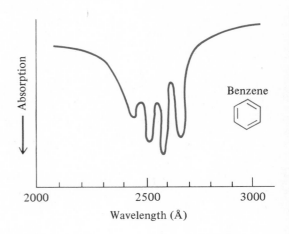

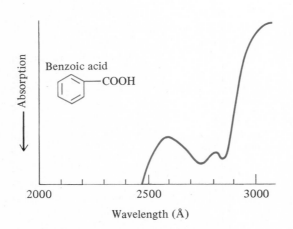

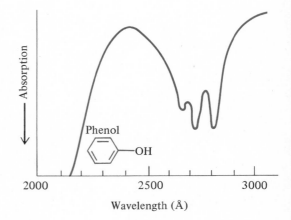

FIGURE 8-25

Some Ultraviolet Absorption Spectra

sponding pi antibonding orbital (π^*). This energy relationship is represented in Figure 8-26.

An electronic excitation (transition from a lower to a higher energy level) may be visualized as the moving of an electron from a lower *occupied* molecular orbital to a higher *unoccupied* molecular orbital. Thus, an electron in a σ, π, or n orbital may undergo a transition to a higher energy σ^* or π^* orbital, provided these antibonding orbitals are not already filled. It is apparent from Figure 8-26 that the six different kinds of possible excitations represent the absorption of different amounts of energy, depending on the difference in energy between the occupied ground-state level and the unoccupied excited level. The six kinds of transitions, arranged according to increasing energy absorption (ΔE) are shown in Table 8-4.

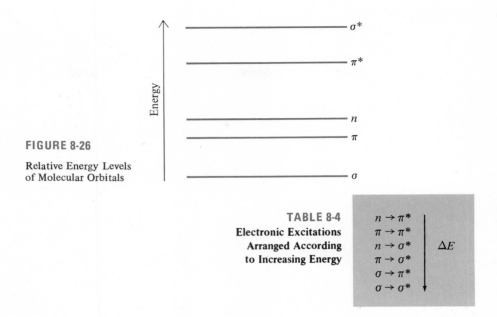

FIGURE 8-26

Relative Energy Levels
of Molecular Orbitals

TABLE 8-4	
Electronic Excitations Arranged According to Increasing Energy	$n \rightarrow \pi^*$ $\pi \rightarrow \pi^*$ $n \rightarrow \sigma^*$ $\pi \rightarrow \sigma^*$ $\sigma \rightarrow \pi^*$ $\sigma \rightarrow \sigma^*$ ΔE

The greater the energy of the electronic transition, the higher the frequency and the shorter the wavelength of the radiation absorbed. Therefore, each type of transition corresponds to absorption in a somewhat different portion of the spectrum. Excitations represented by $\sigma \rightarrow \sigma^*$, $\sigma \rightarrow \pi^*$, and $\pi \rightarrow \sigma^*$ transitions constitute energy changes of such magnitude that they are accompanied by absorptions in the low-wavelength end of the ultraviolet spectrum, the *far ultraviolet* (so named because of its position with respect to the visible). On the other hand, $n \rightarrow \pi^*$, $\pi \rightarrow \pi^*$, and $n \rightarrow \sigma^*$ transitions are of lower energy and correspond to absorptions in the *near-ultraviolet* and *visible* spectrum. Most ultraviolet-visible spectrophotometers are not capable of dealing with wavelengths below approximately 2000 Å. Therefore, in terms of practical applications, one is usually dealing only with those transitions that occur in the near-ultraviolet and visible portions of the spectrum. Specific structural groups that are responsible for absorption in this spectral region are called *chromophores*. A number of chromophores are listed in Table 8-5, together with illustrative compounds, type of excitation, and wavelength of maximum absorption (λ_{max}).

Color.   Another important aspect of the electronic spectrum of a substance is its relationship to the color of that substance. The color that an object appears to be is determined by the wavelengths of light that travel from the object to the retina of the eye. In general, when light falls upon an object, some of it may be absorbed by the object, some may be transmitted through it, and some may be reflected from it. If the object in question is opaque, then essentially no light is transmitted, it is either absorbed or reflected. In the case of translucent objects (including solutions), on the other hand, light is either absorbed or transmitted,

TABLE 8-5
Some Common
Chromophores

Chromophore	Compound	Transition	λ_{max} (Å)
$-\ddot{\text{S}}-$	$C_6H_{13}SH$	$n \rightarrow \sigma^*$	2240
$-\ddot{\text{B}}\text{r}:$	CH_3Br	$n \rightarrow \sigma^*$	2040
$-\ddot{\text{I}}:$	CH_3I	$n \rightarrow \sigma^*$	2580
$\overset{\ddot{\text{O}}:}{\underset{}{\overset{\|}{-\text{C}-}}}$	$\overset{\text{O}}{\overset{\|}{CH_3CCH_3}}$	$n \rightarrow \pi^*$	2790
$\overset{\ddot{\text{S}}:}{\underset{}{\overset{\|}{-\text{C}-}}}$	$\overset{\text{S}}{\overset{\|}{CH_3CCH_3}}$	$n \rightarrow \pi^*$	2080
$>\!\ddot{\text{N}}-$	$(CH_3)_3N$	$n \rightarrow \sigma^*$	2270
$-C\!=\!C\!-\!C\!=\!C-$	$CH_2\!=\!CH\!-\!CH\!=\!CH_2$	$\pi \rightarrow \pi^*$	2200
⬡	C_6H_6	$\pi \rightarrow \pi^*$	2550

with essentially none being reflected. The light that travels from the object to the eye, then, is light that has been either reflected or transmitted by the object; in other words, it is white light minus whatever has been absorbed by the object.

Substances whose electronic spectra consist of absorption bands entirely within the ultraviolet region are without color—they appear white in the solid state, and their solutions are colorless—for they reflect or transmit to the eye all wavelengths of the visible spectrum with equal intensity. On the other hand, a substance that absorbs radiation in the visible region of the spectrum is colored; the light that reaches the eye does not contain all wavelengths of white light in equal intensity. Thus, the color of an object is determined by those wavelength bands that are not absorbed. A solution containing permanganate ion is purple because that ion shows an absorption band in the region of 5000–6000 Å (the green-yellow region); that is, the green portion of the spectrum is absorbed, and the combination of the red and blue regions that are transmitted results in the purple color. Similarly, a red dye transmits the longer wavelengths of the visible spectrum (red) and absorbs the shorter wavelengths (violet, blue, green, and so on).

If the object being viewed is illuminated by light containing wavelengths of only part of the visible spectrum, its color may be different from that when viewed in white light. For example, consider a piece of cloth patterned with red polka dots on a blue background. In white light the dots reflect the red portion of the spectrum and the background reflects the blue portion. If this same cloth is viewed in a room illuminated with blue light, the pattern will appear as black polka dots on a blue background. Since the light striking the cloth does not contain the red wavelengths, these wavelengths cannot be reflected, and the dots appear black. Similarly, if the cloth is viewed under red light, the pattern will appear as red dots on a black background, because there are no wavelengths of the blue portion available for reflection.

What kinds of compounds would one expect to be colored, then? Those whose structures permit electronic transitions of relatively low energy—low enough to correspond to frequencies in the visible portion of the electromagnetic spectrum. In general, this condition is found in molecules that have extended pi bonding (alternating multiple bonds); that is to say, in structures such as the following:

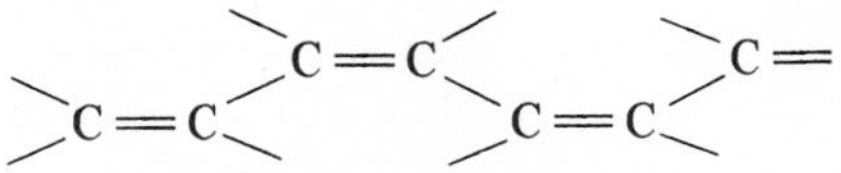

This extended pi bonding results in delocalization of the pi electrons, so that it is "easier" for an electron to move to an unoccupied orbital of higher energy; that is, less energy is required for the transition. As one illustration of the relationship between extended pi bonding and color, consider the class of compounds known as azo dyes. This is an extensive group of yellow, orange, and red compounds, many of which contain within their molecules the structure:

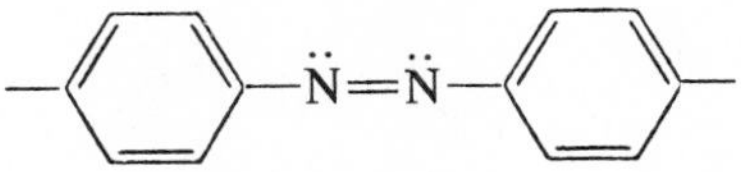

Another type of bonding condition that permits low-energy electronic transitions apparently exists in compounds of the transition elements, for the transition metals form many highly colored ions, while most ions of the representative elements are colorless. Here the color is a result of the unfilled d orbitals. The low-energy transitions that account for color may be rationalized from the crystal field theory as discussed in Chapter 22.

Vibrational Spectra   We have already established that the quanta involved in changes in vibrational energy are considerably smaller than those involved in electronic energy changes. Therefore, vibrational spectra result from absorption of radiant energy of lower frequencies (longer wavelengths) than in the case of electronic spectra. Transitions in vibrational energy correspond to wavelengths in the infrared portion of the spectrum, between approximately 1 μ and 100 μ. (In the infrared spectrum it is customary to express wavelengths in terms of the *micron*, μ, rather than the smaller angstrom unit; 1 μ = 10,000 Å.)

Using instrumentation somewhat similar to that used for measuring electronic spectra, vibrational spectra can be recorded as the amount of infrared radiation transmitted (or absorbed) versus wavelength, or, as is commonly the case, versus wavenumber in cm^{-1} (the reciprocal of the wavelength expressed in centimeters). In most practical applications of infrared spectra, measurement is limited to the spectral region between 2.5 μ and 15 μ.

As in electronic spectra, the absorptions in vibrational spectra appear as bands rather than lines, because a single vibrational energy change is accompanied by a number of rotational energy changes. (Refer to Figure 8-23.)

The vibrations that give rise to infrared absorption are of two kinds: stretching and bending. A *stretching vibration* is a vibration between two atoms along the bond that connects them. This is analogous to two balls attached to a coiled spring that is alternately compressed and extended. Note that the distance between the two atoms is continuously changing because of this vibration, and a "bond length" is therefore really an average value. In a diatomic molecule the stretching vibration can take only one form, but a triatomic molecule possesses two *stretching modes:* a symmetrical stretch, in which both bonds are being shortened or lengthened at the same time, and an unsymmetrical stretch, in which one bond is shortened while the other is lengthened. Figure 8-27 illustrates these two stretching modes.

A *bending vibration* involves a change in bond angle. Thus, reported bond angles, like bond lengths, are also only average values. The various bending modes for a three-atom group are shown in Figure 8-28.

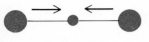

Symmetrical

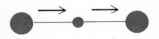

Unsymmetrical

FIGURE 8-27

Stretching Modes
of a Triatomic Molecule

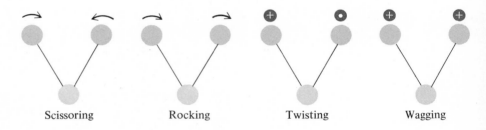

| | Scissoring | Rocking | Twisting | Wagging |

⊕ represents direction above the plane

⊙ represents direction below the plane

FIGURE 8-28

Bending Modes

In order for a molecular vibration to result in absorption of infrared radiation it must cause a rhythmical change in the dipole moment of the molecule. A change in the distribution of charge within the molecule is necessary for interaction between the molecule and the oscillating field of the infrared radiation. Thus, in a molecule such as CO_2, the unsymmetrical stretch gives rise to an absorption band, but the symmetrical stretch does not, for it causes no change in the dipole moment (Figure 8-27).

Using the analogy between atoms connected by a bond and balls connected by a spring, we may apply Hooke's law of simple harmonic oscillators:

$$\nu = \frac{1}{2\pi} \sqrt{\frac{k}{\mu}}$$

where ν is the vibrational frequency, k is the *force constant,* a measure of the stiffness of the spring or the strength of the bond, and μ is the *reduced mass*—the product of the masses of the balls (or atoms) divided by the sum of the masses. It is clear from this relationship that the vibrational frequency (therefore, the wavelength of radiation absorbed) for each mode of vibration is determined by the strength of the bond between atoms and the masses of those atoms. A given bond, then, such as C–H, O–H, or C=O, has approximately the same stretching frequency in different molecules. Moreover, the bending frequencies are nearly the same for any given group of atoms (for example, $-CH_3$, $-C\overset{\diagup O}{\underset{\diagdown H}{}}$, $-NH_2$) in different molecules. While the vibrational frequencies of a given bond or given group of atoms are not precisely the same in all molecules because of the effect of environment (neighboring groups or atoms) within the molecule, rather narrow frequency ranges can be assigned. These assigned frequency ranges, a few of which are listed in Table 8-6, are of great practical value as an aid in deducing the structure of a molecule from its infrared spectrum.

TABLE 8-6
Some Characteristic
Absorption Bands

Group	Vibration	Wavenumber (cm^{-1})	Wavelength (μ)
O–H	Stretching	3700–3200	2.70–3.15
N–H	Stretching	3500–3300	2.86–3.03
C–H	Stretching	3300–2700	3.03–3.70
C=O	Stretching	1900–1640	5.45–6.10
C=C	Stretching	1700–1580	5.90–6.40
N–H	Bending	1650–1490	6.10–6.70
C–H	Bending	1475–1300	6.90–7.70
O–H	Bending	1450–1200	6.85–8.40
C–O	Stretching	1300– 900	7.70–11.2

Every compound has a characteristic infrared spectrum, different from that of every other compound. The infrared spectrum thus serves as a kind of "fingerprint" from which positive identification can be made. Spectra of a few simple compounds are shown in Figure 8-29.

Rotational Spectra

Since differences between rotational energy levels are very small, rotational transitions involve quanta of relatively little energy and absorb radiation in the far infrared and microwave portions of the electromagnetic spectrum.

In order for a molecule to give rise to a rotational spectrum it must have a dipole moment. Thus, molecules such as H_2, N_2, and CO_2 show no absorptions

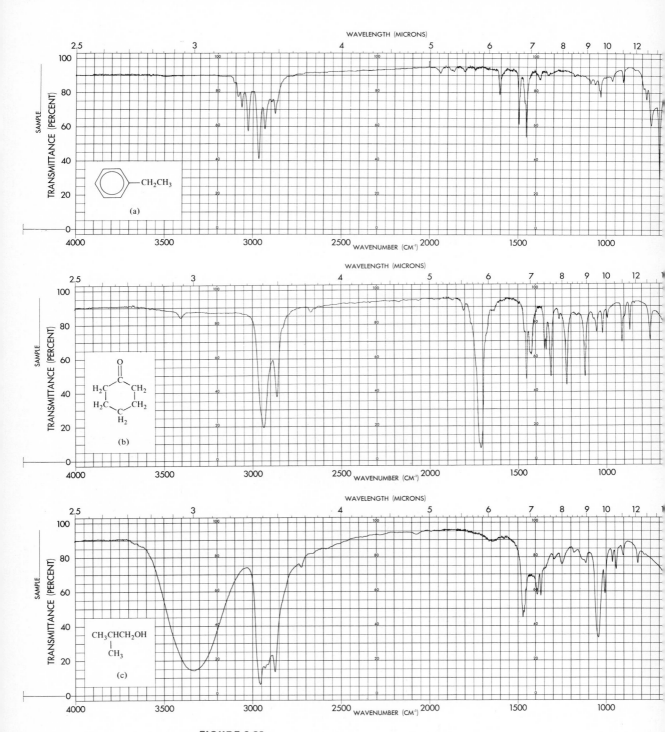

FIGURE 8-29

Some Infrared Spectra

corresponding to changes in rotational energy, while HCl, CO, and similar polar molecules yield definite spectra in this region.

Rotational spectra, which consist of evenly spaced lines, do not have anywhere near the applicability of electronic or vibrational spectra. Their primary usefulness has been in the determination of bond lengths in rather simple molecules, which involves measurement of the spectral line spacings and a rather complex mathematical application of wave mechanics.

SUGGESTED READINGS

Brown, H. C. "Foundations of the Structural Theory." *Journal of Chemical Education,* Vol. 36, No. 3 (March 1959), p. 104.

Gillespie, R. J. "The Valence-Shell Electron-Pair Repulsion (VSEPR) Theory of Directed Valency." *Journal of Chemical Education,* Vol. 40, No. 6 (June 1968), p. 295.

Lambert, J. B., "The Shape of Organic Molecules." *Scientific American,* Vol. 222, No. 1 (January 1970), pp. 58–70.

Pauling, Linus, and Roger Hayward. *The Architecture of Molecules.* San Francisco: W. H. Freeman & Co. 1964.

PROBLEMS

1. Which of the following compounds: (a) are structural isomers; (b) exhibit geometrical isomerism; (c) exhibit optical isomerism?

(a) $CH_3CH_2CH_2CH_2CH_3$ (b) $CH_3CH{=}CHCH_2CH_3$ (c) $CH_3C{\equiv}CH$

(d) 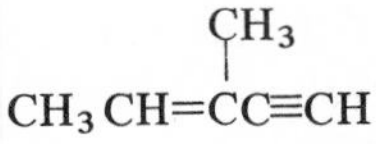

(e) 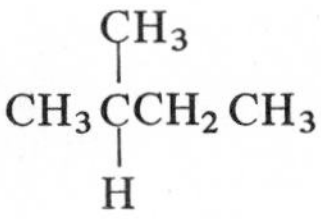

(f) 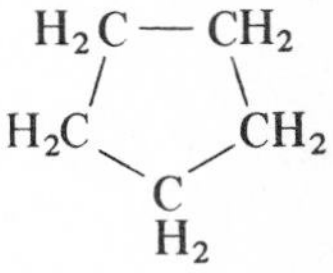

(g) 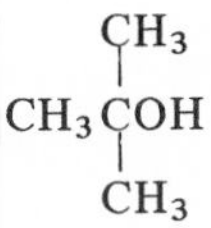

(h) 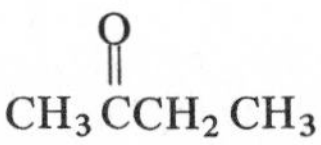

(i) 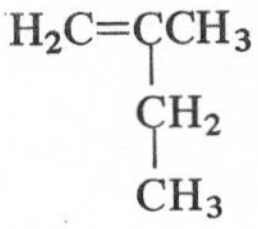

(j) 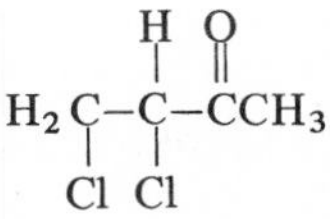

(k) 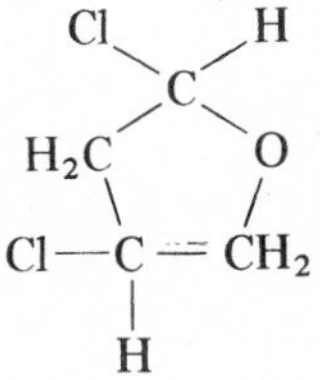

(l) 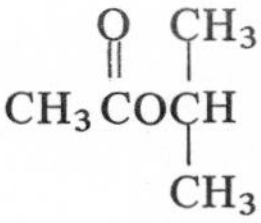

(m) $CH_3OCH_2CH_2CH_3$

2. Account for the difference in bond angles between members of the following pairs:
   (a) NH_3 $(107°)$, NO_3^- $(120°)$ (b) NH_3, PH_3 $(94°)$
   (c) NF_3 $(103°)$, NCl_3 (*ca.* $106°$)   (d) NF_3, BF_3 $(120°)$
   (e) NF_3, ClF_3 $(87°)$ (f) NF_3, NH_3

3. Predict the shape and hybridization at the central atom for each of the following:
   (a) $HgCl_2$ (b) $Si(CH_3)_4$ (c) IF_5
   (d) SF_4 (e) PF_3 (f) BF_4^-
   (g) NO_2^- (h) $B(CH_3)_3$ (i) NO_3^-
   (j) $SnCl_2$ (k) SnF_6^{2-} (l) SO_4^{2-}
   (m) ICl_4^- (n) ClF_3 (o) SO_3^{2-}
   (p) I_3^- (q) ClO_2^- (r) XeF_4
   (s) H_3O^+ (t) NH_2^- (u) NH_4^+
   (v) ClO_3^-

4. The percentage composition of a compound containing only carbon and hydrogen is 85.6% C and 14.4% H. The molecular weight is 56.
   (a) Calculate the empirical and molecular formulas for this compound.
   (b) Write all the reasonable structural formulas that this compound might have and show that the electron dot formula for each obeys the octet rule.
   (c) How many different compounds are represented by these structural formulas?

5. The reaction of fluorine with xenon in a closed nickel container produced a crystalline compound. Analysis showed that 0.295 g of xenon had reacted with 0.173 g of fluorine and that the compound had a molecular weight of 207. Calculate the empirical and molecular formulas of the compound. Write an electron dot formula for the compound and predict its molecular shape.

6. Each of the compounds below has a central atom attached to three, four, five, or six groups. Determine the shape of each and then decide which molecules exhibit geometrical isomerism and which exhibit optical isomerism.
   (a) BBr_2Cl (b) $As(CH_3)ClH$
   (c) $Pt(NH_3)_2Cl_2$ (square planar) (d) PF_2Cl_3
   (e) $SClF_5$

7. Draw all of the possible structural isomers and stereoisomers for compounds with the following molecular formulas:
   (a) $C_2H_2Cl_2$ (b) C_4H_7Cl
   (c) $C_4H_6Cl_2$ (d) C_4H_9Cl
   (e) C_5H_{10}

8. Distinguish carefully between:
   (a) structural isomers and stereoisomers
   (b) molecular formula and structural formula
   (c) geometrical isomers and optical isomers
   (d) translation and rotation

9. For each of the following molecular formulas, draw all possible structural isomers and name them.

   (a) $C_2H_4Cl_2$ (b) C_3H_7Cl
   (c) $C_3H_6Br_2$ (d) $C_6H_{14}O$

10. An aqueous solution of permanganate ion is purple when viewed in white light. What would be the color of this solution in a room illuminated only by blue light? red light? green light?

11. Give the molecular orbital description of the N_2 molecule. According to this description, what is the lowest energy electronic excitation possible for this molecule?

12. On the basis of your answer to problem 24 of Chapter 7, what is the lowest energy electronic excitation possible for CO_2?

13. Alkenes that contain carbon-carbon double bonds at alternate carbons (for example, 1,3-butadiene) behave as though the π-electrons are partially delocalized over the entire molecule. In the valence bond model, this delocalization would be represented by resonance structures such as

$$H_2C=C-C=CH_2 \longleftrightarrow H_2\overset{\oplus}{C}-C=C-\overset{\ominus}{C}H_2 \longleftrightarrow H_2\overset{\ominus}{C}-C=C-\overset{\oplus}{C}H_2$$

   If it is assumed that each π-electron has as its domain the entire length of the molecule, the energy of *molecular orbitals* for the π-electrons can be calculated from the equation derived in Chapter 5 for the energy of an electron confined to a line.

   Using an effective length of 6 Å for a molecule of 1,3-butadiene, calculate the energies of the four lowest energy π-molecular orbitals. Indicate which orbitals are occupied by the four π-electrons and then calculate the energy required for a transition from the occupied orbital of highest energy to the unoccupied orbital of lowest energy. What is the wavelength of the radiation required for this transition? What region of the electromagnetic spectrum is this radiation in? Is 1,3-butadiene colored according to this description?

14. As indicated in Table 8-6, the infrared absorption due to a C–H stretching vibration occurs in the region 3300–2700 cm^{-1}; the C–O stretching vibration occurs at 1300–900 cm^{-1}; and the C=O stretching vibration occurs at 1900–1640 cm^{-1}. Carefully account for the order for the energy of these absorptions: C–H > C–O and C=O > C–O.

15. A substance is believed to be either acetone or methanol. The infrared spectrum of the compound shows absorptions at 3.0, 3.4, 3.5, 7.0, and 9.7 μ. Identify the substance.

16. A liquid has an elemental composition of C, 69.8%; H, 11.6%; and O, 18.6%; and a molecular weight of 86. Its infrared spectrum contains no absorptions above 2950 cm^{-1} and none between 2500 cm^{-1} and 1500 cm^{-1}. Determine a possible structural formula for the compound.

17. A compound with percentage composition C, 63.2%; H, 12.3%; and N, 24.5% has a molecular weight of 57 and exists as geometrical isomers. The

absorption due to the carbon-nitrogen linkage appears at about 1680 cm^{-1} (the C−N single bond gives rise to absorption in the region 1000–1200 cm^{-1}). Determine the molecular formula of the compound and then draw all possible structural formulas. From the data given, determine the structural formula for this compound.

9 *Matter in Bulk I:*
Solids and Liquids

Our study of the structure of matter up to this point has dealt largely with matter on the submicroscopic level. We have examined the structure of the basic particles of matter: atoms, ions, and molecules. Now let us consider how these "building blocks" of matter are held together to form macroscopic specimens—matter in bulk.

In examining the structure of matter in bulk, two basic properties of the particles must be given consideration. First, the particles are in motion and therefore possess kinetic energy (see Chapter 1). The magnitude of their kinetic energy—and therefore their velocity—is dependent on the temperature: the higher the temperature, the greater the velocity of the particles. Second, there are attractive forces between the particles of matter that tend to hold them together. In a sense, then, matter possesses these two opposing influences: kinetic energy, which tends to separate the structural units, and attractive forces, which tend to pull them together. It is the precise relationship between these two opposing factors that determines the physical state in which a particular substance will exist—solid, liquid, or gas.

It is characteristic of gases that they have no definite shape and no definite volume. A gas takes the shape and the volume of the container in which it is confined. Furthermore, the density of a gas is very low and varies greatly with temperature and pressure. These facts suggest that the particles of a gaseous substance are very diffuse and have freedom of motion and that the forces between them are weak.

183

Solids, on the other hand, have both definite shapes and definite volumes, and their densities are high and only slightly affected by temperature and pressure. The structural units of solids, therefore, must be close together, their attractive forces relatively strong, and their motion greatly restricted.

Liquids have characteristics intermediate between gases and solids. A liquid has a definite volume, but its shape is determined by the container. Densities of liquids generally range between those of solids and gases, and although a liquid's density varies more with temperature and pressure than a solid's does, the variation is not nearly so great as it is in a gas. The particles of a liquid, then, may be thought of as having greater freedom of motion than particles in a solid but less than those in a gas. The attractive forces in the liquid state are intermediate between the forces in a solid and those in a gas.

The attractive forces that hold the particles of matter together are all electrical in nature, but they are of several different kinds and of varying strengths. They are described in the discussions of the various states of matter that follow. The remainder of this chapter deals with solids and liquids. The gaseous state and transitions between states are discussed in Chapter 10.

The Solid State

Materials that have the general characteristics of the solid state—rigidity, definite volume, definite shape—are of two structurally different kinds: crystalline solids and amorphous solids. In a *crystalline solid* the structural units are arranged in a regular, repeating, three-dimensional pattern called the crystalline *lattice*. In *amorphous solids*, on the other hand, the units are scattered randomly, without any pattern. Amorphous solids—often called *glasses* because ordinary glass is a common example ("hard candy" and solidified tar are other examples)—are structurally more like liquids than like crystalline solids. In fact, they have the properties of extremely viscous liquids and are usually considered as such. Throughout our discussion of matter in bulk, when we employ the term *solid* we will mean *crystalline solid.*

CRYSTAL STRUCTURE

Much of our knowledge about the internal structure of crystals is the result of measurements involving the interaction of X-rays with crystals. It will be recalled from Chapter 8 that X-rays are high-energy electromagnetic vibrations, and therefore a beam of X-rays has the same general properties as a beam of light. Some understanding of the interaction of X-rays with a crystal lattice can be gained by comparing it with the simple phenomenon of *diffraction* of light through a grating.

If two waves come together so that their maximum amplitudes coincide at

(a) Two waves in phase   (b) Two waves out of phase

FIGURE 9-1

Reinforcement and Interference of Waves

the same point at the same time (that is, the crests of the waves coincide, and so do their troughs), they are said to be *in phase,* and they reinforce each other. This reinforcement results in a wave with greater amplitude but the same wavelength as the original two waves (Figure 9-1a). The increased amplitude corresponds to increased intensity. On the other hand, if two waves combine so that the crest of one coincides with the trough of the other, the waves are *out of phase.* In this case the waves cancel each other, and the intensity of radiation drops to zero (Figure 9-1b). If the two waves are not exactly out of phase, there is still some interference, resulting in partial cancellation and a decrease in intensity of the radiation.

This phenomenon of reinforcement and interference of waves can be demonstrated as shown in Figure 9-2. A beam of monochromatic light is allowed to fall on an opaque barrier containing two narrow slits that are parallel and very close together. Light coming through the slits can be observed on a screen. Each slit acts as a source of light waves that emerge from the slits in phase. Since the distance from point A on the screen to each of the slits is the same, the waves from both slits arrive at point A in phase, causing reinforcement and resulting in a bright line on the screen. At each of the points B, the distances to the two slits differ by exactly one wavelength. Therefore, reinforcement occurs and bright lines are observed at these points. The distances from the slits to each of the points C differ by two wavelengths, again resulting in reinforcement. On the other hand, points on the screen exactly halfway between A and B represent distances to the two slits

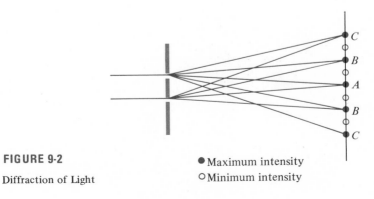

FIGURE 9-2

Diffraction of Light

● Maximum intensity
○ Minimum intensity

that differ by one-half wavelength. Hence these points on the screen remain dark because of interference. Similarly, the points midway between *B* and *C* show darkness, since they represent distances from the slits that differ by one and one-half wavelengths. Extension of this line of reasoning to additional points up and down the screen explains the pattern of alternating light and dark lines observed on the screen. This phenomenon is called *diffraction,* and the pattern observed is referred to as a *diffraction pattern.* (See also the discussion on p. 62.)

Diffraction patterns are produced whenever light is transmitted through (or reflected by) a structure that consists of a regular, repetitive pattern. The two-slit apparatus described above is the simplest example of such a structure. A *diffraction grating* is constructed by producing a large number of parallel, closely spaced, transparent lines on an otherwise opaque surface. The grating functions in the same way as the two-slit apparatus, but the lines obtained are much sharper because of the multiple reinforcements and interferences. In order for the diffraction pattern to be well defined, the spaces between lines should be of the same order of magnitude as the wavelength of the radiation used (about 30,000 lines per inch for visible light).

Because crystals are composed of atoms, ions, or molecules arranged in a regular, repeating pattern, and because the repeat distances are of the same order of magnitude as the wavelengths of X-rays (see Figure 8-22), crystals behave as three-dimensional diffraction gratings for the diffraction of X-rays.

Figure 9-3 represents schematically an apparatus for X-ray diffraction. X-rays are collimated into a narrow beam by passage through a hole in a thick lead shield. The beam is directed onto a crystal of the substance under investigation. As the X-rays penetrate the crystal, they are diffracted through a variety of angles. A photographic film can be used to detect the diffracted radiation, the film being exposed wherever the X-rays strike it.

A representation of an X-ray diffraction pattern is shown in Figure 9-4. The large central spot is caused by the undeflected main beam. The other spots are

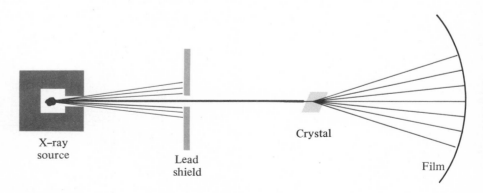

FIGURE 9-3

X-ray Diffraction Apparatus

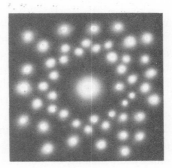

FIGURE 9-4

An X-ray Diffraction Pattern

caused by X-rays that were diffracted through various angles from respective planes of atoms in the crystal. Each crystalline substance gives its own characteristic pattern.

The spacings (*d*) of the planes of the crystal lattice are related to the angle of diffraction (θ) and the wavelength of the X-rays used (λ) by the equation:

$$n\lambda = 2d \sin \theta$$

where *n* is an integer (1, 2, 3, and so on). This equation, named the *Bragg equation* after its discoverer, is useful in analyzing the structure of crystals. Using X-rays of known wavelength, the diffraction angles can be measured, and the interplanar spacings of the crystal can be computed. In this way a complete picture of a crystal lattice can be obtained.

A crystal lattice may be pictured as a pattern of points representing the arrangement of the atoms, ions, or molecules that make up the crystal, and extending in all directions throughout the entire crystal. See Figure 9-5.

In describing a crystal lattice, it is convenient to consider only the smallest section that represents the order of arrangement. If this small section, called the *unit cell,* were moved a distance equal to its own dimensions in various directions, it

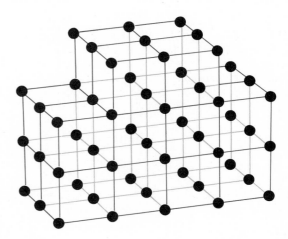

FIGURE 9-5

Representation of a Crystal Lattice

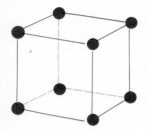

FIGURE 9-6

Unit Cell

would generate the entire lattice. Thus, assuming that the lattice points in Figure 9-5 are identical atoms (or ions or molecules), the unit cell of that lattice is represented by the individual cubic structure shown in Figure 9-6. In other words, the crystal consists of repeating unit cells stacked together, and the external symmetry of the crystal as a whole reflects the arrangement within the unit cell.

When one considers the large number of different crystalline substances in existence, it is perhaps surprising that all crystals can be described by only fourteen different space lattices that can be grouped according to symmetry into only seven different *crystal systems.* The crystal systems may be defined in terms of the relative dimensions of a unit cell along its three axes (*a, b, c*) and the magnitudes of the angles (α, β, γ) between the sides, as shown in Figure 9-7.

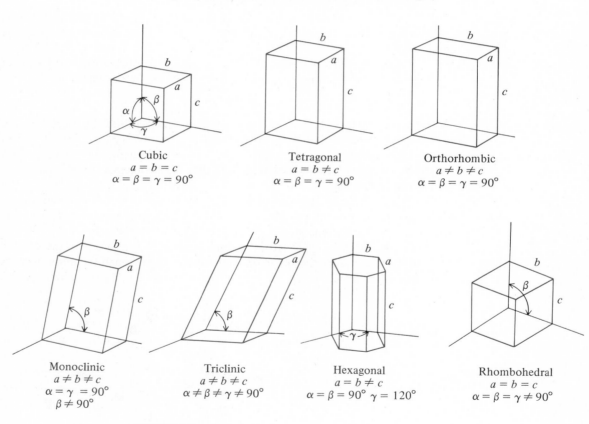

Cubic
$a = b = c$
$\alpha = \beta = \gamma = 90°$

Tetragonal
$a = b \neq c$
$\alpha = \beta = \gamma = 90°$

Orthorhombic
$a \neq b \neq c$
$\alpha = \beta = \gamma = 90°$

Monoclinic
$a \neq b \neq c$
$\alpha = \gamma = 90°$
$\beta \neq 90°$

Triclinic
$a \neq b \neq c$
$\alpha \neq \beta \neq \gamma \neq 90°$

Hexagonal
$a = b \neq c$
$\alpha = \beta = 90° \ \gamma = 120°$

Rhombohedral
$a = b = c$
$\alpha = \beta = \gamma \neq 90°$

FIGURE 9-7

Crystal Systems

That only seven different crystal systems can give rise to fourteen different crystal lattices is due to the fact that lattice points need not occur only at the corners of the unit cell. They may occur also in the center of the unit cell or at the

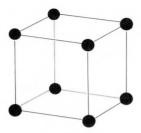
Simple cubic

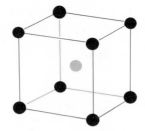
Body–centered cubic

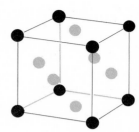
Face–centered cubic

FIGURE 9-8

Cubic Lattices

Simple cubic

Body-centered cubic

Face-centered cubic

FIGURE 9-9

Cubic Lattices Formed
from Identical Spheres

center of the faces of the unit cell. Thus, three different cubic lattices are known, as shown in Figure 9-8: simple cubic, body-centered cubic, and face-centered cubic.

In the foregoing discussion we have visualized lattices as consisting of points connected by imaginary lines, the points representing centers of atoms, ions, or molecules. But atoms are actually space-filling entities, and some further insight into the internal structure of crystals can be gained by picturing crystals as the result of the packing together of spheres. For example, the unit cells of the three cubic lattices are shown as arrangements of identical spheres in Figure 9-9. (Compare with the point lattices of Figure 9-8.)

Of particular interest is the arrangement of spheres in such a way that there is a minimum of empty space. Most metals and many molecular crystals display this *closest-packed* structure. The placing of a single layer of spheres so that there is the least free space between them is illustrated in Figure 9-10a. The spheres in this layer are designated *a*. Obviously, if a second layer of spheres is to be placed on the first layer and a minimum of free space is to be retained, then the second layer must be added by placing spheres over the holes in the first layer. These holes or depressions are identical, but they can be divided into two groups (labeled *b* and *c*), since if a sphere is placed over a hole marked *b*, there is no room for one to be placed over an adjacent hole labeled *c*, and vice versa. It doesn't matter whether we choose to

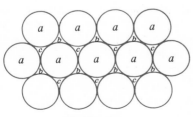
(a) One layer

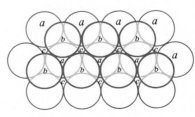
(b) Two layers

FIGURE 9-10

Closest Packing of Spheres

cover the *b* holes or the *c* holes, since the net result is the same. Figure 9-10b shows the placing of a second layer with *b* holes covered.

Now, in adding a third layer of closest-packed spheres, two different arrangements are possible, and these two arrangements lead to two different overall structures. We can place the third-layer spheres in the depressions directly over the first-layer spheres (*a* sites), which results in a structure in which the sequence of layers is continued indefinitely as *ababab*. This is a *hexagonal* close-packed structure (Figure 9-11a). The alternative is to place the third-layer spheres over the holes labeled *c*. This gives a continuous sequence of layers *abcabc* and results in a *cubic* close-packed structure (Figure 9-11b). It will be noted that this structure is a face-centered cubic lattice, as depicted also in Figure 9-9.

(a) Hexagonal close–packing   (b) Cubic close-packing

FIGURE 9-11

TYPES OF CRYSTALS

There are four basically different types of crystalline solids, the distinction between them being based on both the nature of the structural particles and the forces of attraction involved.

Ionic Crystals

FIGURE 9-12

Schematic Representation of a Cross Section of a Portion of an Ionic Crystal

Ionic compounds form crystals in which the units making up the lattice are ions. The forces holding the crystal together are the electrostatic attractions between ions of opposite charge. As illustrated schematically in Figure 9-12, each positive ion is surrounded by negative ions and vice versa. No particular positive ion "belongs to" any particular negative ion; there is no molecule as such. The interionic attractions—sometimes called the *ionic bonds*—are relatively strong, and considerable energy is required to overcome them. Ionic crystals and their forces of attraction have been discussed in detail in Chapter 6.

Metallic Crystals

The properties of metals in the solid state can be accounted for by a rather simple crystal model. The lattice is composed of positive ions, closely packed and embedded in an electron cloud. The ions are derived from neutral atoms by loss of electrons from the outer unfilled shell (valence electrons). These nonlocalized valence electrons constitute the electron cloud—sometimes called the "sea of electrons" or "electron gas"—which belongs to the entire crystal. The crystal is held together by the attraction between the electron cloud and the positive ions. The strength of this attraction—referred to as the *metallic bond*—depends on the size of the ions and the number of valence electrons "contributed" by each atom, and it therefore varies considerably from metal to metal. The structure of a metallic crystal is illustrated schematically in Figure 9-13, using sodium as an example. Since a sodium atom has only one electron in its outermost energy level, each ion in the crystal has a unipositive charge, and the number of electrons in the electron cloud is equal to the number of ions in the crystal.

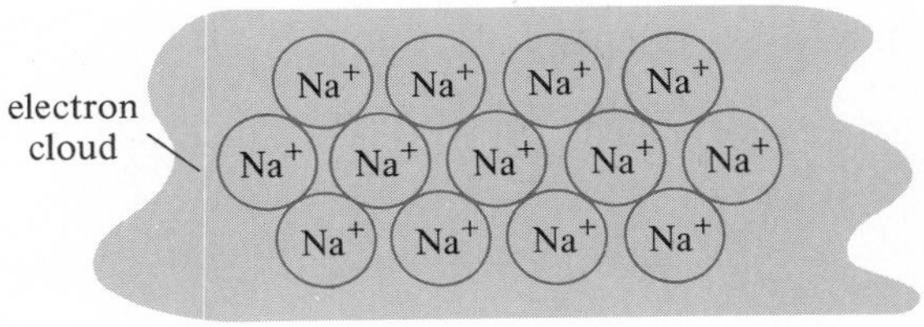

FIGURE 9-13

Single Layer of Ions in a Crystal
of Sodium Metal

While this simple model offers a satisfactory description of most metals, there is evidence that some of the transition metals possess covalent bonding between ions in addition to the attraction between the ions and the electron cloud.

Molecular Crystals   Substances that exist in the form of molecules crystallize in still another type of crystal: molecular crystals. In this type the lattice is composed of individual molecules held together by *intermolecular forces*. The nature of these forces will be discussed in some detail later in this chapter. For the present it will be sufficient to point out that intermolecular forces are generally very weak—much weaker than the ionic bond and, in most cases, weaker than the metallic bond.

As a simple example of this crystal type, consider the elementary substance iodine. Each iodine molecule consists of two iodine atoms bound together by a strong covalent bond, but the molecules are held to each other in the crystal by the weak intermolecular forces. (See Figure 9-14.)

FIGURE 9-14

Schematic Representation
of a Segment of an Iodine Crystal

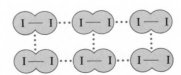

Covalent Network Crystals

The fourth type of crystal is the covalent network crystal. Here the lattice is made up of atoms linked to each other by covalent bonds. No separate molecules exist within the crystal; in fact, the entire crystal is a giant molecule. The classic example of this type of crystal is *diamond,* one of the *allotropes* of carbon.

Allotropy is the existence of two or more different forms of an elementary substance in the same physical state. The various forms themselves are called *allotropes.* Allotropy may occur due to a difference in the number of atoms in a molecule—for example, oxygen (O_2) and ozone (O_3) are allotropes—or a difference in crystal form. Diamond and graphite are two allotropes of carbon; they are elemental carbon existing in different crystal forms.

In diamond each carbon atom is at the center of a tetrahedron and is attached by covalent bonds to four other carbon atoms, which form the corners of that tetrahedron. Each of these four atoms is also the center of a tetrahedron and is attached to four other carbon atoms, and so on, to form a three-dimensional covalent network (Figure 9-15).

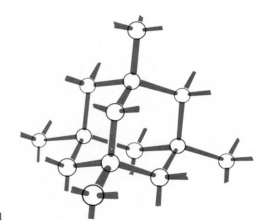

FIGURE 9-15

The Diamond Structure:
A Covalent Network Crystal

Of course, diamond is not the only example of a covalent network crystal. The elementary substances silicon and germanium and one of the allotropes of tin are other examples (see Chapter 21), as is the extremely abundant compound

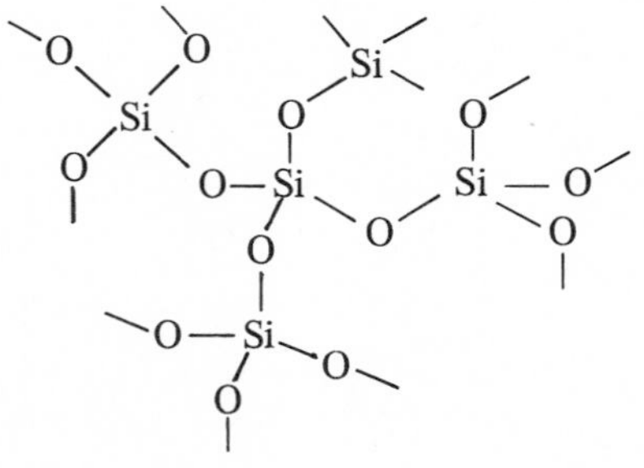

FIGURE 9-16

The Silicon Dioxide Structure

silicon dioxide. The structure of the latter is shown in Figure 9-16. It will be noted that each silicon atom is bonded to four oxygen atoms and each oxygen atom is bonded to two silicon atoms, giving a silicon/oxygen ratio of 1:2 and therefore a formula of SiO_2.

Graphite, the other allotrope of carbon, has a crystal form that is a kind of mixture of molecular and covalent network crystal forms. Each carbon atom is bonded covalently to three other carbon atoms, and each atom has one electron available for pi bonding to the adjacent carbons. This results in planar sheets of atoms arranged in hexagons. The sheets—each of which may be thought of as a giant molecule of indefinite size—are held together by intermolecular forces, as illustrated in Figure 9-17. (The electron dot formula shown in this figure is only one of many possible resonance forms, and since each group of four carbons shares one pi bond, the bond order for each carbon-carbon linkage is 4/3.) The very weak forces between atoms of different layers permit sheets to slide over one another, accounting for graphite's unusual lubricating properties. (When graphite is maintained in a vacuum, however, it loses some of its lubricating ability. This suggests that gases of the air, trapped between the layers, act as molecular ball bearings and allow very facile movement of layers.)

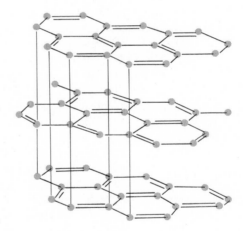

FIGURE 9-17

The Graphite Structure

PROPERTIES OF CRYSTALS

Some properties of solids are dependent upon the strength of the attractive forces within the crystal. Since the four different crystal types represent different kinds of forces, certain generalizations can be made regarding those properties.

Melting Point   Recall that the particles of matter are in constant motion. In a crystalline lattice this motion is greatly restricted and amounts to nothing more than a vibration. This vibratory motion of the particles, which is working against the attractive forces, increases as the temperature is increased. When the temperature gets sufficiently high, the motion of the particles becomes great enough to overcome the attractive forces holding them in the lattice. The particles are then able to move past one another, the lattice structure breaks down, and the substance passes into the liquid state. This process is called *melting,* and the temperature at which it occurs for a particular substance is that substance's *melting point.*

Obviously, the melting point of a solid is related to the lattice forces. In general, the stronger the attractive forces in the crystal, the greater the kinetic energy of the particles must be in order to overcome those forces, and therefore the higher is the melting point.

The covalent bonds in *covalent network crystals* are among the strongest attractive forces found in any of the crystal types. Therefore, crystals of this type have *very high melting points.* (The melting point of diamond is above 4500°C; SiO_2 melts around 1600°C.)

At the other extreme are the *molecular crystals.* Because intermolecular forces are weak, molecular crystals have *low melting points,* rarely over a few hundred degrees. As we shall see, many molecular substances have such low melting points that they exist as liquids, and even gases, at room temperature.

The electrostatic attractions between ions are generally considerably stronger than intermolecular forces. *Ionic crystals,* as a rule, have melting points higher than those of molecular crystals but lower than the melting points of covalent network crystals. To illustrate, the melting points of three common ionic compounds are: NaCl, 800°C; KI, 723°C; $MgSO_4$, 1185°C.

Finally, in *metallic crystals* the strengths of the attractive forces apparently vary widely, for the melting points of metals cover a wide range. Some of the structural and coinage metals are high melting (Fe, 1535°C; Au, 1063°C; Ag, 961°C; Cu, 1083°C), other metals have moderate melting points (Bi, 271°C; Pb, 328°C; Sn 232°C), while a few have melting points even lower than those of many molecular crystals (Na, 98°C; Cs, 28°C; Ga, 30°C).

The alkali metals have particularly low melting points. This fact can be explained with the aid of our metallic crystal model (cations in a "sea" of electrons), for each atom has only one valence electron to contribute to the electron cloud. Thus the density of the cloud is low and the metallic bond is relatively weak. Comparison of the melting points of the alkali metals with those of the alkaline earth metals shows the latter to be considerably higher (Table 9-1). The alkaline earth metals have two valence electrons to contribute, forming dipositive ions, and the metallic bond is therefore stronger.

Heat of Fusion   When a crystalline solid is heated, the temperature of the solid gradually rises until the melting point is reached. Additional heat does not cause a temperature increase. Instead, the solid melts at constant temperature, and only after melting is complete does the temperature begin to rise again. The thermal energy supplied during the melting process is consumed in bringing about the transition from the solid to the liquid state; it is used to overcome the attractive forces holding the particles (ions,

TABLE 9-1 **Melting Points** **of the Alkali Metals** **and Alkaline Earth Metals**	*Alkali Metal*	*mp* $(^{\circ}C)$	*Alkaline* *Earth Metal*	*mp* $(^{\circ}C)$
	Li	179	Be	1283
	Na	98	Mg	650
	K	64	Ca	850
	Rb	39	Sr	770
	Cs	28	Ba	725

molecules, or atoms) of the crystal together. The energy necessary to bring about melting, or *fusion,* is called the *heat of fusion.* (Note that melting is an *endothermic* process.) It is often measured in calories per gram or, more conveniently, in kilocalories per mole.

Obviously, the magnitude of the heat of fusion of a crystal is related to the forces holding the crystal together: the stronger the forces, the higher the heat of fusion. Thus we find heats of fusion following a pattern similar to that of melting points, as they relate to the various crystal types. Covalent network crystals generally have very high heats of fusion (for diamond it is estimated to be 143 kcal/mole), ionic crystals have somewhat lower heats of fusion (for NaCl, 6.8 kcal/mole), and molecular crystals have the lowest (1.44 kcal/mole for ordinary ice). Metals, as might be expected from their melting points, have a range of heats of fusion from low-melting cesium (0.50 kcal/mole) to very high-melting tungsten (8.42 kcal/mole).

Hardness

Another property that varies with crystal type is hardness. The scratching or breaking of a crystal requires the breaking of crystal bonds and disruption of the lattice. Covalent network crystals are very hard because of the large number of strong covalent bonds that must be broken. (Diamond is the hardest naturally occurring substance known.) Most ionic crystals are quite hard also, although they show a greater brittleness and tendency to fracture by cleavage than covalent network crystals do. Molecular crystals are usually soft by comparison.

Metals range from very hard to very soft, but they possess a property not found in the other crystal types. Because the electrons in a metallic crystal are highly mobile and provide a uniform distribution of charge, the positions of the positive ions can be changed without destroying the bonding. Therefore, metallic crystals can be deformed rather easily, and metals possess the properties of malleability and ductility (the ability to be hammered into various shapes and to be drawn into wire, respectively).

Vapor Pressure

Vapor pressure as a property is discussed in detail later in this chapter (p. 203). For now, suffice it to say that vapor pressure is a measure of the tendency of the structural particles of a solid or a liquid to leave the surface and become a gas. (This

TABLE 9-2
**Structural Features
and Relative Properties of Crystal Types**

Crystal Type	Lattice Particles	Attractive Force	Melting Point	Heat of Fusion	Hardness	Vapor Pressure	Electrical Conductivity
Ionic	Positive and negative ions	Electrostatic attraction (ionic bond)	high	high	hard	very low	no
Metallic	Positive ions	Attraction between positive ions and electron cloud (metallic bond)	Range: high to low	Range: high to low	hard to soft	very low	yes
Molecular	Molecules	Intermolecular forces	low	low	soft	relatively high	no
Covalent network	Atoms	Covalent bond	very high	very high	very hard	negligible	no

process is called *evaporation* for liquids, *sublimation* for solids.) With few exceptions, molecular crystals are the only ones that have an appreciable vapor pressure at room temperature, for only intermolecular forces are weak enough to be so easily overcome.

Electrical Conductivity

In order for a substance to conduct electricity, its structure must permit the motion of electrical charges through it. Of the four basic crystal types, only metallic crystals conduct electricity. The highly mobile electrons of the electron cloud are able to move without destroying the crystal lattice. When electrons are forced into one end of a metal wire, they displace electrons of the metallic crystal's electron cloud. These displaced electrons take new positions by displacing neighboring electrons, and so on down the wire until electrons are forced out at the opposite end. In the other three crystal types, the electrons are all too tightly bound to permit delocalized motion.

Ionic substances also conduct electricity, but only in the liquid state. Thus, for example, solid sodium chloride is a nonconductor, but *molten* sodium chloride is a conductor. In the liquid state, the electrical charge can be carried by the motion of ions.

Graphite, the structure of which was described earlier (Figure 9-17), is something of an exception in that it is an electrical conductor. This fact can be rationalized from the graphite structure by the assumption that the pi bonding electrons are not localized between two carbon atoms (as is implied by the double bonds in Figure 9-17) but are free to move throughout the entire sheet. It is the motion of these electrons that gives rise to electrical conductivity. This view is supported by the fact that the degree of conductivity is rather high in a direction parallel to the sheets but is quite low in a direction perpendicular to them.

Structural features and relative properties of the four crystal types are summarized in Table 9-2.

INTERMOLECULAR FORCES

Of all the different substances known to man, most by far exist in the form of molecules. Therefore, the kind of attractive forces with which we most frequently deal are *intermolecular forces*. In our discussion of molecular crystals, these forces were characterized simply as weak electrical attractions between molecules, and discussion of their origin was deferred. Let us now examine in some detail the nature of these forces.

Dipole-Dipole Attractions

The polarity of covalent bonds and the dipole moment of molecules have been discussed in Chapter 7. Recall that in some molecules—because of unequal sharing of electrons and the molecular shape—the electron cloud is not distributed symmetrically about the center of gravity. The result is that even though the molecule as a whole is electrically neutral, there is a charge separation making one end of the molecule slightly positive and the other slightly negative with respect to each other. A molecule of this condition is said to be *polar* and is called a *dipole*.

Because of this charge separation, the negative end of one polar molecule is attracted to the positive end of another, and vice versa. This intermolecular force is called *dipole-dipole attraction.* In the crystalline lattice of a polar molecular substance, the molecules are oriented in a way that maximizes the effect of dipole-dipole attraction, as illustrated in Figure 9-18.

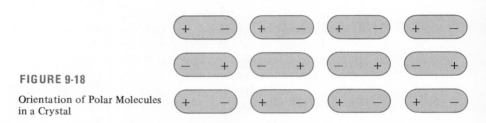

FIGURE 9-18

Orientation of Polar Molecules in a Crystal

Other things being equal, the more polar the molecules, the stronger the attractive force between them, and this strength is reflected in the various properties discussed above. This is illustrated in Table 9-3, where the dipole moments (see p. 105) and melting points of three isoelectronic molecules are compared.

TABLE 9-3
The Effect of Polarity on Melting Point

Molecule	Dipole Moment (D)	Melting Point (°C)
SiH_4	0	−185
PH_3	0.55	−132.5
H_2S	0.94	−82.9

Because dipole-dipole attraction is a relatively strong intermolecular force, crystals of polar molecules generally have higher melting points, higher heats of vaporization, lower vapor pressures, and greater hardness than crystals of nonpolar molecules.

A special kind of intermolecular force that occurs between certain types of polar molecules is called the *hydrogen bond.* This force, which is stronger than ordinary dipole-dipole attractions, is discussed later in this chapter.

Van der Waals Forces

Forces of attraction exist even between nonpolar molecules. The nonpolar gases—O_2, N_2, and even the inert gases—can be liquefied; the nonpolar substances Cl_2 and Br_2 are rather easily solidified, and I_2 is a solid even at room temperature. Thus, intermolecular forces other than dipole-dipole attractions must exist. The existence of these forces was first suggested by the Dutch physicist Johannes van der Waals and are named *van der Waals forces* in his honor.

Van der Waals forces, which are present in all matter, are weaker than dipole-dipole attractions. Their origin may be described as follows. Consider the simple diatomic molecule Cl_2. Because the two atoms are identical, the bond between them is nonpolar and the molecule has no dipole moment. The electrons within the molecule are not in fixed positions but form an electron cloud by their constant motion. Although on the average, with time, the electron cloud is symmetrical within the molecule, at a given instant there may be more electrons around one chlorine nucleus than there are around the other. The result of this distortion, or *polarization,* is that for an *instant* one chlorine atom is negative with respect to the other and the molecule is a *temporary dipole.* A second chlorine molecule, adjacent to the temporary dipole, will be influenced by it, for its electron cloud will be distorted toward the positive end of the first molecule. The first molecule has *induced* a dipole in the second. This arrangement is very fleeting, of course, because the electrons are in motion, but as the electron density of the first molecule moves to the opposite end, the electrons in the second molecule will move in the same direction, reversing the polarization. Thus, a kind of fluctuating dipole is set up. This description of van der Waals forces is illustrated in Figure 9-19. (The Greek letter delta, δ, is used to indicate a partial charge—one of considerably smaller magnitude than the charge of an electron or proton.)

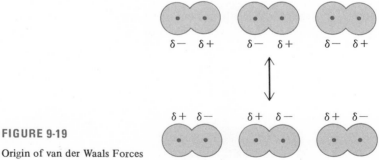

FIGURE 9-19

Origin of van der Waals Forces

On the average, then, over a finite period of time, the electron clouds of all the molecules are symmetrical, but it is the instantaneous polarizations of the molecules that result in attractions between them. The attractions are weak, to be sure, and when molecules are far apart they are negligible, but when molecules are close enough these forces are capable of retaining them in a crystal lattice.

The greater the number of electrons in a molecule, the more easily the electron cloud is polarized and the stronger are the van der Waals forces. Since the number of electrons increases with increasing molecular weight, it follows that the

TABLE 9-4	Halogen	Mol Wt	mp (°C)	bp (°C)
Relationship of				
van der Waals Forces	F_2	38	−218	−188
to Molecular Weight	Cl_2	71	−101	−34
in the Halogens	Br_2	160	−7	59
	I_2	254	114	184

strength of van der Waals forces increases with increasing molecular weight. This relationship is demonstrated elegantly by the nonpolar halogen molecules, wherein the melting points and boiling points—both indications of strength of inter-molecular forces—increase steadily as molecular weight increases (Table 9-4).

Molecular shape is also a factor in determining the strength of van der Waals forces. Other factors being equal, the greater the surface available for close contact between molecules, the stronger van der Waals forces will be. Consider, for example, the three chain isomers of pentane, C_5H_{12} (Figure 9-20). The three molecules have identical molecular weights, yet their boiling points are different, indicating a difference in the strengths of the intermolecular forces. The straight-chain structure of *n*-pentane permits close approach and intimate contact between molecules, giving rise to the strongest forces and therefore the highest boiling point. The branching in isopentane reduces the contact surface between molecules and leads to weaker forces and a lower boiling point. Finally, the neopentane molecule, being almost spherical, offers the least possibility for close contact and has the weakest van der Waals forces and the lowest boiling point of the three. It is a general rule that in chain isomers the more branched the chain, the weaker the van der Waals forces.

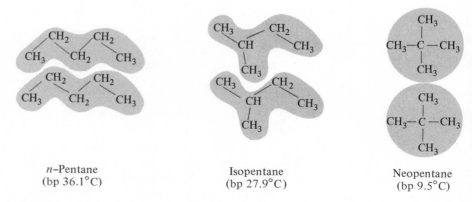

n–Pentane
(bp 36.1°C)

Isopentane
(bp 27.9°C)

Neopentane
(bp 9.5°C)

FIGURE 9-20

Influence of Molecular Shape on Strength of van der Waals Forces

The Liquid State

With few exceptions, substances that exist in the liquid state at ordinary temperatures and pressures are molecular. Ionic, metallic, and covalent bonds are so strong that particles of matter have sufficient kinetic energy to break them only at elevated temperatures. Intermolecular forces, on the other hand, are weak enough to permit many molecular substances to exist as liquids even at very low temperatures.

Our model of the liquid state consists of molecules in motion (and therefore possessing kinetic energy), randomly distributed, and attracted to each other by dipole-dipole attractions and van der Waals forces. Unlike the solid state, in which the molecules are held rigidly in a lattice, the molecules of liquids are free to move with respect to one another within limits determined by the strength of the intermolecular forces. This model enables us to rationalize many of the properties characteristic of the liquid state.

PROPERTIES OF LIQUIDS

Density

As defined earlier, the density of a substance is the mass of a sample of the substance per unit volume. The fact that liquids in general have densities in a range between those of solids and gases means that the average space between molecules in a liquid is greater than in a solid and smaller than in a gas.

For nearly all liquids, an increase in temperature results in a decrease in density. In other words, a liquid expands upon heating; its volume increases and therefore its density decreases. This relationship between temperature and density is easily explained in terms of the kinetic energy of the molecules. As the temperature is raised, the kinetic energy of the molecules is increased. The enhanced vigor of motion decreases the effectiveness of the attractive forces, and the average space between molecules increases.

At a given temperature the densities of different liquids vary widely, and this variation can be correlated with the relative strengths of the intermolecular forces. The stronger the forces, the more closely the molecules are held together and the higher is the density. This relationship is illustrated in Table 9-5. The two hydrocarbons *n*-pentane and *n*-octane are nonpolar, so that the only attractive forces involved are van der Waals forces. The one with the higher molecular weight has the stronger van der Waals forces and therefore the higher density. Of the three, the compound with the highest density is *n*-propyl chloride, even though its molecular weight is only slightly higher than that of *n*-pentane. This may be explained by the

		Density at 20° C
TABLE 9-5 Effect of Intermolecular Forces on the Density of Liquids	*Liquid*	*Mol Wt*	*(g/ml)*
n-C_5H_{12}	72	0.626
n-C_8H_{18}	114	0.703
n-C_3H_7Cl	78	0.890

fact that n-propyl chloride is a polar molecule, so there are dipole-dipole attractions in addition to van der Waals forces.

Viscosity

Because the molecules of a liquid are held only loosely, they are able to move past one another, thus enabling liquids to *flow*. But the tendency to flow, which is expressed in terms of *viscosity*, differs widely for different liquids. *The viscosity of a liquid is its resistance to flow;* the unit of measurement is the *poise* or, more commonly, the *centipoise* (cp). The more readily a liquid flows, the less *viscous* it is and the lower is its viscosity. The fact that viscosity decreases as the temperature is increased is readily explained with our liquid model. Increased molecular motion at higher temperature lessens the effectiveness of the intermolecular forces and permits the liquid to flow more easily. Differences in viscosities of different liquids at the same temperature is simply a matter of differences in strengths of intermolecular forces; the stronger the forces the greater the viscosity. Thus, the high molecular weight hydrocarbons that constitute motor oil give that liquid a much higher viscosity than gasoline, which contains only hydrocarbons of much lower molecular weight.

Surface Tension

Another property of liquids that is related to intermolecular forces is the *surface tension*. This is defined as the *force on a liquid surface opposing the expansion of the surface area*. Liquids tend to present a minimum surface area, and it requires work to expand that surface area. The amount of work required is an expression of the surface tension of the liquid. This property is reflected in the fact that a free-falling drop of liquid takes on a spherical shape, for a sphere has the least surface area of any geometrical shape for a given volume.

As illustrated in Figure 9-21, the molecules in the body of a liquid are subjected to intermolecular forces from all directions, with the result that the *net* force being exerted on any molecule is zero. Molecules on the surface of the liquid, on the other hand, are subjected to forces only from the sides and below, which results in a net force attracting the surface molecules toward the interior of the liquid. This accounts for the tendency of the surface area to be minimized, and thus accounts for the surface tension.

As one would expect, surface tension decreases with increasing temperature, and for different liquids at the same temperature, surface tension increases with increasing strength of intermolecular forces.

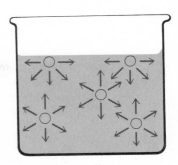

FIGURE 9-21

Surface Tension Caused by Intermolecular Forces

It is the surface tension of liquids that accounts for the familiar phenomenon of *capillarity,* the rise or fall of liquids in a capillary tube. When one places one end of an open tube of very small diameter into a liquid, the liquid either rises or falls in the tube (Figure 9-22). Those liquids that rise do so because their molecules have a stronger attraction for the walls of the capillary tube than they do for each other. Such a liquid is said to "wet" the surface of the tube. This attraction produces a concave surface, increasing the surface area of the liquid. Because of surface tension the liquid rises in the tube until it reaches a height at which the weight of the column of liquid balances the attraction exerted by the tube's surface on the molecules of the liquid. Since the greater the surface tension of the liquid the higher it will rise in the tube, this *capillary rise* can be used to measure surface tension.

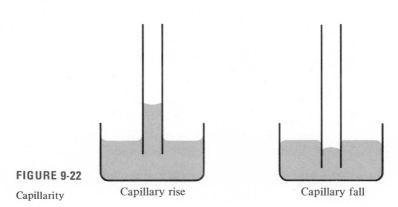

FIGURE 9-22

Capillarity   Capillary rise   Capillary fall

Liquids whose molecules have greater attraction for each other than for the walls of the capillary tube fall instead of rising and produce a convex surface.

Vapor Pressure   In our discussion of the solid state, vapor pressure was described simply as a measure of the tendency of a solid or liquid to become a gas. Let us now examine this property more closely and develop a more specific definition of it.

Consider an open beaker containing a liquid at room temperature. At any given instant the individual molecules of the liquid will have a variety of kinetic energies, although the average kinetic energy will be determined by the temperature. If a molecule on the surface has a kinetic energy sufficiently above the average and is moving in the right direction, it may overcome the intermolecular forces and enter the *vapor* phase above the liquid's surface. (The terms *vapor* and *gas* are often used interchangeably, although *vapor* is usually reserved for the gaseous form of a substance that exists as a liquid at ordinary room conditions.)

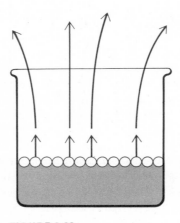

FIGURE 9-23

Evaporation of a Liquid

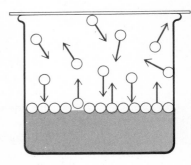

FIGURE 9-24

Evaporation and Condensation

This process, illustrated in Figure 9-23, is called *evaporation* or *vaporization*. Since the container is open, the vapor molecules mingle with the molecules of the air and become distributed throughout the room. Everyone is familiar with the fact that if a glass of water is left standing, the volume of water will gradually diminish.

Now suppose we take the same liquid in the same container at the same temperature, but this time the container is sealed so that vapor molecules cannot escape into the room. The vapor molecules are in motion above the liquid, and during this random motion some will certainly collide with the liquid surface. If when one of these collisions occurs the vapor molecule's kinetic energy is sufficiently low, intermolecular forces may take over, and the molecule becomes part of the liquid again. This process, which is just the opposite of evaporation, is called *condensation* or *liquefaction* (Figure 9-24).

Thus, we have two opposing processes occurring:

$$\text{Liquid} \underset{\text{condensation}}{\overset{\text{evaporation}}{\rightleftharpoons}} \text{Vapor}$$

The *rate* of evaporation of a liquid is dependent upon a number of factors:

1. The nature of the liquid. Even if all conditions are the same, different liquids will not evaporate at the same rate. This is due to the difference in

the strength of the intermolecular forces, which in turn depend on the molecular weight, structure, and degree of polarity of the molecules.

2. Temperature. For any given liquid the rate of evaporation varies with temperature. The higher the temperature, the greater the kinetic energy of the molecules, which means, in turn, the more easily the intermolecular forces will be broken and therefore the greater the rate of evaporation will be.

3. Surface area. Since evaporation is a surface phenomenon, enlarging the surface area increases the rate of evaporation. If a given volume of water is placed in a large, shallow pan and the same volume in a tall, narrow cylinder at the same temperature, the water will evaporate from the pan faster than it will from the cylinder.

It follows, then, that for a given liquid substance in a given container maintained at a given temperature, the rate of evaporation is constant.

On the other hand, the *rate of condensation* depends not only on these same three factors but also on the number of vapor molecules in the phase above the liquid. The greater the number of vapor molecules, the more collisions will occur with the liquid surface, and therefore the higher will be the rate of condensation.

Here we have two opposing processes, one of which is occurring at a constant rate (evaporation). The other (condensation) begins at zero rate, and its rate gradually increases as more and more molecules enter the vapor phase. Eventually, the two rates must become equal; molecules will be leaving the vapor phase just as fast as they are entering it. When this point is reached, both rates will continue to be equal and there will be no net change in the quantity of substance in either the liquid or the vapor phase, provided the temperature is held constant. This situation—two opposing processes occurring at equal rates—is called *dynamic equilibrium.* It must be emphasized that we have not said that evaporation occurs up to a certain point and then stops. On the contrary, although it may *appear* that evaporation has ceased because there is no net change in the amount of liquid present, the two processes will continue to occur *ad infinitum* as long as the conditions of the system are not changed. The concept of *dynamic equilibrium* is a very important one, for it is useful in explaining many chemical phenomena. We shall encounter it frequently throughout the remainder of this book.

In the liquid-vapor equilibrium just described, in addition to colliding with the liquid phase and with each other, the vapor molecules undergo collisions with the walls of the container. In other words, they exert a pressure. It is this pressure that is called vapor pressure. Thus, *vapor pressure* is defined as *the pressure exerted by the molecules of a vapor in equilibrium with its liquid.*

The vapor pressure of any liquid varies with temperature. Suppose we have a liquid and its vapor in equilibrium at 25°C and have measured the vapor pressure. Now let us raise the temperature of the liquid to, say, 50°C. At the higher temperature the kinetic energy of the liquid molecules is increased, so that the intermolecular forces are more easily overcome, and the rate of evaporation increases. The rate of condensation also increases as the number of vapor molecules becomes larger. Eventually the two rates will become equal again and equilibrium will be reestablished at the new temperature (50°C). However, at the higher temperature there are more molecules in the vapor phase than there were at the lower temperature. More vapor molecules means more collisions with the container walls and therefore a higher vapor pressure. Furthermore, at the higher temperature, the vapor molecules themselves have greater kinetic energy, so that even if the number of vapor molecules had not increased, a greater force would be exerted on the container walls. Both factors contribute to an increase in vapor pressure with increasing temperature. This relationship is illustrated in Figure 9-25 with vapor pressure curves (plots of temperature versus vapor pressure) of several common liquids. (A table of vapor pressures of water at various temperatures is given in Appendix 12.)

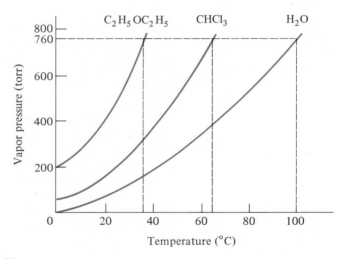

FIGURE 9-25

Vapor Pressure Curves of Ethyl Ether, Chloroform, and Water

Figure 9-25 also demonstrates the relationship between vapor pressure and the strength of the intermolecular forces. At any given temperature, the stronger the intermolecular forces, the lower the vapor pressure. Ethyl ether, which is weakly polar, has a higher vapor pressure than the somewhat more polar (and higher molecular weight) chloroform, and water, which is highly polar, has the lowest vapor pressure of the three.

Heat of Vaporization   During the evaporation of a liquid it is the molecules with high kinetic energy that are escaping from the surface. This lowers the average kinetic energy of the remaining molecules and causes a corresponding decrease in temperature. If the liquid were completely isolated from its surroundings, this decrease in temperature would lower the rate of evaporation until, theoretically, evaporation would cease. Under ordinary circumstances, however, evaporation continues because the liquid takes heat from its surroundings. This phenomenon is easily experienced if one places a few drops of some very *volatile* liquid (one with a high vapor pressure), such as ethyl ether, on one's skin. As the liquid evaporates, there is a sensation of cold at that spot. Another common example of this phenomenon is the coldness felt after a swim: The water on the body evaporates, taking heat from the surface of the skin.

It follows from the foregoing that evaporation is an *endothermic* process. Energy must be supplied to convert a liquid to a gas. This energy, called the *heat of vaporization,* is usually expressed in terms of kilocalories per mole. Like the heat of fusion for solids, the heat of vaporization is a measure of the energy required to overcome attractive forces. Liquids with strong intermolecular forces have correspondingly high heats of vaporization.

The heat of vaporization of any substance varies with temperature. At higher temperatures the molecules have greater kinetic energy, and therefore less energy is required to cause evaporation. In other words, the heat of vaporization decreases as the temperature increases. For example, at one atmosphere of pressure (760 torr) water has a heat of vaporization of 10.6 kcal/mole at 10°C and 9.72 kcal/mole at 100°C.

Boiling Point

While evaporation and boiling are both processes whereby liquids are converted into gases, they differ in several respects. Evaporation is a surface phenomenon, whereas boiling involves the formation of bubbles of vapor within the body of the liquid, which rise to the surface and escape. Furthermore, evaporation occurs spontaneously at any temperature and represents a diffusion of individual molecules into the atmosphere. In boiling, the vapor escapes by pushing back and displacing the gas of the atmosphere above the liquid, and therefore boiling will not occur unless the vapor pressure of the liquid is equal to that of the confining atmosphere. Thus, the *boiling point* is defined as *that temperature at which the vapor pressure of the liquid is equal to the pressure of the atmosphere above the liquid.*

The boiling point of a liquid is not a constant, then, but varies with the pressure of the surrounding atmosphere. Water boils at a lower temperature on a mountain top than it does at sea level, and it may be made to boil at room temperature if the pressure of the atmosphere above it is reduced sufficiently by means of a vacuum pump. The variation of boiling point with pressure may be seen by reference to Figure 9-25. If the pressure of the atmosphere over water were, say, 400 torr, then water would boil at a temperature of 83°C, because at that temperature the vapor pressure of water is 400 torr. On the other hand, the boiling point of water would be about 108°C if the pressure were 1000 torr.

In order to be able to make direct comparisons of boiling points of different liquids, it is necessary to establish a standard pressure as a reference point. By universal agreement this standard pressure is *one atmosphere,* equal to the pressure exerted by a column of mercury 760 mm high. *The temperature at which a liquid* boils when the pressure is one atmosphere is called its *normal boiling point*. The dashed lines in Figure 9-25 show the *normal boiling points* of the three liquids; that is, they show the temperature at which each of the three liquids has a vapor pressure of 760 torr.

Distillation is a process whereby a liquid is converted into a vapor by boiling and the vapor is then condensed back to a liquid by cooling. This process is useful in purifying liquids and in separating substances of widely different boiling points. Figure 9-26 illustrates the separation of water from a mixture of sodium chloride and water. The water vaporizes and is then condensed; the sodium chloride remains behind in the flask.

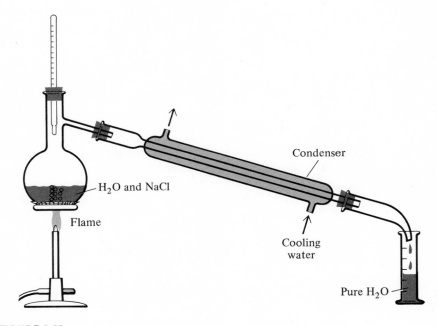

FIGURE 9-26

A Simple Distillation Apparatus

If the liquid to be distilled is unstable at elevated temperatures or if its normal boiling point is extremely high, it may be distilled at a lower temperature—that is, its boiling point may be lowered—by reducing the pressure in the distillation apparatus. This procedure is called a *vacuum distillation*.

Freezing Point   When heat is gradually removed from a liquid, the temperature of the liquid is lowered and the kinetic energy of its molecules decreases. As molecular motion diminishes, the intermolecular forces become more effective, and at a sufficiently low temperature the intermolecular forces "take over." The molecules are "pulled" into a lattice arrangement and *crystallization* occurs. The temperature at which this occurs is called the *freezing point*, and is, of course, identical to the melting point. The exact freezing point of a liquid varies somewhat with pressure, although not nearly so much as the boiling point. The *normal freezing point* is defined as the freezing point under a pressure of one atmosphere.

At the freezing point, a liquid and its solid are in equilibrium. Thus, for example, if a mixture of ice and water is held at a temperature of exactly 0°C and no heat is allowed to enter or leave the system, ice will melt and water will freeze at exactly equal rates, resulting in no net change in the amount of ice or water present. The system is in dynamic equilibrium.

$$\text{Solid} \underset{\text{freezing}}{\overset{\text{melting}}{\rightleftharpoons}} \text{Liquid}$$

During the crystallization of a liquid, the temperature remains constant, and heat must be removed from the system in order for crystallization to continue. Crystallization (freezing) is an *exothermic* process—energy is given off. The amount of energy that is given off is called the *heat of crystallization* and can be expressed in kilocalories per mole. For any given liquid the heat of crystallization is, of course, numerically equal (but of opposite sign) to the heat of fusion, discussed earlier in this chapter.

In our discussion of solids it was noted that molecular crystals possess measurable vapor pressures. Molecules on the surface of a crystal may possess

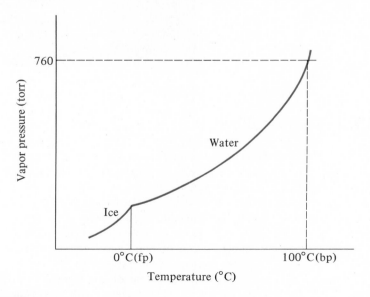

FIGURE 9-27

Vapor Pressure Curves for Ice and Water

sufficient kinetic energy to break away from the lattice and enter the vapor phase. In a closed system, this process of *sublimation* will lead to an equilibrium between solid and vapor. Vapor pressures of solids are defined and measured in the same way as vapor pressures of liquids. The vapor pressure of solids accounts for numerous well-known phenomena. For example, ice will disappear from a sidewalk on a windy day even though the temperature is well below the freezing point. The ice *sublimes,* and the vapor molecules are carried away by the wind. The fact that some solids have odor is an indication of their vapor pressure, for in order for us to smell a substance, molecules of that substance must enter the nasal passages.

At the freezing point the vapor pressure of a solid is equal to the vapor pressure of the liquid. This is illustrated in Figure 9-27, which shows the vapor pressure curves for the solid and liquid states of water. The temperature at which the curves intersect is the freezing (or melting) point.

The effects of the strength of intermolecular forces and temperature on various properties of liquids are summarized in Table 9-6.

TABLE 9-6
Effect of
Intermolecular Forces
and Temperature
on Some Properties of Liquids

Property	Effect of Increasing Temperature	Effect of Increased Intermolecular Forces
Density	Decreases	Increases
Viscosity	Decreases	Increases
Surface tension	Decreases	Increases
Vapor pressure	Increases	Decreases
Heat of vaporization	Decreases	Increases
Boiling point		Increases
Freezing point		Increases

THE HYDROGEN BOND

Certain compounds display properties that suggest the existence of attractive forces far too strong to be explained on the basis of van der Waals forces and dipole-dipole attractions alone. Since the most common compound showing these "strong force" properties is water, let us examine this concept using water as our example. Water has several abnormal properties—abnormal in the sense that they do not follow the trends predicted by consideration of ordinary intermolecular forces. For example, its boiling point and freezing point are both abnormally high. First, let us establish the basis for this statement.

The boiling points of the hydrides of Group IVA (Table 9-7) follow the expected trend: They increase with increasing molecular weight. Since these hydrides are nonpolar compounds, the only forces binding their molecules together

TABLE 9-7
Boiling Points
of the Group IVA
Hydrides

	CH_4	SiH_4	GeH_4	SnH_4
Molecular weight	16	32	76	123
Boiling point	−161	−112	−90	−57

are van der Waals forces, and the direct relationship to molecular weight is predictable.

Although the hydrides of Groups VA, VIA, and VIIA show some degree of polarity, this relationship of boiling point to molecular weight seems to prevail within each group, with the important exception of the first member. Ammonia, water, and hydrogen fluoride do not follow the trend, but have boiling points well above the expected values. The same trends and the same exceptions apply also to the melting points of these hydrides. Table 9-8 illustrates these relationships and shows clearly why we may conclude that water, ammonia, and hydrogen fluoride have abnormally high boiling and melting points.

TABLE 9-8
Normal Boiling Points
and Melting Points
of the Hydrides
of Groups V, VI, and VIIA[a]

Compound	Mol Wt	mp (°C)	bp (°C)
NH_3	17	−77.7	−33.4
PH_3	34	−132.5	−85
AsH_3	78	−113.5	−55
SbH_3	125	−88	−17
H_2O	18	0	100
H_2S	34	−82.9	−59.6
H_2Se	81	−65.7	−41.3
H_2Te	130	−48	−1.8
HF	20	−83	19.4
HCl	36.5	−111	−85
HBr	81	−86	−66.4
HI	128	−50.8	−35.5

[a]Tint blocks highlight exceptions to the trend.

Another abnormal property of water concerns the relationship of density to temperature. For most liquids, as the temperature is lowered the density increases and continues to increase as the liquid undergoes the transition to the solid state. (In other words, liquids contract on solidification; the solid is more dense than the liquid.) This behavior is easily explained on the basis of van der Waals forces and dipole-dipole attractions. Liquid water shows this same behavior—that is, its density increases as the temperature is lowered—but only down to a temperature of approximately 4°C. Further cooling of water below this temperature produces a reversal in the trend, and the density *decreases* as the temperature is lowered. Furthermore, when water crystallizes it expands; ice is *less dense* than liquid water. This particular "abnormal" behavior of water is of extreme importance to life as we know it on this planet. If ice were not less dense than liquid water, then ice crystals forming on lakes and rivers in the winter would fall to the bottom; lakes would freeze from the bottom up. Indeed, even in moderate climates, deep bodies of water might freeze solid and thaw only near the surface in the summertime.

The abnormal properties of water and certain other compounds are explained as follows. Oxygen is a highly electronegative element with two unshared pairs of electrons. The hydrogen atoms are attached to it by means of highly polar bonds. Each hydrogen atom is extremely small, and because it bears a rather large positive partial charge and has no screening electrons it is almost like a bare proton. These conditions permit a strong electrical attraction between a hydrogen atom of one water molecule and the oxygen atom of another molecule. This strong attraction is called a *hydrogen bond*. As a result of the hydrogen bond, liquid water consists of aggregates of molecules rather than entirely separate molecules. This is illustrated in Figure 9-28, where the hydrogen bonds are shown as dashed lines and the covalent bonds as solid lines.

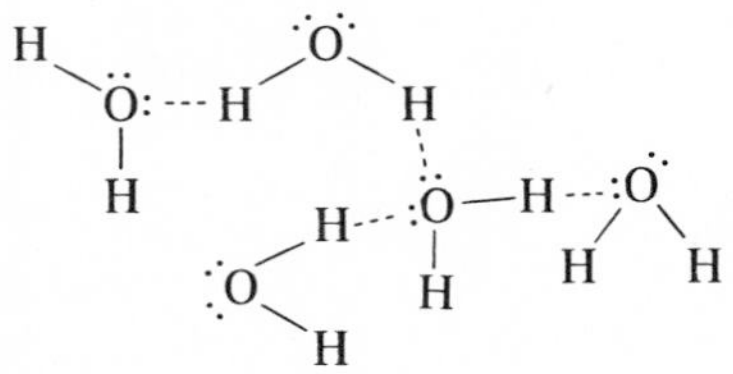

FIGURE 9-28

Hydrogen Bonding in Water

Because the hydrogen bond is considerably stronger than ordinary intermolecular forces, more energy is required to separate hydrogen-bonded molecules than would be required to separate the same molecules if they were not hydrogen bonded. In order for ice to melt and in order for liquid water to boil, the water molecules must gain a higher kinetic energy than would be necessary if the hydrogen bond did not exist. Therefore, both the melting point and the boiling point of water are higher than they would be without hydrogen bonding. However, in H_2S, H_2Se, and H_2Te there is no appreciable hydrogen bonding because sulfur, selenium, and tellurium are not sufficiently electronegative, and these compounds have melting and boiling points that can be explained by ordinary intermolecular forces alone.

The concept of the hydrogen bond also enables us to account for the unusual density-temperature behavior of water. At elevated temperatures the relatively high kinetic energy of the molecules keeps them apart and the amount of hydrogen bonding is small. As the temperature is lowered, the decreased motion of the molecules permits them to approach one another more closely, and hydrogen bond formation occurs to an increasing extent. At 4°C the average distance between molecules is at a minimum. When the temperature is lowered still further, hydrogen bonding becomes so extensive that the aggregates of molecules begin to assume fixed positions, with each oxygen atom bonded tetrahedrally to four hydrogen atoms (two by covalent bonds, two by hydrogen bonds). This results in a molecular orientation in which any given molecule is surrounded by fewer other molecules than would be the case in the absence of hydrogen bonding (illustrated in Figure 9-29). In other words, the total space between molecules increases because of hydrogen bonding, and the liquid expands rather than contracting.

Finally, at 0°C hydrogen bonding is at its maximum and all of the atoms are involved in hydrogen bonding as crystallization occurs. The ice crystal does not consist of separate water molecules attracted to each other by intermolecular forces (as is the case in solid H_2S), but is a network of tetrahedra, with each oxygen atom attached to four hydrogens and each hydrogen atom attached to two oxygens. This arrangement results in hexagonal "open spaces," which accounts for the lower density of ice (Figure 9-30).

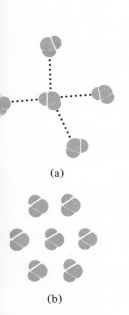

(a)

(b)

FIGURE 9-29

Spacing of Water Molecules (a) with Hydrogen Bonding; (b) without Hydrogen Bonding (Hypothetical)

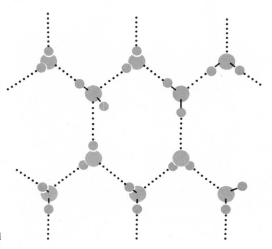

FIGURE 9-30

Structure of the Ice Crystal

Water is not the only substance that exhibits hydrogen bonding. The high melting and boiling points of HF and NH_3 are also explained by the occurrence of hydrogen bonds. Indeed, any molecule in which a hydrogen atom is bonded to a highly electronegative element having at least one pair of unshared electrons is capable of hydrogen bonding. The elements that meet this requirement are fluorine, oxygen, and nitrogen, and, to a small extent, chlorine. (Although nitrogen and chlorine have the same electronegativity, the chlorine atom is considerably larger than the nitrogen atom, and the electron cloud of chlorine is too diffuse to permit strong hydrogen bonding.) Thus, those compounds which undergo hydrogen bonding include amines (RNH_2), alcohols (ROH), and carboxylic acids (RCOOH).

The strength of the hydrogen bond varies with the particular structure of the molecule. Other things being equal, it depends on the electronegativity of the atom to which the hydrogen is covalently bonded, and therefore the strength usually decreases in the order $F > O > N$. The bond energies of hydrogen bonds—the energy necessary to *break* a bond, which is therefore a measure of bond strength—range from about 3 to 10 kcal/mole. This represents a relatively strong bond compared to intermolecular forces (the weakest van der Waals forces have energies of only a few tenths of a kilocalorie per mole) but is only about one-tenth the strength of covalent bonds. For example, in liquid water the bond energy for the hydrogen bond is about 7 kcal/mole, whereas the bond energy for the covalent $O-H$ bonds is 110 kcal/mole.

The effect of hydrogen bonding on the properties of liquids is convincingly demonstrated in Table 9-9 and 9-10, where the properties compared are ones whose magnitudes increase as the attractions between molecules become stronger. Table 9-9 compares water (with two hydrogen atoms available for hydrogen bonding) with methyl alcohol (one hydrogen available for hydrogen bonding) and with methyl ether (no hydrogen bonding). The dipole moments suggest that the strongest forces occur in water and the weakest in methyl ether. However, the differences

TABLE 9-9
Effect of Hydrogen Bonding on Physical Properties: Water, Methyl Alcohol, and Methyl Ether

	H_2O	CH_3OH	$(CH_3)_2O$
Molecular weight	18	32	46
Dipole moment (D)	1.84	1.68	1.30
Melting point (°C)	0	−97.8	−138.5
Boiling point (°C)	100	64.7	−23.7
Heat of vaporization at bp (kcal/mole)	9.72	8.42	5.14
Viscosity (20°C; cp)	1.005	0.597	(gas)
Density (20°C; g/ml)	0.998	0.792	(gas)

TABLE 9-10
Some Physical Properties of *n*-Hexane, 1-Pentanol, and Glycerol

Property	$CH_3(CH_2)_4CH_3$	$CH_3(CH_2)_3CH_2OH$	$\underset{\underset{OH}{\vert}}{CH_2}-\underset{\underset{OH}{\vert}}{CH}-\underset{\underset{OH}{\vert}}{CH_2}$
Molecular weight	86	88	92
Density (20°C; g/ml)	0.659	0.810	1.261
Melting point (°C)	−95.3	−78.9	18.2
Boiling point (°C)	68.7	138.1	290
Viscosity (20°C; cp)	0.33	4.0	1069.0
Surface tension (20°C; dynes/cm)	18.4	23.8	63

in dipole moments are not great and would not be expected to lead to large differences in properties. Furthermore, the trend in molecular weights might be expected to counteract the effect of polarity. The striking differences in properties can be accounted for only by hydrogen bonding.

In Table 9-10, three liquids of nearly the same molecular weight are compared: *n*-hexane, a nonpolar molecule in which the only attractions are van der Waals forces; 1-pentanol, a polar compound capable of dipole-dipole attractions and some hydrogen bonding; and glycerol, a trihydric alcohol capable of undergoing very extensive hydrogen bonding.

Hydrogen bonding may occur between unlike molecules. For example, ketones, which are incapable of hydrogen bonding themselves, will form hydrogen bonds with alcohols (Figure 9-31a); and water molecules will hydrogen bond to alcohol molecules (Figure 9-31b). Furthermore, where the structure will permit it, *intramolecular* hydrogen bonding also takes place, as for example in *o*-methoxyphenol shown in Figure 9-32.

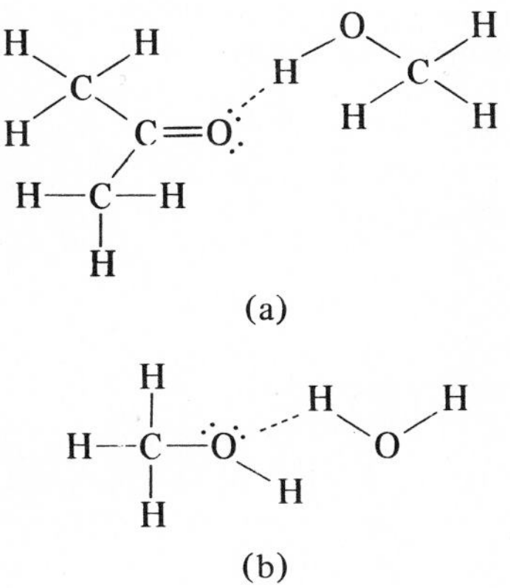

(a)

(b)

FIGURE 9-31

Hydrogen Bonding
between Unlike Molecules

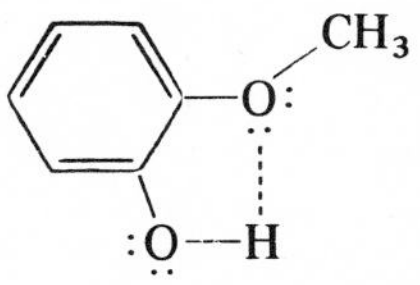

FIGURE 9-32

*Intra*molecular
Hydrogen Bonding

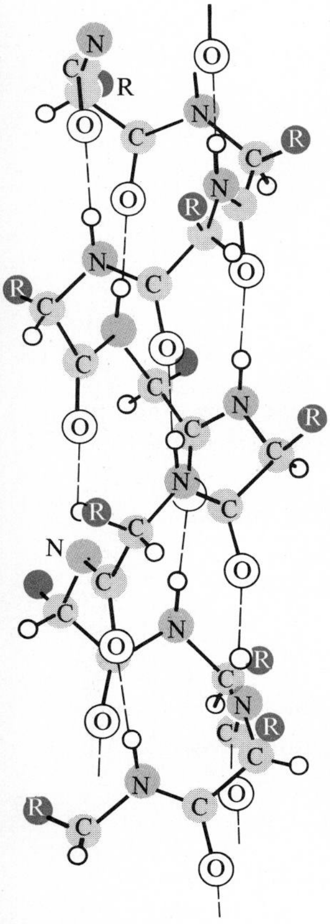

FIGURE 9-33

The α-helix

Intramolecular hydrogen bonding is an important structural determinant in certain *polypeptides*. These compounds contain amino acid units and can be prepared by the condensation of amino acids. For example, a dipeptide would be formed by the reaction

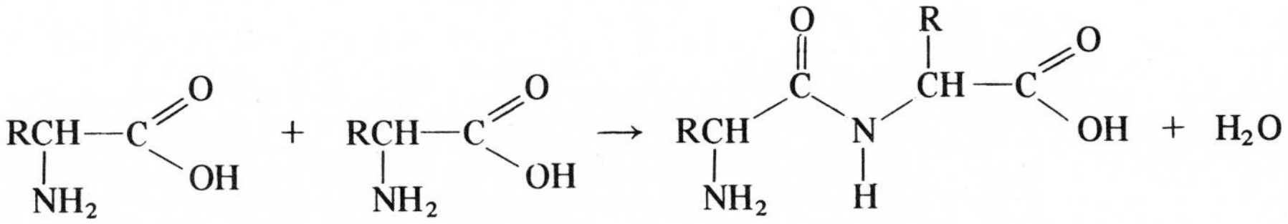

Proteins consist of numerous amino acid units (that is, they are *poly*peptides) and
have the general formula

$$\left[\begin{array}{c} O \\ \| \\ C \end{array} - N - \begin{array}{c} R \\ | \\ C \\ | \\ H \end{array} \right]_n$$

Proteins of the α-keratin class (found in hair, muscle, and elsewhere) have the
helical arrangement shown in Figure 9-33. In this structure the hydrogen atoms
attached to the nitrogens are hydrogen bonded to the oxygens of the carbonyl
groups situated above them.

SUGGESTED READINGS

Mott, N., "The Solid State." *Scientific American,* Vol. 217, No. 3 (September
   1967), pp. 80–89.
Sisler, Harry H. *Electronic Structure, Properties, and the Periodic Law*. New
   York: Reinhold, 1963. Chapter 3.

PROBLEMS

1. Distinguish between the terms in each of the following pairs:
   (a) isotope and allotrope
   (b) ionic crystal and molecular crystal
   (c) heat of fusion and heat of crystallization
   (d) polar bond and polar molecule
   (e) vapor pressure and rate of evaporation

2. Silicon carbide, SiC, is among the hardest substances known and has a
   melting point above 2700°C. Suggest a likely crystal structure for this
   substance.

3. Predict which of the two compounds, $MgCl_2$ and PCl_5, will have
   (a) the higher melting point
   (b) the higher vapor pressure at room temperature
   (c) the higher boiling point
   (d) the greater hardness
   (e) the higher heat of fusion
   (f) the greater electrical conductivity in the liquid state

4. The heat of fusion of silver metal is 2.70 kcal/g-atom. How much heat
   energy is absorbed in melting 10.0 g of silver?

5. The following table lists the dimensions (in angstrom units) and the axial
   angles of the unit cells of six hypothetical crystalline solids. By referring
   to Figure 9-7, assign the proper crystal system to each of the solids.

	a	b	c	α	β	γ
(a)	6.78	11.32	5.69	85°	95°	112°
(b)	3.61	3.61	3.61	90°	90°	90°
(c)	3.45	4.50	5.45	90°	90°	90°

	a	b	c	α	β	γ
(d)	4.90	4.90	5.40	90°	90°	120°
(e)	4.20	7.98	16.50	90°	95°	90°
(f)	4.50	4.50	2.90	90°	90°	90°

6. List those properties which distinguish ionic crystals from molecular crystals and explain the differences.

7. Explain the difference in normal boiling points (given in parentheses in °C) of the compounds in each of the following pairs:
   (a) 1-pentene (30.0°) and 1-heptene (93.3°)
   (b) silicon tetrachloride (57.6°) and phosphorus trichloride (76°)
   (c) methyl acetate (57°) and acetic acid (118°)
   (d) isopropyl alcohol (82.4°) and *n*-propyl alcohol (97.2°)
   (e) acetic acid (118°) and sodium acetate (>500°)
   (f) methylamine (−6.7°) and methyl alcohol (64.7°)

   (g)

   (225°) and   (243°)

8. For each of the pairs of compounds in problem 7, predict which compound has (a) the higher vapor pressure, (b) the greater viscosity, and (c) the greater surface tension, all at 25°C.

9. A closed vessel is partially filled with a liquid. Why is it that the density of the liquid decreases and the density of the vapor increases with an increase in temperature?

10. Explain the fact that when a liquid is in equilibrium with its solid phase, the vapor pressure of the liquid must equal the vapor pressure of the solid.

11. Rationalize the following experimental observations:
    (a) Pressure applied to liquid $TeCl_4$ at a temperature just above its normal freezing point causes the substance to freeze.
    (b) Aluminum chloride has a higher vapor pressure than aluminum fluoride.
    (c) Methyl ether has a higher vapor pressure than ethyl alcohol at room temperature.

12. Explain the difference between the meniscus in an alcohol thermometer and that in a mercury thermometer.

13. Carefully explain the fact that evaporation is an endothermic process.

14. A traveler in the deserts of the Middle East, under a temperature of 135°F, is pleasantly surprised to find that water stored in a porous clay pot is refreshingly cold. Explain.

15. Would it take more or fewer calories to convert 1 g of water at 90°C to vapor at 90°C or 1 g of water at 70°C to vapor at 70°C? Explain.

10 *Matter in Bulk II: Gases and Transitions between States*

The Gaseous State

The gaseous state of matter may be thought of as a highly diffuse extension of the liquid state. Substances that exist in the gaseous state at normal conditions are composed of molecules (or in the case of the *inert gases,* of individual atoms) randomly distributed and moving with relatively high velocities. The average distance between gas molecules is much greater than between molecules in the liquid state, and the attractive influence of neighboring molecules is much smaller. Most substances that exist as gases under ordinary conditions (room temperature and atmospheric pressure) have low molecular weights (therefore, high molecular velocities) and weak intermolecular forces.

The large distances between molecules explains the great variation in the density of a gas with changes in pressure and temperature. When the pressure on a confined gas sample is increased, the molecules are simply forced closer together, the spaces between them decreasing. This results, of course, in increased density. On the other hand, when the temperature is raised, the average velocity of the molecules increases. This causes an increase in the intermolecular distances and decreases the density.

At moderately high temperatures (around room temperature and above) and at moderately low pressures (atmospheric and below), the molecules of most gases are so far apart that two approximations may be made concerning them. First, the volume occupied by the molecules themselves is such a very small fraction of the total volume of the gas that the volumes of the molecules are negligible. In other words, the molecules may be thought of simply as points in space. Second, the attractive forces between molecules are negligible. Under these conditions many of the properties of gases become a function not of the molecules but of the space between molecules. Any gas for which these approximations—negligible molecular volume and negligible intermolecular forces—would be valid under all conditions is called an *ideal gas* or *perfect gas,* and the behavior of this hypothetical gas can be described by several quantitative statements called the ideal gas laws.

THE IDEAL GAS LAWS

The volume occupied by any gas sample is determined by three factors: the pressure, the temperature, and the number of molecules in the sample. The *ideal gas laws* are statements of the relationship of volume to each of these factors.

Variation of Volume with Pressure: Boyle's Law

Gases exert pressure on any surface with which they come into contact. This pressure is interpreted as the result of collisions of the moving molecules with the surface. Since the earth's atmosphere is a mixture of gaseous substances, the atmosphere exerts a pressure on all objects on the earth's surface. This *atmospheric pressure* is approximately 14.7 pounds per square inch, but varies with elevation and fluctuates even in the same location.

The pressure exerted by the atmosphere may be demonstrated as follows. A long glass tube is filled with mercury. The mouth of the completely filled tube is covered, and the tube is inverted into a dish of mercury so that the mouth of the tube is beneath the surface of the mercury in the dish (Figure 10-1). When the cover is removed from the mouth of the tube, some of the mercury runs into the dish, but the mercury within the tube maintains a level well above that in the dish (about 30 inches above it). The column of mercury in the tube is being supported

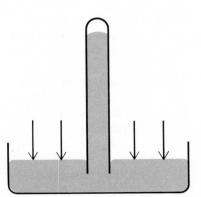

FIGURE 10-1

Atmospheric Pressure
Supporting a Column of Mercury

by the pressure of the atmosphere on the surface of the mercury in the dish. Any other liquid could be used in this demonstration, but the lower the density of the liquid, the higher the column of liquid will be. If water were used, for example, the column would be over 30 feet high. Mercury, with its very high density (13.6 g/ml), is chosen simply because it results in a column of convenient height.

Since the height of the column of mercury is proportional to the pressure supporting it, this affords a convenient way of measuring atmospheric pressure. In chemistry, gas pressures are nearly always expressed in millimeters of mercury (mm Hg). One millimeter of mercury is also called a *torr* (after the Italian physicist Evangelista Torricelli). A device used for measuring atmospheric pressure based on the simple apparatus described above is called a *mercurial barometer. One atmosphere* (atm) is defined as 760 torr (760 mm Hg).

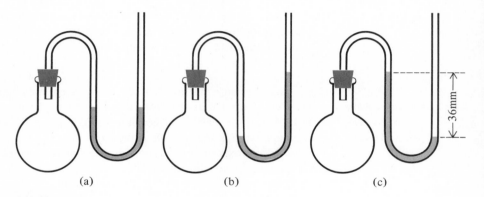

(a)   (b)   (c)

FIGURE 10-2

U-Tube Manometer

The principle employed in the mercurial barometer can also be applied to the measurement of the pressure of a confined gas sample. Figure 10-2 shows a flask containing a gas. The flask is fitted with a U-tube containing mercury and open to the atmosphere at one end. The gas within the flask is exerting pressure on the mercury in one arm of the U-tube, while the other arm is subjected to atmospheric pressure. If the pressure of the gas sample is exactly equal to that of the atmosphere, the mercury stands at the same height in both arms (Figure 10-2a). Figure 10-2b represents the case where the pressure inside the flask is greater than atmospheric pressure, and in Figure 10-2c atmospheric pressure is greater than the pressure of the gas. By measuring the difference in the heights of the two columns and determining the atmospheric pressure with a barometer, one can determine the pressure of the confined gas sample by simple arithmetic. For example, suppose that the mercury columns are positioned as in Figure 10-2c and that the difference in the heights of the columns is 36 mm. Suppose further that atmospheric pressure has been found to be 742 torr. The pressure of the gas in the flask, then, is 742 − 36 = 706 torr. This U-tube device is called a *manometer.*

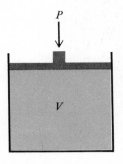
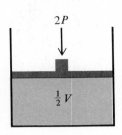
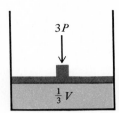

FIGURE 10-3

Illustration of Boyle's Law

The quantitative relationship between the volume and the pressure of a gas sample was first described by Robert Boyle in 1662. Boyle found that *for any given mass of gas the volume varies inversely with the pressure, provided the temperature is held constant.* In other words, if the pressure is doubled, the volume is halved; if the pressure is tripled, the volume is reduced to one-third, and so on. Figure 10-3 shows a sample of a gas confined in a cylinder with a movable piston. The total pressure required to keep the piston in place is equal to the pressure being exerted on the piston by the gas molecules. When the pressure is P, the gas in the cylinder has a volume of V. Now if the pressure is doubled to $2P$ while maintaining the temperature constant, the volume of the gas will be reduced to $\frac{1}{2}V$. Similarly, if the pressure is raised to $3P$, the volume becomes $\frac{1}{3}V$.

This relationship, known as *Boyle's Law,* may be expressed mathematically as

$$V \propto \frac{1}{P}$$

or

$$V = \frac{k}{P}$$

where k is a proportionality constant. Therefore, PV = constant, and for any gas sample that goes from a volume of V_1 and pressure of P_1 to a volume of V_2 at a pressure of P_2 (temperature and number of molecules remaining constant), one may write

$$P_1 V_1 = P_2 V_2$$

Boyle's Law enables us to make quantitative predictions about the effect that a particular pressure change will have on the volume of a gas, and vice versa.

EXAMPLE: If a certain sample of nitrogen has a volume of 250 ml at a pressure of 735 torr, what will the volume of the sample be at a pressure of 1 atm (760 torr)?

Solution 1: Using the relationship $P_1V_1 = P_2V_2$, $P_1 = 735$ torr, $V_1 = 250$ ml, $P_2 = 760$ torr, and $V_2 =$ unknown.

Substituting into the equation, we obtain

$$735 \times 250 = 760 \times V_2$$

Solving for V_2,

$$V_2 = \frac{735 \text{ torr} \times 250 \text{ ml}}{760 \text{ torr}} = 242 \text{ ml}$$

Solution 2: The same results are obtained by a "reasoning" method as follows: The volume of 250 ml is to be "corrected" for a pressure change; that is, the volume must be multiplied by a ratio of the two pressure values. The pressure change is an *increase.* Since pressure and volume vary inversely, the volume must *decrease,* and we must multiply the 250 ml by a "correction factor" *less than one.* The "correction factor" is 735/760. Therefore,

$$250 \text{ ml} \times \frac{735}{760} = 242 \text{ ml}$$

EXAMPLE: If a sample of oxygen has a volume of 800 ml at a pressure of 740 torr and a temperature of $30°C$, what pressure will be required to compress the oxygen sample to 500 ml, keeping the temperature constant?

Solution 1: $P_1 = 740$, $V_1 = 800$, $P_2 =$ unknown, $V_2 = 500$. Therefore,

$$P_2 = \frac{P_1 V_1}{V_2} = \frac{740 \text{ torr} \times 800 \text{ ml}}{500 \text{ ml}} = 1184 \text{ torr}$$

Solution 2: The same result is obtained by reasoning that the "new" pressure is simply the "old" pressure multiplied by the ratio of the two volumes. Since volume and pressure vary inversely, and the volume in this example is decreasing, the pressure must increase. Therefore, the ratio of the volumes must be a factor *greater than one,* namely, 800/500.

$$P_2 = 740 \text{ torr} \times \frac{800}{500} = 1184 \text{ torr}$$

Variation of Volume with Temperature: Charles' Law

Experiments by the French scientist Jacques Charles around 1787 resulted in a statement of the relationship between gas volume and temperature. Confirmation and elaboration of these results was published by another Frenchman, Joseph Gay-Lussac, in 1802. This relationship, usually called *Charles' Law,* resulted from the observation that different gases expand by the same fractional amount for the same rise in temperature if the pressure is held constant. It was found that the volume of a gas increases by 1/273 of its volume at $0°C$ for each degree Celsius that the temperature is raised. If, for example, we had 273 ml of a gas at $0°C$ and some pressure P, then raising the temperature to $1°C$ (keeping the pressure constant) would increase the volume of the gas by a factor of 1/273 or 1 ml, and the volume at 1°C would therefore be 274 ml. Raising the temperature to 2°C would expand the volume to 275 ml; at 10°C the volume would be 283 ml, etc. Naturally, if the temperature of the same gas sample were lowered to −1°C, its volume would contract to 272 ml, and at −10°C the volume would be 263 ml.

This relationship implies that if the temperature were lowered far enough, to −273°C, the volume of a gas would become zero. Of course, any real gas would liquefy before this low temperature was reached, but a hypothetical, nonliquefiable gas would "disappear" at −273°C, suggesting that this temperature represents a degree of coldness beyond which it is impossible to go. This temperature, more precisely −273.15°C, is called *absolute zero* and represents the zero point on the *absolute temperature scale,* also called the *Kelvin* scale after its originator, Lord Kelvin. Each degree on the absolute scale is of the same magnitude as a degree on the Celsius (centigrade) scale. Therefore, absolute temperature (°K) is related to Celsius temperature as follows: °K = °C + 273. These two scales and the Fahrenheit scale are compared in Figure 10-4.

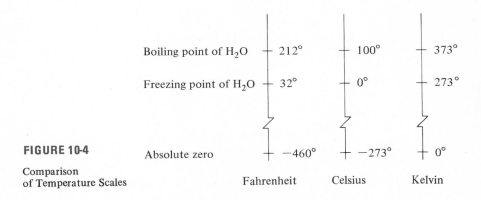

FIGURE 10-4

Comparison
of Temperature Scales

	Fahrenheit	Celsius	Kelvin
Boiling point of H$_2$O	212°	100°	373°
Freezing point of H$_2$O	32°	0°	273°
Absolute zero	−460°	−273°	0°

Charles' Law may be stated as follows: *For any given mass of gas, the volume varies directly as the absolute temperature, provided the pressure remains constant.* Expressed mathematically,

$$V \propto T \quad \text{or} \quad V = kT$$

Therefore,

$$\frac{V}{T} = \text{constant}$$

and

$$\frac{V_1}{T_1} = \frac{V_2}{T_2}$$

Charles' Law has the same usefulness in predicting volume-temperature relationships as Boyle's Law has for volume-pressure relationships.

EXAMPLE: A sample of methane gas has a volume of 1.50 liters at 25°C and 1 atm pressure. What volume will it occupy at 50°C and 1 atm pressure?

Solution 1:

$$\frac{V_1}{T_1} = \frac{V_2}{T_2}$$

$$V_1 = 1.50 \text{ liters}$$

$$T_1 = 25°C + 273 = 298°K$$

$$T_2 = 50°C + 273 = 323°K$$

$$V_2 = \text{unknown}$$

Solving for V_2 and substituting:

$$V_2 = \frac{V_1}{T_1} \times T_2 = \frac{1.50 \text{ liters}}{298°K} \times 323°K = 1.63 \text{ liters}$$

Solution 2: Since volume and temperature vary *directly*, the increase in temperature from 25°C to 50°C must result in an *increase* in volume. Therefore, the initial volume of 1.50 liters must be multiplied by the ratio of the two temperatures, and this ratio must be a factor *greater than one,* namely, 323/298.

$$1.50 \text{ liters} \times \frac{323}{298} = 1.63 \text{ liters}$$

Consideration of the effects described by Boyle's Law and Charles' Law leads to a simple relationship between pressure and temperature. Increasing the temperature tends to expand the volume (Charles' Law); if this expansion of volume is to be prevented, it can only be done by increasing the pressure (Boyle's Law). *At constant volume, the pressure and absolute temperature of a gas vary directly.* This statement is sometimes referred to as *Gay-Lussac's Law.*

EXAMPLE: The gas in a steel bomb exerts a pressure of 1 atm at 33°C. What will the pressure be if the temperature is raised to 100°C? (Note that the volume cannot change unless, of course, the bomb bursts.)

Solution: The temperature increase from 33°C (306°K) to 100°C (373°K) will cause the pressure to be raised by a factor of 373/306.

$$1 \text{ atm} \times \frac{373}{306} = 1.22 \text{ atm}$$

Let us consider now how the volume of a gas is affected by changes in both temperature and pressure. Suppose, for example, we have 100 ml of a gas at 35°C and 740 torr. What will be the volume of the gas sample at standard temperature and pressure? (Standard temperature and pressure, abbreviated STP, is defined as a

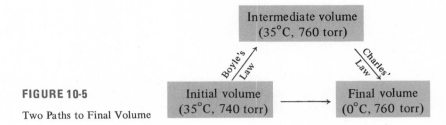

FIGURE 10-5

Two Paths to Final Volume

temperature of 0°C and a pressure of 1 atm.) If we were actually carrying out this change of conditions in the laboratory, we could change temperature and pressure simultaneously to arrive at a final volume. In this case it might appear that the gas laws are not applicable in predicting the new volume, since Boyle's Law assumes constant temperature and Charles' Law assumes constant volume. On the other hand, the same final volume will be reached if we bring about the changes in two steps. First, hold the temperature constant and change the pressure. Second, holding that *new* pressure constant, change the temperature. These alternate paths are illustrated in Figure 10-5. In the two-step process, the first step is a simple Boyle's Law correction of the initial volume, and the second step is a Charles' Law correction of the intermediate volume. The problem can be solved as follows:

1. Holding T constant, change P from 740 torr to 760 torr. Increasing pressure decreases volume.

$$100 \times \frac{740}{760} = \text{intermediate volume}$$

2. Holding P constant, change T from 308°K to 273°K and calculate the effect on the intermediate volume. Decreasing temperature causes a decrease in volume.

$$100 \times \frac{740}{760} \times \frac{273}{308} = 86.3 \text{ ml}$$

Variation of Volume with Number of Molecules: Avogadro's Law

In 1811, experiments in which he measured the volumes of gases reacting or produced in chemical reactions led Amedeo Avogadro to summarize the relationship between the volume of a gas and the number of molecules present. This relationship, sometimes called *Avogadro's Law*, may be stated as follows: *Equal volumes of all gases, at the same temperature and pressure, contain the same number of molecules.* For example, one liter of chlorine contains the same number of molecules as one liter of methane or one liter of nitrogen or one liter of any gas, provided the temperature and pressure are the same in all cases. Furthermore, two liters of chlorine contain twice as many molecules as one liter, if again the temperature and pressure are the same. In other words, according to Avogadro's Law, the volume of a gas is directly proportional to the number of molecules of that gas at constant temperature and pressure. Letting N represent the number of molecules,

$$V \propto N \quad \text{and} \quad V = kN$$

Because we are usually dealing with very large numbers of molecules, it is often more convenient to express the *quantity* of gas in terms of number of *moles* rather than number of molecules. Keeping in mind that a mole of any substance contains the same number of molecules as a mole of any other substance, we may restate Avogadro's Law: The volume of a gas is directly proportional to the *number of moles* of the gas (T and P constant, of course). Therefore,

$$V \propto n \quad \text{and} \quad V = kn$$

where n equals the number of moles.

It follows from this relationship that one mole of *any gas* occupies the same volume as one mole of any other gas at the same temperature and pressure. At *standard conditions* (0°C and 760 torr) one mole of any gas occupies 22.414 liters. This volume, which for most purposes can be rounded off to 22.4 liters, is called the *molar gas volume at standard temperature and pressure* (STP).

Avogadro's Law and the establishment of the molar gas volume enable us to relate the weight of a gas sample to its volume in a number of useful ways. We can calculate the volume occupied by a given weight of a particular gas at STP or—by applying Charles' and Boyle's Laws—at any other temperature and pressure.

EXAMPLE 1: What is the volume of 10.0 g of CO_2 at STP?

Solution: One mole of CO_2 weighs 44.0 g. Therefore,

$$10.0 \text{ g of } CO_2 = \frac{10.0}{44.0} \text{ mole of } CO_2$$

Since 1 mole occupies 22.4 liters at STP,

$$\frac{10.0 \text{ g}}{44.0 \text{ g/mole}} \times \frac{22.4 \text{ liters}}{\text{mole}} = 5.09 \text{ liters at STP}$$

EXAMPLE 2: What is the volume of 5.0 g of CH_4 at 30°C and 735 torr?

Solution: The volume of 5.0 g of CH_4 at STP, following the procedure in Example 1, is

$$\frac{5.0 \text{ g}}{16.0 \text{ g/mole}} \times \frac{22.4 \text{ liters}}{\text{mole}}$$

Correcting this volume for a temperature increase (0°C to 30°C) and a pressure decrease (760 torr to 735 torr):

$$\frac{5.0 \text{ g}}{16.0 \text{ g/mole}} \times \frac{22.4 \text{ liters}}{\text{mole}} \times \frac{303°\text{K}}{273°\text{K}} \times \frac{760 \text{ torr}}{735 \text{ torr}} = 8.0 \text{ liters}$$

The density of any gas can be readily calculated also. Since the weight of a mole and the volume of a mole at STP are known, the density of a gas (in grams per liter) at STP is simply its molecular weight divided by 22.4. The density at any other conditions of temperature and pressure can be found by applying Boyle's and Charles' "corrections" to the density at STP.

EXAMPLE: What is the density of acetylene (C_2H_2) at $100°C$ and 800 torr?

Solution: The molecular weight of C_2H_2 is 26.0; that is, 1 mole weighs 26.0 g. The density of C_2H_2 at STP, then, is

$$\frac{26.0 \text{ g/mole}}{22.4 \text{ liters/mole}}$$

Increasing the temperature from $0°C$ to $100°C$ will *decrease* the density by a factor of 273/373. Increasing the pressure from 760 torr to 800 torr will *increase* the density by a factor of 800/760. Therefore, the density of acetylene at $100°C$ and 800 torr is

$$\frac{26.0 \text{ g/mole}}{22.4 \text{ liters/mole}} \times \frac{273}{373} \times \frac{800}{760} = 0.894 \text{ g/liter}$$

Since one mole of any gas occupies the same volume as one mole of any other gas, the densities of various gases (at any stated temperature and pressure) are in the same ratio to each other as the ratios of their molecular weights. Thus, for example, the density of O_2 (mol wt 32) is sixteen times the density of H_2 (mol wt 2) at the same temperature and pressure; methane (CH_4, mol wt 16) has a density one-half that of O_2, and so on.

Avogadro's Law also permits us to reach certain conclusions regarding the volumes of gaseous substances consumed and produced in chemical reactions. A chemical equation (Chapter 3) tells us nothing about the relative volumes of substances involved in chemical reactions if those substances are in the solid or liquid states; calculations of reacting volumes would require that the densities of the substances be known. However, for gases this is not the case. Because *equal volumes of all gases, at the same temperature and pressure, contain equal numbers of molecules,* and because the coefficients in the equation represent ratios of numbers of molecules, it follows that the coefficients must also represent the ratios of volumes of gaseous substances involved in a chemical reaction. Of course, the volumes must be measured at the same temperature and pressure.

By way of illustration, consider the reaction whereby ammonia is produced by direct combination of nitrogen and hydrogen. The balanced equation for this reaction is

$$N_2(g) + 3H_2(g) \rightarrow 2NH_3(g)$$

This equation tells us that *one volume* of N_2 will combine with *three volumes* of H_2 to yield *two volumes* of NH_3, all volumes being measured at the same temperature and pressure. In other words, 100 liters of N_2 will yield 200 liters of NH_3, or 10 cubic feet of N_2 will react with 30 cubic feet of H_2, and so on.

Furthermore, because of our knowledge of the molar gas volume at STP, it is possible to relate the volume of a gaseous reactant or product to the weights of the other substances taking part in the reaction.

EXAMPLE: When solid potassium chlorate is heated, it decomposes to give oxygen, as follows:

$$2KClO_3(s) \rightarrow 2KCl(s) + 3O_2(g)$$

What volume of oxygen (measured at STP) will be produced from the decomposition of 10.0 g of $KClO_3$?

Solution: Since the formula weight of $KClO_3$ is 122.6,

$$10.0 \text{ g KClO}_3 = \frac{10.0}{122.6} \text{ moles KClO}_3$$

The equation tells us that 2 moles of $KClO_3$ yield 3 moles of O_2. Therefore, the number of moles of O_2 produced is

$$\left(\frac{10.0}{122.6} \times \frac{3}{2} \right) \text{ moles}$$

At STP, each mole of O_2 occupies a volume of 22.4 liters, and therefore the volume of oxygen produced is

$$\frac{10.0}{122.6} \times \tfrac{3}{2} \times 22.4 = 2.74 \text{ liters}$$

The volume of oxygen can be calculated for any other temperature and pressure by application of Boyle's and Charles' Laws to "correct" the 2.74 liters.

THE IDEAL GAS EQUATION

Boyle's Law ($V \propto 1/P$), Charles' Law ($V \propto T$), and Avogadro's Law ($V \propto n$) can be combined into one mathematical statement:

$$V \propto \frac{nT}{P}$$

and this statement can be written as a mathematical equation by use of a proportionality constant, which we will designate as R.

$$V = \frac{RnT}{P}$$

This equation, more commonly written in the form

$$PV = nRT$$

is called the *ideal gas equation* (or the *perfect gas equation*). It describes the relationship between pressure, volume, number of moles, and absolute temperature for an *ideal gas*. If the value of the *ideal gas constant, R,* is known, then, knowing the values of any three of the four variables—P, V, n, T—one can calculate the value of the fourth from this equation.

The ideal gas constant, R, can be evaluated rather simply from our knowledge of the molar gas volume at STP: 1 mole of gas occupies 22.4 liters at STP. In other words, when $n = 1$ mole, $P = 1$ atm, and $T = 273°K$, then $V = 22.4$ liters. Solving the ideal gas equation for R and substituting these values into it gives:

$$R = \frac{PV}{nT} = \frac{1 \text{ atm} \times 22.4 \text{ liters}}{1 \text{ mole} \times 273°K} = 0.0821 \frac{\text{liter-atm}}{\text{mole-}°K}$$

The magnitude and units of R depend on the units in which the variables are expressed. If, for example, we had used 760 torr for P, then the value of R would be 62.4 liter-torr/mole-$°K$. The value of 0.0821 liter-atmospheres per mole-degree Kelvin is the value most commonly used for R.

Now, with R evaluated, the ideal gas equation can be employed to solve a variety of problems by direct substitution of the proper figures. It should be noted, however, that the ideal gas equation does not enable us to solve any "new" kinds of problems. It is simply an alternate route to the solution of problems that can be solved by the application of the individual gas laws using the method illustrated in various earlier examples.

Let us now consider the solution of a number of problems using the ideal gas equation, including several that were solved earlier.

EXAMPLE: A certain gas sample has a volume of 100 ml at $35°C$ and 740 torr. What will be the volume of the sample at STP?

Solution: Using the expression $PV = nRT$, all quantities are known about the original gas except n, the number of moles.

$$P = 740 \text{ torr} = \frac{740}{760} \text{ atm}$$

$$V = 100 \text{ ml} = 0.100 \text{ liter}$$

$$T = 35°C = 308°K$$

$$R = 0.0821 \text{ liter-atm/mole-}°K$$

Substituting these quantities into the equation and solving for n,

$$n = \frac{PV}{RT} = \frac{740}{760} \times \frac{0.100}{0.0821 \times 308} = 0.00385 \text{ mole}$$

Now, changing P and T will change V but will have no effect on n. Therefore, under the final conditions:

$$P = 1 \text{ atm}$$

$$n = 0.00385 \text{ mole}$$

$$T = 0°C = 273°K$$

$$R = 0.0821$$

Solving the ideal gas equation for V and substituting:

$$V = \frac{nRT}{P} = \frac{0.00385 \times 0.0821 \times 273}{1}$$

$$0.0863 \text{ liter} = 86.3 \text{ ml}$$

Compare this with the solution to the same problem on p. 225.

EXAMPLE: What is the volume of 5.0 g of CH_4 at $30°C$ and 735 torr?

Solution:

$$PV = nRT$$

Solving for V:

$$V = \frac{nRT}{P}$$

$$n = \frac{5.0 \text{ g}}{\text{mol wt } CH_4} = \frac{5.0 \text{ g}}{16 \text{ g/mole}} = \frac{5.0}{16} \text{ moles}$$

$$R = 0.0821 \text{ liter-atm/mole-}°K$$

$$T = 30°C = 303°K$$

$$P = 735 \text{ torr} = \frac{735}{760} \text{ atm}$$

Therefore,

$$V = \frac{5.0}{16} \times 0.0821 \times 303 \times \frac{760}{735} = 8.0 \text{ liters}$$

Compare with the solution on p. 226.

EXAMPLE: What is the density of acetylene (C_2H_2) at $100°C$ and 800 torr?

Solution: Densities of gases are usually expressed in terms of grams per liter. Therefore, what we are seeking in this problem is the weight (in grams) of one liter of the gas, which can be expressed as g/V. The relationship between the weight (in grams) of any substance and the number of moles of that substance is

$$n = \frac{\text{grams}}{\text{grams/mole}} = \frac{g}{M}$$

where M is the molecular weight of the substance. Hence, the ideal gas equation may be expressed in the form

$$PV = \frac{g}{M}RT$$

The problem, then, is to find the value of g/V when

$$P = 800 \text{ torr} = \frac{800}{760} \text{ atm}$$

$$R = 0.0821 \text{ liter-atm/mole-}^\circ\text{K}$$

$$T = 100^\circ\text{C} = 373^\circ\text{K}$$

$$M = 26.0 \text{ g/mole (the molecular weight of } C_2H_2)$$

Solving for g/V and substituting:

$$\frac{g}{V} = \frac{PM}{RT} = \frac{800}{760} \times \frac{26.0}{0.0821 \times 373} = 0.894 \text{ g/liter}$$

Compare with the solution on page 227.

EXAMPLE: What weight of N_2 will occupy a volume of 2.5 liters at a pressure of 1 atm and a temperature of 200°C?

Solution:

$$PV = \frac{g}{M}RT$$

$$P = 1 \text{ atm}$$

$$V = 2.5 \text{ liters}$$

$$T = 473^\circ\text{K}$$

$$M = 28.0 \text{ g/mole}$$

$$g = \frac{PVM}{RT} = \frac{1 \times 2.5 \times 28.0}{0.0821 \times 473} = 1.8 \text{ g}$$

One of the major practical applications of the ideal gas laws is in the experimental determination of molecular weights of gases or very volatile liquids. Although a number of different techniques can be used, they all involve the measurement of gas density; that is, the accurate measurement of the weight and the volume of a gas sample and observation of the temperature and pressure at the time of measurement. The general procedure may be illustrated as follows: A glass bulb, of accurately known volume and fitted with a stopcock, is evacuated and weighed. The bulb is then filled with the gas under investigation, the temperature is noted, and the pressure of the gas in the bulb is measured. Finally, the filled bulb is weighed in order to determine the weight of the gas sample. The molecular weight of the gas can be determined by applying the ideal gas equation to the measured data, as in the following illustration:

Weight of bulb + sample	20.3951 g
Weight of evacuated bulb	19.8765 g
Weight of gas sample	.5186 g

Temperature = 25.0°C = 298°K

$$\text{Pressure} = 742 \text{ torr} = \frac{742}{760} \text{ atm}$$

Volume = 203 ml = 0.203 liter

$$PV = \frac{g}{M}RT$$

$$M = \frac{gRT}{PV} = \frac{0.5186 \times 0.0821 \times 298}{(742/760) \times 0.203} = 64.0 \text{ g/mole}$$

DALTON'S LAW OF PARTIAL PRESSURES

In a mixture of gases, each gas exerts the same pressure as it would if it were in the container alone. The pressure exerted by each gaseous constituent is called its *partial pressure,* and the total pressure of the mixture is the sum of the partial pressures. This observation, first published by John Dalton in 1803, is called *Dalton's Law of Partial Pressures.*

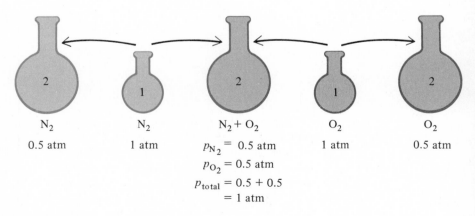

FIGURE 10-6

Illustration of Dalton's Law of Partial Pressures

The significance of these statements may be illustrated as follows (Figure 10-6). Suppose we have two 1-liter containers, one filled with N_2 at 1 atm pressure, and the other filled with O_2 also at 1 atm. Suppose further that we have a completely empty 2-liter flask. Now, if the 1 liter of N_2 is in some way transferred into the empty 2-liter flask, we will then have 2 liters of N_2 at a pressure of 0.5 atm. (The number of molecules has remained constant but the volume of the N_2 has doubled, and therefore the pressure is halved.) In the same way, if we were to transfer the O_2 to the empty 2-liter flask we would have 2 liters of O_2 at 0.5 atm. Now suppose that we transfer the contents of *both* 1-liter flasks into the empty 2-liter flask. We will then have 2 liters of a mixture of N_2 and O_2. The N_2 in the mixture is exerting the same pressure that it would if it were in the flask alone—0.5 atm. Therefore, the *partial pressure* of N_2 (symbolized p_{N_2}) is 0.5 atm. The O_2 in the mixture also has a *partial pressure* (p_{O_2}) of 0.5 atm. The total pressure, then, is 0.5 atm + 0.5 atm = 1 atm.

One very useful application of Dalton's Law is in calculations involving gases collected over water or other liquids. A convenient method of collecting gases produced by chemical reactions is by displacement of water. A delivery tube carries the gas to the collection vessel, which is filled with water and inverted in a pan of water (Figure 10-7). The gas, being much less dense than water, bubbles to the top of the vessel, pushing water out of the vessel into the pan. (Naturally, if the gas in question undergoes a chemical reaction with water, or is very soluble in water, then some other liquid must be used.)

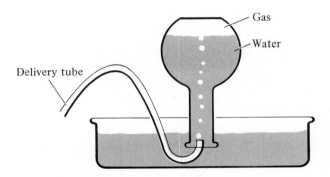

FIGURE 10-7

Collection of a Gas by Displacement of Water

A gas collected over water in this manner is not pure; it is saturated with water vapor. The pressure of the collected sample is not the pressure exerted by the gas itself, but the sum of the partial pressures of the gas and water vapor ($p_{\text{total}} = p_{\text{gas}} + p_{H_2O}$). The partial pressure of the water vapor is dependent on the temperature. In fact, it is the *vapor pressure* of water at that particular temperature. (Recall the discussion of the vapor pressure of liquids, p. 203, and see Appendix 12 for a table of vapor pressures of water at various temperatures.) The application of this concept is illustrated in the following example.

EXAMPLE: 275 ml of oxygen is collected over water at a temperature of $23°C$ and a pressure of 700 torr. What weight of O_2 has been collected?

Solution: We can use expression

$$PV = \frac{g}{M} RT$$

with the values

$$V = 0.275 \text{ liter}$$

$$M = 32 \text{ g/mole}$$

$$T = 296°K$$

However, P in the expression refers to the pressure of oxygen, which in this case is the *partial* pressure of oxygen, p_{O_2}, and 700 torr is the combined pressure of oxygen and water vapor, or

$$p_{O_2} + p_{H_2O} = 700 \text{ torr}$$

The vapor pressure of water at $23°C$ is 21 torr. Therefore,

$$p_{O_2} = 700 - 21 = 679 \text{ torr} = \frac{679}{760} \text{ atm}$$

Solving for g and substituting:

$$g = \frac{PVM}{RT} = \frac{679}{760} \times \frac{0.275 \times 32}{0.0821 \times 296} = 0.324 \text{ g}$$

GRAHAM'S LAW OF DIFFUSION

Diffusion is the ability of a gas to spread out spontaneously and to move through another gas until it completely fills the container. For example, suppose a container of a particularly odorous gas, say hydrogen sulfide, were opened in the front of a crowded lecture hall. Those people in the front row would soon detect it, and gradually the odor would become noticeable to people sitting farther and farther back, until eventually the gas could be smelled in every spot in the room. The H_2S molecules would spontaneously make their way throughout the entire room, moving between and intermingling with the gaseous molecules of the air.

The *rate of diffusion* of a gas at any given temperature and pressure is dependent on its molecular weight. This relationship was investigated by the Scottish chemist Thomas Graham, who in 1833 formulated his *Law of Diffusion of Gases: At the same conditions of temperature and pressure, the rates of diffusion of gases are inversely proportional to the square roots of their molecular weights.* Expressed mathematically, Graham's Law states

$$\frac{r_1}{r_2} = \sqrt{\frac{M_2}{M_1}}$$

where r_1 and r_2 are the rates of diffusion of gas 1 and gas 2, respectively, and M_1 and M_2 are their molecular weights.

Graham's Law provides another method for experimentally determining the molecular weight of a gas. If you measure, at constant temperature and pressure, the rates of diffusion of the gas under investigation and some known gas (O_2 or N_2, for example), you can then insert these rates and the molecular weight of the known gas in the equation and solve for the molecular weight of the "unknown" gas. While this method was of some historical importance, it has very little usefulness today.

Mixtures of gases can sometimes be separated into their constituents by taking advantage of their differing rates of diffusion. The greater the difference in the molecular weights of two gases, the more easily they can be separated by diffusion. An example of an important application of this principle is the separation of the two principal isotopes of uranium, $^{235}_{92}U$ and $^{238}_{92}U$. Since ^{235}U undergoes nuclear fission, it was important to the development of nuclear energy to obtain this isotope in relatively pure form. However, ^{235}U accounts for only about 0.7% of the uranium atoms in nature, almost all the rest being ^{238}U, and the isolation of ^{235}U is extremely difficult. It has been accomplished, however, by converting uranium to uranium hexafluoride (UF_6), which is a gas above $56°C$, and permitting the gas to diffuse through a porous plate. The molecules containing ^{235}U atoms, being of lower mass, diffuse somewhat faster than the molecules containing ^{238}U atoms, so that after a time the gas mixture that has diffused through the plate contains a greater proportion of ^{235}U than that which has not diffused. By allowing part of the ^{235}U-enriched mixture to diffuse through the porous plate again, the proportion of ^{235}U is increased still further. After numerous repetitions of this process, a separation is achieved.

THE KINETIC MOLECULAR THEORY

It must be emphasized that all the laws of gas behavior that have been discussed in the preceding pages were the products of experimentation. They represent statements of fact based on observations of the behavior of matter in bulk. Some of the gas laws were firmly entrenched many years before Dalton's atomic theory, and their authors knew nothing of molecules; all of them predated the concept of intermolecular forces. (In fact, it was the failure of gases to obey the ideal gas laws at high pressures and low temperatures that led van der Waals to suggest the existence of intermolecular forces in the 1870s.) Furthermore, these early workers did not possess the means of achieving either very high pressures or very low temperatures, and the term "ideal" gas could not possibly have occurred to them. They were simply describing the behavior of gases as they knew them.

As more was learned about the *particulate* nature of matter, attempts were begun to explain the macroscopic properties of gases in terms of the particles of which matter is composed. Around the middle of the nineteenth century, there emerged a series of postulates aimed at explaining the behavior of gases as described by the gas laws. These postulates taken together are known as the *kinetic molecular theory*. The major points of this theory follow:

1. Gases are composed of molecules, which are in rapid, random, straight-line motion (translational motion). Because they have mass and motion, molecules possess kinetic energy ($KE = \frac{1}{2}mv^2$).
2. The molecules undergo collisions with the walls of the container and with each other. These collisions are perfectly *elastic;* that is, there is no net loss of kinetic energy during the collision. Although there may be a transfer of kinetic energy between two colliding molecules, their *average* kinetic energy remains the same after the collision as it was before.
3. The molecules are separated by average distances much larger than the molecules themselves. Thus the space occupied by the molecules is negligible compared with the volume of the container.
4. The molecules exert no attractive or repulsive forces on one another.
5. At any given instant, the molecules do not all have the same kinetic energy, but the *average* kinetic energy of the molecules is directly proportional to the absolute temperature. It follows that at any given temperature the average kinetic energy of the molecules is the same for all gases. Since $KE = \frac{1}{2}mv^2$, the higher the molecular weight of a gas the lower its velocity must be.

The kinetic molecular theory provides an interpretation of gas pressure on a molecular level. The constant, rapid motion of the molecules results in countless collisions with the walls of the container during each tiny fraction of a second. Since molecules have mass, these collisions are constantly exerting a *force* on the container's walls, and pressure is simply force per unit of surface area. The pressure exerted by a gas sample, then, will depend on the number of molecules colliding with the container walls and the kinetic energy of the molecules undergoing the collisions (therefore, the temperature).

With this interpretation of pressure and the postulates stated above, one can readily rationalize ideal gas behavior. Consider a gas confined in a container that will permit variable volume; for example, a cylinder fitted with a piston. There is some definite volume of gas, V_1, in the container, consisting of N_1 molecules, and at pressure P_1 and temperature T_1. Now suppose we wish to increase the pressure of the gas to P_2. There are several ways this might be accomplished. We could add more molecules, thus increasing the number of collisions per unit time, or we could raise the temperature, which would raise the average kinetic energy of the molecules, causing more frequent and more forceful collisions. Either of these changes would raise the pressure. The conditions of Boyle's Law, however, require that both the temperature and the number of molecules remain constant. The only other way that the pressure can increase—that is, that the force exerted by the molecules per unit area can increase—is for the volume of the gas to *decrease* to V_2. We will then have more molecules per unit volume, although the total number of molecules has remained constant. Thus, we have accounted for *Boyle's Law:* at constant temperature, for a given number of molecules, pressure and volume vary inversely. See Figure 10-8.

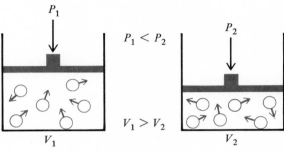

FIGURE 10-8

Rationalization of Boyle's Law

$$V \propto \frac{1}{P} \ (N, T \text{ constant})$$

The explanation of *Charles' Law* follows a similar line of reasoning. Imagine the same volume of gas, V_1, in the same cylinder under the same conditions, P_1, T_1, N_1. Now, without changing the number of molecules, we increase the temperature of the gas to T_2. This causes an increase in the kinetic energy of the molecules so that they make more frequent collisions and their collisions are more forceful. If the volume cannot change, the pressure will necessarily increase. But Charles' Law deals with a constant pressure. How can the pressure remain constant even though the molecular collisions occur with greater force? The surface area over which the collisions occur must become larger; the volume must *increase*. This rationalization

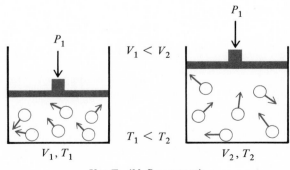

FIGURE 10-9

Rationalization of Charles' Law

$$V \propto T \ (N, P \text{ constant})$$

of Charles' Law is depicted in Figure 10-9, where increased molecular velocity is indicated by longer arrows on the molecules.

Avogadro's Law states that volume and number of molecules are directly proportional at constant pressure and temperature. Starting with N_1 molecules of gas in volume V_1, we increase the number of molecules to N_2. If N_2 molecules now occupy volume V_1, there will be more collisions (increased pressure) unless the temperature is lowered. The only way the number of molecules can be increased without an increase in pressure or a decrease in temperature (or both) is for the volume to increase so that the *number of molecules per unit volume* remains as it was at the start (Figure 10-10).

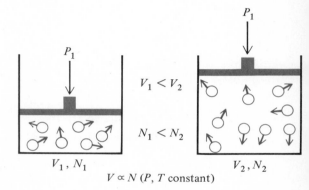

FIGURE 10-10

Rationalization of Avogadro's Law

$V \propto N \ (P, T \text{ constant})$

Dalton's Law of Partial Pressures presents no problems to the kinetic molecular theory. If there are no attractive forces between molecules, then each molecule strikes the container walls with the same frequency and force as it would if no other molecules were present. The pressure exerted by a gas in a mixture is the same as if no other gas were present.

The relationship described in *Graham's Law of Diffusion* is predictable from the kinetic molecular theory. Consider two different gases at the same temperature. The average kinetic energy of the molecules of gas 1 is given by the expression:

$$\text{KE}_1 = \tfrac{1}{2}m_1 v_1{}^2$$

Similarly, the average kinetic energy of the molecules of gas 2 is:

$$\text{KE}_2 = \tfrac{1}{2}m_2 v_2{}^2$$

Now, according to the kinetic molecular theory, the average kinetic energy of the molecules is the same for all gases *at the same temperature*. Therefore,

$$\text{KE}_1 = \text{KE}_2$$

and

$$\tfrac{1}{2}m_1 v_1{}^2 = \tfrac{1}{2}m_2 v_2{}^2$$

Rearranging this expression gives

$$\frac{v_1{}^2}{v_2{}^2} = \frac{m_2}{m_1}$$

or

$$\frac{v_1}{v_2} = \sqrt{\frac{m_2}{m_1}}$$

Since the rate of diffusion of a gas is proportional to the average molecular velocity, the ratio of the molecular velocities of the two gases is directly proportional to the ratio of their rates of diffusion (r_1 and r_2). Therefore,

$$\frac{v_1}{v_2} = \frac{r_1}{r_2}$$

Furthermore, the molecular weight of a substance is directly proportional to the mass of a single molecule of that substance, and therefore

$$\frac{m_2}{m_1} = \frac{M_2}{M_1}$$

We have derived Graham's Law from theoretical considerations:

$$\frac{r_1}{r_2} = \sqrt{\frac{M_2}{M_1}}$$

REAL GASES: DEVIATIONS FROM IDEALITY

The kinetic molecular theory adequately explains the behavior of gases as described by the gas laws with which we have dealt in this chapter. A gas that would adhere to these laws under any conditions, no matter how extreme, is called an *ideal gas*. An ideal gas is purely hypothetical; no such gas really exists. Many real gases do obey the gas laws quite closely when the temperature is relatively high and the pressure relatively low, but at low temperatures and high pressures all gases deviate from ideal behavior.

Suppose, for example, we pose the question: What volume will 0.50 mole of a gas occupy at a pressure of 740 torr and a temperature of 50°C? The ideal gas equation will permit us to calculate an answer to our question. Now, suppose we go into the laboratory and subject 0.50 mole of a gas to a temperature of 50°C and a pressure of 740 torr, and actually measure its volume as accurately as we can. We will find that the *measured* volume and the *calculated* volume agree very closely. The gas is behaving just as the ideal gas laws predict it will behave; it is behaving ideally. Suppose, then, we pose another question about the same gas: What volume will 0.50 mole occupy at a pressure of 100 atm at −100°C? Again, we can calculate the answer using the ideal gas equation. However, if we actually measure the

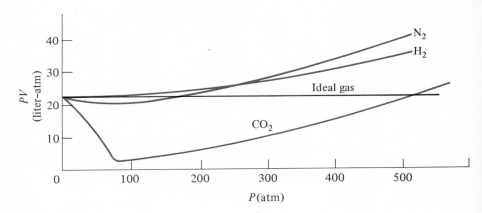

FIGURE 10-11

PV versus P for H_2, N_2, and CO_2 ($T = 273°K; n = 1$ mole)

volume of 0.50 mole of our gas at 100 atmospheres and $-100°C$, the *measured* volume and the *calculated* volume will differ. The actual volume may be less or greater than that predicted by the ideal gas equation, depending on the specific gas used in the experiment. At these conditions of high pressure and low temperature, the gas obviously is not adhering to the gas laws—it is *deviating from ideality*.

The deviation of gases from ideality with increasing pressure may be shown graphically by plotting the product of the pressure and volume (PV) against pressure (P). Assume we have taken one mole of gas at $0°C$, measured its volume at a given pressure, and multiplied these together to get PV. We then increase the pressure to some new value, measure the volume at that pressure, determine PV again, and so on. We can then plot PV as a function of P, as in Figure 10-11.

Since the temperature and number of moles of gas are held constant throughout these measurements, for an ideal gas, $PV = nRT = $ constant $= 22.4$ liter-atm. The plot for an ideal gas is shown as a straight line in Figure 10-11. However, whereas the curves for the three gases show ideal behavior at very low pressures, they deviate drastically as the pressure is increased. It will be noted that hydrogen deviates steadily in the positive direction—the real volume is *larger* than that predicted by the ideal gas laws—but that both nitrogen and carbon dioxide deviate first in a negative direction and then, at higher pressures, in a positive direction.

The effect of temperature on deviation from ideality is illustrated in Figure 10-12, where plots of PV versus P at three different temperatures are given for methane and are compared with the plots for an ideal gas at each temperature (broken lines). Note that as the temperature becomes lower, the extent of deviation increases.

This nonideal behavior of real gases is completely unpredictable from the kinetic molecular theory. Clearly, the theory must contain assumptions that are reasonably valid at "normal" temperatures and pressures but invalid at *low temperatures* and *high pressures*.

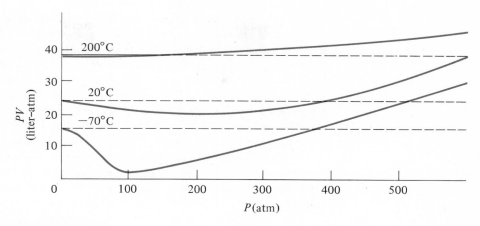

FIGURE 10-12

PV versus *P* for CH_4 at Different Temperatures

One of these assumptions is that the volumes of the molecules themselves are negligible. At conditions of high temperature and low pressure this is a reasonable assumption to make, for under these conditions the average space between molecules is so great that only a tiny fraction of the total volume of a gas sample is occupied by the molecules themselves (Figure 10-13a). However, at high pressure and/or low temperature—conditions that decrease the volume—the molecules are much closer together and their volumes cannot be neglected (Figure 10-13b). The gas laws' predictions concerning the volume of a gas are really predictions concerning the volume of the *space between* molecules. When conditions are such that the volumes of the molecules themselves are *not negligible,* the volume of a real gas is larger than is predicted for an ideal gas, and this results in a positive deviation from ideality in a *PV* plot (Figures 10-11 and 10-12).

A second assumption of the kinetic molecular theory is that there are no attractive forces between molecules. This is, of course, incorrect. As we have seen in the preceding chapter, intermolecular forces—however weak they may be—exist between all molecules. It is reasonable to assume that these forces may be ignored when the temperature is relatively high and the pressure low. Under these condi-

FIGURE 10-13

Effect of Temperature and Pressure
on Negligibility of Molecular Volumes

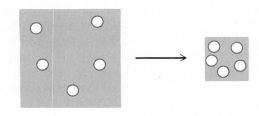

(a) Hight *T*, low *P*   (b) Low *T*, high *P*

tions the molecules are moving so rapidly and the average distance between them is so large that the effect of intermolecular forces may indeed be negligible. However, high pressure and low temperature bring the molecules closer together and reduce their velocity so that intermolecular attractions may no longer be discounted. These attractive forces pull the molecules closer together, reducing the intermolecular spaces to a greater extent than is predicted by the ideal gas laws and giving the real gas a *smaller* volume than an ideal gas would have. This effect appears as a negative deviation on a plot of *PV* versus *P* (Figures 10-11 and 10-12).

Thus we have two factors responsible for the deviation of real gases from ideal behavior. And these two factors are in competition in the sense that they tend to cause deviations in opposite directions. The actual deviation shown by a specific gas is the result of a combination of these factors and depends on the size of the molecules and the strength of the intermolecular forces.

The relationship between deviation, molecular volume, and intermolecular forces can be illustrated by a consideration of the three gases shown in Figure 10-11. The hydrogen molecule, possessing only two electrons, has extremely weak van der Waals forces. (Recall that the strength of van der Waals forces increases as the number of electrons in the molecule increases.) As a result, their effect is not observed at $0°C$, and the only deviation shown must be ascribed to the volumes of the molecules. But the hydrogen molecule is also very small, and the effect of volume would not be expected to be great. This accounts for the fact that of the three gases being compared, hydrogen shows the least deviation. The nitrogen molecule has both a larger volume and stronger van der Waals forces than hydrogen. At pressures up to approximately 100 atmospheres, the forces are more important than the volume, and we see a negative deviation. At still higher pressures, however, the molecules are so close together that the effect of their volumes counteracts the effect of van der Waals forces and the direction of deviation is reversed. The fact that at very high pressures the deviation of N_2 is greater than that of H_2 can be accounted for by the nitrogen molecule's greater size. Carbon dioxide has much stronger intermolecular forces than either of the other two gases and therefore shows the greatest negative deviation, but even this gas at sufficiently high pressures shows the effect of molecular volume.

Figure 10-12 illustrates well the effect of temperature on the importance of van der Waals forces. As the temperature is lowered, the velocity of the molecules decreases. This in turn permits an increase in the effectiveness of the intermolecular attractions and causes a greater decrease in the volume.

An ideal gas could not be liquefied because it would have no intermolecular forces. Any real gas, however, will liquefy if subjected to sufficiently high pressure and sufficiently low temperature. In order for a gas to become a liquid, the kinetic energy of the molecules must be reduced to the point at which the intermolecular forces can "take over." Therefore, for each gas there is a temperature above which it will not liquefy no matter how high a pressure is applied. This temperature is called the *critical temperature*. At temperatures below the critical temperature a gas can be liquefied by application of sufficient pressure, and the lower the temperature the less pressure is required. The pressure necessary to liquefy a gas at its critical temperature is called its *critical pressure*. The critical temperature of a gas is,

Gas	Critical Temperature (°K)	Critical Pressure (atm)
H_2O	647	217.7
NH_3	406	111.5
HCl	224	81.6
O_2	153	49.7
N_2	126	33.5
H_2	33	12.8
He	5	2.3

TABLE 10-1
Some Critical Temperatures and Pressures

of course, related to the strength of its intermolecular forces. The stronger these forces, the more easily the gas can be liquefied and therefore the higher is its critical temperature. In general, polar substances can be expected to have higher critical temperatures than nonpolar ones, and among nonpolar substances—where only van der Waals forces exist—higher molecular weights can be expected to coincide with higher critical temperatures. This relationship is illustrated in Table 10-1.

Transitions between States

Throughout our discussion of matter in bulk we have referred to the processes that occur when a substance changes from one physical state into another. These transitions between states are summarized in Figure 10-14. Transitions that occur in the direction from left to right in the diagram are endothermic—energy is

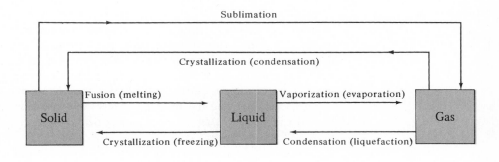

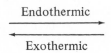

FIGURE 10-14

Transitions between States

absorbed—while those that occur from right to left are exothermic—energy is released.

ENTHALPY

The absorption of evolution or energy accompanying any change (chemical or physical) may be conveniently described in terms of the enthalpy of the system. Each substance at any given set of conditions possesses a certain amount of energy called its *heat content* or *enthalpy*, represented by the symbol *H*. Heat lost or gained during any chemical or physical change represents a change in enthalpy of the system. If heat is absorbed, then the system at its *final* state (after the change) must have a higher enthalpy than it did in its *initial* state (before the change). Conversely, evolution of heat during a change means the enthalpy of the initial state is higher than that of the final state. Thus the *change in enthalpy* (ΔH) is defined as the sum of the enthalpies of all the substances in the final state *minus* the sum of the enthalpies of the substances in the initial state. For example, in the hypothetical process in which substances A and B undergo some sort of change resulting in C and D (represented as A + B → C + D), the enthalpy change for that process is:

$$\Delta H = (H_C + H_D) - (H_A + H_B)$$

Note that a higher enthalpy in the final state than the initial results in ΔH being *positive*, but if the initial state has the higher enthalpy, then ΔH is *negative*. In other words, a positive ΔH represents an endothermic change; a negative ΔH represents an exothermic change. The magnitude of ΔH depends, of course, not only on the particular substances involved in the change but also on the quantities of these substances. Generally, ΔH is expressed in terms of calories (or kilocalories) per mole.

This concept is applicable to all the kinds of changes that matter undergoes (it will be dealt with in connection with chemical reactions in Chapter 13), including transitions between states. The enthalpy changes accompanying the various transitions are assigned names that identify the particular transition. Thus the number of calories per mole involved in the melting of a solid is called the heat of fusion (ΔH_{fus}), and the enthalpy changes of some other transitions of state are heat of sublimation (ΔH_{sub}), heat of vaporization (ΔH_{vap}), heat of crystallization (ΔH_{cryst}), and so forth.

The enthalpy change for the transition from one physical state to another is numerically equal to the enthalpy change for the reverse transition, but of opposite sign. Thus, for example, for the *melting* of ice at 0°C and 1 atm, $\Delta H = +1.44$ kcal/mole; for the *freezing* of water at the same conditions, $\Delta H = -1.44$ kcal/mole.

Because endothermic processes result in an increase in the energy of the system, while exothermic processes lead to a decrease in energy, and because our everyday experiences suggest that any system should tend to go in the direction of lowest energy (water runs downhill, clock springs wind down, hot water cools to the surrounding temperature), there is a temptation to assume that exothermic processes (ΔH is negative) occur spontaneously while endothermic ones (ΔH is positive) do not. This assumption is incorrect. Although many exothermic changes are spontaneous, so are many endothermic ones. A piece of ice melts spontaneously at room temperature—an endothermic change. Solid carbon dioxide (Dry Ice) sublimes spontaneously—another endothermic process. In fact, if spontaneity were determined by enthalpy change alone, all substances would tend toward the solid state, for that is the state with lowest enthalpy.

ENTROPY

While the enthalpy change is certainly a factor in determining spontaneity of a change, it is not the only factor. Spontaneity is also determined by another property of matter called *entropy. Entropy*, represented by the symbol *S, is a measure of the degree of randomness or disorder in a system.* It is a kind of probability factor, for there are many more ways of producing disordered arrangements than there are of producing ordered ones. Suppose one throws a handful of pennies into the air, allowing them to fall back onto a table. The probability that the pennies will fall into some neat arrangement of columns and rows, or into a single vertical pile, or into any other orderly arrangement, is extremely slight. Time after time they will fall in some random, disordered way. What is the probability that all the seedlings produced from a maple tree in the forest will be growing in straight, evenly spaced rows? What is the probability that each of four hands of cards, dealt from a shuffled deck, will contain all the cards of a single suit? In short, disorder is far more probable than order, and so disorder is favored in nature.

Each substance at a given set of conditions possesses a certain entropy—a measure of the degree of disorder in that substance—and *when a substance undergoes a physical or chemical change, there is a change in the entropy of the system.* The entropy change (ΔS)—most commonly expressed in terms of calories per mole per degree K—is defined in a way similar to that for enthalpy change; the entropy change is equal to the sum of the entropies of all the substances in the *final* state *minus* the sum of the entropies of the substances in the *initial* state.

$$A + B \rightarrow C + D$$

$$\Delta S = (S_C + S_D) - (S_A + S_B)$$

Note that when the entropy of the final state is greater than the entropy of the initial state, ΔS is positive; when the reverse is true (when entropy has decreased), ΔS is negative. Since disorder is favored over order, and since ΔS is a measure of the *increase* in disorder, then changes which have a *positive* ΔS are *favored* changes, while those for which ΔS is *negative* are *not favored.*

The application of the concept of entropy to transitions between states is readily apparent. Certainly a crystalline solid is more ordered than either a liquid or a gas, and the gaseous state is the most disordered of the three. Disorder (therefore, entropy) increases from solid to liquid to gas. The changes represented by fusion, vaporization, and sublimation have positive entropy changes, while for crystallization and condensation ΔS is negative. Referring to Figure 10-14, note that each transition occurring from left to right is favored by the entropy change but not favored by the enthalpy change, while transitions from right to left in the figure are favored by the enthalpy change but not favored by the entropy change.

FREE ENERGY

It is the combination of these two factors—enthalpy and entropy—with temperature that determines whether or not a given change will occur spontaneously. The property that represents this combination is called the *free energy,* represented by G. Each substance at a specific set of conditions has a free energy, which is defined as

$$G = H - TS$$

where H and S are the enthalpy and entropy, respectively, and T is the absolute temperature. The free energy, G, may be thought of as the freely *available* energy, and the product of temperature and entropy, TS, as the *unavailable* energy. Thus, the above relationship states that the available energy is less than the enthalpy by the amount TS. The temperature of the system enters into the degree of disorder of the system, for the higher the temperature the greater the molecular motion, and this motion increases the disorder.

The change in free energy (ΔG) that accompanies a physical or chemical change is defined in the same way that ΔH and ΔS are defined. For the process

$$A + B \rightarrow C + D$$

the change in free energy is

$$\Delta G = (G_C + G_D) - (G_A + G_B)$$

It follows that for a change occurring at constant temperature, the free-energy change may be defined as:

$$\Delta G = \Delta H - T\Delta S$$

This is a very significant equation, for it permits us to predict whether or not a process is spontaneous. *A change can occur spontaneously only if there is a decrease in free energy, that is, only if* ΔG *is negative.*

If a process is *favored* by both the enthalpy change and the entropy change, then ΔH is negative and ΔS is positive and, according to our equation, ΔG must be negative, indicating a spontaneous change.

$$\Delta G = (-) - (+) = -$$

Similarly, if a process is not favored by either ΔH or ΔS (ΔH is positive and ΔS negative) then ΔG must be positive and the change in question will not occur spontaneously, but its reverse will.

$$\Delta G = (+) - (-) = +$$

In those cases where a process is favored either by enthalpy *or* entropy but not both, spontaneity is determined by which of the two factors has the greater effect. For example, where a positive ΔS is sufficiently large, it may outweigh a positive ΔH and lead to a spontaneous process. This explains why some endothermic processes are spontaneous (for example, ice melting at room temperature).

When a system is in equilibrium (when a change has the same tendency to occur in both directions), $\Delta G = 0$.

In order to illustrate the application of the free-energy change, let us make some simple calculations regarding the compound benzene. The heat of vaporization (ΔH_{vap}) for benzene is 7.34 kcal/mole, and its entropy of vaporization (ΔS_{vap}) is 20.8 cal/mole-$^{\circ}$K. (The values of these quantities vary somewhat with temperature, but the variations are small enough to permit us to make these illustrative calculations without introducing major errors.)

EXAMPLE: Is benzene a liquid or a gas at 100°C? (That is, does the following process occur spontaneously at 100°C?)

Benzene (liquid) $\rightarrow$ Benzene (gas)

Solution:

$$\Delta G = \Delta H - T\Delta S$$
$$= 7340 \text{ cal/mole} - (373^{\circ}\text{K})(20.8 \text{ cal/mole-}^{\circ}\text{K})$$
$$= 7340 - 7758 = -418 \text{ cal/mole}$$

Therefore, the process *is* spontaneous, and benzene exists as a gas at 100°C.

EXAMPLE: Is benzene a liquid or a gas at 50°C?

Solution:

$$\Delta G = \Delta H - T\Delta S$$
$$= 7340 \text{ cal/mole} - (323^{\circ}\text{K})(20.8 \text{ cal/mole-}^{\circ}\text{K})$$
$$= 7340 - 6718 = 622 \text{ cal/mole}$$

Since ΔG is positive, we may conclude that the vaporization process is not spontaneous, and benzene is a liquid at 50°C.

EXAMPLE: What is the boiling point of benzene?

Solution: At the boiling point the liquid and gaseous states are in equilibrium, and $\Delta G = 0$. The question is, then: At what temperature is ΔG for the vaporization process equal to zero?

$$\Delta G = \Delta H - T\Delta S = 0$$

$$T\Delta S = \Delta H$$

$$T = \frac{\Delta H}{\Delta S} = \frac{7340 \text{ cal/mole}}{20.8 \text{ cal/mole-}^\circ K} = 353^\circ K$$

The boiling point of benzene is $353^\circ K$, or $353 - 273 = 80^\circ C$.

Enthalpy, entropy, and free energy are often called *thermodynamic* properties, and their relationships are embodied in the *Laws of Thermodynamics*. These are discussed in connection with chemical reactions in Chapter 13.

SUGGESTED READINGS

Neville, Roy G. "The Discovery of Boyle's Law, 1661–62." *Journal of Chemical Education,* Vol. 39, No. 7 (July 1962), pp. 356–59.
Kieffer, William F. *The Mole Concept in Chemistry.* New York: Reinhold, 1964. Chapter 2.
Nash, Leonard K. *Stoichiometry.* Reading, Massachusetts: Addison-Wesley, 1966. Chapter 2.

PROBLEMS

1. A sample of carbon dioxide has a volume of 450 ml at $0^\circ C$ and a pressure of 800 torr. What volume will the sample have at STP?

2. If the gas sample in the preceding problem is to be compressed into a volume of 100 ml (keeping the temperature constant), what pressure will be required?

3. One liter of oxygen at STP is heated to $250^\circ C$ while the pressure is held constant. What will the new volume be?

4. A steel bomb having a volume of 500 cm^3 is filled with air at $28^\circ C$ and 742 torr. If the bomb is heated to $100^\circ C$, what pressure will be exerted by the confined gas?

5. What is the density of phosgene gas ($COCl_2$) at STP?

6. What volume is occupied by 50 g of H_2 at a temperature of $25^\circ C$ and a pressure of 740 torr?

7. What is the density of F_2 at $32^\circ C$ and 1 atm?

8. If 10.0 g of methane is confined in a rigid container having a volume of 250 cm^3 at a temperature of $50^\circ C$, what is the pressure exerted by the gas?

9. A sample of carbonoxysulfide (COS) gas occupies a volume of 200 ml at

27°C and a pressure of 750 torr. How many moles of gas are in the sample? What weight of COS is in the sample?

10. What is the temperature required for 25 g of nitrous oxide (N_2O) to occupy a volume of 1.0 liter at a pressure of 2.0 atm?

11. In the complete combustion of 10.0 g of acetylene, according to the equation,

$$2C_2H_2(g) + 5O_2(g) \rightarrow 4CO_2(g) + 2H_2O(g)$$

(a) What volume of CO_2 (measured at STP) is produced?
(b) What volume of O_2 measured at 180°C and 742 torr is consumed?

12. The fermentation of sucrose (cane sugar) to produce ethyl alcohol and carbon dioxide can be represented by the equation

$$C_{12}H_{22}O_{11} + H_2O \rightarrow 4C_2H_5OH + 4CO_2$$

What volume of CO_2 gas (measured at STP) will be liberated during the fermentation of 1 lb of sucrose?

13. A sample of sodium sulfite weighing 2.34 g is treated with excess hydrochloric acid.

$$Na_2SO_3 + 2HCl \rightarrow H_2O + 2NaCl + SO_2 \uparrow$$

What volume of SO_2 gas is produced at 27°C and 730 torr?

14. The following reaction, carried out at high temperature, is an important step in the commercial preparation of nitric acid:

$$4NH_3(g) + 5O_2(g) \rightarrow 4NO(g) + 6H_2O(g)$$

(a) What volume of NO, measured at 1 atm and 1000°C, can be produced theoretically from 50 liters of NH_3 at the same temperature and pressure?
(b) What volume of O_2 at STP will be consumed in reacting with 10 moles of NH_3?
(c) What weight of water is produced from the reaction of 10 liters of NH_3 measured at 150°C and a pressure of 1 atm?
(d) What weight of NH_3 is required to produce 50 liters of NO at 30°C and 742 torr?

15. A sample of a gaseous compound weighing 0.110 g was found to occupy a volume of 51.0 ml when measured at 28°C and 744 torr. Calculate the molecular weight of the compound.

16. A certain gas has a density of 2.37 g/liter at 25°C and 735 torr. What is the molecular weight of the gas?

17. A gaseous compound has a percentage composition of 88.8% C and 11.2% H and has a density of 2.12 g/liter at 31°C and 742 torr. Calculate (a) the empirical formula, (b) the molecular weight, and (c) the molecular formula of the compound.

18. A sample of oxygen collected over water at 30°C and 750 torr has a volume of 222 ml. What volume will the sample occupy *dry* and at STP?

19. If 5.00 g of methane is collected by bubbling it through water at $25°C$ and 758 torr, what is the volume collected?

20. A gas sample weighing 0.523 g has a volume of 456 ml when measured over water at $30°C$ and a pressure of 752 torr. What is the molecular weight of the gas?

21. Passing carbon monoxide gas over powdered nickel metal at $40°C$ gives a volatile compound of the following composition: 34.30% Ni, 28.10% C, and 37.53% O. The density of the gaseous product at this temperature and 750 torr is 6.57 g/liter. Calculate (a) the empirical formula and (b) the molecular formula.

22. A so-called vacuum tube is sealed off at a pressure of 1.0×10^{-5} torr and $100°C$. Calculate the number of remaining molecules of gas per milliliter.

23. A 250-ml flask was evacuated and into it was forced 150 ml of H_2 under a pressure of 750 torr, 75 ml of O_2 under a pressure of 350 torr, and 50 ml of N_2 under a pressure of 250 torr. (The temperature was held constant.) Determine the partial pressure of each of the gases in the 250-ml flask, and the total pressure of the gas mixture.

24. The composition of air, expressed as volume-percent, is 20.95% oxygen, 78.09% nitrogen, and 0.93% argon, and the remaining 0.03% may be considered as CO_2.
    (a) Calculate the apparent molecular weight of air.
    (b) How many moles of argon are present per mole of air?
    (c) Calculate the partial pressure of the CO_2 if the total pressure is 1 atm.

25. List the following gases in order of increasing rate of diffusion at STP: H_2S, CH_4, Cl_2, CO_2, N_2, SO_2.

26. The rate of diffusion of a certain gas was measured in a laboratory apparatus and found to be 10.0 ml/min. Pure oxygen in the same apparatus at the same temperature and pressure diffused at a rate of 14.1 ml/min. Calculate the molecular weight of the gas.

27. Uranium-235 is separated from uranium-238 through the gaseous diffusion of the hexafluorides. Calculate the relative rates of diffusion.

28. Rationalize the following observed facts on the molecular level:
    (a) A sealed glass bottle breaks when heated in an oven.
    (b) When a real gas is allowed to escape through the valve of a compressed gas tank, the gas cools.
    (c) You can drink through a soda straw.
    (d) Gases deviate from ideal behavior at high pressures.
    (e) The critical temperature of ammonia is much higher than that of nitrogen.

29. Rationalize the following facts:
    (a) The critical temperature of argon is $-122°C$, while that of helium is much lower ($-268°C$).
    (b) The critical temperature of dimethyl ether is $126.9°C$, while that of ethyl alcohol is much higher ($243°C$).

30. Explain why it is that at room temperature hydrogen gas cannot be converted to a liquid no matter how high the applied pressure.

31. The heat of fusion of aluminum is 2.6 kcal/g-atom. Calculate the entropy of fusion (ΔS) of aluminum at its melting point (660°C).

32. The heat of fusion of ordinary water is 1440 cal/mole, while the heat of fusion of heavy water (D_2O) is 1550 cal/mole. Assuming the entropy of fusion is the same for D_2O as for H_2O, predict the freezing point of heavy water.

33. The assumption made in the preceding problem is apparently not valid, for the actual freezing point of D_2O is 3.8°C. Calculate the entropy of fusion of H_2O and of D_2O at their freezing points.

11 *Matter in Bulk III: Solutions*

Thus far in our discussion of matter in bulk we have dealt almost exclusively with pure substances. *Mixtures* of substances are also important forms of matter. Indeed, most matter as it occurs in nature is in the form of mixtures; the earth's atmosphere, the oceans, the contents of every living cell are all mixtures. Every chemical reaction carried out by a chemist in the laboratory requires him to deal with a mixture. Of particular interest are those mixtures which consist of a single phase: *homogeneous mixtures,* commonly called *solutions.* (Review the classification of matter in Chapter 1.)

Although in everyday usage the term *solution* brings to mind a solid dissolved in a liquid, the fact is that solutions can exist in any of the three physical states: gas, liquid, or solid. Air, of course, is an example of a gaseous solution, as is moist oxygen or a mixture of iodine vapor and nitrogen. Some *alloys* (mixtures of metals) are solid solutions. For example, brass is a solution of zinc in copper. An amalgam of copper and mercury (the term *amalgam* means any alloy containing mercury) is a solid solution formed from a solid and a liquid. Liquid solutions may be obtained by the mixture of a liquid with a gas, solid, or another liquid.

In describing solutions it is customary to refer to that component which retains its original physical state as the *solvent.* The other components of the solution are called *solutes.* Thus, in moist oxygen, the solvent is oxygen and the solute is water, because the solution is a gas. In the liquid solution formed from salt and water, the solvent is water, the solute is salt. And in a copper-mercury alloy, copper is the solvent, because the solution is in the solid state. In solutions in which all components have the same physical state, the one present in the greatest amount is called the solvent and the other components are called solutes. Thus, in a solution formed from 20 ml of ethyl alcohol and 500 ml of water, water is the solvent, ethyl alcohol the solute.

Liquid solutions are by far the most common, and we shall deal primarily with these in the discussions that follow.

THE SOLUTION PROCESS

The difference between solutions and heterogeneous mixtures lies in the size of the dispersed particles. In a solution these are individual molecules or ions, whereas in heterogeneous mixtures they are larger aggregates. In order to gain some under-standing of how this molecular dispersion comes about, we must examine the solution process on the molecular level.

As an illustration, let us consider the dissolution of a molecular solid (for example, sugar) in a molecular liquid (say, water). Some crystals of the solute are added to the liquid solvent. Three kinds of intermolecular forces must be taken into account in this system. First, there are the intermolecular forces between solute molecules—the forces holding the molecules together in the crystal lattice. Second, there are attractive forces between the molecules of the liquid solvent. Both these types of forces tend to prevent the formation of a solution. The forces within the crystal must be overcome if solute molecules are to leave the crystal and inter-mingle with the solvent molecules, and the forces between solvent molecules must be overcome if solute molecules are to disperse among the solvent molecules. The third type of force at work in the system is the intermolecular attraction that exists between solute and solvent molecules. This third force tends to counteract the other two and bring about dissolution, and the stronger the attractions between solute and solvent molecules, the more easily the solute-solute and solvent-solvent forces can be overcome. It is the overall balance of these three types of forces that determines how readily the solute will dissolve in a particular solvent. If conditions are favorable for dissolution, the surface molecules leave the crystal, enter the liquid phase, and by diffusion become dispersed among the solvent molecules. The next layer of solute molecules becomes the surface layer, and these in turn enter the liquid phase, and so on, as the crystal dissolves.

The *rate* at which this dissolution occurs (the number of solute molecules entering the liquid phase per unit time) is dependent on the nature of the solute and the solvent, for the strengths of the various intermolecular forces are deter-mined, of course, by the particular structure and composition of the substances involved. For a given solute-solvent system, the rate of solution varies with temperature and the surface area of the solute as follows:

Temperature. An increase in temperature has several effects on the system, all of which combine to *increase* the rate of solution. Higher temperatures increase the kinetic energy of the solute molecules, thus lessening the effectiveness of the lattice forces. The kinetic energy of the solvent molecules is also increased at higher temperatures, so that solvent-solvent attractions are more easily overcome. Finally, higher temperatures increase the rate with which solute molecules diffuse through the solvent.

Surface area.   Since dissolution of a solid solute is a surface phenomenon, the greater the surface area of the solid the higher will be its rate of solution. A pulverized solid will dissolve *faster* than the same substance in one large lump, because pulverization increases the surface area. Stirring or agitation of the mixture will also increase the rate of solution, because this action increases the amount of surface in contact with the liquid. Stirring also aids in diffusion of the dissolved molecules.

As the solution process continues, more and more solute molecules enter the liquid phase. These "dissolved" molecules are in random motion along with the solvent molecules. As a result of this motion, solute molecules in the liquid phase undergo collisions with the surface of the crystal. If a collision occurs when a "dissolved" molecule has a sufficiently low kinetic energy, it may be "captured" by the lattice forces at the surface of the crystal and become part of the crystal again. This process, which is the reverse of dissolution, is called crystallization. Thus we have two opposing processes occurring in the system (illustrated in Figure 11-1), which can be represented by the expression

$$\text{Undissolved solute} \rightleftharpoons \text{Dissolved solute}$$

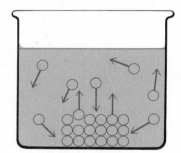

FIGURE 11-1

Dissolution and Crystallization

The *rate* of crystallization is determined by the same factors that determine the rate of solution—nature of solute and solvent, temperature, and surface area of solute—and one additional factor, the number of solute molecules in the liquid phase.

As was indicated above, for a given solute and solvent at a given temperature and with a given surface area, the *rate of solution* is essentially constant with time. The *rate of crystallization,* however, begins at zero and gradually increases because the number of dissolved solute molecules is constantly increasing. Eventually, therefore, the rates of the two opposing processes must become equal, and a state of *dynamic equilibrium* is reached. (Note the similarity to the evaporation of a liquid in a closed system, p. 204). In this state, if the temperature is held constant, solute will continue to dissolve and to crystallize at the same rate, and there will be no net change in the *amount* of solute dissolved. A solution in which this situation exists—that is, one in which *the dissolved solute is in equilibrium with the undissolved solute*—is called a *saturated* solution. When a solution contains *less* dissolved solute than the *equilibrium amount* it is said to be *unsaturated;* if it contains *more* than the equilibrium amount it is called *supersaturated* (see p. 259).

Let us now consider the dissolution of an *ionic* solid in a liquid; for example, sodium chloride in water. In general, our discussion of the dissolution of a molecular solid applies also to an ionic solid, with some modifications necessary because of the difference in the nature of the attractive forces.

Recall that an ionic solid consists of positive and negative ions arranged alternately in the crystal lattice and held together by rather strong electrostatic attractions. Ions in the body of the crystal are surrounded by some number of ions of opposite charge and are therefore subjected to attractions in all directions. The electrostatic attractions on the surface ions, however, are unbalanced, and these are the ions that come into contact with the solvent molecules. Water molecules, being dipoles, are attracted to the surface ions, with their positive ends to the anions and their negative ends to the cations. This attractive force between an ion and a polar molecule is called an *ion-dipole attraction.* It permits the ions to leave the surface of the lattice and become part of the liquid phase. The dissolved ions diffuse through the solution surrounded by their attached water molecules. In this condition the ions are said to be *hydrated,* and the process of their formation is called *hydration.* (These terms are used when the solvent is water. The more general terms *solvated* and *solvation* are used to indicate the attachment of molecules of any solvent.) As in the case with molecular solutes, if sufficient solute is present, an equilibrium will be established between the dissolved and undissolved ionic substance, giving a saturated solution.

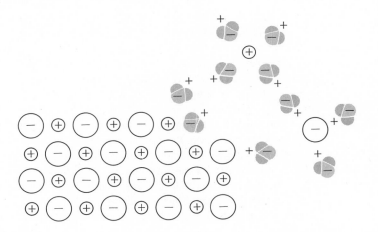

FIGURE 11-2

Dissolution of an Ionic Crystal in Water

SOLUBILITY

The amount of a solute present in the dissolved state in a saturated solution is called the *solubility* of the solute. Solubility is a measure of the *extent* to which one substance will dissolve in another and should not be confused with *rate of solution.* Solubility may be expressed either as the amount of solute in a given amount of solvent (for example, grams of sugar per 100 grams of water) or as the amount of solute in a given amount of solution (for example, grams of sugar per liter of solution). The solubility of a substance is dependent on a number of factors, which we shall treat in turn.

Nature of Solute and Solvent

The solubility of one substance in another varies widely with the structural and compositional nature of the two substances and is related to the three kinds of attractive forces discussed earlier. In order for solution to occur, solvent-solute attractions must overcome solvent-solvent and solute-solute attractions, and it is the balance between these forces that determines the solubility of a given solute in a particular solvent.

In general, solutes tend to be more soluble in solvents to which they are structurally and electrically similar. This is the often-quoted rule of "like dissolves like." More specifically, solubility is likely to be greater if both solute and solvent are polar, or if both are nonpolar, than if one is polar and the other nonpolar. If both solute and solvent are polar (or if hydrogen bonding is possible between them), then the solute-solvent attractions are strong, which enhances solubility. Thus, substances such as sugar, methyl alcohol, and glycerol have high solubilities in water and other polar solvents. On the other hand, sugar is almost completely *insoluble* in such nonpolar liquids as benzene or gasoline (a mixture of hydrocarbons). In this case, the very weak solute-solvent attractions are not strong enough to overcome the strong solute-solute attractions. Similarly, ionic compounds dissolve only in polar solvents, where ion-dipole attractions can occur; their solubility in nonpolar liquids is generally negligible.

A nonpolar substance such as carbon tetrachloride has only a very slight solubility in water. The strong attractive forces between water molecules (solvent-solvent attractions) are much stronger than solvent-solute attractions. On the other hand, carbon tetrachloride shows a high solubility in ethyl ether (a less polar molecule than water and one in which hydrogen bonding does not occur), because the solvent-solvent forces are weak enough to allow carbon tetrachloride molecules to enter between ether molecules.

Covalent network crystals, such as diamond or quartz, are not soluble in any liquid. No potential solute-solvent forces are strong enough to break down the network of covalent bonds.

Temperature

The solubility of gases in water (and most other liquids) usually *decreases* with increasing temperature. An open bottle of carbonated drink goes "flat" quicker when it is warm than it does if it is kept cold; at higher temperatures the dissolved carbon dioxide escapes more rapidly because its solubility is lower. Everyone has observed the formation of small bubbles when water is heated; these are bubbles of air released from the solution as the solubility decreases with increasing temperature.

There is no general rule regarding the effect of temperature on the solubility of liquid or solid solutes in liquid solvents. While it is true that most solids show an *increase* in solubility with increasing temperature, this is certainly not the case for all solids; many have *decreasing* solubilities with increasing temperatures. Furthermore, the extent to which a given temperature change will change the solubility varies greatly between substances. There is no simple equation that expresses solubility-temperature relationships. These must be determined experimentally and are usually presented as solubility curves, in which solubility is plotted against temperature. Figure 11-3 shows the solubility curves of a substance whose solubility increases with increasing temperature (KCl), a substance whose solubility decreases with increasing temperature (Li_2CO_3), and one whose solubility is hardly affected by temperature change (NaCl).

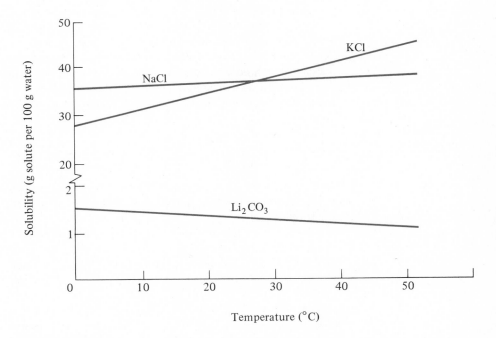

FIGURE 11-3

Solubility Curves of KCl, NaCl, and Li_2CO_3

The effect of temperature change on solubility is related to the energetics of the solution process. The dissolution of a solute in a solvent is accompanied by a change in enthalpy; that is, heat is either absorbed or evolved. This effect is the net result of the energy *required* to break the solute-solute and solvent-solvent attractions and the energy *released* in the formation of solute-solvent attractions. The quantity of energy involved in the change, which is called the *heat of solution* (ΔH_{soln}), is a function not only of the amount of solute but also of the amount of solvent in which it is dissolved. Thus, heats of solution are usually expressed in terms of calories per mole of solute dissolved in a specified number of moles of solvent. One can determine experimentally whether a solution process is endothermic ($\Delta H_{soln} = +$) or exothermic ($\Delta H_{soln} = -$) by dissolving some of the solute in a solution that is already nearly saturated, thereby eliminating any significant change in concentration. If the solution increases in temperature, the process is exothermic, and if it decreases in temperature, the process is endothermic.

From a knowledge of whether a given solution process is endothermic or exothermic, one may predict how the solubility will change with temperature by applying a useful concept known as *Le Chatelier's Principle*. This principle (proposed by Henri Le Chatelier in the 1880s) states that *if a stress is applied to a system in equilibrium, the system will readjust in such a way as to relieve the stress and establish a new equilibrium state.* Since a state of equilibrium exists when two opposing processes are occurring at the same rate, we might restate Le Chatelier's Principle as follows: *if one changes the conditions of a system in equilibrium, that process will be favored which tends to restore the original conditions.*

The solubility of a substance is a measure of the amount of that substance present in a *saturated* solution; that is, the *equilibrium* amount. Therefore, the solubility of a substance may be described as the amount of "dissolved solute" in an equilibrium, represented by one of the following expressions:

(1)   Undissolved solute $\rightleftharpoons$ Dissolved solute + Heat   ($\Delta H = -$)

(2)   Undissolved solute + Heat $\rightleftharpoons$ Dissolved solute   ($\Delta H = +$)

where equation (1) represents an *exothermic* solution process (heat is given off when the solute dissolves) and (2) is an endothermic process (heat is absorbed when the solute dissolves).

Now, *heat* is a form of "stress" referred to by Le Chatelier's Principle. In order to raise the temperature we must supply heat. In so doing we disturb the equilibrium, and, according to Le Chatelier's Principle, that process will be favored which consumes the added heat and tends to return the system to its original conditions. In the exothermic solution process, equation (1), the process to the *left* will be favored (that is, more *undissolved* solute will be formed), and when equilibrium is reestablished at the new, higher temperature there will be *less* dissolved solute than there was at equilibrium at the lower temperature. In other words, the solubility will have *decreased*. On the other hand, in the situation represented by equation (2), raising the temperature will favor the process to the *right* (the process that consumes heat), causing an *increase* in the amount of *dissolved* solute and a *decrease* in the amount of *undissolved*—the solubility will have *increased.* In summary, then, solutes that have exothermic heats of solution (the enthalpy change is negative) show *decreasing solubility with increasing temperature,* while solutes with endothermic heats of solution (the enthalpy change is positive) show *increasing solubility with increasing temperature.*

The change in solubility of a substance with change in temperature may result in a *supersaturated* solution. For example, suppose we prepare a saturated solution of KCl in water at 50°C (the solubility at this temperature is 42.6 g of KCl per 100 g of water). Now we very carefully filter the solution at 50° to remove any undissolved KCl as well as any particles of dust or other foreign matter. The solution is then allowed to cool without any disturbance to room temperature, say 25°C. If we have been very careful in our operations, the solution may cool to 25° without any solute crystallizing out. If so, we have a solution containing 42.6 g of KCl per 100 g of water, even though the solubility of KCl at 25°C is only 35.5 g per 100 g of H_2O. Our solution obviously contains *more* than the equilibrium amount of solute and is therefore *supersaturated.* Such solutions are not stable; if they are agitated or if a crystal of the solute is introduced, crystallization of solute from the solution will quickly occur and equilibrium will be established.

Pressure For solutions of solids or liquids in liquid solvents, the effect of pressure on solubility is negligible. However, *for solutions of gases in liquids, the solubility of a gas is directly proportional to the partial pressure of the gas above the solution.* This relationship is known as Henry's Law (William Henry, 1803). For example, the solubility of hydrogen in water at 25°C and a hydrogen pressure of 1 atm is 1.54 mg per liter of water. Therefore, at the same temperature and a hydrogen pressure of 700 torr, the solubility of hydrogen will be

$$\frac{700}{760} \times 1.54 = 1.42 \text{ mg } H_2 \text{ per liter } H_2O$$

One of the most common applications of this pressure-solubility relationship is in the preparation of "soda water." Carbon dioxide is dissolved in water, bottled, and tightly sealed, all under high pressure. When the bottle cap is removed at ordinary atmospheric pressure, the decreased solubility of the carbon dioxide causes it to escape from the solution, imparting the fizz to the liquid.

Another illustration of this phenomenon can be seen in the "bends," or caisson disease, an ailment sometimes contracted by divers and others who work under high pressures when they experience a decrease in pressure too quickly. Inhalation of air at the high pressure necessary for deep sea diving results in the solution of a large amount of nitrogen in the blood and body tissues—a much larger amount than could dissolve at atmospheric pressure. When the diver returns to the surface, the decrease in pressure causes a lower solubility of nitrogen, which leaves the blood as bubbles. If the decompression takes place sufficiently slowly the bubbles can escape from the body with no ill effects. But if the pressure decrease occurs too suddenly, the rapid formation of nitrogen bubbles within the body can cause intense pain and even death.

THE ENERGETICS OF SOLUTIONS

The solubility of a given solute in a particular solvent—that is, the *extent* to which the solution process occurs—is a function of the free energy change (ΔG) that accompanies the process. Since ΔG is a function of the enthalpy change and the entropy change,

$$\Delta G = \Delta H - T\Delta S$$

both these thermodynamic properties play a role in solubility.

As we have discussed earlier, the enthalpy change (heat of solution) may be either positive or negative, and it represents the net result of the energy required to overcome the solute-solute and solvent-solvent attractions and the energy given off in the formation of solute-solvent attractions. Thus, the dissolution of an ionic solid in water may be thought of as occurring in two imaginary steps. First, the crystal lattice breaks down to form isolated gaseous ions. Using NaCl as an illustration, this step may be represented as:

(1) $NaCl(s) \rightarrow Na^+(g) + Cl^-(g)$ (ΔH is positive; unfavorable)

This will be recognized as the reverse of the equation by which *lattice energy* was earlier defined (see Chapter 6). Since energy is required, the enthalpy change is positive, and the process is not favored. The second step involves the hydration of the gaseous ions, which may be represented by the equation

(2) $Na^+(g) + Cl^-(g) \rightarrow Na^+(aq) + Cl^-(aq)$ (ΔH is negative; favorable)

The enthalpy change for this process—called the hydration energy—is negative and therefore favors the process. (The hydration energy is actually the net of the energy released in the formation of solute-solvent attractions and that absorbed in overcoming solvent-solvent attractions. Because these two energy changes are extremely difficult to determine separately, they are lumped together as hydration energy.)

The addition of these two equations results in an expression that represents the process of the dissolution of solid sodium chloride in water. Therefore, the heat of solution (ΔH_{soln}) is the sum of the enthalpy changes of the two imaginary processes.

(1)	$NaCl(s)$	$\rightarrow Na^+(g) + Cl^-(g)$	$\Delta H =$ lattice energy
(2)	$Na^+(g) + Cl^-(g)$	$\rightarrow Na^+(aq) + Cl^-(aq)$	$\Delta H =$ heat of hydration
(3)	$NaCl(s)$	$\rightarrow Na^+(aq) + Cl^-(aq)$	$\Delta H =$ heat of solution

Since ΔH of step (1) is always positive and ΔH of step (2) is always negative, the sign of the heat of solution depends on which of them is greater. If the heat of hydration is larger than the lattice energy, the solution process will be exothermic and therefore favored. On the other hand, if the lattice energy is larger than the heat of hydration, the solution process will be endothermic and not favored.

The direction of the *entropy* change when a solid is dissolved in a liquid is easily predicted. Certainly, the random dispersion of solute and solvent particles in a solution represents greater disorder (therefore, higher entropy) than either a crystal lattice or a liquid, unmixed. Dissolution of a solid in a liquid, then, usually results in an increase in entropy (ΔS is positive), and this *favors* the process.

Let us apply these considerations to the free energy change ($\Delta G = \Delta H - T\Delta S$). Since in the dissolution of a solid in a liquid ΔS is positive, in order for ΔG to be negative (which means that the dissolution will occur), ΔH must be negative or if ΔH is positive its magnitude must be less than the product $T\Delta S$. If ΔH is a very large positive number (if the solution process is highly endothermic), dissolution will not take place or will occur only to very limited extent.

Similar considerations can be applied to the solution of molecular solids in liquids. The energy required to break up the lattice of molecular crystals is not as great as in the case of ionic crystals, but solvation energy is also lower for the molecular substances.

It is convenient to imagine the dissolution of a nonpolar solute as occurring in two steps. First the solid melts, and then the molecules of the two liquids mix to form a solution, as illustrated in the expressions:

$$
\begin{array}{ll}
\text{solute(s)} \rightarrow \text{solute(l)} & \Delta H_{\text{fus}} = \text{positive, unfavorable} \\
& \Delta S_{\text{fus}} = \text{positive, favorable} \\
\\
\text{solute(l) + solvent(l)} \rightarrow \text{solution(l)} & \Delta H_{\text{mix}} = \text{positive, unfavorable} \\
& \qquad\qquad\quad \text{or negative, favorable} \\
& \Delta S_{\text{mix}} = \text{positive, favorable} \\
\hline
\text{solute(s) + solvent(l)} \rightarrow \text{solution(l)} & \Delta H_{\text{soln}} = \text{positive, unfavorable} \\
& \qquad\qquad\quad \text{or negative, favorable} \\
& \Delta S_{\text{soln}} = \text{positive, favorable}
\end{array}
$$

The melting process is endothermic, and therefore the heat of fusion is positive and *unfavorable* to the solution process. The enthalpy of mixing may be favorable or unfavorable, depending on the relative strengths of the various intermolecular forces involved. On the other hand, both processes lead to greater disorder, and the entropy change is *favorable* to dissolution. Thus, solubility in this case is determined primarily by the magnitude of the heat of mixing (ΔH_{mix}), which in turn is determined by the nature of the solvent, particularly the strength of solvent-solvent attractive forces. This relationship can be illustrated by consideration of the relative solubility of a simple, nonpolar molecular solid, iodine (I_2), in water and in a less polar solvent such as chloroform ($CHCl_3$). The solution process is the same in the two cases, yet although the solubility of I_2 in water is very low (only 0.03 g per 100 g of water at 25°C) it is much higher in $CHCl_3$ (3.1 g per 100 g of water at 25°C). Obviously, the enthalpy change must be more *unfavorable* in the case of water as solvent than in the case of $CHCl_3$, even though the heat of fusion of iodine is identical in both cases. The difference lies in the heats of mixing. Water, with its hydrogen bonding, possesses strong solvent-solvent attractive forces. In order for mixing to occur, these attractions must be "broken" to allow homogeneous dispersion of the I_2 molecules, and a large input of energy is required to do this. Thus, the mixing process is highly endothermic and ΔH_{mix} is very unfavorable. On the other hand, the intermolecular forces between $CHCl_3$ molecules are much weaker than those between water molecules, and less energy is required to "make room for" the I_2 molecules. Hence, when I_2 dissolves in $CHCl_3$, the mixing process is much less endothermic and ΔH_{mix} is not nearly so unfavorable as when water is the solvent. The net result is a more favorable free-energy change (and therefore greater solubility) in the I_2-CHCl_3 system than in the I_2-H_2O system. As we have already seen, under the rule of "like dissolves like," nonpolar solutes are generally more soluble in nonpolar solvents than they are in polar ones.

The reasoning employed in an analysis of the energetics of solution of a molecular solid in a liquid can be applied also to the solution of a *liquid* in a liquid. Since both solute and solvent are liquids, no lattice need be broken and no imaginary change of state need be considered. The solution process is simply one of mixing. The heat of solution is the heat of mixing, just as the entropy of solution is the entropy of mixing. Thus, whether the factor favoring solution, $T\Delta S$, will outweigh an unfavorable heat of mixing depends on just how endothermic the mixing process is, and this in turn depends on the relative strengths of the attractions of solute and solvent molecules toward themselves and each other.

Finally, let us consider briefly the energetics of solution of a gas in a liquid. The two imaginary steps in this case are:

1. Condensation of the gas to a liquid. This is an exothermic process, therefore *favored* by the enthalpy change, but it represents a decrease in disorder and the entropy factor is *unfavorable*.
2. Mixing of the molecules of the two liquids. The entropy change here is *favorable*. However, the enthalpy change may be either favorable or unfavorable, depending on whether the forces of attraction are stronger or weaker in the solution than they are in the two unmixed liquids.

$$(1) \qquad Solute(g) \rightarrow Solute(l) \qquad \begin{matrix} \Delta H_{cond} &= \text{negative, favorable} \\ \Delta S &= \text{negative, unfavorable} \end{matrix}$$

$$(2)\, Solute(l) + Solvent(l) \rightarrow Solution(l) \qquad \begin{matrix} \Delta H_{mix} &= \text{negative, favorable} \\ & \quad \text{or positive, unfavorable;} \\ \Delta S &= \text{positive, favorable} \end{matrix}$$

The entropy change in the liquefaction step is greater than the entropy change of mixing, so that the overall entropy effect in the solution of a gas in a liquid is *unfavorable* to the process. (Recall that for solid and liquid solutes the entropy in all steps is favorable to dissolution.)

When the liquid solvent has strong intermolecular forces (as in water), then the mixing process is highly endothermic (ΔH_{mix} is unfavorable). This, together with the unfavorable entropy factor, accounts for the fact that most gases (except those that undergo chemical reaction with water) have very low solubilities in water.

CONCENTRATIONS OF SOLUTIONS

In dealing with solutions it is necessary that we have some way of describing their compositions. The composition of a *saturated* solution is defined by the solubility at the temperature in question. But the solutions that we more commonly encounter, both in the laboratory and in nature, are *unsaturated* and have variable compositions. The description "a solution of sugar in water" tells us *what* the components are but nothing about *how much* of each component is present. The terms *concentrated* and *dilute* are sometimes used in this regard, but these are merely relative terms; a concentrated sugar solution contains more sugar per given amount of water than a dilute one does.

The composition of a solution is usually described in terms of its *concentration:* an expression of the amount of solute in a given amount of solvent *or* in a given amount of solution. The amounts may be expressed in any units convenient for the purpose. Thus, grams of solute per 100 grams of solvent, grams of solute per kilogram of solution, milliliters of solute per liter of solution, micrograms of solute per milliliter of solution are all acceptable expressions of concentration.

One of the more common expressions of concentration in general use is *weight percentage.* This is the weight of solute per hundred parts by weight of *solution.* Thus, in a 5 wt-% aqueous solution of sodium chloride, there are 5 g of NaCl in every 100 g of solution (and there are 95 g of water in every 100 g of solution). If 10 g of sugar is dissolved in 40 g of water, then the result is a 20 wt-% sugar solution (10 g of sugar in 50 g of solution).

In chemistry it is often more important that we know the composition of a solution in terms of the number of molecules or atoms present than in terms of *weight* of solute. For that reason, a number of methods of expressing concentration in terms of number of moles of solute have been devised and are widely used. Each has its own particular usefulness in various aspects of solution chemistry.

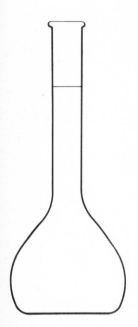

FIGURE 11-4

Volumetric Flask

Molarity

Molarity is defined as *the number of moles of solute per liter of solution.* Thus, for example, a 1.25 *molar* (1.25M) sodium chloride solution contains 1.25 moles of NaCl for each liter of solution. It should be noted that the *amount of solvent* is not involved in the expression of molarity. Dissolving 1.25 moles of NaCl in *one liter of water* is not the same thing as dissolving 1.25 moles of NaCl in *enough water to make one liter of solution.* Moreover, the volume of a liquid solution rarely equals the sum of the volumes of the pure components, but is usually larger or smaller than that sum.

The most convenient way to prepare a solution of specific molarity is to use a volumetric flask (Figure 11-4). The proper amount of solute is placed in the flask, and solvent is then added with thorough mixing until the solution fills the flask just exactly to the calibration mark on the neck. Volumetric flasks are available in a variety of sizes, each with its volume carefully calibrated.

Suppose, for example, we wish to prepare 250 ml of $0.50M$ $CaCl_2$ solution. One mole of $CaCl_2$ weighs approximately 111 g. Therefore, 0.50 mole weighs 55.5 g, and if we wished to make one liter of the solution we would use 55.5 g of $CaCl_2$. However, since we wish to make only 250 ml (0.250 liter), we will need 55.5 × 0.250, or 13.9 g of $CaCl_2$. We place this weight of $CaCl_2$ in a 250-ml volumetric flask and add water slowly to the flask, agitating it thoroughly to dissolve all the solute. Finally, we add enough water to bring the volume just to the calibration mark.

The relationship between molarity and weight of solute is further illustrated in the following examples.

EXAMPLE: What weight of HCl is contained in 350 ml of $0.250M$ HCl solution?

Solution: The weight of 1 mole of HCl (rounded to one decimal place) is 36.5 g. Since the solution is $0.250M$, the weight of HCl in 1 liter of the solution is

$$\text{g of HCl/liter} = 36.5 \text{ g/mole} \times 0.250 \text{ mole/liter}$$

Therefore, in 350 ml there is:

$$\frac{36.5 \text{ g}}{1 \text{ mole}} \times \frac{0.250 \text{ mole}}{1 \text{ liter}} \times \frac{350 \text{ ml}}{1000 \text{ ml/liter}} = 3.19 \text{ g of HCl}$$

EXAMPLE: What weight of chloride ion is contained in 75 ml of $0.200M$ $BaCl_2$ solution?

Solution: Since *one* mole of $BaCl_2$ contains *two* moles of Cl^-, the solution is $0.400M$ in Cl^-. Each mole of Cl^- weighs approximately 35.5 g. Therefore,

$$\frac{0.400 \text{ mole}}{1 \text{ liter}} \times \frac{35.5 \text{ g}}{1 \text{ mole}} \times \frac{75 \text{ ml}}{1000 \text{ ml/liter}} = 1.07 \text{ g } Cl^-$$

EXAMPLE: How does one prepare 500 ml of $0.15M$ $AgNO_3$ solution from solid $AgNO_3$?

Solution: One liter of the solution would contain 0.15 mole of $AgNO_3$, and therefore 500 ml must contain:

$$\left(\frac{500}{1000} \times 0.15 \right) \text{ moles of } AgNO_3$$

One mole of $AgNO_3$ weighs 170 g. Hence the weight of $AgNO_3$ required to prepare the solution in question is:

$$\frac{500 \text{ ml}}{1000 \text{ ml/liter}} \times \frac{0.15 \text{ mole}}{1 \text{ liter}} \times \frac{170 \text{ g}}{1 \text{ mole}} = 12.8 \text{ g}$$

One dissolves 12.8 g of $AgNO_3$ in water in a 500-ml volumetric flask and dilutes to the mark.

EXAMPLE: How would you prepare 500 ml of $0.15M$ $AgNO_3$ solution from a $6.0M$ stock solution of $AgNO_3$?

Solution: The solution we wish to prepare must contain a total of:

$$\left(0.15 \times \frac{500}{1000}\right) \text{ moles of } AgNO_3$$

Since the stock solution contains 6.0 moles of $AgNO_3$ per liter, the question we must ask is, "How many liters of this stock solution contains the number of moles we want to have in our final solution?" This is answered by dividing 6.0 into the number of moles needed:

$$\frac{0.15 \text{ mole}}{1 \text{ liter}} \times \frac{500 \text{ ml}}{1000 \text{ ml/liter}} \times \frac{1 \text{ liter}}{6.0 \text{ mole/liter}} = 0.0125 \text{ liter}$$

$$= 12.5 \text{ ml}$$

Therefore, we place 12.5 ml of the $6.0M$ stock solution in a 500-ml volumetric flask, and dilute with water to the mark. The resulting solution is $0.15M$ $AgNO_3$.

EXAMPLE: What is the molarity of a 20 wt-% NaCl solution which has a density of 1.15 g/ml?

Solution: Since molarity is based on *volume* of solution while percentage is based on *weight* of solution, it is not possible to derive one of these concentration expressions from the other unless the density of the solution is known. Since 1 ml of the solution weighs 1.15 g, 1 liter weighs

$$(1.15 \times 1000) \text{ g}$$

The solution is 20 wt-% NaCl. Therefore, the weight of NaCl in 1 liter of solution is

$$(1.15 \times 1000 \times 0.20) \text{ g of NaCl}$$

To find the number of moles of NaCl in 1 liter (that is, the *molarity*), we need only divide the number of grams of NaCl per liter by the number of grams in one mole. Thus:

$$\frac{1.15 \times 1000 \times 0.20}{58.4} = 3.94M$$

Normality   Another concentration expression, somewhat related to molarity, is *normality:* the number of *equivalents* of solute per liter of solution. Thus $1.2N$ is read 1.2 normal and indicates 1.2 equivalents of solute for every liter of solution. An *equivalent* of any substance is that weight (in grams) of the substance which in a particular chemical reaction is equivalent to (that is, will combine with, or replace, or otherwise react with) one gram-atom of hydrogen. In most cases the equivalent weight of a substance is equal to its formula weight or is some simple fraction (one-half, one-third, and so forth) of its formula weight. Unlike the *mole,* an *equivalent* of a substance is not necessarily a constant amount but depends on the specific reaction in which that substance is taking part. It is therefore inconvenient, to deal further with normality at this time. We shall discuss it in connection with various types of reactions in Part Two.

Molality

Both molarity and normality are expressions of concentration based on the *volume of solution* and are therefore subject to changes in temperature. Furthermore, they do not specify the number of moles of solvent present. For some purposes it is desirable to have an expression of concentration that is not affected by temperature and that indicates the molar ratio of solute to solvent. One such expression in common use is *molality: the number of moles of solute per kilogram of solvent.* Note that although molarity and molality are similar words, they have quite different meanings. Molarity is based on *volume of solution,* molality on *weight of solvent.* A 2.5 molal (abbreviated 2.5*m*) sugar solution contains sugar and water in a ratio such that there are 2.5 moles of sugar for every 1000 g of water. The volume of a solution whose concentration is known only in terms of molality is not predictable unless the density is also known.

EXAMPLE: What is the molality of a solution prepared by dissolving 10.0 g of I_2 in 100 g of CCl_4?

Solution: The formula weight of I_2 is 254. Therefore, 10 g of I_2 is

$$\frac{10}{254} \text{ mole of } I_2$$

This number of moles is the number per 100 g of CCl_4 (the solvent). The number of moles per gram of CCl_4 is

$$\frac{10}{254 \times 100}$$

and the molality—the number of moles per 1000 g of CCl_4—is

$$\frac{10}{254} \times \frac{1000}{100} = 0.39m$$

EXAMPLE: What is the molality of a 20 wt-% solution of glucose ($C_6H_{12}O_6$) in water?

Solution: The solution contains 20 g of glucose per 80 g of water. The molecular weight of glucose is 180; therefore, 20 g is

$$\frac{20}{180} \text{ mole of glucose}$$

If this is the number of moles per 80 g of water, then the number per 1000 g of water must be

$$\frac{20}{180} \times \frac{1000}{80} = 1.4m$$

EXAMPLE: What is the molality of a $2.5M$ H_2SO_4 solution that has a density of 1.18 g/ml?

Solution: One liter of this solution weighs

$$(1000 \times 1.18) \text{ g solution}$$

Since there are 2.5 moles of H_2SO_4 in 1 liter, the weight of H_2SO_4 per liter is

$$(2.5 \times 98) \text{ g } H_2SO_4$$

The weight of water in 1 liter is therefore

$$[(1000 \times 1.18) - (2.5 \times 98)] \text{ g } H_2O$$

or

$$1180 - 245 = 935 \text{ g } H_2O$$

The solution contains, then, 2.5 moles of H_2SO_4 in 935 g of H_2O. Therefore, the number of moles of H_2SO_4 per 1000 g of water is

$$2.5 \times \frac{1000}{935} = 2.67m$$

Mole Fraction Another useful method of expressing concentration is the *mole fraction*. It can be applied to each component of a solution, solvent as well as solutes. *The mole fraction of a component of a solution is the ratio of the number of moles of that component to the total number of moles of all the constituents of the solution.* Thus, in a solution of A dissolved in B, the mole fraction of A is given by

$$\frac{\text{moles of A}}{\text{moles of A + moles of B}} = \text{mole fraction of A}$$

It is obvious, then, that the sum of the mole fractions of all the components of a solution is *one*.

EXAMPLE: What is the mole fraction of glucose ($C_6H_{12}O_6$) in a 20 wt-% solution of glucose in water?

Solution: There are 20 g of glucose for every 80 g of water. The number of moles of glucose in 20 g is:

$$\frac{20}{180} \text{ mole of glucose}$$

and the number of moles of H_2O in 80 g is

$$\frac{80}{18} \text{ moles of water}$$

The mole fraction of glucose, then, is

$$\frac{\dfrac{20}{180}}{\dfrac{20}{180} + \dfrac{80}{18}} = 0.024$$

EXAMPLE: What is the mole fraction of HCl in a 10*m* aqueous solution of HCl?

Solution: A 10*m* solution of HCl contains 10 moles of HCl per 1000 g of water, or 10 moles of HCl per 1000/18 moles of water. Therefore, the mole fraction of HCl is

$$\frac{10}{10 + (1000/18)} = \frac{10}{65.5} = 0.15$$

PROPERTIES OF SOLUTIONS

Solutions have the same general characteristics as their solvents. Thus, solutions in liquid solvents show the typical behavior of liquids; they have measurable vapor pressures, surface tensions, viscosities, and so forth. The magnitudes of the properties—that is, the actual values of the various properties—however, differ between a liquid solution and the pure liquid solvent. This is to be expected, since the molecules in a liquid are in intimate contact, and the molecules of one component will be influenced by those of another.

Of particular interest are four properties of liquid solutions—vapor pressure, boiling point, freezing point, and osmotic pressure—that are dependent only on the *number* of solute particles (molecules or ions) in a given amount of solvent, not on the *nature* of the solute particles. Because these four properties are closely inter-related they are often referred to as the *colligative properties* (from the Latin *colligare*, "to bind together").

Vapor Pressure

When a nonvolatile solute (one that has a negligible vapor pressure) is dissolved in a liquid solvent, the resulting solution has a vapor pressure that is *lower* than that of the pure solvent at the same temperature. This effect can be demonstrated dramatically, as shown in Figure 11-5. Two open beakers, one containing pure water and the other containing a solution of sugar in water, are placed inside a tightly sealed enclosure. After a period of time, it is observed that the volume of pure water has decreased, while the volume of sugar solution has increased. Through the processes of evaporation and condensation, water has been transferred from one beaker to the other, because the vapor pressure of pure water is higher than the vapor pressure of water in the solution. If a similar experiment is performed with two sugar solutions, one more concentrated than the other, it is found that there is a net transfer of water from the less concentrated to the more concentrated solution until the solutions are the same concentration. This indicates that the extent to which the vapor pressure is lowered is dependent on the amount of solute present: the more concentrated the solution, the lower the vapor pressure.

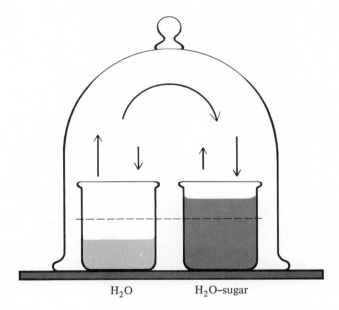

H_2O H_2O–sugar

FIGURE 11-5

Demonstration of Difference in Vapor Pressure between Pure Water and Aqueous Sugar Solution

This vapor pressure lowering can be rationalized on the molecular level as simply a matter of statistics. Since the solute is nonvolatile, only the solvent molecules escape from the solution and enter the vapor phase. Moreover, only *surface* molecules with sufficiently high kinetic energy can escape. The solute molecules simply "take up space" and reduce the proportion of molecules that have kinetic energy high enough to escape from the liquid surface. Thus the effect of solute on vapor pressure is dependent only on the relative number of molecules, not on their nature. (This rationalization is based on the assumption that the solute-solvent attractions are nearly the same magnitude as the attractions in the pure components, and also that the solvent and solute molecules are of comparable size.)

The quantitative relationship between vapor pressure lowering and concentration is embodied in Raoult's Law (Francois-Marie Raoult, 1886): *The vapor pressure (p) of a volatile component of a solution is equal to the product of the mole fraction (X) of that component and the vapor pressure of the pure component (p^0) at the same temperature.*

$$p = p^0 X$$

It is clear from this expression that as the concentration of nonvolatile solute increases, the smaller the mole fraction of solvent (X) becomes, and the lower is the vapor pressure with respect to that of the pure solvent.

The following example illustrates the application of Raoult's Law to predict the vapor pressures of solutions.

EXAMPLE: What is the vapor pressure of a 20 wt-% glucose ($C_6H_{12}O_6$) solution in water at 29°C?

Solution:

$$p^0 = \text{the vapor pressure of water at } 29°C$$

$$= 30.0 \text{ torr (see Appendix 12)}$$

$$X = \text{mole fraction of water in the solution}$$

$$= \frac{\text{number of moles of water}}{\text{total number of moles}}$$

Since there are 20 g of glucose for every 80 g of water, there are 20/180 moles of glucose for every 80/18 moles of water. Therefore,

$$X = \frac{\dfrac{80}{18}}{\dfrac{80}{18} + \dfrac{20}{180}} = 0.976$$

and

$$p = 30.0 \times 0.976 = 29.3 \text{ torr}$$

In the case of a solution in which the solute has an appreciable vapor pressure (for example, the solution of a liquid in a liquid), the application of Raoult's Law becomes somewhat more complex. Each component has a certain vapor pressure, which is less than that of the pure substance at the same temperature, and the vapor pressure of the solution is the sum of the vapor pressures of the components. Suppose we have a solution of two volatile liquids, A and B. According to Raoult's Law, the vapor pressure of component A is given by

$$p_A = p^0{}_A X_A$$

where $p^0{}_A$ is the vapor pressure of pure A at the temperature in question and X_A is the mole fraction of A in the solution. But component B also has a vapor pressure:

$$p_B = p^0{}_B X_B$$

The vapor pressure of the solution, then, is equal to $p_A + p_B$.

> **EXAMPLE:** A solution is prepared by mixing 2.0 moles of benzene and 3.0 moles of carbon tetrachloride. The vapor pressure of pure benzene and carbon tetrachloride at $25°C$ are 95.5 torr and 115.6 torr, respectively. What is the vapor pressure of the solution at $25°C$?
>
> *Solution:* The mole fraction of benzene in the solution is
>
> $$\frac{2.0}{2.0 + 3.0} = 0.40$$
>
> Therefore, the vapor pressure of benzene in the solution is
>
> $$p_{C_6 H_6} = p^0{}_{C_6 H_6} X_{C_6 H_6} = 95.5 \times 0.40 = 38.2 \text{ torr}$$
>
> The mole fraction of carbon tetrachloride is 0.60, and the vapor pressure of CCl_4 is
>
> $$p_{CCl_4} = p^0{}_{CCl_4} X_{CCl_4} = 115.6 \times 0.60 = 69.4 \text{ torr}$$
>
> The vapor pressure of the solution, then, is
>
> $$p_{soln} = p_{C_6 H_6} + p_{CCl_4} = 38.2 + 69.4 = 107.6 \text{ torr}$$

Raoult's Law should not be considered a precise quantitative statement of the behavior of all solutions. A hypothetical solution that would adhere strictly to Raoult's Law under all conditions is called an *ideal solution,* but in actual fact there are no ideal solutions, just as there are no ideal gases. Real solutions deviate from Raoult's Law, some very widely. The relationship expressed in Raoult's Law assumes that the solvent-solute attractions are of the same magnitude as the intermolecular attractions in the pure components. Actually, when they are weaker, the molecules of the component substances can escape more easily from solution than they can from the pure substance, and the solution has a higher vapor pressure than is predicted by Raoult's Law. On the other hand, if the solvent-solute attractions are stronger than the attractions between like molecules, then the molecules cannot escape as readily from the solution as from the pure substances, and the solution's vapor pressure is lower than predicted. In general, solutions of nonvolatile solutes in liquids tend to follow Raoult's Law more closely than solutions with more than one volatile component. Also, as solutions become more dilute, the solvent approaches ideal behavior more closely.

Boiling Point

Since the normal boiling point of a liquid is a function of the liquid's vapor pressure (refer to Figure 9-25), it is to be expected that the vapor pressure lowering caused by dissolution of a solute in a liquid will bring about a change in the liquid's boiling point. This effect is illustrated in Figure 11-6, where the vapor pressure curve for water is plotted along with the vapor pressure curve for a solution of sugar in water. The normal boiling point of a liquid is the temperature at which its vapor pressure is equal to 1 atm (760 torr). Since dissolution of sugar in water lowers the vapor pressure of the water at any given temperature, pure water will have a vapor pressure of 1 atm at a *lower* temperature than the sugar solution will. In other words, a solution of a nonvolatile molecular solute in a liquid has a *higher* boiling point than the pure liquid does. This boiling point elevation is shown as Δt_b in Figure 11-6.

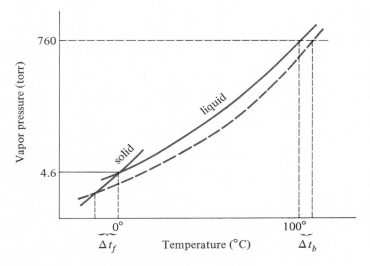

FIGURE 11-6

Vapor Pressure Curves of Pure Water (———) and Sugar Solution (- - -)

    The quantitative relationship between the boiling point elevation and the concentration of solute is revealed by the following hypothetical experiment. A series of solutions of sugar in water, of various molalities, are prepared, and the boiling point of each solution is determined. The results (assuming all solutions behave ideally) are tabulated in Table 11-1, showing that the elevation of the boiling point is directly proportional to the molality. A 1.0m solution has a boiling point of 0.512° higher than that of pure water, a 2.0m solution boils 2 × 0.512°, or

TABLE 11-1 Normal Boiling Points of Ideal Solutions of Various Molalities	*Molality*	*Boiling Point* ($^\circ$C)
	0 (pure H_2O)	100.000
	0.1	100.051
	0.5	100.256
	1.0	100.512
	2.0	101.024
	3.0	101.536

1.024°, higher than water, and so on. If the experiment is repeated using any nonvolatile, nondissociating solute instead of sugar, the results are the same; the boiling point of water is raised by 0.512°C for each mole of solute per 1000 g of water. It is the *number* of molecules of solute, not the nature of those molecules, that determines the boiling point elevation, and we may express this relationship for water solutions as

$$\Delta t_b = 0.512m$$

where m is the molality of the solution.

Other liquids behave the same as water does in this regard, the only difference being that the amount the boiling point is raised for each mole of solute per 1000 g of solvent varies from liquid to liquid. Consequently, the relationship between boiling point elevation and concentration for solutions in general is given by the expression

$$\Delta t_b = K_b m$$

where Δt_b is the number of degrees the boiling point is raised, m is the molality of solute in the solution, and K_b is a constant whose value is characteristic of the particular liquid solvent in question. The value of this constant, called the *molal boiling point constant*, is given in Table 11-2 for several common solvents.

TABLE 11-2 Normal Boiling Points and Molal Boiling Point Constants of Some Liquid Solvents	*Liquid*	*Normal* *Boiling Point* ($^\circ$C)	K_b ($^\circ$C-kg/mole)
	Acetone	56.5	1.67
	Benzene	80.1	2.53
	Carbon tetrachloride	76.8	5.02
	Chloroform	61.2	3.63
	Ethyl alcohol	78.4	1.22
	Water	100.0	0.512

EXAMPLE: Predict the boiling point of a solution made by dissolving 20.0 g of quinone ($C_6H_4O_2$) in 100.0 g of acetone.

Solution: The molecular weight of quinone is 108. The molality of the solution is:

$$\frac{20}{108} \times \frac{1000}{100} = 1.85m$$

The molal boiling point constant for acetone is 1.67.

$$\Delta t_b = K_b m$$

$$\Delta t_b = 1.67 \times 1.85 = 3.09°C$$

Since the boiling point of pure acetone is $56.5°C$, and the boiling point elevation is $3.09°C$, the boiling point of the solution will be

$$56.5 + 3.09 = 59.6°C$$

Freezing Point

Another property of liquids that is related to vapor pressure is the freezing point, for the freezing point is the temperature at which the vapor pressures of the solid and liquid forms of a substance are the same. As shown in Figure 11-6, the vapor pressure curves of the liquid and solid forms of the solvent intersect at the freezing point. However, because the vapor pressure curve of the solution is lower than that of the pure liquid, the intersection occurs at a lower temperature. In other words, dissolving a nonvolatile solute in a liquid causes the resulting solution to have a lower freezing point than that of the pure solvent.

Illustrations of this phenomenon of freezing point lowering are encountered frequently in everyday life. The salt water of the ocean does not freeze at temperatures that freeze bodies of fresh water; we sprinkle salt on ice-covered roads to cause melting; and the water in the cooling systems of automobiles is prevented from freezing by the addition of "antifreeze." These are all examples of aqueous solutions having lower freezing points than pure water.

Experiments show that freezing point depression and concentration are related quantitatively by an expression very similar to that discussed for boiling point elevation. One mole of solute dissolved in one kilogram of solvent lowers the freezing point of that particular solvent by a constant amount called the *molal freezing point constant.* The value of the constant is different for each solvent but is independent of the nature of the solute, provided the solute is nonvolatile and does not dissociate in solution. Table 11-3 lists the molal freezing point constants for a number of solvents.

The expression for freezing point lowering is

$$\Delta t_f = K_f m$$

where Δt_f is the freezing point depression, K_f is the molal freezing point constant, and m is the molality of the solute in the solution. The application of this relationship in predicting freezing points of solutions is illustrated in the following example.

TABLE 11-3	Solvent	Freezing Point (°C)	K_f (°C-kg/mole)
Freezing Points and Molal Freezing Point Constants of Some Solvents	Acetic acid	16.6	3.90
	Benzene	5.5	5.12
	Camphor	179.7	39.7
	Carbon tetrachloride	−22.8	29.8
	Chloroform	−63.5	4.68
	Naphthalene	80.2	6.80
	Water	0.0	1.86

EXAMPLE: The compound most commonly used as an automotive antifreeze is ethylene glycol ($HOCH_2 CH_2 OH$), a water-soluble liquid with relatively low vapor pressure. Assuming that ethylene glycol and water form an ideal solution, what is the freezing point of a 25 wt-% solution of ethylene glycol?

Solution: The solution contains 25 g of ethylene glycol for every 75 g of water. The molecular weight of ethylene glycol is 62, and therefore the molality of the solution is

$$\frac{25}{62} \times \frac{1000}{75} = 5.38m$$

The freezing point depression is

$$\Delta t_f = 1.86 \times 5.38 = 10.0°C$$

The freezing point of the solution, then, is

$$fp = 0.0 - 10.0 = -10.0°C$$

Osmotic Pressure

Membranes of certain materials permit the passage of water molecules but not molecules or ions of solutes dissolved in the water. These *semipermeable* membranes include many natural membranes such as the walls of animal and plant cells and the outer covering of the intestines, as well as some artificial membranes such as parchment paper, cellophane, and certain gelatinous compounds.

Figure 11-7 shows an apparatus in which pure water is separated from a sugar solution by a semipermeable membrane. At the start, both liquid levels are at the same height, but with the passage of time it is observed that the level of sugar solution rises, while the level of the pure water falls. Obviously, there is a transfer of water from the pure solvent, through the membrane, into the solution. This phenomenon is called *osmosis* (from the Greek, meaning "to thrust"). If sugar solutions are placed on both sides of the membrane, with one of the solutions being more concentrated than the other, then water will pass *from the less concentrated into the more concentrated solution.*

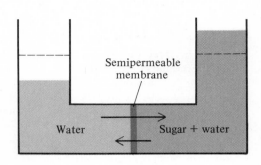

FIGURE 11-7

Apparatus to Demonstrate Osmosis

A very simplified explanation of this process is as follows: Water molecules, in their constant, random motion, collide with the semipermeable membrane on both sides. However, the rate of these collisions is greater on the pure water (or less concentrated) side than on the solution (or more concentrated) side, simply because the concentration of water molecules is greater in pure water than in the solution. Sugar molecules also collide with the membrane, of course, but these collisions are ineffective because the membrane will not permit the passage of the sugar molecules. The net result is that, although water molecules pass through the membrane in both directions, *more* pass through per unit time from the pure water to the solution (left to right in Figure 11-7) than in the opposite direction. Hence, the volume of liquid becomes larger on the right and smaller on the left.

If the open arms of the apparatus in Figure 11-7 were fitted with pistons, then it would be possible to force water through the membrane in either direction by applying sufficient pressure to the surface of the proper liquid. By applying just the right amount of pressure to the solution (the right side), the osmotic flow of water from left to right will be prevented, and the two liquid levels will be maintained at the same height. The amount of pressure necessary to accomplish this is called the *osmotic pressure* of the solution. Osmotic pressure, like the other colligative properties, is dependent on the number of particles of solute, not on their nature. Like vapor pressure, it is a measure of the tendency of the solvent molecules to escape, but in inverse proportion; the lower the escaping tendency, the greater the osmotic pressure.

Quantitative investigations have shown that the osmotic pressure of a solution is proportional to the molal concentration of the solute. For dilute aqueous solutions, however, the molarity and molality are very nearly the same, and the following relationship has been found to apply:

$$\text{Osmotic pressure} = CRT$$

where the osmotic pressure is in atmospheres, C is the concentration in moles per liter, R is the ideal gas constant, and T is the absolute temperature.

Since the walls of living cells are permeable to water and some solutes, but not permeable to others, osmosis is an extremely important process in living organisms. The absorption of soil water by plant roots and the transportation of that water up to the leaves, the passage of nutrients through the intestinal wall into

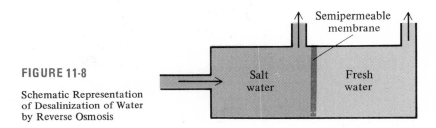

FIGURE 11-8

Schematic Representation
of Desalinization of Water
by Reverse Osmosis

the bloodstream, and the elimination of wastes by passage from the kidney cells into the urine are a few examples of osmosis in life processes.

A potentially valuable application of the phenomenon of osmosis is the production of fresh water from the oceans. If salt water is pumped under pressure into a tank fitted with a semipermeable membrane, water molecules will pass through the membrane, leaving the ions of salt behind (Figure 11-8). Here water is moving from the more concentrated side to the less concentrated side of the membrane because of the high pressure applied to the salt water side—a process referred to as *reverse osmosis*. Although this technique has been successful on a small scale, large-scale desalinization has not yet become practical, primarily because of the lack of suitable large, long-lasting membranes.

MOLECULAR WEIGHT DETERMINATIONS

The colligative properties of solutions provide a means of experimentally determining molecular weights. If a solution is prepared from accurately measured weights of solute and solvent and the colligative properties of that solution are measured, one can then calculate the number of moles of solute present. From the weight of solute and the number of moles of solute, it is a simple matter to calculate the weight of one mole; that is, the molecular weight. While all the colligative properties can be employed for this purpose, the freezing point can usually be measured more conveniently—and often more accurately—than the others, and therefore freezing point depression is the method most commonly used. This is illustrated in the following example.

EXAMPLE: After a sample of compound X weighing 1.286 g was dissolved in 25.00 g of benzene, the freezing point of the resulting solution was found to be $3.27°$C. What is the molecular weight of compound X?

Solution: The freezing point of pure benzene is $5.50°$C. Therefore, the freezing point depression (Δt_f) is $5.50° - 3.27° = 2.23°$C. The molal freezing point constant (K_f) for benzene is 5.12 (see Table 11-3). Therefore, the molality of the solution can be calculated:

$$\Delta t_f = K_f m$$

$$m = \frac{\Delta t_f}{K_f} = \frac{2.23}{5.12} = 0.436m$$

Since the molality is the number of moles of solute per 1000 g of solvent and since the solution used in the measurement contains 1.286 g of solute per 25.00 g of solvent, we may write

$$\frac{1.286}{\text{mol wt}} \times \frac{1000}{25.00} = 0.436$$

Solving for molecular weight (the weight of one mole):

$$\text{Mol wt} = \frac{1.286}{0.436} \times \frac{1000}{25.00} = 118 \text{ g/mole}$$

For molecular weight determinations of compounds of very high molecular weight (for example, proteins), the measurement of osmotic pressure provides a more satisfactory method than the measurement of the other colligative properties. This is because solutions of very low molality provide very small changes in freezing point and boiling point, making these changes difficult to measure accurately.

EXAMPLE: A sample of hemoglobin weighing 1.000 g is dissolved in water to make 50.0 ml of solution. The measured osmotic pressure of this solution is 5.90 torr at 25°C. What is the molecular weight of hemoglobin?

Solution: Denoting osmotic pressure by π, we can write

$$\pi = CRT = \frac{n}{V}RT$$

where n = number of moles of solute and V = volume of solution (in liters). But,

$$n = \frac{\text{grams of solute}}{\text{gram-molecular weight of solute}}$$

and therefore,

$$\pi = \frac{gRT}{(\text{mol wt}) \, V}$$

Solving for molecular weight, we obtain

$$\text{Mol wt} = \frac{gRT}{\pi V}$$

$$= \frac{(1.000 \text{ g}) \left(0.082 \dfrac{\text{liter-atm}}{\text{mole-}°\text{K}} \right) (298°\text{K})}{\left(\dfrac{5.90}{760} \text{ atm} \right) (0.050 \text{ liter})}$$

$$= 63{,}000 \text{ g/mole}$$

The superiority of osmotic pressure over freezing point depression for very high molecular weight substances is easily demonstrated. The aqueous solution of hemoglobin in the example just given is only about 3×10^{-4} molal,

$$\frac{1.000}{63,000} \times \frac{1000}{50} = 0.00032m$$

and this corresponds to a freezing point depression of only $0.00059°C$

$$\Delta t_f = 1.86 \times 0.00032 = 0.00059°C$$

Although such a small temperature change is extremely difficult to measure, a column of mercury 5.9 mm in height is relatively easy to measure.

ELECTROLYTE SOLUTIONS

The structural and electrical characteristics of the water molecule give this liquid the capability of dissolving a large number of ionic and polar covalent substances. This, together with its great abundance, gives water a unique position among liquid solvents, and the nature of *aqueous* solutions deserves special emphasis.

It was recognized as early as the eighteenth century that water-soluble substances could be divided into two categories: those whose aqueous solutions conduct electric current and those whose aqueous solutions do not. The former we call *electrolytes;* the latter, *nonelectrolytes.* These two types of solutes may be distinguished experimentally by use of an apparatus shown schematically in Figure 11-9. Two metallic electrodes are connected to a source of electricity. Some device for detecting the flow of electric current is placed in the circuit. This may be an ammeter or galvanometer, or simply a light bulb or buzzer. When the electrodes are

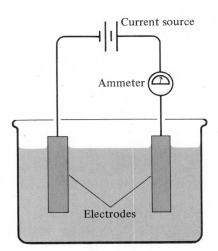

FIGURE 11-9

Apparatus to Investigate
Conductivity of Solutions

suspended in air, no current flows, of course, because the circuit is not completed. If the electrodes are immersed in pure water, there is still no flow of current, indicating that water is not a conductor of electricity. Now, when a solute is dissolved in the water, the resulting solution may or may not be a conductor. If the solute is an electrolyte, then the solution will conduct, as indicated by the flow of current through the circuit. But if the solute is a nonelectrolyte, no current will flow.

The search for an explanation of the conductivity of electrolyte solutions led Svante Arrhenius to formulate his *theory of electrolytic dissociation* in 1887. This theory has been modified considerably over the years as a result of increased knowledge of chemical bonding. Today we recognize that in order for a solution to conduct electricity it must contain ions. Whereas the conductivity of an electric current by a metal is explained by the motion of the loosely bound electrons of the metallic crystal (p. 197), the conductance of current by a solution—called *electrolytic* conductance—is accomplished by the motion of ions. Positive ions (cations) move through the solution toward the negative electrode (cathode), and negative ions (anions) move toward the positive electrode (anode). The phenomenon of electrolytic conductance involves more than just the motion of ions, and this process of *electrolysis* will be discussed in detail in Chapter 17. Nevertheless, the presence of ions that are free to move is essential for the conductance of electric current by a solution. An electrolyte, then, is a substance whose aqueous solution contains ions; solutions of nonelectrolytes do not contain ions.

Classification of Electrolytes

Electrolytes may be divided conveniently into three types of compounds: acids, bases, and salts. (Compounds that do not fit into any of these categories are therefore nonelectrolytes and include hydrocarbons, alcohols, esters, ethers, aldehydes, ketones, etc.) There are a number of useful ways of defining acids, bases, and salts, depending on the purpose at hand, and these various concepts are treated at length in subsequent chapters. For our present purpose, since we are dealing only with aqueous solutions, it will be adequate to use those definitions which have been derived from Arrhenius's theory of electrolytic dissociation.

Acids. Acids are compounds that react with water to produce hydronium ions. These ions, generally symbolized by the formula H_3O^+, may be thought of as hydrated protons. By way of illustration, consider the compound hydrogen chloride, HCl. This is a covalent compound that consists of discrete molecules and exists as a gas at standard temperature and pressure. The bond between the atoms has only about 20% ionic character. However, if HCl is dissolved in water, a reaction occurs, during which the proton is transferred from the HCl molecule to a water molecule. The chlorine atom left after removal of the proton has one more electron than its normal complement and is therefore a chloride ion.

$$H_2O + HCl \rightarrow H_3O^+ + Cl^-$$

Thus, while neither HCl nor H_2O alone conducts current, a solution of HCl in H_2O contains ions and is a conductor. Hydrogen chloride, therefore, is an electrolyte. Sometimes, for convenience, the water is omitted, and the equation representing this reaction is written simply as

$$HCl \rightarrow H^+ + Cl^-$$

There is no harm in using this abbreviated notation, as long as one keeps in mind that it *is* abbreviated.

All acids (refer to the partial listing on p. 92), including the hundreds of known carboxylic acids, react with water in an analogous way, and all acids are therefore electrolytes.

Bases. Compounds that dissolve in water to form hydroxide ions (OH^-) are called *bases*. This includes, of course, all the hydroxides ($NaOH$, $Ca(OH)_2$, and so on).

$$NaOH \rightarrow Na^+ + OH^-$$

$$Ca(OH)_2 \rightarrow Ca^{2+} + 2OH^-$$

But it also includes some other compounds that do not themselves contain hydroxide ions but react with water to form these ions, notably, ammonia and its derivatives the amines.

$$NH_3 + H_2O \rightarrow NH_4^+ + OH^-$$

Salts. When a solution of an acid and a solution of a base are mixed together, a chemical reaction occurs in which the hydronium ion and the hydroxide ion combine to form molecules of water. This reaction is called *neutralization*.

$$H_3O^+ + OH^- \rightarrow 2H_2O$$

The anion of the acid and the cation of the base remain in solution, and if water is removed by evaporation, a compound of these two ions is obtained. This type of compound, which may be thought of as consisting of the anion of an acid and the cation of a base, is called a *salt*. Some neutralizations and the salts that are formed are illustrated in the following "molecular" equations:

$$HNO_3 + NaOH \rightarrow H_2O + NaNO_3$$
acid   base   water   salt

$$H_2SO_4 + KOH \rightarrow H_2O + KHSO_4$$
acid   base   water   salt

$$2CH_3COOH + Ca(OH)_2 \rightarrow 2H_2O + Ca(CH_3COO)_2$$

acid   base   water   salt

Most salts are ionic compounds, and, as we have discussed previously in this chapter, when ionic crystals dissolve in water they are present in the solution as hydrated ions. Salts, therefore, are electrolytes, and their aqueous solutions conduct electricity.

Although they contain neither hydrogen nor hydroxide ions, the oxides of many elements are nevertheless electrolytes, for they undergo reactions with water to produce ions in aqueous solution. In general, the water-soluble oxides of *nonmetallic* elements dissolve in water to produce hydronium ions and may therefore be considered *acids*. This ion production can be illustrated with sulfur dioxide and carbon dioxide, as follows:

$$SO_2 + H_2O \rightarrow H_2SO_3$$

$$H_2SO_3 + H_2O \rightarrow H_3O^+ + HSO_3^-$$

$$CO_2 + H_2O \rightarrow H_2CO_3$$

$$H_2CO_3 + H_2O \rightarrow H_3O^+ + HCO_3^-$$

In fact, many nonmetallic oxides show this relationship to specific acids. These oxides may be thought of as resulting from the removal of water from their corresponding acids, and are therefore called *acid anhydrides*. A number of common acids and their anhydrides are listed in Table 11-4.

TABLE 11-4
Some Common Acids and Their Anhydrides

Acid	− Water	=	Anhydride
H_2SO_3	$- H_2O$	=	SO_2
H_2SO_4	$- H_2O$	=	SO_3
H_2CO_3	$- H_2O$	=	CO_2
$2HNO_2$	$- H_2O$	=	N_2O_3
$2HNO_3$	$- H_2O$	=	N_2O_5
$2HClO_4$	$- H_2O$	=	Cl_2O_7
$2H_3PO_3$	$- 3H_2O$	=	P_2O_3
$2H_3PO_4$	$- 3H_2O$	=	P_2O_5

The water-soluble oxides of *metals,* on the other hand, dissolve in water with the production of hydroxide ions and are thus *bases*.

$$\underline{CaO} + H_2O \rightarrow Ca^{2+} + 2OH^-$$

$$\underline{Na_2O} + H_2O \rightarrow 2Na^+ + 2OH^-$$

For the same reason that oxides of nonmetals are called acid anhydrides, oxides of the metals are called *basic anhydrides;* that is, they may be thought of as resulting from the removal of water from a hydroxide base, as illustrated in Table 11-5.

TABLE 11-5	*Base*	*−*	*Water*	*=*	*Anhydride*
Some Bases and Their Anhydrides	$2NaOH$	−	H_2O	=	Na_2O
	$Ca(OH)_2$	−	H_2O	=	CaO
	$Ba(OH)_2$	−	H_2O	=	BaO
	$Zn(OH)_2$	−	H_2O	=	ZnO
	$2Al(OH)_3$	−	$3H_2O$	=	Al_2O_3

Strength of Electrolytes

Equivalent solutions of all electrolytes are not equally good conductors of electricity. Suppose, for example, we prepare three aqueous solutions; one is $1m$ HCl, another is $1m$ $HC_2H_3O_2$, and the third is $1m$ HCN. Each of these solutions is acidic because the solute reacts with water to produce hydronium ion and an anion according to the following general equation:

$$HX + H_2O \rightarrow H_3O^+ + X^-$$

Moreover, all three of the solutions are of the same concentration. We might expect, therefore, that each of the three will conduct electric current equally well. If, however, we test their conductance in an apparatus, such as the one in Figure 11-8, the ammeter tells us that the HCl solution is a better conductor than the $HC_2H_3O_2$ solution, and the $HC_2H_3O_2$ solution is, in turn, a better conductor than the HCN solution. Since the ability of a solution to conduct electricity is a function of the number of ions present (a dilute solution of HCl, for example, is a poorer conductor than a more concentrated solution of HCl), the difference in conductance of the three solutions suggests that they contain different numbers of ions, in spite of the fact that each was prepared by dissolving 1 mole of solute in 1000 g of water.

In order to understand this difference in conductance, we must realize (1) that the reaction between the acid and water to produce ions is reversible and (2) that, other things being equal, the *rate* of a reaction is proportional to the concentration of the reactants. Thus, if HX can react with H_2O to produce H_3O^+ and X^-, then H_3O^+ and X^- can also react with each other to produce HX and H_2O, and this *reversible* reaction can be represented:

$$HX + H_2O \rightleftharpoons H_3O^+ + X^-$$

When molecules of HX are dissolved in water, their concentration is high at first, and the reaction producing the ions (the one to the right in the above equation) occurs at a relatively high rate. This rate gradually decreases, however, as HX is consumed and its concentration decreases. The rate of the reaction to the left, on the other hand, begins at zero and gradually increases as more and more ions are formed. Eventually the two rates must become equal, and a state of equilibrium exists. Thus, the reaction of an acid with water does not lead to its *complete* conversion into ions. An aqueous solution of HX contains H_3O^+ and X^-, but it also contains *undissociated* molecules of HX.

Equilibria in aqueous solutions will be treated quantitatively in later chapters.

For our present purpose, it is sufficient simply to point out that the extent to which an electrolyte solution contains ions—and therefore the degree to which it conducts electric current—varies widely and depends on the *position* of the equilibrium. In the case of HCl, the equilibrium position lies far in the direction favoring ion production; in $HC_2H_3O_2$ the equilibrium mixture contains many more undissociated molecules of acid and fewer ions; and the HCN equilibrium contains even fewer ions. In other words, the HCl solution is the best conductor because it contains the most ions, and the HCN solution is the poorest conductor because it contains the fewest ions. The solution of $HC_2H_3O_2$ is intermediate. The extent to which an electrolyte yields ions in aqueous solution is referred to as its *strength*. Thus, of the three electrolytes considered in our illustration, HCl is the strongest, $HC_2H_3O_2$ is weaker, and HCN is the weakest.

The relative strengths of electrolytes vary over a wide range, and the quantitative expression of these strengths is discussed in Chapter 15. It is convenient to divide electrolytes qualitatively into two categories: *strong* and *weak*. An electrolyte that is so greatly dissociated into ions in aqueous solution that, for all practical purposes, the dissociation may be considered complete is, by definition, a *strong electrolyte*. All others are *weak electrolytes*. Hence, a solution of the strong electrolyte nitric acid (HNO_3) may be considered to contain only H_2O, H_3O^+, and NO_3^-, no molecules of HNO_3. But the weak electrolyte nitrous acid (HNO_2) contains HNO_2 molecules as well as H_2O, H_3O^+, and NO_2^-.

In recognizing which common electrolytes are strong and which are weak, the following generalizations are useful:

Acids: HCl, HBr, HI, HNO_3, $HClO_4$, and H_2SO_4 are strong electrolytes. Nearly all other acids are weak.

Bases: The hydroxides of the Group IA metals and the Group IIA metals (except Be) are strong electrolytes. Most other hydroxides, NH_3, and amines are weak.

Salts: Nearly all common salts are strong electrolytes. A few (for instance, the halides of Zn, Cd, and Hg(II)) are weak.

COLLIGATIVE PROPERTIES OF ELECTROLYTE SOLUTIONS

The discussion of the colligative properties of solutions earlier in this chapter was confined to those solutions in which the solute remains undissociated so that the number of dissolved particles is directly proportional to the number of moles of solute used. When the solute is an electrolyte, the formation of ions results in a higher concentration of *particles* than the concentration based on number of moles of solute dissolved, and the quantitative application of colligative properties becomes more complex. For example, suppose we prepare a $1m$ solution of the weak electrolyte HX by dissolving exactly 1 mole of HX in exactly 1 kg of water. The solute, upon dissolving, reacts with water to form ions, and a state of equilibrium is reached between the undissociated molecules and the ions:

$$HX + H_2O \rightleftharpoons H_3O^+ + X^-$$

The equilibrium concentration of undissociated HX is less than 1 molal, but for each molecule dissociated, *two* ions have been formed. The concentration of solute particles is the sum of the concentrations of HX, H_3O^+, and X^-, and therefore the molality of *particles* is greater than 1. As a result, the freezing point of this solution will be lowered more than $1.86°C$ and the boiling point will be raised more than $0.512°C$. Likewise, the vapor pressure will be lowered by an amount greater than is predicted by Raoult's Law, and the osmotic pressure will also be greater than is predicted. The extent to which the colligative properties are affected by dissociation depends, of course, on the degree to which dissociation occurs for the specific electrolyte in question; it depends on the strength of the electrolyte. In fact, the degree of dissociation of an electrolyte can be estimated by the measurement of colligative properties of its solutions, as illustrated in the following example.

EXAMPLE: A $1.00m$ solution of the weak acid HX in water has a freezing point of $-2.05°C$. What is the degree of dissociation of the acid in this solution?

Solution: The relationship between the freezing point lowering and concentration for an aqueous solution is:

$$\Delta t_f = 1.86m$$

Since the freezing point of this solution has been lowered $2.05°$, we can calculate the molality of total particles in the solution.

$$2.05 = 1.86m$$

$$m = \frac{2.05}{1.86} = 1.10m \text{ in total particles}$$

Now, if we let x equal the number of moles of acid dissociated at equilibrium, then the molal concentrations of the various particles are as follows:

$$HX + H_2O \rightleftharpoons H_3O^+ + X^-$$
$$(1-x) \qquad\qquad (x) \qquad (x)$$

and the molality of total particles is:

$$(1-x) + x + x = 1 + x$$

Therefore,

$$1 + x = 1.10$$

and

$$x = 1.10 - 1 = 0.10$$

Thus, of the number of moles of HX dissolved in the solution, one-tenth are dissociated at equilibrium. In other words, the electrolyte is 10% dissociated.

On the basis of the foregoing discussion, one might logically conclude that dissolution of a strong electrolyte—one that is known to be 100% in ionic form—

such as NaCl, would have exactly twice the effect on colligative properties that a nondissociated solute has. Since a $1m$ solution of NaCl is $2m$ in ions ($1m$ in Na^+ and $1m$ in Cl^-), one would expect the freezing point to be lowered $2 \times 1.86°$, or $3.72°C$, and the boiling point to be raised $2 \times 0.512°$, or $1.024°C$. Similarly, a salt such as $BaCl_2$, which yields 3 moles of ions per mole of compound, would be expected to lower the freezing point $3 \times 1.86°$, or $5.58°C$, and so forth.

TABLE 11-6 **Freezing Points** **(Actual and Predicted)** **of 0.1m Solutions** **of Some Electrolytes**	*Electrolyte*	*Number of Ions*	*Predicted Freezing Point*	*Actual Freezing Point*
	NaCl	2	−0.372	−0.348
	NaOH	2	−0.372	−0.348
	$MgCl_2$	3	−0.558	−0.494
	K_2SO_4	3	−0.558	−0.432
	$K_3[Fe(CN)_6]$	4	−0.744	−0.530

Experiment reveals, however, that while these strong electrolytes do affect colligative properties to an extent close to that which is predicted on the basis of number of ions, the effect is always less than the whole-number multiple expected. This is demonstrated in Table 11-6, where the experimentally determined freezing points of 0.1 molal aqueous solutions of several strong electrolytes are compared with the predicted freezing points. The same relationship holds for the other colligative properties as well. The differences between the measured colligative properties of strong electrolyte solutions and the predicted whole-number values decrease as the solutions become more dilute, as illustrated in Table 11-7.

TABLE 11-7 **Observed Freezing Point** **Depressions of Some NaCl Solutions** **Compared to Calculated Values**	*Molal Concentration*	*Freezing Point Observed ($°C$)*	*Depressions Calculated ($°C$)*
	1	3.46	3.72
	0.1	0.348	0.372
	0.01	0.0360	0.0372
	0.001	0.00366	0.00372

These deviations shown by ionic solutes may be explained as follows: Because of the electrical charges on ions, each ion tends to be attracted to ions of opposite sign. These interionic attractions restrain the ions somewhat in their freedom of motion, so that they are not as independent of one another as uncharged molecules in solution. The result is that the effectiveness of the ions is decreased, and the solution behaves as if there are fewer particles present. As a solution becomes more dilute, the ions are spread farther and farther apart, so that their influence on one another decreases, and in extremely dilute solutions the ions behave essentially independently. This explanation is a very simplified version of what is known as the *Debye-Hückel Theory*.

COLLOIDS

The statement has been made earlier in this chapter that the fundamental difference between homogeneous mixtures (solutions) and heterogeneous mixtures is in the *size* of the dispersed particles. In a true solution the dispersed particles of solute are individual molecules or ions, and these submicroscopic particles will not "settle out" no matter how long the solution is kept standing. In a heterogeneous mixture, on the other hand, the particles are relatively large aggregates of atoms, visible under a microscope if not to the naked eye, and large enough to be separated by gravity. Thus, for example, if fine sand is shaken with water, the sand particles will remain distributed throughout the water for only a short period of time; they will quickly "settle out." This mixture obviously consists of two distinguishable phases.

There is a type of mixture, intermediate between these two extremes, wherein the dispersed particles are considerably larger than most molecules, yet not large enough to be visible or to be separated by gravity. These mixtures are called *colloidal dispersions* or simply *colloids.*

It is not possible to state precisely the range of particle sizes that represents colloidal dispersions, but this range is generally considered to be approximately between 10 Å and 5000 Å in diameter. Most molecules have diameters of less than 10 Å, and particles larger than 5000 Å can be seen under an ordinary microscope and usually do not remain colloidally dispersed. It should be noted that the size of the particle does not define the makeup of the particle. Colloidal particles may be aggregates of atoms, ions, or molecules; or, in the case of some very large molecules, such as starch, proteins, and other polymers, they may be single molecules. (Hemoglobin, with a molecular weight of more than 60,000 and a diameter of approximately 30 Å, is colloidally dispersed in blood.)

Although colloidal particles are too small to be visible to the naked eye, they are large enough to reflect light, and this fact can sometimes be used to distinguish between colloids and true solutions. When a narrow beam of bright light is directed to pass through a colloidal dispersion, the dispersed particles scatter the light, and by viewing the system at right angles to the beam, one can observe the path of the light beam. This phenomenon has been named the *Tyndall effect,* after its discoverer. A true solution does not display this effect because the dispersed solute particles are too small to reflect light from their surfaces.

Colloids can be classified into a number of different types, the most important of which are *aerosols, emulsions, sols,* and *gels.*

An *aerosol* is a dispersion of either a solid or liquid in a gas. Smoke is a common example of the former; clouds, mists, and fogs are examples of the latter. Many commercial aerosol sprays, such as insecticides, deodorants, and hair sprays, are colloidal dispersions of liquids in air.

Emulsions are colloids in which one liquid is dispersed in another liquid. Milk, cream, and mayonnaise are examples of fats or oils dispersed in water.

A colloidal dispersion of a solid in a liquid is called a *sol.* The dispersed particles may be fragments of the crystalline lattice of the solid—as in colloidal dispersions of gold, sulfur, silver iodide, etc., in water—or they may be single, large molecules such as starch, protein, and gums.

Gels are somewhat different from the other types of colloids in that both phases are continuous. A solid is arranged in a fine network, interlacing a liquid. Jellies, gelatin desserts, and gelatinous precipitates are common examples of gels.

The properties of colloids are more complex and more difficult to quantify than the properties of true solutions. In general, solid-in-liquid colloids have vapor pressures, boiling points, and freezing points which differ very little from those of the pure solvents. The colligative relationships that have been discussed for solutions do not apply to colloidal dispersions.

SUGGESTED READINGS

Nash, Leonard K. *Stoichiometry*. Reading, Massachusetts: Addison-Wesley, 1966. Chapter 4.

PROBLEMS

1. Indicate clearly the difference between the terms in each of the following pairs:
   (a) solute and solvent
   (b) solubility and rate of solution
   (c) saturated solution and unsaturated solution
   (d) molarity and molality
   (e) electrolyte and nonelectrolyte
   (f) strong electrolytes and weak electrolytes
   (g) acid anhydrides and basic anhydrides
   (h) solution and colloid

2. Predict whether each of the following compounds will have greater solubility in water or in benzene.
   (a) potassium nitrate
   (b) ethyl acetate
   (c) ethyl ether
   (d) oxalic acid
   (e) *n*-hexane
   (f) iodine
   (g) chloroform (trichloromethane)

3. From the following thermodynamic data (at $25°C$), calculate the free energy of solution for silver fluoride and for silver chloride at $25°C$. On the basis of your answers, comment on the relative solubilities of the two compounds.

For AgF:	Lattice energy	= 231.1 kcal/mole
	Heat of hydration	= 235.9 kcal/mole
	Entropy of solution	= −4.73 cal/mole-$°K$

For AgCl	Lattice energy	= 219.4 kcal/mole
	Heat of hydration	= 203.6 kcal/mole
	Entropy of solution	= +7.85 cal/mole-$°K$

4. Explain how you would prepare each of the following aqueous solutions:
   (a) 1.00 liter of $0.50M$ KNO_3 solution from solid KNO_3
   (b) 250 ml of $0.25M$ NaCl solution from a $1.0M$ NaCl stock solution
   (c) 500 ml of a solution that is $0.15M$ in Cl^- from solid $CaCl_2 \cdot 2H_2O$
   (d) At least 1 liter of $0.50m$ $NaC_2H_3O_2$ from solid $NaC_2H_3O_2$
   (e) At least $\frac{1}{2}$ liter of a 15 wt-% solution of NaOH from solid NaOH

5. What weight of metal ion is contained in each of the following aqueous solutions?
   (a) 100 ml of $0.20M$ $AgNO_3$
   (b) 20 ml of $0.75M$ $CuSO_4$
   (c) 3.0 liters of $1.2 \times 10^{-3}M$ $MgCl_2$
   (d) 500 ml of a 5.0 wt-% $Hg(NO_3)_2$ solution, which has a density of 1.06 g/ml

6. Commercial concentrated hydrochloric acid contains 36.5 wt-% HCl and has a density of 1.185 g/ml. Express the concentration of this solution in terms of (a) molarity, (b) molality, and (c) mole fraction of HCl.

7. A 9.00 wt-% stock solution of potassium hydroxide has a density of 1.082 g/ml. How would you prepare 1 liter of a $1.0M$ solution of KOH from the stock solution?

8. A solution of $1.86M$ sulfuric acid has a density of 1.115 g/ml. Calculate the molality of the solution.

9. Calculate the percentage composition by weight of nitric acid in a 2.0 molal HNO_3 solution.

10. How many milliliters of 47.0% hydrobromic acid (density 1.50 g/ml) are needed to prepare 1 liter of a $0.156M$ solution?

11. The solubility of oxygen in water at $25°C$ and 1 atm is approximately 0.04 g per liter of solution. What is the molarity of a saturated aqueous solution of oxygen at this temperature and pressure?

12. A solution is prepared by dissolving 25.0 g of methanol (CH_3OH) in 100 g of water. Express the concentration of this solution as (a) weight percent, (b) molality, and (c) mole fraction of methanol.

13. The solubility of $CaSO_4$ in water at $10°C$ is 1.063 g per liter of solution. Calculate the molarity of a saturated $CaSO_4$ solution at $10°C$.

14. A 20.0 wt-% solution of $Na_2S_2O_3$ in water has a density of 1.10 g/ml. Calculate (a) the molarity and (b) the molality of the solution.

15. A mixture of 60.0 g of chloroform and 80.0 g of carbon tetrachloride is prepared. Calculate the mole fraction of each constituent of the resulting solution.

16. At $20°C$ the solubility of lead chloride is 0.99 g per 100 ml of solution. At $100°C$ the solubility is 3.14 g per 100 ml of solution.
   (a) Calculate the molarity of saturated solutions of $PbCl_2$ at the two temperatures.
   (b) Is the solution process endothermic or exothermic?

17. Calculate the vapor pressure at $25°C$ of an aqueous solution containing 100 g of glycerine ($C_3H_8O_3$) in 150 g of water. (Assume glycerine to be nonvolatile.)

18. Methanol is sometimes used as an antifreeze. Calculate (a) the number of grams and (b) the number of moles of methanol that must be added to 10.0 liters of water to lower the freezing point to $-10.0°C$.

19. Explain the fact that freshly boiled water freezes at a higher temperature than unboiled tap water.

20. A solution of 4.785 g of mercury(II) chloride in 60 g of water is found to boil at $100.152°C$. Is mercury(II) chloride dissociated in water solution?

21. A certain nonvolatile, molecular compound has a molecular weight of 75.2. If 1.00 g of this compound is dissolved in 10.00 g of pure benzene, what is the vapor pressure of the solution at $25°C$? (The vapor pressure of pure benzene at $25°C$ is 95.5 torr.)

22. At $23°C$ the vapor pressures of chloroform and carbon tetrachloride are 178 torr and 100 torr, respectively. Calculate the vapor pressure of the solution in problem 15.

23. The vapor pressure of pure benzene at $25°C$ is 95.5 torr. When 10.0 g of compound X is dissolved in 90.0 g of C_6H_6, the vapor pressure of the solution is 94.8 torr. Calculate the molecular weight of compound X.

24. If 50.0 g of a compound of molecular weight 92 is dissolved in 100.0 g of water, what is the normal boiling point of the solution? (See Table 11-2.)

25. When 40.0 g of compound Y is dissolved in 500.0 g of benzene, the resulting solution has a normal boiling point of $81.4°C$. Calculate the molecular weight of compound Y. (See Table 11-2.)

26. What is the freezing point of a $0.50m$ aqueous sugar solution? (See Table 11-3.)

27. A sample of otherwise pure camphor contains an impurity of a solid hydrocarbon, $C_{10}H_8$, to the extent of 1.0 g of the hydrocarbon for every 100 g of camphor. What is the melting point of the camphor sample? (See Table 11-3.)

28. When 1.00 g of compound A is dissolved in 10.00 g of benzene, the resulting solution has a freezing point of $-1.3°C$. Calculate the molecular weight of compound A.

29. When 0.650 g of a certain compound is dissolved in 27.80 g of diphenyl, the freezing point of the solution is $68.44°C$. The freezing point of pure diphenyl is $70.00°C$, and the molal freezing point constant is 8.00. Calculate the molecular weight of the compound.

30. A solution consisting of 0.100 g of a certain compound and 10.00 g of cyclohexane freezes at $1.00°C$. Calculate the molecular weight of the compound. (Cyclohexane has a freezing point of $6.00°C$ and a molal freezing point constant of 20.0.)

31. A compound has a percentage composition of 93.75 wt-% carbon and 6.25 wt-% hydrogen. When 0.320 g of this compound is dissolved in 10.0 g of benzene, the solution has a freezing point of $4.20°C$. Calculate (a) the empirical formula, (b) the molecular weight, and (c) the molecular formula of the compound.

32. A $0.10m$ solution of a certain weak acid freezes at $-0.245°C$. Calculate the percentage dissociation of the weak acid.

33. A $0.50m$ solution of NaCl freezes at $-1.70°C$. What is the *apparent* percentage of dissociation of NaCl?

34. Explain the following experimental observations:
    (a) The freezing point depression per mole of potassium chloride is greater in a 0.010 molal solution than in a 1.0 molal solution.
    (b) The vapor pressure of the solvent over a solution of nonvolatile solute is less than the vapor pressure of the pure solvent.
    (c) When sodium chloride is sprinkled on pavement ice, it causes the ice to melt.

35. Assuming complete dissociation, calculate the moles of calcium chloride needed to lower the freezing point of 10.0 liters of water to $-10.0°C$. Why is $CaCl_2$ not used as an automotive antifreeze?

36. Designate each of the following compounds as strong electrolyte, weak electrolyte, or nonelectrolyte:
    (a) KBr
    (b) NH_4Cl
    (c) CH_3NH_2
    (d) HCN
    (e) CO_2
    (f) $NaC_2H_3O_2$
    (g) HCOOH
    (h) $HClO_4$
    (i) C_6H_{12}
    (j) CaO
    (k) NaOH
    (l) CH_3COCH_3
    (m) CH_3CH_2OH
    (n) SO_2
    (o) $NH_4C_2H_3O_2$
    (p) CH_3OCH_3
    (q) $CH_3CH_2COOCH_3$
    (r) MgI_2
    (s) HF
    (t) CH_3CH_2Cl

37. What volume of $6.0M$ HCl solution is required to prepare 250 ml of Cl_2 at STP according to the following equation:

$$MnO_2 + 4HCl \rightarrow MnCl_2 + Cl_2 + 2H_2O$$

38. On the basis of the equation

$$2Al(s) + 3H_2SO_4(aq) \rightarrow Al_2(SO_4)_3(aq) + 3H_2(g)$$

what volume of $1.0M$ sulfuric acid solution will be required to react completely with 15.0 g of Al?

39. What volume of dry oxygen gas (at STP) is liberated by the complete decomposition of the hydrogen peroxide in 1.0 liter of a 3.0 wt-% aqueous solution of hydrogen peroxide? The density of the solution is 1.009 g/ml.

$$2H_2O_2 \rightarrow 2H_2O + O_2$$

40. Excess zinc metal is added to 100 ml of a $1.21M$ solution of hydrochloric acid. How many liters of wet hydrogen gas are collected over water at $25°C$ and 740 torr?

$$Zn + 2HCl \rightarrow ZnCl_2 + H_2$$

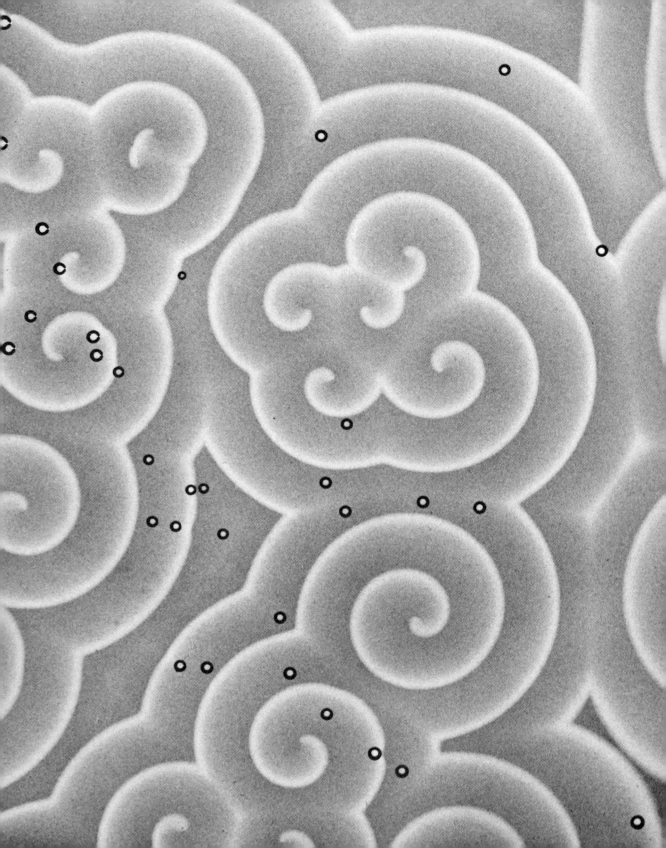

PART TWO

The Reactions of Matter

12 *Rates of Reactions*

A chemical equation is simply a statement that describes the reactants, the products into which they are converted, and the molar ratios pertaining to this conversion. It does not tell us anything about the speed with which the conversion takes place. Yet, knowledge of the speed of a reaction—called the *reaction rate*—is often of great importance if we wish to employ the reaction as a preparative process or to gain an understanding of the mechanism—the theoretical pathway—by which the reaction proceeds.

In general, the rate of any process is measured in terms of quantity per unit time. The rate at which an automobile travels is measured in miles per hour, the rate of flow of water from a pipe may be measured in gallons per minute, and so on. The rate of a chemical reaction is most conveniently measured in terms of the quantity of a reactant consumed or of a product produced per unit of time, and the quantity is most commonly expressed as the number of moles. Thus, for example, in the hypothetical reaction $A + B \rightarrow C + D$, the rate might be expressed as number of moles of A (or of B) consumed per minute or as number of moles of C (or of D) produced per minute.

For any given reaction the rate is greatly influenced by the specific conditions imposed upon it. Furthermore, for most reactions, the rate is not constant with time but is a maximum at the start of the reaction and gradually decreases with time. Therefore, a reaction rate is a measure of the speed with which reactants are converted into products *at a given instant under a given set of conditions*. The study of reaction rates and the factors that influence them is called *chemical kinetics*.

Before discussing the factors that influence the rate of a chemical reaction, we shall find it convenient to divide reactions into two categories: homogeneous and heterogeneous. *Homogeneous reactions* take place in a single phase; the reaction mixture is a solution. Some examples are

$$HCl(aq) + NaOH(aq) \rightarrow NaCl(aq) + H_2O(l)$$

$$N_2(g) + 3H_2(g) \rightarrow 2NH_3(g)$$

In *heterogeneous reactions,* on the other hand, the reaction mixture consists of more than one phase, and reaction occurs at a phase boundary. The following equations illustrate this category:

$$Zn(s) + 2HCl(aq) \rightarrow ZnCl_2(aq) + H_2(g)$$

$$CaO(s) + CO_2(g) \rightarrow CaCO_3(s)$$

FACTORS THAT DETERMINE REACTION RATES

Nature of the Reactants

Since chemical reactions are processes involving the making and breaking of bonds, the rate of a reaction should, logically, depend on the specific bonds formed and broken, and therefore on the specific structure and composition of the reactants. The reaction rates observed for different reactants vary widely from very slow to extremely fast.

Most reactions involving the combination of ions—for example, aqueous acid-base neutralization, or precipitation of an ionic solid from aqueous solution—occur very rapidly. Most nonionic reactions occur at considerably lower rates, requiring minutes or hours to reach completion. The rates of some very slow reactions (for instance, the rusting of iron) might be measured more conveniently in terms of months or years.

Clearly, then, under any given set of conditions, the rate of a chemical reaction is determined to a large extent by the specific nature of the substances taking part in the reaction.

Concentration of the Reactants

For homogeneous chemical reactions the reaction rate depends on the concentration of the reactants. The precise relationship between concentration and rate for any particular reaction is not predictable from the equation and must be determined experimentally. Consider the hypothetical reaction between substances A and B to form C (A + B → C). Suppose we mix together some A and some B in such amounts that their molar concentrations are equal; say, 0.100M. After allowing the reaction to proceed for some definite period of time, we analyze the reaction mixture to determine the concentration of A or B remaining. Or, if it is more convenient, we might analyze for the amount of C formed. From the results of our analysis and the length of time the reaction has proceeded, we have a measure of the reaction rate.

Now suppose we repeat the experiment under identical conditions except that the concentration of A is 0.200*M*, and we then discover the reaction rate to be twice what it was in the first experiment. In other words, doubling the initial concentration of A is found to double the rate. If we use a concentration of A of 0.300*M*, we find that the rate is tripled, and so on. Let us suppose, further, that we repeat the series of experiments, keeping the initial concentration of A constant at 0.100*M* but varying the initial concentration of B—0.100*M*, 0.200*M*, 0.300*M*—and discover that the reaction rate is affected in the same way by changes in the concentration of B as it was by changes in the concentration of A. We have ascertained, then, that for this particular reaction the rate is directly proportional to both the molar concentration of A and the molar concentration of B, and we may write

$$\text{Rate} \propto [A][B]$$

where brackets around the chemical symbol or formula is conventional notation for molar concentration. By use of a proportionality constant, *k,* this expression may be converted into a mathematical equation:

$$\text{Rate} = k[A][B]$$

This equation is called the *rate law* for the particular reaction in question, for it expresses the relationship between reaction rate and concentration of reactants at a specified set of conditions. The constant *k* is called the *specific rate constant* and is characteristic of that reaction at those specified conditions. Once the value of *k* has been determined for a particular reaction, the reaction rate can be predicted for any stated initial concentrations of A and B, provided the specified conditions are maintained. As a further illustration of rate law determination, let us imagine another series of experiments, this time involving reaction between substances X and Y. Let us suppose that our investigation shows that when the concentration of Y is held constant and the concentration of X is varied, the reaction rate varies directly with the concentration of X—when [X] is doubled, the rate is doubled; when [X] is tripled, the rate is tripled, and so on. However, when [X] is held constant and [Y] is varied, we find that doubling [Y] *quadruples* the rate; and when [Y] is tripled, the rate increases by a factor of *nine.* What we have discovered is that the reaction rate is directly proportional to the molar concentration of X, but proportional to the *square* of the molar concentration of Y. This can be expressed as

$$\text{Rate} \propto [X][Y]^2$$

and the rate law for this reaction is

$$\text{Rate} = k[X][Y]^2$$

In general, the rate of a homogeneous reaction is proportional to the product of the molar concentrations of the reactants, each raised to some power. Thus, for the generalized reaction,

$$a\text{A} + b\text{B} + \cdots \rightarrow$$

the rate law is expressed as

$$\text{Rate} = k[A]^m[B]^n \cdots$$

This relationship, first discovered by Cato Guldberg and Peter Waage in 1864, is called the *Law of Mass Action.* The exponents m and n, usually small integers, are often referred to as the *order* of the reaction; m is the order with respect to substance A, n is the order with respect to B, and the sum $m + n$ is the overall order of the reaction. Thus, the hypothetical reaction for which the rate law is

$$\text{Rate} = k[X][Y]^2$$

is first order with respect to X, second order with respect to Y, and third order overall.

It should be emphasized that the values of m and n in our generalized rate equation *cannot* be predicted from the stoichiometry of the chemical equation; they must be determined experimentally. It is an error to assume that the coefficients of the balanced equation are the exponents in the rate law. This is true only in those cases in which the reaction occurs in a *single step.* Most reactions proceed by a mechanism that involves several steps; that is, several successive reactions. Each of these steps proceeds at its own rate, and each rate is dependent on the concentrations of the substances that are the reactants for *that step.* A chemical equation describes the stoichiometry of the *overall* reaction; it implies nothing about the mechanism. Therefore, only in those reactions for which the equation is also a description of the only step that occurs are the coefficients of the equation equal to the exponents in the rate law.

Although the values of m and n are usually integers, they may be fractions or even zero. Again, this is the result of the fact that the overall rate of a reaction is often the summation of a multistep process. The relationship of reaction rate to reaction mechanism is further discussed later in this chapter.

Table 12-1 lists a few simple homogeneous reactions and their experimentally determined rate laws. Note that in some cases the exponents are the same as the coefficients of the equations; in other cases they are not.

TABLE 12-1 **Rate Laws of Some Homogeneous Reactions**	*Reaction*	*Rate Law*
	$2N_2O_5 \rightarrow 4NO_2 + O_2$	$r = k[N_2O_5]$
	$2N_2O \rightarrow 2N_2 + O_2$	$r = k[N_2O]^2$
	$CH_3CHO \rightarrow CH_4 + CO$	$r = k[CH_3CHO]^2$
	$2NO + O_2 \rightarrow 2NO_2$	$r = k[NO]^2[O_2]$
	$F_2 + 2ClO_2 \rightarrow 2FClO_2$	$r = k[F_2][ClO_2]$
	$2H_2 + 2NO \rightarrow 2H_2O + N_2$	$r = k[H_2][NO]^2$

The effect of concentration on reaction rate cannot be stated so simply for heterogeneous reactions. In a heterogeneous reaction the reactants are not components of the same solution, and therefore the relative amounts of reactants present cannot be expressed meaningfully in terms of molarity. Consider, for example, the reaction between metallic zinc and hydrochloric acid:

$$Zn(s) + 2HCl(aq) \rightarrow ZnCl_2(aq) + H_2(g)$$

The amount of HCl in the aqueous acid solution can, of course, be described in terms of molarity, but the zinc metal is not dissolved in the solution and therefore an expression of its concentration is meaningless. Since the reaction can take place only at the boundary between the solid phase and the liquid phase, the area of contact between the two phases influences the reaction rate. The more finely divided the zinc, the greater is its surface area and the faster the reaction will occur. This effect of particle size on reaction rate applies to heterogeneous reactions in general, where one or more of the reactants is a solid.

This effect has been strikingly illustrated by numerous coal mine explosions resulting from the ignition of bituminous dust. A lump of coal requires a high temperature for ignition, and once ignited it burns slowly and quietly. But finely divided coal dust suspended in air may be ignited by a match with explosive violence. Similar disasters have occurred in factories and mills from the ignition of fine dusts of other combustible materials.

Temperature

The results of observations of the effect of temperature on the rate of chemical reactions can be summarized very simply: *increasing the temperature increases the reaction rate,* and *lowering the temperature decreases the rate.* This relationship holds for both endothermic and exothermic reactions. What this means in terms of the rate law is that an increase in temperature increases the specific rate constant k, and a decrease in temperature decreases k.

The magnitude of this effect—like the rate law itself—is not predictable but must be determined experimentally. It varies widely from one reaction to another, and even for a given reaction it varies with temperature range. The effect is generally quite marked, as indicated by the rough rule of thumb often cited: For each $10°C$ rise in temperature the rate is approximately doubled. This is a very rough approximation and should not be relied upon in quantitative determinations.

Catalysis

The rates of some chemical reactions are increased by the presence of certain substances that are themselves not consumed in the overall reactions and therefore do not appear as reactants or products in the chemical equations. These substances are called *catalysts,* and their rate-increasing effect is called *catalysis.* Any reaction whose rate is increased by a catalyst is said to be *catalyzed.*

Several important facts concerning catalysts should be emphasized. First, a catalyst does not *cause* a reaction to occur. If a reaction is thermodynamically impossible at a specified set of conditions, then with or without a catalyst that reaction will not take place. The catalyst merely increases the *rate* of the reaction by providing an alternate path—a different mechanism—by which the reaction can proceed. Second, a catalyst does not speed up a chemical reaction by its mere passive presence; it plays an active role. Although the catalyst is not consumed in the overall reaction and may be recovered unchanged from the reaction mixture, this does not mean that it does not undergo chemical change during some of the steps of the process. The catalyst may be used up in one step and regenerated in a later step, so that it reacts over and over again without undergoing any net change. This is illustrated by a mechanism that has been postulated for the decomposition of hydrogen peroxide, for which iodide ion is a catalyst.

$$H_2O_2 + I^- \rightarrow H_2O + IO^- \qquad \text{(catalyst consumed)}$$

$$H_2O_2 + IO^- \rightarrow H_2O + O_2 + I^- \qquad \text{(catalyst regenerated)}$$

$$\overline{2H_2O_2 \rightarrow 2H_2O + O_2 \qquad \text{(overall reaction)}}$$

A third important aspect of catalysis lies in the fact that whatever effect a given catalyst has on a specific reaction, its effect on the reverse of that reaction is the same. Thus, if a catalyst doubles the rate at which A and B react to form C and D, then that same catalyst will also double the rate at which C and D react to form A and B, given the same set of conditions.

When the catalyst is in the same phase as the reactants and products of the reaction—like the iodide ion in the iodide-catalyzed decomposition of hydrogen peroxide just cited, where all substances are in aqueous solution—the catalysis is said to be *homogeneous.* Catalysis in which the catalyst is in a separate phase from the reactants and the reaction occurs at the catalytic surface is called *heterogeneous* or *surface catalysis;* for example, the gas-phase hydrogenation of an alkene using finely divided platinum as a catalyst.

$$CH_2{=}CH_2 + H_2 \xrightarrow{\text{Pt}} CH_3{-}CH_3$$

In some instances, the presence of a substance in a reaction mixture may bring about a *decrease* in the reaction rate. Such a substance, which is called a *negative catalyst* or *inhibitor,* functions by interfering with the normal mechanism of the uncatalyzed reaction.

Catalysis is a common phenomenon in nature. Nearly all of the multitude of complex chemical reactions that occur in living organisms are catalyzed by highly specific catalysts called *enzymes.* These enzymes, most of which are protein molecules of very high molecular weight, permit vital reactions to occur at high rates in spite of the low concentrations and low temperatures that exist in living cells. The breakdown of complex nutrient molecules (fats, carbohydrates, and proteins) into smaller molecules, as well as the synthesis from these smaller molecules of nucleic acids, vitamins, hormones, lipids, and a host of other vital substances (including enzymes themselves) all involve enzyme-catalyzed reactions. Moreover, unlike most chemical reactions carried out in the laboratory, enzyme-catalyzed reactions give 100 percent yield; there are no side reactions and no unwanted by-products.

THEORY OF REACTION RATES

The foregoing discussion has been largely a reporting of observed facts that describe the influence of various conditions on the rate of chemical reactions. In order to explain this body of facts, let us now develop a model that will permit us to interpret the described behavior on the molecular level.

Molecular Collision Theory

The *collision theory*—an extension of the kinetic molecular theory of gases (see Chapter 10)—holds as its basic assumption that in order for reaction to occur, the reacting particles (molecules, atoms, or ions) must collide. For substances A and B to react with each other, molecules of A must undergo collision with molecules of B in such a way that electrons and atoms are rearranged, leading to the formation of new substances. The greater the frequency of these collisions, the greater the rate of reaction. Therefore, any change of conditions that increases the frequency of collisions will also increase the reaction rate.

Not every collision between reactants results in reaction, however. One can show by calculation from the kinetic molecular theory that the number of collisions between molecules in one liter of gas at STP is of the order of 10^{32} collisions per second. Obviously, if every collision resulted in reaction, all reactions would occur instantaneously. This is not the case, of course, and we must distinguish between *total* number of collisions and number of *effective collisions* (those which result in a reaction). Many reactions have such a low rate that the number of effective collisions per unit time must be a very small fraction of the total number of collisions occurring during that time.

The fact that some collisions are effective while others are not may be explained by considering two factors. The first of these is the *orientation* of the colliding molecules. Since most molecules are not simple spheres, colliding molecules may have a large variety of different positions, relative to each other, at the instant of collision. Many of these positions may be unfavorable for the breaking and making of bonds necessary to the reaction. Only certain orientations, then, can lead to effective collisions. This is illustrated in Figure 12-1 for the simple reaction

$$2AB \rightarrow A_2 + B_2$$

which proceeds in a single step by direct collision of two molecules of AB.

A second, more significant factor in determining the effectiveness of collision is the kinetic energy of the colliding molecules. It will be recalled from our earlier discussion of matter in bulk that the molecules of a gas are in constant linear motion and that, as a result of their velocity and their mass, they possess kinetic energy. At any given instant, the molecules in a gas sample have a wide range of velocities and therefore a wide range of kinetic energies. As a result of random collisions, the instantaneous velocities of some molecules will be very low, or even

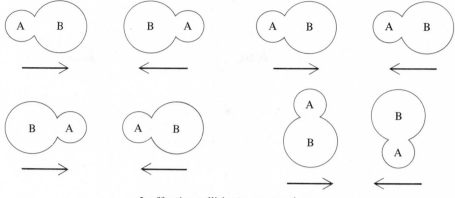

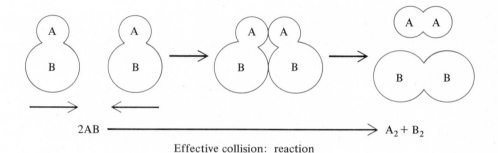

FIGURE 12-1

Collision Orientations between Two Molecules of AB

zero, while other molecules will have very high velocities because of collisions in which momentum was transferred to them. The kinetic energy of any individual molecule is constantly changing with time. However, in a measurable sample of gas, there is such a large number of molecules present that at any one time the number of molecules with a particular velocity will be essentially constant. A very small fraction of the molecules will have extremely high or low kinetic energies, while most of the molecules will have kinetic energies close to some average value. This statistical distribution is shown graphically in Figure 12-2. This curve, called a *Maxwell-Boltzmann distribution curve,* can be predicted theoretically and confirmed experimentally. It will be noted that the curve is not symmetrical, since it is limited to zero at the low-energy end but the high-energy end is limitless. The maximum in the curve represents the most probable kinetic energy; that is, a larger fraction of the molecules possess that kinetic energy than any other. Because the curve is skewed, the average kinetic energy is somewhat higher than the most probable kinetic energy.

In a reaction, then, the collisions between the molecules of the reactants

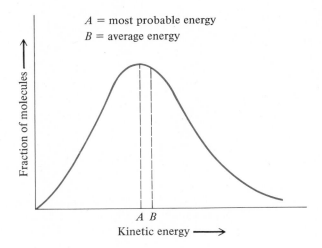

FIGURE 12-2

Maxwell-Boltzmann Distribution Curve

occur with a wide range of total kinetic energies. The negativity of the electron clouds surrounding molecules produces a force of repulsion between molecules. If two molecules collide with low kinetic energy, the repulsion between the electron clouds may cause them to simply rebound without any reaction. For a collision to be *effective,* the colliding molecules must have sufficiently high kinetic energy to overcome the mutual repulsions of the electron clouds. Only then can the molecules penetrate each other far enough to permit the rearrangement of electrons that constitutes the making and breaking of bonds.

For each reaction, then, there is some minimum amount of energy that the colliding molecules must possess in order for the collision to be effective. This minimum amount is called the *activation energy.* As the distribution curve of Figure 12-2 reveals, not all molecules will possess sufficient energy, and only a fraction of the total collisions will be effective.

It should perhaps be pointed out that although our discussion of molecular energy and molecular collisions has been based on the gaseous state, the same considerations apply, in general, to reacting species in liquid solution.

Transition State Theory

An extension of the collision theory, known as the *transition state theory,* is aimed at a more detailed description of the collision process and the relation of this process to the activation energy. The basic postulate of this theory is that an effective collision between reactant molecules results in the formation of an unstable, transitory intermediate species called the *transition state* or *activated complex.* As the two colliding molecules contact each other, some bonds become longer and weaker, and some new bonds begin to be formed. Thus the activated complex may be visualized as being held together by "partial" bonds. Although this activated complex does not represent a substance that can be observed or isolated, it is considered to have many of the same properties as real molecules: molecular weight, bond lengths and angles, energy, and so forth.

The formation of the activated complex and its relationship to the reactants and products of the reaction are illustrated in Figure 12-3 for the simple hypothetical reaction $2AB \rightarrow A_2 + B_2$.

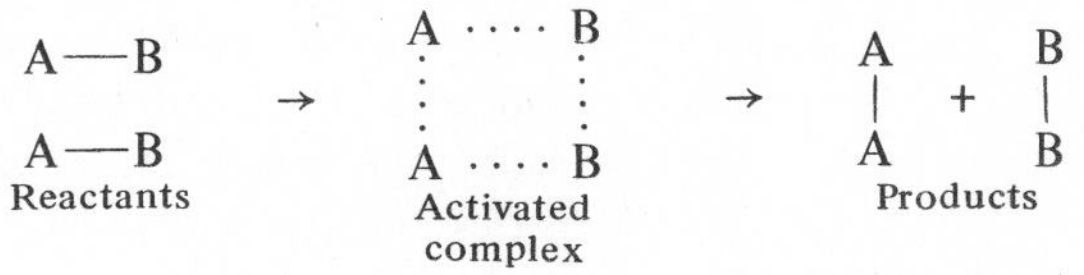

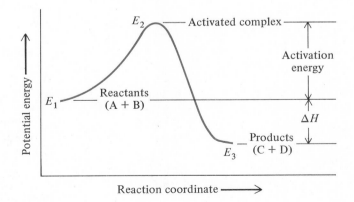

FIGURE 12-3

Schematic Representation of Activated Complex Formation

As its name implies, the activated complex is in a relatively high energy state; it possesses greater energy than either the reactants or the products. This relationship is shown graphically in Figure 12-4, where the potential energy of the system is plotted against the reaction coordinate (a measure of how far the reaction has proceeded). The molecules of the reactants have some average energy, E_1. When collision occurs between reactant molecules, work must be done on the system to overcome the mutual repulsion of the electron clouds. In other words, the potential energy must increase. This is accomplished by conversion of some of the kinetic energy of the colliding molecules into potential energy. In order for a collision to be effective—for the activated complex to be formed—the potential energy must reach E_2. The collision of molecules of low kinetic energy may provide an amount

FIGURE 12-4

Potential Energy Change during a Reaction (Exothermic)

of potential energy somewhere between E_1 and E_2, but not enough to result in formation of the activated complex. These molecules then fly apart unchanged. (In terms of the energy diagram, this corresponds to moving part way up the hill and then sliding back down again.) On the other hand, if the total kinetic energy of the colliding molecules is sufficiently great, the necessary potential energy, E_2, will be reached and the activated complex formed. This potential energy barrier, which must be overcome to form the activated complex—that is, the difference between the potential energy of the reactants (E_1) and that of the activated complex (E_2)—is the *activation energy*. Being unstable, the activated complex "splits apart," forming the products, and the potential energy decreases to E_3. (The system slides down the other side of the hill in the energy diagram.)

The energetics of the *reverse* reaction can also be interpreted in terms of Figure 12-4. Let us suppose that the diagram represents the reaction

$$A + B \rightarrow C + D$$

that is, the reactants are A and B and the products are C and D. Now it is also possible for substances C and D to react to form A and B:

$$C + D \rightarrow A + B$$

In this reverse reaction the activated complex formed is the same as that formed in the forward reaction. For the reverse reaction the reaction coordinate in the diagram is now proceeding from right to left, and energetically the system must move up and over the hill from right to left, so the activation energy is $E_2 - E_3$.

It should be noted that Figure 12-4 represents an *exothermic* reaction, for the products possess less energy than the reactants. The amount of energy *released* in going from the activated complex to the products is greater than the activation energy. The difference between the average energy of the products and the average energy of the reactants ($E_3 - E_1$) is the energy released in the reaction (labeled ΔH

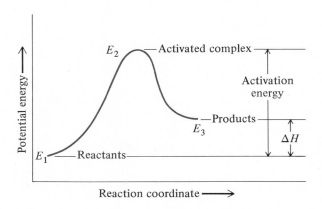

FIGURE 12-5

Potential Energy Diagram for an Endothermic Reaction

in Figure 12-4). The magnitude of ΔH is dependent only on the reactants and the products, not on the pathway, and is therefore not related to either the rate or the activation energy. For comparison, a similar energy diagram of an *endothermic* reaction is given in Figure 12-5. Note that in this case, ΔH represents energy *absorbed* in the reaction, which is also independent of rate and activation energy.

It follows from the foregoing discussion that for any reaction that occurs in a single step (or any single step in a multistep reaction), it is the rate of formation of the activated complex that determines the rate of the reaction. Moreover, other conditions being equal, the lower the activation energy, the higher the fraction of collisions that are effective and the greater the reaction rate.

RATIONALIZATION OF RATE-DETERMINING FACTORS

Taken together, the two theories we have just discussed provide us with a model whereby we can explain satisfactorily many of the known facts regarding reaction rates. Let us now return to the four factors that determine the rate of a chemical reaction—nature of the reactants, concentration of reactants, temperature, and catalysis—and apply our model to the explanation of their effects.

Nature of the Reactants

The fact that different reactions occur at widely different rates even though the same set of conditions is maintained can be explained on the basis of difference in activation energy. The nature of the reactants, of course, determines the nature of the activated complex and therefore the activation energy. In general, very fast reactions have low activation energies; slower reactions have higher activation energies.

Concentration of the Reactants

The effect of concentration on reaction rate can be explained simply on the basis of frequency of collisions. If the number of molecules per unit volume is increased, then the number of molecular collisions occurring per unit time must necessarily increase also.

Let us consider again our simple hypothetical example of the one-step reaction between A and B, for which the rate law is $r = k[A][B]$. The effect of concentration on number of collisions is shown in Figure 12-6. If we assume a volume so small that it contains only one molecule of A and one of B, then only one collision is possible. When the number of molecules of A in the same volume is doubled, the number of A–B collisions is doubled. Similarly, with one molecule of A and two of B, two A–B collisions are possible. In short, the number of collisions per unit time is directly proportional to the concentrations of both A and B, and this is in agreement with the rate law.

It should be emphasized that changing the concentration of a reactant changes the *total* number of collisions per unit time; it does not have any effect on

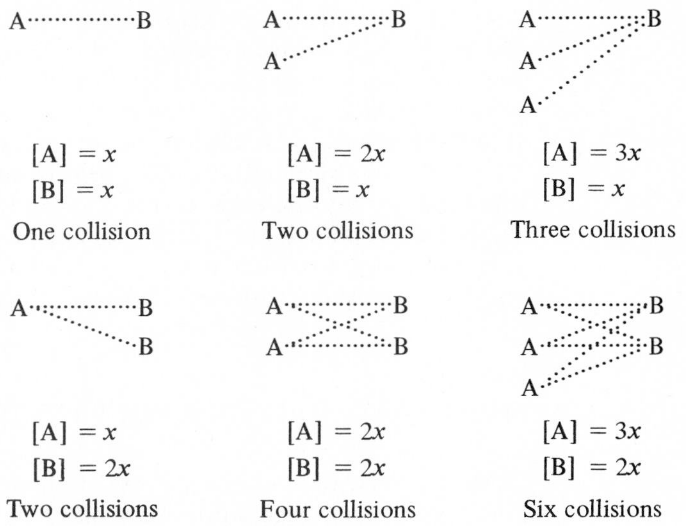

$[A] = x$
$[B] = x$

One collision

$[A] = 2x$
$[B] = x$

Two collisions

$[A] = 3x$
$[B] = x$

Three collisions

$[A] = x$
$[B] = 2x$

Two collisions

$[A] = 2x$
$[B] = 2x$

Four collisions

$[A] = 3x$
$[B] = 2x$

Six collisions

FIGURE 12-6

Chances of Collision between A and B as a Function of Concentration

the *fraction* of the total collisions that result in reaction. The fraction of the collisions that are effective can be changed only by a change in temperature or use of a catalyst (as we shall see later). If the conditions of temperature and catalysis remain constant, then the ratio of effective collisions to total collisions also remains constant. However, since increasing the concentration increases the total number of collisions, it must also increase the *number* of effective collisions per unit time. For example, let us assume that one-tenth of all collisions are effective, so that if there are 1000 collisions, 100 of these will result in reaction. If we now increase the concentration sufficiently to cause 2000 collisions, while keeping other conditions constant, the number of effective collisions will be one-tenth of 2000, or 200. Thus, doubling the total number of collisions also doubles the number of effective collisions.

Temperature In accounting for the increase in reaction rate caused by an increase in temperature, consideration must be given to two different aspects of the effect of temperature on molecular energy. In the first place, since the kinetic energy of a molecule is directly proportional to the temperature, raising the temperature must cause an increase in the average velocity of the molecules, and this in turn must cause an increase in the total number of collisions per unit of time. Thus, we would expect that higher temperature would result in higher rate simply because of the increased frequency of collisions. However, it can be shown by calculation from the kinetic molecular theory that an increase in temperature of 10°C increases the frequency of collision of gas molecules by only about 2 or 3 percent. As was mentioned earlier in this chapter, for many reactions an increase of $10°C$ approximately doubles the rate; that is, it increases it by close to 100 percent. It is clear, then, that the effect of temperature on rate cannot be adequately explained by increased frequency of collision alone.

A more important factor in accounting for the temperature-rate relationship lies in the fact that a rise in temperature increases the fraction of molecules that will have sufficient energy to react. This can be readily understood by consideration of the effect of temperature on the Maxwell-Boltzmann distribution curve.

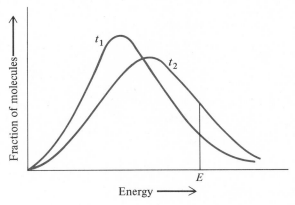

FIGURE 12-7

Energy Distribution of Molecules at Temperatures t_1 and t_2 ($t_2 > t_1$)

Figure 12-7 shows the distribution of molecular energies at two different temperatures, t_1 and t_2, where t_2 is higher. E is the activation energy—the minimum energy necessary to form the activated complex and therefore the minimum necessary to lead to reaction. At both temperatures the number of molecules that possess at least that amount of energy is only a fraction of the total number of molecules in the sample. However, at the higher temperature, t_2, that fraction is larger than at the lower temperature, t_1. In other words, raising the temperature increases the proportion of collisions that are effective. Or, putting it another way, at the higher temperature more of the collisions have sufficient energy to get to the top of the hill in the energy diagram in Figure 12-4.

Catalysis   The role of a catalyst in increasing reaction rate is explained by our model in terms of its effect on the activation energy. A catalyst provides an alternate and "easier" pathway for the reaction to take. This is interpreted as meaning that the catalyst permits the formation of an activated complex of lower energy than the activated complex formed without the catalyst, and hence the catalyst lowers the activation

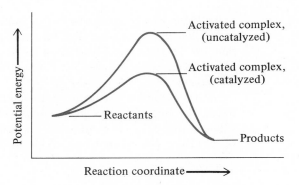

FIGURE 12-8

Potential Energy Diagram Comparing Catalyzed and Uncatalyzed Reactions

energy. If the activation energy is lower, then at any given temperature more collisions have sufficient energy to make it up the hill and to form the activated complex, thereby increasing the rate. This is illustrated schematically in Figure 12-8, with energy diagrams comparing the catalyzed and the uncatalyzed reaction.

PHOTOCHEMICAL REACTIONS

The rate-determining factors that have been discussed are general ones; they apply to all chemical reactions. Another factor that influences the rate of certain reactions is *light*. Some reactions that do not occur at an appreciable rate in the dark will proceed rapidly at the same temperature if the reactants are exposed to light. Reactions of this kind are called *photochemical* reactions.

As we have seen in Chapter 8, light is a form of energy, and when a molecule absorbs light its energy is increased by rotational, vibrational, and electronic excitation. If this energy increase is sufficiently great, bonds within the molecule may break, initiating a chemical reaction.

A rather simple example of a photochemical reaction is the chlorination or bromination of alkanes, which may be illustrated by the reaction of chlorine with methane to produce methyl chloride:

$$CH_4 + Cl_2 \rightarrow CH_3Cl + HCl$$

This reaction does not occur at a measurable rate at room temperature if the reactants are kept in the dark. However, it takes place readily in the presence of ultraviolet light. The mechanism by which the reaction is believed to occur is as follows:

1. A chlorine molecule absorbs light energy ($h\nu$), which causes the bond between the chlorine atoms to be broken in a symmetrical way, so that

each of the chlorine atoms retains one of the electrons of the pair that formed the covalent bond. Thus, each atom possesses an unpaired electron. An atom or group of atoms that has one or more unpaired electrons is called a *free radical*.

$$:\overset{..}{\underset{..}{Cl}}-\overset{..}{\underset{..}{Cl}}: \overset{h\nu}{\longrightarrow} 2:\overset{..}{\underset{..}{Cl}}\cdot$$

2. In a collision between a chlorine free radical and a methane molecule, the methane loses a hydrogen atom to the chlorine, forming hydrogen chloride and a methyl free radical.

$$\underset{\displaystyle H}{\overset{\displaystyle H}{H-\underset{|}{\overset{|}{C}}-H}} + \cdot\overset{..}{\underset{..}{Cl}}: \rightarrow \underset{\displaystyle H}{\overset{\displaystyle H}{H-\underset{|}{\overset{|}{C}}\cdot}} + H-\overset{..}{\underset{..}{Cl}}:$$

3. The methyl free radical can then undergo collision with a chlorine molecule, producing methyl chloride and another chlorine free radical.

$$\underset{\displaystyle H}{\overset{\displaystyle H}{H-\underset{|}{\overset{|}{C}}\cdot}} + :\overset{..}{\underset{..}{Cl}}-\overset{..}{\underset{..}{Cl}}: \rightarrow \underset{\displaystyle H}{\overset{\displaystyle H}{H-\underset{|}{\overset{|}{C}}-\overset{..}{\underset{..}{Cl}}:}} + :\overset{..}{\underset{..}{Cl}}\cdot$$

4. The chlorine free radical produced in step 3 is now available for collision with a methane molecule to produce another methyl free radical, which can, in turn, collide with another chlorine molecule, and so on. In other words, steps 2 and 3 are repeated over and over again. Once the process is initiated by the cleavage of a chlorine molecule through light absorption, it is self-perpetuating. Absorption of one photon can result in the production of many molecules of methyl chloride. A reaction of this kind, in which each step generates a substance that brings about the next step, is called a *chain reaction*.

Photochemical reactions are not rare. Many familiar processes occur through the absorption of light energy. The basis of black-and-white photography is a reaction whereby finely divided silver bromide, on exposure to light, is converted into dark grains of silver metal. Anyone who has observed the fading of a colored fabric on prolonged exposure to sunlight has witnessed a photochemical reaction of the chemical dye. Life on this planet owes its existence to the complex photochemical reaction *photosynthesis*—the conversion of water and carbon dioxide into carbohydrates by green plants, with chlorophyll playing the role of catalyst.

Photochemical reactions are thought to play a major role in air pollution. Nitrogen dioxide is a strong absorber of ultraviolet light and is believed to be the starting point of complex chain reactions that produce smog. The initial step may be represented

$$NO_2 + h\nu \rightarrow NO + O$$

Both NO and O undergo reactions with hydrocarbons in the air to form aldehydes, ketones, and a variety of other undesirable compounds.

REACTION MECHANISMS

As we have seen earlier in this chapter, few chemical reactions occur in a single step. Most reactions involve a series of step-wise processes—sometimes called *elementary processes* to distinguish them from the overall reaction—each of which produces a reactive intermediate substance that then takes part in a subsequent step. This sequence of stepwise processes, which describes the pathway of the reaction, is called the *reaction mechanism.* Just as the equation of a reaction tells us *what* happens, the reaction mechanism tells us *how* it happens.

The theoretical model we have used for discussion of reactions in general can be applied to each elementary process in a reaction sequence. In each separate step the reactants may be visualized as forming an activated complex, with a specific activation energy being required. Our considerations of energy distribution and the effect of temperature apply as well. Moreover, for each step a rate expression can be written that shows the relationship of concentration of reactants to the rate of that particular step. These expressions may be derived directly from the coefficients of the reactants of the individual steps. As an illustration, let us examine the reaction between carbon monoxide and chlorine to produce phosgene ($COCl_2$), a toxic, lung-searing gas that was used as a weapon in World War I. The equation for the reaction is

$$CO + Cl_2 \rightarrow COCl_2$$

and the mechanism that has been proposed for the reaction, along with the derived rate equation for each step, follows:

(1)	$Cl_2 \rightarrow 2Cl$	$r = k_1 [Cl_2]$
(2)	$Cl + CO \rightarrow COCl$	$r = k_2 [Cl][CO]$
(3)	$COCl + Cl_2 \rightarrow COCl_2 + Cl$	$r = k_3 [COCl][Cl_2]$

It should be noted that none of these derived rate equations is the rate law for the overall reaction, which has been determined by experiment to be

$$r = k[Cl_2]^{3/2}[CO]$$

Each elementary process may be considered as the direct result of effective collision between the reacting species (molecules, ions, atoms, free radicals). The number of these species that take part in the collision—that is, that form the activated complex—is referred to as the *molecularity* of the reaction. If only one molecule takes part, the process is said to be *unimolecular; bimolecular* denotes two reacting species; and *termolecular,* three. Thus, step (1) in the mechanism for $COCl_2$ production is a unimolecular process, while steps (2) and (3) are each bimolecular. (Note that *molecularity* is a theoretical concept based on the proposed mechanism, whereas *order* of reaction—discussed earlier—is an experimentally determined quantity.) Most elementary processes are believed to be unimolecular or bimolecular. Termolecular processes are rare, and processes of higher molecularity are never postulated; the probability of more than three species colliding at the same instant is extremely small.

If one of the steps of a multistep reaction has a rate that is quite low compared to the rates of all the other steps, then that slow step will determine the rate of the overall reaction. As a physical analogy, imagine a bucket brigade attempting to put out a fire. The first man in the line is filling buckets with water from a faucet. As each bucket is filled it is handed down the line to be thrown on the fire. Now suppose that the faucet from which the buckets are being filled produces only a feeble stream of water. It takes much longer to fill a bucket than it does to hand the bucket down the line and throw the contents on the fire. The rate at which the fire is being doused is slowed by the bucket-filling step, rather than by any subsequent step in the chain. The bucket-filling step is therefore the rate-determining step, and anything we might do to speed up that one step would increase the rate of the whole process. On the other hand, if the faucet is gushing water at such a high rate that buckets can be filled faster than they can be passed along to the fire, then the filling step is no longer rate determining, and increasing the rate of filling will not in itself increase the rate of the whole process.

As a simple illustration of a rate-determining step, consider the reaction between iodine monochloride and hydrogen,

$$2ICl + H_2 \rightarrow I_2 + 2HCl$$

and assume that this reaction proceeds by a two-step mechanism, as follows:

(1) $ICl + H_2 \rightarrow HI + HCl$

(2) $HI + ICl \rightarrow I_2 + HCl$

(Note that the equation for the overall reaction is the sum of the equations for the two elementary processes.) Now, assume further that step (1) is very slow compared to step (2). This means HI will be consumed as fast as it is formed and that step (1) is the rate-determining step. Therefore, the rate law for the overall reaction is the same as the derived rate expression for step (1):

$$r = k[ICl][H_2]$$

The fact that this expression is indeed the experimentally determined rate law for this reaction lends evidence for the validity of our mechanism. However, it does not *prove* that this is the correct mechanism, for numerous other mechanisms might be postulated that would also satisfy the experimental rate law.

It should not be assumed on the basis of the preceding example that the reaction rate law is always the same as the rate expression for the slow step. Rate laws must be expressed in terms of the reactants of the overall reaction, and in some mechanisms the rate expression for the slow step involves *intermediates*. To illustrate, let us examine again the mechanism for phosgene production proposed on p. 310, where step (3) is believed to be the slow step. The rate expression for this step is $r = k\,[\text{COCl}]\,[\text{Cl}_2]$. Yet this cannot be the rate law for the overall reaction, because COCl is an intermediate formed in the preceding step; it is not one of the original reactants. The fast steps must also be taken into account in this case, and correlation of the rate law with the mechanism is more complex than in the ICl example.

It is not a simple matter to establish a mechanism for most reactions. Some reactions that appear to be very simple have held the attention of research chemists for many years without any satisfactory mechanism being proposed. In fact, detailed mechanisms have been firmly established for relatively few types of reactions. As we have indicated earlier, the rate law for the reaction must be established and must be in agreement with any proposed mechanism, but this information alone is usually not sufficient. Other experimental evidence (for example, identification of intermediates) and often considerable chemical intuition are also required. To be acceptable, a mechanism must be consistent with *all* the known facts about a reaction. But even this is not *proof* that a proposed mechanism is *correct;* it only indicates that it is a likely possibility.

SUGGESTED READING

Phillips, D. C., "The Three-Dimensional Structure of an Enzyme Molecule." *Scientific American,* Vol. 215, No. 5 (November 1966), pp. 78–90.

PROBLEMS

1. Indicate clearly the difference in the meaning of the terms in each of the following pairs:
   (a) homogeneous reaction and heterogeneous reaction
   (b) order of reaction and molecularity
   (c) reaction rate and reaction mechanism
   (d) activated complex and activation energy
   (e) total collisions and effective collisions
   (f) chemical reaction and elementary process

2. Describe the effect of temperature on reaction rate and rationalize this effect on the molecular level.

3. Define *catalysis* and provide a theoretical explanation for this phenomenon.

4. Assuming each of the following reactions occurs directly by a single step, what is the *order* of each reaction?
   (a) $\text{SO}_2\text{Cl}_2 \rightarrow \text{SO}_2 + \text{Cl}_2$ (b) $2\text{N}_2\text{O} \rightarrow 2\text{N}_2 + \text{O}_2$
   (c) $2\text{NO} + \text{O}_2 \rightarrow 2\text{NO}_2$ (d) $2\text{HI} \rightarrow \text{H}_2 + \text{I}_2$

5. Give the *order* for each of the reactions for which the following are the experimental rate laws:
   (a) Rate $= k\,[\text{NO}]^2\,[\text{Cl}_2]$ (b) Rate $= k\,[\text{N}_2\text{O}]$
   (c) Rate $= k\,[\text{N}_2\text{O}_4]$ (d) Rate $= k\,[\text{CO}]\,[\text{Cl}_2]^{3/2}$

6. For the reaction $2\text{A} + \text{B} \rightarrow \text{C}$, the rate of formation of C was measured for a number of different initial concentrations of A and B, with the following results:

[A]	[B]	Rate
0.10M	0.10M	2.0×10^{-3} moles/liter-min
0.20M	0.10M	8.0×10^{-3} moles/liter-min
0.30M	0.10M	1.8×10^{-2} moles/liter-min
0.20M	0.20M	8.0×10^{-3} moles/liter-min
0.30M	0.30M	1.8×10^{-2} moles/liter-min

(a) Write the rate equation for the reaction.
(b) What is the order of the reaction?
(c) Calculate the specific rate constant.
(d) Postulate a two-step mechanism that is in agreement with the data.

7. For the reaction $CO + Cl_2 \rightarrow COCl_2$, the rate of formation of $COCl_2$ at a given temperature, t, is found to be 6.7×10^{-3} moles/liter-min when the concentrations of CO and of Cl_2 are each 0.10M. The rate law for this reaction is

$$r = k[CO][Cl_2]^{3/2}$$

(a) Calculate the numerical value of the rate constant at temperature t.
(b) What is the rate of formation of $COCl_2$ at the same temperature if the concentrations of CO and Cl_2 are each 0.02M?
(c) If the same reaction occurs at a temperature higher than t, will the numerical value of the rate constant be larger or smaller than the answer to part (a)?
(d) If the reaction were run in the presence of a suitable catalyst at temperature t, what effect would this have on the magnitude of the rate constant?

8. The reaction of carbon dioxide with an aqueous solution in which the hydroxide ion concentration is greater than $10^{-4}M$ occurs in two steps:

$$CO_2(aq) + OH^-(aq) \rightarrow HCO_3^-(aq) \qquad \text{slow}$$

$$HCO_3^-(aq) + OH^-(aq) \rightarrow CO_3^{2-}(aq) + H_2O(l) \quad \text{fast}$$

(a) Write a rate equation for the disappearance of CO_2.
(b) The rate constant for the reaction is 8500 liters/mole-sec. If the OH^- concentration is $2.2 \times 10^{-4}M$ and the CO_2 concentration is 0.44 g/liter, calculate the rate of the reaction.

9. Explain why an increase in temperature increases the rate of any reaction independent of whether the reaction is endothermic or exothermic.

10. Explain why iron powder rusts at a much faster rate than an iron nail.

11. Assuming that gases A and B react by the simple collision of A with B, how would an increase in pressure at a constant temperature affect the rate of the reaction?

12. The spontaneous decomposition of a radioactive species (for example, $^{14}_{6}C \rightarrow ^{14}_{7}N + e^-$) is dependent only on the number of $^{14}_{6}C$ nuclei present. What is the order of the reaction? Write a rate equation.

13 *Extent of Reaction*

In addition to the identity of the reactants and products, there are three things that a chemist would like to know about a chemical reaction: its *rate, mechanism,* and *extent.* These factors are particularly important to the chemist who synthesizes compounds. How much of the product will be obtained after how much time and what reaction conditions will maximize the amount of product obtained are vital considerations for the efficiency of many synthetic and industrial processes. The concepts of rate and mechanism have been explored in Chapter 12; in this chapter we will discuss the extent of chemical reactions.

It is essential at the outset of our discussion to distinguish clearly between the rate and extent of reaction. The *rate* of a reaction is a measure of how quickly reactants are converted to products. The *extent of a reaction is a measure of how much product is formed from a certain amount of reactant after the reaction has proceeded as far as it will under a given set of conditions.*

Consider the conversion of *cis*-2-butene to its geometrical isomer *trans*-2-butene at 400°C. When 1.00 mole of the *cis* isomer is introduced into an empty flask at 400°C and a pressure of 400 torr, and the contents of the flask are monitored as a function of time, the data of Table 13-1 are obtained. At the

TABLE 13-1	*Moles* cis	*Moles* trans	*Time*
Extent of Isomerization of 2-Butene	1.00	0.00	0
	0.96	0.04	192 minutes
	0.90	0.10	420 minutes
	0.86	0.14	8 hours
	0.50	0.50	several weeks
	0.50	0.50	several months

beginning of the reaction, time zero, only the *cis* isomer is present; after 420 minutes have elapsed, 0.10 mole of the *cis* isomer has been converted to *trans* isomer; after several weeks, 0.50 mole of the *trans* isomer is present and the composition of the reaction mixture has stabilized. Even after several months, or even years the percentage composition of the mixture will not change; the conversion of the *cis* isomer to the *trans* isomer has proceeded as far as it will under this set of conditions.

If this reaction proceeded 100 percent to completion, 1.00 mole of the *trans* isomer and none of the *cis* isomer would be present. Since 0.50 mole of the *trans* isomer has actually been formed under these conditions, the reaction has gone

$$\frac{0.50}{1.00} \times 100 = 50\%$$

to completion. This is the *extent* of the reaction. A considerable length of time was required to obtain the 50-percent conversion, however, and the *rate* of the reaction could be characterized, therefore, as moderately slow.

Examples of reactions with more extreme extents and rates are given in Table 13-2. The first reaction, the decomposition of nitrogen oxide, occurs at a very slow rate at room temperature. However, the extent of the reaction is high, and if sufficient time—perhaps several years—is allowed, most of the nitrogen oxide will decompose to N_2 and O_2. The reaction of $SiCl_4$ with H_2O also has a high extent, but since its rate is quite fast the virtually total conversion of $SiCl_4$ to SiO_2 occurs within seconds. The decomposition of water has both a low extent and a slow rate of reaction at room temperature. Thus, even if this reaction is allowed to proceed

TABLE 13-2		Rate	Extent
Some Reactions with Extreme Extents and Rates at Room Temperature	$2NO(g) \rightleftharpoons N_2(g) + O_2(g)$	slow	high
	$2H_2O(l) \rightleftharpoons 2H_2(g) + O_2(g)$	slow	low
	$SiCl_4(l) + 2H_2O(l) \rightleftharpoons SiO_2(s) + 4HCl(g)$	fast	high
	$2NaCl(s) + H_2SO_4(l) \rightleftharpoons 2HCl(g) + Na_2SO_4(s)$	fast	low

for an infinite amount of time, only a small percentage of the water will convert to H_2 and O_2. The last reaction in the table, which represents the commercial preparation of HCl, occurs at a fast rate, but only a small percentage of the reactants are converted to products at room temperature.

EQUILIBRIUM

In earlier discussions of phase transitions (Chapter 10) and the solution process (Chapter 11) we have seen that equilibrium is that stage in a physical process at which the rates of two opposing processes are equal. The same is true for a chemical reaction.

In a chemical process the two processes whose rates become equal at equilibrium are the forward and reverse reactions. Only a *reversible* chemical reaction, therefore, can reach a state of equilibrium. That is, equilibrium can be attained only if both the forward and reverse reactions can and do occur after the reaction is initiated. Theoretically, all reactions are reversible when carried out under the proper conditions. However, many reactions have such low rates or extents that for practical purposes they are nonreversible. The decomposition of water at room temperature is an example of such a reaction.

Let us now examine a reversible reaction—for example, the decomposition of nitrogen dioxide carried out at a temperature high enough to insure a moderate rate of reaction. After the decomposition of NO_2 to NO and O_2 has been initiated, NO_2 molecules decompose to NO and O_2 molecules *and* NO and O_2 molecules recombine to form NO_2.

$$2NO_2(g) \rightarrow 2NO(g) + O_2(g) \quad \text{forward reaction}$$

$$2NO(g) + O_2(g) \rightarrow 2NO_2(g) \quad \text{reverse reaction}$$

At the very beginning of the reaction the concentration of NO_2 is large and the concentrations of NO and O_2 are very small. Since the rate of the forward reaction depends upon the concentration of NO_2 and the rate of the reverse reaction depends upon the concentrations of NO and O_2, the rate of the forward reaction will be greater than the rate of the reverse reaction at this point. As the concentrations of NO and O_2 increase and the concentration of NO_2 decreases, the rate of the reverse process will increase and the rate of the forward process will decrease. Eventually the rates will become equal, the concentrations will remain constant, and at this point the system has reached equilibrium. Indeed, equal rates and constant concentrations are the two definitive criteria of equilibrium. Experimentally, however, an equilibrium can be detected only by the second criterion—the constancy of concentrations.

Thus, when the decomposition of NO_2 has reached equilibrium, the amounts of NO and O_2 formed will increase no further (the forward and reverse reactions continue, but at equal rates): the reaction has proceeded as far as it will under the particular set of conditions employed. This is, of course, our definition of extent of reaction, and therefore *the extent of a reversible reaction is determined after it has reached equilibrium.*

It is sometimes desirable to alter the conditions of a reaction to prevent equilibrium from occurring. In the industrial preparation of HCl (the last reaction in Table 13-2), the gaseous HCl is removed from the reaction vessel as soon as it is formed. This prevents the HCl from combining with the other products to form reactants and thus equilibrium cannot be attained. The sodium chloride and sulfuric acid simply continue to react until one or both have been used up, and the maximum possible amount of HCl is therefore obtained. In other words, the reaction is forced to go to completion.

THE EQUILIBRIUM CONSTANT

Let us now examine the isomerization of 2-butene more carefully. Table 13-1 shows that 0.50 mole of the *trans* isomer is obtained at equilibrium when the reaction is begun with 1.00 mole of the *cis* isomer. In order to check the reversibility of the reaction we now introduce 1.00 mole of the *trans* isomer into an empty 1-liter flask at 400° and find that at equilibrium 0.50 mole of the *cis* isomer and 0.50 mole of the *trans* isomer are present. Thus, the reaction is reversible, the equilibrium can be approached from either side of the equation, and the *relative concentrations* of the isomers at equilibrium are independent of how the equilibrium is approached. Indeed, it appears that the ratio of the concentrations of the two isomers may be a constant.

Let us now evaluate the theoretical likelihood of this suggestion. The rate of the forward reaction, conversion of *cis* isomer to *trans* isomer, depends only on the concentration of the *cis* isomer. Since this reaction is first order with respect to the *cis* isomer, this dependence can be expressed as

$$\text{Rate}_{\text{forward}} = k_f[cis]$$

where k_f is the rate constant for the forward reaction. In the same way, the rate of the reverse reaction, conversion of *trans* isomer to *cis* isomer, can be expressed as

$$\text{Rate}_{\text{reverse}} = k_r[trans]$$

At equilibrium, the rate of the forward reaction is equal to the rate of the reverse reaction, and therefore

$$\text{Rate}_{\text{forward}} = \text{Rate}_{\text{reverse}}$$

$$k_f[cis] = k_r[trans]$$

or, rearranging

$$\frac{k_f}{k_r} = \frac{[trans]}{[cis]}$$

Theoretically, then, the ratio of the concentrations of the two isomers *at equilibrium* is equal to a quotient of two constants, which must, of course, be another constant.

$$\frac{k_f}{k_r} = K = \frac{[trans]}{[cis]}$$

Since the rate constants are dependent only on temperature and the presence of catalysts, and since catalysts affect the forward and reverse constants proportionately (see Chapter 12), the *equilibrium constant K* is dependent only on temperature.

The equilibrium constant for the 2-butene isomerization can now be calculated from the previous data on the composition of an equilibrium mixture. Regardless of which side the equilibrium is approached from, the equilibrium composition is 50 percent *cis* isomer and 50 percent *trans* isomer. The equation calls for the use of concentrations, however, and since 0.50 mole of *trans* and 0.50 mole of *cis* are present in a 1-liter flask (both isomers are gases at this temperature), the concentrations are 0.50 mole/liter. Substitution of these concentrations into the equation for the equilibrium constant yields the result that for the isomerization at 400°C, $K = 1.0$.

The considerable importance of the equilibrium expression

$$K = \frac{[trans]}{[cis]}$$

lies in the fact that it describes the composition of any equilibrium mixture of the two isomers at a given temperature. Suppose, for example, that 0.20 mole of the *cis* isomer is isomerized at 400°C in a 1-liter flask. At this temperature the equilibrium constant is 1.0, and so when equilibrium is established the mixture must contain equal concentrations of both isomers. Thus, at equilibrium at 400°C, there will be 0.10 mole of the *cis* isomer and 0.10 mole of the *trans* isomer.

The same result could have been obtained by the use of simple algebra and a knowledge of chemical stoichiometry. Since the amount of the *trans* isomer at equilibrium is unknown, we will represent it by x. Now, every mole of *cis* isomer that reacts produces 1 mole of *trans* isomer, and so the amount of *cis* isomer at equilibrium is the amount introduced into the flask less the amount of *trans* isomer formed. Hence, the amount of the *cis* isomer is $0.20 - x$.

$$[trans] = \frac{x \text{ moles}}{1 \text{ liter}}$$

$$[cis] = \frac{(0.20 - x) \text{ moles}}{1 \text{ liter}}$$

$$1 = \frac{[trans]}{[cis]} = \frac{x}{0.20 - x}$$

$$(1)(0.20 - x) = x$$

$$0.20 - x = x$$

$$2x = 0.20$$

$$x = 0.10 \text{ mole}$$

As another example of the utility of the equilibrium expression, let us calculate the equilibrium concentrations at a temperature at which the equilibrium constant is, say, 4.3. The equilibrium expression

$$K = 4.3 = \frac{[trans]}{[cis]}$$

reveals that at equilibrium there must be 4.3 times as much *trans* isomer as *cis* isomer. If we again begin the reaction with 0.20 mole of the *cis* isomer, the equilibrium amounts (in a 1-liter flask) can be calculated algebraically:

Let x = amount of *trans* isomer at equilibrium.

Then the amount of *cis* isomer at equilibrium is $0.20 - x$.

$$\frac{[trans]}{[cis]} = \frac{x}{0.20 - x} = 4.3$$

$$x = .86 - 4.3x$$

$$5.3x = .86$$

$$x = 0.162 \text{ mole}$$

At equilibrium:

$$[trans] = 0.162 \text{ mole/liter}$$

$$[cis] = 0.20 - 0.162 = 0.038 \text{ mole/liter}$$

As a final example, let us add 0.10 mole of the *trans* isomer to this equilibrium mixture (0.162 mole *trans*, 0.038 mole *cis*) and determine the composition of the mixture after equilibrium has been reestablished. At the very instant the 0.10 mole of *trans* isomer is added to the mixture, the concentration of *trans* isomer is 0.10 + 0.162 = 0.262 mole/liter and the *cis* isomer concentration is still 0.038 mole/liter. At this instant, then, the concentration of the *trans* isomer is *more* than 4.3 greater than the concentration of the *cis* isomer, and the system must now adjust itself to lower the concentration of *trans* isomer to the point where

$$\frac{[trans]}{[cis]} = 4.3$$

This adjustment can only occur by conversion of the *trans* isomer to the *cis* isomer, and when equilibrium is reestablished, the new concentration of the *cis* isomer will be greater than 0.038 mole/liter. The new equilibrium amounts can be calculated by letting x represent the amount of *trans* isomer converted to *cis* isomer after the 0.10 mole of *trans* isomer is added. The amount of *cis* isomer formed during the reestablishment of equilibrium is also x and thus

$$[trans] = (0.262 - x)\text{mole/liter}$$

$$[cis] = (0.038 + x)\text{mole/liter}$$

$$\frac{[0.262 - x]}{[0.038 + x]} = 4.3$$

$$0.262 - x = .163 + 4.3x$$

$$0.099 = 5.3x$$

$$x = 0.019 \text{ mole}$$

At the new equilibrium, then, the concentration of the *trans* isomer is 0.262 − 0.019 = 0.243 mole/liter, the concentration of the *cis* isomer is 0.038 + 0.019 = 0.057 mole/liter, and the concentration of the *trans* isomer is once again 4.3 times as great as the concentration of the *cis* isomer.

The equilibrium expressions for more complex chemical reactions can be generalized as follows. For a reaction of the type

$$a\text{A} + b\text{B} + c\text{C} + \cdots \rightleftharpoons e\text{E} + f\text{F} + g\text{G} + \cdots$$

where A, B, C represent the starting materials; E, F, G are the products; and the lowercase letters *a, b, c,* and so on, are the coefficients of these materials in the balanced equation, the equilibrium expression can be written as

$$K = \frac{[\text{E}]^e [\text{F}]^f [\text{G}]^g \cdots}{[\text{A}]^a [\text{B}]^b [\text{C}]^c \cdots}$$

For example, the equilibrium expression for the decomposition of NO

$$2\text{NO}(g) \rightleftharpoons \text{N}_2(g) + \text{O}_2(g)$$

is

$$K = \frac{[\text{N}_2][\text{O}_2]}{[\text{NO}]^2}$$

and the expression for the reaction

$$\text{N}_2(g) + 3\text{H}_2(g) \rightleftharpoons 2\text{NH}_3(g)$$

is

$$K = \frac{[\text{NH}_3]^2}{[\text{N}_2][\text{H}_2]^3}$$

It is important to be aware of the following features of the equilibrium expression:

1. Unless specified otherwise, equilibrium constants refer to the equation containing the lowest possible whole-number coefficients. For the formation of ammonia, this equation and the corresponding equilibrium expres-

sion and constant are given above. If this equation is multiplied by 2, the equilibrium constant associated with the new equilibrium expression is the old constant raised to the second power.

$$2N_2 + 6H_2 \rightleftharpoons 4NH_3$$

$$K' = \frac{[NH_3]^4}{[N_2]^2[H_2]^6} = K^2$$

where

$$K = \frac{[NH_3]^2}{[N_2][H_2]^3}$$

If the equation is multiplied by $\frac{1}{2}$, the constant K'' is $K^{1/2}$.

$$\tfrac{1}{2}N_2 + \tfrac{3}{2}H_2 \rightleftharpoons NH_3$$

$$K'' = \frac{[NH_3]}{[N_2]^{1/2}[H_2]^{3/2}} = K^{1/2}$$

2. The equilibrium expression always contains the products of the equation *in its numerator*.

$$N_2 + 3H_2 \rightleftharpoons 2NH_3$$

$$K = \frac{[NH_3]^2}{[N_2][H_2]^3} \qquad \text{product}$$

If an equation is reversed, for example,

$$2NH_3 \rightleftharpoons N_2 + 3H_2$$

the equilibrium expression is

$$K''' = \frac{[N_2][H_2]^3}{[NH_3]^2}$$

and K''' is therefore $1/K$, the reciprocal of the previous constant, K.

3. The value of an equilibrium constant that has not yet been experimentally determined for a given chemical reaction can often be calculated from known constants for related reactions. Suppose, for example, that the constant for the reaction

$$A + B \rightleftharpoons D \qquad K_1$$

is to be found, and the constants for the two reactions

$$A + B \rightleftharpoons C \qquad K_2$$

$$C \rightleftharpoons D \qquad K_3$$

are known.

Chemical equations can be manipulated just like mathematical equations, and thus if the last two equations are added, the substance C appears on opposite sides of the equality sign (arrows) and therefore cancels. The sum is the desired equation.

$$A + B \rightleftharpoons \cancel{C}$$
$$+ \quad \underline{\cancel{C} \rightleftharpoons D}$$
$$A + B \rightleftharpoons D$$

Since equilibrium expressions contain the concentrations of substances as products rather than sums, the constant K_1 can be obtained by simply multiplying K_2 by K_3. This can be more readily appreciated if the expressions for each reaction are written out.

$$A + B \rightleftharpoons D \qquad K_1 = \frac{[D]}{[A][B]}$$

$$A + B \rightleftharpoons C \qquad K_2 = \frac{[C]}{[A][B]}$$

$$C \rightleftharpoons D \qquad K_3 = \frac{[D]}{[C]}$$

$$K_2 \times K_3 = \frac{[\cancel{C}]}{[A][B]} \times \frac{[D]}{[\cancel{C}]}$$

$$= \frac{[D]}{[A][B]} = K_1$$

If two or more equations must be *subtracted* to obtain the desired equation, the constant is obtained by *dividing* the constants for the reactions.

4. The equilibrium expression is given in terms of concentrations. When a substance is a gas, its concentration is expressed in either moles per liter (per volume of the reaction vessel) or partial pressure (since under ideal conditions partial pressure is proportional to concentration). When partial pressures are used, the equilibrium constant is referred to as K_p, as opposed to the designation K_c, which is used when moles per liter are the units. When the substance in question is in a liquid solution, its concentration is given in moles per liter of solution.

When the substance is a pure liquid or solid not in solution, its concentration is the number of moles per unit volume of the substance. For example, the concentration of 18 g of H_2O is 1 mole per 18 ml (since 1 g of H_2O occupies 1 ml at room temperature and pressure), or 55.5 moles per liter. The concentration of 100 g of H_2O is exactly the same—55.5 moles/liter. Similarly, the concentration of a solid is the number of moles per unit volume of the solid.

The concentrations of liquids and solids that are part of a heterogeneous mixture are therefore independent of the amount present. More-over, since the densities of liquids and solids are virtually unaffected by changes in pressure, their concentrations are also independent of pressure. At a constant temperature, then, the concentration of a liquid or solid that is not in solution is a constant. For this reason, the concentrations of such substances are generally not included in the equilibrium expression, and the equilibrium constant is adjusted to include these constant concentrations.

The decomposition of barium carbonate in a closed vessel is an example of a heterogeneous equilibrium.

$$BaCO_3(s) \rightleftharpoons BaO(s) + CO_2(g)$$

The equilibrium expression for this reaction could be written as

$$K' = \frac{[BaO(s)][CO_2(g)]}{[BaCO_3(s)]}$$

But the concentrations of the two solids, BaO and $BaCO_3$, are constants and therefore the equilibrium constant can also be written as

$$K = K' \underbrace{\frac{[BaCO_3(s)]}{[BaO(s)]}}_{\text{all constants}} = [CO_2(g)]$$

The final equilibrium expression

$$K = [CO_2(g)]$$

is the simple algebraic statement that at a given temperature the concentration of CO_2 is a constant.

The decomposition of water (Table 13-2) affords an example of a heterogeneous liquid-gas reaction. Here the concentration of the liquid is constant, and the equilibrium expression is

$$2H_2O(l) \rightleftharpoons 2H_2(g) + O_2(g)$$

$$K = [H_2(g)]^2[O_2(g)]$$

5. Just as the ideal gas law, $PV = nRT$, and Raoult's Law, $p = p^0 X$, are expressions of idealized behavior, so is the equilibrium expression. That is, only at low concentrations is the quotient

$$\frac{[E]^e[F]^f \cdots}{[A]^a[B]^b \cdots}$$

a constant, independent of actual amounts and pressure. It has been verified, however, by numerous experimental data that it is a constant at low concentrations and constant temperature.

The actual numerical value of the equilibrium constant can be calculated either from data on equilibrium compositions or from the free-energy change for the reaction. The determination of the free-energy change and other thermodynamic characteristics of a reaction will be considered in a subsequent section of this chapter.

The following examples show the calculation of the equilibrium constant from the composition of the reaction mixture at equilibrium.

EXAMPLE: The decomposition of N_2O_4 to NO_2 is carried out at $8°C$ in chloroform. When equilibrium has been established, 0.20 mole of N_2O_4 and 2.0×10^{-3} mole of NO_2 are present in 2.0 liters of solution. Calculate the equilibrium constant for the reaction at $8°C$.

Solution: The balanced equation for the reaction is

$$N_2O_4 \rightleftharpoons 2NO_2$$

and the corresponding equilibrium expression is

$$K = \frac{[NO_2]^2}{[N_2O_4]}$$

Converting the amounts of N_2O_4 and NO_2 present to concentrations in moles per liter of solution, we obtain

$$[NO_2] = \frac{2.0 \times 10^{-3} \text{ mole}}{2.0 \text{ liters}} = 1.0 \times 10^{-3} \text{ mole/liter}$$

$$[N_2O_4] = \frac{0.20 \text{ mole}}{2.0 \text{ liters}} = 1.0 \times 10^{-1} \text{ mole/liter}$$

Thus,

$$K = \frac{[1.0 \times 10^{-3} \text{ mole/liter}]^2}{[1.0 \times 10^{-1} \text{ mole/liter}]}$$

$$= \frac{1.0 \times 10^{-6} \text{ mole}^2/\text{liter}^2}{1.0 \times 10^{-1} \text{ mole/liter}} = 1.0 \times 10^{-5} \text{ mole/liter}$$

In this case the equilibrium constant has the units of moles per liter. Usually the units are omitted when the constant is reported.

EXAMPLE: When 10 g of PCl_5 is allowed to dissociate into PCl_3 and Cl_2 in a 3-liter flask at $250°C$, 63% of the PCl_5 is converted to products. Calculate the equilibrium constant (K_c) for this reaction at $250°C$.

Solution: The equation and corresponding equilibrium expression are

$$PCl_5(g) \rightleftharpoons PCl_3(g) + Cl_2(g)$$

$$K = \frac{[PCl_3][Cl_2]}{[PCl_5]}$$

The number of moles of PCl_5 at the beginning of the reaction is

$$\frac{10 \text{ g}}{208 \text{ g/mole}} = 0.048 \text{ mole}$$

At equilibrium, 63% of the PCl_5 has been converted to PCl_3 and Cl_2. Therefore, 0.63×0.048 mole, or 0.030 mole of PCl_3 and 0.030 mole of Cl_2 are present at equilibrium. (The stoichiometry of the equation dictates that for every molecule of PCl_5 that reacts, one molecule of PCl_3 and one molecule of Cl_2 are formed.) The amount of PCl_5 remaining at equilibrium is $0.048 - 0.030 = 0.018$ mole. The equilibrium concentrations are

$$[PCl_5] = \frac{0.018 \text{ mole}}{3 \text{ liter}} = 0.0060 \text{ mole/liter}$$

$$[PCl_3] = \frac{0.030 \text{ mole}}{3 \text{ liter}} = 0.010 \text{ mole/liter}$$

$$[Cl_2] = \frac{0.030 \text{ mole}}{3 \text{ liter}} = 0.010 \text{ mole/liter}$$

$$K_c = \frac{(1.0 \times 10^{-2})(1.0 \times 10^{-2})}{6.0 \times 10^{-3}} = 1.7 \times 10^{-2} \text{ mole/liter}$$

EXAMPLE: Ethyl acetate can be prepared according to the equation

$$CH_3COOH + C_2H_5OH \rightleftharpoons CH_3COOC_2H_5 + H_2O$$

At some temperature, 0.10 mole of acetic acid is mixed with 0.10 mole of ethyl alcohol in 1.0 liter of solution (a nonreactive solvent is used) and when equilibrium has been established, 0.060 mole of ethyl acetate is present. Calculate the equilibrium constant.

Solution: Every time a molecule of ethyl acetate is formed in the reaction, a molecule of water is also formed, and therefore 0.060 mole of water must also be present at equilibrium. The 0.060 mole of ethyl acetate and water must be produced by the reaction of 0.060 mole of acetic acid with 0.060 mole of ethyl alcohol. Thus, at equilibrium, 0.040 mole of each starting material remains.

$$[CH_3COOC_2H_5] = [H_2O] = 0.060 \text{ mole/liter}$$

$$[CH_3COOH] = [C_2H_5OH] = 0.040 \text{ mole/liter}$$

$$K = \frac{[CH_3COOC_2H_5][H_2O]}{[CH_3COOH][C_2H_5OH]} = \frac{(0.060)(0.060)}{(0.040)(0.040)}$$

$$K = 2.2$$

EXAMPLE: The commercial preparation of gaseous carbon monoxide is accomplished by heating solid carbon with gaseous carbon dioxide at high temperatures. At some temperature, 0.090 mole of carbon dioxide is mixed in a 5-liter flask with an excess of carbon. When equilibrium is established, 0.040 mole of carbon dioxide remains and the total pressure in the system is 1.3 atm. Calculate K_c and K_p for this process.

Solution: The equation and equilibrium expression for the process are

$$C(s) + CO_2(g) \rightleftharpoons 2CO(g)$$

$$K = \frac{[CO(g)]^2}{[CO_2(g)]}$$

The amount of CO_2 that has been converted to CO is equal to the amount present at the start of the reaction minus the amount present at equilibrium: $0.090 - 0.040 = 0.050$ mole. According to the equation, 2 moles of CO are produced from the conversion of 1 mole of CO_2, and therefore the amount of CO produced is $0.050 \times 2 = 0.10$ mole. The equilibrium concentrations are

$$[CO] = \frac{0.10 \text{ mole}}{5 \text{ liters}} = 0.020 \text{ mole/liter}$$

$$[CO_2] = \frac{0.040 \text{ mole}}{5 \text{ liters}} = 0.0080 \text{ mole/liter}$$

and, in terms of moles per liter, the equilibrium constant is

$$K_c = \frac{(2.0 \times 10^{-2})^2}{8.0 \times 10^{-3}} = 5 \times 10^{-2} \text{ mole/liter}$$

The partial pressure of each gaseous constituent is the mole fraction of the constituent times the *total* pressure. Thus,

$$p_{CO} = (X_{CO})(1.3 \text{ atm}) = \frac{0.10 \text{ mole}}{0.10 \text{ mole} + 0.040 \text{ mole}}(1.3 \text{ atm})$$

$$= 0.93 \text{ atm}$$

$$p_{CO_2} = (X_{CO_2})(1.3 \text{ atm}) = \frac{0.04 \text{ mole}}{0.10 \text{ mole} + 0.040 \text{ mole}}(1.3 \text{ atm})$$

$$= 0.37 \text{ atm}$$

and

$$K_p = \frac{(0.93)^2}{0.37} = 2.3 \text{ atm}$$

THE EQUILIBRIUM CONSTANT AS A MEASURE OF EXTENT

If we know the equilibrium constant for a particular reaction, we can determine its equilibrium composition and consequently the amount of product formed from a given amount of reactant—that is, the extent of the reaction. Also, the *relative* extents of two reactions can often be ascertained by comparing their equilibrium constants.

In order to illustrate the evaluation of relative extents let us examine two simple, generalized reactions. The first reaction can be represented as

$$A \rightleftharpoons B$$

which is a generalized expression for reactions such as the geometrical isomerization considered above. Suppose that for a certain reaction of this $A \rightleftharpoons B$ type, the equilibrium constant is 1.0×10^{-1}. The amount of product obtained in this reaction from, say, 1.0 mole of A can be obtained by the same algebraic procedure used in the discussion of the isomerization (page 318). Accordingly, if x is used to represent the amount of B present at equilibrium, the amount of A remaining at equilibrium will be $1.0 - x$, and inserting these amounts (note that the volume cancels in this case) into the equilibrium expression gives us

$$\frac{x}{1.0 - x} = K = 1.0 \times 10^{-1}$$

$$x = 1.0 \times 10^{-1} - 1.0 \times 10^{-1}x$$

$$1.1x = 1.0 \times 10^{-1}$$

$$x = 0.91 \times 10^{-1} = 0.091 \text{ mole}$$

Hence, from 1.0 mole of A, 0.091 mole of B has been obtained. Total conversion of A to B would have resulted in 1.0 mole of B, and thus the reaction has gone 9.1 percent to completion.

Suppose, now, that for a different reaction of the same type ($A \rightleftharpoons B$) the equilibrium constant is 1.0. In this reaction 50 percent of A is converted to B (see the calculation on page 318). Obviously, the extent of this reaction is greater than that of the reaction for which $K = 1.0 \times 10^{-1}$.

Finally, consider a reaction of different stoichiometry,

$$A \rightleftharpoons 2B$$

for example, the reaction $N_2O_4 \rightleftharpoons 2NO_2$. If we again assume an equilibrium constant of 1.0×10^{-1}, we can calculate the amount of B formed from 1.0 mole of A when the reaction is conducted in a total volume of 1 liter. The stoichiometry of the equation indicates that the conversion of 1 mole of A results in the formation of 2 moles of B, and if we let x represent the amount of A converted, $2x$ is the amount of B formed at equilibrium. Since the total volume is 1.0 liter, the equilibrium concentrations are

$$[A] = 1.0 - x$$

$$[B] = 2x$$

and

$$K = 1.0 \times 10^{-1} = \frac{[B]^2}{[A]} = \frac{(2x)^2}{1.0 - x}$$

or

$$4x^2 = 1.0 \times 10^{-1} - 1.0 \times 10^{-1}x$$

This quadratic equation can be solved with the quadratic formula, which for the equation

$$ax^2 + bx + c = 0$$

is

$$x = \frac{-b \pm \sqrt{b^2 - 4ac}}{2a}$$

Rearranging our equation to the same form results in

$$4x^2 + 1.0 \times 10^{-1}x - 1.0 \times 10^{-1} = 0$$

Thus, $a = 4$, $b = 1.0 \times 10^{-1}$, and $c = -1.0 \times 10^{-1}$. The solution to the equation is then

$$x = \frac{-1.0 \times 10^{-1} \pm \sqrt{(1.0 \times 10^{-1})^2 - 4[4(-1.0 \times 10^{-1})]}}{2(4)}$$

$$= \frac{-1.0 \times 10^{-1} \pm 1.3}{8} = \frac{1.2}{8}, -\frac{1.4}{8}$$

$$= 0.15, -0.175$$

The negative solution cannot be the correct solution to this problem since it would mean that at equilibrium $1.0 - (-0.175)$ or 1.2 mole of A is present, which is more than was present at the beginning of the reaction. The correct solution is therefore 0.15 mole, which is the amount of A that reacted. The amount of B present at equilibrium is twice this amount, of 0.30 mole.

[This problem could also be solved by defining x in a different way, for example as the amount of B formed. The amount of A at equilibrium would then be $(1.0 - \frac{1}{2}x)$. The answer would, of course, be the same in either case.]

Total conversion of 1.0 mole of A would result in 2.0 moles of B, and the extent of this reaction is therefore

$$\frac{0.30}{2.0} \times 100 = 15\%$$

which is greater than the 9.1 percent extent of the A $\rightleftharpoons$ B reaction with the same equilibrium constant. Comparing the equilibrium constants of two reactions will reveal their relative extents, therefore, only when the reactions have the same stoichiometry; that is, when they are both A $\rightleftharpoons$ B, both A + B $\rightleftharpoons$ C + D, both A + B $\rightleftharpoons$ 2C, etc.

THE EFFECT OF CONCENTRATION ON EXTENT

A further complication that arises when equilibrium constants are used as an index of extent of reaction is that extent is also influenced by the concentrations of starting materials and the concentrations of any products that may be present at the beginning of the reaction. The esterification of acetic acid with ethyl alcohol provides an example of the effect of concentration on extent. Let us first calculate the amount of ethyl acetate formed when 0.10 mole of acetic acid is mixed with 0.10 mole of ethyl alcohol in a sufficient amount of nonreactive solvent to give 1 liter of solution. As we have seen on page 325, the equation and equilibrium expression for this reaction are

$$CH_3COOH + C_2H_5OH \rightleftharpoons CH_3COOC_2H_5 + H_2O$$

$$K = \frac{[CH_3COOC_2H_5][H_2O]}{[CH_3COOH][C_2H_5OH]} = 2.2$$

The amounts (which in this case are also the concentrations, because the volume of solution is 1.0 liter) of ethyl acetate and water at equilibrium are unknown, and therefore we define x as the amount of ethyl acetate present at equilibrium. The amount of water at equilibrium must be the same. The amounts of starting materials must be, then, $0.10 - x$. Therefore, we see that

$$K = 2.2 = \frac{(x)(x)}{(0.10-x)(0.10-x)} = \frac{x^2}{0.010 - 0.20x + x^2}$$

$$0.022 - 0.44x + 2.2x^2 = x^2$$

$$1.2x^2 - 0.44x + 0.022 = 0$$

$$x = \frac{0.44 \pm \sqrt{0.194 - 0.106}}{2.4}$$

$$= 0.31, 0.060$$

The value $x = 0.31$ obtained from the quadratic equation cannot be the solution for this particular problem because it would mean that more starting material was converted to product than was originally present. The correct solution is $x = 0.060$, which is, in moles, the amount of ethyl acetate and the amount of water present at equilibrium and also the amount of acetic acid and ethyl alcohol converted to product. The extent of the reaction is then

$$\frac{0.060}{0.10} \times 100 = 60\%$$

Imagine now that the reaction is run a second time at the same temperature, again using 0.10 mole of acetic acid, but with 1.0 mole, rather than 0.10 mole, of ethyl alcohol. The amounts of products at equilibrium can be represented by x; the amount of acetic acid is $0.10 - x$; and the amount of ethyl alcohol is $1.0 - x$.

$$K = 2.2 = \frac{(x)(x)}{(0.10 - x)(1.0 - x)}$$

$$x = 1.9,\ 0.096$$

The correct solution, $x = 0.096$, is the amount of acetic acid converted to products. Since the maximum amount of ethyl acetate (and H_2O) obtainable from this reaction is 0.10 mole, the percent conversion is

$$\frac{0.096}{0.10} \times 100 = 96\%$$

Increasing the concentration of one of the starting materials has therefore increased the extent of reaction. In mathematical terms, an increase in the concentration of the ethyl alcohol results in an increase in the denominator of the equilibrium expression which, in order to maintain a constant numerator/denominator ratio K, must be compensated for by an increase in the numerator.

The synthetic chemist is often concerned with the reaction conditions necessary to obtain a certain amount of product. Suppose now that 0.098 mole of the ethyl acetate is desired but that the amount of acetic acid to be used must still be 0.100 mole. How much ethyl alcohol should be mixed with the acetic acid (in 1 liter of solution) in order to accomplish this goal? In this problem the concentrations of ethyl acetate and water at equilibrium are known—both are 0.098 mole/liter. The concentration of acetic acid at equilibrium is $0.100 - 0.098 = 0.002$ mole/liter. The only unknown equilibrium concentration is that of the ethyl alcohol. Therefore, the expression becomes

$$K = 2.2 = \frac{(0.098)(0.098)}{(0.002)(x)}$$

$$x = \frac{(9.8 \times 10^{-2})^2}{(2 \times 10^{-3})(2.2)} = 2.2 \text{ moles/liter}$$

The concentration of ethyl alcohol *at equilibrium* must be 2.2 moles/liter. This is not, however, the answer to our problem. We want to know how much ethyl alcohol must be mixed with 0.100 mole of acetic acid at the *beginning* of the reaction in order to obtain 0.098 mole of product at equilibrium. We must therefore add to the equilibrium concentration of ethyl alcohol the amount that has been converted to product *during* the reaction. This amount is obviously 0.098 mole, and the amount of alcohol necessary at the beginning of the reaction is 2.2 + 0.098 = 2.3 moles (to two significant figures). [This calculation assumes a constant solution volume of 1 liter.]

EXAMPLE: At a certain temperature, the equilibrium composition of the 2-butene gas-phase isomerization is 0.30 mole *trans* isomer and 0.10 mole *cis* isomer in a 3-liter flask. How much of the *cis* isomer must be added to this mixture in order to obtain a total of 0.60 mole of the *trans* isomer when equilibrium is reestablished?

Solution:

$$cis\text{-2-Butene} \rightleftharpoons trans\text{-2-Butene}$$

$$K = \frac{[trans]}{[cis]} = \frac{0.30 \text{ mole/3 liters}}{0.10 \text{ mole/3 liters}} = 3.0$$

Let x be the concentration of the *cis* isomer at the new equilibrium. The concentration of the *trans* isomer at the new equilibrium is

$$\frac{0.60 \text{ mole}}{3 \text{ liters}} = 0.20 \text{ mole/liter}$$

$$K = \frac{(0.20)}{x} = 3.0$$

$$x = \frac{0.20}{3.0} = 0.067 \text{ mole/liter}$$

The concentration of the *cis* isomer at equilibrium is 0.067 mole/liter. The *amount* of *cis* isomer present at equilibrium is $3 \times 0.067 = 0.20$ mole. During the attainment of the new equilibrium, 0.30 mole of the *trans* isomer was formed, and thus 0.30 mole of the *cis* isomer must have reacted. In order to get 0.20 mole of *cis* at equilibrium after 0.30 mole has reacted (a total of 0.50 mole *cis*), *0.40 mole* of the *cis* isomer must be added to the 0.10 mole already present in the old equilibrium mixture.

Concentrations can also be altered by a change in volume. This change will affect the extent of reaction whenever volumes do not cancel in the equilibrium expression. The formation of hydrogen iodide from hydrogen and iodine is an example of a reaction whose extent is *not* influenced by a change of volume. In the equilibrium expression for this reaction,

$$K = \frac{[HI]^2}{[H_2][I_2]} = \frac{(\text{moles/liter})^2}{(\text{moles/liter})(\text{moles/liter})}$$

the volume cancels. Or, in other words, a change in volume will change all the concentrations, but these changes are proportionately the same in both numerator and denominator.

The decomposition of PCl_5 is a reaction whose extent is influenced by a change in volume. Let us assume, as in the example on page 325, that at equilibrium at 250° there are 0.018 mole of PCl_5, 0.030 mole PCl_3, and 0.030 mole Cl_2 in a 3-liter flask. The equilibrium constant for the reaction at 250° is given on page 325 as 1.7×10^{-2} mole/liter. If the volume of the vessel is now decreased to 1.0 liter, the *instantaneous* new concentrations are $[PCl_5] = 0.018$ mole/liter, $[PCl_3] = 0.030$ mole/liter, $[Cl_2] = 0.030$ mole/liter, and

$$\frac{[PCl_3][Cl_2]}{[PCl_5]} = 5 \times 10^{-2} \text{ mole/liter}$$

which is greater than the equilibrium constant. The system must now change in order to decrease the value of this quotient to its equilibrium value of 1.7×10^{-2} mole/liter. This can be accomplished only if some of the products react to form reactant. The volume change therefore brings about an increase in the amount of reactant and a decrease in the amount of products.

Since a change in volume for a reaction containing one or more gases is also accompanied by a change in pressure, the arguments above also apply to pressure changes brought about by changes in volume.

EFFECT OF TEMPERATURE ON EXTENT

It is an experimentally observed fact that an increase in temperature increases the extent of some reactions but decreases the extent of others. Furthermore, a comparison of the properties of both types of reaction reveals that those reactions whose extents are increased by an increase in temperature absorb heat during reaction (they are *endothermic* reactions), while those reactions whose extents are decreased by an increase in temperature liberate heat during reaction (they are *exothermic* reactions). The change in extent produced by a change in temperature is a result of a change in equilibrium constant.

The reaction of H_2 with I_2 to form HI is an example of an endothermic reaction.

$$\text{heat} + H_2(g) + I_2(g) \rightleftharpoons 2HI(g)$$

The equilibrium constant for this reaction therefore increases with an increase in temperature. The formation of ammonia from N_2 and H_2

$$N_2(g) + 3H_2(g) \rightleftharpoons 2NH_3(g) + \text{heat}$$

is an example of an exothermic reaction, and the equilibrium constant decreases with an increase in temperature.

LE CHATELIER'S PRINCIPLE

The effect of concentration and temperature on the extent of a reaction can be summarized qualitatively by Le Chatelier's Principle, which was introduced in the discussion of solution equilibrium in Chapter 11: *If a change occurs in one of the factors, such as concentration or temperature, under which a system is in equilibrium, the system will tend to adjust itself so as to annul, as far as possible, the effect of that change.*

The decomposition of N_2O_4 to NO_2 provides an apt illustration of the application of Le Chatelier's Principle.

$$\text{heat} + N_2O_4(g) \rightleftharpoons 2NO_2(g)$$

Consider a mixture of N_2O_4 and NO_2 that has reached equilibrium at some temperature.

1. *If N_2O_4 is added,* the concentration of N_2O_4 will instantaneously increase, and the system must adjust itself to relieve this increase. The only way this can be accomplished in a constant volume is for some of the N_2O_4 to react to form more NO_2. The increase in N_2O_4 results in an instantaneous increase in the denominator of the equilibrium expression. In order to maintain the $[NO_2]^2/[N_2O_4]$ ratio at the constant K, the numerator must increase. This can happen only if some N_2O_4 is converted to NO_2.

2. *If the pressure of the system is increased* by decreasing the volume of the container, the mixture will react to reduce the pressure. Since there are twice as many product molecules as reactant molecules, the number of molecules hitting the walls of the container (and therefore the pressure) is reduced if some NO_2 is consumed to produce N_2O_4. The net result of the increase in pressure is, then, a decrease in the *amount* of NO_2 present and an increase in the *amount* of N_2O_4 present.

   In order to analyze the pressure change in terms of the equilibrium expression, assume that x moles of N_2O_4 and y moles of NO_2 are present in an initial volume of 2 liters. The equilibrium constant is therefore

$$K = \frac{(y/2)^2}{(x/2)} = \frac{y^2}{2x}$$

   If the volume is now decreased to 1 liter, the *instantaneous* concentrations change from $\frac{1}{2}x$ and $\frac{1}{2}y$ to x and y. The instantaneous quotient after the volume change is then

$$\frac{(y)^2}{(x)} = \frac{y^2}{x}$$

   which is greater than K. The numerator must therefore decrease and the denominator increase, which can occur only if some NO_2 is converted to N_2O_4.

3. *If the temperature is increased,* the system must react to relieve the increased heat. Since the forward reaction absorbs heat, the stress is relieved by the production of more NO_2. It must be emphasized that the increase of temperature changes (increases) the equilibrium constant.

EXAMPLE: For the reaction

$$\text{heat} + CaCO_3(s) \rightleftharpoons CaO(s) + CO_2(g)$$

predict the effect of the following changes made on the system at equilibrium:

(a) addition of $CaCO_3$
(b) removal of CO_2
(c) increase in volume of reaction vessel
(d) decrease in temperature

Solution (a): Since the concentrations of the two solids are constants, the amount and concentration of CO_2 at equilibrium is independent of the actual amounts of $CaCO_3$ and CaO as long as some of each is present. As dictated by the equilibrium expression,

$$K = [CO_2(g)]$$

the equilibrium *concentration* of CO_2 must be a constant at a given temperature.

Solution (b): If a portion of CO_2 is removed from the equilibrium mixture, this stress (change in concentration of CO_2) must be relieved by the production of more CO_2. Calcium carbonate decomposes, then, to give CaO and CO_2 until the concentration of CO_2 has once again reached its constant equilibrium value.

Solution (c): An increase in the volume of the reaction vessel, which is, of course, the volume occupied by the CO_2, has the same consequence: The instantaneous decrease in pressure is compensated for by reaction of $CaCO_3$ to produce enough CO_2 to bring the pressure and concentration of CO_2 back to their initial values. At the new equilibrium, then, the concentration of CO_2 is the same but the total amount has increased.

Solution (d): A decrease in temperature will remove heat from the equilibrium system, and, since "heat" occurs on the left-hand side of the equation, this stress can be alleviated by conversion of some CO_2 and CaO to $CaCO_3$ and heat. The volume has not changed and therefore the *amount and concentration* of CO_2 have decreased. This is consistent with the decrease in equilibrium constant that must occur when the temperature of an endothermic reaction is decreased.

The Thermochemical Determinants of Extent

Our earlier discussions of phase change and the solution process have disclosed two fundamental laws of nature. One is the drive of physical systems to the state of lowest energy; for example, the simple fact that water flows downhill, not uphill. The other is the drive, the trend, to a state of maximum disorder or randomness; for example, the diffusion of gases throughout their containers. The same two laws also apply to chemical reactions.

The lowest energy state for the simple reaction

$$N_2(g) \rightleftharpoons 2N(g)$$

is the nitrogen molecule (recall the discussion of Chapter 7), because energy is required to dissociate the triple-bonded diatomic nitrogen into unbonded monatomic nitrogen. If this were the only drive of nature, all elemental nitrogen would be present as $N_2(g)$ at even moderately high temperatures. However, the state of maximum disorder for this reaction would be attained if all of the nitrogen were in the form of atoms rather than molecules, because this would produce a greater number of particles and therefore a greater amount of disorder.

These two natural drives tend to force the reaction in opposite directions: the drive to low energy pushes toward the starting material, the drive to maximum disorder pushes to nitrogen atoms, the product. The compromise that is attained is directly related to the establishment of equilibrium and to the magnitude of the equilibrium constant. In the sections that follow, we will examine the thermodynamic expressions of these natural laws and their relationship to the equilibrium constant.

THE THERMOCHEMICAL FUNCTIONS

The Energy Change

Since the greater the difference in energy between the starting materials and products, the greater the push to the low-energy state (just as is true for waterfalls, for example), it becomes essential to be able to express this difference in energy. Mathematically, the difference in energy, ΔE, for a given reaction is equal to the sum of the energies of the products minus the sum of the energies of the reactants.

$$\Delta E = E_{products} - E_{reactants}$$

The energy of each species is the sum of the kinetic and potential energies associated with the electrons, the nucleus, vibration of atoms within the molecule, rotation and translation of the molecule, and intermolecular forces.

As discussed in Chapter 8, by far the major part of the energy of a substance (aside from the nuclear energy, which remains constant during a chemical reaction) is its electronic energy. Moreover, the spacings of the electronic energy levels are so large that at ordinary temperatures virtually all the species are in the ground state. Therefore, the change in energy in a chemical reaction is primarily a reflection of the difference between the ground electronic energy levels of the reactants and products. Clearly, the energy of two molecules or moles of a substance is twice the energy of one molecule or mole, and ΔE is therefore also dependent upon the amount of material in question.

While absolute energies of molecules cannot be measured, differences in energy can. Suppose, for example, in a reaction of the type $A \rightleftharpoons B$, that A has a higher energy than B. According to the law of conservation of energy, the total energy of the system and surroundings must remain constant. (This, incidentally, is a statement of the *First Law of Thermodynamics.*) Thus, the energy that A loses in going to B is given off to the surroundings in the form of heat. The amount of heat given off when, say, one mole of A is completely converted to one mole of B in a closed system is a direct measure of the difference in energy between one mole of A and one mole of B.

Enthalpy Change

Unfortunately, ΔE can be measured directly only if volume is kept constant. Most chemical reactions are performed under conditions of constant pressure, however, which often results in a change in volume. In that case, some of the energy released is used to accomplish the change in volume. For this reason, when a reaction is carried out at constant pressure, it is more convenient to consider its energy change in terms of its heat content, or *enthalpy*. The heat released or absorbed is ΔH, the change in enthalpy for the system, and is defined in the same way as the change in energy: the enthalpy of the products minus the enthalpy of the reactants.

$$\Delta H = H_{products} - H_{reactants}$$

For most reactions, ΔH and ΔE are very similar in magnitude. Even with gas-phase reactions such as

$$N_2O_4(g) \rightleftharpoons 2NO_2(g)$$

in which (if the reactions go to completion and if the pressure is maintained constant) the volume must increase twofold during the reaction, the difference between ΔE and ΔH is only a few kilocalories. Consequently, we will assume (as indeed we have in previous chapters) that there is a negligible numerical difference between ΔE and ΔH. Some thermodynamic quantities—for example, lattice energy, ionization energy, electron affinity, and bond energy—are measured and tabulated as ΔE values. Since we shall henceforth use only ΔH, the enthalpy changes for these reactions will be assumed to be identical to the tabulated ΔE values.

Because of the way in which ΔH is defined, an exothermic reaction will have a negative enthalpy change as shown in Figure 13-1, while an endothermic reaction has a positive enthalpy change. That is, if the products are of lower energy than the reactants, the enthalpy change is negative; if the reactants are of lower energy than the products, the enthalpy change is positive. Indeed, a high extent of reaction is favored by a large negative enthalpy change. Enthalpy changes are measured in calories.

Entropy Change

The amount of disorder or randomness associated with a particular substance is measured in terms of its entropy, S. The change in entropy, ΔS, is an indication of the change in randomness not only in the motion of the molecules but also in electronic, vibrational, and rotational processes. If, for example, only the ground-state vibrational level is populated in a collection of many molecules, this is a low-entropy (highly ordered) situation relative to the situation where a number of vibrational levels are populated by the same collection of molecules. The same is true for the population of the electronic and rotational levels.

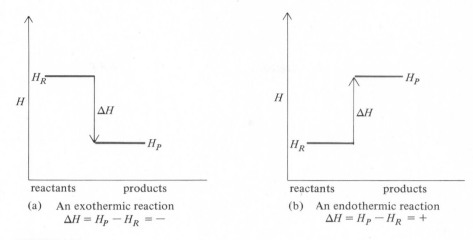

FIGURE 13-1

Enthalpy Changes for Exothermic and Endothermic Reactions

At a given temperature, the more closely spaced the energy levels, the greater the number of levels that can be populated and consequently the greater the entropy. In Chapter 8 the spacings between electronic levels were shown to be greater than those between vibrational levels, which are, in turn, greater than the spacings between rotational levels. Furthermore, the translational levels are so closely spaced as to be almost continuous. Thus, at a given temperature many more translational levels are populated than rotational levels, which are more populated than vibrational levels, and so on; and in general the largest part of the entropy of any collection of molecules, atoms, or ions is due to the population of many translational levels. In other words, the many velocities with which the particles of a substance in the liquid or gaseous state are moving are the major source of its disorder. An increase in temperature provides energy for the population of even more energy levels; the entropy of a substance therefore increases with temperature.

A change in entropy is defined analogously to a change in enthalpy:

$$\Delta S = S_{\text{products}} - S_{\text{reactants}}$$

a positive ΔS indicates an increase in disorder in the reaction, while a negative ΔS denotes an increase in order (Figure 13-2). Entropy changes are given in units of calories per degree.

In a reaction the total entropy of the system (reactants and products) *and* surroundings (reaction vessel and its surroundings) increases as the reaction proceeds. This, a statement of the *Second Law of Thermodynamics,* was combined with the first law in the maxim of the German physicist, R. Clausius: "The energy of the universe remains constant, but the entropy of the universe continually tends toward a maximum." Unfortunately, the difficulties involved in measuring the

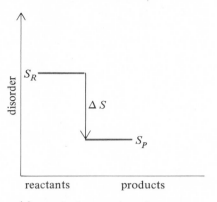

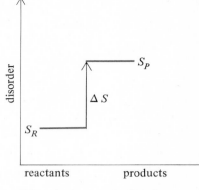

(a)   An increase in order
$$\Delta S = S_P - S_R = -$$

(b)   An increase in disorder
$$\Delta S = S_P - S_R = +$$

FIGURE 13-2

Positive and Negative Entropy Changes

entropy of the system *and* surroundings precludes using this law as the sole index of the progress of a reaction.

Free-Energy Change   The compromise between the drives to low energy and high disorder can be expressed mathematically as

$$G = H - TS$$

where G is the free energy, in calories, and T is the absolute temperature. This arithmetical mixture of enthalpy, entropy, and temperature is the determinant of the extent of a chemical reaction. The optimal mix of low energy and high disorder is attained when the free energy of a system has been minimized. If at any point in a reaction the free energy of the products present is lower than the free energy of the reactants present, the reaction will continue to produce products until the free energy of the system has been minimized. At that point the free energies of the reactants and products are equal; that is, $\Delta G = 0$, and the system has arrived at a state of equilibrium.

The minimization of free energy for the reaction

$$N_2O_4(g) \rightleftharpoons 2NO_2(g)$$

at 25°C is shown graphically in Figure 13-3. One mole of N_2O_4 is arbitrarily assigned a free energy of zero, and relative to this 2 moles of NO_2 have a free energy of +1.2 kcal. The free energy of each compound in the mixture depends upon its concentration. If the reaction is begun with 1 mole of N_2O_4, point A, conversion of N_2O_4 to NO_2 will occur until 0.37 mole of NO_2 has been produced. At this point (C), the mixture of 0.37 mole of NO_2 and 0.815 mole of N_2O_4 has the lowest possible free energy, the free energy of 0.37 mole NO_2 is equal to the free

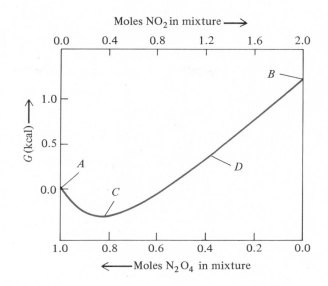

FIGURE 13-3

Variation in Free Energy for the Reaction N_2O_4 (g) $\rightleftharpoons$ 2NO$_2$ (g) at 25° C
SOURCE: William G. Davies, *Introduction to Chemical Thermo-
dynamics.* Philadelphia: W.B. Saunders Co., 1972, Figure 9-4, p. 192.

energy of 0.815 mole of N_2O_4, and the system is at equilibrium. On the other hand, if the reaction is begun with 2.0 moles of NO_2, NO_2 will react to form N_2O_4 until point C is once again reached. At some point between B and C—say, D—the free energy has not been minimized; that is, the free energy of 1.2 moles of NO_2 is greater than that of 0.4 mole of N_2O_4.

In conformity with the definitions of ΔH and ΔS, the change in free energy is

$$\Delta G = G_{\text{products}} - G_{\text{reactants}}$$

The *change* in free energy can also be expressed in terms of the changes in enthalpy and entropy:

$$\Delta G = \Delta H - T\Delta S$$

Thus, for a given reaction, a negative ΔG could be a result of a negative ΔH and a positive ΔS, or a small positive ΔH outweighed by a sufficiently large positive ΔS, or a negative ΔS outweighed by a negative ΔH.

The three thermochemical functions—ΔH, ΔS, and ΔG—have three features in common:

1. Their values do not depend on the mechanism of the reaction; they depend only on the initial and final states of the system (reactants and products). This is analogous to the difference in potential energy of a stream at two points. Since this energy depends only on the difference in

heights of the two points, it is independent of whether the stream takes a circuitous route from A to B or whether it falls directly from A to B (Figure 13-4). The enthalpy change for the reaction $A \to D$ is given by $\Delta H = H_D - H_A$ and is not affected by a possible path for the reaction such as

$$A \to B$$
$$B \to C$$
$$C \to D$$

Only the initial and final states (A and D) are significant to the calculation of ΔH.

FIGURE 13-4

Potential Energy of Stream

An important consequence of this feature is that enthalpy, entropy, and free-energy changes are additive. Suppose, for example, that you have at your disposal the enthalpy changes for the two reactions

(1) $C(s) + \frac{1}{2}O_2(g) \to CO(g)$ $\Delta H_1 = -26.4 \text{ kcal}$

(2) $CO(g) + \frac{1}{2}O_2(g) \to CO_2(g)$ $\Delta H_2 = -67.7 \text{ kcal}$

and that for some reason you need the enthalpy change for the reaction

(3) $C(s) + O_2(g) \to CO_2(g)$ $\Delta H_3 = ?$

Both equations (1) and (2) contain some of the species of equation (3), although neither (1) nor (2) is identical to (3). The sum of equations (1) and (2) is, however, identical to (3), and therefore the sum of the enthalpy changes, ΔH_1 and ΔH_2, will yield the desired ΔH_3.

$$C(s) + \tfrac{1}{2}O_2(g) \rightleftharpoons CO(g) \qquad \Delta H_1$$

$$+ \quad CO(g) + \tfrac{1}{2}O_2(g) \rightleftharpoons CO_2(g) \qquad \Delta H_2$$

$$\overline{\quad C(s) + \quad O_2(g) \rightleftharpoons CO_2(g) \qquad \Delta H_3 = \Delta H_1 + \Delta H_2}$$

2. Their values depend upon the amount of material in question. The free energy of 10 moles of CO_2 is ten times that of 1 mole of CO_2. Likewise, the free-energy change (or the enthalpy change or entropy change) for the conversion of 10 moles of carbon to 10 moles of CO_2 is ten times that for the conversion of 1 mole.

$$C(s) + \quad O_2(g) \rightleftharpoons \quad CO_2(g) \qquad \Delta G_1$$

$$10C(s) + 10O_2(g) \rightleftharpoons 10CO_2(g) \qquad \Delta G_2$$

$$\Delta G_2 = 10\Delta G_1$$

For this reason, the notations for ΔH, ΔS, and ΔG for a reaction often include an indication of how many moles are involved in the reaction. This amount, moreover, is almost always one mole. For example, ΔH for equation (3) shown before is -94.1 kcal per mole of CO_2 formed when the reaction is carried out at $25°C$ and 1 atm. If the amount is not specified, it is assumed that the number of moles involved is represented by the equation written with the lowest possible integral coefficients.

3. Their values are dependent upon temperature. Over short (100-degree) temperature ranges, ΔH and ΔS usually vary only a little, but ΔG, because of the $T\Delta S$ term in the relation $\Delta G = \Delta H - T\Delta S$, is more temperature dependent.

Because of the dependence of ΔH, ΔS, and ΔG on conditions, it is conventional practice to define certain *standard conditions* for the measurement of these parameters as *one atmosphere partial pressure for all gases and one molar concentration for all substances in solution. The standard temperature is usually, but not always, 25°C.* A value measured under these conditions is designated by a superscript zero; for instance, ΔH^0, ΔS^0, ΔG^0. A *standard enthalpy change* of 13.9 kcal for the reaction

$$N_2O_4(g) \rightleftharpoons 2NO_2(g)$$

at $25°C$ indicates that when 2 moles of NO_2 are formed at a pressure of 1 atm from 1 mole of N_2O_4 also at 1 atm, 13.9 kcal of heat are absorbed.

THE RELATIONSHIP BETWEEN ΔG^0 AND K

With standard state defined, we can express a very important relationship between the *standard free-energy change* and the equilibrium constant for a given reaction.

$$\Delta G^0 = -RT \ln K$$

In this expression, R is the ideal gas constant expressed in energy units (1.99 cal/mole-deg), T is the absolute temperature, and $\ln K$ is the natural logarithm of the equilibrium constant. The relationship can also be written in terms of base 10 logarithms by using the factor of 2.303 for conversion of ln to log ($\ln x = 2.303 \log x$).

$$\Delta G^0 = -RT(2.3 \log K)$$

When the reaction involves gases, K must be expressed in partial pressures. (Of course, when the number of moles of gas on both sides of the equation are equal, K_p is identical to K_c.)

The significance of this relation can be appreciated more readily if the equation is written in exponential form

$$K = 10^{-\Delta G^0/2.3RT}$$

Thus, if ΔG^0 is zero, the entire exponent is zero, and since any number raised to the zeroth power is 1, K has the value of 1. If ΔG^0 is positive, the exponent is negative, and K therefore is less than 1. Suppose, for example, that $\Delta G^0 = +1.0$ kcal/mole at 25°C. At that temperature

$$K = 10^{-1000/(2.30)(1.99)(298)}$$

$$= 10^{-1000/1364} = 10^{-0.73}$$

or

$$K = \frac{1}{10^{0.73}}$$

Now, 10 raised to the 0.73 power can be evaluated by finding the antilog of 0.73, which is 5.4. Therefore,

$$K = \frac{1}{5.4} = 0.19$$

On the other hand, if ΔG^0 is negative, K is greater than 1, and the greater the magnitude of the negative ΔG^0 the greater the value of K. Table 13-3 gives values of K for a variety of standard free-energy changes at 25°C. Clearly, the major significance of ΔG^0 is that it can be used to calculate the equilibrium constant.

The standard free-energy change can, in turn, be evaluated in several ways:

K	ΔG^0 (kcal/mole)
TABLE 13-3	
Relationship between	
K and ΔG^0 at 25°C	
10	−1.4
100	−2.8
10^3	−4.1
10^4	−5.5
10^{10}	−14
10^{20}	−28
10^{100}	−140
1	0
0.10	+1.4
0.010	+2.8
10^{-3}	+4.1
10^{-10}	+14
10^{-100}	+140

(a) from a knowledge of the standard enthalpy and entropy changes for a given reaction,

$$\Delta G^0 = \Delta H^0 - T\Delta S^0$$

(b) from the equilibrium constant if it is known, and (c) from the standard electromotive force of an electrochemical reaction, a concept we will discuss in Chapter 17. Since we are concerned primarily with the calculation of K from ΔG^0, we will discuss here the evaluation of the standard enthalpy and entropy changes.

DETERMINATION OF ΔH^0

Standard Heat of Formation

Many enthalpy changes are tabulated as the *standard heat of formation,* ΔH_f^0, *which is defined as the enthalpy change when 1 mole of a substance is formed from its constituent elements, each element being in its stable form at 1 atmosphere pressure and the temperature specified.* Generally, standard heats of formation are measured at 25°C. When 1 mole of NO_2 (g) is formed from N_2 and O_2 at 1 atm and 25°C, 8.1 kcal of heat is absorbed.

$$\tfrac{1}{2}N_2(g) + O_2(g) \rightleftharpoons NO_2(g) \qquad \Delta H^0 = \Delta H_f^0 = 8.1 \text{ kcal/mole}$$

The enthalpy change for this reaction, which is positive because heat is *absorbed,* is the standard heat of formation of NO_2.

The equation representing the formation of liquid methanol from its elements in their stable forms is

$$C_{graphite} + 2H_2(g) + \tfrac{1}{2}O_2(g) \rightarrow CH_3OH(l)$$

The graphite allotropic form of carbon is used in this equation because it is the stable form of carbon at 1 atm and 25°C. The formation of 1 mole of methanol under these conditions liberates 57.02 kcal of heat, and thus ΔH_f^0 for liquid methanol is −57.02 kcal/mole.

When the standard heats of formation of each substance involved in a particular reaction are known, the enthalpy change for the reaction can be calculated. The standard enthalpy change for the $N_2O_4 \rightleftharpoons 2NO_2$ reaction can be determined from the knowledge that ΔH_f^0 (NO_2) = 8.1 kcal/mole and $\Delta H_f^0(N_2O_4)$ = 2.3 kcal/mole. First we write explicitly the equations that represent the formation of these substances from their elements.

$$\tfrac{1}{2}N_2(g) + O_2(g) \rightleftharpoons NO_2(g) \qquad \Delta H_f^0 = 8.1 \text{ kcal/mole}$$

$$N_2(g) + 2O_2(g) \rightleftharpoons N_2O_4(g) \qquad \Delta H_f^0 = 2.3 \text{ kcal/mole}$$

If now the first equation is multiplied by 2 and the second equation is reversed and added to the first, we obtain

$$N_2(g) + 2O_2(g) \rightleftharpoons 2NO_2(g) \qquad \Delta H^0 = 2\Delta H_f^0(NO_2) = 16.2 \text{ kcal/mole}$$

$$\underline{N_2O_4(g) \rightleftharpoons N_2(g) + 2O_2(g) \qquad \Delta H^0 = -\Delta H_f^0(N_2O_4) = -2.3 \text{ kcal/mole}}$$

$$N_2O_4(g) \rightleftharpoons 2NO_2(g) \qquad \Delta H^0 = 2\Delta H_f^0(NO_2) - \Delta H_f^0(N_2O_4)$$

$$= 13.9 \text{ kcal/mole}$$

The elements, N_2 and O_2, cancel, and the resulting sum is the desired equation. The overall ΔH^0 is given by twice the ΔH_f^0 for NO_2 minus the ΔH_f^0 for N_2O_4. (Note that reversal of the second equation changes the sign of the enthalpy change.) This result suggests that ΔH^0 for any reaction can be obtained by simply subtracting the sum of the standard heats of formation of the reactants from the sum of the standard heats of formation of the products after multiplying the ΔH_f^0 for each substance by the coefficient of that substance in the equation for the reaction. Hence, for the reaction

$$a A + b B \rightleftharpoons c C + d D$$

we obtain

$$\Delta H^0 = c\Delta H_f^0(C) + d\Delta H_f^0(D) - [a\Delta H_f^0(A) + b\Delta H_f^0(B)]$$

One other convention is necessary for these calculations; namely, that *the standard heat of formation of any element in its standard state is zero.* For the reaction

$$CO(g) + \tfrac{1}{2}O_2(g) \rightleftharpoons CO_2(g)$$

the enthalpy change is

$$\Delta H^0 = \Delta H_f^0(CO_2) - [\Delta H_f^0(CO) + \tfrac{1}{2}\Delta H_f^0(O_2)]$$

But since $\Delta H_f^0(O_2) = 0$, $\Delta H^0 = \Delta H_f^0(CO_2) - \Delta H_f^0(CO)$.

Standard heats of formation for a variety of compounds can be found in Appendix 7.

EXAMPLE: Trichloromethane decomposes in air to phosgene ($COCl_2$) and hydrogen chloride. Calculate ΔH^0 for this reaction at 25°C.

Solution: The equation for the process must first be written:

$$HCCl_3\,(l) + \tfrac{1}{2}O_2\,(g) \rightleftharpoons \underset{Cl}{\overset{\displaystyle\overset{O}{\|}}{C}}\!\!\diagdown_{Cl}(g) + HCl(g)$$

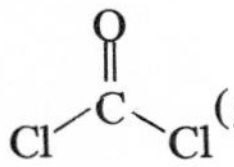

Then,

$$\Delta H^0 = \Delta H_f^0(COCl_2) + \Delta H_f^0(HCl) - \Delta H_f^0(HCCl_3)$$

$$= (-52.8) + (-22.1) - (-31.5)$$

$$= -43.4 \text{ kcal}$$

Care must be taken to specify the physical state of each substance, since the standard heat of formation for a given compound depends on its state—solid, liquid, or gas.

EXAMPLE: Calculate ΔH^0 for the reaction

$$2NaCl(s) + H_2SO_4(l) \rightleftharpoons Na_2SO_4(s) + 2HCl(g)$$

at 25°C using standard heats of formation.

Solution:

$$\Delta H^0 = \Delta H_f^0(Na_2SO_4(s)) + 2\Delta H_f^0(HCl(g))$$
$$-[2\Delta H_f^0(NaCl(s)) + \Delta H_f^0(H_2SO_4(l))]$$

$$\Delta H^0 = -330.9 + 2(-22.1) - [2(-98.2) + (-193.9)]$$

$$= 15.2 \text{ kcal}$$

Heats of Combustion

Enthalpy changes can also be tabulated as *standard heats of combustion,* ΔH_c^0. Since most compounds will react with oxygen at some temperature, many enthalpy changes of this type can be obtained. We will restrict our use of ΔH_c^0 to compounds that contain no elements other than carbon, hydrogen, and oxygen. The heat of combustion can therefore be defined as the *enthalpy change accompanying the complete combustion of a substance into carbon dioxide and water.* These enthalpy changes can be utilized to calculate standard enthalpy changes for reactions, as in the following example.

EXAMPLE: Calculate the standard enthalpy change for conversion of *cis*-2-butene to *trans*-2-butene at 25°C from the following heats of combustion (at 25°C):

$$\Delta H_c^0(cis\text{-2-butene}) = -647.91 \text{ kcal/mole},$$

$$\Delta H_c^0(trans\text{-2-butene}) = -646.96 \text{ kcal/mole}$$

Solution: The reaction for which ΔH^0 is being evaluated is

and the appropriate heats of combustion can be represented by the equations

(1) $cis\text{-}C_4H_8(g) + 6O_2(g) \rightleftharpoons 4CO_2(g) + 4H_2O(l)$ $\Delta H^0 = -647.91 \text{ kcal/mole}$

(2) $trans\text{-}C_4H_8(g) + 6O_2(g) \rightleftharpoons 4CO_2(g) + 4H_2O(l)$ $\Delta H^0 = -646.96 \text{ kcal/mole}$

If equation (2) is now reversed (and its enthalpy change altered in sign) and added to equation (1), the result is

$cis\text{-}C_4H_8(g) + 6O_2(g) \rightleftharpoons 4CO_2(g) + 4H_2O(l)$ $\Delta H^0 = -647.91 \text{ kcal/mole}$

$4CO_2(g) + 4H_2O(l) \rightleftharpoons trans\text{-}C_4H_8(g) + 6O_2(g)$ $\Delta H^0 = +646.96 \text{ kcal/mole}$

_____

$cis\text{-}C_4H_8(g) \rightleftharpoons trans\text{-}C_4H_8(g)$ $\Delta H^0 = -0.95 \text{ kcal/mole}$

Bond Energies

We have previously (Chapter 7) defined the bond energy of a diatomic molecule as the energy required to dissociate a mole of the *gaseous* compound into its constituent *gaseous* atoms. Thus, for example, the bond energy of 57.3 kcal/mole for Cl_2 is the enthalpy change at 25°C for the process

$$Cl_2(g) \rightleftharpoons 2Cl(g)   \Delta H^0 = 57.3 \text{ kcal/mole}$$

For a polyatomic molecule, the term *bond energy* can have several different meanings. For H_2O, the energy required for the dissociation of one O—H bond is 113.4 kcal/mole,

$$H-O-H(g) \rightarrow H(g) + O-H(g)   \Delta H^0 = 113.4 \text{ kcal/mole}$$

while the dissociation of the other O—H bond requires only 100.6 kcal/mole.

$$O-H(g) \rightarrow O(g) + H(g)   \Delta H^0 = 100.6 \text{ kcal/mole}$$

The sum of these two equations represents the *complete* dissociation of $H_2O(g)$ into its constituent atoms, for which 214.0 kcal/mole of energy is required.

$$H-O-H(g) \rightarrow 2H(g) + O(g)   \Delta H^0 = 214.0 \text{ kcal/mole}$$

Since two O—H bonds per molecule have been severed, an *average O—H bond energy for water* can be obtained by dividing 214.0 kcal/mole by 2. The average O–H bond energy in H_2O is, then, 107.0 kcal/mole.

Chemists often require approximate values for ΔH for reactions for which no pertinent thermochemical data are available. This has led to yet another definition of bond energy. If the average bond energy is obtained for a particular bond in each of a series of compounds, these values may then be averaged to obtain a *representative bond energy* value for that bond. For example, the average O–H bond energy in water is 107 kcal/mole, as we have discovered above; the O–H bond energy in methanol, CH_3OH, is 102 kcal/mole; and if the O–H bond energies in a number of compounds are averaged, the value of 110 kcal/mole is obtained. In a similar way, average bond energies have been obtained for other single bonds, such as C–H, N–H, C–Cl, and C–Br, as well as for multiple bonds such as C=C, C=N, and C≡N. Appendix 5 provides a listing of average bond energies.

Use of the exact bond energies for diatomic molecules will produce accurate ΔH^0 values, as illustrated by the reaction

$$N_2 + O_2 \rightleftharpoons 2NO$$

The standard enthalpy change can be calculated by summing the equations (and their respective enthalpy changes).

$$N_2(g) \rightleftharpoons 2N(g) \qquad \Delta H^0 = 225 \text{ kcal}$$

$$O_2(g) \rightleftharpoons 2O(g) \qquad \Delta H^0 = 118 \text{ kcal}$$

$$\underline{2N(g) + 2O(g) \rightleftharpoons 2NO(g) \qquad \Delta H^0 = -300 \text{ kcal}}$$

$$N_2(g) + O_2(g) \rightleftharpoons 2NO(g) \qquad \Delta H^0 = +43 \text{ kcal}$$

The first and second equations are, of course, representative of the dissociation of N_2 and O_2, and the corresponding enthalpy changes are their bond dissociation energies. The third equation is the reverse of the equation for the dissociation of NO, multiplied by 2, and the enthalpy change is the negative of twice the bond energy of NO.

In a mechanical sense, this enthalpy change was calculated by subtracting the sum of the bond energies of the products from the sum of the bond energies of the reactants after multiplying the bond energy of each substance by the coefficient of that substance in the equation in question.

The procedure for obtaining approximate enthalpy changes is similar. Consider the problem of estimating ΔH^0 for the reaction

$$CH_3CH_3(g) + Cl_2(g) \rightleftharpoons CH_3CH_2Cl(g) + HCl(g)$$

The net result of the reaction is the breaking of one C–H bond per ethane molecule and one Cl–Cl bond per Cl_2 molecule, and the formation of one C–Cl bond per chloroethane molecule and one H–Cl bond per HCl molecule. The energy required to break a C–H bond is estimated as the average C–H bond energy, while the energy released when a C–Cl bond is formed is estimated as the average C–Cl bond energy. For the decomposition of Cl_2 and the formation of HCl, exact diatomic bond energies can be used. Hence, addition of the following equations and enthalpy changes produces the desired *estimate* of ΔH^0.

$$CH_3CH_3(g) \rightarrow CH_3CH_2(g) + H(g) \qquad \Delta H^0 \cong 98 \text{ kcal}$$

$$Cl_2(g) \rightarrow 2Cl(g) \qquad \Delta H^0 = 57 \text{ kcal}$$

$$Cl(g) + CH_3CH_2(g) \rightarrow CH_3CH_2Cl(g) \qquad \Delta H^0 \cong -78 \text{ kcal}$$

$$H(g) + Cl(g) \rightarrow HCl(g) \qquad \Delta H^0 = -102 \text{ kcal}$$

$$CH_3CH_3(g) + Cl_2(g) \rightleftharpoons CH_3CH_2Cl(g) + HCl(g) \qquad \Delta H^0 \cong -25 \text{ kcal}$$

The accurate ΔH^0 for this reaction obtained from standard heats of formation is -27.0 kcal/mole, which reveals that the approximate value obtained with average bond energies is not greatly in error.

EXAMPLE: Calculate the standard enthalpy change for the reaction

$$CHCl_3(g) + \tfrac{1}{2}O_2(g) \rightleftharpoons Cl{-}\!\!\overset{\overset{\textstyle O}{\|}}{C}\!\!{-}Cl\,(g) + HCl(g)$$

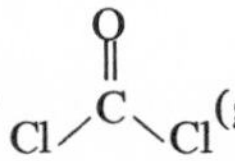

from bond energy data.

Solution: On the left-hand side of the equation, 1 mole of C—Cl bonds, 1 mole of C—H bonds, and $\tfrac{1}{2}$ mole of O_2 bonds are broken:

(1) $\qquad\qquad CHCl_3(g) \rightarrow Cl{-}\!\overset{}{C}\!{-}Cl + H(g) + Cl(g)$

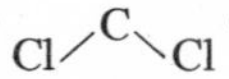

$$\Delta H^0 = 98 \text{ kcal/mole (C–H)} + 78 \text{ kcal/mole (C–Cl)}$$

$$= 176 \text{ kcal}$$

(2) $\qquad\qquad \tfrac{1}{2}O_2(g) \rightarrow O(g) \qquad \Delta H^0 = 59 \text{ kcal}$

On the right-hand side of the equation, 1 mole of C=O bonds and 1 mole of HCl bonds are formed:

(3) 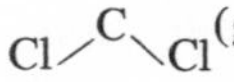 $\quad Cl{-}\!\overset{}{C}\!{-}Cl\,(g) + O(g) \rightarrow Cl{-}\!\!\overset{\overset{\textstyle O}{\|}}{C}\!\!{-}Cl\,(g) \qquad \Delta H^0 = -191 \text{ kcal}$

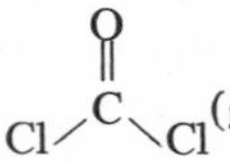

(4) $\qquad\qquad H(g) + Cl(g) \rightarrow HCl(g) \qquad \Delta H^0 = -102 \text{ kcal}$

Addition of the enthalpy changes associated with the four equations results in a ΔH^0 for the overall reaction of -58 kcal.

An alternate way of visualizing the problem is to allow all reactants to break down into their constituent atoms and then re-form these atoms into the products as shown below. Both approaches give the same answer.

$$CHCl_3(g) \rightarrow C(g) + H(g) + 3Cl(g) \qquad \Delta H^0 = 98(C{-}H) + 3 \times 78\ (C{-}Cl)$$

$$= 332 \text{ kcal}$$

$$\tfrac{1}{2}O_2(g) \rightarrow O(g) \qquad\qquad\qquad \Delta H^0 = 59 \text{ kcal}$$

$$C(g) + O(g) + 2Cl(g) \rightarrow \underset{Cl}{\overset{O}{\underset{\diagdown}{\overset{\|}{\overset{\diagup}{C}}}}}_{Cl}(g) \qquad \Delta H^0 = -191(C{=}O) +$$
$$(-2) \times 78(C{-}Cl)$$
$$= -347 \text{ kcal}$$

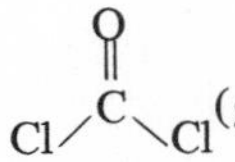

$$\underline{H(g) + Cl(g) \rightarrow HCl(g) \qquad\qquad\qquad \Delta H^0 = -102 \text{ kcal}}$$

$$CHCl_3(g) + \tfrac{1}{2}O_2(g) \rightleftharpoons \underset{Cl}{\overset{O}{\underset{\diagdown}{\overset{\|}{\overset{\diagup}{C}}}}}_{Cl}(g) + HCl(g) \qquad \Delta H^0 = -58 \text{ kcal}$$

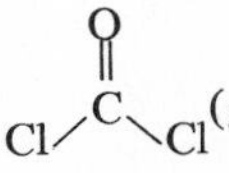

The more accurate enthalpy change for this reaction, derived from heats of formation, is −50.7 kcal.

Other Enthalpy Changes

The standard heat of formation, heat of combustion, and bond energy were emphasized above merely because most heat data are tabulated in these forms. However, any set of reactions with known enthalpy changes can be used to calculate other enthalpy changes. Several examples involving lattice energy, ionization energy, electron affinity, and so on, are given in Part 3 of this book.

DETERMINATION OF ΔS^0

We have previously found the entropy of a substance to be associated with its disorder, this disorder being a result of the number of ways in which its various energy levels (translational, rotational, vibrational, electronic) can be populated. Moreover, the accessibility of energy levels increases as the temperature increases, and therefore the entropy of a substance increases with temperature. This temperature dependence suggests that at absolute zero, any substance should be perfectly ordered; that is, all molecules or ions should occupy the lowest energy state. Numerous experiments have verified this expectation, and the *Third Law of Thermodynamics* states that *the entropy of perfect crystals of all elements and compounds is zero at the absolute zero of temperature.*

As a result of this law, absolute entropies can be determined at any temperature. Entropies evaluated at 25°C are listed in Appendix 7. These entropies can be used to calculate ΔS^0 for a given reaction by subtracting the sum of the entropies of the reactants from the sum of the entropies of the products after multiplying the entropy value for each substance by the coefficient associated with that substance in the equation.

The entropy changes calculated in this way for a number of reactions are

	ΔS^0 (cal/deg) at 25°
TABLE 13-4 Standard Entropy Changes for Selected Reactions	
(1) $PbCO_3(s) \rightarrow PbO(s) + CO_2(g)$	36
(2) $SiCl_4(l) + 2H_2O(l) \rightarrow SiO_2(s) + 4HCl(g)$	98
(3) $\dot{N}_2O_4(g) \rightarrow 2NO_2(g)$	42
(4) $NH_3(g) \rightarrow \frac{1}{2}N_2(g) + \frac{3}{2}H_2(g)$	24
(5) $N_2O_4(g) \rightarrow N_2(g) + 2O_2(g)$	71
(6) $H_2(g) + Cl_2(g) \rightarrow 2HCl(g)$	5
(7) $N_2(g) + O_2(g) \rightarrow 2NO(g)$	6

presented in Table 13-4. Careful analysis indicates that for those reactions, (6) and (7), in which the number of gaseous particles on both sides of the equation is the same, the entropy change is small. For those reactions in which the number of gaseous particles increase, the greater the increase the higher the entropy change. Compare, for example, the ΔS^0 of 98 cal/deg for reaction (2), in which 4 moles of gas are formed from liquids, with the ΔS^0 of 24 cal/deg for reaction (4), in which two moles of gaseous products are formed from 1 mole of gaseous reactants. Indeed, a rough guide is that ΔS^0 is about $25 \Delta n$, where Δn equals the number of moles of gaseous products minus the number of moles of gaseous reactants. These observations can be explained by the much greater freedom of movement (more closely spaced translational levels) of gases as compared to liquids and solids and the fact that an increase in the number of gaseous molecules increases the number of ways in which the translational (and other energy) levels can be populated.

EXAMPLE: Calculate ΔS^0 for the conversion of gaseous N_2O_4 to gaseous NO_2.

Solution: The equation for this conversion is

$$N_2O_4(g) \rightleftharpoons 2NO_2(g)$$

The entropies for $N_2O_4(g)$ and $NO_2(g)$ are obtained from Appendix 7 as 72.7 cal/deg-mole and 57.4 cal/deg-mole, respectively. The standard entropy change is then

$$\Delta S^0 = 2(57.4 \text{ cal/deg-mole}) - 72.7 \text{ cal/deg-mole}$$

$$= 42.1 \text{ cal/deg-mole}$$

DETERMINATION OF ΔG^0

For some substances, standard free energies of formation (ΔG_f^0) have been tabulated. The *standard free energy of formation* is defined as *the free energy change when one mole of a substance is formed from its constituent elements, each element being in its standard state at* 1 *atmosphere pressure and the specified temperature (usually* $25°C$). Standard free energies of formation are entirely analogous, therefore, to standard heats of formation and can be manipulated in exactly the same way. A listing of some ΔG_f^0 values is provided in Appendix 7. The following example indicates that the ΔG_f^0 for any element in its standard state is zero.

EXAMPLE: Calculate ΔG^0 for the following reaction.

$$CH_3OH(l) + 2O_2(g) \rightleftharpoons CO_2(g) + 2H_2O(l)$$

Solution: From Appendix 7, the standard free energies of formation are

$$CH_3OH(l) \quad -39.7 \text{ kcal/mole}$$

$$CO_2(g) \quad -94.3 \text{ kcal/mole}$$

$$H_2O(l) \quad -56.7 \text{ kcal/mole}$$

Then,

$$\Delta G^0 = \Delta G_f^0(CO_2(g)) + 2\Delta G_f^0(H_2O(l)) - \Delta G_f^0(CH_3OH(l))$$

$$= -94.3 + (-113.4) - (-39.7)$$

$$\Delta G^0 = -168.0 \text{ kcal}$$

In many cases, however, standard free energy changes must be calculated from ΔH^0 and ΔS^0 with the equation

$$\Delta G^0 = \Delta H^0 - T\Delta S^0$$

The following examples are illustrative of the kinds of problems that can be solved.

EXAMPLE: Calculate ΔG^0 and K for the isomerization of *cis*-2-butene to *trans*-2-butene.

Solution: We have already calculated ΔH^0 for this process (p. 346). The standard entropy change can be obtained from the absolute entropies of 71.9 cal/deg-mole for the *cis* isomer and 70.9 cal/deg-mole for the *trans* isomer.

Therefore,

$$\Delta S^0 = 70.9 - 71.9 = -1.0 \text{ cal/deg-mole}$$

And from before, p. 346,

$$\Delta H^0 = -0.95 \text{ kcal/mole}$$

Thus,

$$\Delta G^0 = \Delta H^0 - T\Delta S^0$$

$$= -950 \text{ cal/mole} - 298°(-1.0 \text{ cal/deg-mole})$$

$$= -950 + 298 = -652 \text{ cal/mole}$$

$$\Delta G^0 = -0.65 \text{ kcal/mole}$$

Now, from the relation $\Delta G^0 = -RT(2.3 \log K)$,

$$\log K = \frac{-\Delta G^0}{RT(2.3)} = -\frac{-652}{(1.99)(298)(2.3)} = +\frac{652}{1364}$$

$$= +0.48$$

$$K = 3.0 \text{ at } 25°\text{C}$$

EXAMPLE: Determine the feasibility of preparing chlorine monoxide, Cl_2O, by direct combination of the elements.

Solution: The standard free-energy change for the reaction

$$Cl_2(g) + \tfrac{1}{2}O_2(g) \rightarrow Cl_2O(g)$$

can be determined by simply looking up $\Delta G_f^0(Cl_2O)$, since this equation represents the formation of Cl_2O from its elements in their standard states. From Appendix 7 we find that ΔG_f^0 (Cl_2O) is 22.4 kcal/mole. The equilibrium constant calculated from this is 10^{-17}, which means that even under conditions of high pressure (which would favor formation of the product) virtually no product would be formed at 25°C. Of course, since the reaction is endothermic ($\Delta H_f^0 = 18.2$ kcal/mole) higher temperatures will favor the product, but even at 600°C the extent remains small.

Moreover, the positive free energy of formation reveals that at 25°C the compound Cl_2O is unstable relative to its constituent elements and the decomposition

$$Cl_2O(g) \rightarrow Cl_2(g) + \tfrac{1}{2}O_2(g)$$

should occur. In fact, this decomposition does occur but at a slow enough rate so that Cl_2O, once prepared, can be stored for some time.

The preparation can be accomplished by a different route—the reaction

$$HgO(s) + 2Cl_2(g) \rightarrow HgCl_2(s) + Cl_2O(g)$$

The standard free-energy change for this reaction can be obtained from standard free energies of formation and is verification of a sizeable equilibrium constant (10^4) for this reaction.

$$\Delta G^0 = \Delta G^0(\text{Cl}_2\text{O(g)}) + \Delta G_f^0(\text{HgCl}_2(\text{s})) - \Delta G_f^0(\text{HgO(s)})$$

$$= 22.4 + (-44.4) - (-14.0)$$

$$\Delta G^0 = -8.0 \text{ kcal}$$

EXAMPLE: Determine the feasibility of preparing chloroethene (a precursor to the common plastic polyvinyl chloride) from 1,2-dichloroethane. If this process is feasible, what conditions should be used to optimize the extent?

Solution: The equation for the reaction is

$$\text{ClCH}_2\text{CH}_2\text{Cl(g)} \rightarrow \text{H}_2\text{C}{=}\text{CHCl(g)} + \text{HCl(g)}$$

In order to determine the feasibility of the reaction we must, of course, know two things: its rate and its extent. The rate is very difficult to estimate, but the extent can be determined via K and ΔG^0.

    Standard free energies of formation for 1,2-dichloroethane and chloroethene are not given in Appendix 7, and we will have to resort to an approximate calculation of ΔH^0 from average bond energies.

$$\Delta H^0 \cong 83 \text{ (C–C)} + 98 \text{ (C–H)} + 78 \text{ (C–Cl)} - [144 \text{ (C=C)} + 102 \text{ (H–Cl)}]$$

$$\cong +13 \text{ kcal/mole}$$

Since there are twice as many gaseous particles on the right-hand side of the equation as on the left-hand side, the entropy change for the reaction is certainly positive. With the guide given on page 350 we can estimate ΔS^0 as 25 cal/deg-mole. Then,

$$\Delta G^0 = \Delta H^0 - T\Delta S^0$$

$$\cong 13,000 \text{ cal/mole} - (298°)(25 \text{ cal/deg-mole})$$

$$\cong 13,000 \text{ cal/mole} - 7450 \text{ cal/mole}$$

$$\cong 6 \text{ kcal/mole}$$

[The more accurate values obtained from heats and free energies of formation are $\Delta H^0 = 17.3$ kcal/mole, $\Delta S^0 = 34$ cal/deg-mole, $\Delta G^0 = 7.2$ kcal/mole.]

    At 25°C, the reaction clearly has a low equilibrium constant. If we now assume that ΔH^0 and ΔS^0 will remain constant over a temperature range of a couple of hundred degrees, we can calculate the temperature at which $\Delta G^0 = 0$ (and therefore $K = 1$).

$$\Delta G^0 = 0 = \Delta H^0 - T\Delta S^0$$

$$= 13,000 \text{ cal/mole} - T(25 \text{ cal/deg-mole})$$

$$T = \frac{13,000 \text{ cal/mole}}{25 \text{ cal/deg-mole}} = 520°\text{K or } 247°\text{C}$$

[The same calculation based on the more accurate values of ΔH^0 and ΔS^0 produces a temperature of $239°C$.]

Therefore, optimum conditions for this reaction would be a temperature above $200°C$ and relatively low pressure. (The higher temperature will also increase the rate of the reaction.)

SUGGESTED READINGS

Strong, L. E., and W. J. Stratton. *Chemical Energy*. New York: Reinhold, 1965.

Davies, W. G. *Introduction to Chemical Thermodynamics*. Philadelphia: Saunders, 1972.

Sanderson, R. T., "Principles of Chemical Reaction." *Journal of Chemical Education,* Vol. 41, No. 1 (January 1964), pp. 13–22.

PROBLEMS

1. For the reversible exothermic reaction

$$SO_2(g) + \tfrac{1}{2}O_2(g) \rightleftharpoons SO_3(g)$$

determine how the equilibrium constant and the extent of reaction will be affected by each of the following, assuming the system is at equilibrium to begin with.
(a) More $O_2(g)$ is added, the temperature and volume remaining constant.
(b) Some $SO_3(g)$ is removed, temperature and volume remaining constant.
(c) The volume is decreased, the temperature remaining constant.
(d) The temperature is increased, the volume remaining constant.
(e) The temperature is increased, the pressure remaining constant.
(f) A catalyst is added.

2. The endothermic reaction of solid carbon (as coke or coal) with steam at high temperatures is an important route to the industrial fuel called water gas, which is a mixture of carbon monoxide and hydrogen. Assume that the reaction is reversible and at equilibrium. How will the weight of carbon monoxide present be affected by each of the following?
(a) More steam is added, volume and temperature remaining constant.
(b) Some hydrogen is removed at constant volume and temperature.
(c) More solid carbon is added at constant volume and temperature.
(d) The volume is increased at constant temperature.
(e) The temperature is decreased at constant pressure.

3. Solid CuO reacts with gaseous CO to give solid Cu_2O and gaseous CO_2.

$$2CuO(s) + CO(g) \rightleftharpoons Cu_2O(s) + CO_2(g)$$

(a) If the system is at equilibrium, how will the following changes affect the number of moles of CO_2?
(1) More CO is added at constant temperature and pressure.
(2) Some Cu_2O is removed at constant temperature and pressure.
(3) The volume is decreased.
(4) A catalyst is added.

(b) When the reaction has reached equilibrium in a 1-liter flask, 0.3 mole $CuO(s)$, 0.1 mole CO, 0.2 mole $Cu_2O(s)$, and 0.25 mole CO_2 are present. Calculate the equilibrium constant.

4. For the following reaction,

$$CO(g) + Cl_2(g) \rightleftharpoons COCl_2(g)$$

the equilibrium mixture in a 3-liter vessel at a pressure of 2.0 atm was found to contain 21.0 g $COCl_2$, 10.2 g CO, and 18.7 g Cl_2. Calculate the equilibrium constant in terms of both moles per liter and atmospheres.

5. An equilibrium mixture of N_2, H_2, and NH_3 at some high temperature in a 3-liter flask consists of 0.15 mole N_2, 0.10 mole H_2, and 0.05 mole NH_3. What is the equilibrium constant?

6. The equilibrium constant for the reaction

$$PCl_5(g) \rightleftharpoons PCl_3(g) + Cl_2(g)$$

at $250°C$ is 1.7×10^{-2} mole/liter. Calculate the weight of PCl_3 formed when 1.0 mole of PCl_5 is allowed to dissociate in a 1-liter vessel at this temperature.

7. For the reaction

$$CO(g) + H_2O(g) \rightleftharpoons CO_2(g) + H_2(g)$$

at a given temperature, the equilibrium mixture was found to consist of 0.40 mole CO_2, 0.40 mole H_2, 0.20 mole CO, and 0.20 mole H_2O.
(a) Calculate the equilibrium constant.
(b) If 0.50 mole CO and 0.50 mole H_2O are mixed at this temperature, how many moles of CO_2 and H_2 will be present when equilibrium is established?
(c) If 0.50 mole CO_2 and 0.50 mole H_2 are mixed, how many moles of CO will be present at equilibrium?
(d) If equimolar amounts of CO and H_2O are mixed, what is the extent of reaction in terms of percent conversion?
(e) If 0.40 mole CO is mixed with 0.60 mole H_2O, how many moles CO_2 will be present at equilibrium?
(f) How many moles of CO must be mixed with 0.60 mole H_2O in order to obtain 0.35 mole CO_2 at equilibrium?

8. The equilibrium constant (K_c) for the reaction

$$N_2O_4(g) \rightleftharpoons 2NO_2(g)$$

is 2 at some temperature.
(a) Does this constant mean that the number of moles of NO_2 is twice that of N_2O_4 at equilibrium?
(b) How much N_2O_4 must be introduced into a 1-liter flask in order to obtain 0.6 mole NO_2 at equilibrium?
(c) If 1 mole of NO_2 is introduced into a 1-liter flask, how much N_2O_4 is present at equilibrium?
(d) If 1 mole of NO_2 is introduced into a 2-liter flask, how much N_2O_4 is present at equilibrium?

9. Consider the reaction

$$CH_3I + OH^- \rightleftharpoons CH_3OH + I^-$$

for which the equilibrium constant at some temperature is 10.
(a) If 0.5 mole CH_3I is mixed with 0.5 mole OH^- (in 1 liter of solution), how much CH_3OH is obtained?
(b) If 0.5 mole CH_3I is mixed with 1 mole OH^- (in 1 liter of solution), how much CH_3OH is obtained?
(c) In order to obtain 0.6 mole of CH_3OH from 0.8 mole CH_3I, how much OH^- must be mixed with the CH_3I?
(d) 1.0 mole of I^- is added to the equilibrium mixture of part (a). When equilibrium is reestablished, how much CH_3OH is present?

10. Formaldehyde reacts with hydroxide ion to give methanol and formate ion according to the equation:

$$2H_2C{=}O + OH^- \rightleftharpoons CH_3OH + HC\begin{smallmatrix} \nearrow O \\ \searrow O^- \end{smallmatrix}$$

(a) In 1 liter of solution at equilibrium, the following were present: 0.10 mole CH_3OH, 0.60 mole $HC\begin{smallmatrix} \nearrow O \\ \searrow O^- \end{smallmatrix}$, 0.10 mole OH^-, and 0.55 mole $H_2C{=}O$. Calculate the equilibrium constant.
(b) How many moles of OH^- must be mixed with 1.0 mole of $H_2C{=}O$ in order to obtain 0.30 mole of CH_3OH at equilibrium (in 1 liter of solution)?

11. The equilibrium mixture for the reaction

$$CO(g) + H_2O(g) \rightleftharpoons CO_2(g) + H_2(g)$$

at some temperature consists of 0.30 mole CO_2, 0.30 mole H_2, 0.10 mole CO, and 0.20 mole H_2O in a 1-liter vessel.
(a) Calculate the equilibrium constant.
(b) How much CO must be added to this mixture at constant temperature and volume in order to increase the concentration of CO_2 to 0.40 mole/liter?
(c) How much H_2 must be removed at constant temperature and volume in order to increase the concentration of CO_2 to 0.40 mole/liter?

12. For the gas-phase reaction

$$Xe + 2F_2 \rightleftharpoons XeF_4$$

(a) Calculate the equilibrium constant from the observation that at some temperature the extent of reaction is 50% when 0.2 mole Xe and 0.4 mole F_2 have been mixed in an empty 1-liter bulb.
(b) How many moles of F_2 would have to be added to the equilibrium mixture of part (a) in order to increase the conversion of Xe to XeF_4 to 80%?

13. Do the following changes represent an increase or decrease in entropy?
(a) burning propene to carbon dioxide and water
(b) programming a computer

(c) the conversion of a liquid to a solid

(d) breaking a beaker

(e) the rotting of wood

(f) assimilating chemical knowledge

(g) writing a term paper

(h) building a house

14. Carbon tetrachloride appears to be quite inert to water. Silicon tetra-chloride, on the other hand, reacts vigorously with water.

(a) Using standard heats and free energies of formation, calculate the standard enthalpy, entropy, and free-energy changes for the reactions

$$CCl_4 (l) + 2H_2O(l) \rightleftharpoons CO_2 (g) + 4HCl(g)$$

$$SiCl_4 (l) + 2H_2O(l) \rightleftharpoons SiO_2 (s) + 4HCl(g)$$

(b) Why does CCl_4 appear not to react with water?

15. Calculate ΔH^0, ΔS^0, ΔG^0, and K for the gas-phase isomerization of *cis*-1,2-dichloroethene to *trans*-1,2-dichloroethene at $25°C$.

16. Calculate ΔH^0, ΔS^0, ΔG^0, and K for the reaction of glucose, $C_6 H_{12} O_6$, with oxygen in the body to form CO_2 and liquid water.

17. For the following gas-phase reactions

(1) $$N_2 H_4 + O_2 \rightleftharpoons N_2 + 2H_2 O$$

(2) $$H_2 C{=}C(CH_3)_2 + Br_2 \rightleftharpoons H_2 \overset{\overset{\displaystyle Br}{|}}{C}{-}\overset{\overset{\displaystyle Br}{|}}{C}(CH_3)_2$$

(3) $$CH_3 Br + (CH_3)_2 NH \rightleftharpoons (CH_3)_3 N + HBr$$

(4) $$(CH_3)_3 N + CO_2 \rightleftharpoons CH_3 O{-}\overset{\overset{\displaystyle O}{\|}}{C}N(CH_3)_2$$

(a) Calculate ΔH^0 from bond energies (Appendix 5) and standard heats of formation (Appendix 7) where possible.

(b) Estimate ΔS^0 for each reaction and then, where possible, calculate ΔS^0 from absolute entropy values (Appendix 7).

(c) Calculate ΔG^0 at $25°C$ for each reaction from ΔH^0 and ΔS^0 and, where possible, from standard free energies of formation.

18. Which of the following reactions is most likely for the decomposition of $NH_4 NO_3$ at room temperature? Explain!

$$2NH_4 NO_3(s) \rightleftharpoons 2NH_4 NO_2 (s) + O_2 (g)$$

$$NH_4 NO_3(s) \rightleftharpoons N_2 (g) + 2H_2 (g) + \tfrac{3}{2}O_2 (g)$$

$$NH_4 NO_3(s) \rightleftharpoons N_2O(g) + 2H_2O(g)$$

19. Calculate an approximate value for the standard heat of formation of the compound $H_2 N{-}\overset{\overset{\displaystyle H}{|}}{N}{-}NH_2$ in the gaseous state. Is this compound likely to be thermodynamically stable at room temperature? Explain carefully.

20. Structural isomers that differ only in the position of a hydrogen atom are called *tautomers*. Decide on the basis of bond energies which tautomer in each of the following pairs is thermodynamically more stable.

(a) $CH_3 \overset{\overset{O}{\|}}{C}CH_3$, $CH_3 \overset{\overset{OH}{|}}{C}=CH_2$

(b) $CH_3 \overset{\overset{NH}{\|}}{C}-CH_3$, $CH_3 \overset{\overset{NH_2}{|}}{C}=CH_2$

(c) $CH_3 \overset{\overset{O}{\|}}{C}-NH_2$, $CH_3 \overset{\overset{OH}{|}}{C}=NH$

(d) ⟨○⟩—N=N—$\overset{\overset{CH_3}{|}}{\underset{CH_3}{CH}}$,   ⟨○⟩—$\overset{}{\underset{H}{N}}$—N=$\overset{\overset{CH_3}{|}}{\underset{CH_3}{C}}$

21. Which of the following possible gas-phase reactions of methane with CO_2 is thermodynamically more likely to occur at $298°K$? At $1000°K$?

$$CH_4 + CO_2 \rightleftharpoons 2CO + 2H_2$$

$$CH_4 + CO_2 \rightleftharpoons CH_3COOH$$

22. The reaction of methanol with carbon monoxide in the presence of catalysts is an important commercial method of preparing acetic acid.

$$CH_3OH(g) + CO(g) \rightleftharpoons CH_3COOH(g)$$

(a) Calculate ΔH^0 for the reaction from (1) bond energies (Appendix 5), (2) the following heats of combustion, and (3) standard enthalpies of formation (Appendix 7), and compare the values.

$CH_3OH(g)$	$\Delta H_c^0 = -182.6$ kcal/mole
$CO(g)$	$\Delta H_c^0 = -67.7$ kcal/mole
$CH_3COOH(g)$	$\Delta H_c^0 = -220.8$ kcal/mole

(b) Calculate ΔS^0 at $25°$.
(c) Calculate ΔG^0 and K_p at $25°$.

Since the rate of the reaction is low at $25°$, the reaction must be carried out at about $200°C$ in the presence of a rhodium catalyst.

(d) Assume that ΔH^0 and ΔS^0 have the same values at $200°$ as at $25°$, and calculate ΔG^0 and K_p at $200°$.

14 Types of Chemical Reactions I: Ion-Combination Reactions

Perhaps the simplest type of chemical reaction is that which leads to products through a process of the combination of ions. Most commonly, reactions of this type occur in aqueous solution.

It will be recalled from our earlier discussion of solutions (Chapter 11) that in aqueous solutions of ionic compounds and other strong electrolytes the solutes are present in the form of hydrated ions—separated, charged particles that are essentially independent of one another. Thus, an aqueous solution of potassium nitrate is a solution of potassium ions and nitrate ions; and a solution of sodium chloride contains sodium ions and chloride ions. If we mix these two solutions together, no reaction occurs, for the mixture is simply a solution of the four ions.

$$K^+ + NO_3^- + Na^+ + Cl^- \rightarrow \text{no reaction}$$

However, if instead of potassium nitrate we use a solution of *silver* nitrate (Ag^+, NO_3^-) and mix that solution with one of sodium chloride (Na^+, Cl^-), there is an immediate reaction, which results in the formation of a white precipitate of silver chloride. A reaction occurs in this case simply because of the extremely low solubility of silver chloride in water. Upon mixing of the four ions, the solution becomes instantaneously supersaturated with respect to silver chloride. Silver ions and chloride ions come together to form the crystal lattice of AgCl, which precipitates as a solid until a saturated solution is obtained.

The equation for this reaction may be written in the form:

$$AgNO_3 + NaCl \rightarrow \underline{AgCl} + NaNO_3$$

where AgCl is underlined to indicate a precipitate. While this "molecular" equation is acceptable, it does have certain shortcomings. It does not indicate clearly the nature of the reaction; namely, that this is simply the combination of ions to form an insoluble substance. Furthermore, the equation implies that sodium nitrate has been formed as a product, when, in fact, the solution at the completion of the reaction merely contains sodium ions and nitrate ions in exactly the same form in which they were present at the start of the reaction. Only after sufficient water has been removed by evaporation would the compound sodium nitrate be obtained.

It is somewhat clearer, therefore, to describe the reaction by means of an *ionic equation*, which shows more correctly the nature of dissolved strong electrolytes:

$$Ag^+ + NO_3^- + Na^+ + Cl^- \rightarrow \underline{AgCl} + Na^+ + NO_3^-$$

Now it is clear from this ionic equation that neither Na^+ nor NO_3^- have actually taken part in the reaction. The reaction would have been the same if we had used any other soluble chloride salt (KCl, $CaCl_2$, and so on) or any other soluble silver salt. We can refine the ionic equation further by eliminating the nonparticipating ions; that is, by simply removing those substances that appear in the same form on both sides of the equation. The result is called the *net ionic equation:*

$$Ag^+ + Cl^- \rightarrow \underline{AgCl}$$

This equation tells us that silver ion (whatever its source) and chloride ion (whatever its source) combine to form a precipitate of silver chloride.

Some additional examples of precipitation reactions, showing the relationship of "molecular" equation, ionic equation, and net ionic equation follow:

"Molecular": $BaCl_2 + K_2CrO_4$ $\rightarrow \underline{BaCrO_4} + 2KCl$

Ionic: $Ba^{2+} + 2Cl^- + 2K^+ + CrO_4^{2-}$ $\rightarrow \underline{BaCrO_4} + 2K^+ + 2Cl^-$

Net Ionic: $Ba^{2+} + CrO_4^{2-}$ $\rightarrow \underline{BaCrO_4}$

"Molecular": $MgSO_4 + 2NaOH$ $\rightarrow \underline{Mg(OH)_2} + Na_2SO_4$

Ionic: $Mg^{2+} + SO_4^{2-} + 2Na^+ + 2OH^- \rightarrow \underline{Mg(OH)_2} + 2Na^+ + SO_4^{2-}$

Net Ionic: $Mg^{2+} + 2OH^-$ $\rightarrow \underline{Mg(OH)_2}$

DRIVING FORCE OF ION-COMBINATION REACTIONS

The examples of ion-combination reactions just given are reactions that take place because of the formation of a product that is only slightly soluble. The "driving force" of these reactions is the removal of ions from solution by formation of a precipitate. Not all ion-combination reactions result in precipitation, however.

In general, ion-combination reactions occur whenever a product is formed by *removal* of one or more of the ions from solution. *Removal,* as used here, includes not only physical removal, as in precipitate formation, but any "tying up" of ions by combination with other ions or molecules. The "driving force" may be formation of a precipitate, a highly volatile insoluble substance (a gas at normal conditions), a weak electrolyte, or a complex ion.

Precipitate Formation

Several examples of ion-combination reactions in which the driving force is the formation of a precipitate were given above. When the ions of a compound are mixed together in aqueous solution in such concentration that the solubility of that compound is exceeded, the compound precipitates and the solution is saturated with respect to that compound. The term "insoluble" is often used to describe the precipitate. In actual fact, the precipitate is not truly insoluble; it is simply a substance with *low* solubility. The solution above the precipitate does contain some ions of the compound, and these are in equilibrium with the precipitate. The lower the solubility of the precipitate, the greater the extent of the reaction. Precipitation reactions are discussed in greater detail in later sections of this chapter.

Volatile Substance Formation

If a solution containing carbonate ion—say an aqueous solution of sodium carbonate—is mixed with a solution of a strong acid (for example, hydrochloric acid), an effervescence is observed due to the escape of a gas, carbon dioxide. The driving force for the reaction is the formation and escape of the gas. The equations for the reaction may be written

Molecular: $\quad Na_2CO_3 + 2HCl \rightarrow 2NaCl + H_2O + CO_2\uparrow$

Ionic: $\quad 2Na^+ + CO_3^{2-} + 2H^+ + 2Cl^- \rightarrow 2Na^+ + 2Cl^- + H_2O + CO_2\uparrow$

Net Ionic: $\quad CO_3^{2-} + 2H^+ \rightarrow H_2O + CO_2\uparrow$

(Note that we have omitted the water from the hydronium ion, just as we have omitted the water of hydration of all the other ions. We will use the abbreviated H^+ to signify hydronium ion, H_3O^+, throughout this chapter.) That this reaction occurs by a combination of ions is made more obvious by considering it as the formation of the weak electrolyte carbonic acid (H_2CO_3), which then decomposes into water and the acid anhydride CO_2.

$$CO_3^{2-} + 2H^+ \rightleftharpoons H_2CO_3 \rightarrow H_2O + CO_2\uparrow$$

Sulfites and sulfides also react with acids to liberate the gases sulfur dioxide and hydrogen sulfide, respectively.

$$SO_3^{2-} + 2H^+ \rightleftharpoons H_2SO_3 \rightarrow H_2O + SO_2\uparrow$$

$$S^{2-} + 2H^+ \rightarrow H_2S\uparrow$$

In all these examples the gases formed are weak electrolytes, and therefore the reactions would occur even if the products were not gases. Thus, for example, in the reaction of sulfide ion with hydronium ion, the formation of the weak electrolyte H_2S constitutes a removal of ions from solution, and even if H_2S were not a gas the reaction would still take place. Because H_2S is an insoluble gas and escapes from the solution, equilibrium cannot be established and the *extent* of the reaction is increased.

Volatile products from ion-combination reactions are not limited to weak electrolytes. For example, hydrogen chloride can be prepared by reaction of sodium chloride with concentrated sulfuric acid. Heating the reaction mixture causes HCl to be removed as a gas. This is successful, of course, only because HCl is more volatile than H_2SO_4.

$$NaCl + H_2SO_4 \xrightarrow{\text{heat}} NaHSO_4 + HCl$$

$$Cl^- + H^+ \xrightarrow{\text{heat}} HCl \uparrow$$

Similarly, HNO_3 can be prepared from $NaNO_3$ and H_2SO_4, because HNO_3 is more volatile than H_2SO_4.

Weak Electrolyte Formation

Reactions that lead to the formation of a precipitate or a volatile product are usually visible reactions; we know that a reaction has occurred because we see the precipitate or the escaping gas. Not all ion-combination reactions are heterogeneous, however. Some yield products that remain in solution. Although these reactions are not "visible," they are indicated by changes in such measurements as temperature, conductivity, or absorption spectrum. Reactions in which ions combine to form soluble, nonvolatile weak electrolytes are examples of such homogeneous reactions. The driving force is simply the "tying up" of ions through formation of undissociated molecules.

As an illustration, suppose we mix a solution of mercury(II) nitrate with one of sodium chloride. Both these compounds are strong electrolytes, and therefore we are really mixing together the four ions Hg^{2+}, NO_3^-, Na^+, and Cl^-. Now, it will be recalled from our general rules of electrolyte strength (p. 284) that mercury(II) chloride is a weak electrolyte; that is, a solution of $HgCl_2$ contains relatively few Hg^{2+} and Cl^- ions and a relatively large amount of molecules of $HgCl_2$. (An equilibrium exists, of course, between the ions and the molecules.) Therefore, when the two solutions are mixed, Hg^{2+} and Cl^- combine to form molecules of $HgCl_2$, which means that a reaction occurs for which the net ionic equation is

$$Hg^{2+} + 2Cl^- \rightleftharpoons HgCl_2$$

The product, $HgCl_2$, is too soluble to precipitate, and it is not a gas, so there is no visible evidence of reaction. Yet a reaction has occurred, for ions have combined to form a new substance that was not present in either of the reactant solutions.

One of the most common ion-combination reactions leading to weak electrolyte formation is aqueous acid-base neutralization. When an acid solution, such as HCl, is mixed with a solution of a base, for example NaOH, the driving force of the reaction is the combination of hydronium ions with hydroxide ions to form molecules of water.

$$HCl + NaOH \rightarrow NaCl + H_2O$$

$$H^+ + Cl^- + Na^+ + OH^- \rightarrow Na^+ + Cl^- + H_2O$$

$$H^+ + OH^- \rightarrow H_2O$$

As another example of weak electrolyte formation, let us consider what happens when a strong acid is mixed with the salt of a weak acid; for example, HCl mixed with sodium acetate ($NaC_2H_3O_2$). Both the reactants are strong electrolytes, but hydronium ions and acetate ions can combine to form acetic acid, which is a weak electrolyte, and this is the driving force of the reaction.

$$HCl + NaC_2H_3O_2 \rightarrow HC_2H_3O_2 + NaCl$$

$$H^+ + Cl^- + Na^+ + C_2H_3O_2^- \rightarrow HC_2H_3O_2 + Na^+ + Cl^-$$

$$H^+ + C_2H_3O_2^- \rightarrow HC_2H_3O_2$$

Acetic acid, being a weak electrolyte, appears in the equation as a molecule, like $HgCl_2$ and H_2O in the preceding examples.

In order to be able to predict when ion-combination reactions of this kind will occur and what the products will be, one must have some knowledge of which electrolytes are strong and which are weak. Review of the general rules of electrolyte strength in Chapter 11 is recommended at this point.

Complex Ion Formation

Another type of homogeneous ion-combination reaction results in the formation of a stable complex ion. A reactant ion combines with other ions or, in some cases, molecules to form an ion different from that of any of the reactants. For example, let us suppose we mix together solutions containing potassium cyanide, KCN, and cadmium nitrate, $Cd(NO_3)_2$. A reaction occurs because cadmium ions and cyanide ions can combine to form a complex ion, $Cd(CN)_4^{2-}$, as represented by the equations:

$$4KCN + Cd(NO_3)_2 \rightarrow 2KNO_3 + K_2[Cd(CN)_4]$$

$$4K^+ + 4CN^- + Cd^{2+} + 2NO_3^- \rightarrow 2K^+ + 2NO_3^- + 2K^+ + Cd(CN)_4^{2-}$$

$$4CN^- + Cd^{2+} \rightarrow Cd(CN)_4^{2-}$$

The use of sodium thiosulfate ($Na_2S_2O_3$) in the "fixing" process of photographic film development stems from the fact that it forms a complex ion with silver ion and thus removes the unreduced silver bromide from the film surface.

$$AgBr + 2S_2O_3^{2-} \rightarrow Ag(S_2O_3)_2^{3-} + Br^-$$

Complex ions may also be formed by the combination of ions with neutral molecules. For example, Ag^+, Cu^{2+}, Ni^{2+}, and a number of other metal ions form complex ions by combination with molecules of ammonia.

$$Ag^+ + 2NH_3 \rightleftharpoons Ag(NH_3)_2^+$$

$$Cu^{2+} + 4NH_3 \rightleftharpoons Cu(NH_3)_4^{2+}$$

$$Ni^{2+} + 4NH_3 \rightleftharpoons Ni(NH_3)_4^{2+}$$

Further discussion of complex ions, including the nature of the bonding involved, is found in Chapters 16 and 22.

PRECIPITATION REACTIONS

While aqueous reactions involving acids and bases are ion-combination reactions, they are also examples of a more specific type of reaction, which we will designate as *proton-transfer reactions*. We will find it convenient, therefore, to defer any further discussion of acid-base reactions until Chapter 15. Similarly, reactions leading to the formation of complex ions may be classified both as ion-combination reactions and as examples of *electron-sharing reactions*, which are dealt with in detail in Chapter 16. The remainder of the present chapter, therefore, will deal only with those ion-combination reactions that lead to formation of a precipitate.

General Solubility Rules

In order to be able to predict whether mixing certain ions together in aqueous solution will or will not result in a precipitation reaction, it is necessary to have some knowledge of the solubilities of those ionic compounds that might possibly result from the mixture. For example, suppose we are asked, "What reaction, if any, will occur upon mixing a solution of $Pb(NO_3)_2$ with a solution of KCl?" To answer this question we need to know what insoluble compound(s), if any, can be formed by combinations of any of the four ions in the mixture. More specifically, we need to know whether KNO_3 or $PbCl_2$, or both, are insoluble. Given the fact that KNO_3 is soluble but $PbCl_2$ is "insoluble," we can predict that the ionic equation for the reaction that takes place is

$$Pb^{2+} + 2Cl^- \rightleftharpoons \underline{PbCl_2}$$

The solubilities of all the more common ionic compounds have been determined, of course, and can be found in the chemical literature. It is convenient, however, to have some qualitative knowledge about the solubilities of various common compounds in order to make predictions without having to consult tables of solubilities. Table 14-1 provides some useful qualitative rules about the solubilities of some common ionic compounds in water. It should be noted that the rules are general and there are some exceptions. Furthermore, the term "insoluble" should be understood to mean "slightly soluble." As a rough approximation, the term "insoluble," as used in this table, applies to compounds whose solubilities are less than 0.010 mole per liter at 25°C; "soluble" substances are those with solubilities greater than 0.10 mole per liter at 25°C; and a substance with a solubility between 0.010M and 0.10M is referred to as "moderately soluble."

Solubility Product

When a precipitate is formed by the combination of ions in solution, an equilibrium is established between the precipitated compound and the dissolved ions of that

TABLE 14-1
General Rules
of Solubility
in Water

1. All nitrates, chlorates, and acetates are *soluble* except the acetates of Ag and Hg(I), which are moderately soluble.
2. Practically all sodium, potassium, and ammonium salts are *soluble*.
3. All chlorides, bromides, and iodides are *soluble* except those of Ag, Hg(I), and Pb(II).
4. All fluorides are *soluble* except those of Mg, Ca, Sr, Ba, and Pb(II).
5. All sulfates are *soluble* except those of Sr, Ba, and Pb(II), which are insoluble, and those of Ca and Ag, which are moderately soluble.
6. All carbonates, sulfites, phosphates, oxalates, and chromates are *insoluble* except those of Na, K, and NH_4^+.
7. All sulfides are *insoluble* except those of the alkali and alkaline earth metals and NH_4^+.
8. All hydroxides are *insoluble* except those of the alkali metals. The hydroxides of Ca, Sr, and Ba are moderately soluble.

compound. Thus, when silver chloride is precipitated by a reaction of silver ions and chloride ions, the ionic crystals of AgCl are in equilibrium with Ag^+ and Cl^- in solution, and this equilibrium may be represented by the equation

$$Ag^+ + Cl^- \rightleftharpoons \underline{AgCl}$$

As long as the conditions are not changed, AgCl will continue to crystallize from solution (the process to the right) and to dissolve (the process to the left), and these two processes will occur at the same rate so that there is no net change in the amount of solid or dissolved ions. That is, the solution resulting from the reaction is saturated with silver chloride.

It will be recalled from our discussion of the solution process in Chapter 11 that if an ionic solid, such as AgCl, is stirred with water, it will *dissolve* (ions will leave the surface of the crystal), and eventually a saturated solution will be obtained. This saturated solution is a system in equilibrium represented by the equation

$$\underline{AgCl} \rightleftharpoons Ag^+ + Cl^-$$

In short, the same equilibrium is established in the precipitation of a solid from its ions as in the dissolution of the solid in water; it is merely being reached from different starting points. Hence, the relationship between the extent of a precipitation reaction and the solubility of the precipitate formed is clear: the less soluble the precipitate, the greater the extent of the reaction.

As we have seen in the preceding chapter, the *equilibrium constant* of a reaction is a measure of the *extent* of that reaction. In order to make quantitative predictions about precipitation reactions, it will be necessary for us to apply the concept of the equilibrium constant to the equilibrium of saturated solutions.

Let us return to our example of the saturated solution of silver chloride, for which the equilibrium is represented as

$$\underline{AgCl} \rightleftharpoons Ag^+ + Cl^-$$

Now, according to our convention for writing equilibrium constants, we may write

$$\frac{[Ag^+][Cl^-]}{[AgCl]} = K$$

But this is a heterogeneous equilibrium; the AgCl is not dissolved in the solution and its molar concentration is *constant*. Thus, for convenience we may simply incorporate the value of [AgCl] into the equilibrium constant, which lets us write

$$[Ag^+][Cl^-] = K[AgCl] = \text{constant}$$

This new constant, which is simply a modified equilibrium constant for a particular kind of equilibrium, is called the *solubility product constant* and is usually symbolized as K_{sp}. Thus, we write the expression

$$K_{sp} = [Ag^+][Cl^-]$$

which tells us that in any *saturated* aqueous solution of silver chloride—one in which solid AgCl is in equilibrium with its ions—the product of the molar concentration of silver ion and the molar concentration of chloride ion must be constant. If in some way $[Ag^+]$ is reduced, $[Cl^-]$ must increase proportionately, and vice versa. The actual value of the constant varies with temperature, of course, just as the values of all equilibrium constants do.

As further illustrations, let us consider the equilibria involved and the solubility product expressions for some other "insoluble" salts: Ag_2CrO_4, CaF_2, and $Ca_3(PO_4)_2$.

$$\underline{Ag_2CrO_4} \rightleftharpoons 2Ag^+ + CrO_4^{2-} \qquad K_{sp} = [Ag^+]^2[CrO_4^{2-}]$$

$$\underline{CaF_2} \rightleftharpoons Ca^{2+} + 2F^- \qquad K_{sp} = [Ca^{2+}][F^-]^2$$

$$\underline{Ca_3(PO_4)_2} \rightleftharpoons 3Ca^{2+} + 2PO_4^{3-} \qquad K_{sp} = [Ca^{2+}]^3[PO_4^{3-}]^2$$

It will be seen from these illustrations that for an ionic compound with the general formula A_nB_m, we may write

$$\underline{A_nB_m} \rightleftharpoons nA^{m+} + mB^{n-}$$

and

$$K_{sp} = [A^{m+}]^n[B^{n-}]^m$$

The actual values of solubility product constants cannot be assigned on the basis of theoretical considerations but must be determined experimentally. A number of different methods have been used, but one of the simplest, which is applicable to many salts, involves calculation of the constant from the carefully measured solubility of the compound. These calculations, which point out the relationship between solubility and the solubility product constant, are illustrated in the following examples.

EXAMPLE: The solubility of $CaSO_4$ at $10°C$ is found to be 1.063 g per liter of solution. What is the value of K_{sp} for $CaSO_4$ at $10°C$?

Solution: The equation for the equilibrium is

$$CaSO_4 \rightleftharpoons Ca^{2+} + SO_4^{2-}$$

Since equilibrium constants are based on concentrations in terms of molarity, we must first express the solubility as moles per liter. This can be done simply by dividing the solubility in grams per liter by the formula weight of $CaSO_4$.

$$\frac{1.063 \text{ g/liter}}{136.1 \text{ g/mole}} = 7.81 \times 10^{-3} \text{ mole/liter}$$

The equation shows that for each mole of $CaSO_4$ dissolved, there is 1 mole of Ca^{2+} and 1 mole of SO_4^{2-} in solution. Therefore,

$$[Ca^{2+}] = 7.81 \times 10^{-3} \text{ mole/liter}$$

and

$$[SO_4^{2-}] = 7.81 \times 10^{-3} \text{ mole/liter}$$

The value of the solubility product constant, then, is

$$K_{sp} = [Ca^{2+}][SO_4^{2-}]$$
$$= (7.81 \times 10^{-3})(7.81 \times 10^{-3})$$
$$K_{sp} = 6.10 \times 10^{-5}$$

EXAMPLE: BaF_2 has a solubility of 1.326 g/liter at $25°C$. What is K_{sp} for BaF_2 at this temperature?

Solution: The solubility of BaF_2 in moles per liter is

$$\frac{1.326 \text{ g/liter}}{175.3 \text{ g/mole}} = 7.56 \times 10^{-3} \text{ mole/liter}$$

Now the equation for the equilibrium,

$$BaF_2 \rightleftharpoons Ba^{2+} + 2F^-$$

shows that for each mole of BaF_2 dissolved, the solution contains 1 mole of Ba^{2+} but 2 moles of F^-. Therefore, in the saturated solution of BaF_2

$$[Ba^{2+}] = 7.56 \times 10^{-3} \text{ mole/liter}$$

and

$$[F^-] = 2(7.56 \times 10^{-3}) = 1.512 \times 10^{-2} \text{ mole/liter}$$

Substituting these concentrations into the solubility product expression and solving for K_{sp} gives us:

$$K_{sp} = [Ba^{2+}][F^-]^2$$
$$= (7.56 \times 10^{-3})(1.512 \times 10^{-2})^2$$
$$= (7.56 \times 10^{-3})(2.286 \times 10^{-4})$$
$$K_{sp} = 1.73 \times 10^{-6}$$

	Compound	K_{sp}	Compound	K_{sp}
TABLE 14-2 **Selected Solubility** **Product Constants (25°C)**	BaCO₃	8.1×10^{-9}	Mg(OH)₂	1.8×10^{-11}
	BaCrO₄	2.2×10^{-10}	Hg₂Cl₂	2.0×10^{-18}
	CaCO₃	8.7×10^{-9}	AgBr	7.7×10^{-13}
	PbCl₂	1.0×10^{-4}	AgCl	1.8×10^{-10}
	PbI₂	1.4×10^{-8}	Ag₂CrO₄	1.9×10^{-12}
	PbSO₄	1.6×10^{-8}	ZnS	1.0×10^{-23}

The solubility product constants for all the common "insoluble" ionic compounds have been evaluated and are readily available for our use. Table 14-2 lists a few of these values, and a more extensive listing may be found in Appendix 8.

The availability of K_{sp} values makes possible a number of kinds of calculations pertaining to solubility and precipitation phenomena, and several of these will be discussed later in the chapter. One of the obvious applications is the calculation of the solubility of a compound from its solubility product constant, as illustrated in the following examples.

EXAMPLE: Calculate the solubility of silver chloride at 25°C.

Solution: Writing the equilibrium equation and the solubility product expression, and using the value of K_{sp} given in Table 14-2, gives us

$$\underline{AgCl \rightleftharpoons Ag^+ + Cl^-}$$

$$[Ag^+][Cl^-] = K_{sp} = 1.8 \times 10^{-10}$$

If we let s equal the solubility of AgCl at 25°C in moles per liter, then, since for each mole of AgCl dissolved there must be 1 mole of Ag^+ and 1 mole of Cl^- in solution, it follows that

$$s = [Ag^+] = [Cl^-]$$

Therefore,

$$s^2 = [Ag^+][Cl^-] = 1.8 \times 10^{-10}$$

$$s = \sqrt{1.8 \times 10^{-10}} = 1.34 \times 10^{-5} \text{ mole/liter}$$

Thus, a saturated solution of AgCl at 25°C contains 1.34×10^{-5} mole of AgCl per liter of solution. If we wish to express the solubility in terms of grams per liter, we have only to multiply this answer by the weight of 1 mole.

$$1.34 \times 10^{-5} \text{ mole/liter} \times 143 \text{ g/mole} = 1.92 \times 10^{-3} \text{ g/liter}$$

EXAMPLE: What is the solubility of Ag_2CrO_4 at 25°C?

Solution:

$$Ag_2CrO_4 \rightleftharpoons 2Ag^+ + CrO_4^{2-}$$

$$[Ag^+]^2[CrO_4^{2-}] = K_{sp} = 1.9 \times 10^{-12}$$

For each mole of Ag_2CrO_4 dissolved, the solution contains 1 mole of CrO_4^{2-} and 2 moles of Ag^+. Therefore, if we let s equal the solubility,

$$s = [CrO_4^{2-}] = \tfrac{1}{2}[Ag^+]$$

and

$$[Ag^+] = 2[CrO_4^{2-}] = 2s$$

Now, substituting into the solubility product expression, and writing it in terms of s,

$$(2s)^2(s) = K_{sp} = 1.9 \times 10^{-12}$$

$$4s^3 = 1.9 \times 10^{-12}$$

$$s^3 = 4.75 \times 10^{-13}$$

$$s = \sqrt[3]{4.75 \times 10^{-13}} = \sqrt[3]{475 \times 10^{-15}}$$

$$s = 7.8 \times 10^{-5} \text{ mole/liter}$$

Therefore, a saturated solution of Ag_2CrO_4 at 25°C contains 7.8×10^{-5} mole of Ag_2CrO_4 per liter of solution. The concentration of chromate ion in that solution is also 7.8×10^{-5} mole per liter, but the concentration of silver ion is $2(7.8 \times 10^{-5}) = 1.56 \times 10^{-4} M$.

One may be tempted to assume that the smaller the value of K_{sp}, the less soluble the compound. Comparison of the solubilities of AgCl and Ag_2CrO_4 in the preceding examples reveals the fallacy of this assumption. Silver chromate has a smaller K_{sp} (1.9×10^{-12}) than silver chloride (1.8×10^{-10}); yet the molar solubility of Ag_2CrO_4 is more than five times that of AgCl. The direct comparison of relative solubilities with K_{sp} values is valid only for compounds with the same ion ratio, where the ion ratio is the ratio of number of cations to number of anions or vice versa. Thus, referring to Table 14-2, we may correctly predict that $BaCO_3$ is more soluble than $BaCrO_4$, which in turn is more soluble than AgBr, for these are all compounds with an ion ratio of 1:1. Similarly, we know that the solubility of PbI_2 is greater than that of $Mg(OH)_2$, since both have an ion ratio of 1:2; and the solubility of $PbCl_2$ is greater than that of Ag_2CrO_4, since in one the ion ratio is 1:2 and in the other, 2:1.

The Effect of Electrolytes on Solubility

Thus far our discussion of solubility and its relationship to solubility product has dealt only with solubilities of ionic compounds in pure water. Let us now examine the effects on solubility of the presence of other electrolytes in the solution.

Common Ion Effect. Let us imagine that we perform the following experiment: To one liter of water we add some solid $BaSO_4$ and, with stirring, we allow sufficient time for equilibrium to be established (that is, for the solution to become saturated). Then, by some convenient method, we determine how much of the $BaSO_4$ has dissolved. In other words, we measure the solubility of $BaSO_4$ in water. Now we repeat the procedure, except that instead of using water we use an aqueous solution containing sulfate ion—for example, a $0.01M$ solution of Na_2SO_4. We will find that more $BaSO_4$ dissolved in pure water than in the $0.01M$ Na_2SO_4 solution. Similarly, if we determine the solubility of $BaSO_4$ in an aqueous solution of some soluble barium salt (for example, $0.01M$ $BaCl_2$), we will discover that $BaSO_4$ is less soluble in that solution than it is in pure water. In short, the solubility of $BaSO_4$ is decreased by the presence of either Ba^{2+} or SO_4^{2-}.

This phenomenon, which is quite general, is referred to as the *common ion effect: The solubility of an ionic compound is decreased by the presence in the solution of an ion in common with the compound.*

This effect is predictable from Le Chatelier's Principle. The equilibrium in a saturated solution of $BaSO_4$ is expressed by

$$\underline{BaSO_4} \rightleftharpoons Ba^{2+} + SO_4^{2-}$$

and in pure water the concentrations of Ba^{2+} and of SO_4^{2-} must be equal. But if additional sulfate ion is present, then that constitutes an increase in the SO_4^{2-} concentration (a stress on the system), which will favor the process to the left, using up Ba^{2+} and producing more solid $BaSO_4$. Thus, less $BaSO_4$ is dissolved because of the presence of the extra sulfate ion. The same conclusion is reached, of course, by considering the effect of additional barium ions.

If we apply the concept of the solubility product we reach the same qualitative conclusion. According to the expression

$$[Ba^{2+}][SO_4^{2-}] = K_{sp}$$

$[Ba^{2+}]$ times $[SO_4^{2-}]$ must remain constant. Therefore, an increase in $[SO_4^{2-}]$ must mean a decrease in $[Ba^{2+}]$, and vice versa. Furthermore, the solubility product constant enables us to calculate the degree to which the solubility is lowered by a given concentration of common ion. Some examples of these calculations follow.

EXAMPLE: Calculate the solubility (moles per liter) of $BaSO_4$ in (a) water, (b) $0.01M$ K_2SO_4 solution, and (c) $0.02M$ $Ba(NO_3)_2$ solution (K_{sp} for $BaSO_4 = 1.0 \times 10^{-10}$).

Solution: (a) In water, $[Ba^{2+}] = [SO_4^{2-}] = s$. Therefore,

$$[Ba^{2+}][SO_4^{2-}] = 1.0 \times 10^{-10}$$

$$s^2 = 1.0 \times 10^{-10}$$

$$s = 1.0 \times 10^{-5} \text{ mole/liter}$$

(b) For each Ba^{2+} present in this solution, one $BaSO_4$ must have dissolved, and therefore

$$s = [\text{Ba}^{2+}]$$

The SO_4^{2-} present came from two sources: some was present in the 0.01M K_2SO_4 solution, and some entered the solution by dissolving of the $BaSO_4$. Thus,

$$[SO_4^{2-}] = 0.01 + s$$

Now substituting into the solubility product expression:

$$[\text{Ba}^{2+}][SO_4^{2-}] = 1.0 \times 10^{-10}$$

$$s(0.01 + s) = 1.0 \times 10^{-10}$$

This is a quadratic equation, which can be solved by application of the quadratic formula. However, the calculation can be simplified considerably, without introducing any error, by making an approximation. The value of s is quite small compared to 0.01. Even when the solvent is pure water, s has a value of only .00001, and it will be even smaller in this case because of the common ion effect. Hence, within the number of significant figures we have to work with, $0.01 + s$ is equal to 0.01. That is,

$$[SO_4^{2-}] = 0.01 + s \cong 0.01$$

With this approximation, the quadratic equation is avoided, and the solution becomes much simpler.

$$s(0.01) = 1.0 \times 10^{-10}$$

$$s = \frac{1.0 \times 10^{-10}}{1 \times 10^{-2}} = 1.0 \times 10^{-8} \text{ mole/liter}$$

(c) Following the same line of reasoning and making the same approximation as in part (b):

$$s = [SO_4^{2-}]$$

$$[\text{Ba}^{2+}] = 0.02 + s \cong 0.02$$

$$[\text{Ba}^{2+}][SO_4^{2-}] = 1.0 \times 10^{-10}$$

$$s(0.02) = 1.0 \times 10^{-10}$$

$$s = \frac{1.0 \times 10^{-10}}{2 \times 10^{-2}} = 5.0 \times 10^{-9} \text{ mole/liter}$$

EXAMPLE Calculate the solubility of Ag_2CrO_4 in 0.0050M K_2CrO_4 solution (K_{sp} for $Ag_2CrO_4 = 1.9 \times 10^{-12}$).

$$Ag_2CrO_4 \rightleftharpoons 2Ag^+ + CrO_4^{2-}$$

$$[Ag^+] = 2s$$

$$[CrO_4^{2-}] = 5.0 \times 10^{-3} + s \cong 5.0 \times 10^{-3}$$

$$[Ag^+]^2[CrO_4^{2-}] = 1.9 \times 10^{-12}$$

$$(2s)^2(5.0 \times 10^{-3}) = 1.9 \times 10^{-12}$$

$$(2.0 \times 10^{-2})s^2 = 1.9 \times 10^{-12}$$

$$s^2 = \frac{1.9 \times 10^{-12}}{2.0 \times 10^{-2}} = 9.5 \times 10^{-11}$$

$$s = 9.7 \times 10^{-6} \text{ mole/liter}$$

The Effect of Inert Electrolytes. One might expect that the solubility of a slightly soluble ionic compound would not be affected at all by the presence of electrolytes that do not have ions in common with the compound and that do not undergo chemical reaction with the ions of the compound. There is certainly nothing in our treatment of the solubility product concept that would suggest otherwise. Thus, for example, consider the solubility of AgCl in a solution containing KNO_3. There is no way that potassium or nitrate ions can undergo a chemical reaction with either silver or chloride ions, and therefore the presence of KNO_3 will not alter the equilibrium

$$AgCl \rightleftharpoons Ag^+ + Cl^-$$

Hence, it would seem that AgCl should have the same solubility in the KNO_3 solution that it has in pure water.

The fact is, however, that AgCl has a slightly *greater* solubility in KNO_3 solution (or any other solution of an *inert* electrolyte) than it does in water alone. Indeed, the solubility of any slightly soluble salt is increased somewhat by the presence of an inert electrolyte in the solution. This phenomenon is often called the *salt effect* or the *diverse ion effect.* As demonstrated by the experimental values in Table 14-3, the effect is usually not very great, but it increases with increasing concentration of the inert electrolyte.

A theoretical explanation of the salt effect is based on the same considerations used to rationalize the "anomalous" colligative properties of electrolyte solutions (p. 286). Because of their electrical charges, ions in solution are not entirely independent of each other but tend to be attracted to ions of opposite charge. Thus, in our example of a saturated AgCl solution containing KNO_3, the

	KNO_3 Concentration *(molarity)*	Solubility of AgCl *(molarity)*
TABLE 14-3 **Solubility of Silver Chloride in Potassium Nitrate Solutions (25°C)**	0	1.278×10^{-5}
	0.001	1.325×10^{-5}
	0.005	1.385×10^{-5}
	0.01	1.427×10^{-5}

dissolved Ag^+ and Cl^- are "tied up" to some extent by NO_3^- and K^+, respectively, and are not as free to enter into the equilibrium as they otherwise would be. The result is that more ions enter the solution from the solid; that is, the solubility increases slightly. The greater the concentration of the inert electrolyte, the greater is its effectiveness in "tying up" Ag^+ and Cl^-, and the greater is the increase in solubility of AgCl.

The Effect of Acids. The effect of acids (that is, hydronium ion) on the solubility of ionic compounds depends on the particular composition of the compound—more specifically, on the nature of the anion.

Suppose one compares the solubility of AgCl in water with its solubility in a dilute acid solution; for example, $0.01M$ HNO_3. (The acid chosen must, of course, be one that does not have an ion in common with the solute and will not react with it. Thus, in this case we could not use HCl as our acid, nor HBr, since AgBr is itself insoluble.) The solubility will be found to be slightly higher in the acid solution than in water, but the increase can be accounted for entirely by the salt effect. There is no effect on the solubility specifically due to the electrolyte being an *acid*. The same is true for a number of other slightly soluble salts, including, for example, AgBr, PbI_2, and Hg_2Cl_2.

On the other hand, if you compare the solubility in water with the solubility in acid solution of any one of a large number of other ionic compounds (CaF_2 ZnS, BaC_2O_4, $Mg(OH)_2$, $CaCO_3$, to name just a few) you will find the solubility in acid solution considerably greater than in pure water—an increase in solubility too large to be accounted for by the salt effect alone.

What is the difference between these two categories of salts that accounts for the difference in the effect of acids on their solubilities? The compounds of the second group—those that show a marked increase in solubility in the presence of acid—all contain anions that are the anions of weak acids (slightly dissociated electrolytes). Hydronium ion combines with these anions to "tie them up" in molecules, thereby upsetting the equilibrium between undissolved and dissolved solute.

As a specific example, consider the solubility of calcium fluoride, CaF_2. Let us suppose we have prepared a saturated solution of CaF_2 in water. The equilibrium in that solution is

$$CaF_2 \rightleftharpoons Ca^{2+} + 2F^-$$

Now, suppose we add some acid (HCl or HNO_3) to the solution. Because HF is a weak electrolyte, hydronium ions from the acid we have added will combine with fluoride ions from the solution to form molecules of HF, and another equilibrium will be set up in the mixture; namely,

$$H^+ + F^- \rightleftharpoons HF$$

The more acid (H^+) we add, the more HF will be formed and the more F^- will be removed from solution. Now, removal of F^- upsets the heterogeneous equilibrium between CaF_2 and its dissolved ions, causing the process to the right to be favored.

In other words, *more* CaF_2 must dissolve in order to reestablish equilibrium conditions, and the solubility of CaF_2 has been increased. The total effect of the acid in dissolving the CaF_2 can be represented as a single equation that is the sum of the two equilibria involved:

$$\underline{CaF_2 \rightleftharpoons Ca^{2+} + 2F^-}$$

$$\underline{2H^+ + 2F^- \rightleftharpoons 2HF}$$

$$CaF_2 + 2H^+ \rightleftharpoons Ca^{2+} + 2HF$$

The solubilities of all salts containing anions of weak acids are increased by the presence of acid. This includes carbonates, fluorides, oxalates, sulfides, sulfites, nitrites, phosphates, and so forth, and it also includes hydroxides, for their anions (OH^-) combine with hydronium ion to form water molecules.

$$\underline{Mg(OH)_2 \rightleftharpoons Mg^{2+} + 2OH^-}$$

$$\underline{2H^+ + 2OH^- \rightleftharpoons 2H_2O}$$

$$Mg(OH)_2 + 2H^+ \rightleftharpoons Mg^{2+} + 2H_2O$$

It is now clear why the members of our first group of salts—the chlorides, bromides, and iodides—do not show increased solubility in the presence of acids. These are anions of *strong* acids (HCl, HBr, HI). Hydronium ions do not react with them to form molecules, and the equilibrium is not affected. In general, the solubilities of salts containing anions of strong acids are not altered appreciably by the presence of acids.

The extent to which acid increases the solubility of an ionic compound depends both on the solubility product of the compound and on the relative strength of the weak acid formed. The smaller the solubility product, the less the dissolving effect of a given concentration of acid. Thus, for example, it is possible to separate a mixture of ZnS and CuS by using the proper concentration of hydrochloric acid. The ZnS $(K_{sp} = 10^{-23})$ will dissolve, leaving behind the CuS $(K_{sp} = 10^{-45})$. The calculations involved in such a separation are discussed in Chapter 15.

That the strength of the weak acid formed plays a role in determining the extent to which dissolution takes place can be seen by considering the equilibria for the hypothetical salt MX:

(1) $$\underline{MX} \rightleftharpoons M^+ + X^-$$

(2) $$H^+ + X^- \rightleftharpoons HX$$

The weaker the acid HX (that is, the greater its tendency to exist as undissociated molecules), the further to the right will be the equilibrium position in equation (2), and the more X^- will be removed from the solution. The more X^- is removed, the further equation (1) will proceed to the right. Therefore, the weaker the acid formed in the dissolution process, the greater is the dissolving effect of the acid on the salt.

Complex Ion Effect.   Our discussion of the common ion effect established the principle that the solubility of an ionic compound is *decreased* by the presence of an electrolyte having an ion in common with the compound. One might infer from this statement and from the illustrative calculations presented that the greater the concentration of the common ion, the lower the solubility. This is not always the case, however, for some slightly soluble salts react with an excess of their own anions to form complex ions, resulting in an *increase* in solubility.

A striking example of this *complex ion effect* is provided by mercuric iodide. If a solution containing iodide ions is added slowly to a solution of mercury(II) ions, a red precipitate of mercuric iodide (HgI_2) is formed. As addition of the I^- is continued, the amount of precipitate increases, but only up to a point. Addition of I^- beyond that point results in a gradual dissolving of the precipitate, and if the addition is carried far enough, the entire precipitate will redissolve. The equation for the precipitation is

$$Hg^{2+} + 2I^- \rightleftharpoons \underline{HgI_2}$$

and the reaction that is responsible for the redissolving is the formation of the complex ion:

$$\underline{HgI_2} + 2I^- \rightleftharpoons HgI_4^{2-}$$

The equilibrium constant for this second reaction is very large, and HgI_2 dissolves readily in excess I^-. The *minimum* solubility is obtained at the point at which the amount of I^- added is equivalent to the Hg^{2+} present (two I^- for each Hg^{2+}). Addition of excess I^- does not further decrease the solubility through the common ion effect but instead *increases* the solubility through the complex ion effect.

Another example of a complex ion whose formation has a high equilibrium constant is $Ag(CN)_2^-$. As a result, precipitated silver cyanide (AgCN) dissolves readily in an excess of its own anions:

$$Ag^+ + CN^- \rightleftharpoons \underline{AgCN}$$

$$\underline{AgCN} + CN^- \rightleftharpoons Ag(CN)_2^-$$

Both HgI_2 and AgCN are extreme examples of the complex ion effect. There are very few compounds whose solubility is increased so dramatically by an excess of their anions. However, numerous compounds do display the complex ion effect to a lesser extent. In these cases the presence of an excess of anion results in a kind of competition between the common ion effect and the complex ion effect. Usually, when the concentration of excess anion is low, the common ion effect prevails and the solubility is lowered. But as the concentration of anion is raised, the complex ion effect competes more successfully, and at high concentrations the trend is reversed and the solubility increases. This "competition" is revealed in the experimental data of Table 14-4, which show the solubility of AgCl in various concentrations of NaCl. The solubility of AgCl is less in $0.01M$ NaCl than it is in water, but at concentrations of NaCl greater than approximately $0.35M$, AgCl is *more* soluble than it is in water. This increases in solubility is too great to be

	Concentration of NaCl (moles/liter)	*Solubility of AgCl* (moles/liter $\times 10^3$)
TABLE 14-4 **Solubility of AgCl** **in NaCl Solutions**	0	0.013
	0.0039	0.00072
	0.0092	0.00091
	0.036	0.0019
	0.088	0.0036
	0.35	0.017
	0.5	0.028

accounted for by the salt effect (see Table 14-3) and is explained by formation of the complex ion, $AgCl_2^-$.

Several additional examples of the complex ion effect are shown by the following equations.

$$PbCl_2 + Cl^- \rightleftharpoons PbCl_3^-$$

$$CuCN + CN^- \rightleftharpoons Cu(CN)_2^-$$

$$AuBr_3 + Br^- \rightleftharpoons AuBr_4^-$$

$$Zn(OH)_2 + 2OH^- \rightleftharpoons Zn(OH)_4^{2-}$$

Precipitation and the
Solubility Product

The concept of the solubility product is useful not only in calculating the solubilities of slightly soluble salts but also in predicting the concentrations of ions necessary to result in precipitation. If aqueous solutions containing the ions of a slightly soluble salt are mixed together, the salt will not precipitate unless the ions are present in high enough concentration that the product of their concentrations exceeds the solubility product constant. Otherwise, the solution is not saturated with the slightly soluble salt. The following examples are illustrative of this application of the solubility product.

EXAMPLE: Equal volumes of $0.01M$ $Pb(NO_3)_2$ solution and $0.01M$ NaCl solution are mixed together. Will a precipitate of $PbCl_2$ be formed?

Solution: In a saturated solution of $PbCl_2$,

$$[Pb^{2+}][Cl^-]^2 = 1.0 \times 10^{-4}$$

If, in the solution that results from the mixing, $[Pb^{2+}][Cl^-]^2$ is greater than 1.0×10^{-4}, a precipitate will be formed. On the other hand, no precipitate will result if $[Pb^{2+}][Cl^-]^2$ is less than 1.0×10^{-4}.

Since the final solution has a volume twice that of either of the component solutions (this is a valid approximation for dilute solutions), the final concentrations are

$$[Pb^{2+}] = 0.005$$

and

$$[Cl^-] = 0.005$$

Therefore,

$$
\begin{aligned}
[Pb^{2+}][Cl^-]^2 &= (5 \times 10^{-3})(5 \times 10^{-3})^2 \\
&= 5 \times 10^{-3} \times 2.5 \times 10^{-5} \\
&= 1.3 \times 10^{-7}
\end{aligned}
$$

Since $1.3 \times 10^{-7} < 1.0 \times 10^{-4}$, no precipitate will form.

EXAMPLE: What is the minimum concentration of sulfide ion necessary to cause precipitation of ZnS from a $1.0 \times 10^{-3} M$ $Zn(NO_3)_2$ solution?

Solution: Precipitation will not occur unless the S^{2-} concentration is sufficiently large that $[Zn^{2+}][S^{2-}] > 10^{-23}$. Therefore, the S^{2-} concentration required is greater than 10^{-20} mole/liter

$$[S^{2-}] = \frac{10^{-23}}{1.0 \times 10^{-3}} = 10^{-20} \text{ mole/liter}$$

EXAMPLE: What is the maximum weight of solid $CaCl_2$ that can be dissolved in 1 liter of $1.0 \times 10^{-4} M$ Na_2CO_3 solution without any $CaCO_3$ precipitating?

Solution: In a saturated solution of $CaCO_3$,

$$[Ca^{2+}][CO_3^{2-}] = 8.7 \times 10^{-9}$$

and therefore the maximum molar concentration of Ca^{2+} permissible is

$$[Ca^{2+}] = \frac{8.7 \times 10^{-9}}{1.0 \times 10^{-4}} = 8.7 \times 10^{-5} \text{ mole/liter}$$

Thus, the number of moles of $CaCl_2$ that may be dissolved in 1 liter of the solution is 8.7×10^{-5}. This corresponds to a weight of $CaCl_2$ of

$$8.7 \times 10^{-5} \text{ mole} \times 111 \text{ g/mole} = 9.7 \times 10^{-3} \text{ g} = 9.7 \text{ mg}$$

Equivalent Weights in Precipitation Reactions

In connection with our earlier discussion of concentrations of solutions (Chapter 11), an *equivalent* of a substance was defined as *the number of grams of the substance which in a particular chemical reaction is equivalent to one gram-atom of hydrogen.* Emphasis was placed on the fact that the equivalent weight, unlike the formula weight, is not constant but depends on the nature of the reaction in which the substance is to be used.

In precipitation reactions (and in most other ion-combination reactions) the equivalency or "combining power" of an ion is determined by its ionic charge. In ionic form, hydrogen has a charge of *plus one* in aqueous solutions (H^+ or, more correctly, H_3O^+), or of *minus one* in the case of the active metal hydrides (H^-). The ionic charge of unity means that one gram-atom of ionic hydrogen possesses a total ionic charge equal to Avogadro's number. Therefore, the amount of any substance that is equivalent in ion combination reactions to one gram-atom of hydrogen is the weight of that substance that contains Avogadro's number of positive (or of negative) ionic charges. Thus, the equivalent weight of Na^+, K^+, or Ag^+, or of Cl^- or Br^-, is equal in each case to the atomic weight. On the other hand, the equivalent weight of Mg^{2+}, Ca^{2+}, or S^{2-} is one-half of the atomic weight. Similarly, the equivalent weight of an ionic compound may be found by dividing the formula weight by the total positive (or total negative) ionic charge. For example, the equivalent weight of $MgSO_4$ is one-half its formula weight, while the equivalent weights of $Al_2(SO_4)_3$ and Na_3PO_4 are one-sixth and one-third of their formula weights, respectively.

Some further illustrations of this concept and its relationship to normality (the number of equivalents of solute per liter of solution) are given in the following examples.

EXAMPLE: A solution of K_2CrO_4 to be used as a precipitant is prepared by dissolving 10.5 g in enough water to make 1 liter of solution. What is the normality of the solution?

Solution: The formula weight of K_2CrO_4 is 194.2. Since 1 mole of K_2CrO_4 contains 2 moles of K^+ (and 1 of CrO_4^{2-}), the equivalent weight is one-half the formula weight:

$$\frac{194.2 \text{ g/mole}}{2 \text{ equiv/mole}} = 97.1 \text{ g/equiv}$$

The solution contains, therefore,

$$\frac{10.5 \text{ g/liter}}{97.1 \text{ g/equiv}} = 0.108 \text{ equiv/liter}$$

and the normality is $0.108N$.

EXAMPLE: What weight of $Ba(NO_3)_2$ is contained in 50.0 ml of $0.25N$ $Ba(NO_3)_2$ solution?

Solution: The equivalent weight of $Ba(NO_3)_2$ is one-half the formula weight. Therefore, one equivalent weight is

$$\frac{261.4 \text{ g/mole}}{2 \text{ equiv/mole}} = 130.7 \text{ g/equiv}$$

Since the solution contains 0.25 equiv in a liter, 1 liter would contain

$$(0.25 \times 130.7) \text{ g } Ba(NO_3)_2$$

and 50.0 ml contains

$$0.25 \frac{\text{equiv}}{\text{liter}} \times 130.7 \frac{\text{g}}{\text{equiv}} \times \frac{50.0}{1000} \text{liter} = 1.63 \text{ g}$$

PROBLEMS

1. Aqueous solutions of the following pairs of compounds are mixed together. For each pair, state whether the mixture results in a chemical reaction and if so, write the net ionic equation for the reaction.
   (a) $(NH_4)_2 SO_4 + KBr$ (b) $Hg(NO_3)_2 + Na_2 S$
   (c) $HCl + KCN$ (d) $CaCl_2 + (NH_4)_2 CO_3$
   (e) $FeCl_2 + CdSO_4$ (f) $Pb(NO_3)_2 + (NH_4)_2 SO_4$
   (g) $HClO_4 + NaOH$ (h) $Zn(NO_3)_2 + KOH$
   (i) $Na_2 SO_3 + HCl$ (j) $AgClO_3 + CaI_2$
   (k) $Na_2 C_2 O_4 + HBr$ (l) $Na_3 PO_4 + K_2 SO_4$

2. The solubility of $MgC_2 O_4$ at $18°C$ is found to be 1.040 g/liter. Calculate the solubility product constant for $MgC_2 O_4$ at $18°C$.

3. If the solubility of PbF_2 is 0.4655 g/liter at $9°C$, what is the value of K_{sp} for PbF_2 at that temperature?

4. Calculate the solubility (in moles per liter) of each of the following compounds at $25°C$:
   (a) $AgBr$ (b) $Mg(OH)_2$
   (c) ZnS (d) $Li_2 CO_3$ ($K_{sp} = 1.7 \times 10^{-3}$)

5. The solubility product constant for $Mg_3(PO_4)_2$ has a value of 4×10^{-13} at $25°C$. What is the molar concentration of magnesium ion and of phosphate ion in a saturated solution of $Mg_3(PO_4)_2$ at $25°C$?

6. For $MgNH_4 PO_4$ the $K_{sp} = 2.5 \times 10^{-13}$. Calculate the solubility (in grams per liter) of this salt.

7. Calculate the molar solubility of $BaCrO_4$ in each of the following:
   (a) pure water
   (b) $0.050M$ $BaCl_2$ solution
   (c) $0.050M$ $K_2 CrO_4$ solution

8. Calculate the molar solubility of $Mg(OH)_2$ in each of the following:
   (a) pure water
   (b) $0.020M$ $Mg(NO_3)_2$ solution
   (c) $0.020M$ KOH solution

9. To 1 liter of a saturated solution of $PbSO_4$ is added 1.0×10^{-3} mole of $K_2 SO_4$. Calculate the molar concentration of lead ion in the solution before and after the addition of the $K_2 SO_4$.

10. To 100 ml of a saturated solution of CaF_2 is added 0.100 g of NaF. Calculate the final concentration of Ca^{2+} in the solution.

11. Solid $BaSO_4$ is added to 250 ml of $0.025M$ $K_2 SO_4$, and the mixture is stirred until equilibrium is established. What weight of $BaSO_4$ dissolved?

12. Arrange the following in the order of increasing solubility of $CaCO_3$.
   (a) pure water
   (b) $0.02M$ $CaCl_2$ solution
   (c) $0.03M$ $Na_2 CO_3$ solution
   (d) $0.01M$ NaCl solution
   (e) $0.01M$ HCl solution

13. Arrange the following in the order of increasing solubility of PbI_2.
    (a) pure water
    (b) $0.01M$ $Pb(NO_3)_2$
    (c) $0.01M$ NaI
    (d) $0.01M$ HNO_3
    (e) $0.10M$ $NaNO_3$

14. Predict which of the following compounds will dissolve in dilute HNO_3.

    (a) $SrCO_3$ (b) $PbBr_2$
    (c) MgF_2 (d) MnS
    (e) CuI   (f) $Fe(OH)_2$
    (g) Hg_2Cl_2 (h) FeC_2O_4
    (i) $PbCl_2$ (j) $CaCrO_4$

15. When a solution of $Na_2C_2O_4$ is added dropwise to a solution of $Cu(NO_3)_2$, a precipitate forms, which redissolves on further addition of the $Na_2C_2O_4$ solution. Explain.

16. What is the minimum concentration of SO_4^{2-} necessary to cause precipitation of the metal sulfate from each of the following solutions?
    (a) $1.0 \times 10^{-3}M$ $Ba(NO_3)_2$
    (b) $1.0 \times 10^{-3}M$ $Ca(NO_3)_2$
    (c) $1.0 \times 10^{-3}M$ $Pb(NO_3)_2$

17. If equal volumes of the two solutions in each of the following pairs are mixed together, indicate whether or not a precipitate will be formed:
    (a) $1.0 \times 10^{-3}M$ $MgCl_2$ and $1.0 \times 10^{-3}M$ Na_2CO_3
    (b) $1.0 \times 10^{-3}M$ $CaCl_2$ and $1.0 \times 10^{-4}M$ NaF
    (c) $1.0 \times 10^{-5}M$ $Pb(NO_3)_2$ and $1.0 \times 10^{-5}M$ Na_2S
    (d) $1.0 \times 10^{-2}M$ $Ni(NO_3)_2$ and $1.0 \times 10^{-5}M$ Na_2S
    (e) $1.0 \times 10^{-5}M$ $AgNO_3$ and $1.0 \times 10^{-5}M$ KBr

18. What weight of PbI_2 will be formed by the reaction of 100 ml of $0.12N$ $Pb(NO_3)_2$ with 100 ml of $0.12N$ KI?

19. Express the concentration of each of the following solutions in terms of *normality:*

    (a) $0.25M$ $CdSO_4$ (b) $0.25M$ Na_2SO_4
    (c) $0.25M$ $Al_2(SO_4)_3$ (d) $0.25M$ H_2SO_4
    (e) $0.25M$ $CdCl_2$ (f) $0.25M$ $AlCl_3$
    (g) $0.25M$ HCl

20. What volume of $0.10N$ $CaCl_2$ is equivalent to 50 ml of $0.025N$ Na_2CO_3 in the precipitation of $CaCO_3$?

21. The solubility product constants for Ag_2CrO_4 and $Ag_2Cr_2O_7$ are 1.9×10^{-12} and 2.0×10^{-7}, respectively. What effect will the addition of silver ion have on the chromate-dichromate equilibrium?

$$2CrO_4^{2-} + 2H^+ \rightleftharpoons Cr_2O_7^{2-} + H_2O$$

22. When an excess of barium chloride was added to 50.00 ml of a certain aqueous solution of nickel(II) sulfate, 1.346 g of $BaSO_4$ was obtained. What was the molarity of the nickel(II) sulfate solution?

15 *Types of Chemical Reactions II: Proton-Transfer Reactions*

Of the many types of matter that have engaged chemists over the centuries, surely among the most important are those types called acids, bases, and salts. These substances and the theories concerning their nature have played a vital role not only in the development of chemistry but also in the history of civilization. Think, for example, of the impact that saltpeter (KNO_3), a constituent of gunpowder, has had on the political histories of various nations; or of the value of carbonates and hydroxides of potassium and sodium in the ceramics, glass, and soap industries; or yet of the need for acids such as nitric, acetic, and sulfuric as precursors to the products of our vast chemical industry.

Salt as a category of matter probably had its origin in the work of the alchemists on the interrelationships of what they considered to be the four basic elements—fire, water, earth, and air. When certain solids, "earths," were mixed with water, some portion was found to dissolve in the water and was labeled "salt." By the seventeenth century many salts had been discovered and their forms and properties characterized. Moreover, after numerous experiments on the effects of water and that very potent element, fire, on these substances, it was found that some salts could be decomposed into two different substances.

Investigation of the properties of these two new types of matter—acids and bases—led to the conclusion that they were chemical opposites. Acids were known to have a sour taste, to turn blue plant dyes red, and to dissolve many things; while bases have a bitter taste, turn red plant dyes blue, and have a slippery feel. Moreover, the reaction of an acid with a base results in the mutual neutralization of their properties.

381

The similarity of the chemical and physical effects of all known acids suggested the presence of some common "acidifying principle." Since the chemistry of the latter part of the eighteenth century was dominated by experiments on combustion and Lavoisier's demonstration that oxygen is consumed in this process, it is not surprising that Lavoisier and his followers believed this common acidifying principle to be oxygen. Indeed, Lavoisier derived the name *oxygen* from the Greek word for "acid former." According to this theory, acids are binary compounds, one of the elements being oxygen, the other the acid radical. Sulfuric acid, then, was believed to consist of sulfur (the radical) plus oxygen (the acidifying principle), and all of the known acids were classified in this way.

Careful investigations on the composition of hydrochloric acid by Humphry Davy (1778–1829) in the early part of the nineteenth century showed, however, that this acid does not contain oxygen. By 1830 more than ten acids that did not contain oxygen had been discovered and characterized, and it became obvious that the one element that all of the acids had in common was not oxygen but hydrogen.

At the time of its exposition by Davy the new "hydrogen theory" was certainly not universally accepted, however. Indeed, reaction against a new theory is a continual facet of the history of science and is probably just the human psychological counterpart of Le Chatelier's Principle: If a stress is applied to a system in equilibrium, the system reacts so as to annul the effects of that stress. A good example of the reaction against the "hydrogen theory" is provided by the interactions of two excellent chemists and very close friends, Jöns Jacob Berzelius and Justis von Liebig. Liebig, working on the nature of several acids that contain carbon, wrote to Berzelius in November of 1837: "I have let myself in for something which is almost crazy . . . I am dominated by the idea that all organic acids are hydro-acids in a certain sense." And later, in April of 1837: "I have been living in fear and worry over the theory of organic acids that is developing in me . . . If sulphuric acid is really $SO_4 + H_2$, then its saturation capacity [ability to neutralize bases] is dependent only on the hydrogen, which is outside of the radical." Berzelius, in his reply to his friend in 1838, labeled Liebig's theory "absurd" and said, "We must aim to make scientific theories as clear as possible, and to change them as little as possible. . . . You have built up an artificial structure which will topple over at the first scientific blast."

By 1850, the evidence for the ideas of Davy, Liebig, and others was overwhelming, and the "hydrogen theory" was finally accepted. Attention was then directed to the nature of the hydrogen in acids. Since many compounds contain hydrogen, what makes some of them acids while others exhibit none of these characteristics? Svante Arrhenius and Wilhelm Ostwald, in the latter part of the nineteenth century, using Faraday's earlier results on the dissociation of electrolytes (see Chapter 11), showed that acids, bases, and salts are electrolytes and that in water acids dissociate to form hydrogen ions, while bases produce hydroxide ions. Thus, when gaseous hydrogen chloride dissolves in water, molecules of HCl dissociate into hydrogen ions and chloride ions:

$$HCl \rightleftharpoons H^+ + Cl^-$$

On the other hand, methane does not dissociate into ions and therefore is not an acid in water. When solid potassium hydroxide dissolves, it enters the solution as potassium ions and hydroxide ions. Since some bases do not contain hydroxide ions, these substances must produce OH^- ions by reaction with the water; for example,

$$NH_3 + H_2O \rightleftharpoons NH_4^+ + OH^-$$

Moreover, the Arrhenius theory recognized that some acids are strong electrolytes and are therefore completely dissociated into ions in water, while other acids are weak electrolytes and are therefore only partly dissociated into ions. The same is true, of course, for bases.

The Arrhenius concept is clearly limited to the behavior of acids and bases in water, and a number of chemists have since proposed other definitions of acids and bases that incorporate many more compounds and reactions within their confines. One of the most useful of these more general definitions is the one advanced almost simultaneously by the two chemists T. M. Lowry and J. N. Brønsted in 1923. The *Lowry-Brønsted definition* is a logical extension of the Arrhenius concept: *An acid is a substance that can release a proton (hydrogen ion), while a base is a substance that can combine with a proton.* Hence, when gaseous ammonia is dissolved in water, hydroxide ions are formed, because water, an acid, releases a hydrogen ion to the ammonia, a base, which combines with it and forms the ammonium ion.

$$H_2O + NH_3 \rightleftharpoons NH_4^+ + OH^-$$

When the proton leaves the water molecule, a hydroxide ion results. This, then, is a definition independent of the solvent. For example, hydrogen sulfide functions as an acid in releasing a proton to ammonia in liquid ammonia.

$$H_2S + NH_3 \rightleftharpoons HS^- + NH_4^+$$

Another very useful definition, and one that we will explore in greater detail in Chapter 16, was suggested by G. N. Lewis in 1923 but was not widely received until the latter part of the 1930s. The Lewis definition is much more general than either the Arrhenius or Lowry-Brønsted definitions. According to Lewis, *an acid is a species capable of accepting a pair of electrons, while a base is a species that can provide a pair of electrons.* Since the proton can accept a pair of electrons and is therefore a Lewis acid, all Lowry-Brønsted acids are also Lewis acids (likewise for bases). In addition to the proton, however, there are vast numbers of ions and molecules that can also function as Lewis acids.

By far the most all-inclusive definition of acids and bases was provided by a Russian, M. Usanovich, in 1939. According to this definition, an acid is any material that forms salts with bases through neutralization, gives up cations, or combines with anions or electrons. Bases are materials that neutralize acids, give up anions or electrons, or combine with cations. This definition includes not only all substances defined as Arrhenius, Lowry-Brønsted, and Lewis acids and bases, but also all substances involved in oxidation-reduction processes (Chapter 17). Because of this, the definition is little used.

Even today, chemists are attempting to refine and extend their understanding of acid-base phenomena, one of the recent advances being the hard and soft acid and base principle proposed in 1963 by an American professor of chemistry, Ralph Pearson. This concept will be described in Chapter 16.

Before leaving our description of the evolution of our present knowledge of acids and bases, it is perhaps instructive to examine more closely the conditions necessary for the acceptance of a scientific theory. We have already pointed to the reluctance with which a new idea is met. The hesitation on the part of scientists to accept and use a new idea is not all bad, however; the new theory must be shown to be clearly superior to the old.

The acceptance of a new theory often requires more than just the hurdle of this reluctance barrier. As P. Walden, in his study of the historical evolution of the concepts of salts, acids, and bases, puts it: "It is evident that all scientific ideas and theories must have a favorable 'culture medium.' " A part of this favorable "culture medium" is simply timing. For example, a number of chemists in the late 1600s suggested that combustion involved the consumption of something in the air. Not until 1780, when more facts, more inconsistencies, and more minds troubled by the inadequacies of the old model were available, could the "oxygen theory" of Lavoisier be accepted. The time was ripe for a new model.

Another part of the "culture medium" necessary for the growth of an idea is the personality and reputation of the proponent of the idea. Part of the success of Liebig's hydrogen theory of organic acids was due to his reputation and his famous laboratory at Giessen, Germany. Similarly, the Arrhenius theory was promulgated primarily by his student Ostwald, whose laboratory attracted chemists from all over the world. Walden states, "Ostwald's laboratory was a great collector of mental energy, for to it came the young and talented chemists from all over the world to embrace and become followers of new ideas and methods. Indeed, Ostwald himself was a most clever and indefatigable interpreter and promoter of this theory of Arrhenius, and by his unique and exceptional personality made many scientific converts."

PROTON-TRANSFER REACTIONS

Since the Lowry-Brønsted definition of acids and bases encompasses the Arrhenius definition but is not restricted to reactions carried out in water, it is generally the more useful concept, and the remainder of this chapter will be devoted to an exposition of its many features. The definitions

An *acid* is a species capable of releasing a proton

A *base* is a species capable of accepting a proton

lead to the conclusion that a Lowry-Brønsted acid-base reaction involves a *transfer* of protons from the acid to the base. For example, in the reaction of acetic acid with ammonia, a proton is transferred from an acetic acid molecule to an ammonia molecule, resulting in the formation of the ammonium ion and the acetate ion:

$$CH_3COOH + NH_3 \rightleftharpoons CH_3COO^- + NH_4^+$$

The reverse reaction is also a proton-transfer reaction between the acid NH_4^+ and the base CH_3COO^-.

The species formed when an acid loses its proton is called its *conjugate base,*

$$Acid \rightarrow Base + H^+$$

for example,

$$CH_3COOH \rightarrow CH_3COO^- + H^+$$

while the species that results when a base accepts a proton is its *conjugate acid,*

$$Base + H^+ \rightarrow Acid$$

for example,

$$NH_3 + H^+ \rightarrow NH_4^+$$

Every proton transfer reaction is therefore of the type

$$Acid_1 + Base_2 \rightleftharpoons Base_1 + Acid_2$$

where acid$_1$ and base$_1$ and acid$_2$ and base$_2$ are conjugate acid-base pairs.

Although an acid must surely contain hydrogen, not all compounds containing hydrogen are acids. Whether or not a compound containing hydrogen functions as an acid depends upon the strength and polarity of the bond to hydrogen and the ability of the base with which it is reacting to accept the proton. Acetic acid, for example, contains a hydrogen bonded to the very electronegative oxygen. This bond is quite polar, and therefore dissociation into CH_3COO^- and H^+ is moderately easy (the relationship between bond strength and polarity and acidity will be discussed in a subsequent section of this chapter). When acetic acid is mixed with water, protons are released to water molecules.

$$CH_3COOH + H_2O \rightleftharpoons CH_3COO^- + H_3O^+$$

Acetone, on the other hand, contains only hydrogens bonded to carbon, and because of the small electronegativity difference these bonds are not very polar. Acetone is therefore a nonelectrolyte and shows no acidic properties in water. However, if acetone is treated with a stronger base than water—for example, hydroxide ion—the attraction of the hydroxide for the proton is sufficiently great to cause the transfer to occur to some extent.

$$CH_3\overset{\overset{\displaystyle O}{\|}}{C}CH_3 + OH^- \rightleftharpoons CH_3\overset{\overset{\displaystyle O}{\|}}{C}CH_2^- + H_2O$$

Of course, the reaction of acetic acid with hydroxide ion also occurs to a greater extent than the reaction with water.

$$CH_3C\underset{OH}{\overset{O}{<}} + OH^- \rightleftharpoons CH_3C\underset{O^-}{\overset{O}{<}} + H_2O$$

The sole criterion for a base is that it be able to accept a proton. Since the proton is extremely small, the attraction between the base and the proton is more than a simple electrostatic attraction—the extreme smallness of the proton distorts the electron density of the base, and a covalent bond is formed. The proton has no electrons and the base must therefore supply a *pair* of electrons. Any species with a nonbonded pair of electrons is therefore a potential base. Tables 15-1 and 15-2 (pages 388 and 389) list a number of common acids and bases.

A number of acids can clearly provide more than one proton; for example, sulfuric acid, $O_2S(OH)_2$, has two very polar O–H bonds.

$$\underset{O}{\overset{O}{>}}S\underset{OH}{\overset{OH}{<}} \longrightarrow \underset{O}{\overset{O}{>}}S\underset{O^-}{\overset{OH}{<}} + H^+$$

$$\underset{O}{\overset{O}{>}}S\underset{O^-}{\overset{OH}{<}} \longrightarrow \underset{O}{\overset{O}{>}}S\underset{O^-}{\overset{O^-}{<}} + H^+$$

The same is true for acids such as hydrosulfuric, phosphoric, carbonic, oxalic, and so forth. These acids are referred to as *polyprotic* acids. Acids that can provide only one proton per molecule are called *monoprotic acids*.

There are also bases that can accept more than one proton: the carbonate ion, the phosphate ion, the oxalate ion, and so on. These are *polyprotic bases,* while

$$C_2O_4{}^{2-} + H^+ \rightarrow HC_2O_4{}^-$$

$$HC_2O_4{}^- + H^+ \rightarrow H_2C_2O_4$$

those that can accept only one proton per molecule or ion are *monoprotic.* Moreover, some species, for instance H_2O, HS^-, $HCO_3{}^-$, can react as both acids and bases. Hydrogen sulfide ion, for example, has a fairly polar H–S bond, and there are also nonbonded electron pairs on the sulfur. When mixed with a base such as hydroxide ion, it functions as an acid,

$$HS^- + OH^- \rightleftharpoons S^{2-} + H_2O$$

but when mixed with an acid such as hydrochloric acid it behaves as a base.

$$HS^- + HCl \rightleftharpoons Cl^- + H_2S$$

Species that function as both Lowry-Brønsted acids *and* bases are termed *amphiprotic.*

ACID AND BASE STRENGTHS

The extent to which an acid donates a proton to a given base is a measure of its strength as an acid. Conversely, the extent to which a base accepts a proton from a given acid is a measure of its strength as a base. Because of the convenience of working with water, the fact that it is amphiprotic, and the great abundance of aqueous solutions, acid and base strengths are often measured relative to water.

Suppose, for example, that one mole of each of the acids—acetic, nitrous, chlorous, and hydrochloric—were separately mixed with one liter of water. An analysis of each mixture after equilibrium was attained would show that in the acetic acid-water mixture only 0.4 percent of the acetic acid dissociates; that is, that only 0.004 mole of hydronium ions and 0.004 mole of acetate ions are present. In the nitrous acid-water mixture, 2 percent of the nitrous acid molecules dissociate to form 0.02 mole of nitrite ions and 0.02 mole of hydronium ions. In the chlorous acid-water mixture, 10 percent of the acid dissociates; while in the hydrochloric acid mixture, almost 100 percent of the molecules are dissociated into hydronium ions and chloride ions. Thus, HCl is far superior to the other three acids in its ability to release a proton, and the proton-releasing abilities or acid strengths vary in the order

$$HCl > HClO_2 > HNO_2 > HC_2H_3O_2$$

Data from this type of experiment can also be used to calculate the equilibrium constants for dissociation of acids,

$$(1) \qquad HA + H_2O \rightleftharpoons H_3O^+ + A^-$$

(except when the acids are very strong—for example, HCl—or very weak—for example, CH_3OH; see page 401). Since these constants refer to reactions of the same type [given by equation (1)], they are particularly useful as a measure of relative extents of reaction and therefore relative acid strengths. Table 15-1 lists constants for a set of selected acids. The strongest acid listed in the table is, of course, the one with the highest equilibrium constant—$HClO_4$; the weakest acid listed is the one with the lowest constant—C_2H_5OH.

The same approach can be used to determine relative base strengths. For example, if one-mole quantities of the amide ion, ammonia, and the acetate ion are each mixed separately with 1 liter of water, and the amount of base that has accepted a proton from the water is determined after equilibrium is established, the following percent conversions are attained:

$$NH_2^- + H_2O \rightleftharpoons NH_3 + OH^- \qquad 100\% \text{ conversion}$$

$$NH_3 + H_2O \rightleftharpoons NH_4^+ + OH^- \qquad 0.4\% \text{ conversion}$$

$$C_2H_3O_2^- + H_2O \rightleftharpoons HC_2H_3O_2 + OH^- \qquad 2 \times 10^{-3}\% \text{ conversion}$$

TABLE 15-1
Acid Dissociation Constants
in Water at 25°C

$HClO_4$	$> 10^{10}$	HNO_2	4.6×10^{-4}
HI	10^{10}	$p\text{-}NO_2C_6H_4COOH$	3.9×10^{-4}
HBr	10^9	HF	3.5×10^{-4}
HCl	10^8	$m\text{-}ClC_6H_4COOH$	1.5×10^{-4}
H_2SO_4	10^3	H_2Se	2.0×10^{-4}
HNO_3	10^2	$p\text{-}ClC_6H_4COOH$	1.0×10^{-4}
$HClO_3$	10	C_6H_5COOH	6.5×10^{-5}
CF_3COOH	6.0×10^{-1}	$p\text{-}CH_3C_6H_4COOH$	4.2×10^{-5}
CCl_3COOH	2.0×10^{-1}	$p\text{-}CH_3OC_6H_4COOH$	3.4×10^{-5}
$CHCl_2COOH$	3.3×10^{-2}	CH_3COOH	1.8×10^{-5}
$HClO_2$	1.0×10^{-2}	H_2S	1.0×10^{-7}
H_2Te	2.5×10^{-3}	$HClO$	3.0×10^{-8}
$o\text{-}ClC_6H_4COOH$	1.2×10^{-3}		

In other words, in water the amide ion, NH_2^-, is a very strong base, whereas the acetate ion is a rather weak base.

Equilibrium constants for the reaction of selected bases with water are listed in Table 15-2. These reactions are of the following types:

(2a) $$B + H_2O \rightleftharpoons BH^+ + OH^-$$

or

(2b) $$B^- + H_2O \rightleftharpoons BH + OH^-$$

Additional equilibrium constants can be found in Appendix 9.

The data in Tables 15-1 and 15-2 indicate that acidities and basicities vary widely. Even when the bond involved in the dissociation of an acid is the same, say O—H, the equilibrium constants of species containing this bond vary over many orders of magnitude. For example, perchloric acid, O_3ClOH, is a very strong acid with an equilibrium constant of 10^{10}, while hypochlorous acid, $ClOH$, is quite weak and has a constant of 10^{-8}. Or, consider the two structurally similar acids, acetic and trifluoroacetic acids: although both contain the skeleton

$$\overset{\displaystyle\rightarrow}{\underset{\displaystyle\rightarrow}{C}}\!\!-\!C\!\!\overset{\displaystyle\nearrow O}{\underset{\displaystyle\searrow OH}{}}$$

TABLE 15-2	H^-	$>10^{23}$	HS^-	1.0×10^{-7}
Equilibrium Constants for Selected Bases in Water at 25°C	NH_2^-	$\sim10^{21}$	$(C_2H_5)_3P$	5.0×10^{-8}
	OH^-	1	H_2NOH	1.1×10^{-8}
	$(C_2H_5)_3N$	4.0×10^{-4}	C_5H_5N	1.4×10^{-9}
	NH_3	1.8×10^{-5}	CH_3COO^-	5.6×10^{-10}
	CN^-	2.0×10^{-5}	$C_6H_5NH_2$	3.8×10^{-10}
	H_2NNH_2	1.0×10^{-6}	F^-	2.9×10^{-11}

their equilibrium constants differ by 10^5. Clearly, the composition and structure of acids and bases influence their reactivities markedly.

Effect of Structure on Acidity

The relationship between structure and acidity (and basicity) is important not only because of the human desire to understand the workings of nature but also because of the predictions that this understanding allows. We begin our analysis of the effect of structure on acidity (basicity will be considered later) by attempting to determine what factors enhance the ease with which an acid, generalized as HA, dissociates to a proton and its conjugate base A^-.

$$(3) \qquad HA \rightleftharpoons H^+ + A^-$$

If the dissociation occurs in the gas phase or in a solvent that does not solvate either the acid or its conjugate base, the dissociation can be visualized as occurring in three steps: (a) the dissociation of the bond to the hydrogen atom, (b) the loss of an electron from the hydrogen atom, and (c) the gain of an electron by A.

$$(a) \qquad HA \rightleftharpoons H + A$$

$$(b) \qquad H \rightleftharpoons H^+ + e^-$$

$$(c) \qquad A + e^- \rightleftharpoons A^-$$

Since for the most part we shall be interested only in *relative* acidities [step (b) is common to all acid dissociations], only steps (a) and (c) need be considered. If entropy changes are assumed to be similar for a series of acids, the free-energy change and therefore the extent of reaction (3) will depend upon the strength of the H–A bond and the electron affinity of A. The greater the H–A bond energy, the greater the energy required for step (a) and the lower the extent of the reaction. The greater the electron affinity [remember that electron affinity is defined (page 57) as energy *released*], the greater the amount of energy released in step (c) and the greater the extent of reaction. Of course, the *sum* of the bond energy and electron affinity determines the relative extent of reaction (3).

TABLE 15-3
Thermochemical Analysis of the Dissociation of H_2O, HF, and HCl in Water at 25° C

Step	ΔH^0 (kcal/mole)	ΔH^0_{HF} minus $\Delta H^0_{H_2O}$		ΔH^0 (kcal/mole)	ΔH^0_{HF} minus ΔH^0_{HCl}		ΔH^0 (kcal/mole)
1 $H_2O(aq) \rightleftharpoons H_2O(g)$	10	1	$HF(aq) \rightleftharpoons HF(g)$	11	7	$HCl(aq) \rightleftharpoons HCl(g)$	4
2 $H_2O(g) \rightleftharpoons OH(g) + H(g)$	113	22	$HF(g) \rightleftharpoons H(g) + F(g)$	135	33	$HCl(g) \rightleftharpoons H(g) + Cl(g)$	102
3 $H(g) \rightleftharpoons H^+(g) + e^-$	313	—	$H(g) \rightleftharpoons H^+(g) + e^-$	313	—	$H(g) \rightleftharpoons H^+(g) + e^-$	313
4 $OH(g) + e^- \rightleftharpoons OH^-(g)$	−42	−37	$F(g) + e^- \rightleftharpoons F^-(g)$	−79	4	$Cl(g) + e^- \rightleftharpoons Cl^-(g)$	−83
5 $H^+(g) \rightleftharpoons H^+(aq)$	−261	—	$H^+(g) \rightleftharpoons H^+(aq)$	−261	—	$H^+(g) \rightleftharpoons H^+(aq)$	−261
6 $OH^-(g) \rightleftharpoons OH^-(aq)$	−117	−4	$F^-(g) \rightleftharpoons F^-(aq)$	−121	−32	$Cl^-(g) \rightleftharpoons Cl^-(aq)$	−89
$H_2O(aq) \rightleftharpoons H^+(aq) + OH^-(aq)$	+16		$HF(aq) \rightleftharpoons H^+(aq) + F^-(aq)$	−2		$HCl(aq) \rightleftharpoons H^+(aq) + Cl^-(aq)$	−14

$\Delta S^0 = -19$ cal/deg-mole $\qquad\qquad\qquad$ $\Delta S^0 = -23$ cal/deg-mole $\qquad\qquad\qquad$ $\Delta S^0 = -9$ cal/deg-mole

$\Delta G^0 = 22$ kcal/mole $\qquad\qquad\qquad\quad$ $\Delta G^0 = 5$ kcal/mole $\qquad\qquad\qquad\qquad$ $\Delta G^0 = -11$ kcal/mole

$K = 10^{-16}$ $\qquad\qquad\qquad\qquad\qquad\quad$ $K = 10^{-4}$ $\qquad\qquad\qquad\qquad\qquad\qquad$ $K = 10^8$

When reaction (3) is carried out in a solvent, such as water, that can interact with the acid or its conjugate base, two other steps must also be considered—solvation of the acid and its conjugate base. All of the necessary enthalpy changes for the reaction of the binary acids of the three neighboring elements, oxygen, fluorine, and chlorine, with water are known; see Table 15-3.

The first step in Table 15-3 represents the desolvation of the acid into the gaseous phase; the second step is the dissociation of the H—A bond; the third step is the ionization of hydrogen; the fourth step is the acceptance of an electron by gaseous A; and the fifth and sixth steps represent hydration of the ions H^+ and A^-. The sum of all six steps is the equation for dissociation of the acid HA in water. The equation

$$HA(aq) \rightleftharpoons H^+(aq) + A^-(aq)$$

could also be written as

$$HA + H_2O \rightleftharpoons H_3O^+ + A^-$$

to emphasize the proton-transfer aspect of the reaction and also the fact that the proton exists in water firmly attached to a water molecule. Table 15-3 also gives the standard free-energy change and equilibrium constant for each reaction. It is clear that the enthalpy changes do provide a measure of the relative equilibrium constants. (It is important to note, however, the effect of ΔS^0, especially for HF, in making ΔG^0 higher than ΔH^0.)

The dissociation of the three acids can now be compared, step by step, to determine whether the differences in ΔH^0 can be traced to one or more of the individual steps. A comparison of H_2O and HF reveals that the largest differences in ΔH^0 occur for steps 2 and 4. The enthalpy change for step 2, the dissociation of the H—A bond, is lower for H_2O than for HF. If the enthalpy changes for all other steps were the same for both compounds, water would be the stronger acid. The enthalpy changes for the other steps are not the same, however, and the greater electron affinity of fluorine as well as the greater hydration energy of the fluoride ion outweigh the bond energy difference and make HF the stronger acid.

Comparison of the enthalpy changes for the HF and HCl reveals that steps 1, 2, and 4 make the dissociation of HCl more favorable, while step 6 makes dissociation less favorable. Of the first three steps, by far the most important is step 2, the dissociation of the H—A bond. Thus, for the three acids the total ΔH^0 is a rather complex composite of a number of thermochemical steps; but two steps, the bond dissociation and electron affinity, seem to be the most influential.

This result, that the two most important steps for the comparison of acid strengths are bond dissociation and electron affinity, allows the postulation of two guidelines:

1. The relative acidities of two or more acids that contain the acidic proton bonded to a given atom, say oxygen, are determined by the affinity of that given atom in each acid for an electron. For example, the relative acidities of perchloric acid, O_3ClOH, and hypochlorous acid, ClOH, both

of which contain an O—H bond, are determined by the relative affinities of the oxygen of the O_3ClO group and the oxygen of the ClO group for an electron. As we shall see below, the oxygen of the O_3ClO group has a greater affinity for an electron, and perchloric acid is therefore the stronger acid. This rule has its basis in the fact that the H—X bond energies of two acids will be similar when X is the same atom. That is, the bond energies are not very dependent upon the remainder of the acid. The relative acidities will then be controlled by the affinity of the atom X (which is considerably affected by the remainder of the acid molecule) for an electron.

Since the polarity of the H—X bond is determined by the affinity of X for electrons, the ease with which the H—X bond in a given acid dissociates can also be related to the *polarity* of that bond. Hence, the O—H bond in perchloric acid is more polar than the O—H bond in hypochlorous acid, and perchloric acid is therefore the stronger acid.

2. The relative acidities of two acids that contain the acidic proton bonded to different atoms are determined by the relative bond energies and electron affinities (and possibly also relative hydration energies). If the difference in bond energies between the two acids is greater than the difference in electron affinities, the acid with the weaker H—X bond will be the stronger acid, and vice versa. This case is illustrated by the binary acids of the elements of a given periodic group, say Group VI. The H—Te bond for H_2Te is much weaker than the H—S bond in H_2S and thus, in spite of the fact that sulfur is more electronegative and therefore has a greater affinity for an electron, H_2Te is the stronger acid.

We can now use these guidelines to rationalize the relative acidities of the species in Table 15-1. One of the more obvious patterns in these data is that for acids of general structure $O_nQ(OH)_m$—for example O_3ClOH, O_2ClOH, OClOH, ClOH—an increase in the number (n) of oxygens attached to a given central atom (Q) increases the acidity of the acid. Thus, in the series O_nClOH, the acidities vary in the order $O_3ClOH > O_2ClOH > OClOH > ClOH$. Likewise, nitric acid, O_2NOH, is stronger than nitrous acid, ONOH; and sulfuric acid, $O_2S(OH)_2$, is stronger than sulfurous acid, $OS(OH)_2$. Since these acids all have the same X—H bond, the first rule can be applied. Accordingly, nitric acid is stronger than nitrous because the oxygen of the group O_2NO has a greater affinity for an electron than the group ONO. Stated in terms of bond polarity, the O—H bond in O_2NOH is more polar than the O—H bond in ONOH.

This difference in affinity for an electron (and bond polarity) can be explained as follows: In ONOH, two oxygens are attached to nitrogen; the addition of a third oxygen to give O_2NOH removes (because of oxygen's high electronegativity) electron density from the nitrogen. This removal of electron density leaves exposed a greater amount of nuclear positive charge at the nitrogen. This increased "exposed" positive charge in turn attracts electron density from the oxygen of the

O—H bond. The oxygen of this bond is therefore more electron deficient and therefore has a greater affinity for an electron than the corresponding oxygen in ONOH. (Consequently, the O—H bond in O_2NOH is more polar than that in ONOH.) The same reasoning indicates that in the series O_3ClOH, O_2ClOH, $OClOH$, and ClOH, the successive addition of oxygen to the chlorine central atom results in the removal of electron density from the oxygen of the O—H bond, thereby increasing its affinity for electrons.

A variation of the central atom in series of the type $O_nQ(OH)_m$ also affects the acidity. In the series LiOH, H_3COH, ClOH, the electronegativity of the atoms Li, C, and Cl increase in the order Li $<$ C $<$ Cl. Thus, the most polar O—H bond occurs in ClOH and this compound is the strongest acid. In LiOH the electronegativity difference between lithium and oxygen is so large that this linkage is almost totally ionic. Hence, LiOH is a source of hydroxide ions, ClOH is a weak source of hydronium ions, and H_3COH is a source of neither.

The same rule can be applied to the series of substituted acetic acids,

$$F_3CC\overset{O}{\underset{OH}{\diagdown}} , \quad Cl_3CC\overset{O}{\underset{OH}{\diagdown}} , \quad Cl_2CHC\overset{O}{\underset{OH}{\diagdown}} , \quad ClCH_2C\overset{O}{\underset{OH}{\diagdown}} , \quad \text{and} \quad CH_3C\overset{O}{\underset{OH}{\diagdown}}$$

In these compounds the substituent responsible for alteration of the electron density at the oxygen of the O—H group is three bonds removed from the oxygen. Replacement of a hydrogen of the methyl group of acetic acid by the more electronegative chlorine results in a decrease in electron density at the carbon to which it is attached. The decrease in electron density in effect leaves exposed a larger amount of nuclear positive charge. This increase in partial positive charge results in withdrawal of electron density from the carbon of the carbonyl group, which then also obtains a larger partial positive charge. The increased partial positive charge at this carbon finally produces electron withdrawal from the oxygen of the O—H group, thereby giving it a greater affinity for electrons. Thus the electron-withdrawing effect of the electronegative chlorine is passed, bond by bond, through the three intervening bonds to the oxygen. As more hydrogens are replaced by chlorines, more and more density is removed, and the oxygen obtains a greater and greater affinity for electrons (the O—H bond becomes more polar). Hence, the acidities of the chloroacetic acids vary

$$Cl_3CC\overset{O}{\underset{OH}{\diagdown}} > Cl_2CHC\overset{O}{\underset{OH}{\diagdown}} > ClCH_2C\overset{O}{\underset{OH}{\diagdown}}$$

Since fluorine is an even better electron-withdrawing group than chlorine, trifluoroacetic acid is an even stronger acid than trichloroacetic acid. The alteration of electron density at a given atom by an atom or group one or more bonds removed, which results from changes induced in the electron density of intervening bonds, is termed the *inductive effect*.

The acidity of aromatic acids is also affected by remote substituents. Replacement of a hydrogen in benzoic acid by the more electronegative atom chlorine causes an increase in acidity. When the chlorine is on the *para* position,

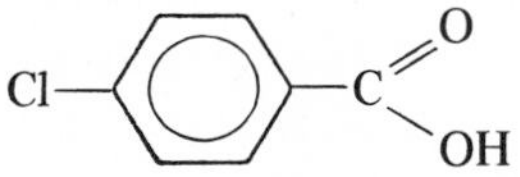

it is six bonds removed from the oxygen, and consequently the difference in acidity between *p*-chlorobenzoic acid and benzoic acid is not nearly so great as the difference between chloroacetic acid and acetic acid, where the chlorine is only three bonds removed. When the chlorine is shifted to the *meta* position it is five bonds from the oxygen, and in the *ortho* position it is only four bonds away. As the chlorine gets closer the acidity increases, and thus for the chlorobenzoic acids the acidity varies:

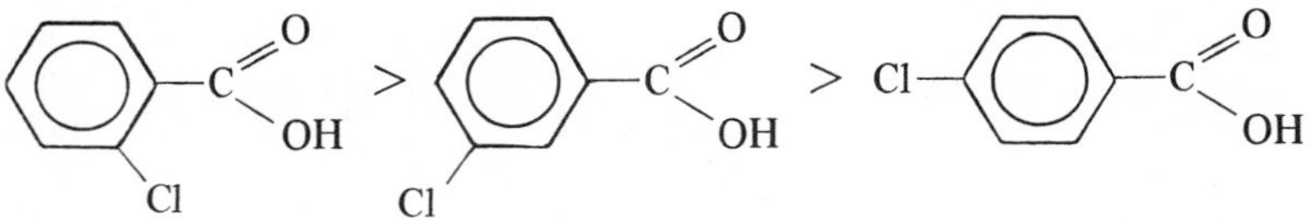

The relative acidities of the binary acids of the Group VI and VII elements are amenable to interpretation with the second rule. That is, the bond energies of the H–X bonds of HF, HCl, HBr, and HI decrease markedly down the group, and this change outweighs the increase in electron affinity up the group. Thus, the acidity orders $HI > HBr > HCl > HF$ and $H_2Te > H_2Se > H_2S > H_2O$ can be ascribed to variations in the H–X bond energies.

Effect of Structure on Basicity

Let us begin our discussion of relative basicities with an almost obvious, but important, generalization: *The stronger the base, the weaker its conjugate acid; the stronger the acid, the weaker its conjugate base.* That is, the greater the basicity of a base, B, the more readily it accepts a proton, and the higher the extent of reaction (4).

(4) $$B + H^+ \rightleftharpoons BH^+$$
   base   conjugate
            acid

The higher the extent of (4), the smaller the degree to which the conjugate acid dissociates back to the base. Tables 15-1 and 15-2 provide a number of examples of this generalization: HCl is a very strong acid, and its conjugate base, Cl^-, is extremely weak; HF is a weaker acid, and its conjugate base, F^-, is stronger than Cl^-, acetic acid is a weaker acid than HF, and its conjugate base, $C_2H_3O_2^-$, is therefore a stronger base than F^-; hydrocyanic acid is weaker than acetic acid, and consequently its conjugate base is stronger than acetate ion.

One of the advantages of this generalization is that it allows an analysis of basicity in terms of the acidity of the conjugate acid. Those characteristics that enhance the strength of an acid (low H—X bond energy and high electron affinity of X) diminish the basicity of its conjugate base. Hence, a strong base will be one whose conjugate acid has a strong H—X bond and whose X atom has a low affinity for electrons. The guidelines for relative basicities, then, are

1. The relative basicities of two bases whose basic atoms (the atoms to which the protons become attached) are the same are determined by the relative affinities of those atoms for an electron: the lower the affinity the stronger the base. For example, ammonia is a stronger base than hydroxylamine, H_2NOH, because substitution of the electronegative OH for a hydrogen increases the affinity of the nitrogen for an electron. Similarly,

   aniline, 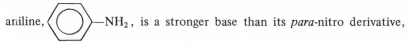—NH_2, is a stronger base than its *para*-nitro derivative,

   which has the more electron-deficient amino nitrogen.

2. The relative basicities of two bases whose basic atoms are different are usually determined by the strength of the bond to those atoms in their conjugate acids: the stronger the bond the stronger the base. For example, triethylamine, $(C_2H_5)_3N$, is a stronger base than triethylphosphine, $(C_2H_5)_3P$, because the N—H bond of the conjugate acid $(C_2H_5)_3NH^+$ is stronger than the P—H bond of the conjugate acid $(C_2H_5)_3PH^+$ (even though nitrogen has a greater affinity for electrons than phosphorus).

EXTENTS OF PROTON-TRANSFER REACTIONS

As indicated in Chapter 12, proton-transfer reactions are generally very rapid processes, with most reactions reaching equilibrium in less than 10^{-3} second.

Although the *extents* of the reactions of acids and bases with water have already been discussed, it is frequently necessary to have at least a qualitative knowledge of the extent of reaction between *any* acid and base. For such a reaction,

$$A \; + \; B \; \rightleftharpoons \; C \; + \; D$$
$$acid_1 \; + \; base_2 \; \rightleftharpoons \; acid_2 + base_1$$

the extent is determined by the relative acidities of the two acids A and C and the relative basicities of the bases B and D. In fact, the reaction can be viewed as a competition between the two sets of acids and bases for the most proton transfers. If A, B, C, and D are present in equal concentrations at the beginning of the reaction, and if A and B are stronger than their conjugates, more protons will be transferred from A to B than from C to D and the equilibrium will lie on the product side ($K > 1$). If the stronger set is the product acid and base, more transfers will occur from C to D than from A to B, and the equilibrium constant will be less than 1.

The determination of which is the stronger acid and which the stronger base can be accomplished by using the relative acidities and basicities listed in Tables 15-1 and 15-2 and the generalization about the strengths of the conjugates (page 394). As a first example, we will consider the reaction between acetic acid and the fluoride ion.

$$HC_2H_3O_2 + F^- \rightleftharpoons HF + C_2H_3O_2^-$$

Table 15-1 shows that HF is a stronger acid than acetic acid, and Table 15-2 shows that the acetate ion is a stronger base than the fluoride ion. There are, therefore, a greater number of proton transfers between HF and the acetate ion than between acetic acid and the fluoride ion, and the extent of reaction is less than 50 percent ($K < 1$).

For the reaction of chlorous acid with the cyanide ion,

$$HClO_2 + CN^- \rightleftharpoons HCN + ClO_2^-$$

the stronger acid is $HClO_2$, and its conjugate base must therefore be weaker than the conjugate base of HCN. The stronger acid and base therefore appear as reactants, and the extent of the reaction is high.

QUANTITATIVE ASPECTS
OF PROTON-TRANSFER EQUILIBRIA IN AQUEOUS SOLUTION

Self-Ionization of Water

Because of the amphiprotic nature of water, protons are transferred between water molecules in liquid water.

$$H_2O + H_2O \rightleftharpoons H_3O^+ + OH^-$$

Since the hydronium ion is a much stronger acid than water, and hydroxide ion is a much stronger base than water, the extent of this reaction is very small. Indeed, at $25°$ its equilibrium constant, designated K_w, has a value of 10^{-14}. The equilibrium expression for this reaction does not contain the concentration of water because so little water is consumed by the reaction that its concentration remains constant.

(5) $$K_w = [H_3O^+][OH^-] = 10^{-14}$$

Both hydronium ions and hydroxide ions, therefore, exist in pure water, and since every proton transfer results in one hydronium ion and one hydroxide ion, their concentrations are equal. The value of this concentration can be calculated by letting x represent the hydronium ion concentration, which must of course also be equal to the hydroxide ion concentration.

$$x = [H_3O^+] = [OH^-]$$

$$K_w = [H_3O^+][OH^-] = 10^{-14}$$

$$= (x)(x) = 10^{-14}$$

$$x = 10^{-7}M$$

The hydronium ion and hydroxide concentrations in pure water are therefore $10^{-7}M$.

Because of the paramount importance of the hydronium and hydroxide ion concentrations in aqueous solution and the inconvenience of writing very small numbers in exponential form, they are frequently expressed as their negative logarithms. The negative logarithm of the hydronium ion concentration is designated pH, while the negative logarithm of the hydroxide concentration is pOH. Indeed, the symbol p can be used to designate the negative logarithm of any variable.

$$pH = -\log[H_3O^+]$$

$$pOH = -\log[OH^-]$$

$$pQ = -\log Q, \quad Q = \text{some variable}$$

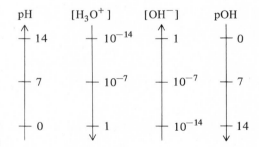

FIGURE 15-1

The pH and pOH Scales

The pH of pure water is then $-\log[10^{-7}] = -(-7) = 7$. The pOH of pure water is, of course, also 7. Figure 15-1 gives a graphic representation of the relationship between pH, pOH, $[H_3O^+]$, and $[OH^-]$. If the negative logarithms of both sides of equation (5) are taken, the useful relationship that pH + pOH = 14 at $25°$ is obtained.

$$K_w = [H_3O^+][OH^-] = 10^{-14}$$

$$-\log[H_3O^+][OH^-] = -\log 10^{-14}$$

$$-\log[H_3O^+] - \log[OH^-] = 14$$

$$pH + pOH = 14$$

Regardless of what is added to water, the relationship between the concentrations of hydronium ions and hydroxide ions as expressed by equation (5) must hold. If a substance is added that produces hydronium ions, the concentration of hydroxide ions will decrease so that the product $[H_3O^+][OH^-]$ remains at its constant value of 10^{-14}. The same effect can be expressed in terms of Le Chatelier's Principle: The addition of hydronium ions is a stress on the system

$$H_2O + H_2O \rightleftharpoons H_3O^+ + OH^-$$

that can only be relieved by the reaction of H_3O^+ with OH^- to form more H_2O.

When hydronium ions dominate an aqueous solution—that is, when $[H_3O^+] > [OH^-]$—the solution is said to be acidic; when $[OH^-] > [H_3O^+]$ the solution is basic or alkaline; when $[H_3O^+] = [OH^-]$ the solution is neutral. Thus, an acidic solution has a pH less than 7.00, a basic solution has a pH greater than 7.00, and a neutral solution has a pH of 7.00.

EXAMPLE: Calculate the pH and pOH of a solution that has a hydronium ion concentration of $5.0 \times 10^{-9}M$.

Solution:

$$[H_3O^+] = 5 \times 10^{-9}M$$

$$pH = -\log(5 \times 10^{-9})$$

$$\log(5 \times 10^{-9}) = 0.70 - 9$$

$$= -8.30$$

$$pH = 8.30$$

$$pOH = 14.00 - 8.30 = 5.70$$

EXAMPLE: Calculate the hydroxide ion concentration in a solution of pH = 4.70.

Solution:

$$pOH = 14.00 - 4.70 = 9.30$$

$$9.30 = -\log[OH^-]$$

$$\log[OH^-] = -9.30$$

$$[OH^-] = \text{antilog}(-9.30)$$

$$= \text{antilog}(-10 + 0.70)$$

$$= 5 \times 10^{-10}$$

The importance of the self-ionization of water to the chemistry of aqueous acid-base reactions is difficult to overemphasize. As a result of the presence of

H_3O^+ and OH^- and the equilibrium involving them, every acid-base reaction that occurs in water can be interpreted in two ways. Consider, for example, the reaction of acetic acid with ammonia in water. The net result of this reaction is the formation of the salt ammonium acetate. Mechanistically, the reaction could take place by transfer of protons from acetic acid molecules directly to ammonia molecules,

$$HC_2H_3O_2 + NH_3 \rightleftharpoons C_2H_3O_2^- + NH_4^+$$

or the acetic acid might first react with water to produce hydronium ions:

$$HC_2H_3O_2 + H_2O \rightleftharpoons C_2H_3O_2^- + H_3O^+$$

while the ammonia reacts to form hydroxide ions:

$$NH_3 + H_2O \rightleftharpoons NH_4^+ + OH^-$$

followed by reaction of OH^- with H_3O^+ to form H_2O:

$$H_3O^+ + OH^- \rightleftharpoons 2H_2O$$

This last reaction, by its consumption of H_3O^+ and OH^-, would serve to force the other two reactions (whose extents are low) nearly to completion. In either case the net result is the same.

Equilibrium Constants for Molecular Acids and Bases

Equilibrium constants for the reaction of molecular acids with water are designated K_a, while constants for molecular bases are designated K_b. For many acids and bases, these constants can be determined by a direct measurement of the hydronium ion concentration, usually with an instrument that determines the pH of the solution.

> **EXAMPLE:** When 0.100 mole of benzoic acid is dissolved in sufficient water to make 1 liter of solution, the pH of the solution is determined to be 2.60. Calculate K_a for benzoic acid.
>
> *Solution:* Benzoic acid reacts with water according to the equation

> and the corresponding equilibrium expression is
>
> $$K_a = \frac{[H_3O^+]\,[C_6H_5COO^-]}{[C_6H_5COOH]}$$

The concentration of water does not appear in this expression. The reaction is carried out in a large excess of water, and the amount of water consumed is so small that the concentration of water remains constant.

Strictly speaking, there are two sources of hydronium ion in this solution: the reaction of the acid with water and the self-ionization of water.

Since the pH of the solution is low, the acid has produced vastly more hydrogen ions than the self-ionization of water (which would be less than 10^{-7} because of suppression by the added hydronium ion), and consequently the concentration of H_3O^+ from the self-ionization of water can be neglected. The concentration of H_3O^+ from the acid can be calculated from the pH.

$$pH = 2.60 = -\log [H_3O^+]$$

$$\log [H_3O^+] = -2.60$$

$$[H_3O^+] = 10^{-2.60} = 10^{-3} + 10^{0.40}$$

$$= 2.5 \times 10^{-3} M$$

According to the equation for the reaction, every molecule of the acid that dissociates produces one benzoate ion and one hydronium ion. Therefore, the concentration of benzoate ion at equilibrium is also $2.5 \times 10^{-3} M$. The concentration of the acid at equilibrium is its original concentration minus the amount consumed during the reaction:

$$0.100 - 0.0025 = 0.0975 M$$

Finally,

$$K_a = \frac{(2.5 \times 10^{-3} M)(2.5 \times 10^{-3} M)}{9.75 \times 10^{-2} M} = 6.4 \times 10^{-5} \text{ mole/liter}$$

When the equilibrium constant is low (10^{-5} and lower), the calculation can usually be simplified by assuming that the equilibrium concentration of the acid is its initial concentration. The validity of such an assumption is totally dependent, however, upon the number of significant figures in the data. Values for pH are usually given to the nearest one-hundredth of a pH unit, which when translated into the hydronium ion concentration usually results in a value for $[H_3O^+]$ of two significant figures. Consequently, a 2 to 5 percent error can generally be tolerated.

EXAMPLE: When 0.10 mole of propanoic acid is mixed with sufficient water to give 1 liter of solution, the pH of the solution is 2.94 when equilibrium has been established. Calculate the K_a for propanoic acid.

Solution: Propanoic acid reacts with water according to the equation

$$CH_3CH_2COOH + H_2O \rightleftharpoons CH_3CH_2COO^- + H_3O^+$$

The equilibrium expression is

$$K_a = \frac{[CH_3CH_2COO^-][H_3O^+]}{[CH_3CH_2COOH]}$$

Assuming that the amount of H_3O^+ from the self-ionization of water is negligible, the concentration of H_3O^+ resulting from the dissociation of the acid is given by

$$pH = 2.94 = -\log [H_3O^+]$$

$$[H_3O^+] = 10^{-3} + 10^{.06}$$

$$= 1.15 \times 10^{-3}M$$

The concentration of propanoate ion is the same, $1.15 \times 10^{-3}M$. The concentration of the acid at equilibrium is $0.10 - 0.00115$, which to two significant figures is $0.10M$.

The equilibrium constant is, then,

$$K_a = \frac{(1.15 \times 10^{-3})^2}{0.10} = 1.3 \times 10^{-5}$$

EXAMPLE: The pH of a solution made by mixing 0.050 mole of diethyl-amine with enough water to make 1 liter of solution is 11.82. Calculate K_b for diethylamine.

Solution: Diethylamine reacts with water according to the equation

$$(C_2H_5)_2NH + H_2O \rightleftharpoons (C_2H_5)_2NH_2^+ + OH^-$$

for which the equilibrium expression is

$$K_b = \frac{[(C_2H_5)_2NH_2^+][OH^-]}{[(C_2H_5)_2NH]}$$

The hydroxide ion concentration can be obtained by first converting pH to pOH with the relationship

$$pH + pOH = 14$$

$$pOH = 14 - 11.82 = 2.18$$

$$-\log[OH^-] = 2.18$$

$$[OH^-] = 10^{-3} + 10^{0.82} = 6.6 \times 10^{-3}$$

Therefore,

$$[(C_2H_5)_2NH_2^+] = [OH^-] = 6.6 \times 10^{-3}M$$

$$[(C_2H_5)_2NH] = 0.050 - 0.0066 = 4.3 \times 10^{-2}M$$

$$K_b = \frac{(6.6 \times 10^{-3})^2}{4.3 \times 10^{-2}} = 1.0 \times 10^{-3}$$

When the acid or base is very strong or very weak, the determination of its equilibrium constant is more difficult. Consider, for example, two very strong acids, such that 99.9 percent of the molecules of the first are dissociated in water whereas the second is 99.99 percent dissociated. The second acid is clearly stronger and has a $K_a = 10^4$ as opposed to $K_a = 10^3$ for the first acid. However, because both are so close to being completely dissociated in water, their strengths are *experimentally indistinguishable;* that is, no experimental technique is capable of distinguishing between the two in water. The strengths of the two can be determined, however, if they are allowed to react in a nonaqueous medium with a base that is not as basic as water—for example, acetic acid.

$$HA + CH_3COOH \rightleftharpoons CH_3COOH_2^+ + A^-$$

In this very weak base, these two strong acids are dissociated to a much smaller extent, and the difference between the two can be measured.

When the acid is very weak, the hydronium ions that result from the self-ionization of water cannot be ignored. That is, both dissociation of the acid and ionization of the water contribute to the total hydronium ion concentration. This situation also complicates the determination of the equilibrium constant, and the K_a for such an acid is often obtained indirectly from the equilibrium constant for the reaction of its moderately strong conjugate base with water.

A *polyprotic* acid can release more than one proton per molecule, and its dissociation very likely occurs in a stepwise fashion. For each step there is a corresponding equilibrium constant: The constant for the removal of one proton from the neutral acid is referred to as K_{a1}, the constant for the removal of a proton from the resulting negatively charged species is K_{a2}, and so on. Thus, for the triprotic acid phosphoric acid, the steps and their corresponding constants are:

$$H_3PO_4 + H_2O \rightleftharpoons H_3O^+ + H_2PO_4^- \qquad K_{a1}$$

$$H_2PO_4^- + H_2O \rightleftharpoons H_3O^+ + HPO_4^{2-} \qquad K_{a2}$$

$$HPO_4^{2-} + H_2O \rightleftharpoons H_3O^+ + PO_4^{3-} \qquad K_{a3}$$

It is generally true that $K_{a1} > K_{a2} > K_{a3}$, and so on.

Once the equilibrium constant for an acid or base is known, it can be used to calculate the concentrations of all the species in a solution of given concentration.

EXAMPLE: Calculate the concentrations of all the species present in a $1.0 \times 10^{-2}M$ solution of *p*-chlorobenzoic acid.

Solution: This acid reacts with water according to the equation

In addition to the four species shown in the equation, there is also OH⁻ from the self-ionization of water. Since the concentration of water remains constant, the concentrations of four species are required.

If x is allowed to represent the equilibrium concentration of H_3O^+ and of the *p*-chlorobenzoate ion, the stoichiometry of the equation dictates that the equilibrium concentration of *p*-chlorobenzoic acid is $1.0 \times 10^{-2} - x$. The K_a as obtained from Table 15-1 is rather large, and it is therefore likely that the x in the equilibrium concentration of the free acid cannot be ignored. When these concentrations are inserted, the equilibrium expression becomes

$$K_a = 1.0 \times 10^{-4} = \frac{[H_3O^+][ClC_6H_4COO^-]}{[ClC_6H_4COOH]}$$

$$= \frac{x^2}{1.0 \times 10^{-2} - x}$$

Solving for x with the quadratic formula, we obtain

$$x^2 + 1.0 \times 10^{-4}x - 1.0 \times 10^{-6} = 0$$

$$x = \frac{-1.0 \times 10^{-4} \pm \sqrt{1.0 \times 10^{-8} + 4.0 \times 10^{-6}}}{2}$$

$$= \frac{-1.0 \times 10^{-4} \pm 2.0 \times 10^{-3}}{2} = 9.5 \times 10^{-4}, -1.05 \times 10^{-3}$$

Thus,

$$[H_3O^+] = 9.5 \times 10^{-4}M$$
$$[ClC_6H_4COO^-] = 9.5 \times 10^{-4}M$$

$$[ClC_6H_4COOH] = 1.0 \times 10^{-2} - 9.5 \times 10^{-4}$$
$$= 9.0 \times 10^{-3}M$$

And, finally, $[OH^-]$ can be obtained from the relationship for K_w:

$$K_w = 10^{-14} = [H_3O^+][OH^-]$$

$$[OH^-] = \frac{10^{-14}}{9.5 \times 10^{-4}} = \frac{10 \times 10^{-15}}{9.5 \times 10^{-4}} = 1.1 \times 10^{-11}M$$

The x in the denominator can usually be ignored unless the equilibrium constant is high or the initial concentration of the acid or base is low. As stated before, the decision of whether to ignore the amount of acid or base that has dissociated in calculating its concentration must be based upon the accuracy desired. Frequently, a quick calculation ignoring the x in the denominator will serve to indicate the approximate percentage error involved.

EXAMPLE: Calculate the concentration of all species in a $1.0 \times 10^{-1}M$ solution of pyridine.

Solution: Pyridine reacts with water according to the equation

The equilibrium constant for this reaction is 1.4×10^{-9}:

$$K_b = 1.4 \times 10^{-9} = \frac{[C_5H_5NH^+][OH^-]}{[C_5H_5N]}$$

If we let x represent the hydroxide ion concentration at equilibrium, we have

$$\frac{(x)(x)}{(1.0 \times 10^{-1}) - x} = 1.4 \times 10^{-9}$$

Assuming that x will be negligibly small relative to 1.0×10^{-1}, we can drop the x in the denominator and solve the equation

$$\frac{x^2}{1.0 \times 10^{-1}} = 1.4 \times 10^{-9}$$

$$x = 1.2 \times 10^{-5}$$

Clearly, x is about one-thousandth as large as 1.0×10^{-1}, and the denominator of the equilibrium expression is indeed 1.0×10^{-1}, to two significant figures. In other words, the solution to the exact equation would give the same answer to two significant figures.

The concentrations are:

$$[OH^-] = 1.2 \times 10^{-5} M$$

$$[C_5H_5NH^+] = 1.2 \times 10^{-5} M$$

$$[C_5H_5N] = 1.0 \times 10^{-1} M$$

$$[H_3O^+] = \frac{10^{-14}}{1.2 \times 10^{-5}} = 8.3 \times 10^{-10} M$$

The calculation of the concentration of all species present in a solution of a polyprotic acid is somewhat more complex. Consider, for example, a $1.0 \times 10^{-1} M$ solution of H_2S. This species reacts with water as a diprotic acid, dissociating in two steps according to the equations:

$$H_2S + H_2O \rightleftharpoons H_3O^+ + HS^- \qquad K_{a1} = 1.0 \times 10^{-7}$$

$$HS^- + H_2O \rightleftharpoons H_3O^+ + S^{2-} \qquad K_{a2} = 1.3 \times 10^{-13}$$

The equilibrium constants for the two steps are quite different, the first being a million times (10^6) as large as the second. Because of this huge difference in extents, the amount of *hydronium ion* resulting from the reaction of H_2S with water can be assumed to result almost totally from the first step. Moreover, the magnitude of K_{a2} reveals that the amount of *hydrogen sulfide ion* consumed in the second step will be negligible compared with the amount produced in the first step. Thus, $[H_3O^+]$ and $[HS^-]$ can be obtained in the usual way from K_{a1}.

$$K_{a1} = 1.0 \times 10^{-7} = \frac{[H_3O^+][HS^-]}{[H_2S]} = \frac{(x)(x)}{1.0 \times 10^{-1} - x}$$

If it is assumed that x is negligible relative to the two significant figures of 1.0×10^{-1}, x can be dropped from the denominator and the equation can be simplified to

$$\frac{x^2}{1.0 \times 10^{-1}} = 1.0 \times 10^{-7}$$

$$x = 1.0 \times 10^{-4} M$$

Hence, $[H_3O^+]$ and $[HS^-]$ in a $1.0 \times 10^{-1} M$ H_2S solution are both $1.0 \times 10^{-4} M$.

The equilibrium concentration of sulfide ion can now be obtained from the expression for K_{a2}.

$$K_{a2} = 1.3 \times 10^{-13} = \frac{[H_3O^+][S^{2-}]}{[HS^-]}$$

The concentration of HS^- is that just calculated, $1.0 \times 10^{-4} M$. The value used for $[H_3O^+]$, however, must reflect the fact that H_3O^+ is produced in both steps. Since the amount of H_3O^+ produced in the second step is assumed to be negligible, the concentration of H_3O^+ in the solution is taken as equal to $[H_3O^+]$ from step 1, $1.0 \times 10^{-4} M$. Hence,

$$K_{a2} = \frac{1.0 \times 10^{-4}[S^{2-}]}{1.0 \times 10^{-4}}$$

and

$$[S^{2-}] = K_{a2} = 1.3 \times 10^{-13} M$$

The concentration of sulfide can also be found by first adding the equations for the two dissociations

$$H_2S + H_2O \rightleftharpoons H_3O^+ + HS^-$$

$$\underline{HS^- + H_2O \rightleftharpoons H_3O^+ + S^{2-}}$$

$$H_2S + 2H_2O \rightleftharpoons 2H_3O^+ + S^{2-}$$

to give an equation representing the overall dissociation to sulfide. The equilibrium constant for this reaction is the product (see p. 322) of K_{a1} and K_{a2}, or 1.3×10^{-20}. Then $[S^{2-}]$ can be obtained from the corresponding equilibrium expression

$$\frac{[H_3O^+]^2[S^{2-}]}{[H_2S]} = K_{a1}K_{a2} = 1.3 \times 10^{-20}$$

This expression can be somewhat misleading, however, because $[H_3O^+]$ is not twice $[S^{2-}]$ (which would imply that all of the HS^- formed in the first step dissociates in the second step), but is instead the value calculated earlier, $1.0 \times 10^{-4} M$. The sulfide concentration is

$$\frac{(1.0 \times 10^{-4})^2[S^{2-}]}{1.0 \times 10^{-1}} = 1.3 \times 10^{-20}$$

$$[S^{2-}] = 1.3 \times 10^{-13} M$$

which is identical with the value obtained in the preceding calculations.

This *overall* expression is particularly valuable in dealing with saturated solutions of H_2S to which an acid has been added. Qualitatively, the effect of adding acid to a saturated solution (approximately $0.1M$) of H_2S is to decrease the concentration of sulfide ion and form more H_2S.

$$S^{2-} + 2H^+ \rightleftharpoons H_2S$$

The initial concentration of sulfide is so low, however, that the increase in concentration of H_2S that results from the addition of any amount of hydronium ion is totally negligible. Hence, in a saturated solution, the concentrations of sulfide and hydronium ions are related by the expression

$$\frac{[H_3O^+]^2[S^{2-}]}{0.1} = 1.3 \times 10^{-20}$$

or

$$[H_3O^+]^2[S^{2-}] = 1.3 \times 10^{-21}$$

For example, the sulfide concentration in a $1.0M$ HCl solution saturated with H_2S is

$$[S^{2-}] = \frac{1.3 \times 10^{-21}}{(1.0)^2} = 1.3 \times 10^{-21}\,M$$

This control of the sulfide ion by the hydronium ion concentration makes possible the analytical separation of certain metal ions. This can be illustrated by the separation of zinc(II) and cadmium(II) through the precipitation of their sulfides. The concentration of sulfide necessary for their precipitation is calculated from their respective solubility product expressions:

$$\underline{ZnS} \rightleftharpoons Zn^{2+}(aq) + S^{2-}(aq) \qquad K_{sp} = [Zn^{2+}][S^{2-}]$$

$$= 1 \times 10^{-23}$$

$$\underline{CdS} \rightleftharpoons Cd^{2+}(aq) + S^{2-}(aq) \qquad K_{sp} = [Cd^{2+}][S^{2-}]$$

$$= 1 \times 10^{-28}$$

If we suppose that a separation is required for a solution that is $0.1M$ in Cd^{2+} and $0.1M$ in Zn^{2+}, the required sulfide concentrations are

$$[Zn^{2+}][S^{2-}] = 1 \times 10^{-23}$$

$$[S^{2-}] = \frac{1 \times 10^{-23}}{10^{-1}} = 1 \times 10^{-22}\,M \text{ for } Zn^{2+}$$

and

$$[Cd^{2+}][S^{2-}] = 1 \times 10^{-28}$$

$$[S^{2-}] = \frac{1 \times 10^{-28}}{10^{-1}} = 1 \times 10^{-27}\,M \text{ for } Cd^{2+}$$

In order to precipitate cadmium but not zinc, a solution with a sulfide concentration of less than $1 \times 10^{-22}\,M$ but greater than $1 \times 10^{-27}\,M$ is required. The concentration of H_3O^+ necessary to produce a sulfide concentration of $1 \times 10^{-22}\,M$ in a saturated H_2S solution is, then,

$$[H_3O^+]^2[S^{2-}] = 1.3 \times 10^{-21}$$

$$[H_3O^+]^2 = \frac{1.3 \times 10^{-21}}{1 \times 10^{-22}} = 13$$

$$[H_3O^+] = \sqrt{13} = 3.6M$$

That is, when the concentration of H_3O^+ is slightly greater than $3.6M$, cadmium sulfide will precipitate and zinc sulfide will not.

Acid–Base Equilibria for Ions

Neutral molecular compounds are certainly not the only substances that behave as acids or bases. Ions also function as acids and bases, and several examples of such ions have already been given. The ammonium ion is a monoprotic acid in water,

$$NH_4^+ + H_2O \rightleftharpoons NH_3 + H_3O^+$$

the acetate ion is a monoprotic base, while the hydrogen carbonate ion can behave as both an acid and base.

$$C_2H_3O_2^- + H_2O \rightleftharpoons HC_2H_3O_2 + OH^-$$

$$HCO_3^- + H_2O \rightleftharpoons H_3O^+ + CO_3^{2-}$$

$$HCO_3^- + H_2O \rightleftharpoons H_2CO_3 + OH^-$$

Since the conjugates of ions are in many cases molecular acids or bases, the relative acidity or basicity of an ion can often be ascertained qualitatively from the K_a or K_b of its conjugate. For example, the chloride ion can theoretically function as a base,

$$Cl^- + H_2O \rightleftharpoons HCl + OH^-$$

but its conjugate acid is one of the very strong acids, and the generalization "the stronger the acid the weaker the conjugate base" would predict that Cl^- should therefore be a very weak base. In fact, the chloride ion shows no basic properties relative to water. In other words, the chloride ion does not affect the pH of an aqueous solution. In general, then, ions that are the conjugate bases of the very strong acids do not function as bases relative to water.

The conjugate bases of the weak acids—for example, acetate ion, fluoride ion, and nitrite ion—are definitely basic relative to water and react to form hydroxide ions. Cyanide ion, whose conjugate is the very weak hydrocyanic acid, is a much stronger base than the fluoride ion, whose conjugate acid is only moderately weak.

The acidity of acidic ionic species can also be predicted from the relative K_b values of their conjugate bases. Thus, the pyridinium ion, $C_5H_5NH^+$, which is the conjugate acid of the weak base pyridine, is a stronger acid than the ammonium ion, NH_4^+, which is the conjugate acid of the stronger base ammonia (see Table 15-2).

A variety of metal cations also act as acids. These cations are generally small and highly charged and are very strongly attracted to water molecules. Indeed, discrete species such as $Al(H_2O)_6^{3+}$ usually result from such interactions. The formation and characteristics of such Lewis acid-base adducts are discussed in Chapter 16. The salient point here is that the bond to the metal cation withdraws enough electron density to increase the polarity of the O–H bonds and thereby increase their acidity. These adducts then donate protons to water, as illustrated by the equation

$$Al(H_2O)_6{}^{3+} + H_2O \rightleftharpoons Al(H_2O)_5(OH)^{2+} + H_3O^+$$

Some common cations that produce hydronium ions in aqueous solutions are Be^{2+}, Al^{3+}, Sn^{4+}, Pb^{4+}, Sn^{2+}, Cr^{3+}, Co^{2+}, Cu^{2+}, Zn^{2+}, Fe^{3+}, and Ni^{2+}. Except for Be^{2+}, the alkali and alkaline earth cations do not react with water.

We are now in a position to evaluate the relative acidities of aqueous solutions of salts. If the salt has a low solubility in water or is a weak electrolyte, it will affect the pH of water very little. If, however, it is soluble and a strong electrolyte, its effect on water can be predicted from our knowledge of the acidity and basicity of its constituent ions. Sodium chloride is a soluble strong electrolyte, but, as we have seen above, neither the sodium ion nor the chloride ion has any acidic or basic properties. An aqueous solution of NaCl is therefore neutral (pH = 7).

Sodium acetate is also a soluble strong electrolyte. The sodium ion does not react with water, but the acetate ion is the conjugate base of the weak acid acetic acid and therefore has some basic properties. The reactions of sodium acetate with water can be summarized as

$$Na^+ + H_2O \rightarrow \text{no reaction}$$

(6) $$C_2H_3O_2{}^- + H_2O \rightleftharpoons HC_2H_3O_2 + OH^-$$

Since hydroxide ions are produced by the reaction of the acetate ion with water, an aqueous solution of sodium acetate is basic. However, a comparison of the acidity of the species in equation (6) shows that the stronger acid and base lie on the product side, and therefore the extent of this reaction is low.

Ammonium chloride provides an example of a salt that contains an acidic cation. The ammonium ion is the conjugate acid of the weak base ammonia and therefore possesses acidic properties. A comparison of the relative acidities of the species of equation (7) shows that the stronger acid and base are on the product side of the equation,

(7) $$NH_4{}^+ + H_2O \rightarrow H_3O^+ + NH_3$$

$$Cl^- + H_2O \rightarrow \text{no reaction}$$

and the extent of this reaction is therefore also low. Since the chloride ion does not react with water, an aqueous solution of NH_4Cl is acidic.

In an aqueous solution of ammonium acetate, both the cation and anion react with water. Whether the solution is acidic or basic depends upon which reaction has the greater extent.

$$NH_4{}^+ + H_2O \rightleftharpoons H_3O^+ + NH_3$$

$$C_2H_3O_2{}^- + H_2O \rightleftharpoons HC_2H_3O_2 + OH^-$$

It now becomes necessary to evaluate the extent of the reaction of an ion with water. While the reaction of, say, the acetate ion with water is fundamentally no different from the reaction of an uncharged species such as ammonia with water, the equilibrium constants for these reactions are often not tabulated. Because of the intimate relationship between the ion and its conjugate, however, the constant is readily evaluated. Consider the conjugate base A^- of the weak acid HA. This base reacts with water according to the equation

$$A^- + H_2O \rightleftharpoons HA + OH^-$$

and it is the equilibrium constant for this reaction that we must evaluate. Since such reactions have traditionally been termed *hydrolyses,* we will label the equilibrium constant K_h, to distinguish it from the K_a for the reaction of the conjugate acid HA with water. If the equation representing the dissociation of HA in water is subtracted from the equation for the self-ionization of water, we obtain

$$H_2O + H_2O \rightleftharpoons H_3O^+ + OH^-$$

$$- [HA + H_2O \rightleftharpoons H_3O^+ + A^-]$$

$$\overline{H_2O + H_2O - HA - H_2O \rightleftharpoons H_3O^+ + OH^- - H_3O^+ - A^-}$$

Upon collecting terms and rearranging the final equation, we obtain the desired equation:

$$A^- + H_2O \rightleftharpoons HA + OH^-$$

The equilibrium constant K_h can therefore be obtained by dividing K_w by K_a.

$$K_h = \frac{K_w}{K_a} = \frac{[H_3O^+][OH^-]}{\dfrac{[H_3O^+][A^-]}{[HA]}} = \frac{[HA][OH^-]}{[A^-]}$$

The same result is obtained for the reaction of an acidic cation, HB^+, with water:

$$HB^+ + H_2O \rightleftharpoons B + H_3O^+$$

$$K_h = \frac{K_w}{K_b}$$

where K_b is the equilibrium constant for the reaction of the uncharged conjugate base with water.

As specific illustrations, let us consider the reactions of the ammonium ion and the acetate ion with water. For the ammonium ion, the equilibrium constant K_h is obtained as

$$NH_4^+ + H_2O \rightleftharpoons NH_3 + H_3O^+$$

$$K_h = \frac{K_w}{K_b} = \frac{10^{-14}}{1.8 \times 10^{-5}} = 5.6 \times 10^{-10}$$

where K_b is the equilibrium constant for the reaction of ammonia with water.

$$NH_3 + H_2O \rightleftharpoons NH_4^+ + OH^- \quad K_b = 1.8 \times 10^{-5}$$

For the acetate ion, the equilibrium constant K_h is obtained from the relationship

$$C_2H_3O_2^- + H_2O \rightleftharpoons HC_2H_3O_2 + OH^-$$

$$K_h = \frac{K_w}{K_a} = \frac{10^{-14}}{1.8 \times 10^{-5}} = 5.6 \times 10^{-10}$$

where K_a is the equilibrium constant for the reaction of acetic acid with water, which coincidentally is numerically identical to K_b for ammonia. Thus, we can conclude that the ammonium ion and the acetate ion react to the same extent with water. Consequently, the concentration of H_3O^+ produced by the hydrolysis of NH_4^- is equal to the concentration of OH^- produced by the hydrolysis of the acetate ion, and an aqueous solution of ammonium acetate is therefore neutral ($[H_3O^+] = [OH^-]$).

> **EXAMPLE:** Determine whether an aqueous solution of NH_4F is acidic, basic, or neutral.
>
> *Solution:* Ammonium fluoride is a soluble strong electrolyte. The ammonium ion reacts with water as follows:
>
> $$NH_4^+ + H_2O \rightleftharpoons NH_3 + H_3O^+$$
>
> The equilibrium constant for this reaction can be evaluated as
>
> $$\frac{K_w}{K_b} = \frac{10^{-14}}{1.8 \times 10^{-5}} = 5.6 \times 10^{-10}$$
>
> The fluoride ion also reacts with water:
>
> $$F^- + H_2O \rightleftharpoons HF + OH^-$$
>
> The constant for this reaction is K_w/K_a, where K_a is the acid dissociation constant for HF ($K_a = 3.5 \times 10^{-4}$).
>
> $$K_h = \frac{10^{-14}}{3.5 \times 10^{-4}} = 2.9 \times 10^{-11}$$
>
> Since the equilibrium constant for the reaction of the ammonium ion with water is larger than that for the reaction of the fluoride ion with water, there are more hydronium ions than hydroxide ions present in an aqueous solution of NH_4F. Thus, the solution is acidic.

The equilibrium constant for hydrolysis can also be used to determine the behavior of an amphiprotic ion in water. The hydrogen carbonate ion can react in both of the following modes:

As an acid,

(8)
$$HCO_3^- + H_2O \rightleftharpoons CO_3^{2-} + H_3O^+$$

As a base,

(9) $$HCO_3^- + H_2O \rightleftharpoons H_2CO_3 + OH^-$$

If the first reaction has a greater extent than the second, the hydrogen carbonate ion serves as a source of hydronium ions. The relative extents of (8) and (9) can be judged from their respective equilibrium constants: The constant for (8) is the second dissociation constant, K_{a2}, for carbonic acid, whereas the constant for (9) is K_w/K_{a1}. To verify that the constant for (9) is indeed K_w/K_{a1} rather than K_w/K_{a2}, let us actually divide the appropriate equilibrium expressions:

$$K_w = [H_3O^+][OH^-]$$

For

$$H_2CO_3 + H_2O \rightleftharpoons H_3O^+ + HCO_3^-,$$

$$K_{a1} = \frac{[H_3O^+][HCO_3^-]}{[H_2CO_3]}$$

$$\frac{K_w}{K_{a1}} = \frac{[H_3O^+][OH^-][H_2CO_3]}{[H_3O^+][HCO_3^-]} = \frac{[OH^-][H_2CO_3]}{[HCO_3^-]}$$

That K_w/K_{a1} rather than K_w/K_{a2} is the correct constant can be recognized at a glance because of the presence of H_2CO_3 in equation (9), which could only be a part of the expression for the *first* dissociation of carbonic acid.

Now, from Appendix 9 we find that K_{a1} for carbonic acid is 4.2×10^{-7}, while K_{a2} is 4.8×10^{-11}. Hence, the equilibrium constants for (8) and (9) are 4.8×10^{-11} and $10^{-14}/(4.2 \times 10^{-7}) = 2.4 \times 10^{-8}$, respectively. Thus, equation (9) has the greater extent of reaction, and the reaction of HCO_3^- with water therefore has the net effect of producing more hydroxide ions than hydronium ions.

EXAMPLE: Determine whether an aqueous solution of sodium hydrogen sulfite is acidic, basic, or neutral.

Solution: Sodium hydrogen sulfite is soluble in water and is a strong electrolyte. In solution there are sodium ions, hydrogen sulfite ions, and the products of any reaction of these ions with water. Since the sodium ion does not react with water, only the two possible reactions of the amphiprotic species HSO_3^- with water need be considered.

(10) $$HSO_3^- + H_2O \rightleftharpoons H_3O^+ + SO_3^{2-}$$

(11) $$HSO_3^- + H_2O \rightleftharpoons H_2SO_3 + OH^-$$

From Appendix 9 we find that for sulfurous acid $K_{a1} = 1.3 \times 10^{-2}$ and $K_{a2} = 5.6 \times 10^{-8}$. The equilibrium constant for equation (10) is then 5.6×10^{-8}, and for equation (11) the constant is

$$\frac{K_w}{K_{a1}} = \frac{10^{-14}}{1.3 \times 10^{-2}} = 7.7 \times 10^{-13}$$

The acid strength of HSO_3^- is therefore greater than its base strength and, as a result, an aqueous solution of $NaHSO_3$ contains an excess of hydronium ions.

Once the equilibrium constant has been evaluated, the concentration of all the species present in a salt solution can be calculated. If both the cation and the anion of the salt react with water, the calculation becomes more complex; this situation will not be discussed here.

EXAMPLE: Determine the concentration of all species present in a 0.10M solution of ammonium chloride.

Solution: The chloride ion does not react with water, but the ammonium ion behaves as an acid, according to the equation

$$NH_4^+ + H_2O \rightleftharpoons NH_3 + H_3O^+$$

The equilibrium constant for this process is

$$\frac{K_w}{K_b} = \frac{10^{-14}}{1.8 \times 10^{-5}} = 5.6 \times 10^{-10}$$

which means that

$$K_h = \frac{[NH_3][H_3O^+]}{[NH_4^+]} = 5.6 \times 10^{-10}$$

The problem is now solved in the usual way, letting $x = [H_3O^+]$. Thus, at equilibrium,

$$[H_3O^+] = x$$

$$[NH_3] = x$$

$$[NH_4^+] = 0.10 - x$$

$$\frac{(x)(x)}{0.10 - x} = 5.6 \times 10^{-10}$$

The constant is low, and the x in the denominator can be ignored.

$$\frac{x^2}{0.10} = 5.6 \times 10^{-10}$$

$$x = 7.5 \times 10^{-6}M$$

Hence, at equilibrium

$$[H_3O^+] = 7.5 \times 10^{-6}M$$

$$[NH_3] = 7.5 \times 10^{-6}M$$

$$[NH_4^+] = 1.0 \times 10^{-1}M$$

$$[OH^-] = \frac{10^{-14}}{7.5 \times 10^{-6}} = 1.3 \times 10^{-9}M$$

$$[Cl^-] = 1.0 \times 10^{-1}M$$

In accordance with Le Chatelier's Principle, the addition of the conjugate base (the common ion) to a weak acid inhibits its dissociation. Similarly, addition of an ion common to a weak base (its conjugate acid) inhibits its production of hydroxide ions. Quantitatively, the effect can be calculated with either the equilibrium expression for the dissociation of the weak acid (or weak base) or the expression for the hydrolysis of the conjugate base (or conjugate acid). Let us consider a weak acid HA to which has been added some of the conjugate base A^-.

$$HA + H_2O \rightleftharpoons H_3O^+ + A^-$$

$$K_a = \frac{[H_3O^+][A^-]}{[HA]}$$

Since the concentration of hydronium ion will be the concentration of HA that has dissociated at equilibrium, we will rearrange this equation as

(12) $$[H_3O^+] = K_a \frac{[HA]}{[A^-]}$$

which emphasizes that the hydronium concentration and therefore the extent of dissociation of HA is a function of the ratio of $[HA]$ to $[A^-]$. The same result can be obtained from the expression for the hydrolysis of A^-.

$$A^- + H_2O \rightleftharpoons HA + OH^-$$

$$K_h = \frac{K_w}{K_a} = \frac{[HA][OH^-]}{[A^-]}$$

If we divide both sides of this equation by $K_w = [H_3O^+][OH^-]$, we obtain

$$\frac{K_w}{K_a K_w} = \frac{[HA][OH^-]}{[A^-][H_3O^+][OH^-]}$$

$$\frac{1}{K_a} = \frac{[HA]}{[A^-][H_3O^+]}$$

$$[H_3O^+] = \frac{K_a[HA]}{[A^-]}$$

which is equivalent to equation (12).

Now let us take the specific case of a $0.10M$ solution of acetic acid to which has been added enough sodium acetate to make the solution $0.10M$ in sodium acetate. The sodium acetate is a strong electrolyte, and this solution contains primarily molecular acetic acid, sodium ions, and acetate ions. Actually there are two sources of acetate ion—the sodium acetate and the dissociation of acetic acid. The concentration of acetate ion resulting from the dissociation of $0.10M$ acetic acid in pure water (no excess acetate) is $1.3 \times 10^{-3}M$:

$$HC_2H_3O_2 + H_2O \rightleftharpoons C_2H_3O_2^- + H_3O^+$$

$$K_a = 1.8 \times 10^{-5} = \frac{x^2}{0.10 - x}$$

$$x = 1.3 \times 10^{-3}$$

and in the presence of acetate ion the dissociation is repressed even further. Thus, the amount of acetate ion from the dissociation of acetic acid is negligible compared to the amount of acetate ion added as sodium acetate. Moreover, the amount of acetate ion consumed by reaction with water to form acetic acid is also negligibly small (see page 410) compared to the 0.10M concentration of the added acetate ion. The equilibrium concentrations can therefore be safely assumed to be

$$[HC_2H_3O_2] = 0.10M$$

$$[C_2H_3O_2^-] = 0.10M$$

and

$$[H_3O^+] = K_a \frac{[HC_2H_3O_2]}{[C_2H_3O_2^-]}$$

$$= 1.8 \times 10^{-5}M \frac{(0.10M)}{(0.10M)}$$

$$= 1.8 \times 10^{-5}M$$

The addition of the common ion has reduced the hydronium ion concentration from $1.3 \times 10^{-3}M$ to $1.8 \times 10^{-5}M$, nearly a hundredfold reduction.

The operation of *buffer solutions*, which are of great importance in many biological and chemical processes, is based upon the common ion effect. A buffer solution consists of nearly equal concentrations of a weak acid and a salt of that acid (its conjugate base) or a weak base and a salt of that base (its conjugate acid). Indeed, the acetic acid-acetate solution described above is a buffer solution. The utility of these solutions lies in their ability to maintain an almost constant hydronium ion concentration when small amounts of even strong acids or bases are added. For example, when a strong acid is added to our acetic acid-acetate solution, the hydronium ion supplied by the strong acid reacts with the acetate ion, thereby converting it to acetic acid.

$$C_2H_3O_2^- + H_3O^+ \rightleftharpoons HC_2H_3O_2 + H_2O$$

If a base is added, the hydroxyl ions formed react with the acetic acid to form acetate ions.

$$HC_2H_3O_2 + OH^- \rightleftharpoons C_2H_3O_2^- + H_2O$$

These additions serve merely to alter the concentration ratio of acetic acid to acetate ion, which in turn affects the hydronium ion concentration via equation (12).

In illustration of the buffering ability of our solution we will calculate (a) the change in hydronium ion concentration that occurs when 1.0×10^{-2} mole of HCl is added to 1.0 liter of water, and (b) the change in hydronium ion concentration that occurs when the same amount of HCl is added to 1.0 liter of water containing 1.0×10^{-1} mole of acetic acid and 1.0×10^{-1} mole of sodium acetate. Since HCl is a strong acid it is completely dissociated in water, and the first solution therefore contains 1.0×10^{-2} mole of H_3O^+ per liter of solution. Pure water has a hydronium ion concentration of 10^{-7}, and the change in $[H_3O^+]$ is from $10^{-7}M$ to $10^{-2}M$, a change of 10^5.

When the same amount of HCl is added to the buffer solution, the hydronium ions react with acetate ions according to the equation

$$H_3O^+ + C_2H_3O_2^- \rightleftharpoons HC_2H_3O_2 + H_2O$$

Since H_3O^+ is a much stronger acid than acetic acid and the acetate ion is a much stronger base than water, this reaction has a very high extent. Thus, the 1.0×10^{-2} mole of H_3O^+ is converted to 1.0×10^{-2} mole of acetic acid. The initial concentrations of acetic acid and acetate ion were $1.0 \times 10^{-1}M$, and when equilibrium is reestablished the concentrations are

$$[HC_2H_3O_2] = 1.0 \times 10^{-1} + 1.0 \times 10^{-2} = 1.1 \times 10^{-1}M$$

$$[C_2H_3O_2^-] = 1.0 \times 10^{-1} - 1.0 \times 10^{-2} = 0.9 \times 10^{-1}M$$

The hydronium ion concentration can now be calculated with equation (12).

$$[H_3O^+] = 1.8 \times 10^{-5} \frac{1.1 \times 10^{-1}}{0.9 \times 10^{-1}}$$

$$= 2.2 \times 10^{-5}M$$

The concentration of hydronium ion in the buffer solution before the HCl was added was $1.8 \times 10^{-5}M$ (see page 414). The change, therefore, is from $1.8 \times 10^{-5}M$ to $2.2 \times 10^{-5}M$, a factor of $1.1/0.9 = 1.2$.

Since the weak acid (or base) and its conjugate base (or conjugate acid) must be present in nearly equal amounts in order to keep the concentration ratio of acid to conjugate base as constant as possible, a given buffer system can operate effectively over only a small pH range. An upper limit on the ratio of acid to conjugate base (or base to conjugate acid) is 10:1. For a given weak acid-conjugate base system this means that buffering can occur between the hydronium ion concentrations of

$$[H_3O^+] = K_a \frac{10}{1} = 10K_a$$

and

$$[H_3O^+] = K_a \frac{1}{10} = \frac{K_a}{10}$$

or, if everything is expressed as negative logarithms ($pK_a = -\log K_a$)

$$-\log[H_3O^+] = -\log K_a \pm \log 10$$

$$pH = pK \pm 1$$

Accordingly, the buffering action for the acetic acid-acetate system can be maintained at any pH between $p(1.8 \times 10^{-5}) - 1$ and $p(1.8 \times 10^{-5}) + 1$ or 4.74 ± 1.

If it is desirable, perhaps for some analytical procedure, to maintain a constant hydronium ion concentration at some other pH, the buffer system can be chosen by simply searching for a weak acid or weak base whose pK_a or pK_b has a value within ±1 of the desired pH.

> **EXAMPLE:** What buffer system could be used to maintain a pH of 9.3?
>
> *Solution:* The desired pH is on the basic side; consequently we need to look for a weak *base* whose pK_b is $14 - 9.3 = 4.7 \pm 1$.
>
> $$pOH = pK_b \pm 1$$
>
> $$pOH = 14 - 9.3 = 4.7$$
>
> $$pK_b = 4.7 \pm 1$$
>
> $$K_b = 2 \times 10^{-4} \text{ to } 2 \times 10^{-6}$$

Table 15-2 reveals that NH_3 is one base with a K_b of approximately 2×10^{-5}. Thus, the buffer solution might be prepared from NH_3 and NH_4Cl. If the pH desired is exactly 9.30, the ratio of NH_3 to NH_4^+ can be determined as follows:

$$[OH^-] = K_b \frac{[NH_3]}{[NH_4^+]}$$

$$[OH^-] = \text{antilog}(-4.70) = 2.0 \times 10^{-5}$$

$$\frac{[OH^-]}{K_b} = \frac{2.0 \times 10^{-5}}{1.8 \times 10^{-5}} = 1.1 = \frac{[NH_3]}{[NH_4^+]}$$

If the solution is made $0.10M$ in NH_3, the concentration of NH_4^+ should be $0.10/1.1 = 0.091M$.

Probably the most important buffer system found in nature is the carbonic acid–hydrogen carbonate–carbonate equilibrium. When carbon dioxide is bubbled into water, a small percentage of it is slowly converted to carbonic acid (recall from Chapter 11 that CO_2 is the anhydride of carbonic acid).

$$CO_2 + H_2O \rightleftharpoons H_2CO_3$$

The carbonic acid can then behave as a normal diprotic acid and dissociate according to the equations

$$H_2CO_3 + H_2O \rightleftharpoons HCO_3^- + H_3O^+ \qquad K_{a1} = 4.2 \times 10^{-7}$$

$$HCO_3^- + H_2O \rightleftharpoons CO_3^{2-} + H_3O^+ \qquad K_{a2} = 4.8 \times 10^{-11}$$

The importance of this buffer system in natural fluids can be illustrated by the equilibria present in seawater and blood plasma. In seawater a buffering action is produced by the equilibrium between absorbed CO_2 and the vast carbonate sediments in the ocean beds. If alkaline materials appear in the water, CO_2 is converted to a hydrogen carbonate ion and is replenished by absorption of atmospheric CO_2 at the surface of the ocean; if acidic materials appear (for example, by volcanic eruptions), they react with carbonate and produce hydrogen carbonate. In this way the pH of the oceans is maintained at a fairly constant value.

In blood plasma the control of pH is vital to the maintenance of life. Since the pH of blood is maintained in a very narrow range, 7.35 to 7.45, plasma is just slightly basic, and the rather strongly basic carbonate ion probably does not play a major role in the buffer. Consequently, the important species in the buffer are CO_2, H_2CO_3, and HCO_3^-. A more detailed description of the mechanism of oxygen and carbon dioxide transportation and the buffering action will be given in Chapter 19.

Acid–Base Titrations

The neutralization of an acid with a base (or vice versa) is one of the oldest and most useful reactions in chemistry. The neutralization reaction, which we have previously defined as the reaction of an acid with a base to form a salt, has been used to determine the rates and extents of many reactions but is most often employed in quantitative analytical procedures. Some important applications are the determination of the percent carbonate in minerals, the percent phosphate in detergents, and the percent nitrogen in blood, proteins, foodstuff, and fertilizers.

The general procedure for the titration of a base with an acid is as follows. The base is added to distilled water in an Erlenmeyer flask, a few drops of an *indicator* solution are added, and the acid is placed in a buret, as shown in Figure 15-2. The concentration of the acid is either known or has been previously determined to at least four significant figures. If the acid can be obtained pure so that the concentration can be accurately known from the weight of acid used, the acid is called a *primary standard*. If a primary standard is not available, the concentration of the acid solution must be obtained by reacting a portion of the solution with a base whose concentration is accurately known. This process is referred to as *standardizing* the acid.

The acid in the buret is then added slowly to the base in the flask until the solution in the flask undergoes a marked change in color. This color change is due to reaction of the indicator with the acid and is the *endpoint* of the titration. If the indicator has been properly selected, this will also be very close to the *equivalence point* of the reaction—the point at which the number of equivalents of acid added from the buret is exactly equal to the number of equivalents of base present in the flask. In proton-exchange reactions, an equivalent is the amount of an acid that can supply one mole of protons or the amount of a base that can accept one mole of protons. Hence the number of equivalents (or moles) of acid can be calculated from the volume added and its concentration. This in turn allows calculation of the weight of base present in the flask. The following examples illustrate the calculations encountered in acid-base titrations.

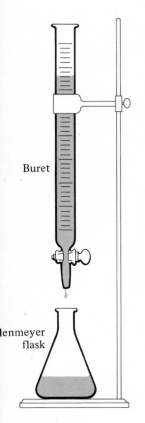

Buret

enmeyer
flask

FIGURE 15-2

Titration

EXAMPLE: A solution of NaOH is standardized by using it to titrate a sample of pure dry potassium hydrogen phthalate. When 0.3200 g of the potassium hydrogen phthalate is used, 32.78 ml of the NaOH solution is required. What are the molarity and the normality of the NaOH solution?

Solution: Potassium hydrogen phthalate is a monoprotic acid and the structure of the anion is

The reaction with OH^- proceeds according to equation (13).

(13)

The number of moles of the acid is obtained by dividing the weight of acid used by its molecular weight (204.23)

$$\text{Moles acid} = \frac{0.3200 \text{ g}}{204.23 \text{ g/mole}} = 1.567 \times 10^{-3}$$

This must also be the number of moles of base added (moles base = 1.567×10^{-3}), and the molarity of the base is simply

$$M = \frac{1.567 \times 10^{-3} \text{ mole}}{3.278 \times 10^{-2} \text{ liter}} = 0.04780$$

Since one mole of OH^- can accept no more than one mole of protons, the number of moles is equal to the number of equivalents, the formula weight is equal to the equivalent weight, and the molarity is equal to its normality (recall that normality is the number of equivalents per liter of solution).

$$N = 0.04780$$

EXAMPLE: What is the percentage of acetic acid in vinegar if a 10.0-g sample requires 35.00 ml of 0.09983M NaOH for titration?

Solution: Acetic acid reacts with NaOH according to the equation

$$HC_2H_3O_2 + OH^- \rightleftharpoons C_2H_3O_2^- + H_2O$$

The number of moles of OH^- used is

$$0.09983 \text{ mole/liter} \times 0.03500 \text{ liter} = 3.494 \times 10^{-3} \text{ mole}$$

which must be the number of moles of acetic acid consumed in the reaction and therefore the number of moles in the sample titrated. The molecular weight of acetic acid is 60.03; the weight of acetic acid in the vinegar sample is therefore

$$60.03 \text{ g/mole} \times 3.494 \times 10^{-3} \text{ mole} = 0.2097 \text{ g}$$

The percentage is

$$\frac{0.2097}{10.0} \times 100 = 2.10\%$$

(Note that the answer can be expressed to only three significant figures.)

EXAMPLE: A 0.4157-g sample containing sodium carbonate and inert impurities requires 37.46 ml of 0.1023M HCl for titration. Calculate the percentage of sodium carbonate in the sample.

Solution: Carbonate ion reacts with hydronium ion according to the equation $CO_3^{2-} + 2H_3O^+ \rightleftharpoons H_2CO_3 + 2H_2O$, from which it is evident that 2 moles of HCl is required for the complete conversion of 1 mole of carbonate ion to carbonic acid. The number of moles of HCl used is

$$0.03746 \text{ liter} \times 0.1023 \text{ mole/liter} = 3.832 \times 10^{-3} \text{ mole}$$

Since 2 moles of HCl is required to react with 1 mole of CO_3^{2-}, the number of moles of sodium carbonate present is half this amount, or

$$\frac{3.832 \times 10^{-3}}{2} \text{ mole} = 1.916 \times 10^{-3} \text{ mole}$$

The formula weight of Na_2CO_3 is 106.0, and the weight of Na_2CO_3 in the sample is therefore

$$1.916 \times 10^{-3} \text{ mole} \times 106.0 \text{ g/mole} = 0.2031 \text{ g}$$

The percentage of Na_2CO_3 in the sample is

$$\frac{0.2031 \text{ g}}{0.4157 \text{ g}} \times 100 = 48.86\%$$

A crucial facet of the acid-base titration is the determination of when the equivalence point is reached. In order to understand more fully the problems involved in this determination, let us now follow the hydronium ion concentration during the titration of 50.0 ml of 0.100M HCl with 0.100M NaOH. At the beginning of the titration, when only 0.100M HCl is present in the flask, the hydronium ion concentration is 0.100M and the pH is 1. Now hydroxide is added to the flask from the buret, and the reaction

$$OH^- + H_3O^+ \rightleftharpoons H_2O + H_2O$$

occurs; this reaction has a very high extent. When 25.0 ml of the hydroxide solution has been added, one-half of the 5.00×10^{-3} mole of HCl originally present has been converted to water, sodium ions, and chloride ions. At this point there is 2.50×10^{-3} mole of hydronium ions in a total of 75.0 ml of solution (this is assuming the volumes—50.0 ml + 25.0 ml—are additive). Thus, halfway through the titration the hydronium ion concentration is

$$\frac{2.50 \times 10^{-3} \text{ mole}}{7.50 \times 10^{-2} \text{ liter}} = 0.0333M$$

and the pH is 1.48.

We continue to titrate and find that when 49.0 ml of hydroxide has been added, only 1.0×10^{-4} mole of hydronium ion remains; the other 4.9×10^{-3} mole has been converted to water by reaction with OH^-. This 1.0×10^{-4} mole is contained in $49 + 50 = 99$ ml of solution, and the H_3O^+ concentration is then

$$\frac{1.0 \times 10^{-4} \text{ mole}}{9.9 \times 10^{-2} \text{ liter}} = 1.0 \times 10^{-3}M$$

The pH of the solution one milliliter away from the equivalence point is therefore 3.0.

At the equivalence point, 50.0 ml of the hydroxide solution has been added to the acid solution, and all of the hydronium ions resulting from HCl have therefore been converted to water. The solution at this point is simply an aqueous solution of sodium chloride.

$$Na^+ + OH^- + H_3O^+ + Cl^- \rightleftharpoons 2H_2O + Na^+ + Cl^-$$

Since neither sodium ions nor chloride ions react with water, the solution is neither acidic nor basic; that is, $[H_3O^+] = [OH^-] = 10^{-7}M$. The pH at the equivalence point is therefore 7.0.

If the titration is continued past the equivalence point, hydroxide ions are simply added to the solution of NaCl, and at one milliliter past the equivalence point, where a total of 51.0 ml of hydroxide solution has been added, the hydroxide ion concentration is 1.0×10^{-3} liter $\times 1.0 \times 10^{-1}$ mole/liter $= 1.0 \times 10^{-4}$ mole in 101 ml of solution, or

$$\frac{1.0 \times 10^{-4} \text{ mole}}{0.101 \text{ liter}} = 1.0 \times 10^{-3}M$$

The pH at this point is therefore 11.0.

Figure 15-3 shows a plot of these five points and others. The titration of any strong acid with any strong base, both at $0.1M$ concentration, will produce exactly

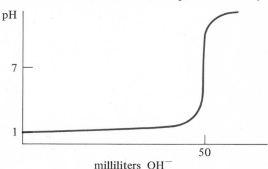

FIGURE 15-3

Curve for the Titration of 50.0 ml
of 0.1M HCl with 0.1M NaOH

the same curve. The one feature of the curve most relevant to our discussion of the determination of the equivalence point is the very fast rate at which the pH changes around this point and the large pH range over which this change occurs.

Theoretically it would be desirable to find some chemical or physical process that would respond to a pH of 7, the exact pH at the equivalence point. However, the very abrupt change of pH near the equivalence point allows use of a process that responds over a wider, more practical range of pH (say 5 to 9) to detect the equivalence point. Chemically, this can be accomplished with a substance that changes color as the pH changes through the critical region.

Such substances, called indicators, are themselves acids or bases. For example, the indicator called phenol red (see structural diagram) can behave as an acid by releasing the proton of the OH group. Thus, when sufficient hydroxide ions are present in the solution, the indicator reacts to form water and its conjugate base [equation (14)]. Since the acid is yellow and the base red, the color of the indicator changes when the acid is converted to its conjugate.

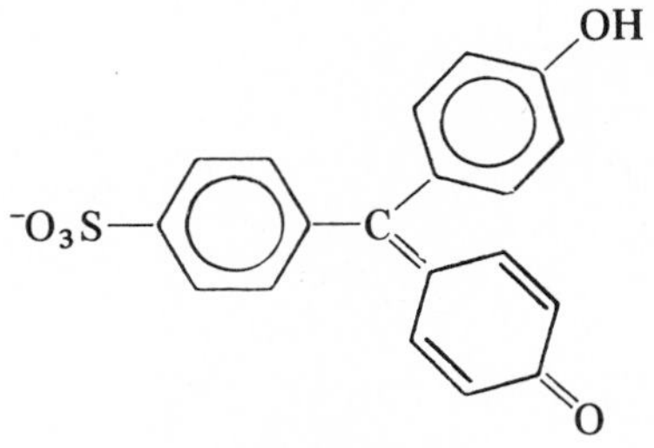

phenol red

(14)

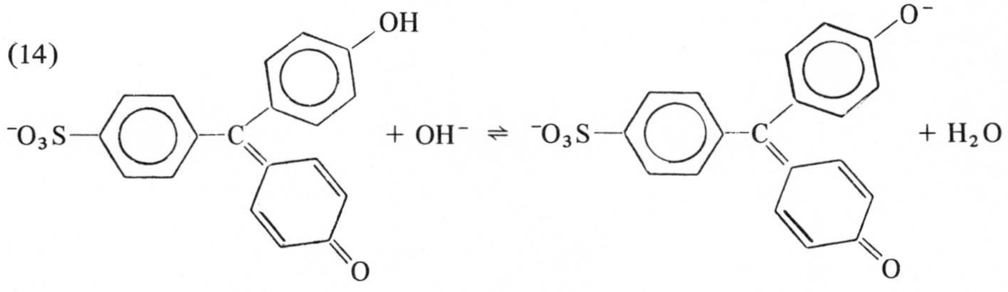

yellow   red

The equilibrium expression for the reaction of the indicator, a weak acid, with water is given by

$$HIn + H_2O \rightleftharpoons H_3O^+ + In^-$$

$$K_a = \frac{[In^-][H_3O^+]}{[HIn]}$$

If this is rearranged to

$$[H_3O^+] = K_a \frac{[HIn]}{[In^-]}$$

the relationship between the color of the solution and its pH becomes more obvious. Thus, when the conversion of the weak acid to its conjugate has occurred to an extent of 50 percent and there are therefore equal amounts of HIn and In⁻, the hydronium ion concentration will be equal to the indicator constant, K_a, and the color of the indicator will be the result of an equal mixture of red and yellow, or orange. When the indicator reaction has occurred to an extent of 9 percent, the ratio of [HIn] to [In⁻] will be approximately 10:1, the hydronium ion concentration will be $10K_a$, and the color of the solution will be yellow with a tinge of red. If the extent goes below 9 percent, the human eye is unable to distinguish further gradations in color; the color at 9 percent and that at, say, 1 percent will both appear to be yellow. The eye is also unable to distinguish between the color produced by a 90 percent extent and that produced by, say, a 99 percent extent. When the indicator reaction has proceeded to 90 percent, the ratio of [HIn] to [In⁻] is approximately 1:10, the hydronium ion concentration is $0.1K_a$ and the color of the solution is red with a tinge of yellow.

Hence, the visible change in indicator color will occur when the hydronium ion concentration is between $10K_a$ and $0.1K_a$ or, in negative logarithmic units, when

$$pH = pK_a \pm 1$$

$$= pK_a$$

Methyl orange, whose structure is illustrated here, is an example of a basic indicator. In acidic solution it is converted from its yellow basic form to its red

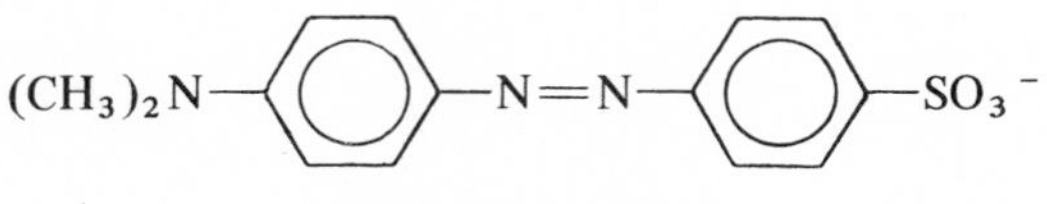

methyl orange

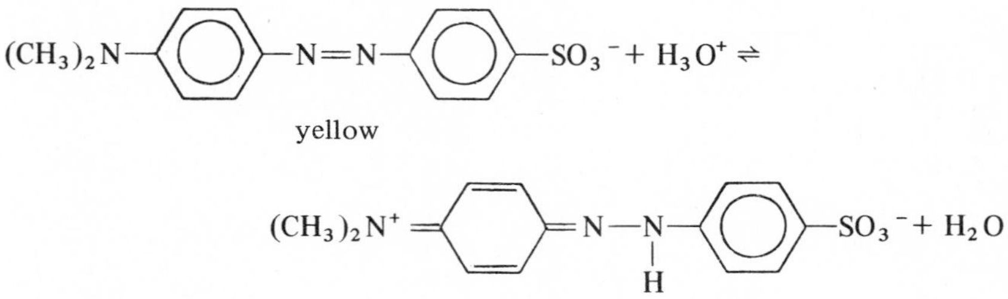

yellow

red

conjugate acid. The expression relating the OH⁻ concentration in the region of color change to the K_b of the *basic* indicator is

$$pOH = pK_b \pm 1$$

Table 15-4 lists some common indicators and the pH range in which the color change occurs. These pH ranges have been experimentally determined and in some cases do not extend over the theoretical range of two pH units.

TABLE 15-4 Some Common Indicators		Acid color	Basic color	pH interval
	methyl orange	red	yellow-orange	3.1–4.4
	methyl red	red	yellow	4.2–6.2
	chlorophenol red	yellow	red	4.8–6.4
	phenol red	yellow	red	6.4–8.0
	cresol purple	yellow	purple	7.4–9.0
	phenolphthalein	colorless	red-violet	8.0–9.8
	alizarine yellow	yellow	violet	10.1–12.0

We can now choose an indicator for our titration of HCl. Since the pH at the equivalence point is 7, an indicator changing in the range of 6 to 8 would be most satisfactory, and certainly phenol red would be an appropriate choice. Since the change of pH is so abrupt near the equivalence point, indicators such as cresol purple, which changes in the pH region of 7.4 to 9, or chlorophenol red, which changes from 4.8 to 6.4, would also be appropriate. Clearly an indicator such as methyl orange, which changes over the range of 3.1 to 4.4, would produce too large an error in the determination of the equivalence point. That is, with methyl orange as indicator the *endpoint* would occur appreciably before the equivalence point was reached and an error would result.

The choice of an indicator becomes more critical in the titration of a weak acid with a strong base or vice versa. To illustrate this point, let us now derive the titration curve for the titration of 50.0 ml of a 0.100M solution of the weak acid acetic acid with the 0.100M solution of sodium hydroxide. At the outset of the titration, when no base has been added, the hydronium ion concentration in the flask is simply that which results from the dissociation of a 0.10M acetic acid solution:

$$HC_2H_3O_2 + H_2O \rightleftharpoons C_2H_3O_2^- + H_3O^+$$

$$x = [H_3O^+] = [C_2H_3O_2^-]$$

$$K_a = 1.8 \times 10^{-5} = \frac{x^2}{0.10 - x}$$

$$x = 1.3 \times 10^{-3}$$

$$pH = 2.89$$

As the hydroxide is added to the flask it reacts with the acetic acid to form acetate ions, and when the titration is halfway completed (when 25 ml of 0.10M NaOH has been added), half of the acetic acid has been converted to acetate ions. This solution is, of course, a buffer system, and since the ratio of the concentrations of weak acid to its conjugate base is 1, the hydronium ion concentration as given by equation (12) is equal to K_a. The pH at this point is then $-\log [1.8 \times 10^{-5}] = 4.74$.

At the equivalence point, the acetic acid has been converted to acetate ions and the only source of acetic acid is the hydrolysis of acetate ions. Hence, in 100 ml of solution there is 0.050 liter $\times$ 0.100 mole/liter = 5×10^{-3} mole of acetate ions and an equal amount of sodium ions. Therefore, the hydronium ion concentration is best calculated by considering the hydrolysis of a $0.0500M$ solution [(5×10^{-3} mole)/0.100 liter = $0.0500M$] of acetate ions.

$$C_2H_3O_2^- + H_2O \rightleftharpoons HC_2H_3O_2 + OH^-$$

$$K_h = \frac{K_w}{K_a} = \frac{10^{-14}}{1.8 \times 10^{-5}} = 5.6 \times 10^{-10}$$

$$5.6 \times 10^{-10} = \frac{[HC_2H_3O_2][OH^-]}{[C_2H_3O_2^-]}$$

$$x = [OH^-] = [HC_2H_3O_2]$$

$$[C_2H_3O_2^-] = 5.00 \times 10^{-2} - x$$

$$5.6 \times 10^{-10} = \frac{x^2}{0.0500 - x}$$

$$x = 5.3 \times 10^{-6} = [OH^-]$$

$$pOH = -\log(5.3 \times 10^{-6}) = 5.28$$

$$pH = 14 - 5.28 = 8.72$$

Thus the pH at the equivalence point is 8.6. Figure 15-4 shows the complete titration curve. Comparison of this curve with that for the titration of a *strong* acid with a strong base reveals that for the weak acid the pH is considerably higher when the region of rapid change in pH occurs around the equivalence point. Moreover, the pH at the equivalence point is higher, and therefore an indicator such as phenolphthalein, which undergoes its color change in the region 8 to 10, would be appropriate. An indicator such as chlorophenol red, which was perfectly adequate

FIGURE 15-4

Curve for the Titration
of 50.0 ml of 0.1M Acetic Acid
with 0.1M NaOH

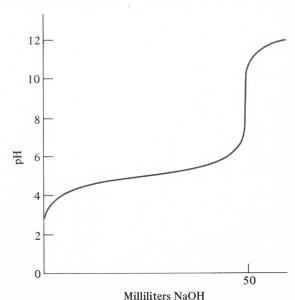

Milliliters NaOH

for the strong acid–strong base titration, is certainly not appropriate for this weak acid–strong base titration; its color change occurs in a pH region that is attained before the equivalence point is reached.

SUGGESTED READINGS

Vanderwerf, C. A. *Acids, Bases, and the Chemistry of the Covalent Bond.* New York: Reinhold, 1961.

Walden, P. *Salts, Acids, and Bases: Electrolytes: Stereochemistry.* New York: McGraw-Hill, 1929. Chapters I and II.

Pauling, L. "Why Is Hydrofluoric Acid a Weak Acid?" *Journal of Chemical Education,* Vol. 33, No. 1 (January 1956), pp. 16–17.

Szabadvary, F., and R. E. Oesper, "Development of the pH Concept." *Journal of Chemical Education,* Vol. 41, No. 2 (February 1964), pp. 105–107.

PROBLEMS

1. For each of the following pairs of acids, predict which is the stronger:

(a) NH_3, PH_3

(b) $H_2C_2O_4$, $HC_2O_4^-$

(c) SiH_4, PH_3

(d) NH_3, H_2O

(e) HCl, HBr

(f)

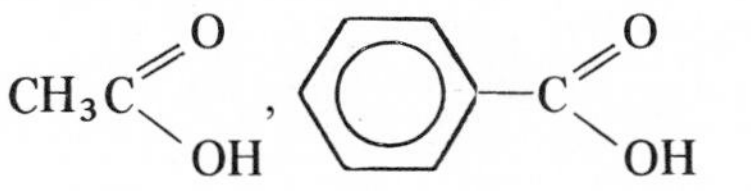

(g)

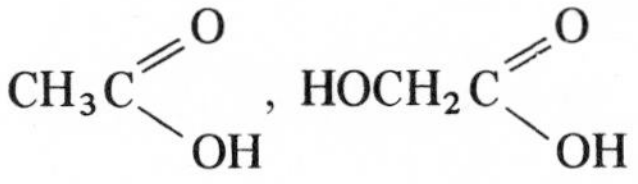

(h)

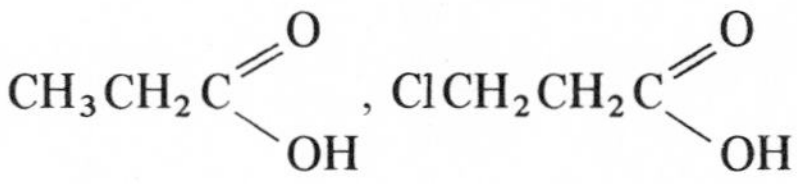

(i)

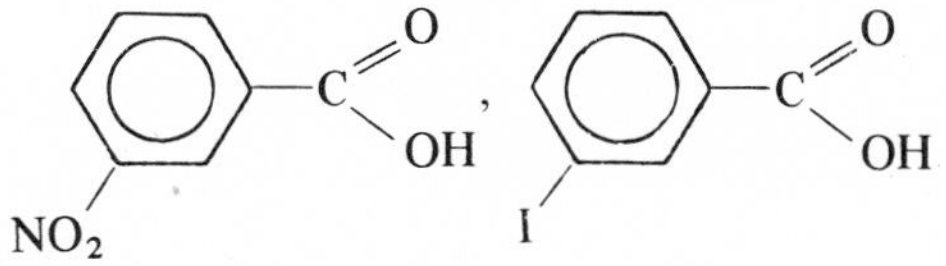

(j)

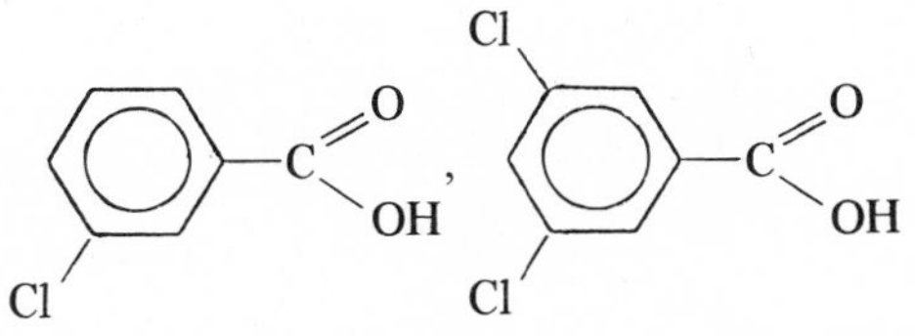

(k)

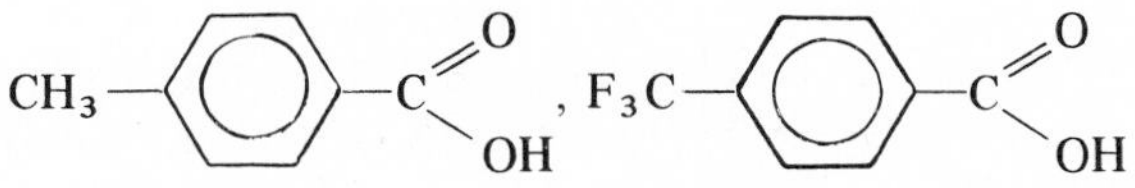

(l) C_2H_5OH, C_2H_5SH
(m) $P(OH)_3$, $As(OH)_3$
(n) $ClOH$, $BrOH$
(o) H_2CO_3, H_2SO_3
(p) IOH, O_2IOH
(q) $NH_4{}^+$, NH_3
(r) H_2S, HS^-

2. For each of the following pairs of bases, predict which is the stronger:
   (a) $B(OH)_3$, $Al(OH)_3$
   (b) BeO, BaO
   (c) $CH_3{}^-$, $NH_2{}^-$
   (d) NH_3, $NH_2{}^-$
   (e) F^-, Cl^-
   (f) S^{2-}, O^{2-}
   (g) $PO_4{}^{3-}$, $HPO_4{}^{2-}$
   (h) OH^-, O^{2-}
   (i) $HONH_2$, $ClNH_2$

   (j) 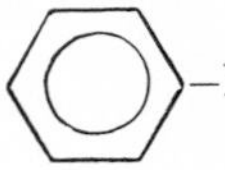 $-NH_2$, CH_3NH_2

   (k) $NaNH_2$, $ClNH_2$

   (l) 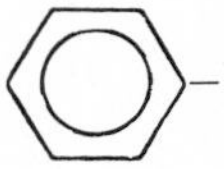 $-NH_2$, $Cl-$ $-NH_2$

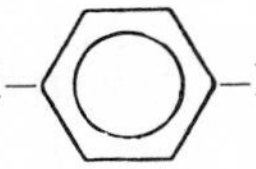

   (m) 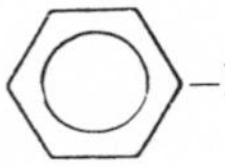 $-NH_2$, 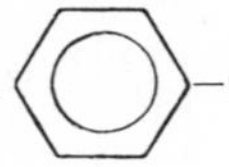 $-OH$ (n) PH_3, AsH_3

   (o) $(C_6H_5)_2NH$, $C_6H_5\overset{\text{H}}{\underset{|}{N}}CH_3$ (p) $HCO_3{}^-$, $HSO_3{}^-$
   (q) C_2H_5OH, C_2H_5SH

3. Calculate the standard enthalpy change for the reaction

   $$HBr(aq) \rightleftharpoons H^+(aq) + Br^-(aq)$$

   and compare the value obtained with that for HCl (Table 15-3). In addition to data found in the appendixes, the following will be necessary.

   $$HBr(aq) \rightarrow HBr(g) \qquad \Delta H^0 = 5.0 \text{ kcal/mole}$$

   $$Br^-(g) \rightarrow Br^-(aq) \qquad \Delta H^0 = -83 \text{ kcal/mole}$$

4. Determine from Appendix 9 whether the methyl group exerts an electron-withdrawing or electron-releasing effect relative to hydrogen.

5. Account for the relative acidities of phenol and methanol. (*Hint:* Refer back to Chapter 7 for the electronic structure of phenol.)

6. For each of the following reactions, predict whether the extent is greater than or less than 50 percent.
   (a) $CH_3COOH + H_2O \rightleftharpoons CH_3COO^- + H_3O^+$
   (b) $HCl + H_2O \rightleftharpoons H_3O^+ + Cl^-$
   (c) $HCOOH + NO_2{}^- \rightleftharpoons HNO_2 + HCO_2{}^-$
   (d) $HNO_2 + HS^- \rightleftharpoons H_2S + NO_2{}^-$
   (e) $H_2CO_3 + CO_3{}^{2-} \rightleftharpoons HCO_3{}^- + HCO_3{}^-$
   (f) $NH_4{}^+ + CH_3COO^- \rightleftharpoons CH_3COOH + NH_3$
   (g) $C_6H_5COO^- + CH_3COOH \rightleftharpoons C_6H_5COOH + CH_3COO^-$
   (h) $C_6H_5OH + OCl^- \rightleftharpoons HOCl + C_6H_5O^-$

7. On page 390, K for H_2O is given as 10^{-16} (a more precise value is 2×10^{-16}), whereas on page 396, K_w for H_2O is given as 10^{-14}. Explain.

8. Calculate the pH of solutions with the following hydronium ion concentrations:
   (a) $1.5 \times 10^{-4}M$ (b) $5.0 \times 10^{-8}M$
   (c) $8 \times 10^{-13} M$

9. Calculate the pH of solutions with the following hydroxide ion concentrations:
   (a) $1.0M$ (b) $6.7 \times 10^{-6}M$
   (c) $4 \times 10^{-10} M$

10. Calculate the hydronium ion concentrations for solutions with the following values for pH:
    (a) 1.53   (b) 3.62
    (c) 11.73

11. A $0.10M$ solution of propanoic acid has a pH of 3.0. Calculate K_a for this acid.

12. When 5.1 g of trimethylacetic acid is dissolved in enough water to make 1 liter of solution, the pH of the solution is 3.2. Calculate the K_a for trimethylacetic acid.

13. A $1.0M$ solution of N-methylaniline has a pH of 9.35. Calculate K_b for this base.

14. Calculate the concentrations of all species in each of the following solutions:
    (a) $0.10M$ formic acid
    (b) $1.0M$ dichloroacetic acid
    (c) $0.010M$ triethylamine
    (d) $0.15M$ aniline ($C_6H_5NH_2$)

15. Calculate the pH of each of the following solutions:
    (a) $0.01M$ HNO_2 (b) $0.10M$ H_2S
    (c) $0.010M$ H_2NOH

16. As a quality control chemist for a pickle processor, you find that the vinegar from a certain supplier is 4 wt-% acetic acid and has a density of 1.004 g/ml. If the process requires vinegar with a pH between 2.4 and 3.4, is this vinegar acceptable?

17. Explain the following observations:
    (a) In general, $K_{a1} > K_{a2} > K_{a3} \cdots$.
    (b) Barium carbonate will dissolve in $1M$ HCl, but barium sulfate will not.
    (c) Copper(II) carbonate will dissolve in $1M$ HCl, but copper(II) sulfide will not.
    (d) When an open bottle of HCl is brought close to an open bottle of aqueous NH_3, a "white smoke" forms.

18. A certain solution is $1 \times 10^{-1}M$ in each of the following ions: Fe^{2+}, Cu^{2+}, Hg^{2+}, Co^{2+}. When the solution is saturated with H_2S, which sulfides will precipitate?

19. Calculate the sulfide ion concentration in a saturated solution of H_2S that is 5.0×10^{-2} molar in HCl. How much copper(II) sulfide would dissolve in 1 liter of this solution?

20. A solution is $1.0 \times 10^{-2}M$ in Cd(II) and $5.0 \times 10^{-2}M$ in Zn(II). If this solution is saturated with H_2S, what hydronium ion concentration is required to allow the precipitation of ZnS to just begin? What concentration of Cd(II) remains in solution at this pH?

21. Calculate the concentrations of all species in each of the following solutions:
    (a) $1.0M$ NH_4Cl (b) $0.01M$ $NaNO_2$
    (c) $0.10M$ KF

22. Determine whether $0.1M$ solutions of each of the following salts will be acidic, basic, or neutral:
    (a) NH_4Cl (b) $(NH_4)_2C_2O_4$
    (c) $KHCO_3$ (d) Rb_3PO_4
    (e) $Ba(NO_3)_2$ (f) K_2SO_3
    (g) NaH_2PO_4 (h) NH_4CN
    (i) $Al(NO_3)_3$ (j) $NaHSO_3$
    (k) $(C_6H_5NH_3)CN$ (l) $Fe_2(SO_4)_3$
    (m) KBr

23. Describe how you would prepare an acetic acid–acetate buffer solution designed to maintain a pH of 5.0.

24. How many grams of NH_4Cl must be added to 100 ml of $0.25M$ NH_3 in order to produce a pH of 8.2?

25. (a) What is the pH of 1 liter of a solution containing 0.10 mole of chloroacetic acid and 0.20 mole of sodium chloroacetate?
    (b) What is the pH of the solution in (a) after the addition of 1.0×10^{-2} mole of HCl? After the addition of 1.0×10^{-2} mole of NaOH?

26. Antacids used to neutralize excess acidity in the stomach generally contain either the water-insoluble hydroxides $Al(OH)_3$ and $Mg(OH)_2$ or sodium hydrogen carbonate. Discuss the relative merits of the two types.

27. Calculate the pH of the solutions resulting from each of the following reactions:
    (a) 50 ml of $0.2M$ HCl + 50 ml $0.2M$ KOH
    (b) 30 ml of $0.1M$ HCl + 30 ml $0.1M$ NH_3
    (c) 25 ml of $0.2M$ HCN + 25 ml $0.2M$ NaOH

28. Calculate the solubility of $Mg(OH)_2$ in an aqueous solution buffered at pH = 12.0.

29. A solution of 3.15 g of an acid dissolved in 100 g of benzene boils at $80.985°C$ (the boiling point of pure benzene is $80.099°C$). A 0.2145-g sample of the same acid requires 45.22 ml of a $0.1054M$ sodium hydroxide solution for complete neutralization. Complete combustion of 1.004 g of the acid produced 0.9817 g CO_2 and 0.2007 g H_2O.
    (a) Calculate the *neutralization equivalent* (the equivalent weight of an acid) of the acid.
    (b) Calculate the molecular weight and molecular formula of the acid.
    (c) Write an electron dot formula for the acid.

30. Complete combustion of 0.3045 g of a monoprotic base produced 0.5955 g of CO_2 and 0.4263 g of H_2O. A 0.1562-g sample, when analyzed for

nitrogen, yielded 43.00 ml of nitrogen gas collected at 25°C and 750 torr. A 0.1891-g sample of the base required 38.06 ml of a 0.1104M HCl solution for complete neutralization. Determine the molecular formula of the base.

31. When 1.00×10^{-2} mole of *p*-chlorobenzoic acid is dissolved in enough water to make 1 liter of solution, the pH of the solution is 3.02.
    (a) Calculate the K_a of *p*-chlorobenzoic acid.
    (b) Calculate the pH of a 1M solution of the acid.
    (c) Calculate the pH of a 0.10M solution of sodium *p*-chlorobenzoate.
    (d) Calculate the concentration of all species in 1 liter of a solution that contains 1.0 mole of *p*-chlorobenzoic acid and 0.10 mole of sodium *p*-chlorobenzoate.

    A 50.0-ml portion of a 0.100M solution of *p*-chlorobenzoic acid is titrated with a 0.100M NaOH solution.

    (e) List all species present at the equivalence point and calculate their concentration.
    (f) Determine the pH at the equivalence point.
    (g) Use Table 15-4 to select a suitable indicator for this titration.
    (h) Assume that methyl orange is used as the indicator for this titration. How many milliliters of NaOH solution will have been added when the endpoint has been reached? (Assume the endpoint to occur in the middle of the indicator range.) What percentage error would this produce?

32. Liquid ammonia is a fairly common nonaqueous medium for acid-base reactions. Since ammonia is similar to water in many of its properties, reactions in liquid ammonia are analogous to their counterparts in water (there are, of course, some important differences). Fundamental to an understanding of reactions in ammonia is its self-ionization, which is strictly analogous to the self-ionization of water.
    (a) Write an equation and equilibrium expression for the self-ionization of ammonia.
    (b) The equilibrium constant for the self-ionization of liquid ammonia is 2×10^{-33} at −50°C. Rationalize the difference between this value and the value for water (ignore the temperature difference) in terms of the strength of H_2O as an acid and as a base relative to the strength of NH_3 as an acid and as a base. (*Hint:* Determine and compare the values of K_a and K_b for H_2O and K_a and K_b for NH_3 in water.)
    (c) What are the ammonia analogs of H_3O^+ and OH^-? Define the ammonia analogs of pH [pH(NH_3)] and pOH, and calculate the pH(NH_3) of liquid ammonia.
    (d) What is the pH(NH_3) of an acidic liquid ammonia solution?
    (e) Calculate the pH(NH_3) of a $10^{-3}M$ NaNH$_2$ liquid ammonia solution and a $10^{-3}M$ NH$_4$Cl liquid ammonia solution.
    (f) Write an equation for a neutralization reaction carried out in liquid ammonia.

    Ammonia is a stronger base than water and therefore can be used to differentiate the acidities of very weak acids. Acetamide, $CH_3\overset{\displaystyle O}{\overset{\|}{C}}NH_2$, is too weak an acid to affect the pH of a water solution but does lose protons in liquid ammonia.

(g) Write an equation and equilibrium expression for the reaction of acetamide with liquid ammonia. Assume the equilibrium constant for this reaction is 10^{-5} and calculate the $pH(NH_3)$ of a $0.1M$ solution of acetamide in liquid ammonia.

(h) Calculate the $pH(NH_3)$ of a $0.1M$ solution of $Na^+CH_3\overset{\displaystyle O}{\overset{\|}{C}}NH^-$ in liquid ammonia.

16 *Types of Chemical Reactions III: Electron-Sharing Reactions*

In the same year in which Lowry and Brønsted delineated the proton-transfer reaction, Gilbert Newton Lewis, one of the great American chemists, provided a more general definition of acids and bases. Lewis's definition, which allows the systematization and understanding of a wide variety of chemical phenomena, is simply:

An *acid* is a substance that can accept a pair of electrons.
A *base* is a substance that can provide a pair of electrons.

A Lewis acid-base reaction, then, occurs when one substance, the base, provides a pair of electrons to share with another substance, the acid. The product of the reaction—the species in which the pair of electrons is shared—is called the acid-base *adduct* or *complex*. Probably the simplest example of such a reaction is

$$(1) \qquad H^+ + :H^- \rightleftharpoons H-H$$

where the hydrogen ion, H^+, is a Lewis acid; the hydride ion, H^-, is a Lewis base; and the product, H_2, is the Lewis adduct. When this reaction is carried out in water or a similar polar solvent, the hydrogen ion is bonded to the solvent, and the reaction can then also be classed as a proton transfer.

$$(2) \qquad H_3O^+ + :H^- \rightleftharpoons H_2 + H_2O$$

The water merely provides the vehicle necessary for the transfer of the proton.

431

Hence, all proton-transfer reactions can also be classified as electron-sharing reactions.

Actually, equations (1) and (2) are examples of two different kinds of electron-sharing reactions. The first, of the type

(3) $A \; + \; B: \; \rightleftharpoons \; A:B$

                       acid   base   adduct

involves the addition of the base B: to the acid A and is therefore called an *addition reaction*. The second, which is of the type

(4) $A:B \; + \; D: \; \rightleftharpoons \; A:D \; + \; B:$

                  adduct   base   adduct   base

is the displacement of the base B: from the adduct A:B by another base D: to produce a new adduct A:D. Such *displacement reactions* can also occur by displacement of one acid by another.

$$A:B \; + \; C \; \rightleftharpoons \; C:B \; + \; A$$

                adduct   acid   adduct   acid

Clearly, in the *base-displacement* reaction, two bases compete to provide the electron pair for the acid; in the *acid*-displacement reaction, two acids compete for the electron pair provided by the base.

ADDITION REACTIONS

The various types of addition reactions can be conveniently categorized in terms of the types of species that function as the acid. Both the Lewis and Lowry-Brønsted definitions require that the base provide a pair of electrons, and there is therefore no difference between the two types of bases. Some common bases are listed in Table 15-2 (Chapter 15, p. 389).

While it is true that the proton is the simplest Lewis acid, the characteristics of Lewis acids are quite different from those of Lowry-Brønsted acids. A Lowry-Brønsted acid must contain a transferrable proton; a Lewis acid must be able to accept a pair of electrons. The most obvious way in which an acid can accept a pair of electrons is to have an empty orbital that can hold the electron pair.

Thus, all cations are potential Lewis acids. For example, the beryllium ion has four empty orbitals in its valence shell—the $2s$ and the three $2p$ orbitals. When this ion is added to water, the following reactions take place.

$$Be^{2+} + H_2O \rightleftharpoons Be(H_2O)^{2+}$$

$$Be(H_2O)^{2+} + H_2O \rightleftharpoons Be(H_2O)_2{}^{2+}$$

$$Be(H_2O)_2{}^{2+} + H_2O \rightleftharpoons Be(H_2O)_3{}^{2+}$$

$$Be(H_2O)_3{}^{2+} + H_2O \rightleftharpoons Be(H_2O)_4{}^{2+}$$

Each reaction is an electron-sharing reaction, and together they portray the stepwise formation of the adduct $Be(H_2O)_4{}^{2+}$. Some of each adduct is present in an aqueous solution of beryllium(II).

The behavior of Be^{2+} shows clearly that more than one molecule of base may complex with a given acid. The number of bases bonded to the acid in a given adduct is called the *coordination number*. In $Be(H_2O)_2{}^{2+}$ the coordination number of Be^{2+} is 2, in $Be(H_2O)_4{}^{2+}$ it is 4. The maximum coordination number for a particular acid-base pair depends upon the number of orbitals available on the acid, the size of the base, the strength of the adduct bond, and other factors. The maximum coordination number of 4 for Be^{2+} is due, of course, to the presence of only four empty orbitals in its valence shell.

The structure of the adducts of ions of the representative elements can be described with the principles developed in Chapters 7 and 8. (Structures of adducts of transition metal cations are more complex, and discussion of their bonding and geometry will be deferred until Chapter 22.) The electron dot formula for the adduct $Be(H_2O)_4{}^{2+}$ is

Since there are four sigma-bonded electron pairs and no nonbonded pairs around the central atom, the geometry of the adduct can be predicted as tetrahedral.

Since the water molecules supply both electrons for the Be—O bonds, the oxygen atoms have *formal* positive charges. As a consequence of this and the strength of the Be—O bonds, sufficient electron density is removed from the oxygen-hydrogen bonds to increase their polarity. This increased polarity results in increased acidity of the hydrogens, and in water, reactions such as the following occur:

$$Be(OH_2)_4{}^{2+} + H_2O \rightleftharpoons [Be(OH_2)_3OH]^+ + H_3O^+$$

$$[Be(OH_2)_3OH]^+ + H_2O \rightleftharpoons Be(OH_2)_2(OH)_2 + H_3O^+$$

These proton-transfer reactions produce hydronium ions; hence, an aqueous solution of Be^{2+} is acidic. If the transfer of a second proton from the adduct occurs to a sufficient extent, the neutral hydroxide, $Be(OH_2)_2(OH)_2$, will precipitate out of the solution.

A number of other metal ions also bond strongly to water. The more common of these ions and the formulas of the adducts (insofar as they are known) are given in Table 16-1. Because of the strong metal ion-oxygen bonds in the adducts, these ions also produce acidic aqueous solutions. A survey of the kinds of ions present in Table 16-1 shows that these are in general rather small, highly

TABLE 16-1 Stable Metal Ion–Water Complexes		
$Be(OH_2)_4^{2+}$	$Ni(OH_2)_6^{2+}$
$Al(OH_2)_6^{3+}$	$Mn(OH_2)_6^{2+}$
$Cr(OH_2)_6^{3+}$	$Sn(OH_2)_6^{4+}$
$Co(OH_2)_6^{2+}$	$Pb(OH_2)_4^{2+}$
$Cu(OH_2)_6^{2+}$	$Zn(OH_2)_4^{2+}$
$Fe(OH_2)_6^{3+}$	

charged species. Such ions are more strongly attracted to the polar water molecules than larger, less highly charged ions.

Bases other than water also complex with metal ions. For example, silver ion reacts with ammonia in liquid ammonia or water to form the adduct $Ag(NH_3)_2^+$.

$$Ag^+ + 2NH_3 \rightleftharpoons Ag(NH_3)_2^+$$

Cadmium ion reacts with the same base to form the adduct $Cd(NH_3)_4^{2+}$,

$$Cd^{2+} + 4NH_3 \rightleftharpoons Cd(NH_3)_4^{2+}$$

while cobalt(II) forms the 6-coordinate complex $Co(NH_3)_6^{2+}$:

$$Co^{2+} + 6NH_3 \rightleftharpoons Co(NH_3)_6^{2+}$$

Complex amines called porphyrins are complexed to Fe^{2+}, Mg^{2+}, and Co^{3+} in biologically vital systems such as hemoglobin, chlorophyll, and vitamin B_{12}. The heme group of hemoglobin is shown in Figure 16-1. These important complexes will be discussed in more detail in Chapter 22.

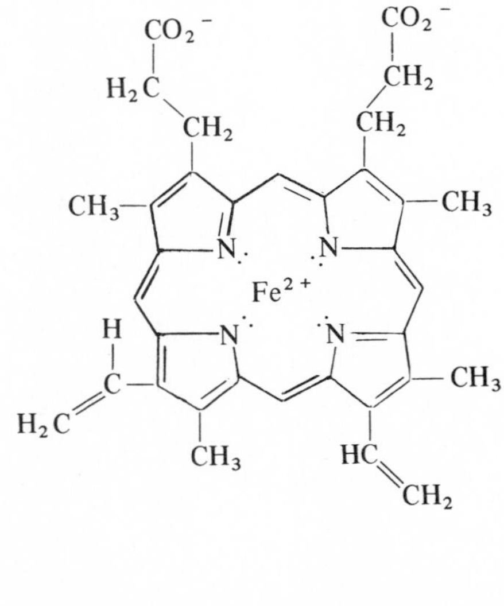

FIGURE 16-1

The Heme Group of Hemoglobin

If the base is an anion, the Lewis acid-base reaction with the cation can also be classified as an ion-combination reaction (see Chapter 14). The addition of sufficient ions to balance the charge of the cation produces a neutral species that may precipitate from aqueous solution. Thus, when a solution of sodium cyanide is added dropwise to an aqueous solution of silver(I), a precipitate of silver cyanide soon forms. If the addition of cyanide is continued until the concentration of cyanide is quite high, the precipitate dissolves because of the formation of the ion $Ag(CN)_2^-$, whose alkali metal salts [for example, $NaAg(CN)_2$] are soluble.

$$Ag^+ + CN^- \rightleftharpoons \underline{AgCN}$$

$$\underline{AgCN} + CN^- \rightarrow Ag(CN)_2^-$$

When the neutral Lewis acid can also undergo a proton-transfer reaction as a *base* it is termed *amphoteric*. (Note that the term *amphiprotic* refers to a species that can behave as both a Lowry-Brønsted acid *and* base, whereas *amphoteric* denotes a species that can function as a Lowry-Brønsted base and a Lewis acid.) Thus, the neutral precipitate AgCN is a Lewis acid because of its further reaction with the base CN^-, but it is also a Lowry-Brønsted base because it can accept a proton to form the weak electrolyte hydrocyanic acid.

$$\underline{AgCN} + H_3O^+ \rightleftharpoons Ag^+(aq) + HCN + H_2O$$

Among the most common amphoteric compounds are the water-insoluble hydroxides listed in Table 16-2. For example, zinc(II) hydroxide dissolves in a sodium hydroxide solution because of the formation of the adduct $Zn(OH)_4^{2-}$

$$\underline{Zn(OH)_2} + 2OH^- \rightleftharpoons Zn(OH)_4^{2-}$$

and, of course, also reacts with acid as a Lowry-Brønsted base:

$$\underline{Zn(OH)_2} + 2H^+ \rightleftharpoons Zn^{2+}(aq) + 2H_2O$$

TABLE 16-2 Some Common Amphoteric Hydroxides	Hydroxide	Reaction as base	Reaction as Lewis acid
	$Zn(OH)_2$	$\underline{Zn(OH)_2} + 2H^+ \rightarrow Zn^{2+}(aq) + 2H_2O$	$\underline{Zn(OH)_2} + 2OH^- \rightarrow Zn(OH)_4^{2-}$
	$Sn(OH)_2$	$\underline{Sn(OH)_2} + 2H^+ \rightarrow Sn^{2+}(aq) + 2H_2O$	$\underline{Sn(OH)_2} + 2OH^- \rightarrow Sn(OH)_4^{2-}$
	$Pb(OH)_2$	$\underline{Pb(OH)_2} + 2H^+ \rightarrow Pb^{2+}(aq) + 2H_2O$	$\underline{Pb(OH)_2} + 2OH^- \rightarrow Pb(OH)_4^{2-}$
	$Sb(OH)_3$	$\underline{Sb(OH)_3} + 3H^+ \rightarrow Sb^{3+}(aq) + 3H_2O$	$\underline{Sb(OH)_3} + OH^- \rightarrow Sb(OH)_4^-$
	$Al(OH)_3$	$\underline{Al(OH)_3} + 3H^+ \rightarrow Al^{3+}(aq) + 3H_2O$	$\underline{Al(OH)_3} + OH^- \rightarrow Al(OH)_4^-$
	$Cr(OH)_3$	$\underline{Cr(OH)_3} + 3H^+ \rightarrow Cr^{3+}(aq) + 3H_2O$	$\underline{Cr(OH)_3} + OH^- \rightarrow Cr(OH)_4^-$
	$Sn(OH)_4$	$\underline{Sn(OH)_4} + 4H^+ \rightarrow Sn^{4+}(aq) + 4H_2O$	$\underline{Sn(OH)_4} + 2OH^- \rightarrow Sn(OH)_6^{2-}$

Some neutral *molecular* species also have empty orbitals, usually on the central atom, and can therefore function as Lewis acids. The electron dot formula for boron trichloride

$$\begin{array}{c} :\ddot{C}l: \\ | \\ B \\ :\underset{\cdot\cdot}{\ddot{C}l} \qquad \ddot{C}l: \end{array}$$

shows only six electrons (three pairs) in the valence shell of boron, and boron can accept two more electrons and thereby fill its octet. Indeed, one of the classical examples of a Lewis acid-base reaction is the formation of an adduct between BCl_3 and NH_3.

$$BCl_3 + :NH_3 \rightleftharpoons Cl_3B:NH_3$$

The lone pair of electrons on the base NH_3 is donated to the acid BCl_3 to form the covalent species $Cl_3B:NH_3$. The electron dot formula of the adduct

$$\begin{array}{ccc} Cl & & H \\ \diagdown \ominus & \oplus \diagup & \\ Cl-B & -N-H \\ \diagup & & \diagdown \\ Cl & & H \end{array}$$

shows four bonds about both boron and nitrogen and suggests sp^3 hybridization and bond angles of approximately 109° at both atoms. It is interesting to note the change in geometry of the acid in this reaction: from planar (sp^2 hybridization at boron) BCl_3 to tetrahedral bonds within the adduct.

Molecules that contain atoms that can expand their octet can also accept a pair of electrons. The central atom of tin tetrachloride has an octet of electrons in its valence shell, but because of the presence of empty d orbitals in the same quantum level it can expand its octet, as in the reaction

$$2Cl^- + SnCl_4 \rightleftharpoons SnCl_6{}^{2-}$$

The hybridization utilized by tin in the adduct is presumably sp^3d^2, and the geometry is octahedral.

Neutral molecules that do not have empty orbitals can also function as Lewis acids if the bonding is such that an electron pair can be accommodated by a rearrangement of the electronic structure of the molecule. The electron dot formula for CO_2, for example, shows a completed octet for each of the atoms, and the molecule therefore has no empty orbitals (although the molecular orbital description of CO_2 indicates the presence of empty antibonding molecular orbitals). Carbon dioxide does react as a Lewis acid, however, with bases such as OH^- or H_2O.

$$H-\ddot{\underset{\cdot\cdot}{O}}:^- + :\ddot{\underset{\cdot\cdot}{O}}=\overset{\delta+}{C}=\ddot{\underset{\cdot\cdot}{O}}: \quad \rightarrow \quad \begin{array}{c} \overset{\cdot\cdot}{\ddot{O}} \\ \diagdown\diagdown \\ C-\ddot{\underset{\cdot\cdot}{O}}:^- \\ | \\ :\ddot{O}: \\ \diagdown \\ H \end{array}$$

The electronegative oxygens remove sufficient electron density from the carbon to give it a partial positive charge that can attract the negative hydroxide ion. As the hydroxide begins to bond to the carbon, one of the carbon-oxygen π bonds breaks to allow formation of the new C—O bond.

The reaction of water with CO_2 can be rationalized in the same way. After the adduct is formed [equation (5)], one of the hydrogens presumably leaves the oxygen, and a hydrogen, either the same one or one from a water molecule or a hydronium ion, attaches itself to the negatively charged oxygen [equation (6)]. The net result is the formation of carbonic acid and is an example of the hydration of an acidic anhydride. The analogous hydration of another acidic anhydride, sulfur trioxide, is shown in equation (7).

(5)

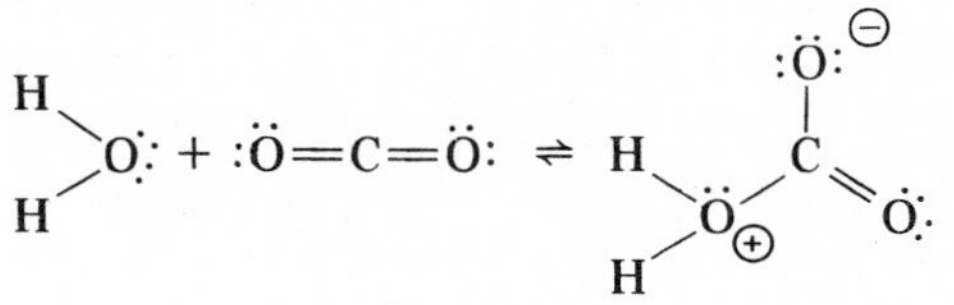

(6)

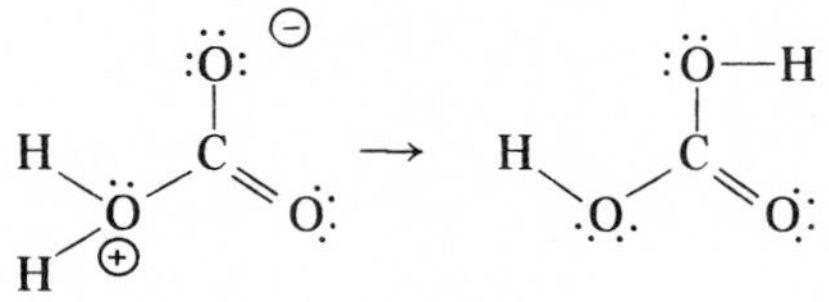

(7)

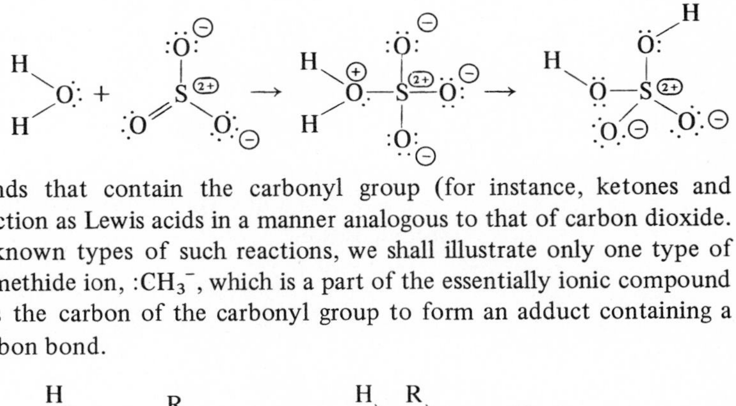

Compounds that contain the carbonyl group (for instance, ketones and aldehydes) function as Lewis acids in a manner analogous to that of carbon dioxide. Of the many known types of such reactions, we shall illustrate only one type of addition. The methide ion, $:CH_3^-$, which is a part of the essentially ionic compound $LiCH_3$, attacks the carbon of the carbonyl group to form an adduct containing a new carbon-carbon bond.

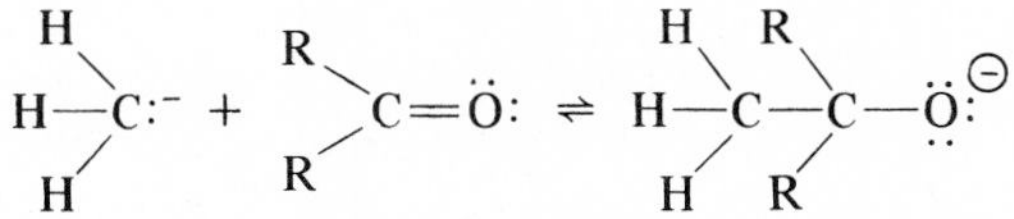

If this adduct is then added to water, it abstracts a proton, and an alcohol results.

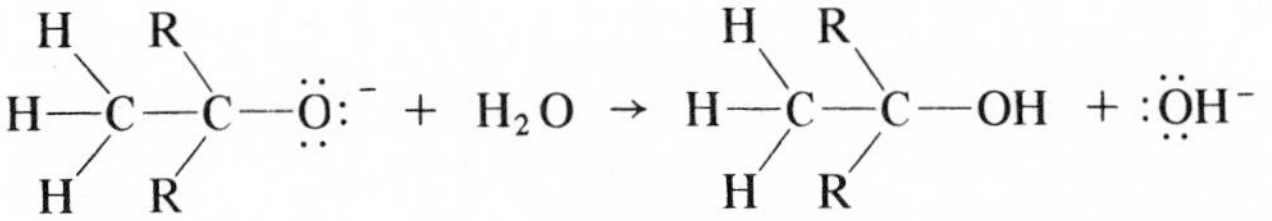

Indeed, this is an important synthetic method for preparing alcohols.

EXTENT OF LEWIS ADDITION REACTIONS

The extents of addition reactions vary considerably. The equilibrium constant for the formation of the AgI_2^- adduct is 10^{22},

$$Ag^+ + 2I^- \rightleftharpoons AgI_2^- \qquad K = 10^{22}$$

whereas the constant for the formation of $PbCl_4^{2-}$ is 10.

$$Pb^{2+} + 4Cl^- \rightleftharpoons PbCl_4^{2-} \qquad K = 10$$

The formation of an adduct containing more than one mole of base per mole of acid is usually visualized as proceeding stepwise. In some cases the intermediate species can be detected and equilibrium constants determined for each. For example, the *overall formation constant* for the complex $Cd(NH_3)_4^{2+}$ is 3×10^7. The sum of the four steps is the equation for the overall formation of the 4-coordinate complex, and the product of the stepwise constants is the overall formation constant.

$$Cd^{2+} + NH_3 \rightleftharpoons Cd(NH_3)^{2+} \qquad K_1 = 5 \times 10^2$$

$$Cd(NH_3)^{2+} + NH_3 \rightleftharpoons Cd(NH_3)_2^{2+} \qquad K_2 = 2 \times 10^2$$

$$Cd(NH_3)_2^{2+} + NH_3 \rightleftharpoons Cd(NH_3)_3^{2+} \qquad K_3 = 30$$

$$\underline{Cd(NH_3)_3^{2+} + NH_3 \rightleftharpoons Cd(NH_3)_4^{2+} \qquad K_4 = 10}$$

$$Cd^{2+} + 4NH_3 \rightleftharpoons Cd(NH_3)_4^{2+} \qquad K = K_1 K_2 K_3 K_4 = 3 \times 10^7$$

Equilibrium constants for complex ions are also tabulated as *instability constants*. For $Cd(NH_3)_4^{2+}$, the instability constant is 3×10^{-8} and refers to the decomposition

$$Cd(NH_3)_4^{2+} \rightleftharpoons Cd^{2+} + 4NH_3 \qquad K_{instab} = 3 \times 10^{-8}$$

Obviously, the instability constant is the reciprocal of the overall formation constant.

Let us now attempt to determine what factors influence the extent of addition reactions. A thorough analysis of the extent of any reaction must include discussion of its entropy change and an examination of the components of its enthalpy change such as bonding in the reactants and products, structural effects on the ground-state energies of the reactants and products, etc. For most reactions, however, such an analysis is difficult, due at least in part to scarcity of the necessary data.

In qualitative terms, addition reactions can be viewed as an attraction between an electron-rich species and an electron-deficient species. This aspect of the reaction can be emphasized by showing the richness of electron density of the base with a partial negative charge, $\delta-$, and the electron deficiency of the acid with a partial positive charge, $\delta+$.

$$B:^{\delta-} + A^{\delta+} \rightleftharpoons B:A$$

Verbally, the same concept can be conveyed by using the term *nucleophile* to refer to the base and the term *electrophile* for the acid. The two terms are derived from the Greek word *philos,* "lover," and mean nucleus (positive charge) lover and electron lover, respectively. As we shall see later, use of these terms is best restricted to discussions of the *rates* of addition and displacement reactions, and we will therefore continue for the present with the more general terms, base and acid.

This view of addition reactions leads to the expectation that for a given acid the extent of reaction should parallel the amount or availability of the electron density of the base, while for a given base the extent should parallel the degree to which the acid is electron deficient. For example, the attraction of a given acid for a negatively charged base, such as OH^-, is certainly greater than its attraction for a neutral species containing the same donor atom, such as H_2O. In other words, not only are the lone pairs of the OH^- more available than those of H_2O, but OH^- also has a *negative charge* and is therefore a stronger base. In the same way, the positively charged Be^{2+} ion is a stronger acid than the neutral species $BeCl_2$.

Let us now determine the generality of this view, or model, of addition reactions. Table 16-3 lists equilibrium constants for the gas-phase interactions of

TABLE 16-3
Equilibrium Constants for the Formation of Adducts with BF_3 and $(CH_3)_3B$

Acid	Base	K
BF_3	$(CH_3)_2O$	6
	$(CH_3)_2S$	0.2
	$(CH_3)_3N$	>15
	$(CH_3)_3P$	15
	$(CH_3)_3As$	2.5
	$(CH_3)_3Sb$	~0
$(CH_3)_3B$	$(CH_3)_3N$	2
	$(CH_3)_3P$	8
	$(CH_3)_3As$	0.6
	$(CH_3)_3Sb$	~0

boron trifluoride and trimethylboron with a variety of bases. The relative extents of the reaction of the bases with a given acid can be assumed to reflect the relative strengths of those bases toward the particular acid. The relative strengths of the bases toward BF_3 are the same as their relative strengths toward the proton. For example, the reaction of $(CH_3)_2O$ with BF_3 has an equilibrium constant greater than that for the reaction of $(CH_3)_2S$ with BF_3, and therefore $(CH_3)_2O$ is a

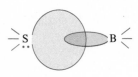

FIGURE 16-2

Rationalization
of the Basicity Order
$(CH_3)_3N > (CH_3)_2O$
$> (CH_3)_2S$ Toward BF_3

stronger base than $(CH_3)_2S$ toward BF_3. The same order of basicity, $O > S$, is observed in proton-transfer reactions, as for example,

$$H^+ + OH^- \rightleftharpoons H_2O \qquad K = 10^{14}$$

$$H^+ + SH^- \rightleftharpoons H_2S \qquad K = 10^7$$

The trends in basicity toward BF_3 can be rationalized crudely on the basis of the model developed here. The greater electronegativity of oxygen relative to nitrogen makes the lone pairs of $(CH_3)_2O$ less available than the lone pair of $(CH_3)_3N{:}$. The same argument applied to $(CH_3)_2O$ relative to $(CH_3)_2S$ should make the sulfur derivative the stronger base. However, the lone pairs of sulfur are in the third quantum level rather than the second and are therefore considerably more diffuse. This diffuseness allows interaction of a smaller percentage of the lone pair with the empty orbital of the electron-deficient boron. These three cases—$(CH_3)_2O$ versus $(CH_3)_3N$ and $(CH_3)_2S$—are shown pictorially in Figure 16-2.

The equilibrium constants for the interaction of the Group V bases with $(CH_3)_3B$ given in Table 16-3 reveal that (a) $(CH_3)_3B$ is a weaker acid toward these bases than BF_3, and (b) $(CH_3)_3P$ is a stronger base toward $(CH_3)_3B$ than $(CH_3)_3N$, while the reverse is true for BF_3. The first observation is easily explained. Since fluorine is more electronegative than carbon; the fluorines in BF_3 remove electron density from the boron, making it more electron deficient, and therefore a better Lewis acid, than the boron in $(CH_3)_3B$. The second observation, that $(CH_3)_3P$ is a stronger base than $(CH_3)_3N$ toward $(CH_3)_3B$, can be rationalized by the argument that bumping of the methyl groups (Figure 16-3) in the amine adduct decreases the stability of the adduct and therefore decreases the extent of the reaction with $(CH_3)_3N$. The phosphorus base has a larger central atom and longer bonds to the methyl groups, which should alleviate this bumping to some extent and thereby make formation of the adduct more favorable.

FIGURE 16-3

Steric Repulsions in $(CH_3)_3BN(CH_3)_3$

Thus, with the possible exception of the interaction of $P(CH_3)_3$ with $B(CH_3)_3$, our model appears to explain the reactivity orders relative to BF_3 and $B(CH_3)_3$, and we can now continue to test its generality with data for some metal ion complexes. Overall formation constants for some 4-coordinate halide complexes

Acid	Base	K for $A^{n+}(aq) + 4X^-(aq) \rightleftharpoons AX_4^{(4-n)-}(aq)$
Fe^{3+}	F^-	10^{15}
	Cl^-	10^{-1}
Hg^{2+}	Cl^-	10^{16}
	Br^-	10^{21}
	I^-	10^{30}
Cd^{2+}	Cl^-	10^3
	Br^-	10^4
	I^-	10^6

TABLE 16-4
Overall Formation Constants for Some 4-Coordinate Halo Complexes

of Fe^{3+}, Hg^{2+}, and Cd^{2+} are listed in Table 16-4. When Fe^{3+} is the Lewis acid, fluoride ion is a stronger base than chloride ion, which is the basicity order expected on the basis of our generalization about availability of the lone pair. This same trend in basicity—in general, $F^- > Cl^- > Br^- > I^-$—is observed when the acid is Cr^{3+}, Al^{3+}, Be^{2+}, or another small and highly charged cation.

The opposite basicity order—$I^- > Br^- > Cl^- > F^-$—is observed for Hg^{2+} and Cd^{2+} (Table 16-4) and also holds for ions such as Ag^+, Cu^+, Pd^{2+}, and Pt^{2+}, which are larger and less highly charged. This order is directly opposite to the one predicted by our model and cannot easily be rationalized. One theory links the greater basicity of I^- to the greater polarizability of its very diffuse lone pairs. This polarizability should result in a stronger covalent bond between the more polarizable metal ions and the heavier halide ions.

Regardless of the reason, it would appear that acids can be classified into two groups: (a) those that bond most strongly to fluoride ion and for which the basicity order $F^- > Cl^- > Br^- > I^-$ holds, and (b) those that bond most strongly to iodide ion and for which the basicity order $I^- > Br^- > Cl^- > F^-$ holds. The acids in the first group also bond more strongly to nitrogen (for example, NH_3) than to phosphorus (for example, PH_3) and more strongly to oxygen [for instance, $(CH_3)_2O$] than to sulfur [for instance, $(CH_3)_2S$]. Those in the second group bond more strongly to the third-period elements, phosphorus and sulfur. While these generalizations have been developed for ions, they also apply to molecular species. Thus, boron trifluoride, which bonds more strongly to $(CH_3)_2O$ and $(CH_3)_3N$ than to $(CH_3)_2S$ and $(CH_3)_3P$, can be classed in the first group, while trimethylboron, since it bonds more strongly to $(CH_3)_3P$ than to $(CH_3)_3N$, falls in the second group (as we shall see later, it is actually a borderline case).

Bases such as I^-, $(CH_3)_2S$, and PH_3, which bond more strongly to the acids in the second group, can be categorized as *soft,* a description intended to portray the polarizable, diffuse character of their electron density. The acids that prefer these bases contain acceptor atoms or ions that are large in size, low in charge, and have unshared electrons in their valence shell. Since such properties result in high polarizability, these acids are also classified as *soft.*

Bases such as F^-, OH^-, $(CH_3)_2O$, and NH_3 contain donor atoms or ions that are highly electronegative and rather unpolarizable. These are *hard* bases, and the acids that prefer them have properties that also allow them to be classified as *hard:* high positive or high partial positive charge and small size. Table 16-5 lists a number of hard and soft acids and bases. Some species have both hard and soft characteristics and are listed as *borderline* cases.

TABLE 16-5
Hard and Soft Acids and Bases

Bases			Acids		
Hard	*Soft*	*Borderline*	*Hard*	*Soft*	*Borderline*
H_2O	R_2S	$C_6H_5NH_2$	H^+	Cu^+Ag^+	$Fe^{2+}, Co^{2+}, Ni^{2+}$
R_2O	PH_3	Br^-	Li^+, Na^+, K^+	Hg^{2+}	$Cu^{2+}, Zn^{2+}, Pb^{2+}$
OH^-	PR_3		$Be^{2+}, Mg^{2+}, Ca^{2+}$	Pt^{2+}	$Sn^{2+}, B(CH_3)_3$
O^{2-}	AsR_3		Sr^{2+}, Mn^{2+}	Cd^{2+}, Pd^{2+}	
NH_3	SbR_3		Al^{3+}, Ga^{3+}	$Ga(CH_3)_3$	
NR_3	I^-		Co^{3+}, Fe^{3+}	$Tl(CH_3)_3$	
F^-	H^-		$BF_3, AlCl_3$		
Cl^-					

With the *hard* and *soft* categories at hand, a great deal of experimental evidence can be summarized by the statement, "Hard acids prefer to bind to hard bases, and soft acids prefer to bind to soft bases." This important generalization, first proposed by Ralph Pearson in 1963, is not a model that attempts to explain why certain acids prefer to bind to certain bases, but rather a simple summary (almost a law) of experimental fact. A reexamination of the data collected in Tables 16-3 and 16-4 shows that the hard acid BF_3 prefers to bond to the hard bases $(CH_3)_2O$ and $(CH_3)_3N$ rather than the soft bases $(CH_3)_2S$ and $(CH_3)_3P$; the borderline acid $(CH_3)_3B$ prefers the soft base $(CH_3)_3P$ over the hard base $(CH_3)_3N$; the hard acid Fe^{3+} bonds more strongly to the hard base F^-, whereas the soft acids Hg^{2+} and Cd^{2+} bond more strongly to the soft base I^-.

The relative abundances of the mineral sources of metals provide an apt illustration of this principle. The hard metal ions occur much more frequently in nature combined with the hard base O^{2-} (that is, as oxides) rather than the softer base S^{2-} (that is, as sulfides), while the soft metal ions occur more frequently as sulfides. For example, the two major sources of aluminum are bauxite, a hydrated oxide of Al^{3+}, and corundum, which is mainly Al_2O_3. There are no major sulfide ores of aluminum. Likewise, the only abundant source of tin is cassiterite, which is primarily SnO_2. The soft metal ions Ag^+ and Hg^{2+}, on the other hand, occur as sulfides, their most abundant ores being argentite, Ag_2S, and cinnabar, HgS.

BASE-DISPLACEMENT REACTIONS

The displacement, or substitution, of one base by another is quite common. Indeed, some of the examples of addition reactions given above are more correctly classified as displacement reactions. The reaction of fluoride ion with the Fe^{3+} ion in aqueous solution certainly proceeds by displacement of complexed water from the adduct $Fe(OH_2)_6^{3+}$

$$Fe(OH_2)_6^{3+} + 6F^- \rightleftharpoons FeF_6^{3-} + 6H_2O$$

Displacement reactions can also occur on the molecular species listed in Table 16-3. Indeed, a convenient way to establish whether an acid should be classified as hard or soft is to run a reaction such as

$$F_3B{:}O(CH_3)_2 + {:}S(CH_3)_2 \rightleftharpoons F_3B{:}S(CH_3)_2 + {:}O(CH_3)_2$$

in which a hard base and a soft base compete for the acid. For this reaction the hard base $O(CH_3)_2$ is preferred by the hard acid BF_3, and the equilibrium constant is less than 1.0. With $GaCl_3$ as the acid, on the other hand, the equilibrium constant for the reaction

$$Cl_3Ga{:}O(CH_3)_2 + {:}S(CH_3)_2 \rightleftharpoons Cl_3Ga{:}S(CH_3)_2 + {:}O(CH_3)_2$$

is greater than 1. Gallium trichloride can be categorized therefore as a soft acid.

One of the most common and useful types of base-displacement reaction occurs at carbon atoms. If methyl iodide is treated with hydroxide ion in a suitable solvent, the iodine is displaced as the iodide ion and methanol is formed:

$$H\ddot{O}{:}^- + ICH_3 \rightleftharpoons HO{-}CH_3 + {:}\ddot{\underset{..}{I}}{:}^-$$

The acid in this reaction is probably best visualized as the CH_3^+ ion, although methyl iodide is certainly not ionic, and, as we shall see below, the mechanism for this reaction does not involve the CH_3^+ ion. This visualization has the advantage of making the analogy to other displacements more obvious: Methyl iodide, then, can be thought of as an adduct of the iodide ion and the CH_3^+ ion.

Numerous other bases can be used to displace the iodide ion in this reaction. Methoxide ion, CH_3O^-, reacts to form dimethyl ether,

$$CH_3\ddot{O}{:}^- + CH_3I \rightleftharpoons CH_3OCH_3 + {:}\ddot{\underset{..}{I}}{:}^-$$

hydrogen sulfide ion produces methanethiol,

$$H\ddot{S}{:}^- + CH_3I \rightleftharpoons CH_3SH + {:}\ddot{\underset{..}{I}}{:}^-$$

and the acetate ion reacts to form the ester methyl acetate:

$$CH_3C \overset{\displaystyle O}{\underset{\ddot{O}:^-}{\big|\big|}} + CH_3I \rightleftharpoons CH_3C \overset{\displaystyle O}{\underset{OCH_3}{\big|\big|}} + :\ddot{I}:^-$$

When ammonia is used as the base, the product adduct is positively charged.

$$H_3N: + CH_3I \rightleftharpoons CH_3NH_3^+ + :\ddot{I}:^-$$

If excess ammonia or some other base is present, a proton transfer from the adduct to the base will occur,

$$CH_3NH_3^+ + :NH_3 \rightleftharpoons CH_3NH_2 + NH_4^+$$

and the reaction can be used as a method for the preparation of amines.

Displacements can also occur at adducts other than methyl iodide: Methoxide ion will react with methyl chloride, methyl bromide, or methyl iodide. In fact, the reaction of alkoxide ions, RO$^-$, with alkyl halides is a convenient general method for preparing ethers.

$$CH_3-\ddot{O}:^- + CH_3Br \rightleftharpoons CH_3OCH_3 + :\ddot{Br}:^-$$

$$CH_3CH_2-\ddot{O}:^- + CH_3Cl \rightleftharpoons CH_3CH_2OCH_3 + :\ddot{Cl}:^-$$

$$CH_3-\ddot{O}:^- + CH_3CH_2Br \rightleftharpoons CH_3CH_2OCH_3 + :\ddot{Br}:^-$$

Compounds containing the acetyl group ($CH_3C{\overset{\nearrow O}{}}$), such as $CH_3C{\overset{\nearrow O}{\underset{\searrow Cl}{}}}$ (which can be visualized as the adduct of $CH_3C{\overset{\nearrow O}{}}$ and $:\ddot{Cl}:^-$), can also be used as adducts for base displacements:

$$:\ddot{O}H^- + CH_3C\overset{\displaystyle O}{\underset{Cl}{\big\langle}} \rightleftharpoons CH_3C\overset{\displaystyle O}{\underset{OH}{\big\langle}} + :\ddot{Cl}:^-$$

$$^-:\ddot{O}CH_3 + CH_3C\overset{\displaystyle O}{\underset{Cl}{\big\langle}} \rightleftharpoons CH_3C\overset{\displaystyle O}{\underset{OCH_3}{\big\langle}} + :\ddot{Cl}:^-$$

Rate of Base Displacements at Carbon

Unlike proton-transfer reactions, the rates of displacement reactions at carbon are generally slow. The rates, rather than extents, of these reactions are therefore of prime importance, and the reactivity of the displacing base is measured in terms of its effect on the rate of the displacement reaction. How readily the base reacts in these terms is referred to as its *nucleophilicity*. Thus, the faster the rate of reaction between the base and the reactant adduct, the greater its nucleophilicity or desire for the acid. This concept of nucleophilicity must be contrasted with the concept of basicity, which refers to reactivity as measured by the *extent* of a reaction. A base can therefore also be called a *nucleophile;* an acid, an *electrophile;* and a base-displacement reaction, a *nucleophilic-displacement* or *nucleophilic-substitution* reaction. The reactant adduct is often referred to as the *substrate*.

$$B: \quad + \quad A:D \quad \rightleftharpoons \quad A:B \quad + \quad D:$$

nucleo-   substrate   product   leaving
phile   nucleophile

Experimentally, it has been determined that the nucleophilicities of various bases toward the adduct methyl iodide (in hydrogen-bonding solvents such as CH_3OH) vary in the order

$$(C_2H_5)_3P > I^- > (C_2H_5)_3N > (CH_3)_2S > NH_3 > Cl^- > F^- > (CH_3)_2O$$

That is, of these bases, triethylphosphine, $(C_2H_5)_3P$, reacts most rapidly with CH_3I, while $(CH_3)_2O$ reacts most slowly. The nucleophilicity orders $(C_2H_5)_3P > (C_2H_5)_3N$, $(CH_3)_2S > (CH_3)_2O$, and $I^- > Br^- > Cl^- > F^-$ are clearly indicative of *soft* acid behavior: the third-row elements phosphorus, sulfur, and chlorine, react more rapidly than their second-row counterparts, nitrogen, oxygen, and fluorine, with the acid portion of the substrate, CH_3^+. These reactivity orders depend somewhat on what solvent the reaction is carried out in, however, and in some solvents the halogen order is actually reversed to $F^- > Cl^- > Br^- > I^-$. For this reason, the acid CH_3^+ is classified as a *borderline* acid. Obviously, the hard and soft acid and base principle can be applied to both the extent *and* rate of chemical reactions.

The nucleophilicity order given above demonstrates that between the neutral hard bases NH_3 and $(CH_3)_2O$ the relative nucleophilicities are determined by the availability of the lone pair of electrons. Thus, as discussed on page 440, the lone pair of NH_3 is more available than that of $(CH_3)_2O$, and NH_3 is therefore the better nucleophile. The availability of the lone pair also determines the relative nucleophilicities of the soft bases—for example, $(C_2H_5)_3P > (CH_3)_2S$.

When acetyl chloride, $CH_3C{\overset{\displaystyle O}{\underset{Cl}{\diagup}}}$, is used as the substrate, the reactivity order is quite different: The hard acids, F^-, OH^-, $(CH_3)_2O$, NH_3, $(C_2H_5)_3N$, and so on, react more rapidly than the soft acids, I^-, SH^-, $(CH_3)_2S$, $(C_2H_5)_3P$, etc. The acid portion of the substrate, $CH_3\overset{\displaystyle O}{\overset{\|}{C}}{}^+$, is therefore a *hard* acid. That $CH_3\overset{\displaystyle O}{\overset{\|}{C}}{}^+$ is *harder* than CH_3^+ is understandable on the basis of the electronegativity of oxygen: The electronegative oxygen removes electron density from the carbon, making it less polarizable than the carbon of the borderline acid CH_3^+.

An altogether different reactivity order is encountered when a tertiary butyl halide is used as the substrate. The nature of the nucleophile does not affect the rate of the reaction at all!

$$(CH_3)_3CX + :B^- \rightleftharpoons (CH_3)_3C:B + X^-$$

This obviously cannot be explained by the hard and soft acids and bases principle, and we must look to the mechanism of the reaction for an explanation.

Not only is the rate of this reaction independent of the nature and concentration of the nucleophile, but it is first order with respect to the concentration of the substrate.

$$r = k[(CH_3)_3 CX]$$

The following two-step mechanism in which the first step is rate determining is consistent with this rate law.

$$(CH_3)_3 CCl \rightarrow (CH_3)_3 C^+ + Cl^- \qquad slow$$

$$(CH_3)_3 C^+ + B:^- \rightarrow (CH_3)_3 C:B \qquad fast$$

The rate-determining step is the dissociation of *tert*-butyl chloride to the $(CH_3)_3 C^+$ ion and Cl^- ion. The rate equation for this step is

$$r = k[(CH_3)_3 CCl]$$

which is in agreement with the experimentally determined rate law for the overall reaction. As soon as the $(CH_3)_3 C^+$ ion is formed in the slow step, it is consumed in the fast step by the nucleophile. This mechanism has been given the abbreviation S_N1, which stands for unimolecular nucleophilic substitution. Unimolecular refers to the molecularity of the rate-determining dissociation of the *tert*-butyl chloride.

A quite different mechanism has been postulated for nucleophilic displacements on methyl halides. Experimentally, the rate of these reactions is proportional to the concentrations of both the substrate and the nucleophile.

$$r = k[CH_3 X][B:]$$

A simple one-step mechanism identical to the overall equation

$$CH_3 X + B:^- \rightleftharpoons CH_3 B + X^-$$

is consistent with the rate equation. This process is bimolecular, and the mechanism is therefore abbreviated as S_N2.

The question now arises: Why should the replacement of the hydrogens in $CH_3 Cl$ by methyl groups to give $(CH_3)_3 CCl$ result in a change in mechanism for displacements at the two substrates? The answer appears to lie at least in part in the structure of the transition state for the S_N2 reaction. Evidence based on the reactions of optically active substrates indicates that the nucleophile attacks the carbon of the substrate at a point directly opposite the leaving group. The structure of the activated complex, where the nucleophile is partially bonded and the leaving group has partially detached itself, is therefore

$$
\begin{array}{ccc}
H & & H \\
\diagdown & & \diagup \\
B \cdots & C & \cdots X \\
& | & \\
& H &
\end{array}
$$

A pictorial time sequence of events is shown in Figure 16-4, along with the energy profile of the reaction. If the same transition state is imagined for *tert*-butyl chloride, the increase in repulsions between the nucleophile and the $C(CH_3)_3$ group and also between the leaving group and the $C(CH_3)_3$ group, due to the greater bulk of a methyl group relative to a hydrogen, are easily visualized. These repulsions

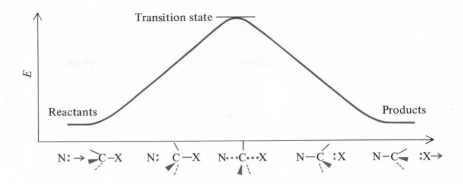

FIGURE 16-4

Reaction Profile for $S_N 2$ Mechanism

increase the energy of the activated complex and, as a result of the increased activation energy, lead to a marked decrease in rate. Instead of proceeding by the relatively unfavorable $S_N 2$ mechanism, the *tert*-butyl substrate utilizes the more favorable $S_N 1$ mechanism.

This model suggests that most primary halides (for example $CH_3 CH_2 Cl$) will undergo nucleophilic substitution via the $S_N 2$ mechanism, while tertiary halides [for example, $(CH_3)_3 CCl$, $(C_6 H_5)_3 CCl$] will react by an $S_N 1$ mechanism. This expectation has been verified by numerous experiments.

ACID-DISPLACEMENT REACTIONS

Whenever a reaction contains two potential Lewis acids, the possibility of an acid-displacement reaction exists. For example, if Zn^{2+} is added to an aqueous solution of the adduct $Cd(NH_3)_4^{2+}$, the following reaction (for which $K = 10^3$) will occur.

$$Zn^{2+} + Cd(NH_3)_4^{2+} \rightleftharpoons Zn(NH_3)_4^{2+} + Cd^{2+}$$

Here the borderline acid Zn^{2+} and the soft acid Cd^{2+} compete for the lone pair provided by the hard base NH_3. The ammonia prefers to bind to the harder acid, in agreement with the hard and soft acids and bases principle.

The destruction of ammonia adducts with acids, a procedure frequently encountered in the qualitative analysis of transition metal salts, can also be thought of as a competition between the acid H^+ and the transition metal cation for the ammonia. The ammonia again prefers the harder acid, the proton.

$$4H^+ + Cu(NH_3)_4^{2+} \rightleftharpoons 4NH_4^+ + Cu^{2+}$$

SUGGESTED READINGS

Vanderwerf, C. A. *Acids, Bases, and the Chemistry of the Covalent Bond.* New York: Reinhold, 1961.

Pearson, R. G., "Hard and Soft Acids and Bases. I." *Journal of Chemical Education,* Vol. 45, No. 9 (September 1968), pp. 581–587.

Pearson, R. G., "Hard and Soft Acids and Bases. II." *Journal of Chemical Education,* Vol. 45, No. 10 (October 1968), pp. 643–648.

PROBLEMS

1. Write an equation showing:
   (a) $Pb(H_2O)_4{}^{2+}$ acting as a Lowry-Brønsted acid
   (b) $Pb(H_2O)_4{}^{2+}$ as a reactant in a base-displacement reaction
   (c) Zn^{2+} acting as a Lewis acid
   (d) $Zn(OH)_2$ reacting as a Lowry-Brønsted base
   (e) $Zn(OH)_2$ reacting as a Lewis acid
   (f) NH_3 reacting as a Lewis base
   (g) NH_3 as a reactant in a base-displacement reaction
   (h) $Cd(NH_3)_4{}^{2+}$ as a reactant in an acid-displacement reaction
   (i) CO_2 acting as a Lewis acid
   (j) $H_2C{=}O$ acting as a Lewis acid
   (k) SO_2 acting as a Lewis acid
   (l) CH_3Br as a reactant in a base-displacement reaction
   (m) CH_3COOH as a reactant in a base-displacement reaction
   (n) Benzene as a reactant in an acid-displacement reaction

2. For each of the following adducts, give a possible preparation and predict the geometry:
   (a) $SnCl_3{}^-$ (b) $SnCl_6{}^-$
   (c) $B(OH)_4{}^-$ (d) $AlF_4{}^-$

3. Distinguish carefully between:
   (a) an addition reaction and a displacement reaction
   (b) a base-displacement reaction and an acid-displacement reaction
   (c) amphiprotic and amphoteric
   (d) the reaction of $Al(OH)_3$ as an acid and its reaction as a base
   (e) basicity and nucleophilicity
   (f) a hard acid and a soft acid
   (g) a nucleophile and an electrophile
   (h) the S_N1 mechanism and the S_N2 mechanism

4. Rationalize the following observations:
   (a) In an attempt to prepare a solution of $SnCl_4$ by dissolving $SnCl_4$ in H_2O, a white precipitate is obtained.
   (b) Silver chloride is insoluble in H_2O but soluble in aqueous NH_3.
   (c) $Mg(OH)_2$ does not dissolve in $1M$ NaOH, but $Al(OH)_3$ does.
   (d) The solubilities of the IIB sulfides vary in the order ZnS > CdS > HgS.
   (e) $B(OH)_3$ is a Lewis acid, while $Al(OH)_3$ is both a Lewis acid and a Lowry-Brønsted base.

5. Present a concise discussion on each of the following topics:
   (a) the mechanism of nucleophilic substitution reactions
   (b) the hard and soft acid-base concept

6. Predict whether the following reactions have equilibrium constants greater than or less than 1.
   (a) $F_3BN(CH_3)_3 + P(CH_3)_3 \rightleftharpoons F_3BP(CH_3)_3 + N(CH_3)_3$
   (b) $F_3BO(CH_3)_2 + N(CH_3)_3 \rightleftharpoons F_3BN(CH_3)_3 + O(CH_3)_2$
   (c) $F_3BN(CH_3)_3 + B(CH_3)_3 \rightleftharpoons (CH_3)_3BN(CH_3)_3 + BF_3$
   (d) $H_3O^+ + H_2S \rightleftharpoons H_3S^+ + H_2O$
   (e) $H_3O^+ + NH_3 \rightleftharpoons NH_4^+ + H_2O$
   (f) $CdCl_4{}^{2-} + 4I^- \rightleftharpoons CdI_4{}^{2-} + 4Cl^-$
   (g) $BF_4{}^- + 4I^- \rightleftharpoons BI_4{}^- + 4F^-$
   (h) $Zn(OH)_4{}^{2-} + Cd^{2+} \rightleftharpoons Cd(OH)_4{}^{2-} + Zn^{2+}$

7. The lithium ion is smaller and therefore less polarizable than the cesium ion. Predict whether the equilibrium constant for the following reaction is greater than or less than 1. Estimate the equilibrium constant at $25°$ by calculating ΔH^0 from the data of Appendix 7.

$$LiF(s) + CsI(s) \rightleftharpoons LiI(s) + CsF(s)$$

8. Determine from the following data whether the trimethylcarbonium ion $(CH_3)_3C^+$ is softer or harder than the methyl ion $CH_3{}^+$.

$$CH_3OH(g) + H_2S(g) \rightarrow CH_3SH(g) + H_2O(g) \quad \Delta H^0 = -10 \text{ kcal/mole}$$

$$(CH_3)_3COH(g) + H_2S(g) \rightarrow (CH_3)_3CSH(g) + H_2O(g) \quad \Delta H^0 = -4 \text{ kcal/mole}$$

9. Hydrogen bonding can be visualized as a Lewis acid-base interaction between a covalently bonded proton (the acid) and a nonbonded pair of electrons (the base). Which kind of base would hydrogen bond better—a hard or a soft base? Provide examples.

10. Predict whether the following common minerals are sulfides or oxides.
    (a) chromite, a chromium(III) ore
    (b) hematite, an iron(III) ore
    (c) pyrolusite, a manganese(IV) ore
    (d) sphalerite, a zinc(II) ore
    (e) rutile, a titanium(IV) ore
    (f) cinnabar, a mercury(II) ore

11. Supply products for the following reactions:
    (a) $CH_3I + {}^-OCH_3 \rightarrow$ (b) $CH_3CH_2Br + {}^-NH_2 \rightarrow$

    (c) 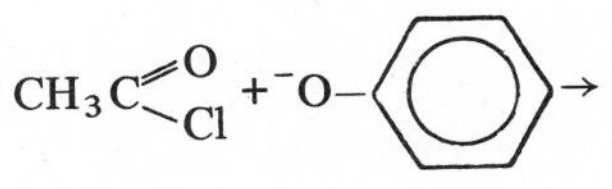 $CH_3C\overset{\diagup O}{\underset{\diagdown Cl}{}} + {}^-O{-}\langle\bigcirc\rangle \rightarrow$ (d) $CH_3C\overset{\diagup O}{\underset{\diagdown OCH_3}{}} + {}^-OH \rightarrow$

    (e) 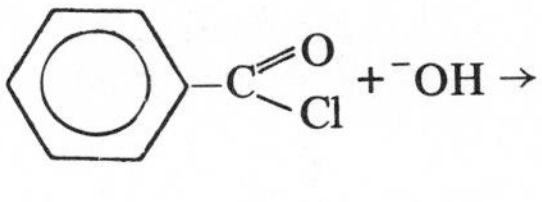 $\langle\bigcirc\rangle{-}C\overset{\diagup O}{\underset{\diagdown Cl}{}} + {}^-OH \rightarrow$

17 Types of Chemical Reactions IV: Electron-Transfer Reactions

The preceding chapters have dealt with chemical reactions involving exchange of ions, exchange of protons, and sharing of electrons. In this chapter we shall consider reactions of a fourth type: those in which electrons are transferred from one reactant to another. Electron-transfer reactions are commonly referred to as *oxidation-reduction* or *redox* reactions.

Originally, the term *oxidation* was applied to a reaction in which a substance combined with oxygen—for example, the oxidation of calcium, in which the calcium is said to be *oxidized*.

$$2Ca + O_2 \rightarrow 2CaO$$

The reverse process—the loss of oxygen by a substance—was called *reduction*. Thus, for example, when mercuric oxide is heated, it decomposes to elemental mercury and oxygen—a reduction, in which HgO is *reduced*.

$$2HgO \rightarrow 2Hg + O_2$$

The meanings of the terms were later broadened to include loss of hydrogen (oxidation) and gain of hydrogen (reduction). Thus, in the hydrogenation of ethylene to ethane, the ethylene is *reduced*.

$$H_2C{=}CH_2 + H_2 \rightarrow H_3C{-}CH_3$$

Considering the fundamental nature of the process in which calcium and oxygen combine, it may be concluded that in order for a calcium atom to become a calcium ion, it must lose electrons, and in order for the atoms of the oxygen molecule to become oxide ions, they must gain electrons.

$$Ca \rightarrow Ca^{2+} + 2e^-$$

$$O_2 + 4e^- \rightarrow 2O^{2-}$$

Similarly, in the reduction of HgO, mercury gains electrons and oxygen loses them.

$$Hg^{2+} + 2e^- \rightarrow Hg$$

$$2O^{2-} \rightarrow O_2 + 4e^-$$

Numerous chemical reactions that do not involve hydrogen or oxygen in any way occur by this same basic process: transfer of electrons. For example, the simple reactions between iron and sulfur, and between sodium and chlorine, are precisely analogous to the *oxidation* of calcium described above.

$$Fe + S \rightarrow FeS$$

$$2Na + Cl_2 \rightarrow 2NaCl$$

The fundamental aspect of these reactions, then, that makes them different from the types of reactions discussed in previous chapters, is not the gain or loss of oxygen or hydrogen, but rather the transfer of electrons. In modern usage, the term *oxidation* means the *loss of electrons; reduction* means *gain of electrons.* Since, in order for a substance to gain electrons, it must gain them from some other substance, which must therefore lose electrons, the two processes occur simultaneously, and the resulting overall transformation is referred to as an *oxidation-reduction reaction.* The reactant that loses the electrons is said to be *oxidized* and is called the *reducing agent* or *reductant;* the reactant that gains the electrons is said to be *reduced* and is called the *oxidizing agent* or *oxidant.* Thus, in the reaction between sodium and chlorine above, sodium is the reductant (it is oxidized) and chlorine is the oxidant (it is reduced).

OXIDATION NUMBERS

Oxidation-reduction processes can be described in terms of a system of useful, although artificial, positive and negative numbers called *oxidation numbers* or *oxidation states.* An oxidation number is assigned to each atom in a compound to represent the electrical charge the atom *would* have *if* the electrons in the compound were assigned in a certain way. It does not represent an actual charge on the atom and should not be confused with the ionic charge.

Oxidation numbers are assigned according to a set of arbitrary rules, as follows:

1. The oxidation number of the atoms in an elementary substance is zero.
2. The oxidation number of monatomic ions is the same as the ionic charge.

3. The oxidation number of hydrogen is +1, except in elemental hydrogen (where it is zero) or in ionic hydrides (where it is −1).
4. The oxidation number of oxygen is −2, except in elemental oxygen (where it is zero) or in peroxides (where it is −1).
5. Covalent binary compounds that contain neither hydrogen nor oxygen are treated as though they were ionic, using rule 2.
6. In a compound the sum of the oxidation numbers must equal zero; in a complex ion the sum must equal the charge of the ion.

With few exceptions, these rules make it possible to assign oxidation numbers to all the atoms in any substance. The following examples illustrate this application. Nitrogen gas, N_2, is an elementary substance; the oxidation number of the nitrogen atoms in the molecule is zero (rule 1), and we indicate this by writing $\overset{0}{N}_2$. $CaCl_2$ is an ionic compound, and the oxidation numbers are the same as the ionic charges (rule 2); that is, +2 for calcium and −1 for chlorine; hence we write $\overset{+2}{Ca}\overset{-1}{Cl}_2$. Note that the sum of one calcium at +2 and two chlorines each at −1 equals zero, satisfying rule 6.

Consider now the oxidation numbers in $HClO_4$. One hydrogen atom with an oxidation number of +1 (rule 3) and four oxygen atoms with oxidation numbers of −2 (rule 4) equals −7. Since the sum of all the numbers must equal zero, we must assign an oxidation number of +7 to the Cl atom: $\overset{+1}{H}\overset{+7}{Cl}\overset{-2}{O}_4$.

Examples of the application of rule 5, where covalent compounds are treated as if they were ionic, include $\overset{+4}{C}\overset{-2}{S}_2$, $\overset{+3}{B}\overset{-1}{F}_3$, $\overset{+4}{C}\overset{-1}{Cl}_4$, and $\overset{+5}{P}\overset{-1}{Cl}_5$.

As further illustrations, a number of formulas are listed below with oxidation numbers assigned in accordance with the rules described.

$$\overset{+1}{N}_2\overset{-2}{O} \qquad \overset{+1}{Na}_2\overset{+4}{C}\overset{-2}{O}_3 \qquad \overset{+6}{S}\overset{-2}{O}_3$$

$$(\overset{+5}{P}\overset{-2}{O}_4)^{3-} \qquad \overset{-2}{C}_2\overset{+1}{H}_6\overset{-2}{O} \qquad \overset{+1}{Na}_2\overset{+2}{S}_2\overset{-2}{O}_3$$

$$(\overset{+6}{U}\overset{-2}{O}_2)^{2+} \qquad (\overset{+3}{C}_2\overset{-2}{O}_4)^{2-} \qquad (\overset{+6}{Cr}_2\overset{-2}{O}_7)^{2-}$$

This system of oxidation numbers provides a method for keeping track of oxidation-reduction processes. First, it enables one quickly to determine whether or not a given chemical reaction is an oxidation-reduction reaction; for if it is, then the oxidation number of at least two different atoms must change—one increasing and the other decreasing. Further, since oxidation is a loss of electrons, and electrons are negative, oxidation results in the oxidation number becoming more positive. On the other hand, reduction, being a gain of electrons, is accompanied by a change in oxidation number in a negative direction. Thus, the oxidant and reductant are easily identified.

Consider the reaction between phosphorus pentachloride and water.

$$\overset{+5}{P}\overset{-1}{Cl}_5 + 4\overset{+1}{H}_2\overset{-2}{O} \rightarrow \overset{+1}{H}_3\overset{+5}{P}\overset{-2}{O}_4 + 5\overset{+1}{H}\overset{-1}{Cl}$$

Superficially this may appear to be a redox reaction, but assignment of oxidation numbers to reactants and products indicates that the reaction results in no change in any of these numbers, and therefore it is not of the redox type. Similarly, one concludes that the conversion of chromate ion to dichromate in the presence of acid is not an oxidation-reduction reaction.

$$2(\overset{+6}{\text{Cr}}\overset{-2}{\text{O}}_4)^{2-} + 2\overset{+1}{\text{H}}^+ \rightarrow (\overset{+6}{\text{Cr}}_2\overset{-2}{\text{O}}_7)^{2-} + \overset{+1}{\text{H}}_2\overset{-2}{\text{O}}$$

On the other hand, the reaction between copper metal and concentrated nitric acid is clearly a redox reaction, as shown by the changes in oxidation numbers.

$$\overset{0}{\text{Cu}} + 4\overset{+1+5-2}{\text{HNO}_3} \rightarrow \overset{+2}{\text{Cu}}(\overset{+5-2}{\text{NO}_3})_2 + 2\overset{+4-2}{\text{NO}_2} + 2\overset{+1}{\text{H}}_2\overset{-2}{\text{O}}$$

Furthermore, it is apparent that copper is the reducing agent (it is oxidized) and nitric acid is the oxidizing agent (it is reduced).

Another practical use to which oxidation numbers can be put is in the balancing of equations. Whereas most equations depicting reactions of the types discussed in previous chapters are easily balanced by simple inspection, many redox reactions lead to equations with rather large coefficients, and balancing them by inspection may be both difficult and time consuming.

A case in point is the reaction between potassium permanganate and oxalic acid in sulfuric acid solution. The skeletal (unbalanced) equation is as follows:

$$\text{KMnO}_4 + \text{H}_2\text{C}_2\text{O}_4 + \text{H}_2\text{SO}_4 \rightarrow \text{K}_2\text{SO}_4 + \text{MnSO}_4 + \text{CO}_2 + \text{H}_2\text{O}$$

Attempts to balance the equation by inspection quickly demonstrate that a balance is not readily achieved. Assigning oxidation numbers to the atoms in the equation, we see that the oxidation number of manganese changes in the reaction from +7 to +2—a decrease of 5.

$$\overset{+7}{\text{Mn}} \overset{-5}{\longrightarrow} \overset{+2}{\text{Mn}}$$

The oxidation number of carbon changes from +3 to +4, and since there must be at least two carbon atoms on each side of the balanced equation (because of the formula of $\text{H}_2\text{C}_2\text{O}_4$), this represents an increase of 2.

$$2\overset{+3}{\text{C}} \overset{+2}{\longrightarrow} 2\overset{+4}{\text{C}}$$

In the overall reaction the net change of oxidation numbers must be zero; that is, the total increase due to oxidation must be equal to the total decrease due to reduction. Therefore, it is necessary to multiply the manganese change by 2 and the carbon change by 5.

$$2(\overset{+7}{\text{Mn}} \overset{-5}{\longrightarrow} \overset{+2}{\text{Mn}}) = 2\overset{+7}{\text{Mn}} \overset{-10}{\longrightarrow} 2\overset{+2}{\text{Mn}}$$

$$5(2\overset{+3}{\text{C}} \overset{-2}{\longrightarrow} 2\overset{+4}{\text{C}}) = 10\overset{+3}{\text{C}} \overset{-10}{\longrightarrow} 10\overset{+4}{\text{C}}$$

Thus, the coefficients of Mn and C in the equation have been fixed, and the equation may be balanced *partially* as follows:

$$2\text{KMnO}_4 + 5\text{H}_2\text{C}_2\text{O}_4 + \text{H}_2\text{SO}_4 \rightarrow \text{K}_2\text{SO}_4 + 2\text{MnSO}_4 + 10\text{CO}_2 + \text{H}_2\text{O}$$

All that remains to be balanced are the H_2SO_4 and H_2O, and these may easily be taken care of by inspection since all the other coefficients are fixed. Three sulfates on the right require three on the left, so we may place a 3 in front of H_2SO_4. This gives a total of 16 hydrogen atoms on the left, and we must place an 8 before H_2O. The equation is now completely balanced.

$$2KMnO_4 + 5H_2C_2O_4 + 3H_2SO_4 \rightarrow K_2SO_4 + 2MnSO_4 + 10CO_2 + 8H_2O$$

As another example of this oxidation state method of balancing equations, consider the following:

Unbalanced equation:

$$H_2S + HNO_3 \rightarrow S + NO + H_2O$$

Changes of oxidation number:

$$\overset{-2}{S} \xrightarrow{+2} \overset{0}{S}$$

$$\overset{+5}{N} \xrightarrow{-3} \overset{+2}{N}$$

Equalize loss and gain:

$$3\overset{-2}{S} \xrightarrow{+6} 3\overset{0}{S}$$

$$2\overset{+5}{N} \xrightarrow{-6} 2\overset{+2}{N}$$

Partially balanced equation:

$$3H_2S + 2HNO_3 \rightarrow 3S + 2NO + H_2O$$

Completely balanced equation:

$$3H_2S + 2HNO_3 \rightarrow 3S + 2NO + 4H_2O$$

HALF-REACTIONS

As was pointed out above, an oxidation-reduction reaction is really the sum of two separate processes, one being oxidation (the loss of electrons), the other reduction (the gain of electrons). Aside from the fact that oxidation and reduction must occur simultaneously, the two processes are independent of each other. Therefore, it is convenient, and quite reasonable, to consider a redox reaction as consisting of two separate *half-reactions*—one involving electron loss, the other electron gain. With this in mind, let us proceed to analyze the reaction represented by the following *unbalanced* molecular equation:

$$K_2Cr_2O_7 + FeSO_4 + H_2SO_4 \rightarrow K_2SO_4 + Cr_2(SO_4)_3 + Fe_2(SO_4)_3 + H_2O$$

Since this reaction occurs in aqueous solution, and since strong electrolytes are involved, the reaction is more appropriately represented by the use of ions.

$$K^+ + Cr_2O_7{}^{2-} + Fe^{2+} + SO_4{}^{2-} + H^+ \rightarrow K^+ + SO_4{}^{2-} + Cr^{3+} + Fe^{3+} + H_2O$$

It is immediately apparent from this *ionic equation* that some of the substances present are not actually involved in the reaction at all; that is, they do not undergo any change whatsoever. This is true of potassium ion and sulfate ion, which may therefore be eliminated, leaving

$$Cr_2O_7{}^{2-} + Fe^{2+} + H^+ \rightarrow Cr^{3+} + Fe^{3+} + H_2O$$

The resulting *net ionic equation* is the best representation of the reaction that occurs, for it shows only those reactants that actually undergo transformation and those products that result from that transformation. This equation may now be separated into two *half-reactions*. Choose any one of the reactants and consider what products must be derived from it. For example, $Cr_2O_7{}^{2-}$ obviously gives rise to Cr^{3+}, and furthermore, one $Cr_2O_7{}^{2-}$ will yield two Cr^{3+}, so we may write:

(1) $$Cr_2O_7{}^{2-} \rightarrow 2Cr^{3+}$$

The oxygen in the $Cr_2O_7{}^{2-}$ may be considered to be converted into water, since that is the only product that contains oxygen, and one $Cr_2O_7{}^{2-}$ supplies enough oxygen for seven H_2O molecules. Therefore, statement (1) can be expanded to:

(2) $$Cr_2O_7{}^{2-} \rightarrow 2Cr^{3+} + 7H_2O$$

But now the 14 hydrogen atoms in the water produced must be accounted for. The source of these is obviously H^+, and statement (2) can be further expanded to:

(3) $$Cr_2O_7{}^{2-} + 14H^+ \rightarrow 2Cr^{3+} + 7H_2O$$

In statement (3) the numbers of each different atom are equal on both sides of the arrow, but the ionic charges are not; the net charge is +12 on the left and +6 on the right. This change in total ionic charge can be accounted for if six electrons are gained in the process. Adding this to statement (3) gives an equation that is balanced in all respects and represents the half-reaction describing the reduction of $Cr_2O_7{}^{2-}$ in acid solution.

(4) $$Cr_2O_7{}^{2-} + 14H^+ + 6e^- \rightarrow 2Cr^{3+} + 7H_2O$$

The other half-reaction can be deduced by the same procedure. Ferrous ion gives rise to ferric ion, and the process involves the loss of one electron. Equation (5), then, is the *oxidation half-reaction.*

(5) $$Fe^{2+} \rightarrow Fe^{3+} + e^-$$

It is obvious from equation (4) that each mole of $Cr_2O_7{}^{2-}$ reduced requires 6 moles of electrons. Equation (5), on the other hand, shows that only 1 mole of electrons is given up for each mole of Fe^{2+} oxidized. Six moles of Fe^{2+} must be oxidized, therefore, to provide enough electrons for the reduction of 1 mole of $Cr_2O_7{}^{2-}$. In other words, before the equations for the two half-reactions can be added together to give the equation for the overall reaction, the number of electrons lost must be made equal to the number gained. This can be accomplished in this instance by multiplying equation (5) by 6. Finally, adding together the two statements yields the balanced net ionic equation for the redox reaction, equation (6).

(4) $$Cr_2O_7^{2-} + 14H^+ + 6e^- \rightarrow 2Cr^{3+} + 7H_2O$$

(5) $$\underline{\qquad\qquad 6(Fe^{2+} \rightarrow Fe^{3+} + e^-) \qquad\qquad}$$

(6) $$Cr_2O_7^{2-} + 14H^+ + 6Fe^{2+} \rightarrow 2Cr^{3+} + 6Fe^{3+} + 7H_2O$$

It will be noted that this procedure for analyzing an oxidation-reduction process results in a balanced net ionic equation. This is, indeed, an alternate method of balancing redox equations and is frequently called the *ion-electron method* to distinguish it from the *oxidation state* method described earlier in this chapter. However, the ability to separate a redox reaction into its two half-reactions is of more importance than simply a means of balancing equations; it is essential to an understanding of the functioning of voltaic and electrolytic cells, which are dealt with later in this chapter.

As another illustration of the application of this procedure, consider the reaction of H_2SO_3 with $KMnO_4$ in acid solution. The unbalanced molecular equation is

$$H_2SO_3 + KMnO_4 + H_2SO_4 \rightarrow MnSO_4 + K_2SO_4 + H_2O$$

Step 1: Write the unbalanced net ionic equation.

$$H_2SO_3 + MnO_4^- + H^+ \rightarrow Mn^{2+} + SO_4^{2-} + H_2O$$

(Note that the weak electrolyte H_2SO_3 is written as a molecule, not as ions.)

Step 2: Deduce the oxidation half-reaction.

(a) $H_2SO_3 \rightarrow SO_4^{2-}$

(b) $H_2SO_3 + H_2O \rightarrow SO_4^{2-}$

(c) $H_2SO_3 + H_2O \rightarrow SO_4^{2-} + 4H^+$

(d) $H_2SO_3 + H_2O \rightarrow SO_4^{2-} + 4H^+ + 2e^-$

Step 3: Deduce the reduction half-reaction.

(a) $MnO_4^- \rightarrow Mn^{2+}$

(b) $MnO_4^- \rightarrow Mn^{2+} + 4H_2O$

(c) $MnO_4^- + 8H^+ \rightarrow Mn^{2+} + 4H_2O$

(d) $MnO_4^- + 8H^+ + 5e^- \rightarrow Mn^{2+} + 4H_2O$

Step 4: Equalize the number of electrons lost and the number gained; that is, multiply the oxidation half-reaction by 5 and the reduction half-reaction by 2.

$$5H_2SO_3 + 5H_2O \rightarrow 5SO_4^{2-} + 20H^+ + 10e^-$$

$$2MnO_4^- + 16H^+ + 10e^- \rightarrow 2Mn^{2+} + 8H_2O$$

Step 5: Add the two half-reactions in Step 4 to obtain the balanced net ionic equation.

$$5H_2SO_3 + 2MnO_4^- \rightarrow 5SO_4^{2-} + 2Mn^{2+} + 4H^+ + 3H_2O$$

It has been seen in these two examples of reactions in acid solution that the half-reactions are arrived at by using H_2O and H^+ as needed in order to write balanced statements. Not all redox reactions take place in acid solution, of course, and if the solution is basic, then the half-reactions must be developed with the proper use of H_2O and OH^-. The reaction of chromite ion with hypochlorite in alkaline solution serves to illustrate:

$$KCrO_2 + KClO + KOH \rightarrow K_2CrO_4 + KCl + H_2O$$

Step 1: The unbalanced net ionic equation is

$$CrO_2^- + ClO^- + OH^- \rightarrow CrO_4^{2-} + Cl^- + H_2O$$

Step 2: Deduce the oxidation half-reaction:

(a) $CrO_2^- \rightarrow CrO_4^{2-}$

(b) $CrO_2^- + 4OH^- \rightarrow CrO_4^{2-} + 2H_2O$

(c) $CrO_2^- + 4OH^- \rightarrow CrO_4^{2-} + 2H_2O + 3e^-$

Step 3: Deduce the reduction half-reaction:

(a) $ClO^- \rightarrow Cl^-$

(b) $ClO^- + H_2O \rightarrow Cl^- + 2OH^-$

(c) $ClO^- + H_2O + 2e^- \rightarrow Cl^- + 2OH^-$

Step 4: Multiply the oxidation half-reaction by 2 and the reduction half-reaction by 3.

$$2CrO_2^- + 8OH^- \rightarrow 2CrO_4^{2-} + 4H_2O + 6e^-$$

$$3ClO^- + 3H_2O + 6e^- \rightarrow 3Cl^- + 6OH^-$$

Step 5: Add.

$$2CrO_2^- + 3ClO^- + 2OH^- \rightarrow 2CrO_4^{2-} + 3Cl^- + H_2O$$

The proper use of H_2O and OH^- in dealing with reactions in basic solution is facilitated if one keeps in mind that two OH^- ions may be divided into one O atom and one H_2O molecule and that one H_2O molecule may be divided into one H atom and one OH^- ion. Thus, for each oxygen atom needed on one side of a half-reaction equation, one adds two OH^- to that side and one H_2O to the other side. Likewise, for each H atom needed on a given side, one may add one H_2O to that side and one OH^- to the other.

REDOX COUPLES

It should be noted that half-reactions are reversible, and the direction in which a given half-reaction goes depends upon whether it is reacting with a stronger or a weaker oxidizing agent. Consider, for example, the half-reaction

$$Cu^{2+} + 2e^- \rightleftharpoons Cu$$

If zinc metal is placed in a solution of Cu^{2+}, the reaction that occurs is

$$Zn + Cu^{2+} \rightarrow Zn^{2+} + Cu$$

Since Zn^{2+} is a weaker oxidizing agent than Cu^{2+}, the Cu^{2+} is reduced and the half-reaction occurs in the direction $Cu^{2+} + 2e^- \rightarrow Cu$. On the other hand, if copper metal is placed in a solution of Ag^+, the reaction is

$$2Ag^+ + Cu \rightarrow 2Ag + Cu^{2+}$$

Silver ion is a stronger oxidizing agent than Cu^{2+}, and therefore Cu is oxidized; the half-reaction for the change in copper occurs in the direction $Cu \rightarrow Cu^{2+} + 2e^-$.

It is convenient, therefore, to speak of the participants in these reversible half-reactions as redox couples—a couple consisting of an oxidized form and the reduced form that results from the gain of electrons by the former. Three redox couples have just been illustrated: Cu^{2+}/Cu, Ag^+/Ag, and Zn^{2+}/Zn. Note that redox couples are somewhat analogous to conjugate acids and bases (p. 385).

ELECTRON TRANSFER IN BIOLOGICAL SYSTEMS

The use of atmospheric oxygen by living organisms involves oxidation-reduction reactions. Although the overall process is a complex series of steps involving enzyme catalysis, the basic concepts developed earlier apply. This is illustrated by the role of cytochrome c in the oxidative chain. Cytochrome c, believed to occur in all animals, plants, and aerobic microorganisms, is a complex molecule in which a protein is bound to the iron-containing heme group (see Figure 16-1). The iron serves as the atom that is reduced and oxidized in the electron-transfer process. Cytochrome c accepts an electron from a molecule that is being oxidized, and the iron is reduced from the +3 to the +2 state. The cytochrome c then transfers an electron to oxygen, reducing the oxygen and regenerating the iron in the +3 state. The two half-reactions and the overall reaction for the latter process may be represented as follows:

$$4[\text{Cyto c}(Fe^{2+}) \rightarrow \text{Cyto c}(Fe^{3+}) + e^-]$$

$$O_2 + 4e^- \rightarrow 2O^{2-}$$

$$\overline{4\text{Cyto c}(Fe^{2+}) + O_2 \rightarrow 4\text{Cyto c}(Fe^{3+}) + 2O^{2-}}$$

The reduced oxygen combines with protons to form water,

$$O^{2-} + 2H^+ \rightarrow H_2O$$

and the overall process releases the energy needed to keep the organism functioning.

VOLTAIC CELLS

That the view of redox reactions described earlier is justified (that a redox reaction is indeed an electron-transfer process and that it may be treated realistically as the sum of two separate half-reactions) may be demonstrated by a rather simple experiment.

Two aqueous solutions are prepared, one containing $K_2Cr_2O_7$ and H_2SO_4, the other containing KI. The first is orange due to $Cr_2O_7^{2-}$, the other is colorless. Now some of the solution containing the $K_2Cr_2O_7$ is poured slowly into a sample of the KI solution. As the pouring is continued, the mixture does not take on an orange color (the $Cr_2O_7^{2-}$ must be transformed) but instead becomes green (some new substance must be produced). In short, a visible reaction occurs on mixing the two solutions. It can be shown by relatively simple chemical tests that the green color is due to the presence of Cr^{3+} and that the mixture also contains I_2. Obviously the reaction that occurred must have been an oxidation-reduction reaction, the $Cr_2O_7^{2-}$ being reduced to Cr^{3+} and the I^- being oxidized to I_2. The equations for the two half-reactions and the overall reaction are as follows:

$$Cr_2O_7^{2-} + 14H^+ + 6e^- \rightarrow 2Cr^{3+} + 7H_2O$$
$$3(2I^- \rightarrow I_2 + 2e^-)$$
$$\overline{Cr_2O_7^{2-} + 14H^+ + 6I^- \rightarrow 2Cr^{3+} + 3I_2 + 7H_2O}$$

Now, the same two solutions, unmixed, are poured into separate beakers and the beakers are placed side by side. Some metallic conductor that will not react with either of the solutions (for instance, a piece of platinum foil) is suspended in each solution, and the two foils are connected with wires through a galvanometer. Finally, a U-tube is filled with a solution of some electrolyte that will not react with either of the solutions in the beakers (dilute K_2SO_4 will suffice), the ends are plugged loosely with glass wool, and the U-tube is inverted so that one arm is dipping into each of the two solutions. The complete apparatus is shown in Figure 17-1.

As soon as the U-tube is immersed in the solutions, the galvanometer needle is deflected, indicating the flow of electric current through it. If the current flow is allowed to continue without any stirring of the two solutions, one observes, after a short period of time, the appearance of the yellow-brown color of iodine around the platinum foil in the right beaker and a change from the orange of $Cr_2O_7^{2-}$ to the green of Cr^{3+} around the foil in the left beaker.

Clearly, the same reaction occurs in this apparatus as occurs when the two solutions are mixed together, the only difference being that in the apparatus the

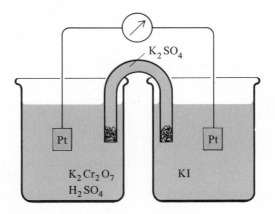

FIGURE 17-1

A Voltaic Cell

electron transfer takes place from one solution to the other externally through the connecting wire. In the beaker on the right, I^- is oxidized to I_2, electrons being given up to the platinum foil.

$$2I^- \rightarrow I_2 + 2e^-$$

The electrons move through the wire and galvanometer to the platinum foil on the left, where they are consumed in the reduction of dichromate ions.

$$Cr_2O_7^{2-} + 14H^+ + 6e^- \rightarrow 2Cr^{3+} + 7H_2O$$

This external flow of electrons, of course, constitutes an electric current.

It will be noted that the color changes (therefore, the reactions) occurred at the surfaces of the metal foils, not at the ends of the U-tube. Yet the U-tube is apparently essential, since the galvanometer indicated no current flow without it. The purpose of the U-tube is to permit the flow of ions to equalize the ionic charge in each solution. Assume that the apparatus is the same as in Figure 17-1, except that the U-tube is not in place. In the right-hand beaker some iodide ions contact the platinum, transfer electrons to it, and are thus oxidized to iodine. The process leaves an excess of potassium ions, giving the solution a positive charge, which prevents the further release of electrons to the platinum. The reduction that occurs in the left beaker leaves an excess of negative ions (sulfate), giving the solution a negative charge that repels electrons and prevents their transfer from the platinum into the solution. In short, without the U-tube, the two reactions would still occur, but to such a slight extent that no measurable current would flow. With the U-tube in place, however, excess sulfate ions on the left can move into it, and on the right sulfate ions move out of it into the solution to balance the charge of the excess potassium ions. Thus, the U-tube provides a pathway for ion flow between the two solutions, preventing the buildup of an ionic charge in either one. Obviously, this equalization of ionic charge can also be accomplished by the flow of potassium ions into the right-hand side of the U-tube and out of the left-hand side. In the actual operating cell, both sulfate and potassium ions migrate.

This apparatus, in which an electric current is produced from a spontaneous oxidation-reduction reaction, is called a *voltaic cell* or a *galvanic cell.* The platinum foils are called *electrodes,* and the U-tube with its electrolyte solution is called a *salt bridge.* Each solution, together with its electrode, is referred to as a *half-cell.* In the half-cell where oxidation occurs, the solution gives up electrons *to* the electrode, and that electrode is therefore designated the *negative* electrode. In the other half-cell, the reduction process takes electrons *from* the electrode, making that electrode the positive one. Electrons flow externally from the negative to the positive electrode. (Thus, in Figure 17-1, the electrode on the right is negative, the one on the left is positive, and electrons flow through the galvanometer from right to left.) The terms *anode* and *cathode* are also used to distinguish between the two electrodes. By definition, *an anode is any electrode at which oxidation occurs, and a cathode is any electrode at which reduction occurs.* Therefore, in a voltaic cell the positive electrode may be called the cathode and the negative electrode may be called the anode.

The specific reaction used in the experiment just described was chosen simply because the color changes that occur make the reaction visible. The experiment is merely an example of a general phenomenon; that is, any redox reaction can be made the basis of a voltaic cell by placing the ingredients of each half-reaction in a separate half-cell and providing suitable electrodes and a suitable salt bridge.

As a further illustration, consider the construction of a voltaic cell based on the reaction of CrO_2^- and ClO^- in basic solution (p. 457). The two half-reactions and the balanced overall reaction are:

Oxidation: $\qquad\qquad CrO_2^- + 4OH^- \rightarrow CrO_4^{2-} + 2H_2O + 3e^-$

Reduction: $\qquad\qquad ClO^- + H_2O + 2e^- \rightarrow Cl^- + 2OH^-$

Overall: $\qquad\qquad 2CrO_2^- + 3ClO^- + 2OH^- \rightarrow 2CrO_4^{2-} + 3Cl^- + H_2O$

In the construction of a voltaic cell from this reaction, one half-cell must contain the materials involved in the oxidation half-reaction, and the other must contain the materials of the reduction half-reaction. Therefore, in one beaker we will place an aqueous solution containing CrO_2^-, CrO_4^{2-}, and OH^-. (Note that CrO_4^{2-} is not a reactant but that it will be formed as soon as any current is drawn from the cell, so that whether we place any CrO_4^{2-} in the solution at the start or not, it will in fact be present as soon as the cell has begun to function.) The other beaker will hold an aqueous solution of ClO^-, Cl^- and OH^-. Platinum will serve as the electrodes, and a solution of Na_2SO_4 or $NaOH$ will be a satisfactory salt bridge. The assembled cell is shown in Figure 17-2.

Since oxidation occurs in the left-hand half-cell, electrons are released to that electrode, move through the wire from left to right and are consumed in the reduction reaction at the right-hand electrode. Therefore, the electrode on the left is negative; the one on the right is positive.

In both the examples cited above, all the substances involved in the reaction—both reactants and products—are in solution, and the electrodes are inert—they do not enter into the reaction itself but serve only as a means of conducting electrons. In redox reactions in which a metal is involved as reactant or product, a piece of

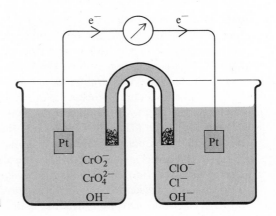

FIGURE 17-2

A Voltaic Cell

that metal may serve as an electrode. Consider, for example, the reaction that occurs when a strip of zinc metal is placed in an aqueous solution of copper(II) sulfate. The zinc dissolves as zinc atoms are oxidized to ions, and copper metal is formed as copper ions are reduced to copper atoms.

$$Zn + Cu^{2+} \rightarrow Zn^{2+} + Cu$$

A voltaic cell based on this reaction can be constructed so that one half-cell consists of a zinc rod dipping into a solution containing zinc ions and the other half-cell consists of a copper rod dipping into a solution of copper ions. As current is drawn from the cell, zinc is oxidized ($Zn \rightarrow Zn^{2+} + 2e^-$) and copper ions are reduced ($Cu^{2+} + 2e^- \rightarrow Cu$). Therefore, the zinc electrode is negative and the copper electrode is positive, and electrons flow externally from zinc to copper. This cell is depicted in Figure 17-3.

There is a conventional shorthand notation for the description of voltaic cells in which a vertical line separates the electrode from the solutes of the half-cell and the salt bridge is designated by a double vertical line. The *negative* electrode (the

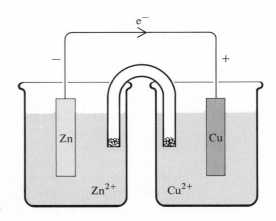

FIGURE 17-3

A Copper-Zinc Voltaic Cell

one at which *oxidation* occurs) is written on the left and the *positive* electrode on the right. This notation is illustrated for the three cells described above.

Figure 17-1: $Pt \mid I^-, I_2 \parallel Cr_2O_7^{2-}, H^+, Cr^{3+} \mid Pt$

Figure 17-2: $Pt \mid CrO_2^-, OH^-, CrO_4^{2-} \parallel ClO^-, Cl^-, OH^- \mid Pt$

Figure 17-3: $Zn \mid Zn^{2+} \parallel Cu^{2+} \mid Cu$

ELECTROMOTIVE FORCE

Any voltaic cell capable of producing an electric current may be thought of as possessing a "driving force" behind the flow of electrons; that is, a force that tends to push the electrons through the external circuit from the negative electrode to the positive electrode. This driving force is called the *electromotive force* (emf) and is commonly expressed in *volts* (V). The emf of a cell may be measured simply by attaching a voltmeter across the two electrodes. (It should be noted, however, that an ordinary voltmeter draws current from the cell, and therefore the emf will change slightly simply as a result of its being measured. For very careful emf measurements, vacuum tube voltmeters or potentiometers are used.)

Since the emf is a measure of the driving force behind the flow of electrons, it is also a measure of the tendency of the cell reaction to occur. The higher the voltage, the greater the tendency of the redox reaction in question to occur.

ELECTRODE POTENTIALS

In order to obtain a clearer understanding of the emf and its relationship to a cell reaction, it is necessary for us to consider the processes occurring at the individual electrodes.

Consider any half-cell, separate and unconnected to anything else; that is, an electrode dipping into a solution containing all the substances involved in a half-reaction. For example, it might be platinum dipping into a solution of $Cr_2O_7^{2-}$, Cr^{3+}, and H^+, it might be platinum dipping into a solution containing I_2 and I^-, or it might be copper dipping into a Cu^{2+} solution.

Now, the solutes in the solution have a certain tendency to remove electrons from the metal electrode, leaving it with less than its normal complement of electrons and making it *positive* with respect to the solution; that is, a potential is developed between the electrode and the solution. This potential is called the *electrode potential* (E), and since its magnitude is a measure of the tendency of the solute system to gain electrons (that is, the tendency for reduction to occur), it may also be called a *reduction potential*. It follows, then, that the stronger an oxidant a given system is, the more readily it is reduced and the larger is the electrode potential.

Now suppose two half-cells with different electrode potentials are connected with a salt bridge and an external circuit so as to produce a voltaic cell. At the electrode with the higher electrode potential (the greater tendency for reduction to occur), reduction will, in fact, take place, causing oxidation to occur at the electrode of lower electrode potential. The emf of the cell, then, is simply the difference between the two electrode potentials. Furthermore, since the emf must be positive, it is equal to the electrode potential at the positive electrode (cathode) *minus* the electrode potential at the negative electrode (anode).

$$\text{emf} = E_{(+)} - E_{(-)}$$

The magnitude of a single electrode potential is dependent upon a number of factors: (1) the particular redox system in question (thus, even if all conditions are kept the same, the electrode potential for the I_2/I^- couple will be different from that for the Cu^{2+}/Cu couple, which will be different from that for the Zn^{2+}/Zn couple, and so on); (2) the temperature; and (3) the relative concentrations of the substances involved in the half-reaction. The relationship between the electrode potential and these factors is expressed by the *Nernst equation* (developed by Walther Nernst in 1889), which is derived from thermodynamic considerations. For the generalized half-reaction

$$a(\text{Ox}) + ne^- \rightleftharpoons b(\text{Red})$$

(the oxidized form plus n electrons yields the reduced form), the Nernst equation states

$$E = E^0 - \frac{RT}{nF} \ln \frac{[\text{Red}]^b}{[\text{Ox}]^a}$$

where E is the electrode potential; E^0 is the *standard electrode potential,* a constant characteristic of the particular half-reaction; R is the gas constant (1.987 cal/°K); T is the absolute temperature; n is the number of electrons lost or gained in the half-reaction; F is the faraday (23,060 cal/V); and $\ln([\text{Red}]/[\text{Ox}])$ is the natural logarithm of the ratio of the molar concentration of the reduced form to that of the oxidized form.

The Nernst equation may be restated in a simplified, more easily used form. Since R and F are constants, they may be combined. Further, if we assume some definite temperature, say 25°C, then T may also be combined with the constant. Finally, since logarithms to the base 10 are more convenient to work with than natural logarithms, we can make that conversion by multiplying by 2.303. The equation then becomes

$$E = E^0 - \frac{0.059}{n} \log \frac{[\text{Red}]^b}{[\text{Ox}]^a}$$

at 25°C; the value of the combined constant has slightly different values at other temperatures, of course.

Some illustrations of the application of the Nernst equation to actual half-reactions follow.

1. $Fe^{3+} + e^- \rightleftharpoons Fe^{2+}$

$$E = E^0 - \frac{0.059}{1} \log \frac{[Fe^{2+}]}{[Fe^{3+}]}$$

2. $Cr_2O_7^{2-} + 14H^+ + 6e^- \rightleftharpoons 2Cr^{3+} + 7H_2O$

$$E = E^0 - \frac{0.059}{6} \log \frac{[Cr^{3+}]^2}{[Cr_2O_7^{2-}][H^+]^{14}}$$

(It will be noted that the $[H_2O]$ does not appear in the Nernst equation, since it is essentially a constant. The logic of this argument has been discussed in previous chapters.)

3. $Zn^{2+} + 2e^- \rightleftharpoons Zn$

$$E = E^0 - \frac{0.059}{2} \log \frac{1}{[Zn^{2+}]}$$

(Note that $[Zn]$ does not appear. Recall the discussion of heterogeneous equilibria in Chapter 13.)

4. $CrO_4^{2-} + 2H_2O + 3e^- \rightleftharpoons CrO_2^- + 4OH^-$

$$E = E^0 - \frac{0.059}{3} \log \frac{[CrO_2^-][OH^-]^4}{[CrO_4^{2-}]}$$

STANDARD ELECTRODE POTENTIALS

As was described above, there is a particular *standard electrode potential (E^0)* associated with each different redox couple. The evaluation of this constant for each couple is considerably useful, for if the values are known they can be used in the Nernst equation to predict the magnitude of electrode potentials for specific solutions and to calculate the emf of various voltaic cells. Furthermore, the E^0 values themselves are measures of relative strengths of oxidizing agents; that is, the stronger the oxidizing agent, the more readily it is reduced, and the larger is the value of the standard electrode potential.

Let us turn our attention, then, to the evaluation of standard electrode potentials.

Returning to the generalized statement of the Nernst equation,

$$E = E^0 - \frac{0.059}{n} \log \frac{[Red]^b}{[Ox]^a}$$

it is apparent that if the concentrations of reduced and oxidized forms are chosen so that the ratio in the logarithmic term is equal to 1, then, since $\log 1 = 0$, E will equal E^0. If both [Red] and [Ox] have a value of 1, then the ratio will equal 1 regardless of the values of the exponents a and b.

To illustrate with a specific example, let us examine the half-reaction

$$Cr_2O_7^{2-} + 14H^+ + 6e^- \rightarrow 2Cr^{3+} + 7H_2O$$

The Nernst equation for the electrode potential for this system is

$$E = E^0 - \frac{0.059}{6} \log \frac{[Cr^{3+}]^2}{[Cr_2O_7^{2-}][H^+]^{14}}$$

Now suppose we prepare a solution in which $[Cr_2O_7^{2-}]$, $[Cr^{3+}]$, and $[H^+]$ are each 1 molar, and we immerse a platinum foil in that solution. Since

$$\frac{(1)^2}{(1)(1)^{14}} = 1 \quad \text{and} \quad \log 1 = 0$$

we obtain $E = E^0$. In other words, the electrode potential developed at the platinum foil will be the *standard* electrode potential, E^0, for the $Cr_2O_7^{2-}/Cr^{3+}$ couple. Therefore, if we can in some way measure the electrode potential, E, of this solution, we will have, in fact, evaluated the constant E^0 for the system in question.

Unfortunately, it is not possible to measure a single electrode potential directly, since there is no complete circuit and no current flow. We can only measure the *difference between* two electrode potentials by connecting them to form a cell and measuring the emf of that cell; as was pointed out in the preceding section, emf $= E_{(+)} - E_{(-)}$.

By way of illustration, let us return to a consideration of the simple cell described in Figure 17-3. Here,

$$E_{(+)} = E^0_{(+)} - \frac{0.059}{2} \log \frac{1}{[Cu^{2+}]}$$

and

$$E_{(-)} = E^0_{(-)} - \frac{0.059}{2} \log \frac{1}{[Zn^{2+}]}$$

Now, if $[Cu^{2+}] = 1M$, and $[Zn^{2+}] = 1M$, then

$$E_{(+)} = E^0_{(+)} \quad \text{and} \quad E_{(-)} = E^0_{(-)}$$

and finally,

$$emf = E^0_{(+)} - E^0_{(-)}$$

Thus, if we construct the cell in question so that one solution is $1M$ in Zn^{2+} and the other is $1M$ in Cu^{2+}, the measured emf will be the *difference* between the E^0 value for the Cu^{2+}/Cu couple and the E^0 value for the Zn^{2+}/Zn couple. When we actually measure the emf of this cell, which may be depicted

$$Zn \mid Zn^{2+}(1M) \parallel Cu^{2+}(1M) \mid Cu$$

we get a value of 1.10 V. Therefore,

$$E^0_{Cu^{2+}/Cu} - E^0_{Zn^{2+}/Zn} = 1.10 \text{ V}$$

Recognizing, then, that we cannot measure the *absolute* values of standard electrode potentials but only differences between them, we find it expedient to develop a system of *relative* standard electrode potentials. Thus, if we choose some particular redox couple and assign it some arbitrary E^0 value, we can measure the E^0 values of other redox couples *relative* to the value assigned the reference couple.

The redox couple that has been chosen to serve as the basis of this relative system may be represented by the half-reaction

$$2H^+ + 2e^- \rightleftharpoons H_2$$

and the E^0 value assigned to it is 0.00 V.

A half-cell for measuring the potential of this H^+/H_2 couple consists of a platinum foil dipping into a solution of acid, with the platinum enclosed inside a tube that permits hydrogen gas to be bubbled over the electrode (Figure 17-4). If

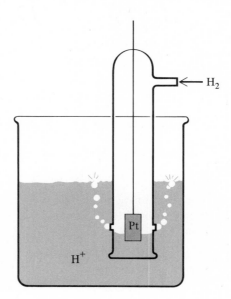

FIGURE 17-4

A Hydrogen Electrode

the hydrogen is admitted under a pressure of exactly 1 atm, then the electrode potential is given by the expression

$$E = E^0 - \frac{0.059}{2} \log \frac{1}{[H^+]^2}$$

Further, if the concentration of H^+ in the solution is exactly $1M$, then the potential at the electrode is the standard potential for the H^+/H_2 couple ($E = E^0$). This potential has been assigned the value of 0.00 V.

The standard electrode potentials for other redox couples can be measured by constructing half-cells from them, connecting these with the standard hydrogen electrode, and measuring the emf of the cell thus produced. Thus, we could measure the E^0 value for the Zn^{2+}/Zn couple as follows. Into a solution one molar in Zn^{2+} is placed a rod of zinc metal. This metal is connected by a wire to the wire of a standard hydrogen electrode (Figure 17-4; $[H^+] = 1$; H_2 at 1 atmosphere pressure). The two solutions are connected by a suitable salt bridge. When the emf of this cell is measured, it is found that the hydrogen electrode is positive, the zinc electrode is negative, and the emf is 0.76 V. Since

$$\text{emf} = E_{(+)} - E_{(-)} = E^0_{H^+/H_2} - E^0_{Zn^{2+}/Zn} = 0.76 \text{ V}$$

and since $E^0_{H^+/H_2}$ has been assigned a value of 0.00,

$$0.00 - E^0_{Zn^{2+}/Zn} = 0.76 \text{ V}$$

and

$$E^0_{Zn^{2+}/Zn} = -0.76 \text{ V}$$

We have thus established that the standard electrode potential for the half-reaction $Zn^{2+} + 2e^- \rightleftharpoons Zn$ is -0.76 V relative to the standard hydrogen electrode.

Note that as current is drawn from the cell, the half-reactions and overall reaction that occur are

$$(+) \quad 2H^+ + 2e^- \rightarrow H_2$$
$$(-) \quad Zn \rightarrow Zn^{2+} + 2e^-$$
$$\overline{\quad Zn + 2H^+ \rightarrow Zn^{2+} + H_2}$$

and that the cell could properly be designated as

$$Zn \mid Zn^{2+}(1M) \parallel H^+(1M), H_2 (1 \text{ atm}) \mid Pt$$

Returning to the zinc-copper cell discussed above, for which the emf was found to be 1.10 V (p. 466), it is now obvious that the E^0 value for the Cu^{2+}/Cu couple is 0.34 V.

$$E^0_{Cu^{2+}/Cu} - (-0.76) = 1.10 \text{ V}$$
$$E^0_{Cu^{2+}/Cu} = 1.10 - 0.76 = 0.34 \text{ V}$$

By the use of procedures similar to the one just described, the standard electrode potentials of a large variety of redox couples have been measured. A few of these are given in Table 17-1; a larger list is found in Appendix 10.

STRENGTHS OF OXIDIZING AGENTS AND THE SPONTANEITY OF REACTIONS

Since the magnitude of E^0 is a measure of the relative tendency for reduction to occur, the relative strengths of oxidizing and reducing agents are predictable from

	Half-Reaction	E^0 (V)
TABLE 17-1 **Some Standard** **Electrode Potentials (25°C)**	$F_2 + 2e^- \rightleftharpoons 2F^-$	+2.87
	$MnO_4^- + 8H^+ + 5e^- \rightleftharpoons Mn^{2+} + 4H_2O$	+1.51
	$Cl_2 + 2e^- \rightleftharpoons 2Cl^-$	+1.36
	$Cr_2O_7^{2-} + 14H^+ + 6e^- \rightleftharpoons 2Cr^{3+} + 7H_2O$	+1.33
	$Br_2 + 2e^- \rightleftharpoons 2Br^-$	+1.06
	$Ag^+ + e^- \rightleftharpoons Ag$	+0.80
	$Fe^{3+} + e^- \rightleftharpoons Fe^{2+}$	+0.77
	$I_2 + 2e^- \rightleftharpoons 2I^-$	+0.54
	$Cu^{2+} + 2e^- \rightleftharpoons Cu$	+0.34
	$2H^+ + 2e^- \rightleftharpoons H_2$	0.00
	$Sn^{2+} + 2e^- \rightleftharpoons Sn$	−0.14
	$Ni^{2+} + 2e^- \rightleftharpoons Ni$	−0.25
	$Fe^{2+} + 2e^- \rightleftharpoons Fe$	−0.44
	$Zn^{2+} + 2e^- \rightleftharpoons Zn$	−0.76
	$Al^{3+} + 3e^- \rightleftharpoons Al$	−1.66
	$Mg^{2+} + 2e^- \rightleftharpoons Mg$	−2.36
	$Na^+ + e^- \rightleftharpoons Na$	−2.71
	$Ba^{2+} + 2e^- \rightleftharpoons Ba$	−2.91
	$K^+ + e^- \rightleftharpoons K$	−2.93
	$Li^+ + e^- \rightleftharpoons Li$	−3.05

the E^0 values. In Table 17-1 the substances on the left of the double arrows are the oxidized forms; therefore, they are the potential oxidizing agents. Since the half-reactions are listed in order of decreasing E^0, it follows that any oxidizing agent is stronger than any that stands below it in the table. Accordingly, F_2 is the strongest oxidant listed, MnO_4^- in acid solution is the next strongest, and so on, with Li^+ being the weakest. On the other hand, the substances to the right of the double arrows are reducing agents, and the order of relative strength as reductants is inverted. That is, if F_2 has a very high tendency to be reduced to F^-, then F^- must have a very low tendency to be oxidized to F_2. Hence, F^- is the weakest reducing agent in the table; Li is the strongest reducing agent, K the second strongest, and so forth.

A logical extension of this relationship between the values of E^0 and the relative strengths of oxidizing and reducing agents is the prediction of whether or not a particular redox reaction will occur spontaneously. It follows from the foregoing discussion that any oxidant in the table will oxidize (be reduced by) any reductant that lies below it. For example, we might ask, will an acidic permanganate solution oxidize chloride ion to chlorine? In other words, will the following reaction occur spontaneously?

$$2MnO_4^- + 16H^+ + 10Cl^- \rightarrow 2Mn^{2+} + 8H_2O + 5Cl_2$$

Note that this equation is arrived at by adding the proper half-reactions after making the number of electrons lost equal to the number gained.

$$2(MnO_4^- + 8H^+ + 5e^- \rightleftharpoons Mn^{2+} + 4H_2O)$$
$$\underline{5(2Cl^- \rightleftharpoons Cl_2 + 2e^-)}$$

Table 17-1 tells us that the E^0 for the reduction of MnO_4^- is 1.51 V, whereas the E^0 for the reduction of Cl_2 is only 1.36 V. Acidic permanganate is a stronger oxidizing agent than Cl_2; therefore, MnO_4^- will be reduced and Cl^- will be oxidized, and the reaction in question *is* spontaneous.

The question of the spontaneity of this reaction may be asked in another way: in terms of the voltaic cell based on the reaction. Will a cell constructed from the two half-cells in question actually generate a flow of current in such a way that the electrode of the MnO_4^- half-cell is positive (reduction occurs) and the electrode in the Cl_2/Cl^- half-cell is negative (oxidation occurs)? In other words, will the cell designated

$$Pt \mid Cl^-(1M), Cl_2(1\ atm) \parallel MnO_4^-(1M), H^+(1M), Mn^{2+}(1M) \mid Pt$$

have an emf that is positive?

$$emf = E_{(+)} - E_{(-)} = 1.51 - 1.36 = +0.15\ V$$

The calculated emf *is* positive; the cell *will* function as written; the reaction in question *is* spontaneous.

Let us consider now the spontaneity of the reaction between acidic dichromate and chloride. Will an acidic dichromate solution oxidize chloride to chlorine? We can write an equation for the reaction in question by combining the proper half-reactions:

$$Cr_2O_7^{2-} + 14H^+ + 6e^- \rightleftharpoons 2Cr^{3+} + 7H_2O$$
$$\underline{3(2Cl^- \rightleftharpoons Cl_2 + 2e^-)}$$
$$Cr_2O_7^{2-} + 14H^+ + 6Cl^- \rightarrow 2Cr^{3+} + 7H_2O + 3Cl_2$$

Referring to Table 17-1, we find that Cl_2 is a stronger oxidizing agent than $Cr_2O_7^{2-}$; therefore, $Cr_2O_7^{2-}$ will not oxidize Cl^-, and the reaction written is *not* spontaneous. In fact, the *reverse* of the reaction *will* occur spontaneously.

Investigation of this reaction in terms of the cell produced from it,

$$Pt \mid Cl^- (1M), Cl_2 \text{ (1 atm)} \parallel Cr_2O_7^{2-} (1M), H^+(1M), Cr^{3+}(1M) \mid Pt$$

leads to the conclusion that this cell would *not* function as written, since the calculated emf is negative.

$$\text{emf} = E_{(+)} - E_{(-)} = 1.33 - 1.36 = -0.03 \text{ V}$$

It must be emphasized that since these predictions of spontaneity are based on the use of *standard* electrode potentials, we are merely predicting whether or not a particular reaction will occur spontaneously at a temperature of 25°C when *all* substances involved in the reaction are in the standard state; that is, when all electrolytes in solution are present at $1M$ concentration and gases are at a partial pressure of 1 atm. Thus, for example, in our prediction that the reaction between acidic dichromate and chloride ion is not spontaneous, we have not said that it is *impossible* to bring about this reaction. In fact, $Cr_2O_7^{2-}$ *can* be made to oxidize Cl^- by proper selection of conditions; for instance, by lowering the partial pressure of Cl_2, removing Cr^{3+} as formed, or using a very high concentration of $Cr_2O_7^{2-}$. In fact, any change of conditions that will raise the electrode potential of the $Cr_2O_7^{2-}/Cr^{3+}$ couple above that of the Cl_2/Cl^- couple will cause the reaction to occur. Furthermore, it should be noted that these predictions are based only on thermodynamic considerations, not on the *kinetics* of the reactions. A prediction of spontaneity indicates only that the reaction is thermodynamically favored; it indicates nothing about the *rate* of the reaction.

STANDARD POTENTIALS AND THE EQUILIBRIUM CONSTANT

In making predictions concerning the spontaneity of redox reactions from values of E^0, as discussed above, what we are really doing, of course, is making a *qualitative* statement about the *equilibrium* position of the reaction. If we predict that a given reaction is spontaneous, we are saying that the equilibrium position lies in favor of the products; conversely, for a nonspontaneous reaction the equilibrium position favors the reactants.

We can relate the standard electrode potentials of the redox couples involved in a reaction to the equilibrium position of that reaction *quantitatively;* that is, we can express the equilibrium constant for a redox reaction as a function of E^0 values. Let us illustrate this relationship with the reaction between acid dichromate and ferrous ion, the balanced equation for which is

$$Cr_2O_7^{2-} + 14H^+ + 6Fe^{2+} \rightarrow 2Cr^{3+} + 6Fe^{3+} + 7H_2O$$

The reduction half-reaction is

$$Cr_2O_7^{2-} + 14H^+ + 6e^- \rightarrow 2Cr^{3+} + 7H_2O$$

and the oxidation half-reaction is

$$Fe^{2+} \rightarrow Fe^{3+} + e^-$$

Now, as we have seen, the driving force behind this reaction is the difference between the two electrode potentials,

$$emf = E_{Cr} - E_{Fe}$$

and the two electrode potentials as expressed by the Nernst equation are

$$E_{Cr} = E_{Cr}^0 - \frac{0.059}{6} \log \frac{[Cr^{3+}]^2}{[Cr_2O_7^{2-}][H^+]^{14}}$$

and

$$E_{Fe} = E_{Fe}^0 - 0.059 \log \frac{[Fe^{2+}]}{[Fe^{3+}]}$$

As the reaction proceeds, $[Cr_2O_7^{2-}]$ and $[H^+]$ decrease, while $[Cr^{3+}]$ increases, making the logarithmic term in the Nernst equation larger, and therefore E_{Cr} becomes smaller. Similarly, $[Fe^{2+}]$ is decreasing and $[Fe^{3+}]$ increasing, causing the ratio to decrease and making E_{Fe} larger. If E_{Cr} is decreasing and E_{Fe} is increasing, these two values must eventually become equal, and the driving force of the reaction (the emf) becomes equal to zero.

$$emf = E_{Cr} - E_{Fe} = 0$$

At this point the system is in equilibrium, since the driving force is the same in both directions.

Note that this same line of reasoning explains why the emf of a voltaic cell gradually decreases as current is drawn from it and will ultimately drop to zero (equilibrium is reached).

In view of the above discussion, at equilibrium,

$$E_{Cr} = E_{Fe}$$

and therefore,

(1) $\quad E_{Cr}^0 - \dfrac{0.059}{6} \log \dfrac{[Cr^{3+}]^2}{[Cr_2O_7^{2-}][H^+]^{14}} = E_{Fe}^0 - 0.059 \log \dfrac{[Fe^{2+}]}{[Fe^{3+}]}$

Rearranging terms, equation (1) becomes

(2) $\quad E_{Cr}^0 - E_{Fe}^0 = \dfrac{0.059}{6} \log \dfrac{[Cr^{3+}]^2}{[Cr_2O_7^{2-}][H^+]^{14}} - 0.059 \log \dfrac{[Fe^{2+}]}{[Fe^{3+}]}$

Now, since

$$0.059 \log \frac{[Fe^{2+}]}{[Fe^{3+}]} = \frac{0.059}{6} \log \frac{[Fe^{2+}]^6}{[Fe^{3+}]^6}$$

we may rewrite (2) as follows:

(3) $\quad E_{Cr}^0 - E_{Fe}^0 = \dfrac{0.059}{6} \log \dfrac{[Cr^{3+}]^2}{[Cr_2O_7^{2-}][H^+]^{14}} - \dfrac{0.059}{6} \log \dfrac{[Fe^{2+}]^6}{[Fe^{3+}]^6}$

Combining terms in equation (3) gives:

(4) $\qquad E_{Cr}^0 - E_{Fe}^0 = \dfrac{0.059}{6} \log \dfrac{[Cr^{3+}]^2[Fe^{3+}]^6}{[Cr_2O_7^{2-}][H^+]^{14}[Fe^{2+}]^6}$

But, for the reaction in question,

$$K = \dfrac{[Cr^{3+}]^2[Fe^{3+}]^6}{[Cr_2O_7^{2-}][H^+]^{14}[Fe^{2+}]^6}$$

and equation (4) becomes

(5) $\qquad\qquad\qquad E_{Cr}^0 - E_{Fe}^0 = \dfrac{0.059}{6} \log K$

and

$$\log K = \dfrac{6[E_{Cr}^0 - E_{Fe}^0]}{0.059}$$

We have derived an expression for the value of the equilibrium constant at 25°C as a function of the values of the E^0s. By following the same derivation procedure for any redox reaction, one can show that the *general* expression for the relationship is

$$\log K = \dfrac{n(E_{red}^0 - E_{ox}^0)}{0.059}$$

where E_{red}^0 is the standard electrode potential of the couple being reduced in the reaction, E_{ox}^0 is the standard electrode potential of the couple being oxidized, and n is the number of electrons transferred in the overall balanced reaction.

Using this formula to calculate the equilibrium constant for the reaction between MnO_4^- and Cl^-, which we have predicted to be spontaneous (p. 470), we get

$$\log K = \dfrac{10(1.51 - 1.36)}{0.059} = \dfrac{10(0.15)}{0.059} = 25.4$$

$$K = 10^{25.4} = 2.5 \times 10^{25}$$

Applying the calculation to the nonspontaneous reaction between $Cr_2O_7^{2-}$ and Cl^- (p. 470) gives

$$\log K = \dfrac{6(1.33 - 1.36)}{0.059} = \dfrac{6(-0.03)}{0.059} = -3.0$$

$$K = 1 \times 10^{-3}$$

Note that the relative magnitudes of these two equilibrium constants agree with our predictions about the spontaneity of the two reactions.

The relationship between standard electrode potentials and the equilibrium constant also enables us to relate standard electrode potentials to the standard free-energy change of the reaction in question. It will be recalled (p. 464) that the factor 0.059 in our simplified form of the Nernst equation is the result of combining several constants and assuming the temperature to be 25°C; namely,

$$(1) \qquad 0.059 = \frac{2.3RT}{F}$$

Substituting this relationship into the equilibrium constant expression derived above,

$$(2) \qquad \log K = \frac{n(E_{red}^0 - E_{ox}^0)}{0.059}$$

we get the expression:

$$(3) \qquad \log K = \frac{n(E_{red}^0 - E_{ox}^0)F}{2.3RT}$$

Now, it will be recalled from Chapter 13 (p. 342) that

$$(4) \qquad \Delta G^0 = -2.3RT \log K$$

Substituting the value for $\log K$ given in equation (3) into equation (4) gives

$$\Delta G^0 = \frac{-2.3RTnF(E_{red}^0 - E_{ox}^0)}{2.3RT}$$

and therefore,

$$(5) \qquad \Delta G^0 = -nF(E_{red}^0 - E_{ox}^0)$$

This relationship may also be expressed as

$$\Delta G^0 = -nF(\text{emf})^0$$

where $(\text{emf})^0$ is the emf of a cell in which the two electrode potentials are the *standard* electrode potentials for the two half-reactions in question.

Thus, we can calculate the value of ΔG^0 for any oxidation-reduction reaction if we know the appropriate E^0 values and the number of electrons (n) transferred in the balanced equation. Some applications of this relationship are discussed in Part 3 of this book.

OXIDATION POTENTIALS

Before concluding our discussion of electrode potentials, it should be made clear that the system of designating electrode potentials is simply a matter of definition. We have here defined the electrode potential as a measure of the tendency of the solutes to *gain* electrons, therefore to be *reduced*. Thus, the electrode potential is a *reduction potential.* This is the definition suggested by the International Union of Pure and Applied Chemistry in 1953. Prior to that time this was only one of two definitions in common use, and, unfortunately, even since that time not all authors have followed the IUPAC suggestion.

Employing the same kind of reasoning as was used on p. 463, let us adapt our definitions to the *other* system of designating electrode potentials. Assume we have a suitable metal foil dipping into a solution containing all the materials of a redox couple. The solutes will have a certain tendency to *give* electrons to the metal, causing it to have *more* than its normal complement of electrons, and thus making it negative with respect to the solution. In other words, a potential is developed between the electrode and the solution. The magnitude of this potential is a measure of the tendency of the couple to *lose* electrons—*be oxidized*—and therefore should properly be called an *oxidation potential.* By this definition, then, the potential is a measure of the tendency of the following process to take place to the *right:*

$$a(\text{Red}) \rightleftharpoons b(\text{Ox}) + n\text{e}$$

Since *oxidation potentials* are measured on the same relative basis as *reduction potentials*—relative to the standard hydrogen electrode potential taken as 0.00 volt—they have the same value but opposite sign. For example, if a given solution has an electrode (reduction) potential of 1.23 V, then its oxidation potential is −1.23 V.

It follows also that when dealing with an *oxidation potential*, that potential as expressed by the Nernst equation becomes

$$E = E^0 - \frac{0.059}{n} \log \frac{[\text{Ox}]^b}{[\text{Red}]^a}$$

Furthermore, under the system of *oxidation potentials*, it is obvious that the larger potential in a voltaic cell is the one at the *negative electrode* (the one at which oxidation occurs), and therefore emf $= E_{(-)} - E_{(+)}$.

Finally, it follows from the foregoing discussion that *standard oxidation potentials* (E^0) have the same value as their corresponding *standard electrode potentials* (*reduction*), but with opposite sign. For example, for the Zn^{2+}/Zn couple,

$$\text{Zn}^{2+} + 2\text{e}^- \rightleftharpoons \text{Zn} \qquad E^0 = -0.76$$

and

$$\text{Zn} \rightleftharpoons \text{Zn}^{2+} + 2\text{e}^- \qquad E^0 = +0.76$$

Table 17-1 can be converted to a table of *standard oxidation potentials* by writing the reverse of each of the half-reactions and changing the signs of all the E^0 values.

Recognizing that either of these two systems can be used as effectively as the other, we shall nevertheless continue to use throughout the remainder of this book the system of designations that was developed earlier in this chapter—when we speak of electrode potentials, we shall mean *reduction* potentials.

CONCENTRATION CELLS

All of the voltaic cells used as illustrations in the preceding discussions have consisted of half-cells involving two different redox couples. Since electrode potentials depend not only on the specific couple involved but also on the concentrations of the electrolytes, it is possible to construct a voltaic cell in which the two half-cells involve the same redox couple, provided the concentrations are different. A cell of this type, called a *concentration cell,* is illustrated by the following simple example.

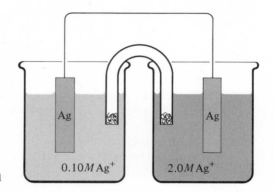

FIGURE 17-5

A Concentration Cell

A piece of silver metal dipping into a 0.10*M* AgNO$_3$ solution serves as one half-cell, and another piece of silver metal dipping into a 2.0*M* AgNO$_3$ solution serves as the other half-cell (Figure 17-5). In both half-cells, the redox couple is

$$Ag^+ + e^- \rightleftharpoons Ag$$

The electrode potential on the *left* in Figure 17-5 may be calculated from the Nernst equation.

$$E = E^0 - 0.059 \log \frac{1}{[Ag^+]}$$

$$= 0.80 - 0.059 \log \frac{1}{0.1}$$

$$= 0.80 - 0.059 \log 10$$

$$E = 0.80 - 0.059(1) = 0.74 \text{ V}$$

Following the same procedure, the electrode potential on the right may be obtained:

$$E = E^0 - 0.059 \log \frac{1}{[Ag^+]}$$

$$= 0.80 - 0.059 \log \frac{1}{2.0}$$

$$= 0.80 - 0.059 \log 0.5$$

$$= 0.80 - 0.059 \, (-.30)$$

$$E = 0.80 - (-.02) = 0.82 \text{ V}$$

Since the right electrode of this concentration cell has the larger electrode potential, it is positive, and the left is negative. The emf of the cell is $0.82 - 0.74 = 0.08$ V. When current is drawn from the cell, the electrode reactions are

Right: $Ag^+ + e^- \rightarrow Ag$

Left: $Ag \rightarrow Ag^+ + e^-$

This cell may be represented by the notation

$$Ag \mid Ag^+(0.10M) \parallel Ag^+(2.0M) \mid Ag$$

SOME VOLTAIC CELLS OF PRACTICAL USEFULNESS

Although, theoretically, voltaic cells can be constructed from a large variety of redox reactions, most of them would not be practical as useful sources of direct current. For instance, cells involving gases in their reactions may be impractical in the matter of space requirements and convenience; cells requiring platinum or other precious metals for electrodes are too expensive; and for many cells, the emf produced is too small for practical applications.

A number of cells, however, have found practical application and are produced commercially. One of the most common is the *lead storage cell,* in which the negative electrode is lead and the positive electrode is a grid of lead packed with lead dioxide. These "plates" are immersed in an aqueous solution of sulfuric acid. When current is drawn from the cell, the electrode reactions are

$$(+) \quad PbO_2\,(s) + SO_4{}^{2-} + 4H^+ + 2e^- \rightarrow PbSO_4\,(s) + 2H_2O$$

$$(-) \qquad\qquad Pb(s) + SO_4{}^{2-} \rightarrow PbSO_4\,(s) + 2e^-$$

The overall reaction involves the conversion of both Pb and PbO_2 into $PbSO_4$, and sulfuric acid is consumed in the process.

$$Pb + PbO_2 + 4H^+ + 2SO_4{}^{2-} \rightarrow 2PbSO_4 + 2H_2O$$

The state of "charge" of the cell can be determined by measuring the amount of H_2SO_4 remaining, and this is most conveniently done simply by measuring the

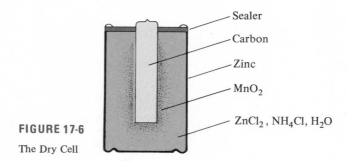

FIGURE 17-6

The Dry Cell

specific gravity of the electrolyte. This cell can be "recharged" by applying an external source of current in a direction opposite to the current produced by the cell, which causes the reverse of these electrode reactions to take place.

The emf of the lead storage cell is approximately 2 V. In its most commonly used form, three or six of these cells are joined in series to produce the common 6- or 12-volt *battery*.

Another cell in common use is the *dry cell*. This usually consists of a zinc container filled with a paste of ammonium chloride, zinc chloride, water, and an inert filler. Imbedded in the center of this paste mixture is a carbon rod surrounded by manganese dioxide (Figure 17-6). The zinc container is the negative electrode (anode), and the carbon rod the positive electrode (cathode). The maximum emf of the dry cell is approximately 1.5 V. The electrode reactions are complex but are usually represented as:

Anode: $Zn \rightarrow Zn^{2+} + 2e^-$

Cathode: $2NH_4^+ + 2MnO_2 + 2e^- \rightarrow Mn_2O_3 + 2NH_3 + H_2O$

FUEL CELLS

Since combustion of natural fuels (coal, oil, gas) is an oxidation-reduction process in which the fuel is oxidized and oxygen is reduced, it is theoretically possible to construct a cell to produce electric current directly from fuel combustion.

In the conventional use of fuel to produce electricity, the heat energy released from the combustion is used to convert water into steam, and the steam is used to drive a turbine, which operates a generator. There is a considerable energy loss in this process, maximum efficiency being only about 40 percent. Direct production of electricity from the oxidation of the fuel in a cell should yield a much higher efficiency, and consequently a great deal of effort in recent years has gone into the development of practical cells based on fuel combustion. These cells, in which the reactants are supplied continuously and the products are removed continuously, are called *fuel cells*.

Unfortunately, the combustion of most fuels requires high temperatures, which introduce practical problems in the operation of such cells: corrosion of the container and electrodes, and reduced efficiency because of the need to continuously supply thermal energy to the cell. Fuels that can be used at low temperature—for example, hydrogen and hydrazine—are too expensive for large-scale use. Thus far they have been used only in situations where cost is of secondary importance, such as in space vehicles.

The operation of a fuel cell using hydrogen as fuel may be illustrated as follows. Hydrogen and oxygen are bubbled through electrodes of porous compressed carbon in which catalysts have been incorporated. The electrodes are immersed in concentrated sodium hydroxide solution.

$$C \mid H_2, OH^-, O_2 \mid C$$

The hydrogen and oxygen are consumed in the reaction, and the water that is formed is removed by evaporation. The reactions which occur are

Anode:	$H_2 + 2OH^- \rightarrow 2H_2O + 2e^-$
Cathode:	$O_2 + 2H_2O + 4e^- \rightarrow 4OH^-$
Overall:	$2H_2 + O_2 \rightarrow 2H_2O$

ELECTROLYTIC CELLS

A previous chapter dealt with the classification of substances as electrolytes or nonelectrolytes (p. 279). It was established that molten ionic compounds (salts) and aqueous solutions of electrolytes (acid, bases, salts) conduct electric current. This conduction, called *electrolytic conduction* to distinguish it from *metallic conduction* (p. 197), is the result of oxidation-reduction processes.

Perhaps the simplest example of this process is the conduction of a molten salt of monatomic ions, such as sodium chloride. Consider some sodium chloride

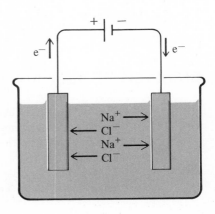

FIGURE 17-7

Electrolysis of Molten Sodium Chloride

that is kept at a temperature above its melting point so that it remains in the liquid state. Two inert electrodes are placed in the melt, and a direct current source (for instance, a battery of voltaic cells) is connected across the electrodes (Figure 17-7). Sodium ions, being positive, are attracted to the negative electrode, and the negative chloride ions are attracted to the positive electrode. But this migration of ions does not in itself account for the completion of the electrical circuit. At the negative electrode, sodium ions accept electrons from the electrode and are thus reduced to sodium atoms. Simultaneously, chloride ions give up electrons to the positive electrode, thus being oxidized to elemental chlorine. The overall process taking place, then, is an oxidation-reduction reaction, consisting of two half-reactions: reduction at the negative electrode and oxidation at the positive electrode.

$$(-) \quad Na^+ + e^- \rightarrow Na$$

$$(+) \qquad 2Cl^- \rightarrow Cl_2 + 2e^-$$

$$\overline{\quad 2Na^+ + 2Cl^- \rightarrow 2Na + Cl_2 \quad}$$

The apparatus described is called an *electrolytic cell,* and the complete process that takes place is called *electrolysis.* The material being *electrolyzed* (in this case molten NaCl) is referred to as the *electrolyte.* During electrolysis, reduction occurs at the negative electrode, which may therefore be called the *cathode;* the positive electrode, then, is the *anode*, since oxidation occurs there. (Note the difference in these designations between an electrolytic cell and a voltaic cell. Recall that in a voltaic cell the anode is the negative electrode and the cathode is the positive one.)

It should be noted that electrons do *not* flow through the electrolyte. They are generated by the direct current source and move through the connecting wire into the cathode, where they are consumed by a reduction process. Other electrons are generated at the anode by the oxidation process and move from there to the source through the wire. The difference between an electrolytic cell and a voltaic cell is now apparent. In a voltaic cell an oxidation-reduction reaction occurs spontaneously to produce an electric current, whereas in an electrolytic cell an electric current is used to force a nonspontaneous reaction to take place.

ELECTROLYSIS OF AQUEOUS SOLUTIONS

The foregoing discussion of electrolysis of molten NaCl also applies in general to the electrolysis of aqueous solutions of electrolytes. The latter are somewhat more complicated, however, because of the presence of water as well as cations and anions.

Let us consider the electrolysis of a dilute solution of NaCl. The electrolytic cell is the same as shown in Figure 17-7, except that dilute aqueous NaCl solution replaces the molten NaCl. At the cathode *two* different reductions are possible: reduction of Na^+ or reduction of water molecules.

$$Na^+ + e^- \rightarrow Na \qquad E^0 = -2.71 \text{ V}$$

$$2H_2O + 2e^- \rightarrow H_2 + 2OH^- \qquad E^0 = -0.83 \text{ V}$$

Comparison of the standard electrode potentials of the two redox couples in question indicates that water is much more readily reduced than sodium ion, and the product at the cathode is hydrogen gas, not sodium metal.

Similarly, at the anode, *two* oxidations are possible: oxidation of Cl^- or of H_2O.

$$2Cl^- \rightarrow Cl_2 + 2e^- \qquad E^0 = -1.36 \text{ V}$$

$$2H_2O \rightarrow O_2 + 4H^+ + 4e^- \qquad E^0 = -1.23 \text{ V}$$

(Note that the E^0 values given here are standard *oxidation* potentials. This is simply a matter of convenience, since we are comparing oxidation processes.) The E^0 values favor oxidation of H_2O rather than Cl^-, but since they differ so little, which of the oxidations will actually occur depends on relative concentration. (Recall that *standard* potentials are based on all concentrations being one molar; the actual electrode potentials are functions of concentration in accordance with the Nernst equation.) Only if the solution of NaCl is dilute will O_2 be the anode product; more concentrated solutions will give only Cl_2.

Thus, the electrode reactions and overall reaction for the electrolysis of aqueous NaCl solution may be summarized as follows:

Very dilute solution:

Cathode: $\qquad 2H_2O + 2e^- \rightarrow H_2 + 2OH^-$

Anode: $\qquad \underline{2H_2O \rightarrow O_2 + 4H^+ + 4e^-}$

Overall: $\qquad 6H_2O \rightarrow 2H_2 + O_2 + 4H^+ + 4OH^-$

More concentrated solution:

Cathode: $\qquad 2H_2O + 2e^- \rightarrow H_2 + 2OH^-$

Anode: $\qquad \underline{2Cl^- \rightarrow Cl_2 + 2e^-}$

Overall: $\qquad 2H_2O + 2Cl^- \rightarrow H_2 + Cl_2 + 2OH^-$

As a further illustration of this "competition" of electrode reactions, let us examine the electrolysis of an aqueous copper(II) sulfate solution using inert electrodes. At the cathode, the two possible reductions are

$$Cu^{2+} + 2e^- \rightarrow Cu \qquad E^0 = +0.34 \text{ V}$$

$$2H_2O + 2e^- \rightarrow H_2 + 2OH^- \qquad E^0 = -0.83 \text{ V}$$

It is obvious that the reduction of Cu^{2+} is favored and will take place rather than the reduction of water.

At the anode, the two possible oxidations and their standard oxidation potentials are

$$2SO_4^{2-} \rightarrow S_2O_8^{2-} + 2e^- \qquad E^0 = -2.01 \text{ V}$$

$$2H_2O \rightarrow O_2 + 4H^+ + 4e^- \qquad E^0 = -1.23 \text{ V}$$

We may conclude that water will be oxidized rather than SO_4^{2-}. The reactions that actually occur, therefore, are

Cathode:	$Cu^{2+} + 2e^- \rightarrow Cu$
Anode:	$2H_2O \rightarrow O_2 + 4H^+ + 4e^-$
Overall:	$2Cu^{2+} + 2H_2O \rightarrow 2Cu + O_2 + 4H^+$

This same kind of "competition" may exist between several ions if more than one electrolyte is present in solution. For example, we may raise the question: In the electrolysis of the solution containing both Cu^{2+} and Ag^+, what will the cathode reaction be? Looking at the standard electrode potentials for the three "competing" reductions,

$$Cu^{2+} + 2e^- \rightarrow Cu \qquad E^0 = +0.34 \text{ V}$$

$$Ag^+ + e^- \rightarrow Ag \qquad E^0 = +0.80 \text{ V}$$

$$2H_2O + 2e^- \rightarrow H_2 + 2OH^- \qquad E^0 = -0.83 \text{ V}$$

we must conclude that the reduction of Ag^+ will take place preferentially. It must be kept in mind that this prediction is based on the assumption that Ag^+ and Cu^{2+} are present in equimolar concentrations. Thus, if we had a solution very concentrated in Cu^{2+} and extremely dilute in Ag^+, copper may, in fact, plate out before silver.

It is seen, then, that the products of an electrolysis depend not only on what electrolytes are present but also on the concentrations in which they are present. Given the exact concentrations of all species present in a solution to be electrolyzed, we can use standard electrode potentials and the Nernst equation to predict accurately what the electrode reactions will be.

However, on the basis of the foregoing discussion, some simple rules of thumb can be derived that enable us to predict qualitatively the electrode reactions in the electrolysis of a solution of a single electrolyte at moderate concentrations.

A. *Cathode reactions:*
   1. If the electrolyte is an acid (HCl, HNO_3, and so forth), then the reaction is

$$2H^+ + 2e^- \rightarrow H_2$$

   2. If the electrolyte is a base or a salt (NaOH, $AgNO_3$, and so on) and
      (a) if the cation is that of a very active metal (Na^+, Ba^{2+}, Mg^{2+}, etc.)—that is, has a very low E^0—then the reaction is

$$2H_2O + 2e^- \rightarrow H_2 + 2OH^-$$

(b) if the cation is that of a less reactive metal (Cu^{2+}, Ag^+, Cd^{2+},)—that is, has a moderate to high E^0—then the reaction is

$$M^{n+} + ne^- \rightarrow M$$

B. *Anode reactions:*
   1. If the electrolyte is a base (NaOH, KOH, and so on), then the reaction is

$$4OH^- \rightarrow O_2 + 2H_2O + 4e^-$$

   2. If the electrolyte is an acid or a salt and
      (a) if the anion is a monatomic ion with a relatively high standard oxidation potential (Br^-, I^-, etc.), then the reaction is, for example,

$$2Br^- \rightarrow Br_2 + 2e^-$$

      (b) if the anion is a complex ion with the central atom in its maximum oxidation state (NO_3^-, SO_4^{2-}, ClO_4^-, etc.), then the reaction is

$$2H_2O \rightarrow O_2 + 4H^+ + 4e^-$$

ELECTROLYSIS WITH ACTIVE ELECTRODES

All of the electrolytic cells discussed thus far have utilized *inert* electrodes—electrodes that do not themselves enter into the reaction but serve only as conductors of electrons, or at most, surfaces on which other metals plate out. If electrolyses are carried out using electrodes of less noble metals, the electrodes themselves may undergo reactions.

In the preceding section (p. 481) the electrolysis of a $CuSO_4$ solution with inert electrodes (Pt) was discussed, the reactions being

Cathode: $Cu^{2+} + 2e^- \rightarrow Cu$

Anode: $2H_2O \rightarrow O_2 + 4H^+ + 4e^-$

Now, suppose a similar electrolysis is carried out with copper electrodes. The cathode reaction is the same as with the inert electrode; that is, copper metal is plated out. However, at the anode an additional reaction is now possible. The "competing" oxidations and their standard oxidation potentials are

$$2SO_4^{2-} \rightarrow S_2O_8^{2-} + 2e^- \qquad E^0 = -2.01$$

$$2H_2O \rightarrow O_2 + 4H^+ + 4e^- \qquad E^0 = -1.23$$

$$Cu \rightarrow Cu^{2+} + 2e^- \qquad E^0 = -0.34$$

Clearly, the tendency for the copper electrode to be oxidized to Cu^{2+} is greater than the tendency for either of the other possible oxidations to occur.

Accordingly, in the electrolysis of $CuSO_4$ solution between copper electrodes, the reactions are

Cathode:	$Cu^{2+} + 2e^- \rightarrow Cu$
Anode:	$Cu \rightarrow Cu^{2+} + 2e^-$
Overall:	$Cu^{2+} + Cu \rightarrow Cu + Cu^{2+}$

Note that there is no net chemical reaction. In the overall process, copper is simply removed from the anode and plated onto the cathode. The concentration of $CuSO_4$ is the solution does not change.

Electrolysis with active electrodes has a number of practical applications, one of which is *electroplating*. Suppose that we wish to plate a covering of silver onto some metallic object. This may be done electrolytically by making the object to be plated the cathode, using a bar of silver as the anode, and using a solution of Ag^+, say, $AgNO_3$ as the electrolyte (Figure 17-8). Silver is removed from the anode by oxidation and plated onto the cathode by reduction.

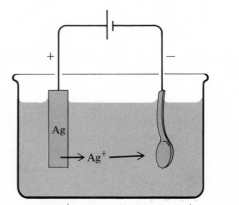

FIGURE 17-8

Electroplating Silver   Anode: $Ag \rightarrow Ag^+ + e^-$ Cathode: $Ag^+ + e^- \rightarrow Ag$

Another application of active electrode electrolysis, *electrorefining,* may be illustrated with the process used to obtain very pure copper. Copper metal produced from its ores contains small amounts of other metals, such as Zn, Fe, Ag, and Au, as impurities. When extremely pure copper is required, these impurities may be removed by employing a bar of the impure copper as the anode in an electrolytic cell. A thin sheet of pure copper serves as the cathode, and a solution of $CuSO_4$ is used as the electrolyte. As electrolysis proceeds, copper is oxidized and goes into solution from the anode along with other more active metals—that is, zinc, iron, and other metals with *high oxidation potentials.* Since they have *low oxidation potentials,* the less active metals, such as silver and gold, are not oxidized; as the impure copper electrode is eroded away, they fall to the bottom of the cell and are later recovered from this "anode mud." Copper plates out at the cathode in pure form, and the other metal ions (Zn^{2+}, Fe^{2+}, for example) remain in solution.

FARADAY'S LAWS

Up to this point, our discussion has been concerned only with the *qualitative* aspects of electrolysis. We have considered what will happen under certain conditions—what the electrode reactions will be—not the extent to which they will occur. Let us now examine the *quantitative* aspect of electrolysis—the relationship between the amount of substance reacting at an electrode and the quantity of electricity used.

The basic unit of *quantity* of electricity generally used in chemistry is the *coulomb.* Passage of one coulomb of electricity through a circuit is equivalent to passage of 6.2418×10^{18} electrons. Rate of flow of electricity (quantity per unit time), called the *current,* is measured in units called *amperes* (amp), with one ampere being defined as one coulomb per second. Thus, the quantity of electricity flowing through a circuit in a given time can be calculated from the current in amperes.

$$\text{Coulombs} = \text{Amperes} \times \text{Seconds}$$

The relationship between quantity of electricity and quantity of substances reacting at the electrodes during electrolysis was described by Michael Faraday in 1833 in two statements now referred to as Faraday's Laws.

Faraday's first law states that *the weight of a chemical substance liberated at an electrode is directly proportional to the amount of current passed through the cell.* This is simply to say that if *one* gram of copper is produced in the electrolysis of a $CuSO_4$ solution by the passage of n coulombs, then $2n$ coulombs will produce *two* grams of copper.

Faraday's second law states that *the weights of different substances produced by the passage of a given quantity of electricity are proportional to the equivalent weights of those substances.* The import of this statement may be illustrated as follows: Suppose we carry out an electrolysis of molten NaCl by passing n coulombs of electricity through the cell, and we determine accurately the weights of Na and of Cl_2 produced. We then carry out an electrolysis of a $CuSO_4$ solution, passing exactly the same quantity of electricity (n coulombs) through that cell, and determine the weight of Cu and O_2 produced. We will find that the weights of Na, Cl_2, Cu, and O_2 produced are in the same ratio to each other as are the equivalent weights of the four substances.

The *equivalent weight* of a substance in an oxidation-reduction reaction is the weight of that substance consumed or produced by the transfer of one mole of electrons (6.023×10^{23} electrons). Thus, the equivalent weight of sodium is the same as its atomic weight, because one gram-atom is oxidized or reduced by one mole of electrons.

$$Na^+ + e^- \rightleftharpoons Na$$

The equivalent weight of copper, on the other hand, is one-half its atomic weight, for *two* moles of electrons are required to oxidize or reduce one gram-atom of copper.

$$Cu^{2+} + 2e^- \rightarrow Cu$$

Since the production of one mole of O_2 (32 g) by oxidation of water requires the loss of *four* moles of electrons,

$$2H_2O \rightarrow O_2 + 4H^+ + 4e^-$$

the equivalent weight of O_2 is *one-fourth* its formula weight, or 8 g.

Now it follows from Faraday's Laws that there is some definite quantity of electricity that will cause the oxidation or reduction of one *equivalent weight* of any substance. That quantity of electricity, equivalent to 6.023×10^{23} electrons, is called one faraday and has been shown to be equal to approximately 96,500 coulombs.

Faraday's Laws enable one to make valuable predictions relating current, time, and quantity of products produced in electrolyses. Some examples follow.

EXAMPLE: A solution of $Ni(NO_3)_2$ is electrolyzed between platinum electrodes using a current of 5.0 amps for 30 minutes. What weight of Ni will be produced at the cathode?

Solution: Since coulombs = amp $\times$ sec

$$5.0 \text{ amp} \times 30 \text{ min} \times 60 \text{ sec/min} = \text{coulombs}$$

and

$$\frac{5.0 \times 30 \times 60}{96,500} = \text{faradays}$$

The reduction half-reaction is

$$Ni^{2+} + 2e^- \rightarrow Ni$$

Therefore the equivalent weight of Ni is

$$\frac{58.71}{2} = 29.36 \text{ g}$$

Since 1 faraday liberates 29.36 g, we must multiply the number of faradays used by 29.36 g:

$$\frac{5.0 \times 30 \times 60}{96,500} \times 29.36 = 2.74 \text{ g Ni}$$

EXAMPLE: In the electrolysis of a sulfuric acid solution using a current of 3.5 amp, how long will it take to liberate 5.0 g of oxygen at the anode?

Solution: The anode reaction is

$$2H_2O \rightarrow O_2 + 4H^+ + 4e^-$$

Therefore the equivalent weight of O_2 is

$$\frac{32}{4} = 8.0 \text{ g}$$

Since the number of equivalents of O_2 produced is 5.0/8.0 and each equivalent requires 96,500 coulombs,

$$\frac{5.0}{8.0} \times 96,500 = \text{number of coulombs required}$$

Since

$$\text{seconds} = \frac{\text{coulombs}}{\text{amperes}}$$

$$\frac{5.0}{8.0} \times \frac{96,500}{3.5} \times \frac{1}{60} = 287 \text{ minutes}$$

EXAMPLE: What current will be required to produce 1.0 g of Na in 10.0 minutes by the electrolysis of molten NaCl?

Solution: Since the equivalent weight of Na is the same as its gram-atomic weight,

$$\text{number of equivalents of Na produced} = \frac{1.0}{23.0}$$

Therefore,

$$\text{number of coulombs required} = \frac{1.0}{23.0} \times 96,500$$

And, because

$$\text{amperes} = \frac{\text{coulombs}}{\text{seconds}}$$

the current required is

$$\frac{1.0}{23.0} \times \frac{96,500}{60 \times 10} = 7.0 \text{ amp}$$

SUGGESTED READINGS

Anson, Fred C. "Common Sources of Confusion; Electrode Sign Conventions." *Journal of Chemical Education,* Vol. 36, No. 8 (August 1959), pp. 394–99.

Young, G. J., and R. B. Rozelle. "Fuel Cells." *Journal of Chemical Education,* Vol. 36, No. 2 (February 1959), pp. 68–73.

Ehl, R. G., and A. J. Ihde. "Faraday's Electrochemical Laws and the Determination of Equivalent Weights." *Journal of Chemical Education,* Vol. 31, No. 5 (May 1954), pp. 226–32.

PROBLEMS

1. Assign oxidation numbers to all the atoms in each of the following:
   (a) $AgNO_3$
   (b) CsF
   (c) SO_3
   (d) C_2H_6
   (e) $SO_4{}^{2-}$
   (f) $C_2H_3O_2{}^-$
   (g) $Mg_2P_2O_7$
   (h) As_4
   (i) $KBrO_3$
   (j) Hg_2Cl_2
   (k) $Cr(OH)_3$
   (l) CaH_2
   (m) $UO_2{}^{2+}$
   (n) $HAsO_2$
   (o) $Na_2S_2O_3$
   (p) $S_2O_8{}^{2-}$
   (q) $Ce(NO_3)_3$
   (r) Sb_2S_3

2. Use oxidation numbers to find the oxidant and the reductant in the following:
   (a) $3CuS + 8HNO_3 \rightarrow 3Cu(NO_3)_2 + 2NO + 3S + 4H_2O$
   (b) $K_2Cr_2O_7 + 14HCl + 2SnCl_2 \rightarrow 2CrCl_3 + 3SnCl_4 + 2KCl + 7H_2O$
   (c) $2CuSO_4 + 4KI \rightarrow 2CuI + I_2 + 2K_2SO_4$
   (d) $2KOH + Cl_2 \rightarrow KCl + KClO + H_2O$
   (e) $AgCl + 2NH_3 \rightarrow Ag(NH_3)_2{}^+ + Cl^-$
   (f) $BaSO_4 + 4C \rightarrow BaS + 4CO$
   (g) $3KClO_3 + 3H_2SO_4 \rightarrow 3KHSO_4 + HClO_4 + 2ClO_2 + H_2O$

3. For each of the following unbalanced "molecular" equations, write (1) the oxidation half-reaction, (2) the reduction half-reaction, and (3) the balanced net ionic equation:
   (a) $Na_2S_2O_3 + I_2 \rightarrow Na_2S_4O_6 + NaI$
   (b) $Ce(SO_4)_2 + FeSO_4 \rightarrow Ce_2(SO_4)_3 + Fe_2(SO_4)_3$
   (c) $FeSO_4 + HNO_3 \rightarrow Fe_2(SO_4)_3 + NO + H_2O$
   (d) $Ag_2S + HNO_3 \rightarrow AgNO_3 + NO + S + H_2O$
   (e) $K_2Cr_2O_7 + H_2S + HCl \rightarrow CrCl_3 + S + KCl + H_2O$
   (f) $NaOH + Cl_2 \rightarrow NaCl + NaClO_3 + H_2O$
   (g) $Fe(OH)_2 + Na_2O_2 + H_2O \rightarrow Fe(OH)_3 + NaOH$
   (h) $Bi_2S_3 + HNO_3 \rightarrow Bi(NO_3)_3 + S + NO + H_2O$
   (i) $H_3AsO_4 + Zn + HNO_3 \rightarrow AsH_3 + Zn(NO_3)_2 + H_2O$
   (j) $CrI_3 + NaOH + Cl_2 \rightarrow Na_2CrO_4 + NaIO_4 + NaCl + H_2O$

4. The following unbalanced skeletal ionic equations represent reactions taking place in acid solution. Convert them to balanced ionic equations, introducing H^+ and H_2O as necessary.
   (a) $Fe^{2+} + NO_3{}^- \rightarrow Fe^{3+} + NO$
   (b) $Cr^{3+} + ClO_3{}^- \rightarrow Cr_2O_7{}^{2-} + ClO_2$

(c) $MnO_4^- + H_2O_2 \rightarrow Mn^{2+} + O_2$
(d) $IO_3^- + I^- \rightarrow I_2$
(e) $Fe^{3+} + SO_3^{2-} \rightarrow Fe^{2+} + SO_4^{2-}$
(f) $\underline{Sb_2S_3} + NO_3^- \rightarrow \underline{Sb_2O_5} + S + NO$
(g) $\underline{As_2S_3} + NO_3^- \rightarrow SO_4^{2-} + H_3AsO_4 + NO$

5. The following unbalanced skeletal equations represent reactions taking place in basic solution. Convert them to balanced ionic equations, introducing OH^- and H_2O as necessary:
(a) $Cl_2 \rightarrow Cl^- + ClO^-$
(b) $\underline{Co(OH)_2} + O_2 \rightarrow \underline{Co(OH)_3}$
(c) $\underline{Al} \rightarrow Al(OH)_4^- + H_2$
(d) $MnO_4^- + S^{2-} \rightarrow \underline{MnS} + S$
(e) $NO_2^- + Al \rightarrow NH_3 + AlO_2^-$

6. Sketch a voltaic cell constructed from each of the following reactions. Label all materials used, indicate the polarity of the cell, and show the direction of electron flow.
(a) $H_2O_2 + 2H^+ + 2Br^- \rightarrow 2H_2O + Br_2$
(b) $Cd + Pb^{2+} \rightarrow Pb + Cd^{2+}$
(c) $Cu + 2Fe^{3+} \rightarrow Cu^{2+} + 2Fe^{2+}$
(d) $Zn + 2H^+ \rightarrow Zn^{2+} + H_2$
(e) $Hg + 2Ce^{4+} \rightarrow Hg^{2+} + 2Ce^{3+}$
(f) $ClO_3^- + 6H^+ + 6I^- \rightarrow Cl^- + 3I_2 + 3H_2O$

7. Sketch the voltaic cell constructed from each of the following pairs of half-cells. Indicate the polarity of the cell and write the electrode reactions and the overall reaction that occur when the cell is producing current.

Half-cell A	*Half-cell B*
(a) $Ni \mid Ni^{2+}(1M)$	$Cd \mid Cd^{2+}(1M)$
(b) $Cu \mid Cu^{2+}(1M)$	$Pt \left\vert \begin{array}{l} NO_3^-(1M),H^+(1M), \\ NO(1\ atm) \end{array} \right.$
(c) $Pt \mid Fe^{3+}(1M),Fe^{2+}(1M)$	$Pt \mid I^-(1M),I_2(1M)$
(d) $Co \mid Co^{2+}(1M)$	$Co \mid Co^{2+}(0.01M)$
(e) $Pt \left\vert \begin{array}{l} H_3PO_4(1M),H^+(1M), \\ H_3PO_3(1M) \end{array} \right.$	$Pt \mid ClO_3^-(1M),H^+(1M),Cl^-(1M)$

8. Using the half-reactions in Table 17-1, write balanced equations for five reactions that *will occur spontaneously* and for five reactions that *will not occur spontaneously*.

9. Calculate the equilibrium constant for each of the reactions in problem 6.

10. Calculate the equilibrium constant for the reaction represented by each of the equations written in answer to problem 8.

11. Calculate the *standard emf* for each of the following voltaic cells:

    (a) $Pt \mid Cr^{2+}, Cr^{3+} \parallel Cu^{2+} \mid Cu$

    (b) $Fe \mid Fe^{2+} \parallel Br_2, Br^- \mid Pt$

    (c) $Ag \mid Ag^+ \parallel Cr_2O_7{}^{2-}, H^+, Cr^{3+} \mid Pt$

    (d) $Co \mid Co^{2+} \parallel Co^{3+}, Co^{2+} \mid Pt$

    (e) $Pb \mid PbSO_4 \mid SO_4{}^{2-} \parallel SO_4{}^{2-}, H^+ \mid PbSO_4 \mid PbO_2$

12. After a solution of tin(II) chloride in dilute HCl has been prepared, solid tin metal is added to prevent the tin(II) from being oxidized by the air to tin(IV). Explain.

13. Is iron(III) ion reduced to iron(II) ion by iron metal?

14. Account for the following observations: A sample of pure solid barium sulfite dissolves in hydrochloric acid. However, when the resulting solution is allowed to stand for a period of time, a white precipitate slowly forms.

15. Aqueous hydrazine (N_2H_4) is a strong enough reducing agent in basic solution to convert oxygen gas into hydrogen peroxide; nitrogen gas is also formed in the reaction. Write a balanced equation for the reaction.

16. A major commercial process for production of nitric acid involves the following steps: (a) ammonia and oxygen react at high temperature in the presence of a platinum catalyst to yield nitric oxide (NO) and water; (b) the NO reacts with excess oxygen to form NO_2; and (c) the NO_2 then reacts with water to yield nitric acid and NO (which is then recycled). Write balanced equations for this process.

17. In preparing an acidic solution of potassium permanganate, would you use sulfuric acid or hydrochloric acid? Explain your choice.

18. In the electrolysis of an aqueous solution of each of the following using platinum electrodes, write an equation for the anode reaction, the cathode reaction, and the overall reaction.

    (a) H_3PO_4 (b) $NaC_2H_3O_2$ (c) $CdSO_4$
    (d) KOH (e) $CuBr_2$ (f) $Mg(ClO_4)_2$
    (g) CaI_2 (h) $Hg(NO_3)_2$

19. What weight of chromium will be plated from a solution of $Cr(NO_3)_3$ by a current of 6.0 amp in 2.0 hours?

20. How long will it take to liberate exactly 1 liter of O_2 (measured at STP) by electrolysis of a sulfuric acid solution using a current of 2.5 amp?

21. What is the current required to liberate 1 mole of H_2 from an acid solution in 30 min?

22. A certain alloy of silver and zinc is 70.91 percent silver. If a 1.203-g sample of the alloy is dissolved, and the solution is electrolyzed between platinum electrodes, how many coulombs are required to completely plate out the metals?

23. In the electrolysis of a saturated solution of $KHSO_4$ using platinum electrodes and a high current, a gas is evolved at the cathode and a white solid precipitates in the vicinity of the anode. Analysis of the dried solid gave results of 28.77 percent K, 23.65 percent S, and 47.58 percent O; formula weight = 270.

(a) What is the formula of the white solid?

(b) Write equations for the electrode reactions occurring during the electrolysis.

PART THREE

The Chemistry
of the Elements

18 *The Origin of the Elements*

Of the 105 known elements, 90 occur naturally in large enough quantities to isolate and identify. Technetium (atomic number 43), promethium (atomic number 61), and the transuranium elements from neptunium (atomic number 93) through element 105 are synthetic elements; that is, they have been synthesized in the laboratory by nuclear reactions. All isotopes of the synthetic elements, as well as all isotopes of those elements with a mass number greater than 209 and an atomic

TABLE 18-1
Partial List of Relative Abundance[a]
of the Elements on the Earth and Sun

Element	Abundance	
	Earth	Sun
H	—[b]	10.5
He	—[b]	*ca.* 9.5
Li	1.7	−0.7
Be	−0.1	−0.4
Na	4.8	4.7
Mg	6.0	6.0
Al	5.0	4.9
Si	6.0	6.0
Ca	4.9	4.8
Fe	6.0	6.1
Cu	2.8	2.7

[a]The abundance is expressed as the logarithm of the number of atoms of each element on a scale in which the number of silicon atoms equals 10^6.
[b]Estimated values for H and He are not considered to be reliable; however, values for both are significantly lower than those for the sun.

number greater than 83, are *radioactive* and spontaneously decompose to stable nuclei, emitting either alpha particles (helium nuclei) or beta particles (electrons) along with gamma radiation (high-energy electromagnetic radiation).

A comparison of the elemental composition of the earth with that of the sun (Table 18-1) shows a remarkable similarity in the relative abundance of many elements. A significant difference is apparent in the lighter elements: Beryllium and lithium are more abundant on earth, but hydrogen and helium (which form the major part of the sun) are much less abundant. The universe as a whole is thought to be composed almost entirely of hydrogen and helium, in about a 3:1 ratio. The heavier elements account for less than 2 percent of the total mass. The inner planets, Mars, earth, Venus, and Mercury, apparently have compositions that are about the same. On the other hand, it is believed that the massive outer planets have retained the lighter gases and have a composition essentially the same as that of the sun.

In order to understand the genesis of the chemical elements as well as the differences in composition within the solar system, we need to know something about the nuclear processes that are thought to be responsible for the formation of many of these elements.

Nuclear Phenomena

NUCLEAR FORCES

In Chapters 2 and 5 we found that, according to the atomic theory, matter is composed of atoms, which constitute the smallest particles of an element. The atom itself contains a small, dense nucleus that carries a positive charge. Surrounding the nucleus in quantized energy levels are electrons, which are negatively charged, the number of electrons in each atom being equal to the positive charge of the nucleus. This electron cloud occupies about 10^{12} times as much space as the nucleus.

The nucleus is made up of neutrons and protons (both of which are also called *nucleons*), each with a mass number of one. Whereas the proton carries a unit positive charge, the neutron carries no charge and therefore contributes to the mass but not the charge of the nucleus. The atomic number, Z, is equal to the number of protons in the nucleus; the mass number, A, is equal to the number of nucleons—the sum of the protons and neutrons. In an atom of hydrogen, the simplest element, one proton is surrounded by an electron, whose mass is about 1/1840 that of the proton. Therefore, almost the total mass of an atom is concentrated in the nucleus.

A few other particles that are involved in nuclear reactions warrant mention here. In addition to the electron of negative charge (denoted e^-), there is also a particle of the same mass but of opposite charge, the *positron* (e^+). Positive, negative, and neutral *mesons,* particles with masses intermediate between those of electrons and protons, have also been identified. Some are apparently involved in the binding force that holds the nucleus together. The *neutrino* (ν), which is emitted in certain nuclear reactions, is a particle of very small mass (less than 0.00002 amu) and no charge.

In chemical reactions, the atomic numbers and mass numbers of the atoms are conserved. In nuclear reactions, mass is not conserved and neither, necessarily, is the atomic number. In fact, there is usually a change in either mass number or atomic number, or in both. However, *the total number of nucleons and the total charge are conserved.* Therefore, equations for nuclear reactions can be balanced just like those for chemical reactions. In the equations that follow, the number of nucleons in the particle (its mass number) is designated by a superscript, while the charge on the particle is given in a subscript. Thus, the proton is symbolized as $^1_1$H, the neutron as $^1_0$n, the electron as $^0_{-1}$e, the deuterium nucleus as $^2_1$H, and the alpha particle as $^4_2$He, while gamma radiation is abbreviated as γ. Equations for nuclear reactions are balanced when the sums of both the superscripts and subscripts of both sides are equal.

$$^{13}_{6}C + ^{1}_{1}H \rightarrow\ ^{14}_{7}N + \gamma$$

$$^{9}_{4}Be + ^{4}_{2}He \rightarrow\ ^{12}_{6}C + ^{1}_{0}n$$

$$^{32}_{16}S + ^{1}_{0}n \rightarrow\ ^{32}_{15}P + ^{1}_{1}H$$

$$^{56}_{26}Fe + ^{2}_{1}H \rightarrow\ ^{4}_{2}He + ^{54}_{25}Mn$$

Since nuclear processes release amounts of energy 10^5 to 10^8 times as great as the amounts involved in chemical processes, nuclear forces must be very strong. For example, 68.3 kcal of energy is released in the formation of one mole of water from its elements in their standard states, while 6.5×10^8 kcal of energy is released by the hypothetical formation of one gram-atom of helium nuclei from their constituent components (two neutrons and two protons per nucleus). In contrast to gravitational and electrostatic forces, nuclear forces operate over only very short distances (about 2×10^{-13} cm) and involve neutron-neutron, neutron-proton, and proton-proton interactions. In addition to the nuclear attractive energy, there is also electrostatic repulsion between the protons. The existence of a large number of stable nuclei indicates that nuclear attractive forces overcome what must be very strong repulsions between protons at these short distances and is further evidence of the great strength of these nuclear forces.

In the hypothetical synthesis of a helium nucleus from its component particles (two protons and two neutrons), the large amount of energy released comes from the conversion of mass into energy according to the familiar Einstein relationship:

$$E = mc^2$$

where m is mass and c is the velocity of light. That is, the helium nucleus has a mass of 4.00150 amu, which is less than the sum of the masses of two protons (2 $\times$

1.00728 amu) and two neutrons (2 $\times$ 1.00867 amu) by an amount equal to 0.0304 amu.

$$\begin{aligned}
{}^4_2\text{He} &= 4.00150 \text{ amu} \\
-2{}^1_1\text{H} &= -2(1.00728) \text{ amu} \\
\underline{-2{}^1_0\text{n}} &= \underline{-2(1.00867) \text{ amu}} \\
\text{Mass loss} &= 0.03040 \text{ amu}
\end{aligned}$$

From the Einstein relationship it can be shown that 1 amu is equivalent to 1.493×10^{-3} erg or 3.57×10^{-11} cal of energy. Since one atomic mass unit equals 1.661×10^{-24} g, and the speed of light is 2.998×10^{10} cm/sec,

$$E = (1.661 \times 10^{-24} \text{ g})(2.998 \times 10^{10} \text{ cm/sec})^2$$

$$= 1.493 \times 10^{-3} \text{ g-cm}^2/\text{sec}^2 \text{ or ergs}$$

or, since 1 erg = 2.39×10^{-8} cal,

$$E = 3.57 \times 10^{-11} \text{ cal}$$

A more convenient unit for the large amount of energy involved is the *electron-volt* (eV), which is defined as *the energy required to raise one electron through a potential difference of one volt.* One electron-volt is equivalent to 1.602×10^{-12} erg, and a million electron-volts (MeV) is therefore 1.602×10^{-6} erg. Therefore, we can express the energy equivalent of one atomic mass unit as

$$\frac{1.493 \times 10^{-3} \text{ erg/amu}}{1.602 \times 10^{-6} \text{ erg/MeV}} = 931 \text{ MeV/amu}$$

A mass loss of 0.0304 amu results, then, in 931 MeV $\times$ 0.0304 amu, or 28.30 MeV of energy per nucleus. The difference in energy between a nucleus and its isolated component particles is known as the *binding energy.* It is more useful to express the binding energy in MeV/nucleon. For the ${}^4_2$He nucleus, which has four nucleons, this is 28.30/4 or 7.08 MeV/nucleon.

When binding energy per nucleon is plotted against mass number for known nuclei (Figure 18-1), the curve obtained is very useful, for it reveals important information about the relative stability of these nuclei. The curve rises rapidly to a maximum in the vicinity of mass number 58 (that is, at iron and its close neighbors) and then slowly falls off with increasing mass number. Therefore, iron and the elements directly around it have the highest binding energy per nucleon and are the most stable elements; that is, more mass is lost (more energy released) in their formation from their constituent nucleons than in the formation of other nuclei.

Why does the universe, then, consist primarily of hydrogen and helium, which are certainly highly unstable in comparison with isotopes of iron, cobalt, and nickel? According to the binding energy curve, iron and its near neighbors might be expected to make up a significant percentage of the total universe. However, just as in chemical reactions, the rate as well as the energetics of a process must be considered. Thus, whereas hydrogen and helium are energetically unstable relative

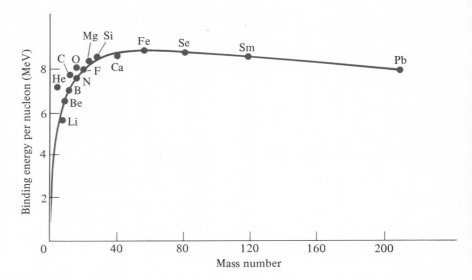

FIGURE 18-1

Nuclear Binding Energy Curve

to heavier nuclei, the rate of their fusion to form these heavier nuclei is very low. Indeed, temperatures of at least 10^7 °C are required before there is any perceptible reaction rate.

In order to understand this high barrier to reaction, let us consider two nuclear reactions, one in which a neutron is being added to a nucleus and the second in which a proton is being added. In the case of the neutron, which carries no charge, there is no repulsive force and thus no energy barrier to overcome. The only requirement, then, is that the neutron bullet must be fired very accurately, so that it comes close enough to the nucleus for the short-range nuclear forces to capture it. If the aim is successful and the neutron enters the nucleus, energy is usually released.

In the case of the proton, however, another dimension is added to the reaction. We are now firing a positively charged bullet toward a positively charged target. The closer the bullet approaches the target, the greater the repulsive force becomes, and a high potential energy barrier must be overcome before the proton can get close enough for nuclear forces to prevail. Therefore, in addition to the accuracy of the shot, the proton bullet must also travel at a high enough velocity to overcome this repulsive force. If the proton is not energetic enough, it is simply deflected before it comes within range of the nuclear attractive force. If the conditions of great accuracy and high velocity are met and the proton is captured by the nucleus, energy is usually released.

Because the fusion of nuclei to give heavier nuclei involves positively charged particles, the electrostatic repulsion is very great. Moreover, as the charges of the nuclei increase, the repulsions increase greatly, and higher and higher temperatures (particle velocities) are needed to overcome the energy barrier to reaction. These two-body collision reactions that result in a new nucleus are second order, and the rate of reaction depends upon the concentration of the reacting species. Therefore, an increase in temperature and an increase in concentration both increase the rate of particle-particle reactions.

NEUTRON-PROTON RATIO

Another indicator of the stability of a nucleus is its neutron-proton ratio, N/P. A plot of the number of neutrons against the number of protons in the nuclei of the stable isotopes of the elements (Figure 18-2) shows that for the first 20 elements,

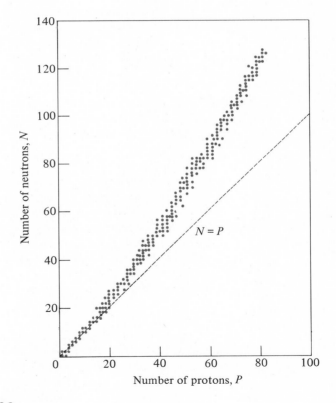

FIGURE 18-2

Plot of Neutrons versus Protons in Stable Isotopes of Naturally Occurring Nuclei
SOURCE: T. Moeller, *Inorganic Chemistry*. New York: John Wiley & Sons, 1952. Reprinted by permission.

the trend is for the most abundant stable nuclei to have a neutron-to-proton ratio near unity. Beyond mass number 40 the stable nuclei become enriched with neutrons. Apparently the proton-proton repulsion, which increases with increasing atomic number, requires increasing numbers of neutrons as a counterbalance. Even so, the most abundant stable nuclei have N/P ratios of less than 1.2, and no stable nucleus has a ratio greater than 1.6. The electrostatic repulsion is so great that beyond an atomic number of 83 even additional neutrons cannot stabilize the nucleus, and, as noted earlier, all such nuclei spontaneously decompose by either alpha or beta emissions.

Nuclei with N/P ratios *higher* than the stable region on the curve are radioactive. They can achieve stability by two different pathways. The first, the emission of a neutron, has the direct effect of decreasing the excessive neutron content.

$$\ce{^{87}_{36}Kr} \rightarrow \ce{^{86}_{36}Kr} + \ce{^{1}_{0}n}$$

However, neutron emission is relatively rare and is found most commonly in fission fragments where N/P ratios are much higher than the stable region.

Most of the natural and synthetic radioisotopes reach stability through the second pathway, emission of a beta particle and an associated neutrino.

$$\ce{^{40}_{19}K} \rightarrow \ce{^{40}_{20}Ca} + \ce{^{0}_{-1}e} + \nu$$

This has the net effect of increasing the charge by 1 while the mass number remains the same. The reaction can be considered the transformation of a neutron into a proton, an electron, and a neutrino,

$$\ce{^{1}_{0}n} \rightarrow \ce{^{1}_{1}H} + \ce{^{0}_{-1}e} + \nu$$

thus significantly decreasing the N/P ratio. Many fission fragments undergo successive beta emissions before reaching a stable region of the curve.

Nuclei with N/P ratios *below* the stable region of the curve have three possible pathways for achieving stability. The first, the spontaneous emission of a proton, is seldom observed. The second, the emission of a positron and a neutrino, has the opposite effect of beta emission and decreases the charge by 1 with no change in mass number. This process can be considered to be the change of a proton to a neutron and a positron, thus increasing the neutron-proton ratio.

$$\ce{^{1}_{1}H} \rightarrow \ce{^{1}_{0}n} + \ce{^{0}_{+1}e} + \nu$$

Positron emission is found only with synthetic nuclei.

The third pathway is the capture of an orbital electron—most probably a $1s$ electron—and is therefore commonly called K capture. (Recall that the first quantum level is called the K shell.) This has the same net effect as positron emission and can also be considered as the transformation of a proton into a neutron, accompanied by the emission of a neutrino.

$$\ce{^{1}_{1}H} + \ce{^{0}_{-1}e} \rightarrow \ce{^{1}_{0}n} + \nu$$

Most examples of this spontaneous reaction are also found in synthetic nuclei.

	Type	Number of Stable Nuclei
TABLE 18-2		
Distribution of Stable Nuclei	Z even, N even	163
	Z even, N odd	56
	Z odd, N even	50
	Z odd, N odd	5

$$\ce{^{55}_{26}Fe} \xrightarrow{K \text{ capture}} \ce{^{55}_{25}Mn} + \nu$$

A careful examination of Figure 18-2 also reveals that nuclei with even atomic numbers are more numerous than those with odd atomic numbers (Table 18-2). Elements with odd atomic numbers commonly exist as a single nuclide and never have more than two stable isotopes. (This is illustrated by the halogens, where stable isotopes are fluorine-19, chlorine-35 and -37, bromine-79 and -81, and iodine-127.) Nuclei with an even number of neutrons are also more common than those with an odd number of neutrons. Over 85 percent of the elements in the earth's crust consist of nuclei with an even mass number; and for a given element, isotopes with an even mass number are almost always more abundant than those with odd mass numbers. With only a few exceptions, nuclei of even mass numbers have even atomic numbers and therefore an even number of neutrons. Four of the five exceptions where Z and N are odd—$^2_1$H, $^6_3$Li, $^{10}_5$B, and $^{14}_7$N—are light nuclei where apparently there are strong interactions between equal numbers of neutrons and protons.

The very successful correlation of chemical properties with orbital electron distributions has led to similar attempts to explain nuclear properties by postulating energy levels for nucleons. The nucleus has proven to be much more complex, however, and our understanding of it is still quite empirical. According to one model, there are specific quantum states in the nucleus analogous to the quantized energy levels of electron distribution. The major quantum levels contain, in order of increasing energy, the following numbers of protons, neutrons, or both protons and neutrons:

$$2, \ 8, \ 20, \ 28, \ 50, \ 82, \ 126$$

These "magic numbers" are completely analogous to the closed quantum-level structures of the inert gases. The numbers have been derived from empirical observations of the following type: Elements with an exceptionally high number of stable isotopes include tin ($Z = 50$), which has ten isotopes, and calcium ($Z = 20$) and nickel ($Z = 28$), each of which has five isotopes. Helium-4 ($Z = 2$, $N = 2$) is much more stable (see binding-energy curve) than other nuclei of light mass such as helium-3 and lithium-7. For both lead-208 and bismuth-209, which have the highest mass of the stable nuclei, $N = 126$. Also, lead ($Z = 82$) is the stable end product of the three naturally occurring radioactive series (which are discussed in the next section). In fact, nuclei at or near these numbers are more abundant than surrounding nuclei.

Each nuclear orbital can contain two nucleons of opposite spin. Since neutrons and protons are different particles, either two neutrons or two protons of opposite spin can occupy an orbital. If Z and N are both even, all the protons and neutrons are paired, and as a result, the nucleus possesses zero spin. If either Z and N is odd, the nucleus possesses a spin equivalent to the odd particle. If both Z and N are odd, the spin possessed by the nucleus is more difficult to evaluate.

NATURAL RADIOACTIVITY

Within two years of Henri Becquerel's accidental discovery of radioactivity in 1896, Marie and Pierre Curie isolated radium and polonium from the mineral pitchblende (essentially U_3O_8). Isolation of other heavy, radioactive elements followed in rapid order. After some years of chemical confusion, it was recognized that the isotopes of these heavy nuclei fitted into three naturally occurring radioactive series, with each series having a long-lived parent. The radioactive series for $^{238}_{92}U$, as presented schematically in Figure 18-3, begins with a loss of an alpha particle from uranium-

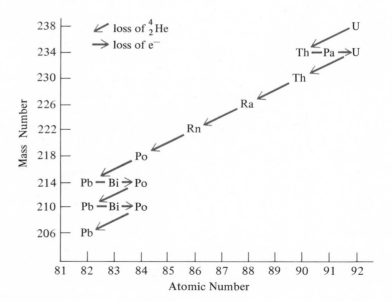

FIGURE 18-3

Uranium-238 Disintegration Chain
SOURCE: T. Moeller, Inorganic Chemistry. New York: John Wiley & Sons, 1952. Reprinted by permission.

	Series	Parent	Half-Life of Parent (years)	Stable End Product
TABLE 18-3 **Parents of** **Disintegration Series**	$4n$	$^{232}_{90}\text{Th}$	1.39×10^{10}	$^{208}_{82}\text{Pb}$
	$4n + 1$	$^{237}_{93}\text{Np}$	2.25×10^{6}	$^{209}_{83}\text{Bi}$
	$4n + 2$	$^{238}_{92}\text{U}$	4.51×10^{9}	$^{206}_{82}\text{Pb}$
	$4n + 3$	$^{235}_{92}\text{U}$	7.07×10^{8}	$^{207}_{82}\text{Pb}$

238 to form thorium-234:

$$^{238}_{92}\text{U} \rightarrow \ ^{234}_{90}\text{Th} + ^{4}_{2}\text{He}$$

The daughter nucleus, thorium-234, rapidly decomposes to protactinium-234 by loss of a beta particle.

$$^{234}_{90}\text{Th} \rightarrow \ ^{234}_{91}\text{Pa} + ^{0}_{-1}\text{e} + \nu$$

The decay process continues with another beta emission to yield uranium-234, and, after subsequent steps in which either an alpha particle or a beta particle is lost, the end product of the series is the stable lead-206 isotope. Divergent decay paths for a few of the daughters are not indicated, because these paths subsequently merge with the main decay path and only one end product is formed.

The parent of a radioactive series is obviously the longest-lived member of the series. Uranium-238 has a half-life (the time required for half the nuclei to decompose) of 4.5×10^{9} years. The three naturally occurring radioactive decay series, along with a fourth whose parent is the synthetic element neptunium-237, fit a pattern in which the mass number of each member of the series divided by 4 gives remainders of 0, 1, 2, and 3, symbolized as $4n$, $4n + 1$, and so on (see Table 18-3). (Whether any significance can be attached to this pattern is not known. The symmetry obtained by finding the fourth series is at least esthetically pleasing.) The three parents of the *naturally occurring* series can account for all of the naturally occurring radioactive isotopes of the heavier elements. All isotopes of a mass number greater than 209 and all isotopes with an atomic number greater than 83 are radioactive and must have been present in the material out of which the earth formed. Whether neptunium-237 was present in the newly formed earth cannot be answered, because, with a half-life of 2.25×10^{6} years, all of it would have disappeared within five billion (5×10^{9}) years, the estimated age of the earth. The fact that there is no naturally occurring technetium and promethium can also be understood, in that all known isotopes of these elements have considerably shorter half-lives than 10^{9} years.

Unlike chemical reactions, nuclear decay processes are independent of their environment and depend only on the number of nuclei present; that is, their rates of reaction do not change with changes in conditions such as temperature and pressure. If the decay process in a radioactive series with a long-lived parent has been occurring for a long enough time and there has been no separation of products, a steady state is achieved in which the rate of disappearance of the parent is equal to the rate of formation of the end product, or, for that matter, the rate of disappearance of any of the daughters. Such a condition obtains for the three naturally occurring series, and the ratio of lead-206 to uranium-238 in uranium minerals can be used to calculate the age of the earth. This ratio is found to be slightly less than unity in uranium minerals contained in the oldest known terrestrial rocks. If the ratio were unity, it would indicate that half of the original uranium-238 had disappeared since the rock had formed, and the age of the rock would be equal to the half-life of uranium-238, or 4.5×10^9 years. Since the observed value is slightly less than unity, Precambrian rocks are judged to be at least three billion years old. The one major assumption necessary in such calculations is that no lead was present when these rocks formed. The absence of $^{204}_{82}Pb$, which is not derived from a decay series, is evidence of the validity of this assumption. Since leaching or other geological processes may have altered the composition during the years, the ratio of $^{206}_{82}Pb$ to $^{207}_{82}Pb$ in uranium minerals is probably a more accurate indicator, and values of about five billion years have been obtained from this ratio. Therefore, the earth is probably at least this old.

Naturally occurring radioactive elements are not limited to the heavier nuclei. The small number of lighter radioactive elements that are found on the earth are characterized by long half-lives (tritium, ^3_1H, is an exception, with a half-life of 12.4 years) and *soft emissions:* emissions in which the particles emitted are of low energy and are consequently not harmful to living cells. A radioactive isotope of great significance to us is carbon-14. With an N/P ratio lying above the stable region (see Figure 18-2), it decomposes to nitrogen-14 by beta emission.

$$^{14}_6C \rightarrow {}^{14}_7N + {}^{\ 0}_{-1}e + \nu$$

Carbon-14 is being continuously produced in the atmosphere by bombardment of nitrogen-14 atoms with neutrons from cosmic radiation.

$$^{14}_7N + {}^1_0n \rightarrow {}^{14}_6C + {}^1_1H$$

The carbon-14 thus produced is rapidly oxidized and becomes a part of the atmospheric carbon dioxide, where a steady-state concentration is established. The ratio of carbon-14 atoms to carbon-12 atoms in atmospheric CO_2 is about $1:10^{12}$. Because green plants consume atmospheric CO_2 in the photosynthetic process, carbon compounds in plant tissues have this same $^{14}C/^{12}C$ ratio. Furthermore, the ratio remains the same in the tissues of animals that eat the plants. In other words, the ratio of radioactive carbon-14 to nonradioactive carbon-12 remains essentially constant throughout the living cycle. On the other hand, when carbon compounds are removed from the living cycle—when a plant or animal dies—the intake of carbon-14 ceases, but the radioactive disintegration of carbon-14 continues, with the result that the $^{14}C/^{12}C$ ratio decreases with time.

This phenomenon forms the basis of a method introduced by W. F. Libby and co-workers in the 1950s for determining the age of carbonaceous matter. A comparison of the radioactivity of the carbon in a piece of cloth from an ancient Egyptian tomb or in a fragment of wood from the remnants of a campfire uncovered in an archeological dig with that of the steady-state radioactivity gives the age of the material. Unfortunately, the half-life of carbon-14 is only about 5700 years, so radiocarbon dating can be of no value in determining the age of coal or petroleum deposits that were laid down so long ago that no carbon-14 remains. The method is, however, very useful in establishing dates of more recent events. The reliability of radiocarbon dating has been checked against well-documented archeological dates. Perhaps the most dramatic test was the age of core wood in a California giant sequoia, where the tree rings gave a value of 2928 ± 10 years compared with a radiocarbon value of 2900 years.

The large amounts of carbon dioxide emitted into the atmosphere from the combustion of fossil fuels will render this technique less useful in the future; the steady-state concentration of carbon-14 in the living cycle is being diluted by this inactive carbon. Future archeologists will need to depend on some other isotopic dating technique.

FISSION AND FUSION

The binding-energy curve of Figure 18-1 suggests that the fragmentation of heavy nuclei into smaller particles of higher binding energy should result in the conversion of mass to energy. Such a reaction, discovered by Otto Hahn and Fritz Strassman in 1939, is called *fission*. Uranium-235 can be fissioned by thermal neutrons (neutrons of low energy) to form uranium-236, a very unstable isotope, which spontaneously splits into two fragments of roughly equal size, plus two or three neutrons.

$$^{235}_{92}U + ^{1}_{0}n \rightarrow \ ^{236}_{92}U \rightarrow \ ^{139}_{56}Ba + ^{94}_{36}Kr + 3^{1}_{0}n$$

There are many different ways in which the fission of uranium-235 can occur, and most of the nuclei produced have masses grouped around mass numbers 95 and 138. The average energy yield is about 200 MeV for each $^{235}_{92}U$ fissioned. This quantity of energy is more than two million times as great as the energy produced by the chemical combustion of the same mass of coal.

Because two to three neutrons are produced per fission, this reaction can be self-sustaining if at least one of the neutrons is captured by another $^{235}_{92}U$ nucleus. If a sample of uranium-235 is large enough so that essentially all of the neutrons formed cause fission in other uranium nuclei, a *chain reaction* occurs that is propagated so rapidly that within a few microseconds a nuclear explosion takes place. The mass required for this to happen is called the *critical mass*. The fission (atom) bomb consists of two subcritical masses of either $^{235}_{92}U$ or $^{239}_{94}Pu$ (also fissionable by slow neutrons) with some device for firing them together within a few microseconds. The chain reaction is initiated by neutrons derived from spontaneous fission. A piece of uranium-235 in large excess over the critical mass does not increase the force of the explosion significantly, since the explosion occurs before the chain can expand to include all of the fissionable material.

In order for the energy produced in this way to be usable, the rate of fission must be controlled. A nuclear reactor employs slightly more than the critical mass of fissionable material, but it is spread out and mixed with a moderator—graphite or heavy water—which slows down the neutrons. Rods of cadmium or some other suitable material are inserted to soak up those neutrons in excess of the number needed to keep the reactor going at the desired rate. The heat generated by the reaction is absorbed by water, liquid sodium, air, or other suitable coolants. The heated liquid or gas is circulated through heating coils, which in turn produce steam to run generators.

Considerable energy is presently being produced by controlled fission reactors, and much larger quantities will be produced in the near future. The major problems associated with nuclear reactors are the disposal of radioactive by-products and the dispersal of the huge amount of heat produced. The nuclei formed in fission reactions have high N/P ratios; most of them have short half-lives and emit beta particles of very high energy. Radioactive isotopes with relatively short half-lives have found some use as tracer elements in both industry and medicine. However, this use accounts for only a very small fraction of the total production. As yet, no completely satisfactory solution has been found for their safe disposal.

The procurement of the raw materials for nuclear reactors also presents problems. Although uranium is a fairly abundant element, most ores presently mined are low in uranium content. Moreover, fissionable $^{235}_{92}U$ accounts for only 0.71 percent of the isotopic composition; the remainder is almost all $^{238}_{92}U$. Fortunately, $^{238}_{92}U$ can be converted to fissionable $^{239}_{94}Pu$ by the capture of a neutron, followed by two successive beta emissions.

$$^{238}_{92}U + ^{1}_{0}n \rightarrow ^{239}_{92}U$$

$$^{239}_{92}U \rightarrow ^{239}_{93}Np + ^{0}_{-1}e$$

$$^{239}_{93}Np \rightarrow ^{239}_{94}Pu + ^{0}_{-1}e$$

A major objective of reactor scientists is to design an efficient and safe *breeder reactor*. The current design utilizes $^{239}_{94}Pu$ to provide the neutron necessary for the conversion of uranium-238 to plutonium-239. In this process, slightly more than one fissionable atom of plutonium-239 is produced for each plutonium-239 consumed in sustaining the nuclear pile. Such a reactor increases tremendously the available fissionable materials. By a similar neutron-capture process, thorium-232 can be converted to uranium-232, which also can be fissioned by slow neutrons.

It is evident from the binding-energy curve that energy can also be produced by the *fusion* of the lightest nuclei into heavier nuclei. Although helium-4 has an extraordinarily high binding energy for a light nucleus, the direct preparation of this nucleus from neutrons and protons is not practical. Neutrons have a short half-life, and the only source of large quantities is a fission reactor. Another disadvantage of fusion is the very high temperature required. Of the few successful fusions that have been carried out in the laboratory on a very small scale, one of the most promising is the production of helium-3 from deuterium.

$$^{2}_{1}H + ^{2}_{1}H \rightarrow ^{3}_{2}He + ^{1}_{0}n$$

Not only does this reaction occur at a rapid rate, it also requires the lowest temperature of any known fusion process—about 10^7 degrees.

In the fusion (hydrogen) bomb, the required temperature is derived from a fission bomb. The fusion bomb consists of a central core containing a fission bomb surrounded by a casing of fusionable material. The fission bomb explodes first and imparts a high enough temperature to the outer casing to cause some fusion to take place. Obviously, there is no critical mass, as in the fission bomb, and the force of the explosion is limited only by the size of the bomb. The engineering problem of packing gaseous deuterium densely enough for fusion to occur was solved rather ingeniously by using lithium deuteride, a salt-like solid.

$$[{}^7_3\text{Li}]^+[{}^2_1\text{H}]^-$$

In addition to the energy obtained from the deuterium fusion, more energy is produced by the reaction of lithium-7 with deuterium to produce helium-4 nuclei.

$$^7_3\text{Li} + {}^2_1\text{H} \rightarrow 2{}^4_2\text{He} + {}^1_0\text{n}$$

If the release of the vast amount of energy from the fusion process could be suitably controlled, man would gain an almost unlimited source of useful energy. The great abundance of water on earth and the fact that deuterium forms a significant part of the total hydrogen in water ensure a plentiful supply of raw materials. Furthermore, this process would not present the environmental problem of radioactive waste disposal that accompanies the production of energy by fission reactions. However, the development of a self-sustaining, controlled fusion reactor presents enormous technical problems, not the least of which is that of reaching and maintaining the required temperature. At a temperature of 10^7 degrees, no substance remains a solid to serve as a container for the reaction.

INDUCED NUCLEAR TRANSFORMATIONS

Probably the most common induced nuclear transformations are particle-particle reactions in which, in addition to the product nucleus, a lighter particle such as a proton or neutron is formed. In 1919, Ernest Rutherford, discovered the first particle-particle reaction by bombarding nitrogen gas with alpha particles to form oxygen and hydrogen.

$$^{14}_7\text{N} + {}^4_2\text{He} \rightarrow {}^{17}_8\text{O} + {}^1_1\text{H}$$

James Chadwick, a student of Rutherford, discovered the neutron by bombardment of beryllium with alpha particles.

$$^9_4\text{Be} + {}^4_2\text{He} \rightarrow {}^{12}_6\text{C} + {}^1_0\text{n}$$

The synthetic elements have also been prepared by similar reactions; for example, the reactions for production of berkelium and californium are

$$^{241}_{95}\text{Am} + {}^4_2\text{He} \rightarrow {}^{243}_{97}\text{Bk} + 2{}^1_0\text{n}$$

$$^{238}_{92}U + ^{12}_{6}C \rightarrow ^{244}_{98}Cf + 6^{1}_{0}n$$

Carbon-14, which is used extensively in metabolic studies, is produced in the laboratory by the same reaction that is believed to account for its natural presence in the atmosphere (p. 504).

$$^{14}_{7}N + ^{1}_{0}n \rightarrow ^{14}_{6}C + ^{1}_{1}H$$

Neutron-capture reactions are also important: The production of plutonium-239 and the fission of uranium-235 are examples given previously. Cobalt-60, an energetic gamma emitter that has replaced radium in the radiation therapy of some malignant tumors, is prepared by the addition of a neutron to cobalt-59.

High-energy photons can also be used to induce nuclear transformations: For example, a gamma reaction can probably account for the very small abundance of beryllium on the earth.

$$^{9}_{4}Be + \gamma \rightarrow ^{8}_{4}Be + ^{1}_{0}n$$

Beryllium-8 is very unstable and spontaneously breaks up into two $^{4}_{2}He$ nuclei (α particles) in less than 3×10^{-16} seconds.

Genesis of the Elements

Having acquired a background of some of the principles of nuclear reactions, we are now able to proceed with an investigation of the evolution of the chemical elements. Any model for this evolution must account for a universe that is about 98 percent hydrogen and helium. Moreover, it must also furnish some viable process for the formation of the heavier elements. Cosmologists are in essential agreement that the overwhelming abundance of both hydrogen and helium suggests a relatively simple origin of the universe. Therefore, the logical starting point in a discussion of the chemical elements is the birth of the universe.

BIRTH OF THE UNIVERSE

The Big Bang theory is the most widely accepted model used to explain the origin of the universe. According to the original version of this theory, all the elements were formed within about 30 minutes after the explosion of an infinitely hot and dense primordial matter. This model accounts for the fact that the elemental composition of the universe remains approximately constant. It is also in harmony with the observation that the universe is expanding, apparently still moving outward from its birthplace.

A modified and more acceptable version of the Big Bang theory suggests that most of the helium present in the universe, but only traces of the heavier elements, were created at this early epoch. A major portion of the elements with an atomic number greater than 2 were formed at a later date. As we shall see, these elements were probably formed in the nuclear reactions that occur in stellar material.

Astronomers have found that stars vary widely in mass, size, surface temperatures, and other characteristics. For example, their absolute luminosity (or radiant energy emitted), which is dependent on the surface temperature and area, ranges from about 10^{-4} to 10^6 times that of the sun. In like fashion, the temperature of the deep interiors of stars may vary from that of a cold, dead star up to 10^9 degrees, and perhaps even higher.

Even amid this great diversity among the stars of the universe, it has been possible to separate many of them into a few major classifications. By far the greatest number of stars (probably about 90 percent) fall under the *main sequence* category, of which the sun is a member. Main sequence stars are relatively stable and differ from each other mainly in luminosity and temperature. It is believed that most main sequence stars will evolve into *giants* with greatly increased diameters and generally lower surface temperatures. About 1 percent of the stars fall into the giant category. Probably most giants evolve to the *white dwarf* phase, which accounts for about 9 percent of the stars. More important are the more massive stars that evolve to the rare *supernova* category, where most of the heavier element production is thought to occur.

Nuclear fusion is considered to be a major source of the energy output of many stars, and it is likely that the heavier elements have been produced by fusion and other nuclear reactions in these nuclear furnaces. Consequently, it is necessary to examine briefly the different stages in the evolution of a star in order to follow the formation of the elements through these nuclear processes.

Birth of a Star

Scientists now generally believe that each star had its beginnings when a whirling cloud of rather tenuous gas, mostly hydrogen and helium and with a diameter of about a light year, separated from a huge parent body of gas and slowly started to coalesce. As the central portion of this whirling gas increased in density, the outer particles were pulled toward the center of gravity with ever-increasing velocities. This coalescence resulted in an increase in the temperature of the gas, and the star was born when the gas in the central portion became incandescent. As this gravitational process continued, the temperature required for the initiation of nuclear fusion—about 10^7 degrees—was attained. At this point in time—the beginning of nuclear fusion—the star evolved to the main sequence category. For our sun, this all probably happened more than five billion years ago.

Main Sequence Phase

In this beginning phase, it is impossible to form nuclei by the fusion of neutrons and protons, for the simple reason that the concentration of neutrons is too low for an effective reaction rate. The production of helium-4 from hydrogen, the main source of energy for stars in the main sequence can, however, occur by two other routes. In the first, high-energy protons fuse to form a deuterium nucleus, a positron, and a neutrino:

$$_1^1 H + _1^1 H \rightarrow _1^2 H + _{+1}^0 e + \nu$$

The deuterium reacts further with another proton to yield a helium-3 nucleus plus gamma radiation.

$$_1^2 H + _1^1 H \rightarrow _2^3 He + \gamma$$

Then two helium-3 nuclei fuse to form a helium-4 nucleus and two very high-energy protons.

$$_2^3 He + _2^3 He \rightarrow _2^4 He + 2 _1^1 H$$

Although most of the energy yield is derived from the last fusion reaction, the other two reactions are also exoergic (energy is given off). In addition to these sources of energy, the positrons formed are annihilated by electrons to give more radiant energy.

$$_{+1}^0 e + _{-1}^0 e \rightarrow 2h\nu$$

The second process is the carbon-nitrogen cycle, in which the following sequence of reactions takes place:

$$_1^1 H + _6^{12} C \rightarrow _7^{13} N + \gamma$$

$$_7^{13} N \rightarrow _6^{13} C + _{+1}^0 e + \nu$$

$$_1^1 H + _6^{13} C \rightarrow _7^{14} N + \gamma$$

$$_1^1 H + _7^{14} N \rightarrow _8^{15} O + \gamma$$

$$_8^{15} O \rightarrow _7^{15} N + _{+1}^0 e + \nu$$

$$\underline{_1^1 H + _7^{15} N \rightarrow _6^{12} C + _2^4 He + \gamma}$$

$$4 _1^1 H \rightarrow _2^4 He + 2 _{+1}^0 e + 2\nu + \gamma$$

Clearly, the net result of this sequence is the fusion of four protons to form a helium nucleus along with two positrons, two neutrinos, and, of course, gamma radiation. Carbon-12 is simply a catalyst and is reformed in the last step of the sequence. The two positrons formed are rapidly annihilated by two electrons. This sequence of reactions requires a temperature of about 2×10^7 °C and a total time of 6.5×10^6 years per cycle, the time required for the slowest step, the first one. Clearly, the overall results of the two routes are the same, only the pathways differ. The energy yield of each of the routes is approximately 30 MeV.

This, then, is the present state of our sun, a naturally occurring thermo-nuclear reactor that has been "burning" hydrogen nuclei to form helium-4 for about the past 5×10^9 years. Probably both the proton-proton route and the carbon-nitrogen route contribute to the fusion process. In any case, both routes consume protons and form helium. Consequently, solar-type stars continually become enriched in helium in their deep interiors. It is estimated that the mass of the burning core in the sun is still about 30 percent hydrogen, and that it therefore can continue as a main sequence star for another five billion years.

When the core of a star changes to almost pure helium, the energy production in the core ceases, but hydrogen fusion continues on the surface. According to one model, when the helium core becomes about one-tenth of the star's mass it collapses inward, furnishing a large amount of gravitational energy, which increases the rate of proton fusion on the core's surface. This increase in energy is so great, according to the model, that all of the excess cannot be lost and the star increases in diameter. At this point, then, the star evolves into the giant phase.

Giant Phase

The very small percentage of observed stars in the universe that have characteristics of the giant category suggests that the *red giant* phase covers a relatively short time span. Apparently, the hydrogen rapidly becomes exhausted through fusion in the shell surrounding the helium core, and a further gravitational collapse occurs near the end of this phase, which raises the temperature to around 10^8 degrees. At this temperature and high concentration of helium, two helium nuclei can fuse at a significant rate to form the highly unstable beryllium-8,

$$\ce{^4_2He + ^4_2He \rightarrow ^8_4Be}$$

most of which immediately decomposes back to the starting materials. However, small amounts of this unstable isotope are probably converted to carbon-12 by fusion with another alpha particle,

$$\ce{^8_4Be + ^4_2He \rightarrow ^{12}_6C}$$

Some of the ^{12}C then reacts with another helium nucleus to give ^{16}O. Additional nuclear reactions may also occur, resulting in the formation of some other light elements.

For the sun, the expansion that will probably occur during the transition to its giant phase may be great enough to swallow up Mercury and Venus and may even extend as far as the earth's orbit. This increase in diameter from the present 865 thousand miles to a projected 94.5 million miles or more makes obvious the name *giant phase*. With this greatly expanded radius, the gravitational force holding the outermost layers would be relatively weak, and it is possible that minor disturbances on the surface could result in the expulsion of material that would exceed escape velocity and be lost to outer space.

White Dwarf Phase

It seems probable that the ultimate fate of most stars (including the sun) is a continuation of the nuclear processes just discussed until all of the usable nuclear fuel becomes exhausted. With only gravitational contraction remaining as an energy source, a star slowly shrinks to a body composed of bare nuclei and electrons with a density of about 10^5 to 10^7 g/cm^3. At this point, called the white dwarf phase, the gravitational contraction is finished, and the star slowly cools as the kinetic energy of the nuclei is converted to radiant energy and lost to space. This extremely dense white dwarf has a very high temperature, and the loss of energy is apparently an exceedingly slow process. (It is doubtful if the universe is old enough for any star to have reached the low temperature of outer space.)

Supernova Phase

For stars considerably more massive than, say, three times the mass of the sun, a more glamorous end (among other possibilities) may be postulated. Because of the huge mass involved in the contraction of the inner helium core, mind-staggering amounts of gravitational energy are converted to kinetic energy. As a result, the temperature could easily exceed the energy barrier required for the fusion of carbon nuclei to magnesium and of oxygen nuclei to silicon.

Temperatures and densities probably do not increase, however, to the point where these nuclei can fuse to form the iron group. Nevertheless, highly energetic collisions occur, which result not in fusion but in *spallation reactions,* in which small particles are chipped off the heavier nuclei, yielding, for example, helium nuclei, deuterium nuclei, protons, and neutrons. By successive additions of helium nuclei, elements up to and including the iron group may be formed. In addition, neutron-addition reactions probably occur with the iron group elements, which, as was discussed earlier, can lead to elements of higher atomic number. For example,

$$^{58}_{26}\text{Fe} + ^{1}_{0}\text{n} \rightarrow ^{59}_{26}\text{Fe}$$

followed by

$$^{59}_{26}\text{Fe} \rightarrow ^{59}_{27}\text{Co} + ^{0}_{-1}\text{e} + \nu$$

Therefore it is possible that the temperature and neutron density in this stage are favorable for the formation of elements up through lead.

All of this may happen at such an accelerated rate that what had been a controlled thermonuclear reactor becomes a thermonuclear bomb, resulting in that most spectacular astronomical catastrophe, a *supernova,* where even uranium, thorium, and elements of higher atomic number are formed. During such an explosion it is believed that most of the elements of higher atomic number than neon are formed. It has also been suggested that matter spewed out in this fashion is the major origin of cosmic rays.

On the other hand, with stars that are even more massive, the explosion may be combined with an *implosion* leaving behind a very small, dense body of almost pure neutrons, appropriately called a *neutron* star. The process does not necessarily stop at this stage. Recent data from sophisticated X-ray detecting satellites have been interpreted as confirming the postulated existence of *black holes* in space. The enormous gravitational field of the neutron star presumably continues to pull the star into itself until it literally disappears from sight. Mass has not disappeared, but the gravitational pull is so concentrated that even light and other forms of electromagnetic radiation are trapped. Therefore, it is not possible to directly observe a black hole with either optical or radio telescopes. The X-ray emissions measured are believed to have their origin in the excited states of atoms being pulled into a black hole.

Naturally, many questions are left unanswered by this introductory treatment of the origin of the elements. Even with a more extensive discussion there is no satisfactory explanation for the relative abundance of all of the elements and their isotopic composition. In summary, then,

1. Hydrogen and most of the helium were produced at the birth of the universe.
2. Formation of a star by gravitational attraction produces temperatures up to about 10^7 degrees, the threshold of nuclear fusion reactions.
3. Most newly born stars evolve to the main sequence stage, in which by some fusion process hydrogen is converted to helium-4.
4. Most main sequence stars evolve to the giant phase, where helium fusion begins forming carbon-12 and oxygen-16. Probably other nuclear reactions occur to form some of the other light elements.
5. For more massive stars, fusion of carbon-12 and oxygen-16 forms magnesium-24 and silicon-32, respectively. Other helium additions probably form elements up to the iron group; and neutron-addition reactions with the iron group, followed by beta emissions, can form elements up to and including lead.
6. A possible end for these massive stars is the *supernova* phase. This thermonuclear explosion probably accounts for the production of a large percentage of elements heavier than neon. Such explosions are thought to be the source of cosmic rays whose violent interactions with interstellar gas appear to play an important role in the production of some of the lighter elements; that is, lithium, beryllium, and boron.

If we accept the view of most cosmologists, the big bang occurred about 12 to 14 billion years ago. In addition, if we take the age of the solar system as about five billion years, our sun must be classified as a middle-generation star. Accordingly, innumerable numbers of stars were formed, ran their inevitable course, and died long before our sun started to glow. Moreover, massive stars formed supernovae and spewed forth the stardust which, along with the primeval hydrogen and helium, formed the gas and dust clouds out of which our solar system was born.

On at least one planet revolving around this middle-generation star, life formed and still exists. This living matter is composed of many elements, although it is based mainly on carbon, nitrogen, oxygen, and hydrogen, which are four of the most abundant elements in the universe. The carbon, nitrogen, and oxygen, according to the theories we have just discussed, were formed by nuclear combinations at a time so distant as to be inconceivable. The hydrogen in the water we drank in this morning's coffee may actually have been primal. Because life is based primarily on these four common elements, it may not be unique to this planet. Although we are not concerned here primarily with the origin or distribution of living matter, the formation of the precursors to life from these elements continues to be an active area of scientific investigation. In the section that follows we discuss some of the theories of the formation of chemical compounds on prebiotic earth.

CHEMICAL EVOLUTION

Whereas a large part of Darwin's theory of evolution is well documented with factual observations, facts are not available for the chemical evolution that preceded biological evolution. Even if the abiogenic (noncellular) synthesis of molecules that can be considered precursors to living matter still occurs, it is doubtful that this phenomenon is discoverable, since compounds produced in this way would be exposed to instant metabolism by bacteria.

Serious study of the formation of the chemical precursors to life did not begin until the 1920s. Within this early period both A. I. Oparin, a Russian biochemist, and J. B. S. Haldane, a British biochemist, published papers on the possible origin of life from organic molecules that had formed during the prebiotic period. There is a gap of over half a century between Darwin's early work and the early investigations into the origin of life. Moreover, little laboratory research was undertaken before the 1950s.

It must be remembered, however, that our knowledge of the chemistry of carbon compounds did not begin to develop sufficiently until the latter part of the nineteenth century. Indeed, contemporary ideas about the composition of the atom, chemical bonding, nuclear reactions, and the extent and rate of reaction were just emerging in the early part of the twentieth century. Research in stellar evolution, geochemistry, and other related areas of science were still in their infancy. The Rutherford atom, the Bohr atom, and the valence bond concept all contributed to our understanding of the world around us.

With all the data and theoretical ideas that are currently available to contemporary scientists, it is hardly surprising that chemical evolution has become a fruitful and rapidly expanding research field. The advent of the electronic, computerized space age, the acquisition of data from the planets and their satellites, and stellar observations that are no longer hindered by the atmosphere give promise of an early resolution of some of the many problems in this field.

We begin our discussion of the precursors of biological evolution with the formation of the earth and its early environment.

Formation of the Earth

One theory of the formation of the earth postulates that the whirling gases, which were to form the sun, contracted to approximately the present diameter of the solar system and took the form of a relatively flat disk, with most of the mass concentrated in the hot center. In a few regions within the more tenuous, colder outer part of the disk, small versions of the large spinning disk of gas apparently coalesced to form the planets. (A repetition of this process on a much smaller scale may account for the extensive satellite systems of the two most massive planets, Jupiter and Saturn.) Since they presumably formed from the colder outer gases in which nuclear reactions were not possible, the planets should have an elemental composition essentially the same as that of the sun. Indeed, the outer planets do consist primarily of hydrogen, helium, and other light gases.

How, then, does this theory account for the much smaller abundance of hydrogen and helium on *earth?* One possibility is that during this period of its formation, the sun hurled out huge solar flares that were hot enough to sweep the lighter elements out of the region in which the inner planets were formed. It is entirely possible, then, that the inner planets were formed from matter that had already lost most of the hydrogen and helium. Under these conditions, it is assumed that the other light elements were chemically combined and not lost. For example, carbon may have been present as carbides or elemental carbon; nitrogen as nitrides or more likely in ammonium ion; oxygen as oxides, hydroxides, and in silicate minerals; fluorine and chlorine as halides; and so on. Hydrogen may have been retained in the form of metallic hydrides, but most of it was probably present as physically entrapped water and as water of hydration. This explanation also accounts for the relative deficiency on earth of the elements neon and argon in comparison with their cosmic abundance. A second possibility is that the earth was formed by the slow aggregation of solar material, followed later by the removal of most of the hydrogen, helium, and lighter, chemically uncombined elements by intense solar radiation.

Whether the earth formed primarily from material deficient in certain elements or whether these elements were lost after its formation cannot be answered. But the answer is not required in order to explain the deficiency in hydrogen and helium. Under present conditions, molecules of hydrogen and helium in the upper atmosphere of the earth possess kinetic energies high enough to allow them to escape. Moreover, conditions on the infant earth were probably such that the rate of escape of these two light molecules was considerably greater. Neither the earth nor the other inner planets have high enough gravitational fields to maintain either hydrogen or helium in their atmospheres.

In any event, during the millions of years that passed, the earth remained in a constant state of turmoil. As it radiated energy into space, it cooled, and a thin crust of rock eventually formed on the surface. Intense volcanic activity along with other forces must have repeatedly broken this surface. It is believed that such constant breaking and reforming of the crust caused a chemical differentiation between molten phases and solid phases, which would account for the fact that the earth's outer crust contains a great abundance of the lighter iron-poor silicate minerals. The heavier, iron-rich minerals constitute the denser rocks that separate the surface crust from the inner core. This material has enough structure to pass seismic waves, but at these great depths the high temperatures and pressures give it some degree of plasticity. The very dense core of the earth appears to be iron or nickel-iron and is at least in part in the liquid state.

Chemical Evolution

The age of the oldest known rocks in the earth's crust is estimated to be 4.5 to 4.8 billion years. Estimates of the date of the appearance of the first living organisms (unicellular plants) place the event at approximately 3.3 billion years ago. Thus, there was a period of perhaps 1.5 billion years during which the chemical precursors of life evolved.

What were the conditions that might account for the synthesis of these compounds—amino acids and sugars, for example? If the theories developed in the preceding paragraphs are even approximately correct, the earth at this stage could not have had an atmosphere—the light gases were lost. The fact that volcanoes still spew out gases that are mainly water vapor and carbon dioxide, with much smaller amounts of SO_2, H_2S, H_2, N_2, HF, HCl and traces of ammonia and methane, suggests that the early atmosphere may have had a volcanic origin.

As the surface of the earth cooled and the earth was degassed of its water of crystallization, an atmosphere of superheated steam would have built up. With further cooling, torrential rains may have poured down, eventually forming pockets of hot surface waters. Some of this water could have percolated through cracks in the surface and reacted with carbides to form low molecular weight hydrocarbons, principally methane.

Perhaps a more likely source of methane is the direct combination of carbon with hydrogen. At this early period (before any chemical differentiation had occurred), free iron was probably still present in surface rocks, and the reaction of iron with water could have furnished the necessary hydrogen. The reaction of either carbon dioxide or carbon monoxide with hydrogen may also have been the source of methane gas, but the postulation of a large amount of hydrogen in the primeval gases argues against the presence of carbon dioxide.

The presence of ammonia in the primitive atmosphere may have been the result of the hydrolysis of ammonium salts, but more likely it was due to the direct combination of nitrogen and hydrogen.

It seems plausible, therefore, that the primitive atmosphere of the earth was a *reducing* mixture of gases, consisting primarily of H_2O, NH_3, and CH_4, with smaller amounts of H_2 and N_2 and traces of other volcanic gases. Little or no O_2 was present. Some rather convincing arguments can be made for the primitive atmosphere having been reducing rather than oxidizing:

1. The matter out of which the earth formed was mostly hydrogen, an excellent reducing agent. (Harold Urey and others have pointed out that both Jupiter and Saturn have reducing atmospheres containing ammonia and methane.)
2. Sedimentary rocks of the Precambrian period contain *ferrous* iron and *not ferric*. Therefore, they were deposited under a reducing environment.
3. Meteorites contain iron either as metallic iron or ferrous iron. Carbon is present as elemental carbon, carbides, and hydrocarbons. The recent reports of trace amounts of amino acids in meteorites is still under dispute. Phosphorus, if present, is found as phosphides.
4. It has been postulated that if photosynthesis ceased, all of the elemental oxygen remaining would be used up in a measurable span of time by reaction with the not fully oxidized igneous rocks.

Having postulated an atmosphere that lacked oxygen and perhaps carbon dioxide, let us consider the kinds of energy available. Without an ozone layer to stop high-energy photons, the earth was under a constant bombardment of ultraviolet radiation from the sun and cosmic rays from outer space. Conditions were such that robust and frequent electrical storms must have kept the skies streaked with lightning. Radioactive decay furnished alpha particles, beta particles, and gamma radiation. It seems reasonable to assume that no significant chemical evolution occurred either at extremely high temperatures or before the hydrosphere (which now covers 70 percent of the earth's surface) had formed. Thus, the average temperature was probably close to that of contemporary earth. There were, however, pockets of higher temperatures in the neighborhood of volcanoes and wherever there existed concentrated deposits of radioactive elements.

Armed with this information, scientists have performed many laboratory experiments in which reducing atmospheres of methane, ammonia, water, and, in some cases, hydrogen, of varying composition have been bombarded with ultraviolet radiation, electrical discharges, alpha particles, electron beams, and gamma radiation either singly or in various combinations. These experiments are usually carried out at room temperature, although some very significant higher temperature research has been reported. Results, to say the least, have been startling. From these reactions have been isolated such significant molecules as amino acids, pyrimidines, purines, sugars, and fatty acids, all of which are essential building blocks of cellular material.

Since Darwinian evolution most likely had its point of origin in the primordial oceans, the more complex chemical evolution also must have taken place in the oceans—the so-called primordial soup. These terrestrial aqueous solutions contain elements essential to living matter (for example, iron and calcium) that are not found in the atmosphere. At the same time, the solvent ocean brought together molecules that had been formed at different locales under different conditions. Many experiments have been carried out in aqueous media, using previously formed simple molecules in dilute solutions. Products reported from these experiments have included sugars and fatty acids, metal complexes of porphyrins essential for photosynthesis, peptides, and protein-like molecules. For example, adenine, an important constituent of DNA and RNA, has been prepared in very small amounts by shooting an electron beam into a mixture of methane, ammonia, water, and hydrogen. Further reaction of adenine in solution with polyphosphate esters gave identifiable yields of adenosine phosphate, as well as adenosine diphosphate (ADP) and triphosphate (ATP). Adenosine triphosphate is very important biologically as both a repository and a source of energy for biological processes.

In addition to its function as a reaction medium, the ocean also served as a solvent in which significant concentrations of the intermediates necessary for the formation of more complex molecules could build up. Ionizing radiation is important in producing the ions and free radicals that served as reaction intermediates. For example, ultraviolet radiation of the proper wavelength splits water into hydrogen atoms and hydroxyl free radicals, from which both hydrogen gas and hydrogen peroxide can form.

$$H-O-H + h\nu \rightarrow H\cdot + \cdot OH$$

$$H\cdot + H\cdot \rightarrow H_2$$

$$\cdot OH + \cdot OH \rightarrow H_2O_2$$

This is a possible source of hydrogen gas in the primitive atmosphere and can also account for loss of hydrogen from the earth. Ultraviolet radiation also destroys the more complex organic molecules that were formed by similar processes. However, in the ocean these molecules could be adsorbed on the surfaces of clay particles that slowly settled to the bottom, and since ultraviolet radiation cannot penetrate water to great depths, they would have been protected from photolytic decomposition.

The formation of the precursors to life, although thermodynamically favored, probably proceeded at a very slow rate. Of course, there was time for these reactions, but there was just as much time available for photolytic and thermal decomposition. The clay particles may have functioned as catalysts. Other possible catalysts were high molecular weight hydrocarbons, which, taking into account their presence in meteorites, could easily have been part of the primordial soup.

The oxygen and carbon dioxide of the earth's *present* atmosphere were probably derived from the process of photosynthesis carried out by the newly formed algae in the seas. The buildup in oxygen led to the formation of the ozone layer that now protects us from the high-energy bombardments so destructive to living matter—the same bombardments that were apparently vital to the spontaneous generation of life.

Our understanding of chemical evolution can never be complete, but it is gradually increasing. Information gained from continued exploration of our near neighbors in space—the moon and the planets—should prove helpful in increasing this understanding as well as providing a more adequate explanation of the origin of the solar system.

SUGGESTED READINGS

Abell, George. *Exploration of the Universe.* 2nd ed. New York: Holt, Rinehart and Winston, 1969.

Hahn, O. "The Discovery of Fission." *Scientific American,* Vol. 198, No. 2 (February 1958), pp. 76–84.

Landis, J. W. "Fusion Power: Hallmark of the 21st Century." *Journal of Chemical Education,* Vol. 50, No. 10 (October 1973), pp. 658–662.

Lemmon, Richard M. "Chemical Evolution." *Chemical Reviews,* Vol. 70, No. 1 (February 1970), pp. 95–109.

Miller, Stanley L. "Production of Some Organic Compounds Under Possible Primitive Earth Conditions." *Journal of the American Chemical Society,* Vol. 77, No. 9 (May 1955), pp. 2351–61.

Mueller, Robert E. "Chemistry in Planetology." *Journal of Chemical Education,* Vol. 42, No. 6 (June 1965), pp. 294–301.

Seaborg, Glenn T. "Some Recollections on Early Nuclear Age Chemistry." *Journal of Chemical Education,* Vol. 45, No. 5 (May 1968), pp. 278–89.

Selbin, J. "The Origin of the Elements, I." *Journal of Chemical Education,* Vol. 50, No. 5 (May 1973), pp. 306–310.

Selbin, J. "The Origin of the Elements, II." *Journal of Chemical Education,* Vol. 50, No. 6 (June 1973), pp. 380–387.

Swartout, John A. "Critical Chemical Problems in the Development of Nuclear Reactors." *Journal of Chemical Education,* Vol. 45, No. 5 (May 1968), pp. 304–10.

Wildeman, T. R. "The Automobile and Air Pollution." *Journal of Chemical Education,* Vol. 51, No. 5 (May 1974), pp. 290–294.

PROBLEMS

1. Complete the following nuclear equations.

   (a) $^{23}_{11}\text{Na} + ^{1}_{0}\text{n} \rightarrow \gamma +$ (b) $^{24}_{12}\text{Mg} + ^{1}_{0}\text{n} \rightarrow ^{1}_{1}\text{H} +$

   (c) $^{35}_{17}\text{Cl} + ^{1}_{0}\text{n} \rightarrow ^{4}_{2}\text{He} +$ (d) $^{14}_{7}\text{N} + ^{4}_{2}\text{He} \rightarrow ^{1}_{1}\text{H} +$

   (e) $^{9}_{4}\text{Be} + ^{4}_{2}\text{He} \rightarrow ^{1}_{0}\text{n} +$ (f) $^{10}_{5}\text{B} + ^{1}_{1}\text{H} \rightarrow ^{4}_{2}\text{He} +$

   (g) $^{63}_{29}\text{Cu} + ^{1}_{1}\text{H} \rightarrow ^{1}_{0}\text{n} +$

2. Potassium-40 is unstable and spontaneously decomposes by either K capture or β^- emission. Write nuclear equations for each mode of decomposition.

3. Calculate the binding energy in MeV per nucleon for the nuclide $^{6}_{3}\text{Li}$. Atomic masses are: electron, 0.000548597 amu; neutron, 1.008665 amu; hydrogen, 1.007825 amu; $^{6}_{3}\text{Li} = 6.01347$ amu.

4. Calculate the energy in ergs and in kilocalories for the formation of one gram-atom of helium by the reaction

$$^{2}_{1}\text{H} + ^{3}_{1}\text{H} \rightarrow ^{4}_{2}\text{He} + ^{1}_{0}\text{n}$$

   Atomic masses are: hydrogen-2, 2.0140 amu; hydrogen-3, 3.01605 amu; and helium-4, 4.00260 amu. See also problem 3.

5. The representative fission reaction

$$^{235}_{92}\text{U} + ^{1}_{0}\text{n} \rightarrow ^{92}_{36}\text{Kr} + ^{141}_{56}\text{Ba} + 3^{1}_{0}\text{n}$$

   is reported to release about 175 MeV of energy per nucleus fissioned. Calculate the quantity of energy released by the fission of one gram of uranium-235. State your answer in kilocalories and in kilojoules (see Appendix 1).

6. Are any of the products of the nuclear reactions in problem 1 radioactive? If so, write decay reactions. Use an atomic weight table as a reference point.

7. Iron has four stable isotopes: ^{54}Fe, ^{56}Fe, ^{57}Fe, and ^{58}Fe. The isotopes ^{53}Fe and ^{59}Fe are radioactive. What type of particle would you expect each of these isotopes to emit? Write nuclear equations for the process.

8. Uranium-235 undergoes the following successive decays: alpha, beta, alpha, beta, alpha, alpha, alpha, alpha, beta, alpha, beta to lead-207. Construct a chart of the changes similar to Figure 18-3.

9. Uranium and radium are chemically extracted from a sample of pitch-blende, and the separated uranium is carefully converted to pure uranyl nitrate. The lead residue is chemically separated and converted to lead nitrate.
   (a) Five years later, what impurities are present in the uranyl nitrate?
   (b) How does the lead nitrate differ from commercially available lead nitrate?

10. Account for the fact that fission products are radioactive and for the most part are β^- emitters.

11. How many tons of coal producing 14,622 Btu/lb would be required to furnish the same amount of energy as the answer to problem 5? (One British thermal unit equals 251.98 calories.)

12. According to the reaction in problem 4, what quantity of tritium is needed to produce the same amount of energy as the answer to problem 5?

13. List the advantages and disadvantages of producing energy
   (a) by fission processes
   (b) from fossil fuels

14. Why are thorium ores liable to be as important in the future as uranium ores are today?

15. Stars vary in composition, temperature, luminosity, and mass. Account for these differences.

16. According to one model, the sun was formed from interstellar gases. How did it form, and how did it reach its present temperature?

17. If the earth formed from solid particles, account for the molten phase it appears to have passed through and for the present molten center.

18. It is stated that water was entrapped by hydrated minerals in the formation of the earth. Do you think that this was present as molecular water, as in $BaCl_2 \cdot 2H_2O$?

19. Explain the fact that the four isotopes ^{16}O, ^{24}Mg, ^{28}Si, and ^{40}Ca constitute about 78 percent by weight of the earth's crust.

20. The reaction of surface iron with hot water or steam

$$Fe + H_2O \rightarrow FeO + H_2$$

probably accounts for most of the hydrogen gas that appears to have been present in the primeval atmosphere. Calculate the ΔG for this reaction with water (use values at standard conditions). Would you expect the reaction to be more favorable at a higher temperature? Explain.

21. Natural gas produced from some gas wells contains a significant quantity of helium, and this is the source of commercially obtained helium gas. If it is true that most of the helium present when the earth formed was lost to outer space, account for this source of helium gas.

22. A lunar experiment conducted during an Apollo mission suggests that a large part of the oxygen in the atmosphere is actually formed in the upper atmosphere from the decomposition of water vapor rather than from photosynthesis.

(a) Write equation(s) to illustrate this process.

(b) Would it be feasible to recycle water by this process to furnish H_2 and O_2 for the fuel cells used in spaceships?

23. Recently, astronomers claimed to have found evidence that ice particles exist on four of the moons of Jupiter. They conclude that discovery of water increases the possibility of finding life on these moons. Explain.

24. (a) Solid carbon dioxide apparently makes up the polar "ice caps" of Mars. How can it exist at the very low pressure of the Martian atmosphere?

(b) An Apollo mission reported passing through solid ice particles left in space by previous Apollo missions. Explain the existence of ice at the almost zero pressure of outer space.

25. Isotopes can be of great value in deducing the possible pathways of a reaction. For example, in a plant, photosynthesis can be represented by the following overall reaction:

$$6CO_2 + 6H_2O + \text{light energy} \rightarrow C_6H_{12}O_6 + 6O_2$$
$$\text{glucose}$$

Algae grown in water containing some oxygen-18 ($H_2{}^{18}O$) evolve oxygen of the same isotopic concentration as the oxygen in the water. However, when the algae in water containing only oxygen-16 were furnished carbon dioxide that contained oxygen-18, no oxygen-18 was present in the evolved oxygen gas. What did this tell researchers about the above process?

19 *Hydrogen*

Hydrogen, with mass number 1 and atomic number 1, is the first element in the periodic chart and has the simplest electronic configuration of all the elements, $1s^1$. With only one electron and only one orbital at low enough energy for covalent bonding, it superficially can be considered as either a member of the alkalies (because of its one valence electron) or of the halogens (because of the possibility of forming a negative ion). However, with the exception of the formation of the negative ion, the chemistry of hydrogen bears little resemblance to either of these two families.

ABUNDANCE AND PROPERTIES

Hydrogen is the only stable element without any neutrons in the nucleus, and hydrogen atoms can be considered the building blocks out of which the other elements are formed. According to the Big Bang theory of the universe (Chapter 18), hydrogen may be thought of as the original element. Although about 70 percent of the universe is composed of hydrogen, it accounts for only 0.87 percent of the earth's mass and is most commonly found in combination with oxygen. However, since water covers about 70 percent of the earth's surface, hydrogen, as a natural resource, is easily accessible. Table 19-1 lists some of the properties of hydrogen.

TABLE 19-1 Properties of Hydrogen		
Ionization energy		313 (kcal/mole)
Electron affinity		17 (kcal/mole)
Electrode potential		0.00 V
Covalent radius		0.4 Å
Dissociation energy, H_2		103 (kcal/mole)
Bond length, H_2		0.74 Å
ΔH hydration, H^+ gaseous ion		−268 (kcal/mole)

Actually, there are three isotopes of hydrogen: ordinary hydrogen, $_1^1H$, which accounts for 99.985 percent of the naturally occurring element; heavy hydrogen or deuterium, $_1^2H$ (or D); and tritium, $_1^3H$ (or T), a radioactive isotope with a half-life of approximately 12 years. Since isotopes generally have very small differences in mass, their chemical and physical properties are virtually identical. The deuterium atom, however, is twice as heavy as an ordinary hydrogen atom, and this difference results in a higher boiling point (101.4°C) for heavy water, D_2O. This difference in boiling point between ordinary water and heavy water is great enough that the Dead Sea, which has entrapped water for eons with no outlets other than through evaporation, has a higher ratio of D_2O/H_2O than the average isotopic ratio.

Hydrogen exists in the elemental form as diatomic molecules and is the lightest known gas. The molecule is, of course, nonpolar, and with only two electrons per molecule, van der Waals forces are extremely weak. Consequently, the boiling point of hydrogen is lower than that of the inert gas neon. Other consequences of these weak forces are the low critical temperature (−240°C) of hydrogen and the fact that it obeys the ideal gas law under all but the most extreme conditions of temperature and pressure.

While the *inter*molecular forces in the elementary substance are very weak, the *intra*molecular forces are strong. The hydrogen-hydrogen sigma bond is stronger than the halogen-halogen sigma bonds and is approximately as strong as the bond in hydrogen chloride. Clearly, a pair of electrons pulls the two protons very close together (as witness also the short bond length). As a direct result of this strong hydrogen-hydrogen bond, molecular hydrogen is a relatively unreactive substance at ambient conditions.

Some of the properties of hydrogen *atoms* also deserve comment. The ionization energy is very high, higher, in fact, than that for the inert gas xenon. This is not too surprising, since the loss of the only electron leaves behind a bare nucleus, the proton. The strong attraction of the proton for the electron is also illustrated by the small size of the hydrogen atom. The electron affinity, on the other hand, is relatively low. Only 17 kcal of energy are released when one mole of electrons is accepted by a mole of gaseous hydrogen atoms, whereas 79 kcal are released upon capture of a mole of electrons by fluorine atoms.

Very pure hydrogen is prepared by the electrolysis of water to which some electrolyte such as sodium chloride has been added. Continuous electrolysis results in an enrichment of heavy water in the electrolytic bath. This is the common method used for the isolation of both deuterium oxide and, by subsequent electrolysis, deuterium. Most hydrogen used commercially is prepared by less expensive methods, such as the reaction of steam with iron.

$$3Fe + 4H_2O \rightarrow Fe_3O_4 + 4H_2$$

Large quantities of hydrogen are also produced by the reaction of steam with hydrocarbons. The reaction of propane is representative.

$$C_3H_8 + 3H_2O \rightarrow 3CO + 7H_2$$

In the laboratory, small amounts of hydrogen are readily obtained by the reaction of an active metal with water,

$$Ca + 2H_2O \rightarrow Ca(OH)_2 + H_2$$

or the reaction of certain metals with aqueous acid solutions.

$$Zn + 2H_3O^+ \rightarrow Zn^{2+} + 2H_2O + H_2$$

COMPOUNDS

The compounds of hydrogen can be divided into three major types:

1. Anionic or saltlike hydrides
2. Interstitial or metallic hydrides
3. Covalent hydrides

Before discussing these hydrogen compounds, it is essential to clarify the *cationic* chemistry of hydrogen. Because of its high ionization energy, hydrogen does not form ionic compounds with even the two most electronegative elements, oxygen and fluorine. The hydrogen cation is a bare nucleus with a radius approximately $1/10^5$ the size of the hydrogen atom and has only transitory existence in chemical reactions. With such a high charge density, the proton either strips electrons from surrounding matter or, at the very least, takes a share of a nonbonded pair of electrons. Therefore, compounds in which hydrogen exists as a simple cation are not known.

On the other hand, the hydrogen cation plays an important role in solution. Numerous hydrogen-containing compounds react as acids in solvating media to produce solvated protons. These solvated species, for example H_3O^+ and NH_4^+, do not, however, contain discrete hydrogen ions. Instead, the protons are covalently bonded to the central atom, and the entire species carries the positive charge. The extensive formation of covalent bonds is reflected in the amount of heat liberated when gaseous hydrogen ions are allowed to react with liquid water. This heat of hydration is -268 kcal/mole, compared with the heat of hydration of the smallest singly charged metal ion (Li^+, -123 kcal/mole), where the interactions of the cation with water are primarily ion-dipole forces.

Anionic Hydrogen

Hydrogen reacts with the most active metals of low ionization energies, the alkalies and alkaline earths, to form compounds containing the hydride ion (H^-). (According to the rules of nomenclature, only those compounds which contain hydrogen in a negative oxidation state should be named hydrides. In actual practice, the term *hydride* is used for all binary hydrogen compounds and is so used here.) These active metal hydrides are white, crystalline, high-melting solids. In the electrolysis of fused lithium hydride, hydrogen gas is liberated at the anode. Since the anode process is an oxidation reaction, the formation of hydrogen gas at the anode confirms the negative charge on the hydrogen.

$$H^- \rightarrow \tfrac{1}{2}H_2 + e^-$$

Both lithium and calcium hydride are formed in good yields by direct combination of the metals with hydrogen gas at elevated temperatures.

$$Ca + H_2 \rightarrow CaH_2$$

Yields with the other metals are much poorer, and higher temperatures are required for their formation. The heats of formation of the alkali and alkaline earth hydrides become less negative with increasing cation size, a trend that will be discussed in Chapter 20.

Chemically, the metal hydrides are powerful reducing agents and react with water to form hydrogen gas and the metal hydroxide.

$$LiH + H_2O \rightarrow LiOH + H_2$$

The heat generated is sufficient to ignite the liberated hydrogen gas; consequently, most of the hydrides burn in moist air. Although the above reaction is undoubtedly an oxidation-reduction reaction, it is also an acid-base reaction. The hydride ion is a much stronger base than the hydroxide ion, and the extent of reaction is therefore high.

$$H^- + H_2O \rightarrow H_2 + OH^-$$

Interstitial Hydrides

Hydrogen gas reacts reversibly with some of the transition metals to form a series of compounds, many of which, although of constant composition, are nonstoichiometric in terms of whole number ratios. Some examples are the hydrides of titanium, vanadium, and zirconium: $TiH_{1.73}$, $VH_{0.6}$, and $ZrH_{1.9}$. These substances can be considered solutions of hydrogen in the solid metal in which hydrogen atoms occupy tetrahedral "holes" in the lattice. That there is some chemical interaction between the metal and hydrogen, however, is evidenced by slight changes in metallic properties. The lattice, although generally not significantly distorted, is expanded, and the interstitial hydrides are therefore less dense than the metals themselves. These hydrides also have lower ductilities, greater hardness, and greater electrical resistance than the corresponding metals. Hydrogen, then, is not simply dissolved in the metal but appears to be bonded to the metal lattice in some fashion. One model describes these compounds as *interstitial alloys* in which the hydrogen, having lost its valence electron, occupies interstices in the metal lattice as a proton, thereby simulating metallic bonding. These metallic hydrides are formed only with the transition elements (and some inner transition elements) that have partially filled *d* orbitals. Nonmetals with small covalent radii—boron, carbon, and nitrogen—form analogous nonstoichiometric compounds with transition metals.

Covalent Hydrides

Molecular binary hydrogen compounds of the nonmetals are either gases or liquids at standard conditions. Table 19-2 lists the boiling points and heats of formation of the simple binary hydrides of families IVA through VIIA. The boiling points of the covalent hydrides increase down the families, a trend that can be rationalized by the increase in intermolecular forces due to an increasing number of electrons. This is most obvious in the carbon family, where the molecules have no dipole moments and only van der Waals forces are operative. In the other families, the electronegativities of the central elements of the heavier molecules are not much different from hydrogen, and their dipole moments are consequently small. Thus, the dipole-dipole forces are outweighed by the larger van der Waals forces. The boiling points of ammonia, water, and hydrogen fluoride are anomalous; the hydrogen bonding that accounts for this was discussed in Chapter 9.

TABLE 19-2
Boiling Points (°C)
and Heats of Formation (kcal/mole)
at 25°C for Molecular Hydrides

	MH_4	MH_3	MH_2	MH
M	C	N	O	F
bp	−161	−33	100	19
ΔH_f^0	−18	−11	−68	−64
M	Si	P	S	Cl
bp	−112	−85	−60	−85
ΔH_f^0	−15	2	−5	−22
M	Ge	As	Se	Br
bp	−90	−55	−41	−67
ΔH_f^0	22	41	20	−9
M	Sn	Sb	Te	I
bp	−57	−17	−2	−35
ΔH_f^0	39	34	37	6

The heats of formation of the covalent hydrides become less negative within a family as atomic number increases, and the formation of some of the heavier members is actually endothermic. A thermochemical analysis of the formation of hydrogen sulfide shows that its heat of formation is equal to the dissociation energy of $H_2(g)$ plus the heat of formation of gaseous sulfur atoms minus twice the H—S bond energy.

$$H_2(g) \rightarrow 2H(g) \qquad \Delta H^0 = \text{bond energy of } H_2$$

$$S(s) \rightarrow S(g) \qquad \Delta H^0 = \Delta H_f^0 \, [S(g)]$$

$$2H(g) + S(g) \rightarrow H_2 S(g) \qquad \Delta H^0 = -(2 \times \text{H—S bond energy})$$

$$\overline{H_2(g) + S(s) \rightarrow H_2 S(g) \quad \Delta H_f^0 \, [H_2 S(g)]}$$

Therefore, for the generalized compound MH_n, the heat of formation varies with the heat of formation of the gaseous M atom and with the M—H bond energy (the heat of formation of hydrogen atoms is common to all). Since the heats of formation of gaseous atoms decrease or remain approximately constant down a family, the decrease in the negative value of the heat of formation is due to the decrease in the M—H bond energy with increasing atomic number of M (Table 19-3). This trend in bond energy can be attributed to (a) the increasing size of M, which results in increased M—H bond lengths; and (b) the greater diffuseness of the bonding orbitals on M as M gets larger, which produces poorer orbital overlap and thus weaker bonds.

As suggested by their negative heats of formation, some of the hydrides can be prepared by direct combination of their constituent elements. Others are prepared by the reaction of a nonmetal halide with an active metal hydride,

$$4LiH + SiCl_4 \rightarrow SiH_4 + 4LiCl$$

TABLE 19-3
Heats of Formation of Gaseous Atoms and M—H Bond Energies (kcal/mole) at 25°C

M	C	N	O	F
ΔH_f^0 (M(g))	172	113	59	18
M—H bond energy	98	92	110	135
M	Si	P	S	Cl
ΔH_f^0 (M(g))	109	75	66	29
M—H bond energy	76	77	87	102
M	Ge	As	Se	Br
ΔH_f^0 (M(g))	78	60	49	27
M—H bond energy	69	59	66	87
M	Sn	Sb	Te	I
ΔH_f^0 (M(g))	72	61	47	25
M—H bond energy	60	61	57	70

or by the hydrolysis of a salt containing a nonmetal anion:

$$Ca_3P_2 + 6H_2O \rightarrow 2PH_3 + 3Ca(OH)_2$$

Probably the most important covalent hydrides are water and ammonia. The properties of water will be discussed in detail later in this chapter. Ammonia and its salts have been known for centuries. Sal ammoniac, NH_4Cl, first appeared in the writings of alchemists in the ninth century, and by the 1600s it was well known that ammonia resulted from the reaction of sal ammoniac with caustic lime (CaO). By the early 1900s it was known that certain bacteria convert atmospheric nitrogen to ammonia, a process referred to as *nitrogen fixation*. In 1905, Fritz Haber, a German chemist, set out to synthetically "fix" nitrogen. Because of the central role of nitrogen derivatives in explosives such as trinitrotoluene (TNT), the Haber process soon became the keystone of the German munitions industry.

Let us now examine the thermochemistry of this process. The standard free energy of formation of ammonia, which is the ΔG^0 for the Haber process

$$\tfrac{1}{2}N_2\,(g) + \tfrac{3}{2}H_2\,(g) \rightleftharpoons NH_3\,(g)$$

is -3.86 kcal/mole at $25°C$. This free-energy change is related to the equilibrium constant by the expression (Chapter 13):

$$\log K = -\frac{\Delta G^0}{RT(2.3)}$$

$$\log K = -\frac{(-3860 \text{ cal/mole})}{(1.99 \text{ cal/deg-mole})(298°)(2.3)} = 2.8$$

$$K = 6.3 \times 10^2$$

The equilibrium constant at $25°C$ is therefore 6.3×10^2, and the extent of the reaction of N_2 with H_2 should be quite high. Experimentally, when N_2 and H_2 are mixed at room temperature, no ammonia is formed. Clearly, the *rate* of the reaction must be extremely slow.

To increase the rate, it is necessary to increase the temperature at which the reaction is carried out; at $400°C$ in the presence of a catalyst, the rate is high enough for equilibrium to be reached within a reasonable amount of time. Since the reaction is exothermic, an increase in temperature decreases the extent of reaction, and at $400°C$ the standard free energy of formation of NH_3 is $+6.55$ kcal/mole. Hence, the equilibrium constant at $400°C$ is small.

$$\log K = -\frac{6550 \text{ cal/mole}}{(1.99 \text{ cal/deg-mole})(700°)(2.3)} = -2.0$$

$$K = 1.0 \times 10^{-2}$$

Since, according to the equation for the reaction, the number of moles of gaseous reactants is greater than the number of moles of gaseous product, a commercially acceptable extent of reaction can be obtained by increasing the pressure (Le

Chatelier's Principle). When nitrogen and hydrogen are mixed in a mole ratio of 1:3 at 400°C and 300 atm, the extent of reaction is greater than 50 percent in spite of the unfavorable equilibrium constant.

Essentially the same reaction is utilized by living organisms to fix nitrogen. These organisms—algae and various kinds of bacteria—do not need to resort to high temperatures in order to increase the rate of the reaction. Instead, the enzyme nitrogenase serves as an extremely effective catalyst and allows the reaction to proceed at room temperature. Despite some rather concerted efforts to unravel the mechanism by which this enzyme performs its rate-increasing function, its workings still remain a mystery.

Since nitrogen-fixing bacteria are associated with alfalfa, soybeans, peas, and certain other plants, these crops are especially valuable in maintaining the nitrogen content of soil. The ammonia produced by these organisms is oxidized by various soil bacteria to nitrate ion, which is then utilized by plants in their formation of proteins, nucleic acids, and other molecules. If the nitrogen-fixing bacteria are not present, plant nutrition depends upon nitrogen added to the soil as nitrate, urea, or natural fertilizers.

The nitrogen within decaying plants is converted back to diatomic elemental nitrogen by yet other bacteria. The biological utilization of nitrogen is therefore a very efficient cyclical process.

Borderline Hydrides

A small number of elements form hydrogen derivatives with properties intermediate between those of the simple covalent and ionic hydrides. Beryllium and magnesium form solid hydrides that decompose at low temperatures. These compounds are best formulated as linear polymers with hydrogen bridges. In the formula below, n is some large number and denotes repetition of the unit inside the brackets.

The small beryllium(II) and magnesium(II) "ions" have high charge densities and tend toward more covalency. In this process they also tend toward a filled valency shell and achieve it through polymerization. The nature of this hydrogen-bridge bond is best described with the boron hydrides, which also fill their valence shell through hydrogen bridging.

A boron hydride was apparently first prepared in 1881 by the reaction of acids with magnesium boride, Mg_3B_2. An incorrect analysis of the gas produced by this reaction resulted in the formula of BH_3 for this compound. Alfred Stock, a German chemist, later developed techniques for handling gases in very small amounts and was able to isolate a number of different boron hydrides from this same reaction. Among these hydrides were the compounds B_4H_{10} (the most abundant product), B_5H_9, B_6H_{10}, and $B_{10}H_{14}$. The simpler compound B_2H_6 was then prepared by heating B_4H_{10}. The simplest member of the family, BH_3, was not found by Stock and indeed has never been isolated.

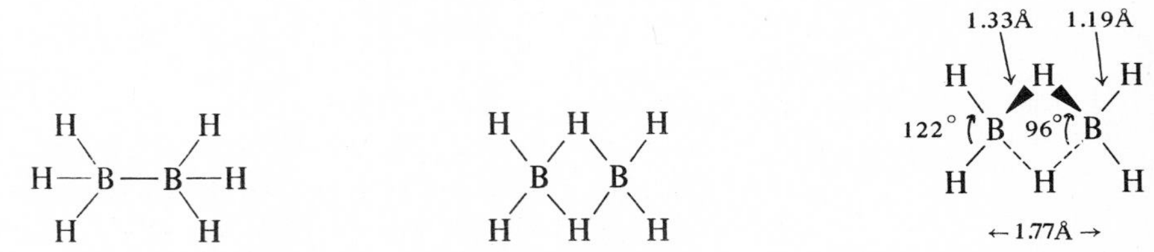

H   H
H—B—B—H
H   H

(a) ethane structure

H H H
B   B
H H H

(b) bridges structure

1.33Å   1.19Å
H H H
122° B 96° B
H H H
← 1.77Å →

(c) actual structure

FIGURE 19-1

Structural Possibilities for Diborane

The nature of the structure and bonding of the simplest known hydride, B_2H_6, has been the topic of many chemical investigations. Until the 1920s the structure of diborane, B_2H_6, was generally believed to be similar to that of ethane, as shown in Figure 19-1a. In 1921, a bridge structure was proposed, but this received little attention until the 1940s, when chemists began to study the compound with various spectroscopic techniques. The infrared spectrum of diborane was found to bear no resemblance to the spectrum of its supposed structural analog, ethane, and the ultraviolet spectrum was found to be rather similar to that of ethene. Finally, a careful electron diffraction analysis showed that the structure was the bridge form proposed earlier (the structural parameters are given in Figure 19-1c). Thus, the two borons and four terminal hydrogens lie in a common plane, with two hydrogens situated between the boron atoms, one above and one below this plane.

This structure is similar to the dimeric structure of aluminum chloride (Chapter 21) but differs from it in one very important aspect: There are too few valence electrons in B_2H_6 to permit all eight B–H linkages to be normal two-electron bonds. Each boron atom has three valence electrons, each hydrogen has one, and there are therefore a total of $(2 \times 3) + (6 \times 1) = 12$ electrons to distribute over eight bonds. Since the terminal B–H linkages are shorter than the bridge linkages, it is reasonable to assume that the end B–H bonds are normal two-electron bonds. This leaves four electrons to distribute over four bonds, and the bridge B–H linkages can therefore be thought of as one-electron bonds. In the terminology of the valence bond model, this is equivalent to a hybrid of the following resonance contributors:

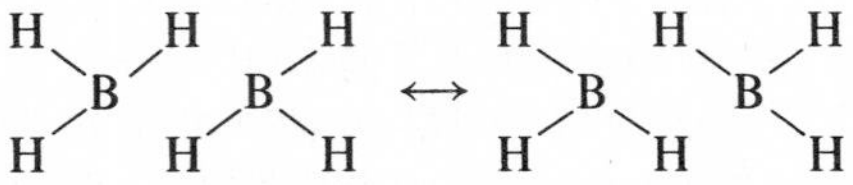

H H H   H H H
B B   ↔   B B
H H H   H H H

The molecular orbital model assumes the existence of three-center molecular orbitals that include the two boron atoms and the bridging hydrogens. A rather useful description that conveys some of the flavor of the molecular orbital view can be generated by visualizing an ethenelike structure for the two borons and four terminal hydrogens. That is, the two boron atoms would be linked by a double bond and would therefore have formal negative charges.

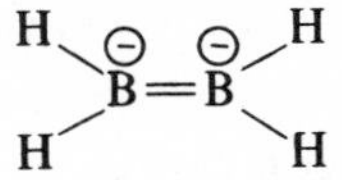

If two protons are imbedded, then, in the π-electron density above and below the plane containing the six atoms, the following electronic structure results:

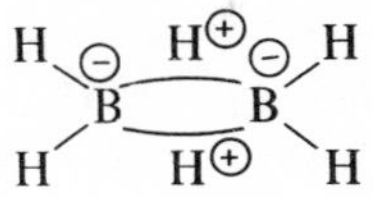

It is interesting to note that if the two "imbedded" protons could be forced into the boron nuclei, ethene would indeed result.

HYDROGEN AS A FUEL

The replacement of fossil fuels (coal, oil, and natural gas) by hydrogen gas as a source of energy has recently received considerable attention. Since the reaction of hydrogen with oxygen releases 58 kcal/mole,

$$H_2(g) + \tfrac{1}{2}O_2(g) \rightleftharpoons H_2O(g) \qquad \Delta H^0 = -58 \text{ kcal/mole}$$

hydrogen could be burned in air or it could be used more indirectly in a fuel cell (p. 478). Since water is the product of both processes, the use of hydrogen as an energy source does not involve the production of the pollutants that accompany the combustion of fossil fuels.

As mentioned earlier, hydrogen can be obtained from the reaction of metals with steam, the destructive distillation of bituminous coal, the electrolysis of water, and the reaction of steam with hydrocarbons. The electrolysis of water is probably the most satisfactory source of hydrogen, because the combustion to water and its subsequent electrolysis constitute an efficient cyclical process. The energy necessary for the electrolysis process would presumably be supplied by nuclear power plants. Nuclear reactors would then be the ultimate source of energy in an economy based on hydrogen as a fuel.

Although mixtures of hydrogen and oxygen are potentially explosive, the use of hydrogen is probably not much more hazardous than the use of natural gas. Hydrogen, mixed with other gases, has been used in the past as a fuel and is currently present in varying concentrations in natural gas. Distribution systems have handled *water gas* (a 50-50 mixture of carbon monoxide and hydrogen obtained by treating either coke or anthracite coal at white-hot temperatures with steam in the absence of air) with no major difficulties. Indeed, water gas was a common fuel for home consumption before the advent of the more efficient natural gas.

In terms of energy production, hydrogen is less efficient than natural gas. The combustion of one mole of hydrogen releases about one-third as much energy as the combustion of one mole of methane, a major constituent of natural gas. Hence, about three volumes of hydrogen are required to equal one volume of methane in energy produced.

$$H_2 + \tfrac{1}{2}O_2 \rightleftharpoons H_2O(g) \qquad\qquad \Delta H^0 = -58 \text{ kcal/mole}$$

$$CH_4 + 2O_2 \rightleftharpoons CO_2 + 2H_2O(g) \qquad \Delta H^0 = -192 \text{ kcal/mole}$$

Storage of hydrogen underground is not as simple a problem as the storage of natural gas, however. Hydrogen permeates other matter more readily than the larger methane molecules and is more readily lost by diffusion. Nevertheless, some of the underground storage facilities now used for natural gas could serve as storage places for hydrogen gas. Both cryogenic (low-temperature) storage and transportation of hydrogen and other gases are common in our present technology.

In addition to its use as a fuel for the generation of electricity, hydrogen should also be suitable as a motor fuel in internal combustion and jet engines. Other possible uses of hydrogen include reduction of ores to the metallic state, conversion of carbon dioxide to methane, and conversion of oil shale—which like coal is hydrogen-deficient—to petrochemicals. In addition, the large amounts of oxygen that would be produced in the electrolysis of water could be used in sewage treatment plants, in hydrogen combustion instead of air (which would eliminate the formation of oxides of nitrogen), and in the basic oxygen furnaces used for the purification of molten iron. Moreover, the deuterium oxide resulting from the electrolysis of water is required in nuclear plants (Chapter 18).

Among other compounds being given serious consideration as energy sources are molecular hydrides. Pilot plants are currently producing methane from the destructive distillation of coal. Diborane and silane have also been suggested as possible fuels of the future. As indicated by the following enthalpy changes, these compounds would be excellent fuels in terms of their output of energy.

$$B_2H_6(g) + 3O_2(g) \rightarrow B_2O_3(s) + 3H_2O(g) \qquad \Delta H^0 = -483 \text{ kcal}$$

$$SiH_4(g) + 3O_2(g) \rightarrow SiO_2(s) + 2H_2O(g) \qquad \Delta H^0 = -306 \text{ kcal}$$

For example, higher molecular weight solid boranes have been extensively researched as solid rocket fuels. However, their chemical production is more complicated than that of simple hydrogen. Consequently, of the molecular hydrides, probably only methane will play a significant role as a fuel in the future. Everything considered, hydrogen appears to be a reasonable future fuel and would seem to be a natural solution to both our energy and pollution problems.

WATER

Water is so common to our environment that we tend to forget what an extraordinary substance it is. The fact that the solid phase, ice, floats in liquid water does not seem to be unusual. Yet, for nearly all other substances the solid phase is more dense than the liquid phase. Numerous other properties of water are also rather unique: It has a higher heat of fusion and a higher heat of vaporization per gram than almost any other substance; it has a higher surface tension than other molecular compounds, a higher heat capacity than most other solids and liquids, and a higher conductance of heat than other liquids. For a compound of such low molecular weight, water is liquid over a wide temperature range—100 centigrade degrees. With its abnormally high dielectric constant, water is also an excellent solvent for ionic compounds. Most of these properties have been accounted for in terms of strong intermolecular hydrogen bonding in Chapter 9.

Rainwater   Rainwater is the only natural source of relatively pure water. It does contain some dissolved gases, absorbed by contact with the atmosphere. Because carbon dioxide is fairly abundant in the atmosphere and is also somewhat soluble in water, the most common of these dissolved gases in CO_2.

Traces of oxides of nitrogen and nitric acid are also present in rainwater. Nitrogen and oxygen combine very slowly at ambient temperatures, and mixtures of the two gases can be kept for years with no noticeable change (as in our atmosphere). Moreover, their reaction to form nitric oxide is endothermic, and the equilibrium greatly favors the reactants.

$$N_2 + O_2 \rightleftharpoons 2NO \qquad \Delta H \text{ is positive}$$

At high temperatures, however, the equilibrium is shifted in favor of the product, and the rate of reaction also increases significantly. Such a condition of high temperature exists when lightning passes through the atmosphere, and over the ages much nitrogen has fallen to earth as nitrate ion—a necessary ingredient for plant life. (The nitric oxide formed is rapidly oxidized to nitrogen dioxide, NO_2, which reacts with water in the air to form nitric acid and nitrous acid.) This formation of nitric oxide occurs to a much greater extent at the high temperatures realized in internal combustion engines and, to a more limited extent, whenever fuel is burned with air as the source of oxygen. Although conditions necessary for the formation of smog are not completely understood, the oxides of nitrogen are known to be necessary ingredients, and their increasing concentrations in the atmosphere pose a potential danger to the future of both plant and animal life.

Recently, increasing concentrations of SO_2 have also been found in rainwater. The combustion of fossil fuels containing sulfur produces gaseous SO_2, which dissolves in the atmospheric water vapor.

Rainwater is an important agent in both the physical and chemical breakdown of rock, and this weathering process accounts for most of the dissolved ions found in fresh water. For example, feldspar, the most abundant mineral, is converted to clay, silica, and potassium hydrogen carbonate by reaction with water containing carbon dioxide.

$$2KAlSi_3O_8(s) + 3H_2O(l) + 2CO_2(aq) \rightarrow$$

$$H_4Al_2Si_2O_9(s) + 4SiO_2(aq) + 2K^+(aq) + 2HCO_3^-(aq)$$

This is the principal source of the small amounts of both potassium ion and silica that are transported by streams and rivers into the oceans.

Dissolved carbon dioxide is responsible for most of the acidity of surface and ground waters.

$$CO_2(aq) + H_2O \rightleftharpoons H_2CO_3(aq)$$

Slightly acidic water passing through limestone (mainly $CaCO_3$) regions dissolves some of the limestone through formation of the more soluble hydrogen carbonate.

$$H_2CO_3(aq) + CaCO_3(s) \rightleftharpoons Ca^{2+}(aq) + 2HCO_3^-(aq)$$

In certain regions, water can be highly acidic. For example, acid mine drainage occurs when water passes over or through coal beds that contain significant amounts of sulfur. This sulfur is oxidized to sulfur dioxide, which reacts with water to form sulfurous acid (remember that SO_2 is the anhydride of H_2SO_3). The sulfurous acid is then further oxidized to sulfuric acid.

Fresh Water and Seawater

Fresh water, then, contains varying concentrations of cations and anions, the nature of which is dependent upon the soil and rock composition of the drainage area. Calcium and magnesium ions are the most common impurities in streams and rivers; sodium, potassium, iron, and other cations are found in smaller concentrations. The anions most commonly found in fresh water are hydrogen carbonate and sulfate, with smaller concentrations of chloride, nitrate, and other anions. For millions of years, rivers have carried these dissolved salts into the sea. Evaporation, followed by cloud formation and then rainfall, produces a cycle that leads to a steady buildup of dissolved salts in seawater (the present salinity of seawater is about 35,000 parts per million). Sodium chloride accounts for most of these dissolved salts. Actually, four cations and four anions account for almost 100 percent of the dissolved material in surface seawater. In order of decreasing concentration the cations are Na^+, Mg^{2+}, Ca^{2+}, and K^+; while the anions are Cl^-, SO_4^{2-}, HCO_3^-, and Br^-. Other species, such as NO_3^- and PO_4^{3-}, are present in much lower concentrations. In fact, trace amounts of nearly all the other elements are found in seawater.

Many of the ions transported to the ocean by fresh water are removed by either chemical or biological processes. For example, the myriad mollusks, which require calcium ion for the formation of their calcium carbonate shells, constantly remove calcium ion from seawater. Since little sodium ion or chloride ion is removed by these processes, the concentrations of these ions increase very slowly with time.

Hard Water

Most of the water that we use is obtained from natural springs, wells, and rivers. Water that contains calcium and magnesium ions is known as *hard water,* since these polyvalent cations form an insoluble precipitate with soap.

Soaps are sodium salts of long-chain carboxylic acids; for example, sodium stearate:

$$\left[CH_3(CH_2)_{16}C \begin{matrix} \diagup O \\ \diagdown O \end{matrix} \right]^- , \ Na^+$$

The cleansing power of soap is dependent upon the dual nature of the carboxylate salt. The carboxylate group at one end of the carbon chain is polar and dissolves in water; that is, it behaves like a simple electrolyte. The long carbon chain, however, is nonpolar and repels water; that is, it behaves like a typical high-molecular-weight hydrocarbon. Since most fibers used for clothing and most dirt surfaces are nonpolar, they are not "wet" by water alone. However, the hydrocarbon part of the soap molecule (which resides in the interface between the water and these nonpolar surfaces) "wets" the fabric and disperses (dissolves) the oil and dirt into very fine particles, which are then washed away with the water. Since calcium and magnesium ions form insoluble precipitates (soap scum) with these long-chain carboxylate ions, no cleansing occurs until enough soap has been added to precipitate the polyvalent cations. Therefore, the harder the water the more soap is required for efficient cleansing.

Soaps are prepared by boiling animal fats with a solution of sodium hydroxide. An animal fat is an ester derived from glycerol (1,2,3-trihydroxypropane) and long-chain carboxylic acids; for example,

$$\begin{matrix} & & O \\ & & \parallel \\ H_2C&-O-&C-C_{17}H_{35} \\ | & & \\ & & O \\ & & \parallel \\ HC&-O-&C-C_{15}H_{31} \\ | & & \\ & & O \\ & & \parallel \\ H_2C&-O-&C-C_{17}H_{35} \end{matrix}$$

This ester is hydrolyzed by the hot sodium hydroxide solution to yield glycerol

$$\begin{matrix} H_2C&-OH \\ | \\ HC&-OH \\ | \\ H_2C&-OH \end{matrix}$$

and the sodium salts of the three carboxylic acids (fatty acids). This process is called *saponification.*

Detergents are synthetic molecules that are similar in structure to soap but contain a sulfonate group in place of the carboxy group as the solubilizing center. A representative detergent is

$$\left[CH_3(CH_2)_{10} - \overset{\displaystyle H}{\underset{\displaystyle H}{C}} - \langle\bigcirc\rangle - \overset{\displaystyle O}{\underset{\displaystyle O}{S}} - O \right]^- , \ Na^+$$

Since the calcium and magnesium salts of these compounds are more soluble than the salts formed with soap, they are more efficient cleansing agents in hard water. Even with detergents, however, the polyvalent cations interfere to some extent with the cleansing action. The effect of these ions can be further alleviated by the addition of sodium tripolyphosphate, $Na_5P_3O_{10}$. The tripolyphosphate anion forms soluble complexes with calcium, magnesium, and many other poly-valent cations, thereby preventing their interference.

The phosphate polymer is readily produced from simple, inexpensive starting materials, and, of greater importance, phosphate is a necessary substance for plant and animal growth. However, the phosphates from detergents, human and animal wastes, and the excessive use of fertilizers have caused the explosive growth of blue-green algae in some lakes, sluggish streams, and rivers. This overfertilization or *eutrophication* causes excessive algae growth and results in the depletion of the oxygen content of these waters, thereby killing off the more desirable aquatic life. The net result is similar to the slow natural aging of lakes.

Solutions to the phosphate problem are available. For example, the phos-phates could be precipitated from solution during the sewage treatment process. Indeed, recovery of phosphate may be essential, since phosphorus is necessary for life and is not a very abundant element.

An alternative solution to the problem associated with hard water is to *soften* the water before using it; that is, remove the polyvalent cations that interfere with the efficient use of soaps and detergents. Hard water that contains hydrogen carbonate as the anion is known as *temporary hard water* and can be softened simply by boiling; calcium and magnesium are removed by precipitation of their carbonates.

$$Ca^{2+} + 2HCO_3^- \xrightarrow{\text{heat}} \underline{CaCO_3} + H_2O + CO_2 \uparrow$$

Hard water containing the chloride or sulfate ion is not changed by boiling and is known as *permanent hard water*. Both cations and anions can be removed by distillation, but while this is an important source of pure water for laboratory work, it is too expensive to be used for softening water on a large scale.

For home consumption, hard water is commonly softened by passing it through zeolites (see Chapter 21) or synthetic ion-exchange resins. Ion-exchange resins consist of rigid cross-linked polymers that are relatively porous materials. A cation-exchange resin contains weakly acidic groups ($-COOH$, $-SO_3H$, or $-OH$) and can be represented as

By percolating a saturated solution of sodium chloride through the resin, the ionizable hydrogens of the carboxylic acid groups can be replaced by sodium ions.

$$(RCOOH)_{resin} + Na^+(aq) \rightleftharpoons ([RCOO]^-, Na^+)_{resin} + H^+(aq)$$

This is the form of the resin used in a water softener. Since the higher the charge on a cation the stronger its affinity for the resin, passing hard water through the resin removes calcium, magnesium, iron, and other polyvalent cations.

$$([RCOO^-], 2Na^+)_{resin} + Ca^{2+}(aq) \rightleftharpoons ([RCOO^-]_2, Ca^{2+})_{resin} + 2Na^+(aq)$$

Thus, the polyvalent cations in the water are replaced by sodium ions. Although the equilibrium constant for this reaction is significantly greater than unity, the process can be reversed by increasing the sodium ion concentration. Thus, passing saturated sodium chloride solution through the resin regenerates the sodium resin, and the polyvalent cations are flushed out of the system. This is an efficient process because the resin can be used over and over again and sodium chloride is inexpensive.

Chemists have also synthesized anion-exchange resins in which amino groups are the active sites and can be represented as

$$[R-\overset{\displaystyle H}{\underset{\displaystyle H}{N}}-H]^+, OH^-$$

Anions are more strongly attracted to the resin than the hydroxide ion, and the exchange of anions that occurs when a chloride solution is passed through the resin can be represented as

$$([R-\overset{\displaystyle H}{\underset{\displaystyle H}{N}}-H]^+, OH^-)_{resin} + Cl^-(aq) \rightleftharpoons ([R-\overset{\displaystyle H}{\underset{\displaystyle H}{N}}-H]^+, Cl^-)_{resin} + OH^-(aq)$$

Hence, if one first percolates hard water through the acid form of a cation-exchange resin, cations are entrapped by the resin and hydrogen ions replace them in solution. On passing this acid solution through an anion-exchange resin, hydroxide ions replace the solution anions, the hydrogen ions react with the hydroxide ions, and the result is deionized water.

$$H^+ + OH^- \rightleftharpoons H_2O$$

Much "distilled" water is presently produced in this way.

Body Fluids

Approximately 70 percent of the human body is water. Most of this is intracellular water (70 percent); the remainder is divided between interstitial cellular water (24 percent) and blood plasma, which contains the remaining 6 percent of the total. The salinity of body fluids is remarkably similar to that of the ocean. The principal cations present in body fluids, then, are Na^+, K^+, Ca^{2+}, and Mg^{2+}, with sodium and potassium being the predominant species. Potassium ion is almost exclusively found in intracellular fluids, and sodium ion in extracellular fluids. This is most interesting, since chemically there is little difference between sodium and potassium ions and those slight differences that exist are mostly caused by the difference in ionic radii. The larger size of the potassium ion is apparently an important factor in biological selectivity.

The major anions found in body fluids are Cl^-, HCO_3^-, $H_2PO_4^-$, HPO_4^{2-}, and SO_4^{2-}. The chloride and hydrogen carbonate ions are the predominant species in extracellular fluids, while the intracellular fluids contain, in decreasing order of concentration, phosphate ($H_2PO_4^-$ and HPO_4^{2-}), sulfate, and hydrogen carbonate ions. The distribution of water and its electrolyte composition and pH are of vital importance in body functions. For example, good health tolerates only a small variation in the pH of the blood (about 7.35 to 7.45), although life can be sustained within the limits of pH 7.0 to 7.8. The major buffer system is H_2CO_3/HCO_3^-, and it is intimately tied up with the transportation of both carbon dioxide and oxygen by the blood.

As we saw in Chapter 10, oxygen is not very soluble in water, and only a very small percentage of the body's oxygen requirements are met by dissolved oxygen in the blood plasma. Most of it is carried reversibly by the hemoglobin (Chapter 22).

$$Hb \; + \; O_2 \; \rightleftharpoons \; HbO_2$$

hemoglobin   oxyhemoglobin

The globin part of hemoglobin contains a basic group that can add a hydrogen ion. The protonated hemoglobin, HbH^+, does not have as strong a capacity for adding oxygen as does plain hemoglobin. To understand the buffer system in the blood, let us look at one of the main functions of the blood—that of transporting oxygen to muscle tissues and carbon dioxide to the lungs.

When the venous blood enters the lungs, it contains high concentrations of hydrogen carbonate ion and protonated hemoglobin with only a small amount of oxyhemoglobin. The partial pressure of oxygen in the inhaled air is much higher than in the blood, and oxygen diffuses through the capillary walls of the lungs into the blood cells. There it is reversibly bonded to the protonated hemoglobin, pushing the equilibrium to the right in accordance with Le Chatelier's Principle.

$$HbH^+ + O_2 \rightleftharpoons HbO_2 + H^+$$

The released protons combine with the hydrogen carbonate ion to form water and carbon dioxide.

$$H^+ + HCO_3^- \rightleftharpoons H_2O + CO_2$$

The partial pressure of carbon dioxide is much higher in the blood than in the inhaled air; CO_2 passes through the capillary walls and is exhaled.

The oxygen-rich blood is transported via the arteries to capillaries of muscle tissue, where the partial pressure of oxygen in the blood is now much higher than that of the interstitial fluids. Thus, the oxygen passes through the capillary walls into the interstitial fluids and then into the cells, where it bonds to myoglobin (Chapter 22). At the same time there occurs a reverse flow of the carbon dioxide that was formed by the oxidative metabolism. As the CO_2 concentration increases in the blood plasma, it increases the hydrogen ion concentration, which in turn results in protonation of the hemoglobin and the consequent release of more oxygen to the cells. The rate of attainment of equilibrium between gaseous carbon dioxide and carbonic acid is very slow. However, the enzyme carbonic anhydrase, which is present in the blood, catalyzes this reaction. The blood that is rich in CO_2 is then transported back to the lungs, where the cycle starts over again. This is an abbreviated description of the maintenance of a constant pH in the blood. A more complete discussion would include the hydrogen phosphate and dihydrogen phosphate ions, the formation of carbamine hemoglobin, and—of very great importance—the action of the kidneys, which play a major role in maintaining both a proper water balance in the body and the proper concentration of electrolytes.

SUGGESTED READINGS

Sanderson, R. T. "Principles of Hydrogen Chemistry." *Journal of Chemical Education*, Vol. 41, No. 6 (June 1964), pp. 331–333.

Chane, K. E. "Chemical Reactions and the Composition of Sea Water." *Journal of Chemical Education,* Vol. 48, No. 3 (March 1971), pp. 148–151.

Hammond, A. L., W. D. Metz, and T. H. Maugh, II. *Energy and the Future.* Washington, D.C.: American Association for the Advancement of Science, 1973.

PROBLEMS

1. Contrast the hydrogen compounds of the extreme members of the third period—NaH and HCl—in terms of
   (a) the nature of the bonding
   (b) the type of solid formed
   (c) melting and boiling points
   (d) electrical conductivity of liquid
   (e) the reaction with water
   (f) acidic or basic properties
   (g) an analysis of their standard heats of formation

2. Both hydrogen and oxygen are diatomic gases. The bond strength in O_2 is greater than that in H_2, yet O_2 is much more reactive than H_2. Account for the greater reactivity of oxygen by calculating and analyzing the standard heats of formation of Li_2O and LiH and those of CO_2 and CH_4.

3. For the hydrogen compounds of the Group VI elements, determine and rationalize the trends in (a) standard heats of formation, (b) bond energies, (c) bond angles, (d) boiling points, (e) acidities, and (f) basicities.

4. For the compounds CH_4, NH_3, H_2O, and HF, determine and rationalize the trends in (a) bond energies, (b) dipole moments, (c) bond angles, (d) boiling points, and (e) acidities.

5. Account for the difference between the standard heat of formation of solid NaCl (−98.2 kcal/mole) and that of solid NaH (−13.7 kcal/mole).

6. Complete the following equations:
   (a) $K + H_2$
   (b) $Na + H_2O$
   (c) $NaH + H_2O$
   (d) $LiH + SnCl_4$
   (e) $Li_3N + H_2O$

7. Liquid ammonia has been called a water-similar solvent. Write equations for:
   (a) the self-ionization of liquid ammonia
   (b) the reaction of acetic acid with liquid ammonia
   (c) an acid-base reaction in liquid ammonia
   Why are ionic compounds much less soluble in liquid ammonia than in water?

8. A chemist hoping to prepare borane, BH_3, reacted boron trichloride with hydrogen gas at an elevated temperature. The major product of the reaction was a gas for which the following data were collected:
   Elemental composition: boron, 78.26 percent; hydrogen, 21.70 percent.
   Gas density at 22°C and 740 torr: 1.11 g/liter.
   What is the molecular formula and structure of the compound that was isolated?

9. Assuming a 40-percent conversion of heat energy to electrical energy, how much hydrogen would have to be burned to furnish the energy needed for a 100-watt bulb for 24 hours? (1 watt = 14 cal/min)

10. Gasoline is a mixture of many straight-chain, branched, cyclic, and aromatic hydrocarbons. Assume that it is equivalent to *n*-heptane (density = 0.68 g/ml) and that a 100-percent combustion to carbon dioxide and water vapor occurs in the automobile engine. How much hydrogen would have to be burned to supply the same amount of energy as that resulting from one gallon of gasoline?

11. Discuss the role water has played in the distribution of elements on the earth. Account for the fact that NaCl is found in huge mineral deposits.

12. Account for the following:
    (a) The rate of evaporation of water from the ocean is slower than the rate of evaporation from Lake Erie.
    (b) In many regions of the world a scum forms on the inside surface of containers in which water is boiled. This surface scale dissolves in dilute acids. The water is also "softer" after boiling.
    (c) Many limestone caves contain stalactite and stalagmite formations.

13. Assume limestone to be pure calcium carbonate, and calculate the molarity of calcium ion in ground water saturated with limestone. If 1000 liters of this water is passed through a water softener, how much sodium chloride is required to replace all of the calcium ions?

14. The following table lists some physical properties of water and ammonia.

	H_2O	NH_3
Melting point	0°C	−78°C
Boiling point	100°C	−33°C
Heat of fusion	1.4 kcal/mole	1.3 kcal/mole
Heat of vaporization	10.5 kcal/mole	5.6 kcal/mole
Density of liquid	1.000 g/ml at 4°C	0.677 g/ml at −33°C

Examine the data and rationalize any differences between water and ammonia.

20 *The Representative Elements I: The s-Fillers*

The elements of the first two groups of the periodic chart have only s electrons in their valence shells. These elements have the largest atomic radii and the lowest ionization energies and electronegativities within a given period and therefore have the most metallic characteristics. They have metallic elemental forms and form ionic compounds.

Group I: The Alkalies

The alkali metals, which constitute the first family of the periodic chart, have the outer electron distribution $(n - 1)s^2(n - 1)p^6 ns^1$ ($1s^2 2s^1$ for lithium). The chemical and physical properties of the elements are quite similar; those differences that do exist primarily reflect the variation in atomic radius. In many ways the chemistry of this group of elements is more predictable than that of any other family.

THE ELEMENTS

Important properties of the Group I elements are listed in Table 20-1.

541

	Li	Na	K	Rb	Cs
TABLE 20-1 **Some Properties of the Group I Elements**					
Density (g/cc)	0.53	0.97	0.86	1.53	1.90
Melting point (°C)	179	98	64	39	28
Boiling point (°C)	1331	892	766	701	690
Ionization energy (kcal/mole), 1st	124	119	100	96	90
2nd	1744	1091	734	634	579
E^0 (volts)	−3.05	−2.71	−2.93	−2.99	−3.02
Heat of sublimation (kcal/mole)	37	26	22	20	19
Heat of hydration, gaseous ion (kcal/mole)	−123	−97	−77	−70	−63
Atomic radius (Å)	1.23	1.57	2.03	2.16	2.35
M^+ crystal radius (Å)	0.74	1.02	1.38	1.49	1.70

Properties

Density. An alkali metal is the first member of a period and has the largest atomic radius and lowest mass of any member of the period. As a direct result of this low mass and large radius, the alkali metals have the lowest densities of the metallic elements, with lithium being the lightest solid element. With increasing atomic number within the family, the mass increases at a faster rate than the atomic radius (all the metals have the same crystal structure), and the density therefore increases from lithium to cesium. Potassium is an exception, with a slightly lower density than sodium.

Hardness, Heat of Sublimation, Melting and Boiling Points. In our discussion of metallic bonding in Chapter 9, we found that, according to the generally accepted model, the metallic lattice consists of closely packed spherical cations "floating in a sea" of quite mobile electrons. That is, the valence electrons can be considered as residing in metal orbitals, which, somewhat like molecular orbitals in covalent compounds, are not directly associated with any specific atom.

Although this model does not adequately explain all the properties of metals, it would appear that the greater the number of valence electrons, the stronger the metallic bond. For example, the hardness, melting point, and heat of sublimation increase from sodium to aluminum. Thus, the alkalies have the lowest melting points and the lowest heats of sublimation and are the softest metals.

If the valence electrons are the "glue" that holds the lattice together, then spreading them through a larger volume should weaken the metallic bond. Consequently, the hardness, heat of sublimation, and melting point all decrease with increasing atomic size from lithium to cesium. The boiling points are much higher than the melting points and follow the same order, decreasing with increasing atomic size.

Ionization Energy.   Since the atomic size decreases with increasing atomic number within a given period, the outermost electrons must experience a progressively larger effective nuclear charge from left to right across the period. Hence, the first member of each period, an alkali metal, has the lowest ionization energy.

Although the nuclear charge increases down the alkali family, the presence of electrons in progressively higher quantum levels more than compensates for this. That is, the increased nuclear charge is more effectively shielded by the inner electron core, and the ionization energies decrease monotonically down the family.

The second ionization energy of the alkali metals is very large; for example, the second ionization energy of sodium is about nine times as great as its first ionization energy. This can be attributed to (a) the removal of the second electron from a much smaller *positively charged* species rather than from the larger neutral atom and (b) the fact that the second electron is being removed from an inert-gas type ion which, as we have seen in Chapter 5, has the special stability of the filled quantum level. This high second ionization energy limits the chemistry of the alkali metals to that of the +1 oxidation state.

Hydration Energy.   The interaction of a cation with a water molecule can be thought of as a simple ion-dipole attraction. Both the charge and the radius of the cation determine the strength of this interaction: the higher the charge and the smaller the radius, the greater the ion-dipole interaction. It is more convenient to express these two variables as a charge-to-radius ratio; thus, the greater the charge-to-radius ratio, the greater the force of interaction. This can be illustrated by a comparison of the hydration energies of the lithium(I) ion (-123 kcal/mole) and the magnesium(II) ion (-460 kcal/mole). The radii of these two ions are nearly the same, but the magnesium ion has twice the charge.

For ions of like charge within a family, the smaller the radius of the ion the closer the approach of the water molecule, and thus the greater the ion-dipole interaction. Hence, both the interaction and the energy released in the hydration process decrease with increasing atomic number of the alkali metal.

$$M^+(g) + H_2O \rightleftharpoons M^+(aq) \qquad \Delta H^0 = \text{hydration energy}$$

The alkali metal cations have a relatively low charge-to-radius ratio, and except for the small lithium cation, which has four water molecules bonded to it in a tetrahedral arrangement, no stoichiometrically definite hydrated species exist.

Bonding   The loss of the ns^1 electron results in an ion with an inert-gas structure and a radius much smaller than the atom from which it is derived. These Group I ions have spherically symmetrical electron density and because of their small size are relatively nonpolarizable. Because of the single positive charge and consequent small charge-to-radius ratios, these ions do not, however, polarize anions in the crystal lattice. Therefore, compounds of the alkali metals are best described by the ionic model, and their crystal lattices are composed of spherical, nondistortable cations and anions.

The lithium ion is much smaller than the other cations, and it does distort the electron density of large anions. (Recall that we have discussed this polarization of the anion by a cation under Fajans' Rules in Chapter 7.) Therefore, some of the compounds of lithium have some covalent character.

Occurrence and Preparation

Because most of the compounds of the alkali metals are very soluble in water, the only significant mineral deposits are found in evaporated inland seas. There, selective crystallization has produced relatively pure deposits of sodium and potassium chlorides, along with smaller amounts of other halides, sulfates, nitrates, and carbonates. Both the Dead Sea and Great Salt Lake are saturated with alkali salts from which K_2SO_4, Na_2SO_4, and other compounds are isolated. Of course, the oceans are rich in dissolved sodium and potassium salts.

Lithium, rubidium, and cesium are found in the earth's crust in highly insoluble silicate minerals.

Because the alkali metals are powerful reducing agents, the chemical reduction of their salts to the metals themselves is not favored. Instead, the metals are produced by electrolysis of their molten salts. For example, sodium is prepared by electrolysis of molten sodium chloride, to which calcium chloride is added to lower the melting point. The E^0 value of calcium is slightly lower than that of sodium, and sodium metal is selectively produced at the cathode with chlorine gas as a by-product at the anode.

$$2NaCl(l) \xrightarrow{\text{electrolysis}} 2Na(l) + Cl_2(g)$$

Reactivity toward Water

The alkali metals are very active elements that exothermically react with water to yield hydrogen gas, hydroxide ions, and M^+ ions.

$$2M + 2H_2O = 2M^+ + 2OH^- + H_2$$

The reactivity of the metals increases with increasing atomic number and decreasing ionization energy. Lithium reacts rather quietly with water. The hydrogen liberated in the sodium reaction may burn, while the reaction of potassium, rubidium, and cesium is quite violent and the liberated hydrogen bursts into flame. At first glance the relatively slow rate of the reaction of lithium with water may seem rather surprising, since the electrode potential (Table 20-1) for lithium is about the same as that for cesium. It is important to remember, however, that the rate of a chemical reaction is not necessarily related to the extent of the reaction.

That the extent of the reaction of lithium with water is indeed high can be verified by calculation of the equilibrium constant with the relationship developed in Chapter 17,

$$\log K = \frac{n(E^0_{red} - E^0_{ox})}{0.059}$$

When the electrode potentials for the pertinent half-reactions are substituted above, an equilibrium constant of 10^{75} is obtained.

$$2H_2O + 2e^- \rightarrow H_2 + 2OH^- \qquad E^0 = -0.83$$

$$2Li^+ + 2e^- \rightarrow 2Li \qquad E^0 = -3.05$$

$$\log K = \frac{2[-0.83 - (-3.05)]}{0.059} = 75$$

$$K = 1 \times 10^{75}$$

The same calculation for the reaction of water with sodium, the alkali metal with the least negative standard reduction potential, gives an equilibrium constant of 10^{64}. Beyond doubt, all of the reactions go to completion, that of lithium simply taking a longer time.

COMPOUNDS

Oxides, Peroxides, Superoxides

All of the metals burn in air to give, preferentially, lithium *oxide* (Li_2O); sodium *peroxide* (Na_2O_2); and potassium, rubidium, and cesium *superoxide* (MO_2).

$$2Li + \tfrac{1}{2}O_2 \rightarrow Li_2O$$

$$2Na + O_2 \rightarrow Na_2O_2$$

$$K + O_2 \rightarrow KO_2$$

The preparation of the oxides other than lithium oxide is accomplished by the reaction of the free metal with the metal nitrate, as shown for potassium.

$$10K + 2KNO_3 \rightarrow 6K_2O + N_2$$

The oxide ion is a much stronger base than the hydroxide ion and is completely converted to the hydroxide ion in water.

$$M_2O + H_2O \rightarrow 2M^+ + 2OH^-$$

The hydroxides are strong bases and can be isolated as crystalline compounds.

Close followers of space travel will remember how important lithium hydroxide was as an air purifier to the returning astronauts of the ill-fated Apollo 13 flight.

$$2LiOH + CO_2 \rightarrow Li_2CO_3 + H_2O$$

The ability to absorb carbon dioxide, discussed in Chapter 16, is a common reaction for all strong hydroxy bases. Lithium hydroxide is also used in submarines as an air purifier. Although more expensive than the more common sodium and potassium hydroxides, lithium hydroxide does not deliquesce (absorb enough water to form a solution) like the other hydroxides, and the technical problems in using a solid are obviously simpler. Lithium hydroxide also has the lowest formula weight, so that the greatest number of moles are present in a given weight of material.

The peroxides and superoxides are strong oxidizing agents, while the peroxide ion is also a reducing agent. A molecular orbital description of the peroxide and superoxide ions has already been given in Chapter 7. Superoxides enter into an unusual and useful reaction with carbon dioxide. For example, potassium super-oxide, an orange, paramagnetic compound, has been used by mountain climbers and in the Russian space program as a source of oxygen gas. In the presence of catalytic amounts of $CuCl_2$, it reacts with the CO_2 in expired air to form potassium carbonate and liberates three moles of oxygen gas for every four moles of KO_2.

$$4KO_2 + 2CO_2 \xrightarrow{CuCl_2} 2K_2CO_3 + 3O_2$$

Group II: The Alkaline Earth Metals

The alkaline earths constitute the Group II family of the periodic chart and have an outer electron distribution of $(n-1)s^2(n-1)p^6ns^2$ ($1s^22s^2$ for beryllium). Calcium, strontium, and barium form a nearly perfect triad whose properties, like those of the alkalies, vary systematically with an increase in size. The behavior of the much smaller beryllium is significantly different. The properties of magnesium lie in between.

THE ELEMENTS

Some properties of the elements are listed in Table 20-2.

Rationalization of Properties

Physical Properties. The atomic radii of the Group II metals are much smaller than the radii of the Group I metals. These smaller radii together with the increased mass give the Group II metals larger densities than the alkali metals. With two valence electrons to contribute to the bonding in the metallic state, they are also much harder and have higher melting points, boiling points, and heats of sublimation. However, within the family these physical properties do not show the smooth periodic variations observed with the alkali metals. One of the most important reasons for this lack of periodicity with the Group II elements is that the crystal structure of the metallic state changes down the family. However, for calcium, strontium, and barium, the rate of change of the atomic radius is small, and the physical properties of these elements generally follow the trends discussed for the alkalies.

TABLE 20-2
Properties of
Group II Elements

	Be	Mg	Ca	Sr	Ba
Density (g/cc)	1.84	1.75	1.55	2.6	3.75
Melting point ($^\circ$C)	ca. 1300	650	851	770	725
Boiling point ($^\circ$C)	ca. 1500	1100	ca. 1300	ca. 1300	ca. 1500
Ionization energy (kcal/mole), 1st	215	176	141	131	120
2nd	420	347	274	254	231
3rd	3548	1848	1181	—	—
E^0 (volts)	−1.85	−2.36	−2.87	−2.89	−2.91
Heat of sublimation (kcal/mole)	77	36	46	39	42
Heat of hydration, gaseous ion (kcal/mole)	−591	−460	−395	−355	−305
Atomic radius (Å)	0.89	1.36	1.74	1.91	1.98
M^{2+} crystal radius (Å)	0.3	0.72	1.00	1.16	1.36

Reactivity. Beryllium is a hard, brittle metal that is relatively inert chemically. Magnesium metal is more reactive, but once protected by its insoluble oxide coating it resists further reaction with either air or water. Magnesium is a relatively abundant element and is used as a structural material, where its chemical inertness and light weight give it special advantages. Beryllium is too rare to be used economically as a structural material. However, it is alloyed in small amounts with copper as a hardener and also acts as an antioxidant. Since it stops X-rays to a lesser extent than any other suitable structural material, it is used for windows in X-ray apparatus.

Calcium, strontium, and barium are bright, shiny metals that are rapidly tarnished when exposed to air. They are strong reducing agents and combine directly with most nonmetals. Their E^0 values are about the same as those of the alkali metals. The extents of reaction with water are similar to those of the alkali metals, but the rates are less and increase from calcium to barium.

$$M + 2H_2O \rightleftharpoons M^{2+} + H_2 + 2OH^-$$

Bonding. The M^{2+} ions of Group II are isoelectronic with the M^+ ions of the alkali metals of the same period. However, because of the increased nuclear charge, the dipositive ions are considerably smaller. The larger charge and the smaller size of the M^{2+} ions produce a larger charge-to-radius ratio, so that these cations are even less distortable than those of Group I. This large charge-to-radius ratio results, however, in greater polarization of anions. The bonding in compounds of beryllium is characteristically covalent—even BeF_2 is not completely dissociated in aqueous solution. With the large increase in ionic radius from beryllium to magnesium, the charge-to-radius ratio changes drastically, so that, with the exception of some compounds of magnesium, the ionic model best describes the bonding in compounds of the other elements of the group.

Occurrence and Preparation. The Group II elements are found in widely dispersed mineral deposits and as dissolved salts in the ocean. Beryllium is present in the earth's crust in highly insoluble silicate minerals. Magnesium is found extensively in deposits of dolomitic rock—a limestone rock that has a high magnesium content. Calcium, the fifth most abundant element on the earth, is found in massive deposits of limestone (mainly calcium carbonate), dolomite, and gypsum ($CaSO_4 \cdot 2H_2O$). Strontium and barium are found primarily as the insoluble carbonates and sulfates.

The alkaline earth metals, like the alkali metals, can be prepared by electrolysis of their molten chlorides.

Properties Common to Groups I and II

ELECTRODE POTENTIALS

As discussed in Chapter 17, the standard electromotive force for an electron-transfer reaction is related to the standard free-energy change by the equation

$$\Delta G^0 = -nF \, (\text{emf})^0$$

where n is the number of electrons transferred in the reaction and F is the Faraday constant. Hence, a thermodynamic rationalization of trends in electrode potentials involves analysis of their parent ΔG^0 values. However, because of the fact that enthalpy changes are more easily understood and also more accessible than free-energy changes, our analysis will be based on standard enthalpy changes. The assumption, then, is that the entropy changes are not very different for a series of similar half-reactions.

The standard electrode potentials for the alkali metals decrease from sodium to cesium. That is, the aqueous sodium ion is more easily reduced than the cesium

$$M^+(aq) + e^- \rightleftharpoons M(s)$$

ion, and the enthalpy and free-energy change for the reduction of $Na^+(aq)$ are lower than those for the reduction of $Cs^+(aq)$. Our analysis will consist, then, of identifying the thermochemical steps that lead to a lower standard enthalpy and free-energy change for sodium. The appropriate steps are the following:

(1) $$M^+(aq) \rightleftharpoons M^+(g)$$

(2) $$M^+(g) + e^- \rightleftharpoons M(g)$$

(3) $$\underline{M(g) \rightleftharpoons M(s)}$$

$$M^+(aq) + e^- \rightleftharpoons M(s)$$

From Table 20-1 it is apparent that the heats of hydration become less negative from sodium to cesium, and therefore the enthalpy change for the first step is greatest for sodium. This, then, cannot be the dominant or controlling step. Table 20-1 also shows that the ionization energies decrease from sodium to cesium. Therefore, more energy is released in step (2) when M = Na. The enthalpy change for step (3) is also most favorable for sodium, as indicated by the values for the heats of sublimation. Thus the second and third thermochemical steps lead to the observed decrease in electrode potential.

Although the trend in electrode potentials is monotonic from sodium to cesium, the value for lithium is anomalously low. This can be attributed to the large negative heat of hydration of the gaseous lithium ion, which results in a very high enthalpy change for the first step. The high ΔH^0 for this step outweighs the ΔH^0 values for the second and third steps (which are lower for lithium than the other alkali metals) and produces a very unfavorable enthalpy change for the reduction of the aqueous lithium ion.

In view of the anomalous behavior of lithium, it is perhaps unexpected to find that the electrode potentials for the Group II elements decrease smoothly from beryllium to barium. Although the charge-to-radius ratio is quite large for the beryllium(II) ion, and it is thus very strongly solvated, the energy of hydration is not great enough to compensate for the much higher first and second ionization energies. The heat of sublimation is also significantly larger for beryllium than for the other Group II elements. Hence, the electrode potentials of the Group II elements decrease with increasing atomic number.

OXIDATION STATES FOR GROUPS I AND II

It was stated earlier that there is only one stable oxidation state for each of the *s*-filler groups: +1 for the alkalies and +2 for the alkaline earth metals. That is, all attempts to prepare M^{2+} ions for the alkalies have been unsuccessful, and M^{3+} ions have never been observed for the Group II elements. However, there is some evidence for the +1 oxidation state for beryllium and magnesium.

Let us now investigate the thermodynamic feasibility of preparing the +2 ion for the alkali metals and the +3 and +1 ions for the alkaline earth metals. We will first calculate the standard enthalpy change for the preparation of NaF_2, where sodium is present in the lattice as the Na^{2+} ion. This small cation provides a high lattice energy, which is the driving force of the reaction. The compound LiF_2 would seem to be more favorable than NaF_2 because the smaller size of the cation would give an even greater lattice energy. However, the second ionization potential of lithium is about 60 percent greater than that of sodium, so the size advantage would be more than lost in the extra energy required to remove the second electron. Fluorine has been chosen for the anion because it is the strongest oxidizing agent known, and it is the higher oxidation state that we wish to prepare.

The component thermochemical steps for the formation of NaF_2 from its constituent elements in their standard states, $Na(s) + F_2(g) = NaF_2(s)$ are:

$Na(s) = Na(g)$	$\Delta H_1^0 = \Delta H_{sub}^0 = 26.0 \text{ kcal/mole}$
$Na(g) = Na^+(g) + e^-$	$\Delta H_2^0 = \text{(Ionization energy)}_1 = 118.5 \text{ kcal/mole}$
$Na^+(g) = Na^{2+}(g) + e^-$	$\Delta H_3^0 = \text{(Ionization energy)}_2 = 1091 \text{ kcal/mole}$
$F_2(g) = 2F(g)$	$\Delta H_4^0 = \text{Bond energy} = 37 \text{ kcal/mole}$
$2F(g) + 2e^- = 2F^-(g)$	$\Delta H_5^0 = 2(-\text{Electron affinity}) = -158 \text{ kcal/mole}$
$Na^{2+}(g) + 2F^-(g) = NaF_2(s)$	$\Delta H_6^0 = -\text{Lattice energy} = -695 \text{ kcal/mole}$
$Na(s) + F_2(g) = NaF_2(s)$	$\Delta H_f^0 = 419 \text{ kcal/mole}$

All the corresponding enthalpy changes are known except for the lattice energy, step 6. This can be approximated with the lattice energy of MgF_2 because the Na^{2+} ion would probably be about the same size as the Mg^{2+} ion (this also assumes a similar crystal structure). The sum of the enthalpy changes is the standard heat of formation and is a very large positive number. Since the entropy change for the reaction is negative, the standard free-energy change is even more positive. Clearly, the formation of NaF_2 from the elements is extremely unfavorable. Only if a thermodynamically favorable synthesis were found and if the *rate* of decomposition to the elements were low could NaF_2 be prepared.

Oxygen is the other element most likely to produce the higher oxidation state, and the same type of calculation for the formation of NaO results in a heat of formation of 570 kcal/mole. For both compounds, the large second ionization energy is not nearly balanced by the lattice energy, and the +2 oxidation state therefore seems highly unlikely.

A similar calculation for the heat of formation of MgF_3 would show that the +3 oxidation state is also unlikely for the alkaline earths because of the extraordinarily high third ionization energy.

We now turn to the +1 oxidation state for the Group II metals. The relatively large second ionization potential for these metals suggests that this state may be stable. We will choose MgI as a favorable candidate, because iodine, the weakest oxidizing agent of the halides, will tend to favor the lower oxidation state of the metal.

From the summation of the component steps and using the lattice energy of

NaI to approximate the lattice energy of MgI, we obtain a heat of formation that is very slightly endothermic.

$Mg(s) = Mg(g)$	$\Delta H_1^0 = \Delta H_{sub}^0 = 35.9$ kcal/mole
$Mg(g) = Mg^+(g) + e^-$	$\Delta H_2^0 =$ Ionization energy $= 176.3$ kcal/mole
$\frac{1}{2}I_2(s) = I(g)$	$\Delta H_3^0 = \Delta H_f^0(I(g)) = 25.3$ kcal/mole
$I(g) + e^- = I^-(g)$	$\Delta H_4^0 = -$Electron affinity $= -71$ kcal/mole
$Mg^+(g) + I^-(g) = MgI(s)$	$\Delta H_5^0 = -$Lattice energy $= -165$ kcal/mole
$Mg(g) + \frac{1}{2}I_2(s) = MgI(s)$	$\Delta H_f^0 = 1.5$ kcal/mole

Since the heat of formation of magnesium(I) iodide is only slightly endothermic, let us try the same calculation for a compound with a considerably higher lattice energy. Magnesium(I) chloride fulfills this objective. The calculated heat of formation for MgCl is -26 kcal/mole, which is certainly favorable. However, the heat of formation of magnesium(II) chloride is -153 kcal/mole. If perchance we were able to isolate MgCl, its disproportionation (the formation of species of higher and lower oxidation states from a species of intermediate oxidation state) at standard conditions would be energetically most favorable.

$$2MgCl(s) = MgCl_2(s) + Mg(s) \qquad \Delta H^0 = -101 \text{ kcal/mole}$$

Only if a relatively large activation energy were needed for this conversion would we expect to isolate MgCl under these conditions. Thus, the possibility of compounds containing the alkaline earths in the +1 oxidation state is not ruled out, but it is unlikely that they will be thermodynamically stable relative to the +2 state.

SOLUBILITY TRENDS

Although chemists have not been able to devise a simple, overall treatment of solubility, and while it is not possible to predict the degree of solubility of ionic compounds in water, the trends in the solubilities of many of the compounds of Groups I and II are sufficiently regular to permit their thermodynamic rationalization. This is at least partly a result of the hard-sphere ionic nature of the bonding in these compounds.

In our discussion of the solubility of ionic compounds in Chapter 11, we found that the process of dissolution can be separated into two energies, the lattice energy and the hydration energy. The destruction of a lattice is an *endoergic* process; that is, it takes energy to separate cations and anions. The hydration of a cation or anion is an *exoergic* process; that is, energy is given off by the interaction of cations and anions with polar water molecules. The heat of solution is simply the difference between these two energies and is positive if the lattice energy is greater than the hydration energy and negative if it is less.

From consideration of only the lattice energy, we would expect that the smaller the lattice energy the greater the solubility of an ionic compound. Thus, the order of solubility of the alkali fluorides, in which the lattice energy decreases from LiF to CsF, should be

$$LiF < NaF < KF < RbF < CsF$$

which agrees with the experimental solubilities.

However, the reverse solubility order is observed for the iodides, with cesium iodide being the least soluble.

$$LiI > NaI > KI > RbI > CsI$$

This order can be explained by the hydration energy of the cation (since the anion is the same, its hydration energy remains constant), which becomes less negative down the family.

Thus, it is the degree of change of each of the two opposing forces that is the important factor in the comparison of the solubilities of salts. With salts of small anions, the magnitude of the change of the hydration energy of the cation (Table 20-1) is less than the magnitude of the change of the lattice energy (Table 20-3), and solubility therefore *increases* with increased cation size within a given group. With large anions the magnitude of the change of the hydration energy is greater than the magnitude of the change of the lattice energy, and solubility therefore *decreases* with increased cation size within a given group.

TABLE 20-3 Lattice Energy of Alkali Halides (kcal/mole)	F^-	Cl^-	Br^-	I^-
Li^+	241	198	189	175
Na^+	216	184	176	165
K^+	192	167	161	151
Rb^+	184	162	155	147
Cs^+	171	153	148	140

The same interplay of forces is responsible for the solubility trends of the lithium halides

$$LiF < LiCl < LiBr < LiI$$

and the cesium halides,

$$CsF > CsCl > CsBr > CsI$$

where with the small lithium cation, the lattice energy is the significant factor, while with the large cesium cation, the hydration energy of the anion becomes more important. The hydration of the halide ions becomes less exothermic, in the order $F^- > Cl^- > Br^- > I^-$, which is the solubility order for the cesium halides.

The above trends are illustrated by the data of Table 20-4, which show that the heats of solution become more negative from LiF to CsF and that the opposite is true for the bromides and iodides. It appears that the radius of the chloride ion is such that neither force dominates, and no clear trend is evident.

	F^-	Cl^-	Br^-	I^-
Li^+	+1	−9	−12	−15
Na^+	0	+1	+1	−2
K^+	−4	+4	+5	+5
Rb^+	−6	+4	+6	+7
Cs^+	−9	+5	+7	+9

TABLE 20-4
Heats of Solution of Alkali Halides (kcal/mole)

The solubilities of many compounds of the alkaline earths are also governed by the size of their ions. For small anions, the lattice energy is the controlling factor. For example, the solubilities of the hydroxides increase from magnesium to barium:

$$Mg(OH)_2 < Ca(OH)_2 < Sr(OH)_2 < Ba(OH)_2$$

The solubilities of the oxides parallel this order, but the solubilities of the fluorides do not vary in a monotonic fashion. With compounds containing large anions, the hydration energies are the dominant factor. For example, the solubilities of the sulfates vary in the order

$$MgSO_4 > CaSO_4 > SrSO_4 > BaSO_4$$

The higher charge of the alkaline earth cation increases the lattice energy to a greater extent than the hydration energy, so that in general, species isoelectronic with the alkali metals are less soluble. For example, magnesium oxide is less soluble than sodium oxide, barium sulfate is much less soluble than cesium sulfate, and so on.

From the foregoing, we can make the following generalizations:

1. Within a family of metals, salts of large anions almost invariably decrease in solubility with increasing cation size; that is, the hydration energy is the controlling factor.
2. Within a family of metals, salts of small anions generally increase in solubility with increasing cation size. The lattice energy is the controlling factor.

3. For a given metal ion, the halide salts generally decrease in solubility with increasing anion size, while for small lithium the reverse is true.
4. For isoelectronic cations, the alkali salt is generally more soluble.

In the discussion above, we have used the enthalpy change alone as an index to solubility. It is, of course, the change in free energy from which the feasibility of a process may be predicted. For example, the dissolution of ammonium chloride in water is so endothermic that the temperature of the solution may fall to $0°C$ or less. Yet this salt is very soluble in water. Actually, there is a large increase in entropy in going from the highly ordered solid NH_4Cl to the solution phase. The entropy increase is large enough to more than balance the enthalpy change, and the net effect is a favorable free-energy change. Usually, however, the trend in the heat of solution parallels the free-energy change for the solution process.

UNIQUE BEHAVIOR OF FIRST-ROW ELEMENTS

We conclude our discussion of the s-fillers with a brief description of how lithium and beryllium, as well as the other first-row elements of the periodic chart, differ from the other elements in their groups. In Part One of this book we noted some unique features of the first-row elements. For example, the number of covalent bonds formed by these elements is limited to four (the octet rule), and the bond in F_2 is abnormally weak. These observations and, as we shall see, most of the unique aspects of the first-row elements, can be explained by the small size of these elements relative to the others in their groups and the fact that their bonding involves the second quantum shell which contains only s and p orbitals.

The compounds of lithium and beryllium differ from those of the other members of Groups I and II in their covalent character. This difference in bonding is easily rationalized with Fajans' model (p. 106): the smaller and more highly charged the "cation," the greater the distortion of the electron density of the "anion" and the greater the covalent character. For example, a number of the halides of lithium and beryllium are soluble in nonpolar solvents. In the vapor phase, $BeCl_2$ is a linear molecule with sp hybridization, while in the solid phase the octet of beryllium is filled by electrons donated by bridging chlorines in a chain polymer. In this structure,

$$\left\{ \begin{array}{c} \text{Cl} \\ \text{Be} \\ \text{Cl} \end{array} \underset{\text{Cl}}{\overset{\text{Cl}}{\diagup}} \text{Be} \underset{\text{Cl}}{\overset{\text{Cl}}{\diagup}} \text{Be} \underset{\text{Cl}}{\overset{\text{Cl}}{\diagup}} \right\}_n$$

each chlorine formally contributes three electrons to two Be—Cl bonds. Beryllium also obtains an octet of electrons by forming complex ions such as the tetrafluoroberylate(II) ion, BeF_4^{2-}, or the aquo complex, $Be(OH_2)_4^{2+}$.

In Group III, boron differs from the other members in the same way: The halides of boron are considerably more covalent. Boron trifluoride is a molecular species with a low boiling point and high solubility in nonpolar solvents, whereas aluminum trifluoride is a high melting solid of low solubility in such solvents.

The first-row elements also differ from their congeners in other ways. The vast majority of double and triple bonds are formed between the first members of Groups IV, V, and VI—carbon, nitrogen, and oxygen. Most of the compounds that exhibit hydrogen bonding have hydrogen attached to one of the first-row elements—nitrogen, oxygen, or fluorine. All of these phenomena and many others have their origin primarily in the small size of the first-row elements.

PROBLEMS

1. Sodium chloride is often used as an example of a typical ionic compound. Use your knowledge of the ionic model to contrast NaCl with other ionic compounds.
   (a) Predict the order for the standard heats of formation of NaCl, KCl, and RbCl.
   (b) Rationalize the fact that the ΔH_f^0 of $MgCl_2$ is more negative than that of NaCl.
   (c) Explain why halite, NaCl, is slightly harder than sylvite, KCl.
   (d) Explain why NaCl and CsCl have different crystal structures but NaCl and MgO have the same structures.
   (e) Predict which has the more negative lattice energy, NaCl or CsCl.
   (f) Predict which has the higher melting point, NaCl or $MgCl_2$.

2. Explain the following experimental observations.
   (a) Lithium carbonate becomes less soluble in water as the temperature is increased, but potassium carbonate becomes more soluble as the temperature is increased.
   (b) Calcium chloride is deliquescent, whereas barium chloride is not.
   (c) Beryllium chloride has a much lower boiling point than barium chloride.
   (d) Lithium chloride is more soluble than sodium chloride in ether.
   (e) Many ammonium salts have about the same solubilities as their analogous potassium salts.

3. Explain the following:
   (a) Calcium chloride is used to melt ice on walks and streets.
   (b) Magnesium is a very reactive metal, yet it is used as a structural material.
   (c) Calcium chloride scattered over a dusty street settles the dust.
   (d) Alkali metals are kept under a solvent such as toluene out of contact with air.
   (e) Slaked lime, $Ca(OH)_2$, is often lightly dusted on plants to repel certain insects. Lime, CaO, is never used.

4. Answer the following, giving reasons for your choice in each case.

   Which has the largest radius?
   (a) Na, K, or Rb
   (b) K^+ or Ca^{2+}

Which is paramagnetic?

(c) K_2O, K_2O_2, or KO_2

Which is most soluble in water?

(d) $CaSO_4$, $SrSO_4$, or $BaSO_4$

(e) CaF_2, SrF_2, or BaF_2

(f) NaF or MgF_2

Which hydrolyzes to the greatest extent?

(g) $BeCl_2$, $MgCl_2$, or $CaCl_2$

Which has the highest hydration energy?

(h) Li^+, Na^+, or K^+

Which has the greatest lattice energy?

(i) MgO, CaO, or SrO

Which has the highest heat of formation?

(j) NaF, NaCl, or NaBr

Which is the best reducing agent?

(k) Ca, Sr, or Ba

Which salt produces the least favorable entropy change upon dissolving in water?

(l) LiCl, NaCl, or KCl

(m) NaF, NaCl, or NaI

5. An acceptable name for BaO_2 is barium peroxide, whereas PbO_2 is called lead dioxide. Explain the difference between the two compounds.

6. Calculate the standard heats of formation of CuCl(s), $CuCl_2$(s), MgCl(s), and $MgCl_2$(s). For MgCl(s) use the lattice energy of NaCl; for the others use the following lattice energy values: CuCl, 237 kcal/mole; $CuCl_2$, 668 kcal/mole; $MgCl_2$, 600 kcal/mole. Why is CuCl a stable species, whereas MgCl is not?

7. Calculate the standard heats of formation of MgS(s) and CaS(s), and compare them with the values for MgO and CaO given in Appendix 7. (See Table 6-7 for lattice energies.)

8. Rationalize carefully the choice of sodium and fluorine as reagents most likely to result in a compound containing a Group I 2+ ion.

9. What properties should an anion have in order to form a compound containing the Mg^+ ion?

10. Compare the E^0 values for

$$Cu^+ + e^- \rightarrow Cu$$

$$Cu^{2+} + 2e^- \rightarrow Cu$$

and determine whether ionization energies or hydration energies control their relative values.

11. Calculate ΔH^0 for the dissolution of CaF_2.

$$\Delta H^0_{hyd} \text{ for } F^- = -121 \text{ kcal/mole}$$

$$\text{Lattice energy for } CaF_2 = 624 \text{ kcal/mole}$$

12. A white precipitate forms when a $1M$ solution of sodium sulfide is added to a $1M$ solution of magnesium nitrate. The separated precipitate dissolves in hydrochloric acid without any evolution of hydrogen sulfide. If the white precipitate is burned in air, no fumes of sulfur dioxide are given off. Identify the precipitate and use equations to explain its formation.

13. Using data from Appendix 7, would you expect that the following reaction in a suitable solvent would be favorable?

$$2Li + CaH_2 = Ca + 2LiH$$

14. Predict the formulas of the compounds that the element astatine forms with hydrogen, sodium, and carbon, and predict their boiling points, standard heats of formation, and reactivities toward water.

15. Write equations for:
    (a) the reaction of rubidium with air
    (b) the decomposition of calcium carbonate with heat
    (c) the reaction of barium oxide with water
    (d) the preparation of magnesium from magnesium chloride
    (e) the reaction of potassium with water

21 *The Representative Elements II: The p-Fillers*

The elements of Groups IIIA through 0 contain p as well as s electrons in their valence shells. These elements have smaller atomic radii and higher ionization energies than the s-fillers of the same period, and therefore display primarily nonmetallic characteristics. For example, only the elements with the larger atomic radii on the left-hand side of the p-filler block have metallic elemental forms; only the very electronegative elements on the right-hand side form compounds containing the monatomic ions; the other elements form covalent compounds.

We begin our discussion of the p-fillers with the elements of Group VII—the halogens. These elements are to some extent the p-filler counterparts of the Group I s-fillers: both exhibit fairly regular periodic trends, both form ions of unit charge, and so on. Then we shall examine the main features of the Group IV elements, and, since many of the trends in the chemistry of these elements also appear in the neighboring p-filler groups, we shall include some of the pertinent features of these groups.

Group VII: The Halogens

The halogens occur just before the inert gases and have an outer electron distribution of $ns^2 np^5$. These elements show a strong family relationship, with the same smooth chemical variation with increasing size that is found for the alkali metals, the group just following the inert gases.

THE ELEMENTS

The halogens are very reactive nonmetals and are never found in nature in the elemental form. They exist as diatomic molecules, varying from the two gases, colorless fluorine and pale green chlorine, to reddish-brown liquid bromine and black, solid iodine. Table 21-1 lists some of the chemical and physical properties of the elements.

TABLE 21-1 Some Properties of the Group VII Elements	F	Cl	Br	I
Melting point (°C)	−218	−101	−7	114
Boiling point (°C)	−188	−35	59	184
Ionization energy (kcal/mole, 25°C)	402	300	273	241
Electron affinity (kcal/mole)	79	83	78	71
E^0, $X_2 + 2e^- \rightleftharpoons 2X^-$ (volts)	2.87	1.36	1.06	0.54
ΔH^0 of hydration, $X^-(g)$ (kcal/mole)	−121	−88	−80	−70
Covalent radii (Å)	0.64	0.99	1.14	1.33
Ionic radii (Å)	1.33	1.81	1.96	2.20
Dissociation energy, X_2 (kcal/mole)	37	57	46	36

Rationalization of Properties

Melting and Boiling Points. The very low melting points of the halogens, compared with the much higher melting points of the alkalies, are indicative of a different type of crystal. The halogens are strongly electronegative elements and exist in their elemental forms as simple diatomic molecules, where the formation of a sigma bond between the two halogen atoms satisfies the octet rule.

$$:\ddot{X}—\ddot{X}:$$

The crystal lattice, then, is composed of diatomic molecules held together by relatively weak van der Waals forces. Since these forces increase as the number of electrons increases, both the melting points and the boiling points of the elemental forms increase with increasing atomic number from fluorine to iodine.

Electron Affinity and the X—X Bond Energy. The electron affinity—the energy released when an electron is added to a neutral, gaseous atom, $X(g) + e^- \rightarrow X^-(g)$—decreases from chlorine to iodine, while fluorine has a value between chlorine and bromine. This intermediate value for fluorine is rather surprising, since the very small fluorine, the most electronegative element, would be expected to have the greatest affinity for an electron. The addition of an electron to the small and compact $2p$ sublevel apparently results in considerable electron-electron repulsions, which are minimized in the much larger atoms of chlorine, bromine, and iodine. Support for this explanation is found in the oxygen and nitrogen families, where both oxygen and nitrogen have lower electron affinities than the larger sulfur and phosphorus.

The bond energies of the diatomic molecules also decrease from chlorine to iodine, following the increasing bond length. As was pointed out earlier, the F—F bond energy is lower than the Cl—Cl and Br—Br bond energies. Yet the length of the F—F bond is certainly shorter than that for the other halogen molecules.

The effect that nonbonding electrons may have on molecular shape has already been discussed in Chapters 7 and 8. However, we have so far ignored any influence that nonbonding electrons may have on bond strength. Since each fluorine in F_2 has three nonbonded pairs of electrons and since, moreover, fluorine is very small, it has been suggested that the nonbonding electrons on adjacent atoms are close enough to produce a significant amount of repulsion, thereby producing a weaker bond. Certainly, if it can be argued that pi bonds are formed by the overlap of p_y and p_z orbitals to give the N≡N triple bond, then *filled* p_y and p_z orbitals should lead to repulsion in the bonding of two small atoms. This repulsion appears to be nonexistent or at least insignificant in the much larger diatomic chlorine molecule.

Reliable data on some aspects of fluorine chemistry have been acquired only recently. Before it was possible to experimentally measure the electron affinity of fluorine, a value of 95 was assigned because it agreed with the periodic trend for the electron affinities of the other halogens. This high value for fluorine resulted in a higher F—F bond energy (as determined from a thermodynamic analysis of the type shown on p. 550), which also seemed more likely from periodic considerations. Although experimental results were slowly accumulated which suggested that both the electron affinity and the F—F bond energy were too high, some time and many arguments were required before these new values were generally accepted. The resistance to the experimentally determined value arose from the belief that electron affinity should fit into a monotonic periodic order. While it is certainly true that the concept of periodicity is a very useful model, it does not apply to all properties of all matter.

Ionization Energies and X^+ Ions. The highest first ionization energies of the halogens virtually exclude the possibility of the formation of a +1 cation, X^+. However, as we have already seen, the hydration energy for the ion in solution or the lattice energy for the ion in a crystalline solid must also be considered in determining the feasibility of formation of a cation. The removal of a p electron from an outer $ns^2 np^5$ electron configuration would not be expected to give a cation with a much smaller radius than that of the atom from which it is derived. (The significant difference in size of the cation relative to the atom in the alkali and alkaline earth metals is a result of the removal of one or more s electrons and consequent loss of *all* the electrons in a principal quantum level). Therefore, neither the hydration energy nor the lattice energy would outweigh the ionization energy of approximately 402 kcal/mole for fluorine. The existence of an F^+ ion is therefore unlikely in either the solution or the solid phase.

Since the ionization energies decrease significantly with increasing atomic size, the feasibility of an X^+ ion increases as one moves down the family. For example, the first ionization energy of iodine, 241 kcal/mole, is only slightly greater than that of some metals (zinc, 217 kcal/mole). Indeed, there are a few compounds of iodine that appear to contain an I^+ ion. There is also some evidence for the participation of +1 halogen species as reaction intermediates in solution. Nevertheless, cationic species play only a minor role in the chemistry of the halogens.

Electrode Potentials. The standard reduction potentials decrease down the family, with fluorine, the most electronegative element, being the strongest oxidizing agent known. An analysis of the component thermodynamic steps involved in the reduction of the halogen shows that the lower electron affinity of fluorine is more than balanced by the much higher hydration energy and the lower heat of formation of the gaseous atom.

$$\tfrac{1}{2}X_2(g) \rightarrow X(g) \qquad \Delta H^0 = \text{heat of formation of } X(g)$$

$$X(g) + e^- \rightarrow X^-(g) \qquad \Delta H^0 = -(\text{electron affinity})$$

$$\underline{X^-(g) \rightarrow X^-(aq) \qquad \Delta H^0 = \text{hydration energy of } X^-(g)}$$

$$\tfrac{1}{2}X_2(g) + e^- \rightarrow X^-(aq) \qquad \Delta H^0 \text{ of reduction half-reaction}$$

Therefore, the standard electrode potential of fluorine is the most positive, and there is a monotonic decrease with increasing size within the family.

Occurrence and Preparation

Fluorine and chlorine are relatively abundant elements. The earth's crust is about 0.08 percent fluorine and 0.06 percent chlorine. Bromine is less plentiful (about 0.002 percent), and iodine can be classified as a rare element. This order of abundance is fairly common within a family, the lighter elements generally being more abundant, and it is in harmony with the discussion in Chapter 18 on the origin of the chemical elements. The sparsity of naturally occurring isotopes (fluorine and iodine have one each, and chlorine and bromine have two each) is also consistent with the observation in Chapter 18 on the relative instability of nuclei with odd atomic numbers.

Because the halogens are very reactive, they are never found in their elemental forms in nature. Considering their high electronegativities, it is not surprising to find that they occur almost exclusively as metal halides. Since almost all their salts are water soluble, ground waters over millions of years have carried large quantities of halide ions into the sea. Seawater contains about 19 g of chloride ion per kilogram, by far the most plentiful element in the sea. The abundance of the other halogens in seawater is much less: 65 ppm for bromine and 1.4 and 0.05 ppm for iodine and fluorine, respectively. The great abundance of calcium ion carried into the sea by ground waters and the insolubility of calcium fluoride result in the most plentiful halogen in the earth's crust having the smallest concentration in seawater.

The elemental form of fluorine, the most electronegative element known, cannot be obtained by a chemical reaction, but is prepared by the electrolysis of molten KHF_2. Because of its great reactivity it was the last of the halogens to be isolated. The industrial use of fluorine has paralleled an increase in the technical knowledge necessary for the handling of very reactive materials. It is used primarily for the production of fluorocarbons, such as CCl_2F_2, a nonpoisonous gas commonly used as a refrigerant, and Teflon, a polymeric substance containing repeating $+CF_2+$ units (see page 580).

However, fluorine is perhaps best known as an essential element in the formation of decay-resistant tooth enamel. This phenomenon was first discovered from data showing a lower incidence of tooth decay in children living in regions where ground waters contained fluoride ion. Today many communities, under the supervision of health departments, add about 1 ppm fluoride ion as sodium fluoride to drinking water as a tooth decay preventive. Stannous fluoride and other fluorides are also added to many toothpastes.

Chlorine is produced by either the electrolysis of molten sodium chloride or, more commonly, by the electrolysis of a concentrated aqueous solution of sodium chloride (see Chapter 20). Before World War II, the major industrial uses of chlorine were for the purification of drinking water and as a bleaching agent in paper pulp mills. Today the manufacture of chlorinated hydrocarbons, such as carbon tetrachloride, chloroform ($CHCl_3$), and especially 1,2-dichloroethane ($ClCH_2CH_2Cl$), an important intermediate in the production of polyvinyl chloride, accounts for better than 50 percent of the greatly increased consumption of chlorine.

Chlorine, a more powerful oxidizing agent than bromine, is used to oxidize bromide ion in acidified seawater to free bromine.

$$Cl_2 + 2Br^- \rightarrow Br_2 + 2Cl^-$$

Bromine is used for the production of 1,2-dibromoethane (ethylene bromide), $BrCH_2CH_2Br$, which is an additive in gasolines containing tetraethyl lead. The ethylene bromide forms volatile lead(II) bromide in the combustion chamber, thereby preventing the formation of a lead coating in the cylinders. Since the lead(II) bromide is emitted in the exhaust vapors, extensive studies have been conducted on its role as a pollutant. Bromine is also important in the production of silver bromide, which is used in photographic emulsions.

The primary commercial source of iodine in the United States is the iodide found in oil-well brines. The free element is produced by reaction with chlorine.

$$Cl_2(g) + 2I^-(aq) \rightarrow 2Cl^-(aq) + I_2(s)$$

Another important source is in the form of iodate ion, found mainly as an impurity in Chilean nitrate deposits. The iodate is reduced with sodium hydrogen sulfite, and the crude iodine obtained is purified by sublimation.

$$2IO_3^-(aq) + 5HSO_3^-(aq) \rightarrow I_2(s) + 5SO_4^{2-}(aq) + 3H^+(aq) + H_2O(l)$$

An alcoholic solution of I_2 and KI (tincture of iodine) finds use as an antiseptic, and elemental iodine is used in the preparation of certain dyes and drugs. Iodine plays an important role in the growth and metabolism of human beings. It is contained in the hormone thyroxine, which is produced in the thyroid gland. A deficiency in the diet may lead to subnormal growth or the development of goiter, an enlargement of the thyroid gland.

THE COMPOUNDS

Halides

Table 21-2 lists the fluorides and chlorides of the second-period and third-period elements. Several of the elements form more than one chloride or fluoride; for example, both PCl_3 and PCl_5 are well-known compounds. Only the higher oxidation state compounds are shown in the table. As indicated by the shading in the table, the halides on the right are covalent, while those on the extreme left are definitely ionic. It is not quite as easy, however, to identify the type of bonding in some of the Group II and Group III halides.

The structure of aluminum fluoride, in which aluminum is octahedrally surrounded by six fluorine atoms, is that of a typical ionic lattice, and the high melting point is consistent with this interpretation. The chemical inertness of aluminum fluoride compared with the other halides of aluminum is, however, more consistent with a three-dimensional covalent network, which would also be ex-

TABLE 21-2
Fluorides and Chlorides
of Second and Third Periods

LiF	BeF_2	BF_3	CF_4	NF_3	OF_2
NaF	MgF_2	AlF_3	SiF_4	PF_5	SF_6
LiCl	$BeCl_2$	BCl_3	CCl_4	NCl_3	Cl_2O
NaCl	$MgCl_2$	$AlCl_3$	$SiCl_4$	PCl_5	SCl_4

☐ ionic   ☐ molecular   ☐ intermediate

pected to have a high melting point. Similar problems are found in attempting to elucidate the bonding for BeF_2. Even though both compounds are very poor conductors in the molten phase, the ionic versions appear to be preferable.

Aluminum chloride exists as a dimer in the vapor phase and in solution in a nonpolar solvent such as benzene. This dimer is composed of two $AlCl_4$ tetrahedra

sharing an edge through two chlorines in common. On the basis of this and the low melting point of aluminum chloride, it is tempting to conclude that the *solid* is molecular in nature. However, the existence of molecules in one phase does not necessarily mean that the substance is molecular in all phases. For example, we found in Chapter 20 that while beryllium chloride is monomeric in the vapor state, it forms a linear polymer in the solid state. In fact, the structure of aluminum chloride in the solid state is a layer lattice, in which each aluminum is surrounded by six chlorines in a slightly distorted octahedron. Although this kind of lattice is found with some ionic compounds, the structure of crystalline aluminum chloride is probably best described as intermediate between ionic and molecular.

Aluminum bromide and aluminum iodide are also dimeric compounds in the vapor phase and in solution in noncoordinating solvents. In contrast to $AlCl_3$, aluminum bromide and iodide in the solid state are also molecular in nature and consist of discrete units of the dimer.

Thus the structure and bonding of the halides can be explained on the basis of either Fajans' model or electronegativity differences (see p. 106). In the aluminum halides the change from ionic structure (AlF_3) to molecular structure ($AlBr_3$, AlI_3) can be related to the increasing size of the anion or to a decreasing difference in the electronegativities of the atoms.

In the series NaCl to $SiCl_4$, the increased charge-to-radius ratio of the cation accounts for the transition from the ionic lattice of sodium chloride to the molecular lattice of silicon tetrachloride. The effect of cation size and electronegativity differences can also be found in the series BCl_3 to $TlCl_3$, where BCl_3 has a typical molecular lattice while $TlCl_3$ has an ionic lattice.

		$NaX(s)$	$MgX_2(s)$	$AlX_3(s)$	SiX_4	PX_3
TABLE 21-3 Standard Heats of Formation of Halides of the Third Period (kcal/mole)	F	−136	−263.5	−311	−370(g)	
	Cl	−98	−153	−166	−153(l)	−81(l)
	Br	−86	−124	−126	−95(l)	−47.5(l)
	I	−69	−86	−75	−32(s)	−11(s)

Heats of Formation.   Table 21-3 lists the standard heats of formation of the halogen compounds of the third-period elements, sodium through phosphorus. For each of these elements the heats of formation become less negative from the fluoride to the iodide. This can be attributed to either the decrease in lattice energy for the ionic halides or the decrease in bond energy for the covalent halides with increasing size of the halogen.

Table 21-4 gives the heats of formation of the chlorides of several families of the representative elements. There is a steady increase in the absolute values of the heats of formation for the chlorides of family IIA from beryllium(II) to barium(II). Although there is a decrease in lattice energy with increasing cation size, the decrease in the energy required to form the gaseous ion (sublimation energy plus ionization energy) more than compensates for this. Therefore, the heats of formation become more negative down the family for these typical ionic chlorides. The principal deviation from this trend, which holds for the ionic halides of families IA and IIA, is found with the fluorides of the alkalies. With these compounds the small size of the fluoride ion causes the lattice energies to fall off rapidly with increasing cation size. Therefore, the heats of formation of the alkali fluorides become less negative from lithium fluoride to cesium fluoride.

TABLE 21-4
Standard Heats of Formation
of the Chlorides
of Groups IIA, IIIA, IVA, and VA
(kcal/mole)

IIA	IIIA	IVA	VA
$BeCl_2$ (s) −122	BCl_3 (g) −94.5	CCl_4 (l) −32	NCl_3 (g) −
$MgCl_2$ (s) −153	$AlCl_3$ (s) −166	$SiCl_4$ (l) −153	PCl_3 (l) −81
$CaCl_2$ (s) −190	$GaCl_3$ (s) −125	$GeCl_4$ (l) −130	$AsCl_3$ (l) −80
$SrCl_2$ (s) −198	$InCl_3$ (s) −128	$SnCl_4$ (l) −130	$SbCl_3$ (s) −91
$BaCl_2$ (s) −206	$TlCl_3$ (s) −84	$PbCl_4$ (l) −	$BiCl_3$ (s) −91

The heats of formation of the halides of the other families listed in Table 21-4 do not show any simple discernible trends.

Reaction with Water.   If one single property could be used to characterize the *covalent* halides, it would be their extensive reaction with water. Since the covalent fluorides are very insoluble and do not react with water at room temperature (or their reactions are more complex), the discussion will be limited to the chlorides, bromides, and iodides.

The halides of the more electronegative elements—boron, silicon, and phosphorus—are completely and irreversibly hydrolyzed to boric acid, hydrated silica, and phosphorous acid, respectively. The hydrolysis of carbon tetrachloride will be discussed later in this chapter.

$$BX_3 + 3H_2O \rightarrow H_3BO_3 + 3HX$$

$$SiX_4 + 4H_2O \rightarrow SiO_2 \cdot 2H_2O + 4HX$$

$$PX_3 + 3H_2O \rightarrow H_3PO_3 + 3HX$$

The hydrolyses of the halides of the other elements of Groups III and V have limited extents. The hydrolyses of both antimony and bismuth trihalides

$$SbX_3 + H_2O \rightleftharpoons \underline{SbOX} + 2HX$$

proceed only as far as the formation of the very insoluble oxyhalides. The reaction of tin(IV) halides with water is complicated by numerous intermediates and is very sensitive to the amount of water added. One of the end products can best be represented as the hydrated tin(IV) oxide, $SnO_2 \cdot xH_2O$.

Hydrogen Halides   The hydrogen halides are gases at standard conditions. With the exception of hydrogen fluoride, which is a weak electrolyte, they are strong acids whose strengths increase in the order HCl < HBr < HI. The reasons for the weakness of hydrogen fluoride as a proton donor and the order of acidity have already been discussed in Chapter 15.

The standard heats and free energies of formation of all hydrogen halides except HI are negative, as shown in Table 21-5, and they all can be prepared by direct combination of the halogens with hydrogen. The equilibrium constants in Table 21-5 for the formation of all but hydrogen iodide are certainly highly favorable. While fluorine reacts violently with hydrogen, the rate of reaction of chlorine is very slow in the dark. However, a mixture of chlorine and hydrogen reacts with explosive violence upon exposure to sunlight. This is an example of a

TABLE 21-5
Thermochemical Data
for HX

	HF	*HCl*	*HBr*	*HI*
Bond energy (kcal/mole)	135	102	87	70
Standard heat of formation (kcal/mole)	−64	−22	−9	6
Standard free energy of formation (kcal/mole)	−65	−23	−13	0.3
K (25°C)[a]	10^{48}	10^{17}	10^9	6×10^{-1}

[a]Equilibrium constant for $\frac{1}{2}H_2 + \frac{1}{2}X_2 \rightleftharpoons HX$.

photon-induced reaction in which photons of the proper energy first split the chlorine molecule into atoms.

$$Cl_2 + h\nu \rightarrow 2Cl$$

The chlorine atoms then react with hydrogen molecules to form HCl and hydrogen atoms.

$$Cl + H_2 \rightarrow HCl + H$$

The hydrogen in turn reacts with chlorine molecules to produce HCl and chlorine atoms.

$$H + Cl_2 \rightarrow HCl + Cl$$

The chlorine atoms formed react with more hydrogen, and thus a very rapid self-propagating chain reaction occurs.

The rate of reaction of hydrogen with bromine is extremely slow. However, both chlorine and bromine can be burned in hydrogen catalytically at controlled rates. Although the reaction is not very favorable, hydrogen iodide can also be produced by direct combination of the elements.

Hydrogen fluoride and hydrogen chloride are much more conveniently prepared by the treatment of readily available calcium fluoride and sodium chloride with hot concentrated sulfuric acid. The hydrogen halides are more volatile than sulfuric acid and are removed from the reaction as they are formed, thereby driving the reaction to completion.

$$CaF_2 + H_2SO_4 \rightarrow CaSO_4 + 2HF\uparrow$$

$$NaCl + H_2SO_4 \rightarrow NaHSO_4 + HCl\uparrow$$

The reaction with calcium fluoride cannot be carried out in glass vessels, because the liberated hydrogen fluoride reacts with silicon dioxide in glass.

$$4HF(g) + SiO_2(s) \rightarrow SiF_4(g) + 2H_2O(l)$$

The relatively strong Si—F bonds and the formation of SiF_4 gas provide the driving force for this reaction.

Since both bromide ion and iodide ion are strong enough reducing agents to be oxidized by sulfuric acid, the method described above for the preparation of HF and HCl is not suitable for the production of HBr and HI.

$$2Br^- + 4H^+ + SO_4^{2-} \rightarrow Br_2 + SO_2 + 2H_2O$$

$$8I^- + 10H^+ + SO_4^{2-} \rightarrow 4I_2 + H_2S + 4H_2O$$

Although phosphoric acid, a nonoxidizing acid, can be used in place of sulfuric acid, the significant decomposition of hydrogen iodide at elevated temperatures gives a highly contaminated product. Hydrogen bromide is conveniently prepared by adding liquid bromine dropwise to moistened red phosphorus. The phosphorus tribromide that is first formed is immediately hydrolyzed to phosphorous acid and hydrogen bromide.

$$2P + 3Br_2 \rightarrow 2PBr_3$$

$$PBr_3 + 3H_2O \rightarrow H_3PO_3 + 3HBr$$

The analogous hydrolysis of phosphorus triodide can be used for the production of hydrogen iodide.

Oxides and Oxyacids   Although the halogens are very reactive elements, they do not combine directly with oxygen. With the exception of I_2O_5, all the binary oxygen compounds (such as OF_2, Cl_2O, ClO_2) have positive heats of formation, and most are thermally unstable. The systematic periodic trends that are so clearly defined with most of the halide compounds are not observed with either the oxides or the oxyacids, in which the halogens have a positive oxidation state.

Chlorine, bromine, and iodine all react with water to a limited extent to form the *hypohalous* acids.

$$X_2 + H_2O \rightleftharpoons H^+ + X^- + HOX$$

The equilibrium constants for this reaction decrease from about 10^{-4} for chlorine to 10^{-13} for iodine. This is an oxidation-reduction reaction in which the halogen plays the role of both the oxidizing agent and the reducing agent. The hypohalous acids are strong oxidizing agents whose strengths decrease in the order HOCl > HOBr > HOI. This order can be rationalized on the basis of the electronegativity of the halogen: the more electropositive the element the more stable the positive oxidation state. (Recently, hypofluorous acid has also been isolated from the reaction of F_2 with water.)

Chlorine, bromine, and iodine react in cold basic solutions to yield the halide ion and the hypohalite ion.

$$X_2 + 2OH^- \rightarrow X^- + OX^- + H_2O$$

This reaction of the halogens with a base is as characteristic of the behavior of nonmetals as the reaction of the Group I elements with an acid is of the behavior of metals. If the reaction is carried out with *hot* basic solutions, halide and halate ions are produced by the autooxidation-reduction of the hypohalite ion.

$$3OX^- \xrightarrow{\text{heat}} 2X^- + XO_3^-$$

Of the *halous* acids, HXO_2, only chlorous acid is known. Of the three *halic* acids, HXO_3, the most stable is iodic acid, a white solid. The heats of formation of the three halic acids (they are all exothermic) become less negative in the following order: $HIO_3 > HClO_3 > HBrO_3$. They are all strong oxidizing agents and decrease in oxidizing strength in the reverse order: $HBrO_3 > HClO_3 > HIO_3$.

Three *perhalic* acids have been isolated—$HClO_4$, $HBrO_4$, and HIO_4. Perbromic acid was first prepared in 1968 and is the strongest oxidizing agent of the three. The perchlorate ion and the periodate ion are not as strong oxidizing agents as the chlorate and iodate ions, respectively. Since the radius of iodine is much larger than that of chlorine, iodine is able to coordinate more oxygen atoms and is also found as H_5IO_6 in addition to HIO_4.

There was a sense of excitement in the field of chemistry in June 1962 after the first report of the preparation of a noble gas compound. Historically, for decades after the discovery of the noble gases in the 1890s, repeated efforts to prepare noble gas compounds ended in failure. *Rare, noble,* and *inert,* all words used to characterize these elements, suggested that unusual and exotic conditions would be necessary to form chemical compounds. The "stable octet," which so strongly influenced earlier bonding concepts, was used to argue that these elements would undoubtedly be chemically inert. Now, after the fact, as is so often true, it is not surprising to find that some of the fluorides of the heavier noble gases are relatively easy to prepare. Thermodynamic calculations indicate that the formation of the fluorides of xenon is exothermic and that the formation of the chlorides and oxides should be endothermic.

As is true with many discoveries, serendipity played a major role. N. Bartlett, a Canadian chemist, was investigating the reaction of platinum metal with fluorine gas, using glass apparatus. One of the products of the reaction was an orange-red solid which, after many difficulties in purification, was found to have the composition PtO_2F_6. This was an unexpected product. By the direct interaction of molecular oxygen with PtF_6, the same orange solid was obtained which, after X-ray and other analyses, was best formulated as O_2^+, PtF_6^-. (The bonding in the O_2^+ ion was discussed in Chapter 7.)

In August of the same year, Bartlett reported that upon consideration of the fact that the first ionization energy of xenon is slightly less than the energy required to remove an electron from O_2, it seemed reasonable to suppose that xenon might also be oxidized by the hexafluoride of platinum. From the reaction of xenon gas with PtF_6, he obtained an orange-yellow solid that had a negligible vapor pressure and was insoluble in carbon tetrachloride. The platinum in this compound was found to have an oxidation state of +5, and the compound was formulated as Xe^+, PtF_6^-.

This first report of a compound of a noble gas suggested that the direct interaction of fluorine with xenon under mild conditions should be successful. Chemists at Argonne National Laboratory reported in August of 1962 that five volumes of fluorine gas plus one volume of xenon gas, heated together in a closed nickel vessel at 400°C and then rapidly cooled, gave crystals of XeF_4. In rapid order XeF_2 and XeF_6 were prepared by other researchers by direct combination under relatively mild conditions.

These xenon fluorides are stable at room temperature but react readily with water. Structurally, XeF_2, with two sigma bonds and three lone pairs on the central xenon, is a linear molecule like the isoelectronic ICl_2^- ion; XeF_4 is a square coplanar molecule like the isoelectronic ICl_4^- ion; and XeF_6 appears to be a distorted octahedron. The structures of XeF_2 and XeF_4 can be rationalized by either the molecular orbital model or the valence bond model. Neither theory adequately explains the shape of the XeF_6 molecule.

A variety of inert gas compounds are now known, and more will almost certainly be synthesized in the future. Known compounds include krypton difluoride, radon fluorides, and oxides and oxyfluorides of xenon.

The Group IV Elements

The elements of Group IV are unique in a number of respects. Carbon, the first member of the group, forms the skeleton of the organic world. Silicon, in combination with oxygen, has the same role in the matter that constitutes the earth's crust. Tin has the largest number of stable isotopes of any element, while lead is the end product of a number of natural radioactivity decay series.

The nuclear stability of tin and lead has been commented on in Chapter 18. Trends in chemical and physical properties of the Group IV elements and their parallels in the other *p*-filler groups will be discussed here.

THE ELEMENTS

Several of the Group IV elements provide excellent examples of allotropic elemental forms. Elemental carbon exists in the diamond and graphite forms, the structures of which have been discussed in Chapter 9. These allotropes afford an interesting illustration of kinetic versus thermodynamic control of stability. The following heats of combustion can be used to calculate the standard enthalpy change for the diamond-graphite conversion.

$$
\begin{array}{ll}
C_{diamond} + O_2 \rightarrow CO_2 & \Delta H^0 = -94.505 \text{ kcal/mole} \\
-(C_{graphite} + O_2 \rightarrow CO_2) & -(\Delta H^0 = -94.052 \text{ kcal/mole}) \\
\hline
C_{diamond} \rightleftharpoons C_{graphite} & \Delta H^0 = -0.453 \text{ kcal/mole}
\end{array}
$$

The absolute entropies at 25°C of graphite and diamond are, respectively, 1.4 cal/mole-deg and 0.6 cal/mole-deg. The entropy change for the conversion is, then, 0.8 cal/mole-deg. The standard free-energy change is

$$\Delta G^0 = \Delta H^0 - T\Delta S^0$$

$$\Delta G^0 = -453 - 298(0.8)$$

$$= -0.7 \text{ kcal/mole}$$

and the conversion of diamond to graphite is definitely favored at room temperature and pressure. Of course, diamonds do not spontaneously convert to graphite (fortunately for the jewelry trade), and their stability must therefore be attributed to a very slow rate of conversion. In view of the extensive bond breaking and rearrangements that would have to occur throughout the crystal, this very slow rate is hardly surprising.

The opposite conversion, graphite → diamond, has for obvious reasons received considerable attention. The chemical literature contains a number of early reports of success, all of which have since been shown to be incorrect or fraudulent. More recently, synthetic diamonds have been prepared at very high temperatures and pressures in the presence of transition metals, which serve as catalysts for the conversion.

There is only one elemental form of silicon—the diamond form, in which each silicon is bonded to four other silicon atoms in a giant three-dimensional network "molecule." The same is true of germanium.

There are three allotropic forms of tin: grey tin, which has the diamond structure and is stable below room temperature; white tin, which has a metallic crystal lattice and is the stable form at room temperature; and brittle tin, which also has a metallic lattice. The two metallic allotropes differ in the extent of close-packing of the tin atoms. Lead, the final member of the group, has only a metallic elemental form.

The trend from nonmetallic to metallic character as the group is descended is in agreement with our previous generalization that a low electronegativity or ionization energy is characteristic of a metal. The elements silicon and germanium are often classed as *metalloids* because a number of their properties, such as luster and conductivity, are intermediate between those characteristic of nonmetals and metals.

The same trend is observed in the elemental forms of Groups III, V, and VI. The Group III elements have lower electronegativities than their Group IV analogs, and consequently all except boron are metals. The heavier Group V elements—arsenic, antimony, and bismuth—have high electrical conductivities and are generally considered to be metals even though their structures are similar to the layer structure of the black allotrope of phosphorus (which is definitely a nonmetal). Of the Group VI elements, only radioactive polonium has the electrical characteristics of a true metal. Thus, the metallic character of the elemental forms parallels electronegativity and increases from top to bottom and from right to left on the periodic chart.

The melting points, boiling points, and heats of formation of gaseous atoms (heats of atomization) of each of the Group IV elements at 25°C are given in Table 21-6. The heats of atomization are a direct measure of the energy required to break the crystal lattices into the constituent gaseous atoms. The atomization of carbon, silicon, and germanium requires the destruction of the covalent bonds within the network crystals, and the decrease in the heats of atomization from carbon to germanium reflects a parallel decrease in the element-element bond dissociation energies, $D(M-M)$ (also shown in Table 21-6). The atomization of tin and lead provides a measure of the energy necessary to break up the metallic lattices of these elements. The decrease in the heats of atomization from tin to lead can be

TABLE 21-6		*C*	*Si*	*Ge*	*Sn*	*Pb*
Some Properties	Melting point (°C)	>3700	1412	937	232	328
of the Group IV Elements	Boiling point (°C)	3627	2727	2827	2687	1751
	ΔH_f^0 [M(g)] (kcal/mole)	172	109	78	72	47
	Bond energy, D(M—M) (kcal/mole)	83	46	38	39	

attributed to an increase in size of the atoms of the lattice. The same effect of size on the metallic lattice energy was noted for the alkali metals (Chapter 20).

With only a few exceptions, the melting points and boiling points of the elements follow the trend in heat of atomization.

Isolation   Carbon occurs naturally as diamond, graphite, and (mixed with other materials) coal. It can be prepared in the laboratory by reduction of CO_2 with hot magnesium.

$$2Mg + CO_2 \rightarrow C + 2MgO$$

The metals and metalloids are generally prepared by reduction of the oxides with carbon at high temperatures. For the reaction

$$C + SiO_2 \rightarrow Si + CO_2$$

the standard enthalpy and free-energy changes can be calculated from standard heats and free energies of formation (Appendix 7) as 111 kcal/mole and 98 kcal/mole, respectively. The large positive ΔG^0 shows clearly the very unfavorable nature of the process at 25°C.

The entropy change for this reduction can also be determined from the relation

$$\Delta G^0 = \Delta H^0 - T\Delta S^0$$

and is 44 cal/deg-mole. The temperature above which the process becomes favorable can be calculated by assuming that ΔS^0 and ΔH^0 are temperature independent and by taking as favorable that temperature at which the standard free-energy change is zero.

$$\Delta G^0 = 0 = \Delta H^0 - T\Delta S^0 = 111,300 \text{ cal} - T(44 \text{ cal/deg-mole})$$

$$T = 2600°K \text{ or } 2.3 \times 10^3 °C$$

Even though ΔS^0 and ΔH^0 are almost certainly not temperature independent over this considerable range of temperature, the calculation still suffices as an estimate of the very high temperature required. The reduction of SiO_2 is, in fact, generally performed in an electric furnace, where an electric arc passed between carbon electrodes produces temperatures of about $1700°C$.

Of the Group III, V, and VI elements, only three—oxygen, nitrogen, and sulfur—occur naturally in their elemental forms. The elemental forms of the others are prepared by chemically or electrolytically converting the naturally occurring source to the element. Thus, since the most common Group VI sources contain the elements in negative oxidation states, they are oxidized to obtain the free element. The Group V sources, on the other hand, contain the elements in their positive oxidation states (for example, phosphates, antimony sulfide) and must consequently undergo reduction in order to secure the free element. The Group III sources also contain the elements in their positive oxidation states. Boron is obtained by chemical reduction of the oxide B_2O_3; aluminum is prepared by electrolysis of the ore bauxite ($Al_2O_3 \cdot xH_2O$) in molten cryolite, Na_3AlF_6; gallium, indium, and thallium are obtained by electrolytic reduction of their salts in aqueous solution.

COMPOUNDS

The valence electron configuration ns^2np^2, as well as the intermediate values for the ionization potentials and electronegativities of the Group IV elements, suggest that, contrary to the trends observed for the Group I and Group VII elements, the existence of inert gas–type ions such as Si^{4-} and Si^{4+} is quite unlikely. The only possible exceptions to this occur with the alkali or alkaline earth carbides, silicides, or germanides, such as Be_2C and Mg_2Si. Even in these compounds, however, the bonding appears to be appreciably covalent.

Thus, instead of bonding ionically by loss or gain of electrons, the Group IV elements utilize the valence s and p electrons to form covalent bonds with other elements. In most Group IV compounds all four valence electrons are used to form four covalent bonds. In addition to these compounds, in which the element is in the +4 oxidation state, there are also compounds with the Group IV element in the +2 oxidation state. These compounds can be thought of as containing +2 ions or as covalent, with only two electrons from the Group IV element being used to form two covalent bonds. The +2 oxidation state is stable relative to the +4 state only for tin and lead.

In order to establish the nature of the bonding—whether covalent or ionic—in compounds containing the Group IV element in the +2 oxidation state, let us examine some properties of the lead(II) halides (Table 21-7). The melting point and boiling point of the fluoride are the highest, and if these properties are used as an estimate of ionicity, the fluoride would appear to be the most ionic. Predictions of ionic character based on Fajans' rules or electronegativity differences are in agreement with this observation.

TABLE 21-7
Some Properties
of the Lead(II) Halides

	PbF_2	$PbCl_2$	$PbBr_2$	PbI_2
Melting point (°C)	855	498	373	ca. 402
Boiling point (°C)	1290	954	916	ca. 954
ΔH_f^0 (kcal/mole)	−158.5	−86	−66	−42

The question of just how ionic PbF_2 really is can be answered, at least approximately, by comparing its experimentally derived lattice energy (580 kcal/mole) with the lattice energy calculated by assuming 100 percent ionic character (563 kcal/mole). The latter value was obtained with the Coulomb expression derived in Chapter 6, using the Madelung constant for the PbF_2 fluorite-type structure. Because of the approximations (for example, the 10-percent reduction for repulsions) employed in the calculations, the agreement between the calculated and experimental lattice energies can certainly be considered satisfactory and is therefore evidence for a high degree of ionic character in PbF_2. As stated earlier, this is probably the most ionic of the lead(II) halides. Since the lead(II) cation is larger than the tin(II) ion, Fajans' rules suggest that PbF_2 is also more ionic than any of the tin(II) halides. In general, then, except for some Sn(II) and Pb(II) compounds, the vast majority of Group IV bonds are covalent.

Compounds of the Group IV elements exhibit a number of trends that are important in the chemistry of neighboring groups as well. We shall now examine these trends.

Stability of the
Lower Oxidation
State

Except for species that exist only under unusual conditions (for example, SiO at high temperatures in the gas phase), the +2 oxidation state is quite unstable for silicon. That is, there are no compounds stable at room temperature and pressure that contain silicon to which an oxidation state of +2 can be assigned. For germanium also, the +2 oxidation state is of little importance, although a few compounds such as GeS and GeF_2 are relatively stable and appear to be ionic. As noted above, for tin and lead the +2 oxidation state is rather stable. There appears, then, to be an increase in the stability of the +2 oxidation state as the group is descended from silicon to lead. Because of the uniqueness of carbon, we will reserve comment on it until later.

This trend is verified by the enthalpy changes given in Table 21-8 for the decomposition of the +4 oxidation state to the +2 oxidation state. For the oxides, the enthalpy change decreases from silicon to lead, indicating an increase in stability of the +2 state *relative* to the +4 state as the group is descended. The data for the chlorides show that energetically the formation of $PbCl_2$ (s) is favored. The free-energy change for this reaction is also negative, and $PbCl_4$ (l) does indeed decompose to $PbCl_2$ (s) at room temperature.

TABLE 21-8
Relative Stability
of the Oxidation States

Group IV	ΔH^0 (kcal/mole)
$CO_2(g) \rightarrow CO(g) + \frac{1}{2}O_2(g)$	68
$SiO_2(s) \rightarrow SiO(g) + \frac{1}{2}O_2(g)$	179
$GeO_2(s) \rightarrow GeO(g) + \frac{1}{2}O_2(g)$	95.5
$SnO_2(s) \rightarrow SnO(s) + \frac{1}{2}O_2(g)$	70
$PbO_2(s) \rightarrow PbO(s) + \frac{1}{2}O_2(g)$	14
$SnCl_4(l) \rightarrow SnCl_2(s) + Cl_2(g)$	47
$PbCl_4(l) \rightarrow PbCl_2(s) + Cl_2(g)$	−7
Group III	ΔH^0 (kcal/mole)
$GaCl_3(s) \rightarrow GaCl(g) + Cl_2(g)$	106
$InCl_3(s) \rightarrow InCl(s) + Cl_2(g)$	84
$TlCl_3(s) \rightarrow TlCl(s) + Cl_2(g)$	35
Group V	ΔH^0 (kcal/mole)
$PCl_5(s) \rightarrow PCl_3(l) + Cl_2(g)$	30
$SbCl_5(l) \rightarrow SbCl_3(s) + Cl_2(g)$	13.5

Table 21-8 also includes values for several Group III and Group V oxidation state changes, and the same trend is obvious in these groups. Although not evident from the data of Table 21-8, the +1 oxidation state is the stable state for thallium (but not for indium), as indicated by the standard electrode potential for the half-reaction:

$$Tl^{3+}(aq) + 2e^- \rightleftharpoons Tl^+(aq) \qquad E^0 = 1.25 \text{ V}$$

In Group V the +3 oxidation state is the stable state for bismuth. The only known pentahalide of bismuth is BiF_5.

Thus, the lower oxidation state increases in stability relative to the higher oxidation state as Groups III, IV, and V are descended, and in fact becomes the stable state for the bottom member of each group. The reasons for these trends are not well understood, and we will not attempt an explanation.

Catenation

Catenation (from the Latin *catena,* meaning "chain") is a term that refers to covalently bonded chains of atoms of the same element. We have seen previously that in the diamond forms of carbon, silicon, germanium, and tin, there is a very extensive proliferation of element-element linkages. In most substances, however, the extent of catenation is quite limited. The element carbon is most unusual in its ability to form very long chains of carbon atoms. These carbon-carbon linkages may consist of single bonds, as in the alkanes, or of multiple bonds, as in aromatic compounds, alkenes, and alkynes. As illustrations of the extent of catenation possible with carbon, consider the following:

1. Chloroethene, also called vinyl chloride, reacts with itself in the presence of peroxides to form a long-chain polymer, generally termed polyvinyl chloride. This polymer contains from 2000 to 5000 carbon atoms per chain.

$$n\text{H}_2\text{C}=\text{CHCl} \rightarrow \begin{bmatrix} & \text{H} & \text{H} \\ & | & | \\ -&\text{C}-\text{C}&- \\ & | & | \\ & \text{H} & \text{Cl} \end{bmatrix}_n$$

The formula $+\text{CH}_2\text{CHCl}+_n$ indicates that n $-\text{CH}_2\text{CHCl}-$ units are linked together by carbon-carbon bonds. The nature of the groups terminating the chain is indeterminate. These rubberlike polymeric materials are used commercially as floor tile, plastic tubing, wire insulation, plastic coverings for raincoats, and so forth.

2. 1,2-Benzopyrene, a constituent of coal tar that is believed to be responsible for skin cancers among workers in the coal tar industries, has the structure

Here, 20 carbons are linked in a series of fused benzene rings.

Catenation of silicon is much more limited. Silanes, analogs of the alkanes, are known that contain chains of up to 30 silicon atoms. Catenation is even more restricted with germanium and tin; and for lead, only the dilead compounds [for example, $(\text{C}_6\text{H}_5)_3\text{Pb}-\text{Pb}(\text{C}_6\text{H}_5)_3$] are stable.

The decrease in the extent of catenation down Group IV can be rationalized by a careful examination of the bond energies given in Table 21-9. First, the M–M bond energies decrease down the group, with a large difference between the C–C bond energy and the Si–Si bond energy. The element-element linkages therefore

TABLE 21-9
Average Group IV
Bond Energies (kcal/mole)

Linkage	M = C	M = Si	M = Ge	M = Sn
M–M	83	46	38	39
M–H	98	76	69	60
M–C	83	76	57	46
M–N	73	ca. 80	ca. 60	–
M–O	85	106	ca. 85	–
M–F	116	142	113	–
M–Cl	78	97	83	77
M–Br	68	76	66	65
M–S	65	61	–	–

become weaker from carbon to lead, as would be expected because of the increase in size of the atoms in that order. Second, the C–C linkage is one of the strongest bonds formed by carbon, only the C–H, C–O, and C–F bonds being stronger, while for the other Group IV elements the reverse is true. The Si–Si bond is the least stable of the silicon linkages presented in Table 21-9. Thus, in reactions that are thermodynamically controlled, bonds between silicon and other elements will be formed rather than silicon-silicon bonds.

The trend in catenation for Groups V and VI is strikingly different. In these groups, the first element—N or O—shows little tendency to form M–M chains. The second element of each group catenates to the greatest extent, and catenation drops off down the remainder of each group. An explanation for this anomaly can be gleaned from the bond energies in Table 21-10. Obviously, the trend in catenation follows the trend in element-element bond energies. The low F–F single-bond energy was discussed earlier in this chapter, and the rationalization for the N–N and O–O bonds (for example, in $H_2N–NH_2$ and HO–OH) is similar.

TABLE 21-10 Group V and VI Bond Energies (kcal/mole)	Group V		Group VI	
	N–N	40	O–O	34
	P–P	50	S–S	54
	As–As	35	Se–Se	44
	Sb–Sb	29	Te–Te	33
	Bi–Bi	25		

Multiple Bonding

In addition to its unexcelled ability to catenate, carbon is also unique in the stability and number of multiply bonded structures it forms. We have previously discussed the bonding in alkenes (C=C bonds), alkynes (C≡C bonds), and ketones, aldehydes, and carboxylic acids (C=O bonds). In addition, there are numerous carbon-nitrogen linkages that contain multiple carbon-nitrogen bonds.

There are, however, no known compounds stable under ambient conditions (room temperature and pressure) that contain silicon doubly or triply bonded to itself or another element, despite the efforts of many chemists to prepare such compounds. There are also no stable compounds containing germanium, tin, or lead multiply bonded to themselves or to other elements.

In an attempt to rationalize the absence of such bonds for the members of Group IV other than carbon, consider the thermodynamic stability of two structures possible for compounds with the empirical formula R_2MO. Such a compound could contain an M=O double bond as in formula A (for instance, propanone, where M = C, R = CH_3), or it could contain M–O single bonds as depicted in structure B (which might be part of a ring structure such as in formula C). The thermodynamic stability of A versus B will depend primarily on the strength of an M=O bond as opposed to the strength of *two* M–O single bonds. The carbon-oxygen double bond in ketones has a bond energy of about 191 kcal/mole, whereas the carbon-oxygen single-bond energy in structures of type B is about 86 kcal/mole. Thus, structure A is about 19 kcal more stable [191 − (2 × 86) = 19 kcal/mole] than structure B, and of course A is indeed the structure of ketones such as propanone. When M = Si, however, the reverse is true. The best estimate for the bond energy of the silicon-oxygen double bond, based on gaseous SiO, is 190 kcal/mole, which is less than twice the Si—O single-bond energy of 106 kcal/mole. For silicon, then, structure B is considerably more stable, and cyclic structures such as C (or polymeric structures, see p. 587) are indeed the structures adopted by compounds with the empirical formula $(CH_3)_2 SiO$.

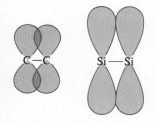

FIGURE 21-1

Pi Overlap in the C—C and Si—Si Bonds

From the foregoing, the basic difference between carbon and silicon appears to be that for carbon the strength of the double bond is equal to or greater than twice that of the single bond, while the opposite is true for silicon. This is apparently true even for Si=C or Si=Si bonds. The rationalization most often offered for this is that because of the greater Si—Si (for example) bond length relative to C—C and the greater diffuseness of the silicon $3p$ orbitals relative to the carbon $2p$ orbitals, the pi overlap between two silicons will be less effective than that between two carbons. This is represented pictorially in Figure 21-1. This hypothesis is not borne out by quantum-mechanical calculations, however, and the reason for the weakness of pi bonds to silicon (and the heavier members of the group) remains a matter for speculation.

The Halides

Table 21-11 summarizes the boiling points and standard heats of formation of the simple binary tetrahalides of the Group IV elements. The large negative heats of formation indicate that the tetrahalides can be prepared by direct reaction of the elements. For example, tin tetrachloride can be synthesized by passing gaseous Cl_2 over molten tin. Once initiated, the reaction is quite exothermic.

Most of the tetrahalides react with water according to the equation

$$MCl_4 + 2H_2O \rightleftharpoons MO_2 + 4HCl$$

Silicon tetrachloride, for example, hydrolyzes very vigorously to give hydrated SiO_2. Carbon tetrachloride, however, appears to be quite inert to water. The standard free-energy changes for these two reactions can be calculated from the known standard free energies of formation.

$$CCl_4 (l) + 2H_2O(l) \rightleftharpoons CO_2 (g) + 4HCl(aq)$$

$$\Delta G^0 = -94.3 + 4(-31.35) - [-15.0 + 2(-56.71)] = -91.3 \text{ kcal/mole}$$

TABLE 21-11
Boiling Points[a] (°C)
and Standard Heats of
Formation (kcal/mole)
of the Group IV
Tetrahalides, MX_4

M	C		Si		Ge		Sn		Pb	
X	B.P.	ΔH_f^0	B.P.	ΔH_f^0	B.P.	ΔH_f^0	B.P.	ΔH_f^0	B.P.	ΔH_f^0
F	−128	−223	−96	−370	−37	—	705(s)	—	—	−222
Cl	76	−32	57	−153	86	−130	114	−130	150	—
Br	d.	—	155	−95	186	—	203	−97	—	—
I	d.	—	290	−32	ca. 348	—	346	—	—	—

[a]s = sublimes; d = decomposes

$$SiCl_4(l) + 2H_2O(l) \rightleftharpoons SiO_2(s) + 4HCl(aq)$$

$$\Delta G^0 = -192.4 + 4(-31.35) - [136.9 + 2(-56.7)] = -67.5 \text{ kcal/mole}$$

Surprisingly, the free-energy change for the hydrolysis of CCl_4 is more negative than that for $SiCl_4$ and corresponds to an equilibrium constant of 10^{65}. The apparent inertness of CCl_4 must be attributed, therefore, to a very slow rate of reaction with water.

Although the exact mechanisms for the hydrolyses of CCl_4 and $SiCl_4$ are not known, the difference in rate of reaction has generally been ascribed to the presence of empty d orbitals in the valence shell of silicon. These orbitals presumably allow the attacking water molecules to bond rather strongly to the silicon atom in the activated complex and thereby lower the activation energy for the reaction. Since carbon has no comparable orbitals in its valence shell, this type of stabilization is impossible for CCl_4. This hypothesis assumes, of course, that the interaction of the water molecules with the tetrahalide, rather than, say, dissociation of M—Cl bonds, determines the rate of reaction.

The tin(II) and lead(II) halides also react with water. Of the variety of products formed, the most abundant is the hydroxide. For example,

$$SnCl_2(s) + 2H_2O \rightleftharpoons Sn(OH)_2(s) + 2HCl(aq)$$

A wide variety of Group IV halides have achieved commercial importance. In 1938, R. J. Plunkett, a du Pont chemist, was experimenting with the gas tetrafluoroethylene, which was routinely stored in metal cylinders under pressure. Upon opening the valve of one particular cylinder, Plunkett found that no gas was released. Since the cylinder weighed more than it had when empty and therefore should have contained some tetrafluoroethylene, he proceeded to systematically dismantle the cylinder. Having finally sawed the cylinder in half, Plunkett discovered it was partially filled with a white powder. This was the serendipitous discovery of Teflon. The white powder was later found to consist of long chains of CF_2 groups; it is polymerized tetrafluoroethylene.

The polymerization of the monomer, $F_2C=CF_2$, can be viewed as proceeding by initial opening of the π-bond to give $F_2\dot{C}-\dot{C}F_2$. These species could then react with one another as follows:

$$
\begin{array}{c}
F_2C-\dot{C}F_2 \\
\dot{} \\
F_2C-\dot{C}F_2 \\
\dot{} \\
F_2C-\dot{C}F_2
\end{array}
\longrightarrow
\quad
\begin{array}{c}
F \quad \left[\begin{array}{c} F \end{array}\right] \quad F \quad F \\
| \quad\quad | \quad\quad | \quad | \\
-C-\!\!-\!\!-C-\!\!-\!\!-C-C- \\
| \quad\quad | \quad\quad | \quad | \\
F \quad \left[\begin{array}{c} F \end{array}\right]_n F \quad F
\end{array}
$$

The considerable chemical and thermal stability of the polymers reflect the great strength of the C—F bond. (Compare the energy of the C—F bond with the energies of other C—X linkages in Table 21-9.)

Many of the chlorocarbon compounds are toxic and find heavy use as insecticides, germicides, and herbicides. Probably the best known pesticide is 1,1,1-trichloro-2,2-bis(*p*-chlorophenyl)ethane, also named dichlorodiphenyltrichloroethane (DDT). While this compound has undoubtedly saved thousands of lives by reducing insect-carried diseases such as malaria and typhus, it is now believed to be responsible for a number of harmful ecological changes. The compound is not easily degraded (either by biological action or reaction with soil and water) to nontoxic substances and can remain toxic for up to 20 years. In 1970 the United States government issued a series of regulations designed to restrict the use of a number of the "hard" (persistent) pesticides. 2,4-Dichlorophenoxyacetic acid (2,4-D) and 2,4,5-trichlorophenoxyacetic acid (2,4,5-T) are weed killers— probably the two most frequently used by the homeowner. The trichloro derivative has been extensively employed as a defoliating agent. Recent evidence suggests, however, that 2,4,5-T, or a chlorine-containing impurity, may cause birth defects, and therefore the use of this herbicide has also been restricted. Fortunately, neither compound is as long-lasting as DDT; 2,4-D has a lifetime of 1 to 18 months.

DDT

2,4-D

2,4,5-T

The Oxides

Of the Group IV oxides, the carbon oxides are most unique. Both carbon monoxide and carbon dioxide are gases at room temperature, while the other Group IV oxides are all solids with high melting points. This difference in melting point and boiling point is a result of the formation of pi bonds in the carbon compounds. In carbon dioxide, carbon's valence of 4 and oxygen's valence of 2 are satisfied by valence bond structures such as O=C=O, and the compound is therefore molecular. Since the carbon dioxide molecule is linear and has no dipole moment, the intermolecular forces are solely van der Waals forces. The pi bonding is, however, unfavorable for silicon, and in silicon dioxide the valences of both atoms must be satisfied by single bonds. Silicon dioxide is therefore a three-dimensional covalent network solid in which each silicon is bound to four oxygens and each oxygen to two silicons, as shown in Figure 9-16 (p. 192). As we have seen in Chapter 9, such a structure has a very high melting point.

Both carbon monoxide and carbon dioxide are formed when carbon is burned in air. The temperature and mole ratio of carbon to oxygen determine the relative amounts of the two products. Since carbon monoxide is isoelectronic with diatomic nitrogen, a bond order of 3 for the C–O linkage would be expected, and the data of Table 21-12 substantiate this expectation. The data also reveal how very similar the physical properties of two isoelectronic species can be. Carbon monoxide and diatomic nitrogen also share some chemical similarities. Both are relatively inert, but carbon monoxide does behave as a Lewis base toward certain metals. The coordination chemistry and physiological effects of CO will be discussed in Chapter 22. The bonding in CO_2 has been discussed in Chapter 7.

	CO	N_2
Bond length (Å)	1.13	1.10
Bond energy (kcal/mole)	256	225
Dipole moment (D)	0.13	0
Boiling point (°C)	−191.5	−195.8
Melting point (°C)	−205.1	−210.0
Density of liquid (g/ml)	0.793	0.796
Viscosity (micropoise), 20°C	163	166
Critical temperature (°C)	−122	−127
Critical pressure (atm)	35	33

TABLE 21-12 Some Properties of CO and N_2

Carbon dioxide is essential to a number of biological processes, such as respiration, photosynthesis, and the regulation of the pH of blood. In the laboratory, CO_2 is sometimes prepared by the action of an acid on a hydrogen carbonate or carbonate.

$$HCO_3^- + H^+ \rightarrow CO_2 + H_2O$$

$$CO_3^{2-} + 2H^+ \rightarrow CO_2 + H_2O$$

This is also the reaction that occurs in some fire extinguishers and when sodium hydrogen carbonate is used in baking.

Silicon dioxide occurs naturally in such forms as quartz, tridymite, cristobalite, flint, opal, jasper, agate, and sand. The first three substances are pure SiO_2, while the others contain impurities of some sort. The general structure of SiO_2 was given above: Each silicon atom is tetrahedrally surrounded by oxygen atoms, and the SiO_4 tetrahedra are linked through oxygen atoms at the corners of the tetrahedra. The three forms of pure SiO_2 differ from one another in the way in which the SiO_4 tetrahedra are linked. Quartz is the stable form up to 870°C, tridymite from 870°C to 1470°C, and cristobalite from 1470°C to 1710°C, the melting point. The transition from one form to another is extremely slow, and thus tridymite and cristobalite do exist at room temperature even though they are thermodynamically unstable relative to quartz at room temperature. This polymorphism (*polys* meaning "many," *morphe* meaning "form") is quite valuable to the geologist, because the particular form of SiO_2 found at a given site is indicative of the temperature prevailing during rock formation.

Carbon dioxide reacts as a Lewis acid with water to form the weak electrolyte, carbonic acid.

$$CO_2 + H_2O \rightleftharpoons H_2CO_3$$

Silicon dioxide reacts with hydroxides to form salts (silicates) and also is acidic in aqueous solution ("silicic acid"). Neither CO_2 nor SiO_2 has any basic properties. Germanium dioxide is not appreciably acidic or basic. Stannic oxide is amphoteric: It reacts with both acids and bases. Hence, the basicity of the oxides increases as the group is descended, and this can be attributed to an increase in the ionic character of the metal-oxygen bond, as discussed in Chapter 15.

The Silicates

The naturally occurring silicates are the most common rock-forming minerals. They are also of considerable commercial importance: Asbestos and mica are used as insulators, granite is a building stone, the various clays are the basis of bricks, pottery, and china, and minerals such as beryl, topaz, and garnet are valued as gemstones.

The composition of the silicates is often variable; for example, the formula of the mineral olivine can be given as Mg_2SiO_4, but in many forms of olivine some of the magnesium is replaced by iron or other metals.

Structurally, the silicates can be thought of as comprising metal cations and anions that consist of silicon atoms covalently and tetrahedrally bonded to four oxygen atoms. The most fundamental difference between the various silicates is the structure of the anion. The anion can exist as a single SiO_4^{4-} ion, as two SiO_4 tetrahedra linked through one oxygen atom, as tetrahedra linked through two of the oxygens of each tetrahedron, or as tetrahedra linked through three of the oxygen atoms of each tetrahedron.

Single SiO_4^{4-} Ions. A few silicates are known to contain discrete SiO_4^{4-} ions. Zircon, $ZrSiO_4$, is an example of such an *orthosilicate*. While the binding between the zirconium and the anion (actually the oxygens of the anion) is surely somewhat ionic in nature, the compound should not be regarded as totally ionic, as implied by the description Zr^{4+}, SiO_4^{4-}.

Two Tetrahedra Linked through One Oxygen.   There are only a few minerals that contain the *pyrosilicate* ion, $Si_2O_7^{6-}$, and these are rare.

Tetrahedra Linked through Two Oxygens of Each.   The *metasilicate* anions can be either cyclic or chainlike. In beryl, $Be_3Al_2Si_6O_{18}$, the anion is cyclic (see diagram). The beryllium and aluminum "ions" (the cation-oxygen bonding is

$$Si_6O_{18}{}^{12-}$$

almost certainly not totally ionic) are located between the anions in sufficient number to make the compound electrically neutral. Beryl is better known as the gem aquamarine when it is pale blue, and as emerald when it is green. The colors of these gems are due to the presence of trace amounts of iron and chromium.

In the pyroxene minerals the anion is an infinite chain of tetrahedra linked as

The repeating unit of the anion is SiO_3^{2-} and this, therefore, is the empirical formula of the pyroxene anions. An example of a mineral in the pyroxene group is spodumene, $LiAl(SiO_3)_2$, a lithium ore. In the amphibole minerals, two such infinite chains are cross-linked at every other tetrahedron. The chains are held together in parallel fashion by the cations, and it would seem reasonable that such compounds should cleave preferentially along the anion chain. This hypothesis is certainly substantiated by the long flexible fibers of the asbestos minerals (Figure 21-2a), most of which are members of the amphibole family.

(a) tremolite   (b) muscovite

FIGURE 21-2

Some Silicate Minerals

amphibole cross-linking

Tetrahedra Linked through Three Oxygens of Each. Anions of this type have two-dimensional sheet structures of the following type:

Minerals containing these sheet anions might be expected to cleave into sheets, and this is confirmed by the sheetlike nature (Figure 21-2b) of the mica family of minerals. The clay minerals also contain sheetlike anions.

Three-dimensional framework anions. The logical extension of the foregoing structural categories is the bonding of every oxygen of each tetrahedron to the silicon of another tetrahedron. This, of course, produces the structure of SiO_2. If some of the silicons are replaced by aluminum atoms, the framework must become negatively charged (aluminum requires one less electron than silicon for neutrality). This is the case for a variety of minerals such as the feldspars (the most common rock-forming minerals) and the zeolites. Interspersed within the framework are cations of the proper charge and number to make the substance electrically neutral. In orthoclase, a member of the feldspar group, one-fourth of the silicon atoms have been replaced by aluminum. Potassium serves as the cation for this material, and the chemical composition can be expressed as $K[(AlO_2)(SiO_2)_3]$.

The silicate glasses also contain three-dimensional networks of silicon-oxygen linkages. Actually, the term *glass* refers to a state of matter, but as generally used means ordinary window glass, which is made by fusing together sand, limestone, sodium carbonate, and potassium carbonate in a furnace lined with alumina (Al_2O_3) blocks. During the fusion, some of the silica is converted to sodium silicate, the carbonates are converted to oxides, and the liberated carbon dioxide serves to stir the molten mass. Once the operation of the furnace is begun, it is maintained by continuously feeding in the raw materials at one end and removing the molten glass at the other. This continues until the alumina blocks, which are slightly soluble in molten SiO_2, have been completely eroded.

Window glass contains about 80 percent SiO_2, 8 percent CaO, 10 percent Na_2O, and 2 percent K_2O. The sodium and potassium oxides extend the liquid range of the glass and make it easier to "work." When glass with a low coefficient of expansion—for example, laboratory glassware—is required, boric oxide, B_2O_3, is added.

Several other elements combine with oxygen to form oxyanions that are structurally related to the silicates. The orthophosphate ion is $PO_4{}^{3-}$, while sodium tripolyphosphate ($Na_5P_3O_{10}$) contains the anion chain

$$\begin{array}{ccccc} & O & & O & & O \\ & \| & & \| & & \| \\ {}^-O{-}P{-}O{-}&P{-}O{-}&P{-}O^- \\ & | & & | & & | \\ & O{-} & & O{-} & & O{-} \end{array}$$

Metaphosphates can contain either rings such as

or infinite chains.

When alkyl or aryl groups are added to the oxygens of the phosphates, the result is a phosphate ester. Several of these phosphate esters are extremely important in biological systems. For example, the nucleic acid DNA contains phosphate

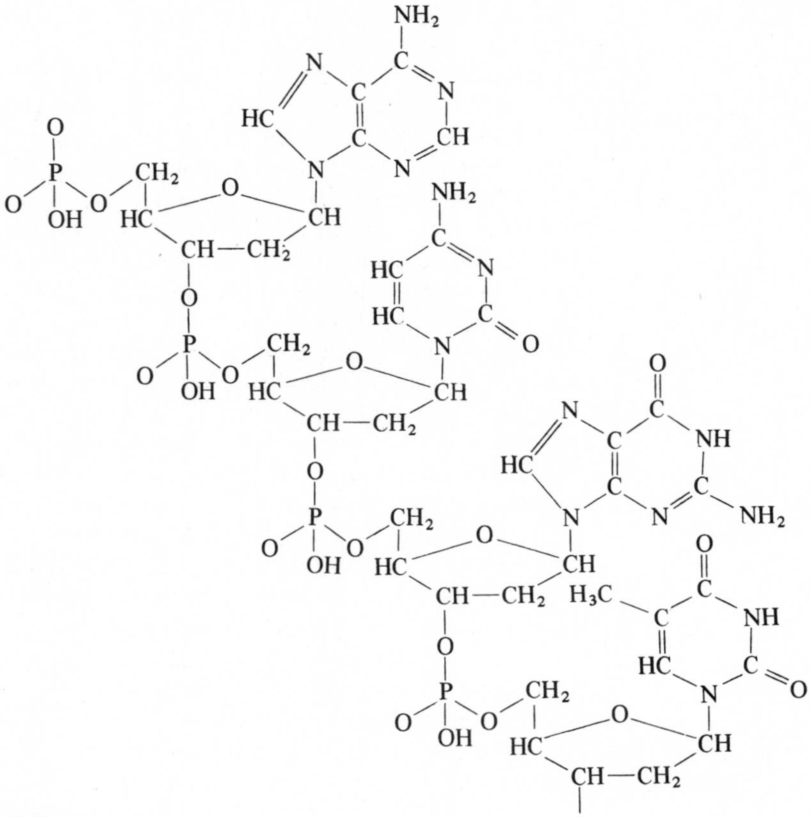

FIGURE 21-3

Portion of the DNA Chain

groups linked together by deoxyribose sugar groups, whereas the phosphate groups in RNA are linked by ribose sugar groups. As illustrated by Figure 21-3, which shows a portion of the DNA chain, each sugar group is also linked to a cyclic nitrogen base.

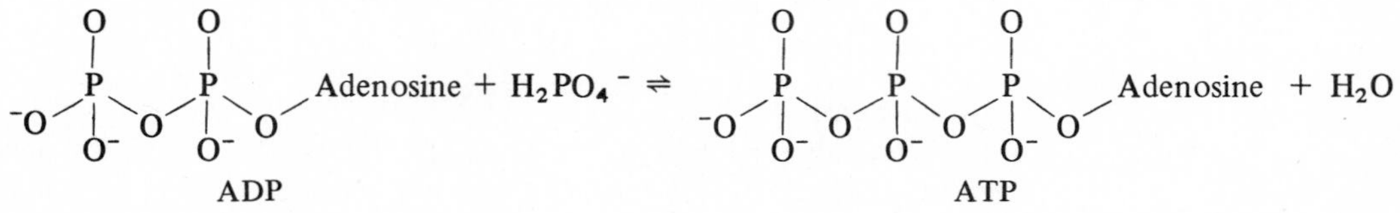

Phosphate esters are also important in storing and using energy from metabolic processes. The compounds adenosine triphosphate (ATP) and adenosine diphosphate (ADP) work together in a cycle for this purpose. ATP can donate a phosphate group and in the process gives off about 10 kcal of energy per mole. This energy can be used to drive reactions such as the activation of glucose prior to oxidation to CO_2 and H_2O. On the other hand, when energy is being generated, as in the degradation of glucose, this energy can be stored by converting ADP and phosphate to ATP.

Silicones   When two of the oxygen atoms attached to each silicon of an infinite-chain (pyroxene-type) silicate are replaced by methyl groups, the resulting electrically neutral compound is a silicone polymer. When the silicone polymer has a fairly low molecular weight, it is an oil and has a good many advantages over the usual hydrocarbon oils: It is very stable to heat and suffers only a small change in viscosity with temperature; it does not become as viscous as hydrocarbon oils at low temperatures, nor does its viscosity decrease as much at high temperatures.

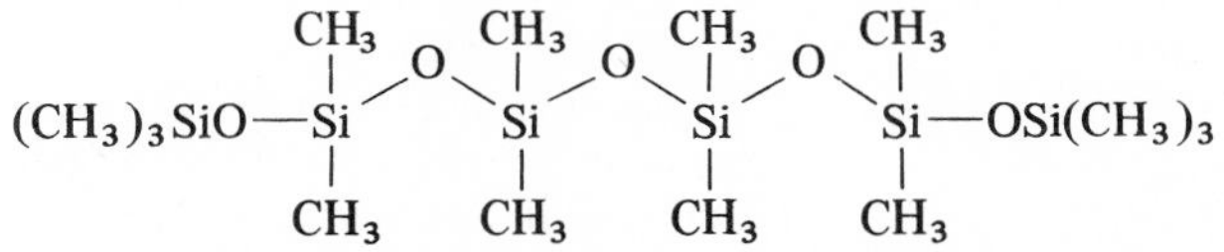

One of the constituents of a silicone oil

When the polymer has a high molecular weight and therefore many

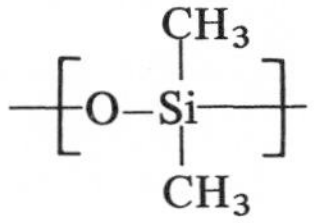

units, it is a rubber. It has a number of advantages over natural rubber: It is virtually unaffected by high and low temperatures and is chemically quite inert. These properties make it very useful in the aircraft and space industry as a sealing material for windows, engine parts, and so forth.

SUGGESTED READINGS

Sanderson, R. T. "Principles of Halogen Chemistry." *Journal of Chemical Education,* Vol. 41, No. 7 (July 1964), pp. 361–366.

Ward, R. "Would Mendeleev Have Predicted the Existence of XeF_4?" *Journal of Chemical Education,* Vol. 40, No. 5 (May 1963), pp. 277–279.

Hall, H. T. "The Synthesis of Diamond." *Journal of Chemical Education,* Vol. 38, No. 10 (October 1961), pp. 484–489.

Rochow, E. G. *The Metalloids.* Lexington, Mass.: D. C. Heath, 1966.

PROBLEMS

1. How, in general, do the *p*-filler elements differ from the *s*-filler elements within the same period in terms of:
   (a) elemental forms
   (b) ionization energies

(c) electron affinities

(d) size of atoms

(e) electrode potentials

(f) common oxidation states

(g) bonding in compounds formed with chlorine

(h) volatility of compounds formed with chlorine

(i) reaction with water

(j) reaction of their chlorine compounds with water

Support your answers with specific examples.

2. How do the elements of Group IV differ from those elements of Group VII within the same period in terms of:

(a) elemental forms

(b) boiling point of their elemental forms

(c) ionization energies

(d) size of atoms

(e) reaction with water

(f) strength of their bonds to hydrogen

(g) reaction of their hydrogen compounds with water

(h) bonding in their compounds with oxygen

Support your answers with specific examples.

3. Account for the following observations:

(a) The strength of the oxyacids of chlorine vary in the order $HClO_4 >$ $HClO_3 > HClO_2 > HClO$.

(b) The halogens decrease in oxidizing strength in the order $F_2 > Cl_2 >$ $Br_2 > I_2$.

(c) Fluorine is very useful in preparing the highest oxidation state of an element.

(d) Hydrofluoric acid cannot be stored in glass bottles.

(e) Hydrochloric acid is a common laboratory reagent, hydroiodic acid is not.

(f) Hydrofluoric acid is a weak acid, the other hydrohalic acids are strong acids.

(g) The dissociation energy of F_2 is less than that of Cl_2.

4. Draw electron dot structures, state the hybridization at the central atom, and give molecular shapes and approximate bond angles for each of the following:

(a) XeF_4 (b) $HClO_3$ (c) BrF_5

(d) ClF_3 (e) Cl_2O (f) ICl_4^-

(g) I_3^- (h) ClO_2^-

5. The H—F bond is said to be greater than 50 percent ionic, yet H—F is considered to be a molecular species. Explain.

6. The standard heat of formation of $XeF_6(g)$ is reported to be -70.4 kcal/mole. Calculate the average Xe—F bond energy in XeF_6.

7. Ammonium perchlorate is isomorphous with (has the same crystalline structure as) perchloric acid monohydrate. Comment on the significance of this fact.

8. The cyano-group behaves in some compounds like a halogen atom. Indeed, cyanogen gas, $(CN)_2$, can be classified as a pseudo-halogen. For example, it reacts with an aqueous sodium hydroxide solution in the same way that chlorine reacts with base. Write an equation for this reaction.

9. Potassium chlorate decomposes to potassium chloride and oxygen at elevated temperatures. Calculate ΔH^0, ΔS^0, and ΔG^0 at $25°C$. Then assume ΔH^0 and ΔS^0 remain constant and calculate an approximate ΔG^0 at $400°C$. Rationalize the value obtained for ΔS^0.

10. Write equations for the reaction of
    (a) hydrogen with bromine
    (b) phosphorous trichloride with water
    (c) chlorine with water
    (d) chlorine with aqueous sodium bromide
    (e) sulfur with fluorine
    (f) bromine with sodium hydroxide solution

11. Make the following predictions for the Group IV elements, giving a careful rationalization for each prediction.
    (a) Which element is most metallic?
    (b) Which element has the highest heat of atomization?
    (c) Which element has the lowest boiling point?
    (d) Which element forms the strongest bond to hydrogen?
    (e) For which element is the +2 oxidation state most stable relative to the +4 oxidation state?
    (f) Which element catenates to the greatest extent?
    (g) Which element forms the most acidic oxide?
    (h) Which hydroxide, $Sn(OH)_2$ or $Sn(OH)_4$, is the stronger base?
    (i) Which chloride should be the hardest Lewis acid, $SiCl_4$, $SnCl_4$, or $SnCl_2$?

12. Use the valence bond model to describe the bonding in SnF_6^{2-}, $SnBr_2$, CO, CO_2, and $Si(CH_3)_4$.

13. Carefully account for the following observations:
    (a) Carbon dioxide is a gas, but silicon dioxide is a solid.
    (b) Graphite has a large electrical conductance parallel to the two-dimensional sheets and almost no conductance perpendicular to the sheets.

14. Write equations for the following reactions:
    (a) $Pb + Cl_2 \rightarrow$ (b) $Si + Cl_2 \rightarrow$
    (c) $SiCl_4 + H_2O \rightarrow$ (d) $Mg + SiO_2 \rightarrow$
    (e) $SnO + O_2 \rightarrow$ (f) $CO_2 + H_2O \rightarrow$
    (g) $CO_2 + OH^- \rightarrow$ (h) $SiO_2 + OH^- \rightarrow$
    (i) $SnO_2 + H^+ \rightarrow$ (j) $SnO_2 + OH^- \rightarrow$

15. Referring to Table 21-13, on page 590, answer the following questions.
    (a) Provide a concise explanation for
        1) the decrease in boiling points of the elements from aluminum to thallium.
        2) the trend in the value of $\Delta H_{hydration}$ of the +3 ions from gallium to thallium.

	B	Al	Ga	In	Tl
TABLE 21-13 **Some Properties of the Elements of Group III**					
Ionization energy (kcal/mole)					
1st	191	138	138	133	141
2nd	580	434	473	435	471
3rd	874	656	708	646	687
Covalent radius (Å)	0.81	1.25	1.25	1.50	1.55
Melting point (°K)	2300	932	312	429	577
Boiling point (°K)	4000	2700	2500	2300	1740
ΔH_f^0(atom) (kcal/mole)	133	78	65	58	43
Ionic radius, M^{3+} (Å)	—	0.53	0.62	0.80	0.88
$\Delta H_{hydration}$, M^{3+} (kcal/mole)	—	−1121	−1124	−994	−984
E^0 ($M^{3+} \rightarrow M$) (V)	—	−1.66	−0.52	−0.34	+0.72

    3) the fact that the standard electrode potential of thallium is higher than that of indium.

(b) Assuming they both have the same crystal structure, which salt, AlF_3 or GaF_3, should be more soluble in water? Why?

(c) Calculate the standard enthalpy change for the reaction

$$Ga(s) \rightarrow Ga^{3+}(aq) + 3e^-$$

(d) Would gallium be likely to react with a $1M$ HCl solution according to the equation below? Why?

$$Ga(s) + 3H^+(aq) \rightarrow Ga^{3+}(aq) + \tfrac{3}{2}H_2(g)$$

16. The boiling points and standard heats of formation of the boron halides are given below.

	bp (°C)	ΔH_f^0 (kcal/mole)
BF_3	−101	−270
BCl_3	12	−96
BBr_3	91	−45
BI_3	210	—

(a) Provide a concise explanation for the trend in boiling points.

(b) Provide a concise explanation for the trend in ΔH_f^0.

17. Sulfur tetrafluoride reacts very readily with water according to the equation

$$SF_4(g) + 2H_2O(l) \rightleftharpoons SO_2(g) + 4HF(g)$$

Sulfur hexafluoride, on the other hand, is very inert to water and most chemicals. Calculate ΔH^0 and ΔG^0 for the reaction

$$SF_6(g) + 3H_2O(l) \rightleftharpoons SO_3(g) + 6HF(g)$$

and explain why SF_6 is unreactive toward water.

18. Use the valence bond model to describe the bonding in PO_4^{3-}, NO, N_2O_4, SF_4, SF_6, SO_3, BF_3, and BF_4^-.

19. The ΔH_f^0 for liquid hydrazine (H_2NNH_2) is 22.8 kcal/mole. Rationalize this endothermic heat of formation. Would you expect the standard free energy of formation to be favorable?

20. A compound of nitrogen and hydrogen is 12.6 wt-% hydrogen. Its density is the same as oxygen gas measured under the same conditions. Calculate the molecular formula of the compound and draw a valence bond structure.

21. Write equations for the following reactions:

(a) $BCl_3 + H_2O \rightarrow$

(b) $Al(OH)_3 + OH^- \rightarrow$

(c) $B(OH)_3 + OH^- \rightarrow$

(d) $In_2O_3 + H^+ \rightarrow$

(e) $N_2 + H_2 \rightarrow$

(f) $MgS + H^+ \rightarrow$

(g) $Mg_3B_2 + H^+ \rightarrow$

(h) $Mg_2Si + H^+ \rightarrow$

22 *Transition Elements: The d-Fillers*

The transition elements—so named because they separate the more electropositive elements, the *s*-fillers, from the more electronegative elements, the *p*-fillers—have, with few exceptions, an $(n-1)s^2(n-1)p^6(n-1)d^{1\ to\ 9}ns^2$ outer electron distribution. The transition elements are all metals and range in abundance from plentiful iron and titanium to the very rare rhenium. Since we will discuss only some of the important aspects of their chemical behavior, especially how they differ from the representative elements, we will restrict ourselves for the most part to the elements of the first transition period: scandium through zinc. Strictly speaking, copper and zinc, whose outer electron arrangements are $3d^{10}\,4s^1$ and $3d^{10}\,4s^2$, respectively, are not transition metals. However, copper(II) with a d^9 electron configuration is certainly a transition metal ion, and zinc(II) is in many ways chemically similar to the *d*-fillers.

Under the rubric of transition metals, we find the more common heavy metals. Of the seven metals known to ancient man (in probable order of discovery: gold, silver, copper, tin, lead, iron, and mercury), five are transition metals. Copper was used by the Sumerians as early as 3000 B.C. for making tools. The combination of copper with tin produced bronze, the first man-made alloy, which was culturally so important to the development of ancient man that archeologists have designated a period of prehistory as the Bronze Age. This was superseded by the Iron Age, and iron remains to this day our most commonly used metal. Molybdenum, chromium, and manganese are all important constituents of different kinds of stainless steels. Titanium is useful in the production of jet engines and in space technology, where lightness and strength are both important. Platinum, which was considered worthless when it was first discovered in placer deposits (ore-bearing sand or gravel deposited by water) in Colombia, is useful for its inertness. It is an important catalyst in the cracking of petroleum and in the control of exhaust emissions in the internal combustion engine.

The transition elements also include gold, silver, and copper, generally known as the coinage metals, although they have become too valuable to be used extensively in modern coins. Silver is now used primarily in photography and in the manufacture of jewelry. Copper is more important as a conductor of electricity and is also used in water pipes and other articles where its chemical inertness is important. Gold, the most glamorous metal of all, is chemically very inert and can also be beaten into very thin sheets and drawn into microscopically fine threads. Along with silver and copper, it is an excellent conductor of electricity and thus finds use as electrical contacts in miniature circuits. However, while it is more useful for its chemical properties, gold remains in most peoples' minds the only metal with true inherent value.

THE ELEMENTS

The transition metals show an irregular decrease in size with increasing nuclear charge within a period. Although there is a size increase down the families with increasing atomic number, the incremental increase between the second and third transition periods is very small. Indeed, hafnium, even though it contains 32 more electrons, is actually slightly smaller than zirconium, and niobium and tantalum are about the same size. This small difference in atomic size is a direct result of the filling of the inner $4f$ level of the lanthanide elements. The f electrons are poor shielders of the nuclear charge, consequently there is a significant decrease in atomic radii from lanthanum to lutetium. Because zirconium and hafnium have identical outer electron arrangements and are of similar size, they have very similar chemical properties. (The same is true of niobium and tantalum.) In fact, the chemical separation of zirconium and hafnium (hafnium is generally found with zirconium minerals) posed one of the great challenges of the past, and even today this separation is an involved process.

In general, the transition elements are harder and have higher melting and boiling points and higher heats of vaporization than the representative metals. The greater hardness and higher heats of vaporization probably mean that d electrons in addition to the outer s electrons are involved in the metallic bonding. Since the nuclear charge is more poorly shielded by d electrons than by s or p electrons, ionization energies of the transition elements are also much higher than those of the alkali and alkaline earth metals. As a consequence, the oxidation potentials are much lower than those of the s fillers, and thus the reactivity of the elements is relatively low.

COMPOUNDS

Since the number of electrons that need to be lost or gained to form simple inert gas-type ions is very large, much of the chemistry of the first transition series involves the loss of the $4s$ electrons to give the +2 ion. The scandium family's chemistry is exclusively that of the +3 oxidation state, and therefore these ions have no d electrons present. Since their chemical behavior more nearly conforms to that of the lanthanide elements, the scandium family will not be discussed here.

Although most of the first-row transition elements exist in the +2 oxidation state, it is not necessarily the most stable state. Indeed, many of the elements exist in a variety of oxidation states. For example, manganese exists in all of the oxidation states from +2 through +7, as illustrated by the compounds:

$$\overset{+2}{Mn}O \quad \overset{+3}{Mn_2O_3} \quad \overset{+4}{Mn}O_2 \quad Na_3\overset{+5}{Mn}O_4 \quad K_2\overset{+6}{Mn}O_4 \quad K\overset{+7}{Mn}O_4$$

Note that the oxidation state of manganese (and other transition elements) changes in units of one, while the common oxidation states of the p-fillers change in units of two. Figure 22-1 lists the highest common oxidation state (above the element) and the most stable oxidation state in aqueous solution (below the element). As indicated in the figure, the +2 aqueous ion becomes increasingly more common with increasing atomic number, and the maximum oxidation state of +7 is reached with manganese.

3	4	5	6	7	3	3	2	2	2
Sc	Ti	V	Cr	Mn	Fe	Co	Ni	Cu	Zn
3	4	4,5	3	2	3	2	2	2	2

FIGURE 22-1

Oxidation States

Within a given transition family, the higher oxidation states become more common with the heavier elements. For example, the chromate ion, CrO_4^{2-}, is a strong oxidizing agent and is therefore easily converted to Cr^{3+}, while the analogous anion WO_4^{2-} is the most stable species for the bottom member of the chromium family. Similarly, fluorine, which tends to bring out the highest oxidation state of a metal, forms FeF_3 with iron but produces OsF_6 with the heaviest member of the iron family.

If any single word could be used to characterize the compounds of the transition elements, it might well be "color." Almost all their compounds are colored, and the colors of dilute aqueous solutions of their simple cations often provide an easy means of identification. For example, in aqueous solution, copper(II) is blue, nickel(II) is green, cobalt(II) is pink, chromium(III) is violet, manganese(II) is faint pink, iron(III) is yellow-brown, and zinc(II) is colorless.

Colors of dilute solutions of transition metal cations are sensitive to changes in the chemical environment of the cation. For example, the pale blue of a solution of copper(II) nitrate changes to green and then yellow-green upon the addition of concentrated hydrochloric acid. This change in color results from the formation of new solution species as some of the water molecules directly surrounding the copper ion are displaced by chloride ions. In like fashion, the rose-pink of a dilute solution of cobalt(II) nitrate changes to deep blue upon the addition of excess thiocyanate ion, and the yellow-brown aqueous solution of the iron(III) ion becomes colorless upon the addition of fluoride ion. This change in color with a change in the environment is characteristic of elements that have partially filled d orbitals and shows that d orbitals are in some fashion involved in the bonding. Thus, both scandium(III), an inert-gas-type ion with no d electrons, and the pseudo inert-gas-type ions zinc(II) and copper(I), which have filled d orbitals, form colorless cations in aqueous solution and when combined with most simple colorless anions.

The presence of d electrons also results in a greater polarizability for the transition metal ions. The much stronger hydration of these ions relative to the s-filler cations is apparently due to their polarizability combined with their greater charge-to-radius ratio. The increased strength of hydration results in considerable hydrolysis of the cations in aqueous solution. For example, the hexahydrate of iron(III) is a strong acid and is almost completely dissociated to the red-brown $[Fe(OH)(H_2O)_5]^{2+}$:

$$[Fe(H_2O)_6]^{3+} + H_2O \rightleftharpoons [Fe(OH)(H_2O)_5]^{2+} + H_3O^+$$

Coordination Compounds

The fact that the color of transition metal cations depends on their chemical environment can be directly related to another outstanding characteristic of these elements: their ability to form coordination compounds. As early as 1798, it was discovered that cobalt salts combined with ammonia to form highly colored "complex compounds." The complexity of these cobalt compounds, along with that of the solid hydrates of transition metal compounds and other "complex compounds," posed a major problem for chemists in the nineteenth century; their structure and bonding were not understood until near the end of the century.

The second half of the nineteenth century saw an expansion in the chemistry of carbon that can best be described as explosive. The elucidation of the tetrahedral bonding of the carbon atom, the bonding of carbon to itself, the simple valence rules for carbon, geometric and optical isomerism, and the structure of the benzene ring all laid the foundation for this rapid expansion in molecular chemistry.

The more complex chemistry of the transition metals languished during this period. The simple valence rules of carbon could not explain the apparent anomaly of the reaction of a "valence-satisfied" compound like $PtCl_2$ with a "valence-satisfied" molecule like NH_3 to give the compound $PtCl_2 \cdot 2NH_3$ or, depending upon the method of preparation, $PtCl_2 \cdot 4NH_3$. It was not until the work of Alfred Werner, a Swiss chemist (1866–1919), that problems of this kind were clarified.

WERNER'S THEORY

The essence of Werner's theory is simply that transition metals exhibit two different kinds of valence.

1. The *Hauptvalenz* or *principal valence* depends upon the charge on the cation and is satisfied by the equal but opposite charge of the anions. For example,

$$Cu^{2+}, 2Cl^- \quad \text{or} \quad Fe^{3+}, 3Cl^-$$

where the double positive charge of the copper cation and the triple positive charge of the iron cation are satisfied, respectively, by the two and three negatively charged chloride ions to give neutral species.

2. The *Nebenvalenz* or *secondary valence* is nonionizable and therefore a directed covalent valence. It is satisfied by either neutral molecules or negative ions. Each metal has a characteristic secondary valence, which is independent of the primary valence. For example, that of cobalt(III) is always 6. This is now called the *coordination number* of a metal, and it has been found to vary with the size and charge of the cation.

Figure 22-2 is a plot of conductivity data for a series of platinum(IV) complexes that Werner used in an early paper in support of his model.

The conductivity of complex E, $[Pt(NH_3)_2Cl_4]$, is essentially zero, and that

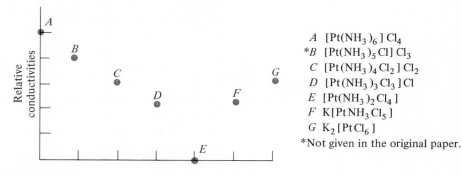

A $[Pt(NH_3)_6]Cl_4$
*B $[Pt(NH_3)_5Cl]Cl_3$
C $[Pt(NH_3)_4Cl_2]Cl_2$
D $[Pt(NH_3)_3Cl_3]Cl$
E $[Pt(NH_3)_2Cl_4]$
F $K[PtNH_3Cl_5]$
G $K_2[PtCl_6]$
*Not given in the original paper.

FIGURE 22-2

Graphical Representation of Conductivities of $0.1M$ Aqueous Solutions of Some Platinum(IV) Complexes compound is therefore not dissociated in aqueous solution. The principal valence of +4 for platinum is satisfied by the four covalently bonded chloride "ions." The secondary valence, which is the number of nonionizable groups attached to the central metal ion, must in this case be 6 because of the two ammonia molecules and four chloride "ions." Complex *D* dissociates into two ions, the $Pt(NH_3)_3Cl_3^+$ ion and the Cl^- ion. In this compound the principal valence is satisfied by the one chloride ion and the three covalently bonded chloride "ions," while the secondary valence of 6 is satisfied by three ammonia molecules and the three covalently bonded chloride ions.

Complex *G* dissociates to three ions—two potassium ions and one $PtCl_6^{2-}$ ion. The principal valence of platinum in this complex is satisfied by four covalently bonded chloride ions, while the secondary valence is satisfied by the six covalently bonded chloride ions. Thus, this series of platinum(IV) complexes illustrates the fulfillment of the secondary valence with all variations from six coordinated ammonias to six coordinated chloride ions. The number of moles of ions per mole of compound in the aqueous solution varies from five in complex *A* to none in complex *E* to three in complex *G*.

As a further example of the application of Werner's theory, consider the ammonia complexes of cobalt(III) chloride, which are listed in Table 22-1 with

TABLE 22-1
Ammonia Complexes of Cobalt(III)

Compound	Color	Number of Chloride Ions Precipitated
$CoCl_3 \cdot 6NH_3$	orange-yellow	3
$CoCl_3 \cdot 5NH_3 \cdot H_2O$	rose	3
$CoCl_3 \cdot 5NH_3$	purple	2
$CoCl_3 \cdot 4NH_3$	green	1
$CoCl_3 \cdot 4NH_3$	violet	1

their colors and the number of chloride ions precipitated at room temperature by the addition of a silver nitrate solution. Since the precipitation of chloride indicates an ionizable chloride, the cobalt(III) compounds in Table 22-1 can be formulated as follows:

$CoCl_3 \cdot 6NH_3$ is $[Co(NH_3)_6]^{3+}, 3Cl^-$

$CoCl_3 \cdot 5NH_3 \cdot H_2O$ is $[Co(NH_3)_5 H_2O]^{3+}, 3Cl^-$

$CoCl_3 \cdot 5NH_3$ is $[Co(NH_3)_5 Cl]^{2+}, 2Cl^-$

$CoCl_3 \cdot 4NH_3$ is $[Co(NH_3)_4 Cl_2]^+, Cl^-$

In the first complex the principal valence of cobalt is satisfied by the three chloride ions; the secondary valence is satisfied by the six coordinated ammonia molecules. In the last compound, the principal valence is still satisfied by three chlorines, only one of which is "ionic." The secondary valence is satisfied by two coordinated chloride "ions" and four ammonia molecules.

Werner elaborated upon his secondary directed valence by postulating three possible arrangements of the six groups around the central metal cation in cobalt(III) and platinum(IV) complexes: planar hexagonal, trigonal prismatic, and octahedral. These arrangements are shown in Figure 22-3. When all the coordinated groups are the same, as in $Co(NH_3)_6^{3+}$, it is difficult to determine which geometric form is adopted by a particular compound. When there are two different coordinated groups, as in $Co(NH_3)_4Cl_2^+$, geometrical isomers are possible for each form. For example, if the cation $Co(NH_3)_4Cl_2^+$ exists in the planar hexagonal form, the two chlorines could be situated relative to one another in three different ways: adjacent, one corner removed, or opposite. Or, in terms of Figure 22-3, if one chlorine is at position 1, the other could be at position 2, 3, or 4 (the ammonia molecules occupy the other positions, of course). These three geometric isomers are chemically and physically distinct, and if the cation adopts the planar hexagonal form, three isomers should be isolable.

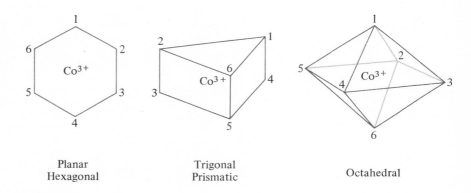

Planar Hexagonal   Trigonal Prismatic   Octahedral

FIGURE 22-3

Possible Distribution of Groups for Coordination Number 6

There are also three possible isomers for the trigonal prismatic form (chlorine occupation of, say, the 1,2, the 1,3, or the 1,4 positions), but only two isomers for the octahedral form (chlorine occupation of, say, the 1,2 or 1,6 positions). The geometrical isomers for the octahedral form are distinguished in Figure 22-4 as *cis* (chlorines adjacent) and *trans* (chlorines opposite). It is important to realize that all corners of the octahedron are identical, and therefore the *trans* isomer, for example, can be generated by 1,6, 2,4, or 3,5 substitution in the octahedron shown in Figure 22-3.

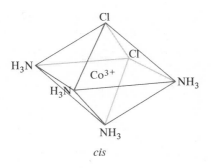
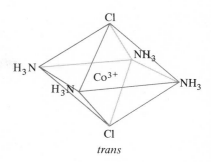

cis   *trans*

FIGURE 22-4

Geometric Isomers of $[Co(NH_3)_4Cl_2]^+$

Since Werner was able to isolate only two isomers for $[Co(NH_3)_4Cl_2]Cl$, he argued for the octahedral symmetry. (It must be noted that the inability of a chemist to isolate a third isomer is not proof of its nonexistence.) Then, through a series of cleverly conceived substitution reactions, Werner and his students concluded that the green form (see Table 22-1) was the *trans* isomer and the purple form the *cis* isomer.

Cis and trans isomers were also isolated for the analogous compound with ethylenediamine (abbreviated en) in place of ammonia molecules, $[Co(en)_2Cl_2]Cl$ (Figure 22-5). Werner then realized that the *cis* isomer should have a nonsuperimposable mirror image and would therefore be optically active. The resolution of the purple form into two optical isomers lent credence to the octahedral arrangement. These assignments were later verified by more sophisticated physical techniques.

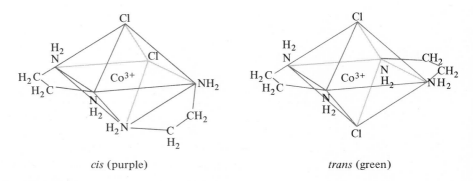

cis (purple)   *trans* (green)

FIGURE 22-5

Geometric Isomers of $[Co(en)_2Cl_2]^+$

The determination of the geometric form adopted by the four-coordinate Pt(II) compounds was somewhat simpler. The two possible geometric forms—

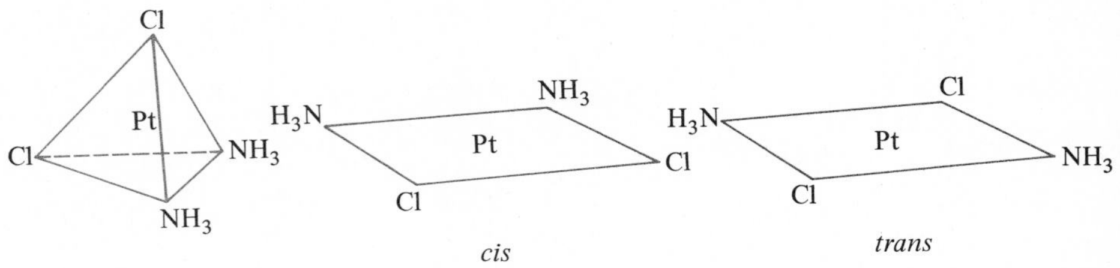

FIGURE 22-6

Tetrahedral and Square Coplanar Geometric Isomers for $Pt(NH_3)_2Cl_2$

tetrahedral and square coplanar—are shown in Figure 22-6. For the compound $Pt(NH_3)_2Cl_2$, only one isomer is possible for the tetrahedral form, but two isomers are possible for the square coplanar form. The isolation of two isomers was evidence for the latter form.

While the octahedral form is the only common arrangement for 6-coordination, both the tetrahedral and square coplanar forms are common for 4-coordination.

Although Werner's concepts appear obvious to us today, his ideas were not readily accepted by his contemporaries, and he spent some thirty years working to perfect and support his theory. He did gain recognition within his lifetime, however, and received the Nobel Prize in chemistry in 1913. It was not until the Lewis concept of the coordinate covalent bond and the extensive work of Pauling and others with the valence bond model that the genius of Werner could be fully recognized. He had in a sense anticipated these more sophisticated treatments of the chemical bond, and his concepts, with the minor modifications required by our ever-changing ideas, are still valid.

NOMENCLATURE

Before discussing more contemporary bonding models for coordination compounds, it will be helpful to learn the basic rules of nomenclature. Let us begin with a few conventions and definitions. The coordination entity is enclosed in brackets; for example, $[Co(NH_3)_6]^{2+}$, $[Fe(CN)_6]^{4-}$, and $[CoCl_4]^{2-}$. The atom, ion, or molecule attached to the central metal is called a *ligand*. Ligands that can be attached in only one position are called *monodentate* (from the Latin *dentatus*, meaning "tooth") ligands; those that contain more than one possible point of attachment are called *multidentate* ligands. The word *chelate* (from the Greek for crab's claw) is also used to represent a ligand with more than one seat of attachment to the central metal. Some examples of multidentate ligands are illustrated here: two bidentate ligands, ethylenediamine and the glycinate ion; and the tridentate anion of *o,o'*-dihydroxyazobenzene.

ethylenediamine

glycinate ion

anion of
o,o'-dihydroxyazobenzene

The rules for naming coordination compounds are:

1. The cation is named first, followed by the anion, in agreement with the rules of nomenclature for salts.
2. Negative ligands have the suffix -o added to the anion stem: Cl^-, chloro; SO_4^{2-}, sulfato; CO_3^{2-}, carbonato; NH_2^-, amido; $C_2O_4^{2-}$, oxalato. Neutral ligands simply use the name of the compound itself: $P(CH_3)_3$, trimethylphosphine; $H_3NCH_2CH_2NH_2$, ethylenediamine; $O(CH_3)_2$, dimethyl ether. However, coordinated water is named *aquo;* ammonia, *ammine;* and CO, *carbonyl.* Positive ligands are given an -ium ending: NO^+, nitronium. The usual prefixes of di-, tri-, tetra-, and so on, are used for more than one ligand of the same kind. For more complicated ligands where some confusion may be caused by the simple prefixes, bis-, tris-, and tetrakis- (for 2, 3, and 4) are used instead.
3. The complex *cation* is named as follows: The ligands are named in the order negative, neutral, positive, followed by the name of the central metal with its oxidation state in parentheses. Roman numerals are used (Arabic 0 for zero) to represent the oxidation state. For example,

$[Cu(NH_3)_4]^{2+}$	tetraamminecopper(II) ion
$[Ni(H_2O)_6]^{2+}$	hexaaquonickel(II) ion
$[Cr(H_2O)_5Cl]^{2+}$	chloropentaaquochromium(III) ion
$[Co(en)_2Br_2]^+$	dibromobis(ethylenediamine)cobalt(III) ion

4. A *neutral* complex is given the same name as the complex cation. For example,

$[Pt(NH_3)_2Cl_4]$	tetrachlorodiammineplatinum(IV)
$[Pt(NH_3)_2ClBr]$	bromochlorodiammineplatinum(II)

Note that negative groups are named in alphabetical order. This convention is commonly followed for both negative and neutral groups.

5. A complex anion is named in the same way as the complex cation, but with an -ate ending attached to the metal stem. For example,

$[CoCl_4]^{2-}$	tetrachlorocobaltate(II) ion
$[Ni(CN)_4]^{2-}$	tetracyanonickelate(II) ion
$[Fe(CN)_6]^{3-}$	hexacyanoferrate(III) ion

LEWIS AND VALENCE-BOND BONDING MODELS

Groups that behave as ligands have at least one lone pair of electrons. Therefore, Werner's secondary valence concept can be visualized as the formation of Lewis adducts with each monodentate ligand (the base) supplying an electron pair to the metal ion (the acid). Hence, the Lewis model and its wave-mechanical analog, the valence bond model, were the first theories used to describe the bonding in coordination compounds.

Although the valence bond model is not nearly so useful today in describing the chemical behavior of these compounds, coordination chemistry owes a huge debt to its early applications. The bonding interactions derived from this theory are readily visualized, and some knowledge of the valence bond model is useful in understanding other more sophisticated bonding models.

In the valence bond theory, the metal cation furnishes a set of empty hybridized orbitals, while the ligand furnishes the electron pair for bond formation. For the first transition series, the d orbitals in the third principal quantum level ($n = 3$) and the s, p, and d orbitals in the fourth quantum level can all be used to form hybridized orbitals. When the $3d$ orbitals are used, the complex is called an *inner orbital complex,* and when $4d$ orbitals are used, the complex is called an *outer orbital complex.* Table 22-2 lists the hybridization used by the metal cation and the spatial arrangement of bonds in the coordination sphere for 2-, 4-, and 6-coordination (even coordination numbers are much more common than odd ones).

TABLE 22-2
Hybridization and Shape for Common Even Coordination Numbers

Coordination Number	Hybridization	Shape
2	sp	linear
4	sp^3	tetrahedral
4	dsp^2	square planar
6	d^2sp^3	octahedral

The application of the valence bond model to complex ions can be illustrated with the following ions:

In the diamminesilver(I) ion, $[Ag(NH_3)_2]^+$, the hybridization at the central ion is sp, and these orbitals overlap with the sp^3 hybrid orbital on each ammonia. The resulting geometry is linear.

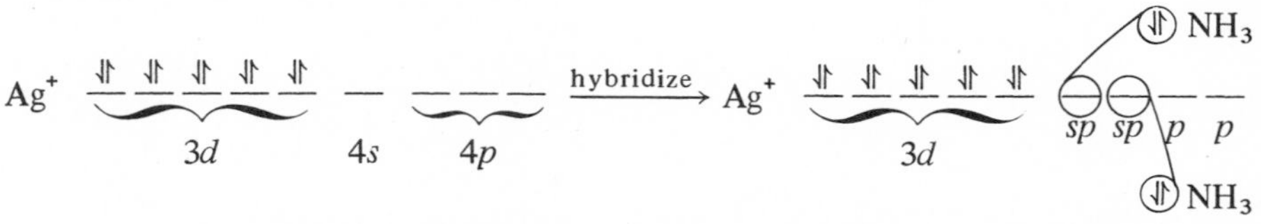

For the tetraamminezinc(II) ion, $[Zn(NH_3)_4]^{2+}$, the hybridization is sp^3, and the resulting geometry is tetrahedral. (Vertical arrows represent cation electrons, and ○ represents shared electrons from the ligand.)

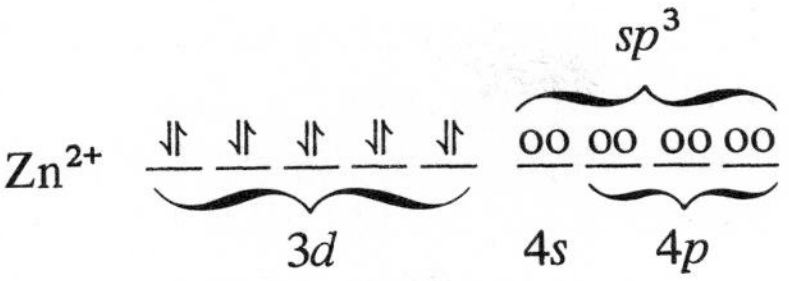

$$sp^3$$

Zn²⁺   ⥮ ⥮ ⥮ ⥮ ⥮   oo oo oo oo

      3d   4s   4p

For the hexaamminenickel(II) ion, $[Ni(NH_3)_6]^{2+}$, six empty orbitals are required and the outer d orbitals are used. The hybridization is sp^3d^2, and the resulting geometry is octahedral.

$$sp^3d^2$$

Ni²⁺   ⥮ ⥮ ⥮ ↑ ↑   oo oo oo oo oo oo   __ __ __ __

      3d   4s   4p   4d

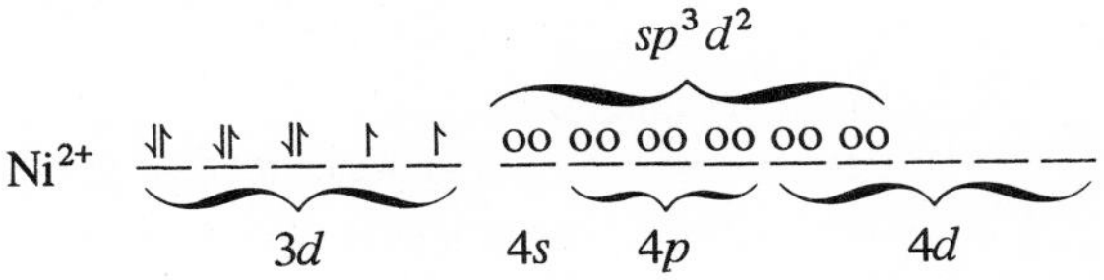

In the hexacyanochromate(III) ion, $[Cr(CN)_6]^{3-}$, two inner d orbitals are available. The hybridization is d^2sp^3, and the geometry is octahedral.

$$d^2sp^3$$

Cr³⁺   ↑ ↑ ↑ oo oo oo oo oo oo

      3d   4s   4p

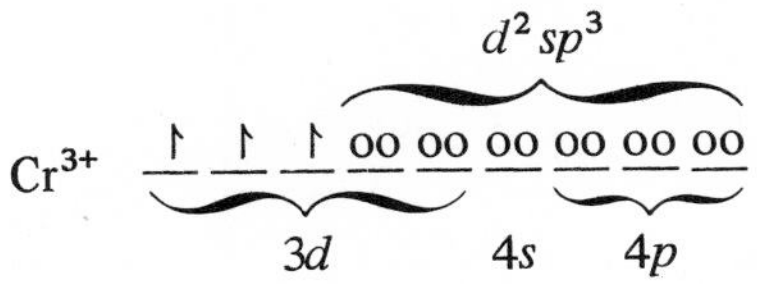

For the cobalt(III) ion, the electron distribution is

Co³⁺   ⥮ ↑ ↑ ↑ ↑   __   __ __ __

      3d   4s   4p

The measured paramagnetism (see p. 608) for the hexafluorocobaltate(III) ion, $[CoF_6]^{3-}$, corresponds to four unpaired electrons. Therefore, in this species cobalt(III) utilizes its $4d$ orbitals to form sp^3d^2 hybrids.

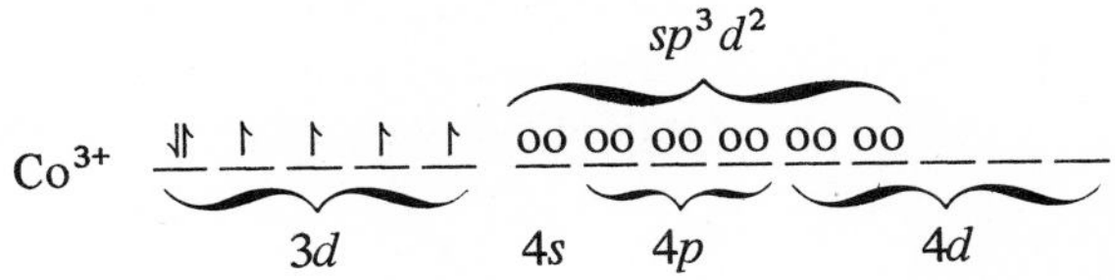

$$sp^3d^2$$

Co³⁺   ⥮ ↑ ↑ ↑ ↑   oo oo oo oo oo oo   __ __ __ __

      3d   4s   4p   4d

The hexaamminecobalt(III) complex, on the other hand, is diamagnetic. Apparently, in this complex the metal d electrons are paired, and the inner $3d$ orbitals are hybridized with the $4s$ and $4p$ orbitals.

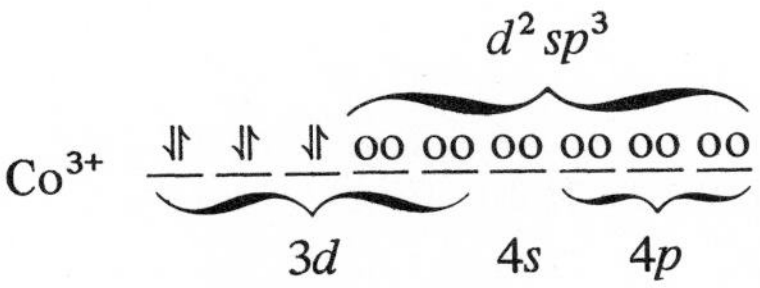

$$d^2sp^3$$

Co³⁺   ⥮ ⥮ ⥮ oo oo oo oo oo oo

      3d   4s   4p

Of course, the pairing of electrons requires energy (recall that electrons of the same spin repel one another less than those of opposite spin), but this energy is more than compensated for by the energy released from the formation of bonds to the lower energy $3d$ orbitals rather than $4d$ orbitals.

For coordination compounds the valence bond model has been almost completely replaced in recent years by the crystal field theory. One of the principal reasons for this was that the visible spectra of coordination compounds could not be rationalized with the valence bond model. Nor could it explain certain thermodynamic effects such as the trends in hydration energies for transition metal cations or the trends in formation constants for coordination compounds. In addition, although valence bond theory appeared to adequately explain the magnetic properties of many compounds, it did not provide a basis for a more quantitative treatment of paramagnetic compounds.

THE CRYSTAL FIELD MODEL

FIGURE 22-7

Octahedral Environment of Cation

The crystal field model is based on the influence of the electrical environment of the central metal ion on the energies of its d orbitals. Consider first the d orbitals of, for example, the Fe^{3+} ion, which has an electron configuration of $1s^2 2s^2 2p^6 3s^2 3p^6 3d^5$. If this ion were completely isolated from all electrical influences, the d orbitals would all have the same energy.

Consider now an octahedral complex of the type $MX_6{}^{n-}$, for example, $FeF_6{}^{3-}$. Even though the bonds between M and X are at least partially covalent, the crystal field model assumes that both M and X are present as ions. Thus, this complex could be represented as in Figure 22-7, where the apices of the octahedron are centered on the x, y, z axes.

The electrostatic field of the six anions will affect the electronic energy of the cation (the energy of its atomic orbitals) in two ways: First, the electrons of the cation will certainly be repelled by the six negatively charged anions. This electrostatic repulsion results in an increase in energy of all the cation electrons relative to the energy of the isolated cation. This is shown in Figure 22-8 for the d electrons

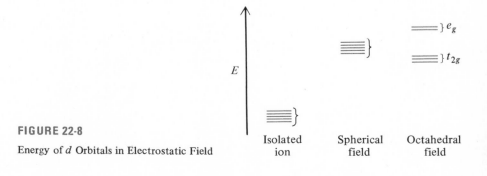

FIGURE 22-8

Energy of d Orbitals in Electrostatic Field

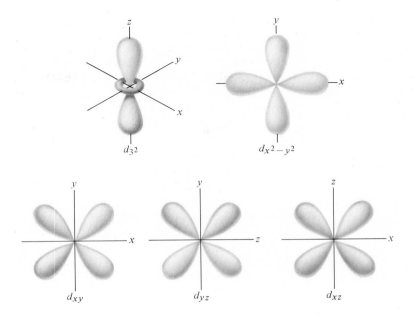

FIGURE 22-9

Orientation of d Orbitals

only. Second, a careful analysis of the orientations of the d orbitals relative to the anions reveals that the energies of the five d orbitals are affected differently. That is, if the cation were located in a spherically symmetrical electrostatic environment (a total of six negative charges smeared evenly over the surface of a sphere surrounding the cation), the energies of the d orbitals would be higher than the energies of the isolated ion, but the energies would also be degenerate (the same). In the octahedral field, however, the energies of the d orbitals will be different.

In order to understand the origin of this second phenomenon, we must review the orientation of the five d orbitals. Figure 22-9 reveals that two of the d orbitals, the d_{z^2} and $d_{x^2-y^2}$, have their maximum density areas directed along the x, y, and z axes. Now, the anions are also located on these axes (Figure 22-7), and thus the electrostatic repulsions between anions and these orbitals will be quite strong. The other three orbitals have their maximum density areas directed *between* the x, y, and z axes, and are therefore further from the anions than the d_{z^2} and $d_{x^2-y^2}$ orbitals and consequently less strongly repelled.

Keeping in mind that the greater the repulsion the higher the energy, it should now be clear that two of the d orbitals will have higher energies than the other three. Although it is not obvious from Figure 22-9, the d_{z^2} and $d_{x^2-y^2}$ orbitals have the same energy, and the d_{xy}, d_{xz}, and d_{yz} are also a degenerate set of orbitals. The energies of the two sets of d orbitals are shown in Figure 22-8. The higher energy set is usually referred to as the e_g set, while the lower energy orbitals are termed the t_{2g} orbitals.

The difference in energy between the two sets of orbitals is commonly termed Δ. Since the total energy of the two e_g orbitals and the three t_{2g} orbitals must be the same as the energy of the five orbitals in a spherical field, the e_g orbitals are destabilized by 0.6Δ and the t_{2g} are stabilized by 0.4Δ. Therefore, each electron in the e_g level is 0.6Δ higher in energy and each electron in the t_{2g} level decreases in energy by 0.4Δ. The total gain in stability of the d electrons over that of the d electrons in a spherical field is known as the *crystal field stabilization energy* (CFSE).

We have thus far observed the effect of the octahedral electrostatic environment on the energies of the d orbitals. Let us now turn our attention to the distribution of electrons in these orbitals.

The distribution of electrons in a set of degenerate orbitals is controlled by two wave-mechanical principles: the Pauli exclusion principle, which permits no more than two electrons in any orbital, and Hund's first rule, which allows only single occupancy of each orbital until the degenerate set of orbitals has been half-filled. For example, the electron configuration of an isolated Cr^{2+} ion, in which all five d orbitals are degenerate, is

$$\underline{\uparrow}\ \underline{\uparrow}\ \underline{\uparrow}\ \underline{\uparrow}\ \underline{}$$

In CrF_2, where the Cr^{2+} ion is in an octahedral environment, the degeneracy of the five d orbitals is removed and the distribution of electrons is not as clear-cut. The problem becomes more obvious when the theoretical basis for Hund's first rule is understood. The placement of two electrons in the same orbital, according to Hund's first rule, is energetically less favorable than placement in two orbitals of equal energies. The energy required to place the electrons in the same orbital—that is, to pair the electrons—is presumably due to the repulsion of electrons occupying roughly the same volume of space.

Now, for the d electrons of transition metal cations, the pairing energy, E_p, is of the same general magnitude as Δ, the difference in energy of the e_g and t_{2g} orbitals. If Δ is greater than the pairing energy, the electron distribution for Cr^{2+} will be

$$\underline{}\ \underline{}\quad e_g$$

$$\underline{\uparrow\downarrow}\ \underline{\uparrow}\ \underline{\uparrow}\quad t_{2g}$$

because it takes less energy to pair the fourth electron than to place it in the e_g orbitals. If Δ is less than the pairing energy, the total energy of the ion is minimized by placing the electron in the e_g orbital.

$$\underline{\uparrow}\ \underline{}\quad e_g$$

$$\underline{\uparrow}\ \underline{\uparrow}\ \underline{\uparrow}\quad t_{2g}$$

The latter is actually the case for CrF_2.

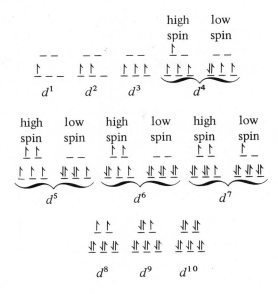

high   low
spin   spin

FIGURE 22-10

Occupancy of *d* Orbitals
in Octahedral Field

Electron distributions for both cases, $\Delta > E_p$ and $\Delta < E_p$, are given in Figure 22-10 for one through ten *d* electrons. As is obvious from this figure, the distributions for 1, 2, 3, 8, 9, and 10 *d* electrons are independent of Δ. For a given *d* electron system, say a d^4 system, the distribution with the greater number of unpaired electrons (when $\Delta < E_p$) is often referred to as the *high-spin* configuration, while the $\Delta > E_p$ situation is called the *low-spin* configuration. High-spin coordination compounds are equivalent to the outer orbital complexes of the valence bond model and involve ligands with low splitting energy (small Δ). Low-spin or inner orbital complexes involve ligands with a high splitting energy (large Δ).

Extensions of the crystal field theory that incorporate covalent bonding situations—a mixture of the molecular orbital model and the crystal field model—are known as *ligand field theory*.

The principal value of crystal field theory is in the prediction or rationalization of the properties of transition metal compounds that depend upon electron configuration and energies. Among these are magnetic properties and color, which will be discussed below.

Magnetic Properties   It is well known that a moving electrical charge has an associated magnetic field. Since an electron in an atom, according to the particle model, moves in an orbit about the nucleus and also spins about its own axis, both motions produce a magnetic field. The magnetic field due to the spin of the electron is usually more important than the orbital contribution. In a substance in which there are no unpaired electrons, equal numbers of electrons spinning in opposite directions produce magnetic fields that cancel. Such a substance is said to be *diamagnetic* and is very slightly repelled by an applied magnetic field. When an atom or ion contains more electrons of one spin, the magnetic fields do not cancel, and the substance is attracted to a magnetic field. Such a substance is termed *paramagnetic;* the extent of its paramagnetism depends upon the number of unpaired electrons. Paramagnetism is expressed in terms of magnetic moment, which is measured in a unit called the *Bohr magneton* (BM). When only the spin contribution to magnetic moment is considered, the moment can be approximated by the equation

$$\mu_{BM} = \sqrt{n(n+2)}$$

where μ_{BM} is the magnetic moment expressed in Bohr magnetons and n is the number of unpaired electrons. The orbital contribution to the magnetic moment varies with different ions and geometries, and for most coordination compounds the measured moment is greater than that calculated by the "spin-only formula." Table 22-3 lists the approximate range of some experimentally measured moments, along with calculated moments.

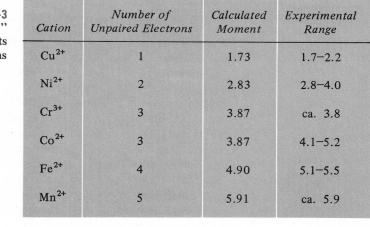

TABLE 22-3
Experimental and "Spin-Only" Calculated Magnetic Moments in Bohr Magnetons

Cation	Number of Unpaired Electrons	Calculated Moment	Experimental Range
Cu^{2+}	1	1.73	1.7–2.2
Ni^{2+}	2	2.83	2.8–4.0
Cr^{3+}	3	3.87	ca. 3.8
Co^{2+}	3	3.87	4.1–5.2
Fe^{2+}	4	4.90	5.1–5.5
Mn^{2+}	5	5.91	ca. 5.9

Balance

Sample

Magnet

FIGURE 22-11

Measuring Magnetic Properties

Experimentally, magnetic properties are measured by weighing the substance in a magnetic field (Figure 22-11). If the substance is paramagnetic, it is attracted to the external magnetic field and consequently weighs more in the external magnetic field than outside the field. If the substance has no unpaired electrons, it is very slightly repelled by the magnetic field. This repulsion is usually several orders of magnitude lower than the attractive force of unpaired electrons and can be ignored except for the most exact measurements.

As an example of how the experimentally determined magnetic moment can be utilized, consider the two cobalt(III) complexes $Co(NH_3)_6^{3+}$ and CoF_6^{3-}. The magnetic moments of these complexes measured from their salts, $Co(NH_3)_6Cl_3$ and K_3CoF_6, are 0 and 5.3 BM, respectively. Since the first complex is diamagnetic, the spins of all the electrons must be paired. The fluoro complex, however, has a magnetic moment that corresponds to four unpaired electrons (see Table 22-3). Both complexes are octahedral, and therefore the electron configuration of the d^6 cobalt(III) in the complexes must be:

$$— —$$

$$\uparrow\quad\uparrow$$

$$\downuparrow\quad\downuparrow\quad\downuparrow$$

$$\downuparrow\quad\uparrow\quad\uparrow$$

$$Co(NH_3)_6{}^{3+} \qquad\qquad CoF_6{}^{3-}$$

The difference in the electron configuration of the two complexes can be rationalized with the crystal field model. The separation of the e_g and t_{2g} levels in the ammine complex is large enough to force the electrons to pair rather than occupy the e_g orbitals. In the fluoro complex, the orbitals are closer, and the energy required to occupy the e_g orbitals is less than the energy necessary to pair the electrons. Thus, $Co(NH_3)_6{}^{3+}$ is a low-spin and $CoF_6{}^{3-}$ a high-spin complex (recall also the valence bond description on p. 603).

Color

More than one reference has been made previously to the fact that transition metal compounds show a wide variety of colors. Perhaps the most significant contribution of the crystal field theory is its explanation of the origin of the visible spectra of coordinated transition metal cations. Figure 22-12 presents the visible absorption spectrum of the hexaaquotitanium(III) ion in aqueous solution. The maximum absorption occurs at a wavelength of 5000 Å. The hydrated titanium(III) ion is purple. This means that light is absorbed by this solution species in the green and yellow regions, and the resulting color is a result of a mixture of those photons that are not absorbed. The crystal field theory explanation is that the single d electron of $[Ti(H_2O)_6]^{3+}$, which in the octahedral field resides in a t_{2g} orbital, is excited to

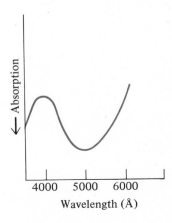

FIGURE 22-12

Visible Absorption Spectrum for $[Ti(H_2O)_6]^{3+}$

the e_g level. The energy for this electronic transition is about 57 kcal/mole and is equivalent to Δ, the difference in energy of the t_{2g} and e_g orbitals.

Absorption spectra for species with more than one d electron are considerably more complicated. For example, the hexaaquovanadium(III) ion, a d^2 species, in aqueous solution shows three absorption bands due to electron interactions. However, the same principle applies, and the color of a transition metal cation with colorless ligands is determined by the splitting energy. The magnitude of Δ can be directly determined from experimentally measured absorption curves.

Ligands can be arranged in order of their increasing crystal field strength. For the abbreviated list of ligands,

$$I^- < Br^- < Cl^- < F^- < OH^- < H_2O < NH_3 < CO < CN^-$$

a coordinated iodide ion splits the d orbitals to the smallest extent (Δ is small), and the CN^- ion splits the d orbitals to the greatest extent (Δ is large). This order, except for minor variations between adjacent pairs, remains the same for different metal ions.

Although a complete quantitative explanation of the position of a ligand in the series is beyond the scope of this book, some major effects can be noted. The increasing crystal field strength from the largest anion, I^-, to the smallest anion, F^-, clearly indicates that as the cation-anion distance becomes shorter, the crystal field splitting increases. Whether the ligand is negatively charged or neutral is in itself not the sole determinant of the position of a ligand in the series. In addition to size and charge, the polarizability of the ligand is also important and probably accounts for the greater effect of the neutral water molecule relative to the negatively charged hydroxide ion. Ammonia, a stronger Lewis base than water, is also more polarizable than water and produces a larger Δ value.

This order of increasing crystal field strength as it relates to changes in color for a given transition metal ion is called the *spectrochemical series*. As we move through the series from iodide ion to cyanide ion, the Δ values increase and the energy required to promote an electron from the t_{2g} level to the e_g level increases. Thus, the light absorbed by the coordinated entities ranges from the red end of the spectrum to the violet end, and therefore the color of the complexes varies from purple to red. For example, the hexafluoroferrate(III) ion, $[FeF_6]^{3-}$, is colorless. The crystal field splitting is very small, and the photons needed to bring about an electron jump are of low energy in the near-infrared region beyond the visible range. The hexaaquoiron(III) ion, $[Fe(H_2O)_6]^{3+}$, is a pale lavender. This corresponds to an absorption of photons in the green and yellow region of the spectrum, and the resulting color is a mixture of the photons from the white light that is not absorbed. Thus, the greater crystal field of H_2O increases the energy of the electronic transition, and the photons required appear in the visible region of the electromagnetic spectrum.

The color of transition metal complexes can also be a clue to their structures. Most tetrahedral complexes of cobalt(II) are deep blue. This characteristic color is used as a qualitative identification for cobalt(II): The addition of excess thiocyanate ion to aqueous cobalt(II) gives the deep blue tetrathiocyanato complex, $[Co(SCN)_4]^{2-}$. Most octahedral complexes of nickel(II) are green, blue, or purple; tetrahedral complexes are blue; and square planar nickel(II) complexes are red, yellow, or brown.

THE CHELATE EFFECT

A comparison of the equilibrium constant for the formation of the hexaammine-nickel(II) complex in aqueous solution

$$Ni^{2+} + 6NH_3 \rightleftharpoons [Ni(NH_3)_6]^{2+} \quad K = 10^8$$

with the formation constant for the *tris*(ethylenediamine)nickel(II) complex under the same conditions shows that the bidentate ligand

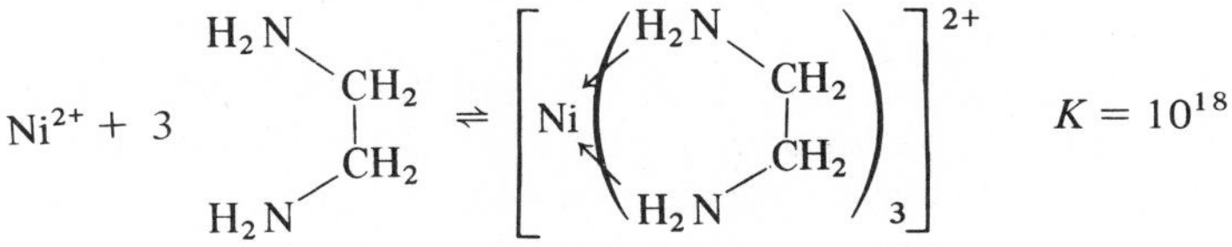

forms a more stable complex than the monodentate ligand. This increased stability is known as the *chelate effect*.

In order to understand this chelate effect, we need to remember that the equilibrium constant is related to the standard free energy by the equation

$$\Delta G^0 = -RT \ln K$$

and that

$$\Delta G^0 = \Delta H^0 - T\Delta S^0$$

Therefore, the equilibrium constant is directly related to ΔH^0 and ΔS^0; also, the greater the negative value for ΔH^0 and the greater the positive value for ΔS^0, the larger the value of K in the above reactions. The ΔH^0 of each of these reactions reflects the difference between water-nickel bonds and ammonia-nickel bonds or amine-nickel bonds.

$$[Ni(H_2O)_6]^{2+} + 6NH_3 \rightarrow [Ni(NH_3)_6]^{2+} + 6H_2O \quad \Delta H^0 = -26 \text{ kcal}$$

$$[Ni(H_2O)_6]^{2+} + 3en \rightarrow [Ni(en)_3]^{2+} + 6H_2O \quad \Delta H^0 = -28 \text{ kcal}$$

The increase in the negative value for the ΔH^0 of the ethylenediamine reaction over that of the ammonia reaction shows that the $Ni-NH_2CH_2CH_2NH_2$ bond is slightly stronger than the $Ni-NH_3$ bond. This increase in bond energy, however, is not large enough to account for the very large difference between the equilibrium constants. Hence, the major effect must be a very favorable entropy change in the formation of the ethylenediamine complex.

As is readily apparent in the above reactions, the formation of the ammonia complex leads to no change in the number of particles—six ammonia molecules simply replace six water molecules in the coordination sphere of the nickel(II) ion. In the formation of the ethylenediamine complex, however, three ethylenediamine ligands replace six water molecules, yielding a net increase of three particles. Therefore, there is a significant increase in randomness—four particles form seven particles—in the formation of the ethylenediamine complex.

This chelate effect is common to all multidentate ligands where steric factors are not important. For the most part, only five- and six-membered chelate rings are favored. In harmony with the above explanation, a nickel(II) complex with diethylenetriamine, $NH_2CH_2CH_2NHCH_2CH_2NH_2$, which is a tridentate chelating agent, has a larger overall formation constant than that of the ethylenediamine complex.

COORDINATION COMPOUNDS IN NATURE

Perhaps the two coordination compounds most important to us are chlorophyll and hemoglobin. Figure 22-13 depicts the structure of heme, which contains the basic nucleus of four pyrrole rings interconnected by CH groups. Each pyrrole is bonded through the nitrogen to an Fe^{2+} cation. Chlorophyll is a slightly modified version of this structure, containing magnesium(II) in place of iron(II) and with different groups attached to the basic nucleus. The magnesium ion is in an octahedral environment and is attached to plant proteins in the two axial positions.

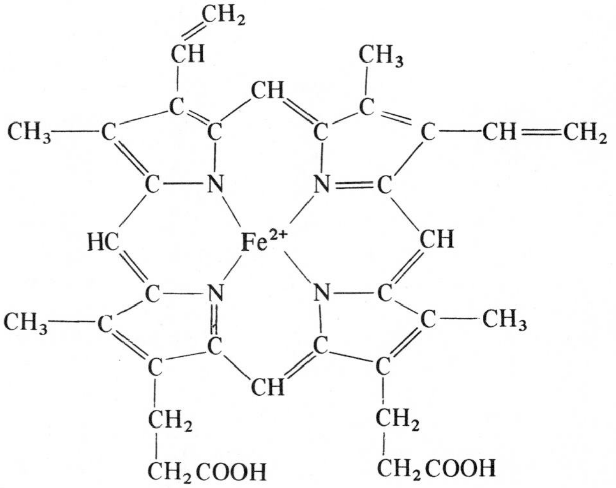

FIGURE 22-13

Heme

The role that chlorophyll plays in plant metabolism has been extensively researched and is well characterized. Consider the photosynthesis of a simple sugar such as glucose:

$$6CO_2 + 6H_2O \xrightarrow[\text{chlorophyll}]{\text{sunlight}} C_6H_{12}O_6 + 6O_2$$

Obviously, this representative synthesis does not occur in a single step but rather through a series of stepwise reactions, the first of which is the splitting of water into oxygen and hydrogen ions. The 680 kcal of energy required to form one mole of glucose is furnished by the sunlight absorbed by the chlorophyll (the components of green light are not absorbed). When an electron of a chlorophyll molecule is raised to a higher energy level, an activated molecule is formed. As these electrons fall back to ground energy level, the energy released brings about the decomposition of water, which only occurs during daylight hours. Much of the energy is stored chemically in the plant, and many of the other steps in photosynthesis occur in the dark.

In hemoglobin, which consists of the protein globin attached to four heme groups, the iron(II) is also in an octahedral environment. Each iron of a heme nucleus is bonded to a nitrogen of the globin, leaving the sixth position available for the bonding of molecular oxygen. The heme groups are protected by the globin in some fashion so that the iron remains in the +2 oxidation state. (The iron in heme that has been separated from the protein is rapidly oxidized to the +3 oxidation state and loses its ability to bond to molecular oxygen.) The oxygen transported by the hemoglobin is transferred to the myoglobin in the muscle tissue, where it is stored until needed by the cell.

Myoglobin contains the same heme nucleus as hemoglobin. It differs in that it is attached to a different protein and has only one heme per molecule. It also adds oxygen reversibly while maintaining iron in a +2 oxidation state. The oxygen stored by the myoglobin is passed on to cytochrome c when needed for oxidation of food.

Cytochromes, found in both plants and animals, are pigments that contain a modified heme nucleus and participate as catalysts in cellular oxidation-reduction reactions. They do not bond to molecular oxygen, but rather serve as electron conductors in oxidation reactions in which the coordinated iron goes through a +2/+3 cycle. For example, cytochrome c, an enzyme, is the catalyst for the conversion of oxygen to water, one of the steps in the oxidation of blood sugar (glucose) to carbon dioxide and water (see p. 458).

Enzymes are truly remarkable compounds. They function as very specific catalysts—a specific catalyst for a single specific reaction—which allow the complicated biological processes (primarily the oxidation of food, which furnishes energy, and the synthesis of cellular materials, which uses up energy) to occur at the relatively mild conditions that exist within a cell. Certainly, one of the most significant reactions that occurs through enzymatic catalysis is the formation of ammonia from the chemically inert nitrogen molecule by soil bacteria (see p. 529).

A number of enzymes contain metals such as magnesium, zinc, iron, manganese, copper, and molybdenum. Enzymes are high molecular weight proteins and therefore contain many amino acid coordination sites, and the metal ions serve as coordination centers between the enzyme and the substance undergoing a chemical change—called the *substrate*. The metal ion changes the electron distribution in the substrate and lowers the activation energy for the desired reaction.

A number of metal ions can also inhibit the catalytic effect of an enzyme. Heavy metal cations, such as silver(I) or mercury(II), coordinate more strongly with the enzyme than the metal ion required for enzymatic action. Therefore, the heavy metal cations act as *poisons* by tying up the coordination sites. On the other hand, a strong ligand such as cyanide ion coordinates with the metal ion that is necessary for the catalytic behavior of the enzyme and effectively removes it from solution. Carbon monoxide is a very effective poison, not because of its interference with enzymatic reaction, but because it coordinates so strongly with heme that it cannot be replaced by oxygen and thus causes oxygen deficiency.

Of special interest is vitamin B_{12}, the only known vitamin that contains a metal ion. This compound is related to heme but is considerably more complicated and contains cobalt(III) instead of iron(II). Vitamin B_{12}, first discovered in 1948 and only recently synthesized, is one of the essential vitamins for higher animals, and its complete metabolic function is as yet not understood. It is known that vitamin B_{12} is needed for normal blood formation and is necessary for the synthesis of DNA.

Metallurgy

Since the transition metals in their elemental forms are so valuable to our society, the extraction of metals from their ores is of utmost importance. A mineral deposit (or mixture of minerals) that is concentrated enough to make it economically feasible to recover the desired metal is known as an *ore*. Whether or not the recovery of a metal is economically worthwhile depends, among other things, upon the richness and extent of the mineral deposits, the state of the technology, and the type of mineral or minerals. A gold-bearing rock that contains as little as 11 parts per million gold can be mined with a profit, whereas the recovery of copper metal from an ore containing about 10,000 parts per million copper may be a marginal operation.

The impurities associated with the desired minerals—usually clay and silicate minerals—are called *gangue* (pronounced "gang"). This gangue may contain other valuable minerals in small concentrations, but their separation may be economically unfavorable because of technological problems or the market demand. Gangue is often stockpiled for possible future use. For example, some uranium ores contain significant concentrations of lanthanide elements. The development of a scandium-europium-containing red phosphor for color television tubes led to a surge in demand for very pure scandium and europium and some of the other lanthanide metals, thus making the recovery of these metals from the gangue of the uranium operation economically worthwhile. In some cases the minerals in the gangue can be extremely important to the success of the mining venture. For example, the recovery of the small amounts of silver associated with lead minerals can be the decisive factor as to whether the mining of the mundane metal lead is a marginal operation or an economically sound venture.

The recovery of a metal from an ore consists of three major steps. The first step is the separation of the desired mineral or minerals from the gangue. In some cases, ores are found that contain little or no impurities. For example, extensive deposits are found of almost pure bauxite, $Al_2O_3 \cdot xH_2O$; rock salt, $NaCl$; gypsum, $CaSO_4 \cdot 2H_2O$; and hematite, Fe_2O_3; these ores can be used with little or no purification. Most ores, however, contain large quantities of impurities, primarily clay and silicate minerals. Separation of the gangue is often accomplished by taking advantage of differences in physical properties. The use of some magnetic device to separate magnetite, Fe_3O_4, from its nonmagnetic impurities is an obvious approach. Gold panning of placer deposits is accomplished by swirling the sands in a shallow pan under running water. The less dense sand and gravel are washed away, while the denser gold flakes remain behind in the pan.

A flotation method is very commonly used for the separation of sulfide minerals from the gangue. Advantage is taken of the fact that oil preferentially wets sulfide minerals, while water wets the clay and silicate impurities. The finely crushed ore is first mixed with oil and then agitated with water containing soap or some other frothing material while air is bubbled through the mixture. The air bubbles adhere to the oily surface of the sulfide minerals, which then rise to the surface, and the gangue settles to the bottom. The surface froth containing the sulfide minerals can be readily skimmed or poured off.

Chemical methods of separation are also used. This is especially true for valuable metals whose minerals are generally not found in high concentrations. Uranium is separated from poor quality ores by a leaching process. The crushed ore is treated with dilute sulfuric acid, which forms a sulfate complex with the uranium, $[UO_2(SO_4)_2]^{2-}$. This solution is passed through an anion-exchange resin, which retains the complex anion of uranium. By flowing a dilute nitric acid solution through the resin, the anion is desorbed, and evaporation of the solution yields relatively pure uranyl nitrate, $UO_2(NO_3)_2 \cdot 6H_2O$.

The second step is the reduction of the metallic element to the free metal. A few metals are sometimes found in the metallic state (gold, silver, copper, and the platinum metals), but most metals occur in chemical combination. The most common minerals are oxides, sulfides, carbonates, and silicates. The sulfides are commonly roasted in air before attempting a chemical reduction.

$$2ZnS + 3O_2 \xrightarrow{\Delta} 2ZnO + 2SO_2$$

$$2PbS + 3O_2 \xrightarrow{\Delta} 2PbO + 2SO_2$$

Mercuric sulfide is one of the few minerals that decompose to the metallic state during the roasting process.

$$HgS + O_2 \xrightarrow{\Delta} Hg + SO_2$$

Thus, mercury can be separated by distillation directly from the roasting process without any preliminary separation being required. The sulfur dioxide produced by the roasting of a sulfide ore is now recognized as a valuable by-product and can be converted to either sulfuric acid or free sulfur, depending upon economic considerations.

In choosing the method to be used for the reduction of the ore to the metallic state, numerous considerations need to be taken into account. Most important to a choice of the most economical method are the thermodynamics of the possible reduction processes. Very active metals are usually prepared by electrolytic techniques rather than by chemical reduction. We have already mentioned the production of the alkali and alkaline earth metals by electrolysis of their molten salts (Chapter 20). Aluminum metal is produced by the electrolysis of a molten mixture of Al_2O_3, Na_3AlF_6, and CaF_2.

Oxides of many of the less active metals are reduced by treating them at high temperatures with carbon in a blast furnace. Inexpensive hydrogen gas (Chapter 19) could be substituted for many of these reductions. Since titanium metal forms a stable carbide and also combines with both oxygen and nitrogen at high temperatures, it cannot be produced from its common ore rutile, TiO_2, by this simple process. The titanium dioxide is first converted to the tetrachloride, which is then reduced by magnesium metal in an argon atmosphere. This involved process is in part responsible for the slow commercial development of this extremely useful metal.

The third step is the purification of the metal to the degree necessary for its intended use. For relatively inactive metals, electrolysis of an aqueous solution is commonly used for the final purification. Crude copper metal is purified in this way in a cell constructed of a pure copper cathode, an impure copper anode, and a solution of copper(II) sulfate. Distillation processes for low-boiling metals are also common. Cadmium metal obtained as a by-product of zinc recovery is separated from the zinc metal by distillation. Mercury, magnesium, and zinc are also purified by distillation. Pure zirconium, hafnium, and beryllium are obtained by the *van Arkel process,* in which a volatile iodide of the metal is passed over tungsten wires at temperatures up to $1800°C$. The iodide decomposes, depositing the metal on the tungsten wires. In the *Mond process,* carbon monoxide gas is passed over crude nickel metal at $60°C$, forming the volatile carbonyl complex of nickel, $Ni(CO)_4$. This gaseous mixture is then passed over pure nickel pellets heated to $200°C$, where the complex decomposes, depositing pure nickel on the pellets, and the carbon monoxide is recycled. Neither iron nor cobalt, the two common impurities, form carbonyl complexes under these mild conditions.

Of great importance in recent years is the method of *zone refining.* A high-intensity heater—a circular narrow coil of wire—is placed around the end of a bar of impure metal and very slowly moved toward the other end of the bar. Only a very narrow band of impure metal is melted at any one time, and crystals of pure metal form at the trailing edge of the molten zone. The impurities remaining in the molten band are carried to the other end of the bar. Thus, advantage is taken of the fact that the pure metal melts at a higher temperature than the impure metal. (Recall the discussion on freezing-point lowering in Chapter 11.) This method produces metals of extremely high purity and is used to produce both germanium and silicon for transistors.

Because iron is probably our most important metal, let us consider in some detail the recovery of iron from its ores. Most iron ore is found in huge deposits of hematite (Fe_2O_3), limonite ($Fe_2O_3 \cdot H_2O$), siderite ($FeCO_3$), or magnetite (Fe_3O_4), which require little or no ore concentration and consequently can be used directly from the mine. The ore is reduced to the metal in a smelting furnace constructed of steel and lined with silica bricks. A mixture of the crushed ore, limestone, sand, and coke is put into the furnace. Hot dry air, preheated to about 600°C, is blown into the bottom of the furnace. This air partially oxidizes the coke to carbon monoxide.

$$C + \tfrac{1}{2}O_2 \rightarrow CO + \text{heat}$$

The heat given off by this reaction is enough to raise the temperature near the bottom of the furnace to greater than 1200°C. The temperature near the top of the furnace where the feedstock is continuously introduced is about 650°C. Within this temperature range, both direct and indirect reduction of the iron ore occurs. Direct reduction does not start until around 850°C. Above 1000°C, most of the reduction is by the direct reaction of coke with the iron ore.

$$3C + 2Fe_2O_3 \rightarrow 4Fe + 3CO_2 - \text{heat}$$

Indirect reduction (primarily by carbon monoxide) occurs at lower temperatures and reaches its maximum between 800 and 825°C.

$$3CO + Fe_2O_3 \rightarrow 2Fe + 3CO_2 + \text{heat}$$

Since the indirect reduction with carbon monoxide is slightly exothermic, while that of direct reduction with coke is highly endothermic, there is a considerable saving of fuel if most of the iron ore can be reduced by the indirect method. The rate of ore reduction is also considerably faster with the indirect method. Injection of carbon monoxide gas or even hydrogen gas (reduction with hydrogen is slightly endothermic) into the blast furnace is being studied.

The gangue present reacts according to the following simplified equations:

$$CaCO_3 \rightarrow CaO + CO_2$$

$$CaO + SiO_2 \rightarrow CaSiO_3$$

$$CaO + Al_2O_3 \rightarrow Ca(AlO_2)_2$$

$$MgO + SiO_2 \rightarrow MgSiO_3$$

to form a fusible mixture called *slag,* which floats on the top of the molten iron. The iron is withdrawn at the bottom of the furnace and is directly transported in insulated cars from there to the steelmaking furnaces. The slag is drawn off from a separate tap and allowed to cool, forming a glassy, rocklike material that is mostly calcium silicate. In the past, the huge quantities of slag produced were considered to be of no value except for land fill, but one of its contemporary uses is in the production of Portland cement.

The pig iron from the blast furnace is contaminated with carbon and silicon. Depending on the kind of ore, it also contains sulfur, phosphorus, and metallic impurities. The *Bessemer process,* which is rapid but does not remove all of the impurities, and the *open-hearth process,* which is very slow but produces pure iron, used to be the two principal methods used for purifying pig iron. Now, however, both of these processes are being replaced by the *basic oxygen process,* in which a stream of oxygen gas is played over the surface of the molten pig iron in a vessel lined with calcium oxide bricks. The total time to produce as pure an iron as that obtained from the open-hearth process is 10 to 20 minutes. The following equations represent the reactions of the impurities with oxygen:

$$C + O_2 \rightarrow CO_2$$

$$Si + O_2 \rightarrow SiO_2$$

$$CaO + SiO_2 \rightarrow CaSiO_3$$

$$S + O_2 \rightarrow SO_2$$

The small amount of slag that forms on the surface is readily removed. Most of the iron obtained by this process is alloyed with various materials to produce the many different kinds of steel.

SUGGESTED READINGS

Quagliano, J. V., and L. M. Vallarino. *Coordination Chemistry.* Lexington, Mass.: Heath, 1969.

Calvin, M. "The Nurture of Creative Science and the Men Who Make It. The Photosynthesis Story." *Journal of Chemical Education,* Vol. 35, No. 9 (September 1958), pp. 428–432.

Moeller, T. "Periodicity and the Lanthanides and Actinides." *Journal of Chemical Education,* Vol. 47, No. 6 (June 1970), pp. 417–423.

PROBLEMS

1. Supply names for the following:
   (a) $K_3[Co(NO_2)_6]$ (b) $[Co(en)_2Br_2]Br$
   (c) $[Cr(NH_3)_5SO_4]Cl$ (d) $K_4[Ni(CN)_4]$
   (e) $Na[AuCl_4]$ (f) $[Co(NH_3)_4CO_3]NO_3$
   (g) $[Cr(NH_3)_4Cl(H_2O)]Cl_2$ (h) $[Ni(CO)_4]$
   (i) $K_4[Fe(CN)_6]$ (j) $K[Cr(C_2O_4)_2(H_2O)_2]$

2. Write formulas (structural where necessary) for
   (a) hydroxopentaaquoiron(III) ion
   (b) potassium hexacyanoferrate(III)
   (c) tetrathiocyanatocobaltate(II) ion
   (d) chloronitrotetraaminecobalt(III) chloride
   (e) potassium tris(oxalato)aluminate(III)
   (f) *trans*-dichlorotetraamminecobalt(III) chloride

3. Show possible geometric isomers for
   (a) $[Pt(NO_2)(C_5H_5N)(NH_2CH_3)(NH_3)]Cl$
   (b) $[Pd(NH_3)_2Cl_4]$
   (c) $[Co(NH_3)_3(NO_2)_3]$

4. Which of the compounds in Problems 1 and 2 exhibit geometric isomerism?

5. For each of the following octahedrally coordinated transition metal cations
   (1) Describe the bonding with the valence bond model.
   (2) Show the d electron distribution in the crystal field model.
   (3) Calculate the crystal field stabilization energy.
   (4) Calculate the magnetic susceptibility in Bohr magnetons, neglecting the orbital contribution.

(a)	$[Mn(H_2O)_6]^{2+}$	high spin—outer orbital
(b)	$[Co(NH_3)_6]^{2+}$	high spin—outer orbital
(c)	$[Co(en)_3]^{3+}$	low spin—inner orbital
(d)	$[Cr(NH_3)_6]^{3+}$	inner orbital
(e)	$[Fe(CN)_6]^{3-}$	low spin—inner orbital
(f)	$[Ni(NH_3)_6]^{2+}$	outer orbital

6. Using octahedrally coordinated iron(II), show d electron distributions for both the low-spin and the high-spin species. Discuss the factors that determine whether a complex is low spin or high spin.

7. The magnetic moment of $K_4[Cr(CN)_6]$ is about 3.2 BM and the one for $K_4[Cr(SCN)_6]$ is about 5.0 BM.
   (a) How does the valence bond model account for these differences?
   (b) How does the crystal field model account for these differences?

8. Lattice energies (in kcal/mole) are given below for the difluorides and the divalent oxides of the metals from calcium to zinc. Plot these values against increasing atomic numbers of the metals. Account for the curves obtained.

M	MF_2	MO	M	MF_2	MO
Ca	624	828	Fe	696	938
Ti	657	928	Co	708	954
V	672	936	Ni	728	974
Cr	688	—	Cu	727	—
Mn	662	911	Zn	710	964

9. Elements other than transition metals form coordination compounds, although usually to a lesser degree.
   (a) Account for the generally observed stability order for the alkaline earths:

   $$Be^{2+} > Mg^{2+} > Ca^{2+} > Sr^{2+} > Ba^{2+}$$

   (b) Account for the fact that these cations coordinate almost exclusively with bidentate or polydentate chelating ligands.

10. Assuming that the yet-to-be synthesized element 110 will be stable enough to determine its chemical properties, predict:
    (a) its highest oxidation state
    (b) its lowest oxidation state
    (c) the physical state of the element
    (d) the formula of a compound with chlorine
    (e) If it forms coordination compounds, give the most likely shape for both 4- and 6-coordination and determine whether high-spin or low-spin complexes would be most common.

11. Show that the overall formation constants for the ethylenediamine complexes of manganese through zinc are consistent with the crystal field model.

$$M^{2+} + 3en \rightleftharpoons [M(en)_3]^{2+}$$

Metal	Mn	Fe	Co	Ni	Cu	Zn
$\log K_{overall}$	5.67	9.52	13.82	18.06	18.60	12.09

12. Analogous complexes for second- and third-period transition metals are more commonly low spin than those for first-period transition metals. Explain.

13. The Δ octahedral value for +3 cations is higher than the Δ value for +2 cations. Explain.

14. Calculate the crystal field stabilization energies for the electron distributions in Figure 22-10.

15. Explain why the chemical behavior of hafnium so closely parallels that of zirconium.

16. Three different compounds of chromium(III) of identical composition—19.44 percent chromium, 39.83 percent chlorine, 40.60 percent water—have been isolated. Given the following information, draw structural formulas for A, B, and C.

    Compound A is violet. Dissolved in water, it behaves as a uni-trivalent electrolyte (M^{3+}, $3X^-$), and all of the chlorine precipitates instantly upon adding excess silver nitrate solution.

    Compound B is green. Dissolved in water, it behaves as a uni-divalent electrolyte (M^{2+}, $2X^-$), and two-thirds of the chlorine precipitates instantly upon adding excess silver nitrate solution. A solution of B very slowly turns violet and has the same visible absorption spectrum as A.

    Compound C is green. Dissolved in water, it behaves as a uni-univalent electrolyte (M^+ X^-), and one-third of the chlorine precipitates instantly upon the addition of excess silver nitrate solution. This solution also very slowly turns violet and has the same visible absorption spectrum as A.

17. Cobaltous ion forms a more stable complex with heme than ferrous ion. Can you suggest a reason why the evolutionary process ended with iron(II) in hemoglobin rather than cobalt(II)?

18. You have isolated the compound $2KCl \cdot MgCl_2$ from a solution that contained potassium chloride and magnesium chloride. Suggest experimental methods to show whether or not it is a coordination compound.

Constants and Conversion Factors

Constants

Mass of electron	9.11×10^{-28} g
Mass of proton	1.672×10^{-24} g
Mass of neutron	1.675×10^{-24} g
Charge on electron	4.80×10^{-10} esu
	1.60×10^{-19} coul
Velocity of light	3.00×10^{10} cm/sec
Planck's constant	6.626×10^{-27} erg-sec
Ideal gas constant	1.99 cal/mole-deg
	0.0821 liter-atm/mole-deg
Faraday constant	96,487 coulomb/equiv
	23.1 kcal/volt-equiv
Molar gas volume (STP)	22.41 liters
Avogadro's number	6.023×10^{23} particles/mole

Conversion Factors

1 atomic mass unit (amu)	=	1.6604×10^{-24} g
1 kilogram (kg)	=	2.2 lb
1 pound (lb)	=	453.5 g
1 Å	=	10^{-8} cm
1 micron (μ)	=	10^{-6} m
1 nanometer (nm)	=	10^{-9} m
1 inch	=	2.54 cm
1 liter	=	1.057 qt
1 coulomb	=	3.00×10^{9} esu
1 esu^2/cm	=	1 erg
1 kcal	=	4.184×10^{10} ergs
1 joule	=	10^{7} ergs
$\ln x$	=	$2.303 \log x$

The following prefixes can be used with a given unit to designate a multiple of that unit. For example,

$$1 \text{ kilogram} = 1000 \text{ grams}$$
$$1 \text{ millimeter} = 1 \times 10^{-3} \text{ meter}$$
$$1 \text{ microliter} = 1 \times 10^{-6} \text{ liter}$$

Prefix	*Multiple*
mega	10^6
kilo	10^3
hecto	10^2
deca	10
deci	10^{-1}
centi	10^{-2}
milli	10^{-3}
micro	10^{-6}
nano	10^{-9}

Atomic and Ionic Radii (Å)

Ag	1.34	Cl	0.99	I	1.33	O	0.66	Si	1.17
Ag^+	1.15	Cl^-	1.81	I^-	2.20	O^{2-}	1.40	Sn	1.40
Al	1.25	Co	1.16	In	1.50	Os	1.26	Sn^{2+}	0.93
Al^{3+}	0.53	Co^{2+}	0.7	In^{3+}	0.80	P	1.10	Sr	1.91
Ar	0.95[a]	Co^{3+}	0.6	Ir	1.27	Pb	1.54	Sr^{2+}	1.16
As	1.21	Cr	1.18	K	2.03	Pb^{2+}	1.18	Te	1.37
As^{3+}	0.58	Cr^{3+}	0.62	K^+	1.38	Pd	1.28	Te^{2-}	2.21
Au	1.34	Cs	2.35	Kr	1.10	Pd^{2+}	0.80	Th	1.65
Au^+	1.37	Cs^+	1.70	La	1.69	Po	1.53	Ti	1.32
B	0.81	Cu	1.17	La^{3+}	1.05	Pt	1.30	Ti^{2+}	0.86
Ba	1.98	Cu^+	0.96	Li	1.23	Pt^{2+}	0.80	Ti^{3+}	0.67
Ba^{2+}	1.36	Cu^{2+}	0.73	Li^+	0.74	Rb	2.16	Tl	1.55
Be	0.89	F	0.64	Lu	1.56	Rb^+	1.49	Tl^+	1.50
Be^{2+}	0.3	F^-	1.33	Lu^{3+}	0.85	Rh	1.25	Tl^{3+}	0.88
Bi	ca. 1.5	Fe	1.17	Mg	1.36	Rn	1.45[a]	U	1.42
Bi^{3+}	1.02	Fe^{2+}	0.7	Mg^{2+}	0.72	Ru	1.25	V	1.22
Br	1.14	Fe^{3+}	0.6	Mn	1.17	S	1.04	V^{2+}	0.79
Br^-	1.96	Ga	1.25	Mn^{2+}	0.7	S^{2-}	1.84	W	1.30
C	0.77	Ga^{3+}	0.62	N	0.70	Sb	1.41	Xe	1.30
Ca	1.74	Ge	1.22	Na	1.57	Sb^{3+}	0.76	Y	1.62
Ca^{2+}	1.00	H	0.37	Na^+	1.02	Sc	1.44	Y^{3+}	0.89
Cd	1.41	He	0.5[a]	Ne	0.6[a]	Sc^{3+}	0.75	Zn	1.25
Cd^{2+}	0.95	Hf	1.44	Ni	1.15	Se	1.17	Zn^{2+}	0.75
Ce	1.65	Hg	1.44	Ni^{2+}	0.70	Se^{2-}	1.98	Zr	1.45
Ce^{3+}	1.01	Hg^{2+}	1.02						

[a] Extrapolated from neighboring elements.

Ionization Energies
of Gaseous Atoms (kcal/mole)

Atomic Number	Element	1	2	3	4
1	H	313.5			
2	He	566.9	1254		
3	Li	124.3	1744	2823	
4	Be	214.9	419.9	3548	5020
5	B	191.3	580.0	874.5	5980
6	C	259.6	562.2	1104	1487
7	N	335.1	682.8	1094	1786
8	O	314.0	810.6	1267	1785
9	F	401.8	806.7	1445	2012
10	Ne	497.2	947.2	1500	2241
11	Na	118.5	1091	1652	2280
12	Mg	176.3	346.6	1848	2521
13	Al	138.0	434.1	655.9	2767
14	Si	187.9	376.8	771.7	1041
15	P	254	453.2	695.5	1184
16	S	238.9	540	807	1091
17	Cl	300.0	548.9	920.2	1230
18	Ar	363.4	637.0	943.3	1379
19	K	100.1	733.6	1100	1405
20	Ca	140.9	273.8	1181	1550
21	Sc	151.3	297.3	570.8	1700
22	Ti	158	314.3	649.0	997.2
23	V	155	328	685	1100
24	Cr	156.0	380.3	713.8	1140
25	Mn	171.4	360.7	777.0	
26	Fe	182	373.2	706.7	
27	Co	181	393.2	772.4	
28	Ni	176.0	418.6	810.9	
29	Cu	178.1	467.9	849.4	
30	Zn	216.6	414.2	915.6	
31	Ga	138	473.0	708.0	1480
32	Ge	183	367.4	789.0	1050
33	As	226	466	653	1160
34	Se	225	496	738	989
35	Br	273.0	498	838	

Atomic Number	Element	1	2	3	4
36	Kr	322.8	566.4	851	
37	Rb	96.31	634	920	
38	Sr	131.3	254.3	1005	1300
39	Y	147	282.1	473	
40	Zr	158	302.8	530.0	791.8
41	Nb	158.7	330.3	579.8	883
42	Mo	164	372.5	625.7	1070
43	Tc	168	351.9		
44	Ru	169.8	386.5	656.4	
45	Rh	172	416.7	716.1	
46	Pd	192	447.9	759.2	
47	Ag	174.7	495.4	803.1	
48	Cd	207.4	389.9	864.2	
49	In	133.4	435.0	646.5	1250
50	Sn	169.3	337.4	703.2	939.1
51	Sb	199.2	380	583	1020
52	Te	208	429	720	880
53	I	241.1	440.3		
54	Xe	279.7	489	740	
55	Cs	89.78	579		
56	Ba	120.2	230.7		
57	La	129	263.6	442.1	
72	Hf	160	344		
73	Ta	182	374		
74	W	184	408		
75	Re	182	383		
76	Os	200	390		
77	Ir	200			
78	Pt	210	428.0		
79	Au	213	473		
80	Hg	240.5	432.5	789	
81	Tl	140.8	470.9	687	1170
82	Pb	171.0	346.6	736.4	975.9
83	Bi	168.1	384.7	589.5	1040
84	Po	194			
85	At				
86	Rn	247.8			
87	Fr				
88	Ra	121.7	234.0		
89	Ac	160	279		

Electron Affinities (kcal/mole)

H	17.3		S	48
He	<0		$S^- \to S^{2-}$	−141
Li	14		Cl	83
B	7[a]		K	16[a]
C	29		Ca	<0
N	−14[a]		Fe	13[a]
O	34		Co	22[a]
$O^- \to O^{2-}$	−187		Cu	42[a]
F	79		Zn	−21[a]
Na	12[a]		Br	78
Mg	−12[a]		Rb	7
Al	12[a]		I	71
Si	32[a]		Cs	7
P	18[a]			

[a] Estimated values; not experimentally verified.

Bond Energies and Bond Lengths[a]

Bond	Bond Energy (kcal/mole)	Bond Length (Å)
Hydrogen Compounds		
H–H	103.25	0.74
H–F	135	0.92
H–Cl	102.3	1.27
H–Br	86.6	1.41
H–I	70.4	1.61
H–O	110	0.96
H–S	87	1.34
H–Se	ca. 66	1.46
H–Te	ca. 57	1.7
H–N	92	1.01
H–P	77	1.44
H–As	59	1.52
H–C	98	1.09
H–Si	76	1.48
H–Ge	69	1.53
H–Sn	60	1.70
H–B	ca. 93	1.19
Group III		
B–F	146	
B–Cl	109	1.75
B–Br	90	
Group IV		
C–F	116	1.35
C–Cl	78	1.77
C–Br	68	1.94
C–I	51	2.14
C–O	85.5	1.43
C=O	191	1.20
C≡O	256	1.13
C–S	65	1.82
C=S	137	1.60

[a] Abridged from Table F.1, *Inorganic Chemistry: Principles of Structure and Reactivity* by James E. Huheey (New York: Harper and Row, 1972), pp. 694-702.

Bond	Bond Energy (kcal/mole)	Bond Length (Å)
C—N	73	1.47
C=N	147	1.29
C≡N	212	1.16
C—C	83	1.54
C=C	144	1.34
C≡C	200	1.20
C—Si	76	1.85
C—Ge	57	1.95
C—Sn	46	2.16
C—Pb	31	2.30
Si—F	142	1.60
Si—Cl	97	2.02
Si—Br	76	2.15
Si—I	56	2.43
Si—O	106	1.63
Si—Si	46	2.33
Ge—Cl	83	2.10
Ge—Ge	38	2.41
Sn—Cl ($SnCl_4$)	77	2.33
Sn—Cl ($SnCl_2$)	92	2.42
Pb—Cl ($PbCl_4$)	58	
Pb—Cl ($PbCl_2$)	73	2.42
Group V		
N—F	68	1.36
N—Cl	72	1.75
N—O	48	1.40
N=O	145	1.21
N—N	ca. 40	1.45
N=N	100	1.25
N≡N	225.1	1.10
P—F	117	1.54
P—Cl	78	2.03
P—O	ca. 80	1.63
P—P	ca. 50	
As—Cl	77	2.16
As—O	72	1.78
As—As	35	2.43
Sb—Cl ($SbCl_3$)	75	2.32
Sb—Cl ($SbCl_5$)	59	
Sb—Sb	ca. 29	

Bond	Bond Energy (kcal/mole)	Bond Length (Å)
Group VI		
O—F	45	1.42
O—O	34	1.48
O=O	118.0	1.21
S—F	68	1.56
S—S	54	2.05
Group VII		
F—F	37.0	1.42
Cl—Cl	57.3	1.99
Br—Br	45.5	2.28
I—F (IF)	66	1.91
I—F (IF_3)	65	
I—F (IF_5)	64	1.75, 1.86
I—Cl (ICl)	50	2.32
I—Br (IBr)	42	
I—I	35.6	2.67
Group 0		
Xe—F (XeF_2)	31	2.00
Xe—F (XeF_4)	31	1.95
Xe—F (XeF_6)	30	1.90

Molecular Parameters for Selected Simple Molecules[a]

No. Valence Electrons	Molecule	Shape	Bond Angle	Bond Length (Å)
		TRIATOMIC MOLECULES		
8	H_2O	bent	104.5°	0.96
8	H_2S	bent	92°	1.33
8	H_2Se	bent	ca. 90°	1.49
8	H_2Te	bent	ca. 90°	1.69
10	HCN	linear	180°	CH, 1.06; CN, 1.16
12	HNO	bent		NO, 1.21
16	CO_2	linear	180°	1.16
16	CS_2	linear	180°	1.55
16	OCS	linear	180°	OC, 1.16; CS, 1.56
16	OCSe	linear	180°	OC, 1.16; CSe, 1.71
16	ClCN	linear	180°	ClC, 1.63; CN, 1.16
16	N_2O	linear	180°	NN, 1.13; NO, 1.19
16	NCO^-	linear	180°	NC, 1.18; CO, 1.18
16	NCS^-	linear	180°	NC, 1.25; CS, 1.59
16	N_3^-	linear	180°	1.15
16	ZnI_2	linear	180°	2.42
16	CdI_2	linear	180°	2.60
16	HgI_2	linear	180°	2.61
17	NO_2	bent	134°	1.19
18	SO_2	bent	119°	1.43
18	O_3	bent	117°	1.28
18	NO_2^-	bent	115°	1.24
18	FNO	bent	110°	FN, 1.52; NO, 1.13
18	ClNO	bent	116°	ClN, 1.95; NO, 1.14
18	$SnCl_2$	bent	ca. 95°	
19	ClO_2	bent	116.5°	1.49
20	F_2O	bent	101.5°	1.38
20	Cl_2O	bent	111°	1.70
22	I_3^-	linear	180°	

[a] Structure of neutral molecules generally determined in gas phase; ions in solid phase. Adapted from A. D. Walsh, "The Shape of Simple Molecules," *Progress in Stereochemistry,* Vol. 1, W. Kline, ed., Butterworths, London, 1954.

No. Valence Electrons	Molecule	Shape	Bond Angle	Bond Length (Å)

TETRATOMIC MOLECULES

AB$_3$ types

No. Valence Electrons	Molecule	Shape	Bond Angle	Bond Length (Å)
8	NH_3	pyramidal	107°	1.01
8	PH_3	pyramidal	93.5°	1.42
8	AsH_3	pyramidal	92°	1.53
8	SbH_3	pyramidal	91.5°	1.73
24	CO_3^{2-}	planar	120°	1.31
24	NO_3^-	planar	120°	1.22
24	SO_3	planar	120°	1.43
24	BO_3^{3-}	planar	120°	1.29
24	BF_3	planar	120°	1.30
24	BCl_3	planar	120°	1.72
26	NF_3	pyramidal	102°	1.37
26	PF_3	pyramidal	104°	1.55
26	PCl_3	pyramidal	100°	2.04
26	PBr_3	pyramidal	101.5°	2.18
26	PI_3	pyramidal	102°	2.43
26	$AsCl_3$	pyramidal	98°	2.16
28	ClF_3	T-shape	87°	Two Cl–F bonds = 1.70; One Cl–F bond = 1.60

B$_2$AC types

No. Valence Electrons	Molecule	Shape	Bond Angle	Bond Length (Å)
12	H_2CO	planar	∠ HCH, 118°	CH, 1.12; CO, 1.21
24	O_2NCl	planar	∠ ONO, 130.5°	NO, 1.20; NCl, 1.84
24	Cl_2CO	planar	∠ ClCCl, 112.5°	CCl, 1.74; CO, 1.18
24	Cl_2CS	planar	∠ ClCCl, 116°	CCl, 170; CS, 1.63
26	Cl_2SO	pyramidal	∠ ClSCl, 114° ∠ ClSO, 106°	SCl, 2.07; SO, 1.45

Others

No. Valence Electrons	Molecule	Shape	Bond Angle	Bond Length (Å)
10	C_2H_2	linear	180°	CH, 1.06; CC, 1.20
14	H_2O_2	nonlinear, nonplanar	∠ HOO, 97°	OO, 1.48
16	HN_3	N$_3$ group linear	∠ HNN, 112.5°	HN, 1.02; N_1N_2, 1.24 N_2N_3, 1.13
16	HCCCl	linear	180°	CH, 1.05; CC, 1.21 CCl, 1.63
16	HNCO	NCO group linear	∠ HNC, 128°	HN, 0.99; NC, 1.21; CO, 1.17
18	NCCN	linear	180°	CC, 1.37; CN, 1.16
18	HCO_2^-	planar	∠ OCO, 125°	CO, 1.26

No. Valence Electrons	Molecule	Shape	Bond Angle	Bond Length (Å)
			OTHERS	
16	H_2CCO	planar, CCO linear	∠ HCH, 122°	CH, 1.08; CC, 1.32; CO, 1.16
24	$(CH_3)_2CO$		∠ CCO, 120°	CO, 1.22; CC, 1.55
32	$POCl_3$	irregular tetrahedron	∠ ClPCl, 103.5°	PCl, 1.99; PO, 1.45
32	SO_2Cl_2	tetrahedron	∠ OSO, 119.5° ∠ ClSCl, 111°	SO, 1.43; SCl, 1.99
34	N_2O_4	planar	∠ ONO, 126°	NO, 1.17; NN, 1.64

Standard Enthalpies and Free Energies
of Formation and Absolute Entropies at 25°C

Substance	ΔH_f^0 (kcal/mole)	ΔG_f^0 (kcal/mole)	S^0 (cal/deg-mole)
H_2 (g)	0.0	0.0	31.2
H(g)	52.1	48.6	27.4
Group I			
Li(s)	0.0	0.0	6.7
Li(g)	37.1	29.2	33.1
$Li_2 O$(s)	−142.4	−134	
$Li_2 O_2$ (s)	−151.7	−135	
LiH(s)	−21.6	−16.7	5.9
LiCl(s)	−97.7	−91.7	ca. 13.2
LiF(s)	−146.3	−139.6	8.6
LiI(s)	−64.8		
Na(s)	0.0	0.0	12.2
Na(g)	26.0	18.7	36.7
NaH(s)	−13.7		
NaF(s)	−136.0	−129.3	14.0
NaCl(s)	−98.2	−91.8	17.3
NaBr(s)	−86.0	−83.1	
NaI(s)	−68.8		
$Na_2 SO_4$(s)	−330.9	−302.8	35.7
$Na_2 SO_4 \cdot 10H_2 O$(s)	−1033.5	−870.9	141.7
$NaNO_3$(s)	−111.5	−87.4	27.8
$NaNO_2$(s)	−85.9		
$Na_2 CO_3$(s)	−270.3	−250.4	32.5
$Na_2 O$(s)	−99.4	−90.0	17.4
$Na_2 O_2$ (s)	−120.6	−102.8	ca. 16.0
NaO_2(s)	−62.3	−52.3	
$NaC_2 H_3 O_2$(s)	−169.8		
NaOH(s)	−102.0	−90.1	ca. 12.5
K(s)	0.0	0.0	15.2
K(g)	21.5	14.6	38.3
KF(s)	−134.5	−127.4	15.9
KCl(s)	−104.2	−97.6	19.8
$KClO_3$(s)	−93.5	−69.3	34.2
$K_2 O$(s)	−86.4	−76.2	
$K_2 O_2$(s)	−118	−100	

Substance	ΔH_f^0 (kcal/mole)	ΔG_f^0 (kcal/mole)	S^0 (cal/deg-mole)
$KO_2(s)$	−67.6	−59.4	
$KOH(s)$	−101.8	−89.5	
$KNO_3(s)$	−117.8	−94.0	31.8
$KClO_4(s)$	−103.6	−72.7	36.1
$Rb(s)$	0.0	0.0	16.6
$Rb(g)$	20.5	13.5	40.6
$RbF(s)$	−131.3		
$RbCl(s)$	−102.9		
$Rb_2O(s)$	−78.9	−70	
$Rb_2O_2(s)$	−101.7	−84	
$RbO_2(s)$	−63		
$Cs(s)$	0.0	0.0	19.8
$Cs(g)$	18.8	12.2	41.9
$CsF(s)$	−126.9		
$CsCl(s)$	−103.5		
$CsI(s)$	−80.5	−79.7	
$Cs_2O(s)$	−75.9	−78	
$Cs_2O_2(s)$	−96.2	−86	
$CsO_2(s)$	−62		
Group II			
$Be(s)$	0.0	0.0	2.3
$Be(g)$	76.6	67.6	32.5
$BeCl_2(s)$	−122.3		
$Mg(s)$	0.0	0.0	7.8
$Mg(g)$	35.9	27.6	35.5
$MgF_2(s)$	−263.5	−250.8	13.7
$MgCl_2(s)$	−153.4	−141.6	21.4
$MgO(s)$	−143.8	−136.1	6.4
$MgSO_4(s)$	−305.5	−280.5	21.9
$MgSO_4 \cdot 7H_2O(s)$	−808.7		
$Mg(NO_3)_2$	−188.7	−140.6	39.2
$Ca(s)$	0.0	0.0	9.95
$Ca(g)$	46.0	40.0	37.0
$CaH_2(s)$	−45.1	−35.8	10
$CaO(s)$	−151.9	−144.4	9.5
$Ca(OH)_2(s)$	−235.8	−214.3	18.2
$CaF_2(s)$	−290.3	−277.7	16.5
$CaCl_2(s)$	−190.0	−179.3	27.2
$CaCO_3(s)$	−288.5	−269.8	22.2
$Sr(s)$	0.0	0.0	13.0
$Sr(g)$	39.2	26.3	39.3
$SrCl_2(s)$	−198.0	−186.7	−28

Substance	ΔH_f^0 (kcal/mole)	ΔG_f^0 (kcal/mole)	S^0 (cal/deg-mole)
Ba(s)	0.0	0.0	16
Ba(g)	42.0	34.6	40.7
$BaCl_2$(s)	−205.6	−193.8	30
Group III			
B(s)	0.0	0.0	1.6
B_2O_3(s)	−302.0	−283.0	12.9
B_2H_6(g)	7.5	19.8	55.7
BCl_3(g)	−96.3	−92.7	69.3
Al(s)	0.0	0.0	6.8
Al_2O_3(s)	−399.1	−376.8	12.2
$AlCl_3$(s)	−168.6	−150.6	26.4
Group IV			
C(s, graphite)	0.0	0.0	1.4
C(s, diamond)	0.5	0.7	0.6
C(g)	171.7	160.8	37.8
CH_4(g)	−17.9	−12.1	44.5
C_2H_2(g)	54.2	50.0	48.0
C_2H_4(g)	12.5	16.3	52.5
C_2H_6(g)	−20.2	−7.9	54.9
$CH_3CH_2CH_3$(g)	−24.8	−5.6	64.5
$CH_3CH_2CH_2CH_3$(g)	−30.15	−4.1	74.1
C_7H_{16} (g, heptane)	−44.9	1.9	102.3
$CH_3-\underset{\underset{CH_3}{\mid}}{\overset{\overset{H}{\mid}}{C}}-CH_3$(g)	−32.15	−5.0	70.4
cyclopropane: $H_2C\overset{\overset{H_2}{C}}{\diagdown\diagup}CH_2$(g)	12.7	24.95	56.75
cyclobutane: $H_2C{-}CH_2 / H_2C{-}CH_2$(g)	6.4	26.3	63.4
cyclohexane (g)	−29.4	7.6	71.3
$H_2C{=}CHCH_2CH_3$(g)	0.0	17.0	73.0
cis-$CH_3CH{=}CHCH_3$(g)	−1.7	15.7	71.9
trans-$CH_3CH{=}CHCH_3$(g)	−2.7	15.05	70.9
cis-$ClCH{=}CHCl$(g)	0.45	5.8	69.2
trans-$ClCH{=}CHCl$(g)	1.0	6.35	69.3

Substance	ΔH_f^0 (kcal/mole)	ΔG_f^0 (kcal/mole)	S^0 (cal/deg-mole)
$C_6H_6(g)$	19.8	31.0	64.3
$C_6H_6(l)$	11.7	29.7	41.4
$C_6H_5CH_3(g)$	11.95	29.2	76.6
$C_6H_5CH_3(l)$	2.9	27.2	52.8
$CH_3Br(g)$	−9.0	−6.7	58.75
$CH_3I(g)$	3.3	3.7	60.7
$CH_3I(l)$	−3.3	3.6	38.9
$CH_3CH_2Cl(g)$	−25.1	−12.7	65.9
$(CH_3)_3CBr(g)$	−32.0	−6.7	79.3
$CHCl_3(l)$	−31.5	−17.1	48.5
$CHCl_3(g)$	−24.2	−16.4	70.7
$CH_3CHBrCHBrCH_3(g)$	−24.4	−2.85	94.4
$CH_3CH_2OCH_2CH_3(g)$	−60.3	−29.2	81.9
$CH_3CH_2OCH_2CH_3(l)$	−66.8	−29.4	60.5
$CH_3OH(g)$	−48.1	−38.8	57.3
$CH_3OH(l)$	−57.0	−39.7	30.3
$CH_3CH_2OH(l)$	−66.2	−41.6	38.4
$(CH_3)_3COH(g)$	−77.9	−45.7	78.0
$(CH_3)_3COH(l)$	−85.9	−44.2	46.2
$C_6H_{12}O_6$ (s, glucose)	−304.6	−217.6	50.7
$CH_2O(g)$	−27.7	−26.3	52.3
$CH_3CHO(g)$	−39.8	−31.9	63.2
$CH_3COCH_3(l)$	−59.3	−37.1	47.9
$HCOOH(g)$	−90.5	−83.9	59.45
$CH_3COOH(g)$	−103.9	−90.0	67.5
$CH_3COOH(l)$	−115.7	−93.1	38.2
$(COOH)_2(s)$	−198.4	−167.6	28.7
$CH_3COOCH_3(l)$	−105.5		
$CH_3NH_2(g)$	−5.5	7.7	58.0
$(CH_3CH_2)_2NH(g)$	−17.3	17.2	84.2
$(CH_3CH_2)_2NH(l)$	−24.8		
$C_6H_5NH_2(l)$	7.4	35.6	45.7
$CF_4(g)$	−223.0	−212.3	62.5
$CCl_4(g)$	−24.0	−13.9	74.1
$CCl_4(l)$	−31.75	−15.0	51.7
$Cl_2CO(g)$	−52.8	−49.4	67.8
$CO(g)$	−26.4	−32.8	47.3
$CO_2(g)$	−94.05	−94.3	51.1
$OCS(g)$	−33.1	−39.6	55.3
$CS_2(g)$	28.0	16.0	56.8
$HCN(g)$	31.2	28.7	48.2
$Si(s)$	0.0	0.0	4.5
$Si(g)$	108.9		

Substance	ΔH_f^0 (kcal/mole)	ΔG_f^0 (kcal/mole)	S^0 (cal/deg-mole)
$SiH_4(g)$	−14.8	−9.4	48.7
$SiCl_4(g)$	−145.7	−136.2	79.2
$SiCl_4(l)$	−153.0	−136.9	57.2
$SiO_2(s)$	−205.4	−192.4	10.0
$Ge(s)$	0.0	0.0	10.1
$Ge(g)$	78.4	69.5	40.1
$GeCl_4(l)$	−130		
$Sn(s)$	0.0	0.0	12.3
$Sn(g)$	72	64	40
$SnO(s)$	−68.4	−61.5	13.5
$SnO_2(s)$	−138.8	−124.2	12.5
$SnCl_2(s)$	−83.6		
$SnCl_4(l)$	−130.3	−113.3	61.8
Group V			
$N_2(g)$	0.0	0.0	45.8
$NH_3(g)$	−10.9	−3.9	46.0
$N_2H_4(g)$	22.75	38.0	57.0
$NO(g)$	21.6	20.7	50.35
$NO_2(g)$	8.1	12.4	57.35
$N_2O(g)$	19.5	24.8	52.6
$N_2O_4(g)$	2.3	23.5	72.7
$NH_4Cl(s)$	−75.4	−48.7	22.6
$NH_4NO_3(s)$	−87.3		
$NH_4NO_2(s)$	−63.1		
$HNO_3(l)$	−41.4	−19.1	37.2
$P(s, white)$	0.0	0.0	10.6
$P(s, red)$	−4.4	−3.3	7.0
$P(s, black)$	−10.3		
$P(g)$	75.2	66.7	39.0
$PH_3(g)$	2.2	4.4	50.2
$P_4O_{10}(s)$	−720.0		
$PCl_3(g)$	−66.6	−61.8	74.5
$PCl_5(g)$	−88.7	−70.9	84.3
$As(s)$	0.0	0.0	8.4
$AsH_3(g)$	41.0		
Group VI			
$O_2(g)$	0.0	0.0	49.0
$O_3(g)$	34.0	39.1	56.8
$H_2O(g)$	−57.8	−54.6	45.1
$H_2O(l)$	−68.3	−56.7	16.7
$H_2O_2(l)$	−44.8	−27.2	

Substance	ΔH_f^0 (kcal/mole)	ΔG_f^0 (kcal/mole)	S^0 (cal/deg·mole)
$OCl_2(g)$	18.2	22.4	63.7
S(s, rhombic)	0.0	0.0	7.6
S(g)	66.4	56.7	40.1
$H_2S(g)$	−4.8	−7.9	49.2
$SF_6(g)$	−262	−237	69.5
$SO_2(g)$	−70.8	−71.8	59.4
$SO_3(g)$	−94.5	−88.5	61.2
$H_2SO_4(l)$	−193.9		
Se(s)	0.0	0.0	10.1
Se(g)	49.4	39.8	42.2
$H_2Se(g)$	20.5	17.0	52.9
$SeO_2(s)$	−56.4		
Te(s)	0.0	0.0	11.9
Te(g)	46.5	37.0	43.6
$H_2Te(g)$	36.9	33.1	56.0
Group VII			
$F_2(g)$	0.0	0.0	48.6
HF(g)	−64.2	−64.7	41.5
$Cl_2(g)$	0.0	0.0	53.3
HCl(g)	−22.1	−22.8	44.6
$Br_2(l)$	0.0	0.0	36.4
$Br_2(g)$	7.3	0.8	58.6
HBr(g)	−8.7	−12.7	47.4
BrF(g)	−11.0	−14.7	54.8
$BrF_3(g)$	−75	−69	69.9
$I_2(s)$	0.0	0.0	27.9
$I_2(g)$	14.9	4.6	62.3
HI(g)	6.2	0.3	49.3
$IF_5(g)$	−195.1	−178.3	78.7
$ICl_3(s)$	−21.1	−5.4	41.1
Transition Metals			
Hg(l)	0.0	0.0	18.5
$HgCl_2(s)$	−55.0	−44.4	ca. 34.5
$Hg_2Cl_2(s)$	−63.3	−50.35	46.8
HgO(s)	−21.6	−14.0	17.2
Cu(s)	0.0	0.0	8.0
Cu(g)	81.5		
CuO(s)	−37.6	−31.0	10.2
$CuSO_4(s)$	−184.0	−158.2	27.1
Ag(s)	0.0	0.0	10.2
AgCl(s)	−30.4	−26.2	23.0

Substance	ΔH_f^0 (kcal/mole)	ΔG_f^0 (kcal/mole)	S^0 (cal/deg-mole)
$AgNO_3(s)$	−29.4	−7.7	33.7
$Fe(s)$	0.0	0.0	6.5
$Fe(g)$	99.8	88.9	43.1
$FeO(s)$	−63.7	−58.4	12.9
$Fe_2O_3(s)$	−196.5	−177.1	21.5
$Ti(s)$	0.0	0.0	7.3
$Ti(g)$	112.6	101.9	43.1
$TiCl_4(l)$	−191.7		
$TiO_2(s)$	−225.8	−212.5	12.0

Solubility Product Constants (K_{sp})

Bromides

$PbBr_2$	4.6×10^{-6}
Hg_2Br_2	1.3×10^{-22}
$AgBr$	7.7×10^{-13}

Carbonates

$BaCO_3$	8.1×10^{-9}
$CdCO_3$	2.5×10^{-14}
$CaCO_3$	8.7×10^{-9}
$CuCO_3$	1.3×10^{-10}
$FeCO_3$	5×10^{-11}
$PbCO_3$	1.6×10^{-13}
$MgCO_3$	8×10^{-11}
$MnCO_3$	9×10^{-11}
$NiCO_3$	1.3×10^{-7}
Ag_2CO_3	1.3×10^{-11}
$SrCO_3$	1.3×10^{-9}
$ZnCO_3$	1.0×10^{-7}

Chlorides

$PbCl_2$	1.0×10^{-4}
Hg_2Cl_2	2.0×10^{-18}
$AgCl$	1.8×10^{-10}

Chromates

$PbCrO_4$	1.7×10^{-14}
Ag_2CrO_4	1.9×10^{-12}
$BaCrO_4$	2.2×10^{-10}

Cyanides

$Hg_2(CN)_2$	5×10^{-40}
$AgCN$	1.6×10^{-14}

Fluorides

BaF_2	1.6×10^{-6}
CaF_2	4×10^{-11}
LiF	5×10^{-3}
PbF_2	4×10^{-8}
MgF_2	6×10^{-9}
SrF_2	3×10^{-9}

Hydroxides

$Al(OH)_3$	3.7×10^{-15}
$Cd(OH)_2$	2×10^{-14}
$Ca(OH)_2$	6×10^{-6}
$Cr(OH)_3$	7×10^{-31}
$Co(OH)_3$	1×10^{-43}
$Co(OH)_2$	3×10^{-16}
$Cu(OH)_2$	2×10^{-19}
$Fe(OH)_3$	6×10^{-38}
$Fe(OH)_2$	2×10^{-15}
$Pb(OH)_2$	4×10^{-15}
$Mg(OH)_2$	1.8×10^{-11}
$Mn(OH)_2$	2×10^{-13}
$Ni(OH)_2$	2×10^{-16}
$Zn(OH)_2$	5×10^{-17}

Iodides

PbI_2	1.4×10^{-8}
Hg_2I_2	5×10^{-29}
AgI	8×10^{-17}

Oxalates

BaC_2O_4	1.6×10^{-7}
CaC_2O_4	1.9×10^{-9}
MgC_2O_4	9.6×10^{-7}
SrC_2O_4	6×10^{-8}
$Ag_2C_2O_4$	8.9×10^{-12}
PbC_2O_4	3×10^{-11}

Phosphates

Ag_3PO_4	2×10^{-21}
$Ca_3(PO_4)_2$	1×10^{-25}
$Mg_3(PO_4)_2$	4×10^{-13}

Sulfates

$BaSO_4$	1.0×10^{-10}
$CaSO_4$	2×10^{-5}
$PbSO_4$	1.6×10^{-8}
Hg_2SO_4	7×10^{-7}
Ag_2SO_4	6×10^{-5}
$SrSO_4$	3×10^{-7}

Sulfides

CdS	1×10^{-28}
CoS	1×10^{-22}
CuS	1×10^{-45}
FeS	1×10^{-18}
PbS	1×10^{-28}
MnS	1×10^{-11}
HgS	1×10^{-54}
NiS	1×10^{-22}
Ag_2S	1×10^{-50}
SnS	1×10^{-28}
ZnS	1×10^{-23}

Acid and Base Dissociation Constants in Water at 25°C

Acid Constants

Compound	K_a	Compound	K_a
HF	3.5×10^{-4}	o-FC$_6$H$_4$COOH	5.4×10^{-4}
HOCl	3.0×10^{-8}	m-FC$_6$H$_4$COOH	1.4×10^{-4}
HClO$_2$	1.0×10^{-2}	p-FC$_6$H$_4$COOH	7.2×10^{-5}
H$_2$S	$1.0 \times 10^{-7}, 1.3 \times 10^{-13}$	o-ClC$_6$H$_4$COOH	1.2×10^{-3}
H$_2$SO$_3$	$1.3 \times 10^{-2}, 5.6 \times 10^{-8}$	m-ClC$_6$H$_4$COOH	1.5×10^{-4}
C$_2$H$_5$SH	2.5×10^{-11}	p-ClC$_6$H$_4$COOH	1.0×10^{-4}
H$_2$Se	2.0×10^{-4}	o-NO$_2$C$_6$H$_4$COOH	7.0×10^{-3}
H$_2$Te	2.5×10^{-3}	m-NO$_2$C$_6$H$_4$COOH	3.4×10^{-4}
HNO$_2$	4.6×10^{-4}	p-NO$_2$C$_6$H$_4$COOH	3.9×10^{-4}
H$_3$PO$_4$	$7.5 \times 10^{-3}, 6.2 \times 10^{-8},$ 2.2×10^{-13}	p-CH$_3$C$_6$H$_4$COOH	4.2×10^{-5}
HCN	4.9×10^{-10}	p-CH$_3$OC$_6$H$_4$COOH	3.4×10^{-5}
H$_2$CO$_3$	$4.2 \times 10^{-7}, 4.8 \times 10^{-11}$	p-H$_2$NC$_6$H$_4$COOH	1.2×10^{-5}
H$_2$C$_2$O$_4$	$5.9 \times 10^{-2}, 6.4 \times 10^{-5}$		
HOOCCH$_2$COOH	$1.5 \times 10^{-3}, 2.0 \times 10^{-6}$	—COOH ~COOH	$1.3 \times 10^{-3},$ 3.9×10^{-6}
HCOOH	1.8×10^{-4}		
CH$_3$COOH	1.8×10^{-5}	C$_6$H$_5$OH	1.3×10^{-10}
FCH$_2$COOH	2.2×10^{-3}	o-NO$_2$C$_6$H$_4$OH	6.8×10^{-8}
F$_2$CHCOOH	6.0×10^{-2}	m-NO$_2$C$_6$H$_4$OH	5.3×10^{-9}
F$_3$CCOOH	6.0×10^{-1}	p-NO$_2$C$_6$H$_4$OH	7.0×10^{-8}
ClCH$_2$COOH	1.4×10^{-3}		
Cl$_2$CHCOOH	3.3×10^{-2}		
Cl$_3$CCOOH	2.0×10^{-1}		
BrCH$_2$COOH	1.4×10^{-3}		
ICH$_2$COOH	7.6×10^{-4}		
CH$_3$CH$_2$COOH	1.3×10^{-5}		
C$_6$H$_5$COOH	6.5×10^{-5}		

Base Constants

Compound	K_b	Compound	K_b
NH$_3$	1.8×10^{-5}	C$_6$H$_5$NH$_2$	3.8×10^{-10}
H$_2$NNH$_2$	1.0×10^{-6}	C$_6$H$_5$N(CH$_3$)$_2$	1.0×10^{-9}

Compound	K_b	Compound	K_b
H_2NOH	1.1×10^{-8}	$o\text{-}FC_6H_4NH_2$	1.6×10^{-11}
$(CH_3)_3N$	6.5×10^{-5}	$m\text{-}FC_6H_4NH_2$	2.5×10^{-11}
$(C_2H_5)_3N$	4.0×10^{-4}	$p\text{-}FC_6H_4NH_2$	4.5×10^{-10}
$(C_2H_5)_2NH$	1.3×10^{-3}	$o\text{-}NO_2C_6H_4NH_2$	2.0×10^{-14}
$C_2H_5NH_2$	5.6×10^{-4}	$m\text{-}NO_2C_6H_4NH_2$	2.5×10^{-12}
$C_6H_5CH_2NH_2$	2.0×10^{-5}	$p\text{-}CH_3C_6H_4NH_2$	1.0×10^{-9}
		$p\text{-}CH_3OC_6H_4NH_2$	2.2×10^{-9}

$$\begin{array}{c} H_2 \\ H_2C\text{---}C \\ | \qquad\quad NH \\ H_2C\text{---}C \\ H_2 \end{array} \qquad 2.0 \times 10^{-3}$$

$\qquad 1.4 \times 10^{-9}$

$(C_2H_5)_3P$	5.0×10^{-8}
$(C_2H_5)_3As$	5.0×10^{-12}

Standard Electrode Potentials

Half-Reaction	E^0 (V)
$F_2 + 2H^+ + 2e^- \rightleftharpoons 2HF$	+3.06
$F_2 + 2e^- \rightleftharpoons 2F^-$	2.87
$S_2O_8^{2-} + 2e^- \rightleftharpoons 2SO_4^{2-}$	2.01
$Co^{3+} + e^- \rightleftharpoons Co^{2+}$	1.82
$H_2O_2 + 2H^+ + 2e^- \rightleftharpoons 2H_2O$	1.77
$MnO_4^- + 4H^+ + 3e^- \rightleftharpoons MnO_2 + H_2O$	1.70
$PbO_2 + SO_4^{2-} + 4H^+ + 2e^- \rightleftharpoons PbSO_4 + 2H_2O$	1.69
$2HClO + 2H^+ + 2e^- \rightleftharpoons Cl_2 + 2H_2O$	1.63
$2HBrO + 2H^+ + 2e^- \rightleftharpoons Br_2 + 2H_2O$	1.59
$MnO_4^- + 8H^+ + 5e^- \rightleftharpoons Mn^{2+} + 4H_2O$	1.51
$2HIO + 2H^+ + 2e^- \rightleftharpoons I_2 + 2H_2O$	1.45
$ClO_3^- + 6H^+ + 6e^- \rightleftharpoons Cl^- + 3H_2O$	1.45
$Ce^{4+} + e^- \rightleftharpoons Ce^{3+}$	1.44
$Cl_2 + 2e^- \rightleftharpoons 2Cl^-$	1.36
$Cr_2O_7^{2-} + 14H^+ + 6e^- \rightleftharpoons 2Cr^{3+} + 7H_2O$	1.33
$O_2 + 4H^+ + 4e^- \rightleftharpoons 2H_2O$	1.23
$Br_2 + 2e^- \rightleftharpoons 2Br^-$	1.06
$HNO_2 + H^+ + e^- \rightleftharpoons NO + H_2O$	1.00
$NO_3^- + 4H^+ + 3e^- \rightleftharpoons NO + 2H_2O$	0.96
$NO_3^- + 3H^+ + 2e^- \rightleftharpoons HNO_2 + H_2O$	0.94
$2Hg^{2+} + 2e^- \rightleftharpoons Hg_2^{2+}$	0.92
$HO_2^- + H_2O + 2e^- \rightleftharpoons 3OH^-$	0.88
$Hg^{2+} + 2e^- \rightleftharpoons Hg$	0.85
$2NO_3^- + 4H^+ + 2e^- \rightleftharpoons N_2O_4 + 6H_2O$	0.80
$Ag^+ + e^- \rightleftharpoons Ag$	0.80
$Hg_2^{2+} + 2e^- \rightleftharpoons 2Hg$	0.79
$Fe^{3+} + e^- \rightleftharpoons Fe^{2+}$	0.77
$MnO_4^- + 2H_2O + 2e^- \rightleftharpoons MnO_2 + 4OH^-$	0.60
$I_2 + 2e^- \rightleftharpoons 2I^-$	0.54
$Cu^+ + e^- \rightleftharpoons Cu$	0.52
$O_2 + 2H_2O + 4e^- \rightleftharpoons 4OH^-$	0.40
$Ag(NH_3)_2^+ + e^- \rightleftharpoons Ag + 2NH_3$	0.37
$ClO_4^- + H_2O + 2e^- \rightleftharpoons ClO_3^- + 2OH^-$	0.36
$Cu^{2+} + 2e^- \rightleftharpoons Cu$	0.34
$AgCl + e^- \rightleftharpoons Ag + Cl^-$	0.22
$HCHO + 2H^+ + 2e^- \rightleftharpoons CH_3OH$	0.19
$SO_4^{2-} + 4H^+ + 2e^- \rightleftharpoons H_2SO_3 + H_2O$	0.17
$Cu^{2+} + e^- \rightleftharpoons Cu^+$	0.15
$Sn^{4+} + 2e^- \rightleftharpoons Sn^{2+}$	0.15
$NO_3^- + H_2O + 2e^- \rightleftharpoons NO_2^- + 2OH^-$	0.01
$2H^+ + 2e^- \rightleftharpoons H_2$	0.00
$O_2 + H_2O + 2e^- \rightleftharpoons HO_2^- + OH^-$	−0.08
$Pb^{2+} + 2e^- \rightleftharpoons Pb$	−0.13
$Sn^{2+} + 2e^- \rightleftharpoons Sn$	−0.14
$Ni^{2+} + 2e^- \rightleftharpoons Ni$	−0.25
$H_3PO_4 + 2H^+ + 2e^- \rightleftharpoons H_3PO_3 + H_2O$	−0.28
$Co^{2+} + 2e^- \rightleftharpoons Co$	−0.28
$Ag(CN)_2^- + e^- \rightleftharpoons Ag + 2CN^-$	−0.31
$PbSO_4 + 2e^- \rightleftharpoons Pb + SO_4^{2-}$	−0.36
$Cd^{2+} + 2e^- \rightleftharpoons Cd$	−0.40
$Cr^{3+} + e^- \rightleftharpoons Cr^{2+}$	−0.41
$Fe^{2+} + 2e^- \rightleftharpoons Fe$	−0.44
$S + 2e^- \rightleftharpoons S^{2-}$	−0.48
$2CO_2(g) + 2H^+ + 2e^- \rightleftharpoons H_2C_2O_4$	−0.49
$Cr^{3+} + 3e^- \rightarrow Cr$	−0.74
$Zn^{2+} + 2e^- \rightarrow Zn$	−0.76
$2H_2O + 2e^- \rightleftharpoons H_2 + 2OH^-$	−0.83
$Se + 2e^- \rightleftharpoons Se^{2-}$	−0.92
$SO_4^{2-} + H_2O + 2e^- \rightleftharpoons SO_3^{2-} + 2OH^-$	−0.93
$Mn^{2+} + 2e^- \rightleftharpoons Mn$	−1.18
$Al^{3+} + 3e^- \rightleftharpoons Al$	−1.66
$Be^{2+} + 2e^- \rightleftharpoons Be$	−1.85
$Al(OH)_4^- + 3e^- \rightleftharpoons Al + 4OH^-$	−2.35
$Mg^{2+} + 2e^- \rightleftharpoons Mg$	−2.36
$Na^+ + e^- \rightleftharpoons Na$	−2.71
$Ca^{2+} + 2e^- \rightleftharpoons Ca$	−2.87
$Sr^{2+} + 2e^- \rightleftharpoons Sr$	−2.89
$Ba^{2+} + 2e^- \rightleftharpoons Ba$	−2.91
$K^+ + e^- \rightleftharpoons K$	−2.93
$Rb^+ + e^- \rightleftharpoons Rb$	−2.99
$Cs^+ + e^- \rightleftharpoons Cs$	−3.02
$Li^+ + e^- \rightleftharpoons Li$	−3.05

Overall Formation Constants of Coordination Compounds

Halo Complexes

AlF_6^{3-}	1×10^{20}
$AgCl_2^-$	3×10^5
$AgCl_4^{3-}$	5×10^6
$HgCl_4^{2-}$	5.0×10^{15}
$AgBr_4^{3-}$	8×10^9
$CdBr_4^{2-}$	6.0×10^3
$HgBr_4^{2-}$	1.0×10^{21}
$PbBr_4^{2-}$	1×10^4
AgI_4^{3-}	1×10^{14}
CdI_4^{2-}	2.5×10^6
HgI_4^{2-}	1.9×10^{30}
PbI_4^{2-}	1.6×10^6

Cyano Complexes

$Ni(CN)_4^{2-}$	1.0×10^{22}
$Fe(CN)_6^{3-}$	1.0×10^{31}
$Fe(CN)_6^{4-}$	1.0×10^{24}
$Ag(CN)_2^-$	1.0×10^{20}
$Cd(CN)_4^{2-}$	5.9×10^{18}
$Hg(CN)_4^{2-}$	3.0×10^{41}
$Zn(CN)_4^{2-}$	1.0×10^{17}

Ammine Complexes

$Co(NH_3)_6^{2+}$	5.0×10^4
$Co(NH_3)_6^{3+}$	4.6×10^{33}
$Ni(NH_3)_6^{2+}$	2.0×10^8
$Cu(NH_3)_4^{2+}$	1.1×10^{13}
$Ag(NH_3)_2^+$	1.6×10^7
$Zn(NH_3)_4^{2+}$	2.8×10^9
$Cd(NH_3)_4^{2+}$	3.0×10^7

Others

$Fe(C_2O_4)_3^{3-}$	4.0×10^{21}
$Al(C_2O_4)_3^{3-}$	2.0×10^{16}
HgS_2^{2-}	3.5×10^{54}
$Fe(NCS)^{2+}$	1.1×10^3
$Hg(NCS)_4^{2-}$	7.8×10^{21}
$Pb(C_2H_3O_2)_4^{2-}$	1.0×10^3
$Cu(en)_2^{2+}$	4×10^{19}
$Ni(en)_3^{2+}$	3×10^{18}

Vapor Pressure of Water (mm Hg) from 0 to 100°C

T	p	T	p	T	p	T	p
0	4.6	26	25.2	51	97.2	76	301.4
1	4.9	27	26.7	52	102.1	77	314.1
2	5.3	28	28.4	53	107.2	78	327.3
3	5.7	29	30.0	54	112.5	79	341.0
4	6.1	30	31.8	55	118.0	80	355.1
5	6.5	31	33.7	56	123.8	81	369.7
6	7.0	32	35.7	57	129.8	82	384.9
7	7.5	33	37.7	58	136.1	83	400.6
8	8.1	34	39.9	59	142.6	84	416.8
9	8.6	35	42.2	60	149.4	85	433.6
10	9.2	36	44.6	61	156.4	86	450.9
11	9.8	37	47.1	62	163.8	87	468.7
12	10.5	38	49.7	63	171.4	88	487.1
13	11.2	39	52.4	64	179.3	89	506.1
14	12.0	40	55.3	65	187.5	90	525.8
15	12.8	41	58.3	66	196.1	91	546.1
16	13.6	42	61.5	67	205.0	92	567.0
17	14.5	43	64.8	68	214.2	93	588.6
18	15.5	44	68.3	69	223.7	94	611.0
19	16.5	45	71.9	70	233.7	95	634.0
20	17.5	46	75.7	71	243.9	96	658.0
21	18.7	47	79.6	72	254.6	97	682.0
22	19.8	48	83.7	73	265.7	98	707.3
23	21.1	49	88.0	74	277.2	99	733.2
24	22.4	50	92.5	75	289.1	100	760.0
25	23.8						

Four-Place Common Logarithms

N	0	1	2	3	4	5	6	7	8	9	Proportional Parts								
											1	2	3	4	5	6	7	8	9
10	0000	0043	0086	0128	0170	0212	0253	0294	0334	0374	4	8	12	17	21	25	29	33	37
11	0414	0453	0492	0531	0569	0607	0645	0682	0719	0755	4	8	11	15	19	23	26	30	34
12	0792	0828	0864	0899	0934	0969	1004	1038	1072	1106	3	7	10	14	17	21	24	28	31
13	1139	1173	1206	1239	1271	1303	1335	1367	1399	1430	3	6	10	13	16	19	23	26	29
14	1461	1492	1523	1553	1584	1614	1644	1673	1703	1732	3	6	9	12	15	18	21	24	27
15	1761	1790	1818	1847	1875	1903	1931	1959	1987	2014	3	6	8	11	14	17	20	22	25
16	2041	2068	2095	2122	2148	2175	2201	2227	2253	2279	3	5	8	11	13	16	18	21	24
17	2304	2330	2355	2380	2405	2430	2455	2480	2504	2529	2	5	7	10	12	15	17	20	22
18	2553	2577	2601	2625	2648	2672	2695	2718	2742	2765	2	5	7	9	12	14	16	19	21
19	2788	2810	2833	2856	2878	2900	2923	2945	2967	2989	2	4	7	9	11	13	16	18	20
20	3010	3032	3054	3075	3096	3118	3139	3160	3181	3201	2	4	6	8	11	13	15	17	19
21	3222	3243	3263	3284	3304	3324	3345	3365	3385	3404	2	4	6	8	10	12	14	16	18
22	3424	3444	3464	3483	3502	3522	3541	3560	3579	3598	2	4	6	8	10	12	14	15	17
23	3617	3636	3655	3674	3692	3711	3729	3747	3766	3784	2	4	6	7	9	11	13	15	17
24	3802	3820	3838	3856	3874	3892	3909	3927	3945	3962	2	4	5	7	9	11	12	14	16
25	3979	3997	4014	4031	4048	4065	4082	4099	4116	4133	2	3	5	7	9	10	12	14	15
26	4150	4166	4183	4200	4216	4232	4249	4265	4281	4298	2	3	5	7	8	10	11	13	15
27	4314	4330	4346	4362	4378	4393	4409	4425	4440	4456	2	3	5	6	8	9	11	13	14
28	4472	4487	4502	4518	4533	4548	4564	4579	4594	4609	2	3	5	6	8	9	11	12	14
29	4624	4639	4654	4669	4683	4698	4713	4728	4742	4757	1	3	4	6	7	9	10	12	13
30	4771	4786	4800	4814	4829	4843	4857	4871	4886	4900	1	3	4	6	7	9	10	11	13
31	4914	4928	4942	4955	4969	4983	4997	5011	5024	5038	1	3	4	6	7	8	10	11	12
32	5051	5065	5079	5092	5105	5119	5132	5145	5159	5172	1	3	4	5	7	8	9	11	12
33	5185	5198	5211	5224	5237	5250	5263	5276	5289	5302	1	3	4	5	6	8	9	10	12
34	5315	5328	5340	5353	5366	5378	5391	5403	5416	5428	1	3	4	5	6	8	9	10	11
35	5441	5453	5465	5478	5490	5502	5514	5527	5539	5551	1	2	4	5	6	7	9	10	11
36	5563	5575	5587	5599	5611	5623	5635	5647	5658	5670	1	2	4	5	6	7	8	10	11
37	5682	5694	5705	5717	5729	5740	5752	5763	5775	5786	1	2	3	5	6	7	8	9	10
38	5798	5809	5821	5832	5843	5855	5866	5877	5888	5899	1	2	3	5	6	7	8	9	10
39	5911	5922	5933	5944	5955	5966	5977	5988	5999	6010	1	2	3	4	5	7	8	9	10
40	6021	6031	6042	6053	6064	6075	6085	6096	6107	6117	1	2	3	4	5	6	8	9	10
41	6128	6138	6149	6160	6170	6180	6191	6201	6212	6222	1	2	3	4	5	6	7	8	9
42	6232	6243	6253	6263	6274	6284	6294	6304	6314	6325	1	2	3	4	5	6	7	8	9
43	6335	6345	6355	6365	6375	6385	6395	6405	6415	6425	1	2	3	4	5	6	7	8	9
44	6435	6444	6454	6464	6474	6484	6493	6503	6513	6522	1	2	3	4	5	6	7	8	9
45	6532	6542	6551	6561	6571	6580	6590	6599	6609	6618	1	2	3	4	5	6	7	8	9
46	6628	6637	6646	6656	6665	6675	6684	6693	6702	6712	1	2	3	4	5	6	7	7	8
47	6721	6730	6739	6749	6758	6767	6776	6785	6794	6803	1	2	3	4	5	5	6	7	8
48	6812	6821	6830	6839	6848	6857	6866	6875	6884	6893	1	2	3	4	4	5	6	7	8
49	6902	6911	6920	6928	6937	6946	6955	6964	6972	6981	1	2	3	4	4	5	6	7	8
50	6990	6998	7007	7016	7024	7033	7042	7050	7059	7067	1	2	3	3	4	5	6	7	8
51	7076	7084	7093	7101	7110	7118	7126	7135	7143	7152	1	2	3	3	4	5	6	7	8
52	7160	7168	7177	7185	7193	7202	7210	7218	7226	7235	1	2	2	3	4	5	6	7	7
53	7243	7251	7259	7267	7275	7284	7292	7300	7308	7316	1	2	2	3	4	5	6	6	7
54	7324	7332	7340	7348	7356	7364	7372	7380	7388	7396	1	2	2	3	4	5	6	6	7
N	0	1	2	3	4	5	6	7	8	9	1	2	3	4	5	6	7	8	9

N	0	1	2	3	4	5	6	7	8	9	Proportional Parts								
											1	2	3	4	5	6	7	8	9
55	7404	7412	7419	7427	7435	7443	7451	7459	7466	7474	1	2	2	3	4	5	5	6	7
56	7482	7490	7497	7505	7513	7520	7528	7536	7543	7551	1	2	2	3	4	5	5	6	7
57	7559	7566	7574	7582	7589	7597	7604	7612	7619	7627	1	2	2	3	4	5	5	6	7
58	7634	7642	7649	7657	7664	7672	7679	7686	7694	7701	1	1	2	3	4	4	5	6	7
59	7709	7716	7723	7731	7738	7745	7752	7760	7767	7774	1	1	2	3	4	4	5	6	7
60	7782	7789	7796	7803	7810	7818	7825	7832	7839	7846	1	1	2	3	4	4	5	6	6
61	7853	7860	7868	7875	7882	7889	7896	7903	7910	7917	1	1	2	3	4	4	5	6	6
62	7924	7931	7938	7945	7952	7959	7966	7973	7980	7987	1	1	2	3	3	4	5	6	6
63	7993	8000	8007	8014	8021	8028	8035	8041	8048	8055	1	1	2	3	3	4	5	5	6
64	8062	8069	8075	8082	8089	8096	8102	8109	8116	8122	1	1	2	3	3	4	5	5	6
65	8129	8136	8142	8149	8156	8162	8169	8176	8182	8189	1	1	2	3	3	4	5	5	6
66	8195	8202	8209	8215	8222	8228	8235	8241	8248	8254	1	1	2	3	3	4	5	5	6
67	8261	8267	8274	8280	8287	8293	8299	8306	8312	8319	1	1	2	3	3	4	5	5	6
68	8325	8331	8338	8344	8351	8357	8363	8370	8376	8382	1	1	2	3	3	4	4	5	6
69	8388	8395	8401	8407	8414	8420	8426	8432	8439	8445	1	1	2	2	3	4	4	5	6
70	8451	8457	8463	8470	8476	8482	8488	8494	8500	8506	1	1	2	2	3	4	4	5	6
71	8513	8519	8525	8531	8537	8543	8549	8555	8561	8567	1	1	2	2	3	4	4	5	5
72	8573	8579	8585	8591	8597	8603	8609	8615	8621	8627	1	1	2	2	3	4	4	5	5
73	8633	8639	8645	8651	8657	8663	8669	8675	8681	8686	1	1	2	2	3	4	4	5	5
74	8692	8698	8704	8710	8716	8722	8727	8733	8739	8745	1	1	2	2	3	4	4	5	5
75	8751	8756	8762	8768	8774	8779	8785	8791	8797	8802	1	1	2	2	3	3	4	5	5
76	8808	8814	8820	8825	8831	8837	8842	8848	8854	8859	1	1	2	2	3	3	4	5	5
77	8865	8871	8876	8882	8887	8893	8899	8904	8910	8915	1	1	2	2	3	3	4	4	5
78	8921	8927	8932	8938	8943	8949	8954	8960	8965	8971	1	1	2	2	3	3	4	4	5
79	8976	8982	8987	8993	8998	9004	9009	9015	9020	9025	1	1	2	2	3	3	4	4	5
80	9031	9036	9042	9047	9053	9058	9063	9069	9074	9079	1	1	2	2	3	3	4	4	5
81	9085	9090	9096	9101	9106	9112	9117	9122	9128	9133	1	1	2	2	3	3	4	4	5
82	9138	9143	9149	9154	9159	9165	9170	9175	9180	9186	1	1	2	2	3	3	4	4	5
83	9191	9196	9201	9206	9212	9217	9222	9227	9232	9238	1	1	2	2	3	3	4	4	5
84	9243	9248	9253	9258	9263	9269	9274	9279	9284	9289	1	1	2	2	3	3	4	4	5
85	9294	9299	9304	9309	9315	9320	9325	9330	9335	9340	1	1	2	2	3	3	4	4	5
86	9345	9350	9355	9360	9365	9370	9375	9380	9385	9390	1	1	2	2	3	3	4	4	5
87	9395	9400	9405	9410	9415	9420	9425	9430	9435	9440	0	1	1	2	2	3	3	4	4
88	9445	9450	9455	9460	9465	9469	9474	9479	9484	9489	0	1	1	2	2	3	3	4	4
89	9494	9499	9504	9509	9513	9518	9523	9528	9533	9538	0	1	1	2	2	3	3	4	4
90	9542	9547	9552	9557	9562	9566	9571	9576	9581	9586	0	1	1	2	2	3	3	4	4
91	9590	9595	9600	9605	9609	9614	9619	9624	9628	9633	0	1	1	2	2	3	3	4	4
92	9638	9643	9647	9652	9657	9661	9666	9671	9675	9680	0	1	1	2	2	3	3	4	4
93	9685	9689	9694	9699	9703	9708	9713	9717	9722	9727	0	1	1	2	2	3	3	4	4
94	9731	9736	9741	9745	9750	9754	9759	9763	9768	9773	0	1	1	2	2	3	3	4	4
95	9777	9782	9786	9791	9795	9800	9805	9809	9814	9818	0	1	1	2	2	3	3	4	4
96	9823	9827	9832	9836	9841	9845	9850	9854	9859	9863	0	1	1	2	2	3	3	4	4
97	9868	9872	9877	9881	9886	9890	9894	9899	9903	9908	0	1	1	2	2	3	3	4	4
98	9912	9917	9921	9926	9930	9934	9939	9943	9948	9952	0	1	1	2	2	3	3	4	4
99	9956	9961	9965	9969	9974	9978	9983	9987	9991	9996	0	1	1	2	2	3	3	3	4
N	0	1	2	3	4	5	6	7	8	9	1	2	3	4	5	6	7	8	9

Answers to Selected Problems

CHAPTER 1

2. 2.01×10^{-30} lb/electron
3. (a) 305 dimes (b) 1.68 lb
5. $-40°$
7. tin cube weighs 26.3 g more than lead cube
9. 8.37 ml
10. (a) 1.61 km (d) 100 nm (f) 2.37 liter (g) 28,344 mg
11. 277 cal
13. 431 kJ/mole

CHAPTER 2

2. $^{31}_{15}$P = 15 protons, 16 neutrons; $^{206}_{82}$Pb = 82 protons, 124 neutrons
3. (a) 3.06×10^{21} atoms (c) 1.36×10^{23} atoms
4. (a) 4.66×10^{-23} g/atom
6. same
8. one billion lead atoms
11. 0.237 g-atom

CHAPTER 3

1. (a) 202.3 g
2. (a) 7.99% Li, 92.01% Br (c) 49.49% Fe, 50.51% F (d) 56.58% K, 8.69% C, 34.73% O
3. (a) NO (c) Fe_3O_4 (e) C_5H_5N
4. (a) 1.26 moles (c) 15.1 g (e) 1.26 g-atoms
6. (a) B_2H_6
8. (a) 0.77 mole (b) 6.92 g (c) 5.79×10^{23} molecules
10. 61.9 g
12. (a) 88 kg (c) 270 g
14. 49 kg
16. 24.8 g, 82.7%
18. 115 g
20. 86%

CHAPTER 5

1. (a) 3×10^{15} sec^{-1} (b) 1.99×10^{-11} erg (c) 234 kcal
6. 6.02×10^{-11} erg; 24.1×10^{-11} erg; 54.2×10^{-11} erg

7. -21.7×10^{-12} erg; -5.43×10^{-12} erg; -2.41×10^{-12} erg
14. $1s, 2s, 2p, 3s, 3p, 3d, 4s, 4p$
17. (b) $1s^2 2s^2 2p^3$
   (e) $1s^2 2s^2 2p^6 3s^2 3p^6$
   (f) $1s^2 2s^2 2p^6 3s^2 3p^6 3d^6 4s^2$
   (k) $1s^2 2s^2 2p^6 3s^2 3p^6 3d^{10} 4s^2 4p^6 4d^{10} 4f^{14} 5s^2 5p^6 5d^{10} 5f^3 6s^2 6p^6 6d^1 7s^2$
   (n) $1s^2 2s^2 2p^6 3s^2 3p^6 3d^{10} 4s^2 4p^6 4d^{10} 5s^2 5p^6$

CHAPTER 6

3. (a) chlorous acid (c) nitrous acid (e) sodium hydrogen carbonate
   (g) nickel oxide (i) silver cyanide (k) calcium phosphate
   (m) aluminum nitrate
4. (b) $CsClO_3$ (e) H_2S (g) $Cu_3(PO_4)_2$ (j) $Cr_2(SO_4)_3$ (m) In_2O_3
   (u) $Ba(H_2PO_4)_2$
6. -5.96×10^{-11} erg
8. 816 kcal/mole
10. (a) NaCl structure (c) zinc blende structure

CHAPTER 7

1. (b) methylbutane; isopentane
   (c) 2,2-dibromopropane
   (e) cyclohexane
   (g) 2-methyl-1-butene
   (i) 3-methyl-1,3-pentadiene
   (j) 1-chloro-4-fluorobenzene; *p*-chlorofluorobenzene
   (m) 2-methyl-2-propanol; *t*-butyl alcohol
   (n) 2-pentanone; methyl *n*-propyl ketone
   (q) sodium benzoate
   (r) propyl ethanoate; propyl acetate
   (x) 2-aminopropane; isopropylamine

2. (c) 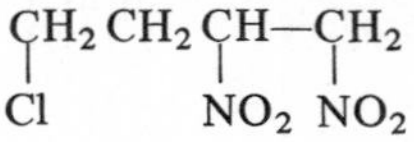 $\underset{\underset{Cl}{|}}{CH_2} \ CH_2 \ \underset{\underset{NO_2}{|}}{CH} - \underset{\underset{NO_2}{|}}{CH_2}$ (o) 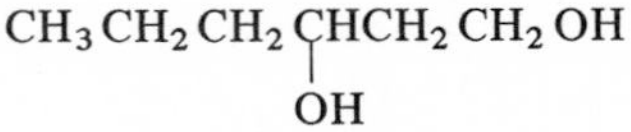 $CH_3 CH_2 CH_2 \underset{\underset{OH}{|}}{CH} CH_2 CH_2 OH$

   (f) 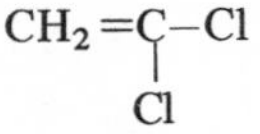 $CH_2 = \underset{\underset{Cl}{|}}{C} - Cl$

   (p)

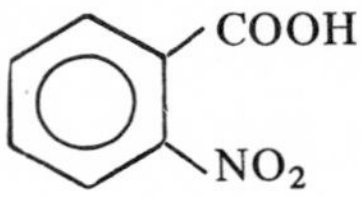

   (h)

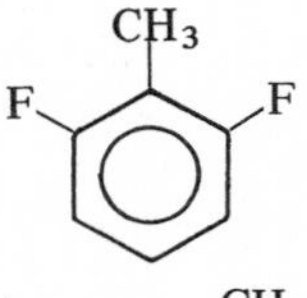

   (r) 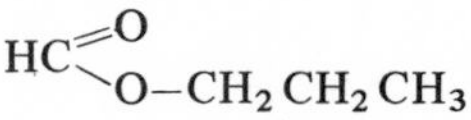 $HC \overset{O}{\underset{O-CH_2 CH_2 CH_3}{\diagup\kern-0.5em\diagdown}}$

   (l) 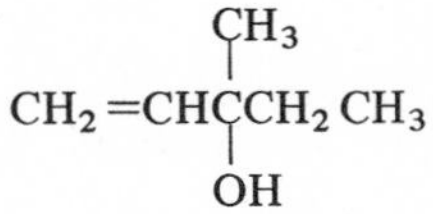 $CH_2 = CH \underset{\underset{OH}{|}}{\overset{\overset{CH_3}{|}}{C}} CH_2 CH_3$ (v) 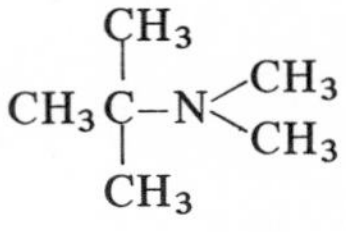 $CH_3 \underset{\underset{CH_3}{|}}{\overset{\overset{CH_3}{|}}{C}} - N \diagdown_{CH_3}^{CH_3}$

3. (b) dinitrogen oxide (nitrous oxide) (e) carbon disulfide (g) sulfur hexafluoride
4. (a) LiBr (d) CF_4 (f) $SnCl_4$ (j) $FeCl_2$

5.

(a)

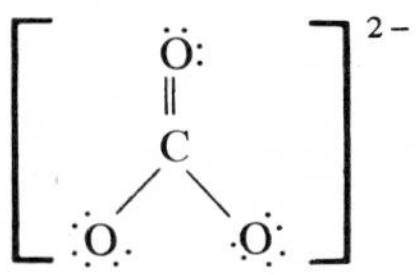

(d)

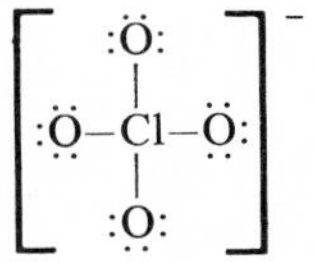

(g)

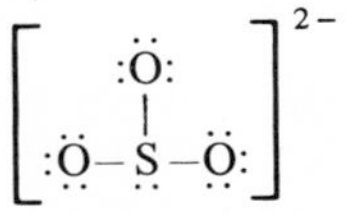

(k)

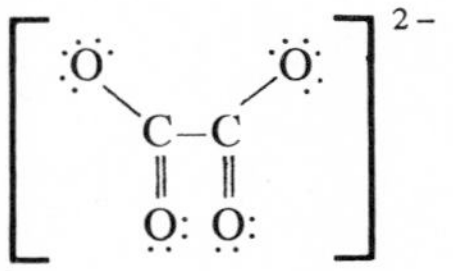

(p)

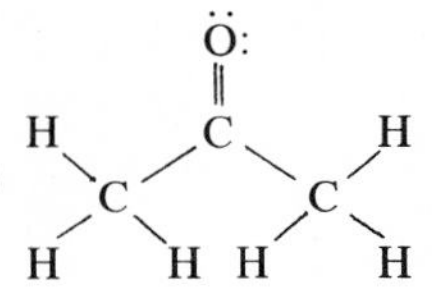

(v)

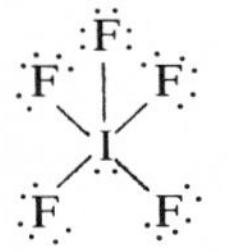

(z)

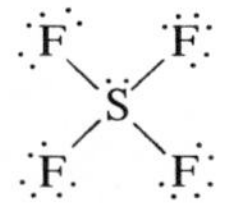

12. $[:\ddot{N}=C=\ddot{N}:]^{2-}$; CO_2; linear

16.

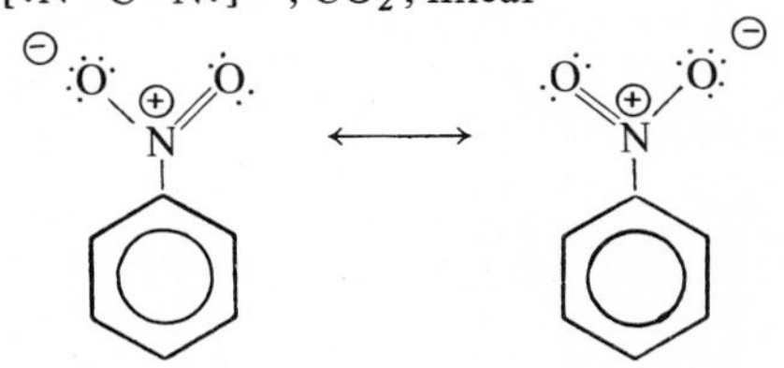

18.

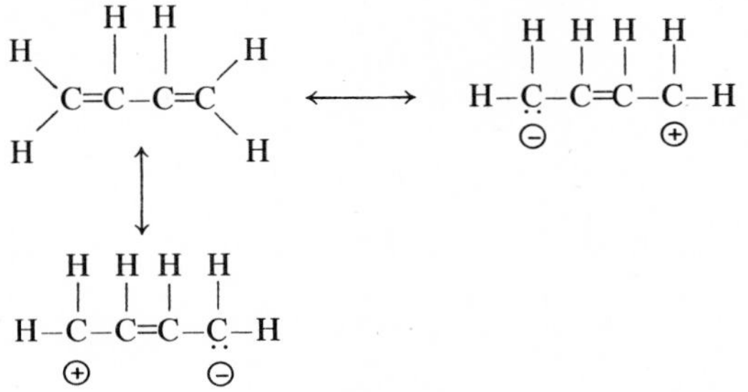

20. (c) $(\sigma 1s)^2 (\sigma^* 1s)^2 (\sigma 2s)^2 (\sigma^* 2s)^2$; bond order = 0; unstable
   (g) $(\sigma 1s)^2 (\sigma^* 1s)^2 (\sigma 2s)^2 (\sigma^* 2s)^2 (\pi 2p)^2 (\pi 2p_z)^2 (\sigma 2p_x)^2 (\pi^* 2p_y)^2 (\pi^* 2p_z)^1$;
   bond order = 1.5
   (j) $(\sigma 1s)^2 (\sigma^* 1s)^2 (\sigma 2s)^2 (\sigma^* 2s)^2 (\pi 2p)^2 (\pi 2p_z)^2 (\sigma 2p_x)^2$; bond order = 3

CHAPTER 8

1. (a) structural isomers: a and e; b, f, and i; g and m; j and k
   (b) geometrical isomerism: b, d, k
   (c) optical isomerism: j, k
3. (a) linear, sp (b) tetrahedral, sp^3 (c) square pyramid, $sp^3 d^2$ (d) see-saw, $sp^3 d$ (e) trigonal pyramid, sp^3 (g) bent, sp^2 (i) trigonal planar, sp^2 (k) octahedral, $sp^3 d^2$ (m) square planar, $sp^3 d^2$ (n) T-shaped, $sp^3 d$ (p) linear, $sp^3 d$ (q) bent, sp^3 (u) tetrahedral, sp^3
5. XeF_4
6. (b) $As(CH_3)ClH$: trigonal pyramid, optical isomerism; (d) $PF_2 Cl_3$: trigonal bipyramid, geometrical isomerism
9. (c) $C_3 H_6 Br_2$: 1,1-dibromopropane; 1,2-dibromopropane; 1,3-dibromopropane; 2,2-dibromopropane
   (d) $C_6 H_{14}O$: 32 structural isomers
10. blue; red; black
11. $(\sigma 1s)^2 (\sigma^* 1s)^2 (\sigma 2s)^2 (\sigma^* 2s)^2 (\pi 2p_y)^2 (\pi 2p_z)^2 (\sigma 2p_x)^2$
15. methanol
16.

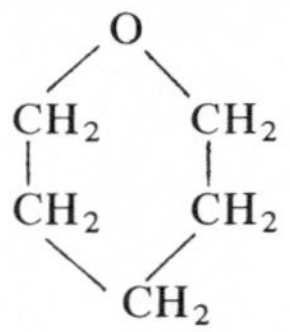

```
          O
        /   \
     CH₂     CH₂
      |       |
     CH₂     CH₂
        \   /
         CH₂
```

CHAPTER 9

3. (a) $MgCl_2$ (b) PCl_5 (c) $MgCl_2$ (d) $MgCl_2$ (e) $MgCl_2$ (f) $MgCl_2$
4. 250 cal
5. (a) triclinic (c) orthorhombic (e) monoclinic
8. 7a: (a) 1-pentene (b) 1-heptene (c) 1-heptene
   7c: (a) methyl acetate (b) acetic acid (c) acetic acid
   7f: (a) methylamine (b) methyl alcohol (c) methyl alcohol

CHAPTER 10

1. 474 ml
3. 1.92 liters
5. 4.42 g/liters
6. 628 liters
8. 66.3 atm
9. 0.00801 mole; 0.481 g
11. (a) 17.2 liters (b) 36.6 liters
13. 476 ml
15. 54.4
17. (a) $C_2 H_3$ (b) 54.2 (c) $C_4 H_6$

19. 7.90 liters
21. (a) $Ni(CO)_4$
23. p_{H_2} 450 torr; p_{O_2} 105 torr; p_{N_2} 50 torr
26. 63.6
31. 2.79 cal/mole-$^{\circ}$K
32. 21°C

CHAPTER 11

3. AgF, $\Delta G = -3.39$ kcal/mole; AgCl, $\Delta G = +13.46$ kcal/mole
4. (a) Dissolve 50.6 g KNO_3 and dilute to 1.00 liter.
   (c) Dissolve 5.51 g $CaCl_2 \cdot 2H_2O$ and dilute to 500 ml.
   (e) Dissolve 88.2 g NaOH in 500 g H_2O.
5. (a) 2.16 g Ag^+ (d) 16.3 g Hg_2^{2+}
6. (a) 11.9M (b) 15.7m (c) 0.221 mole fraction
8. 1.99m
10. 17.9 ml
12. (a) 20.0 wt-% (b) 7.81m (c) 0.123 mole fraction
15. 0.492 mole fraction $CHCl_3$; 0.508 mole fraction CCl_4
17. 21.1 torr
18. (a) 1722 g (b) 53.8 moles
20. no
22. 138 torr
23. 1.2×10^3
25. 156
29. 120
31. (a) C_5H_4 (b) 126 (c) $C_{10}H_8$
32. 32.0%
36. (a) strong (c) weak (e) weak (g) weak (i) non (k) strong (m) non (o) strong (q) non (s) weak
37. 7.4 ml
39. 9.97 liters

CHAPTER 12

4. (a) 1st order (b) 3rd order
5. (a) 3rd order (d) 5/2 order
6. (c) 0.20 liter/mole-min
7. (a) 2.1 liter$^{3/2}$/mole$^{3/2}$-min (b) 1.2×10^{-4} mole/liter-min
8. (b) 1.9×10^{-2} mole/liter-sec

CHAPTER 13

3. (a) (1) increase (2) no effect (3) no effect (4) no effect
6. 17 g

7. (a) 4.0 (b) 0.333 mole CO_2, 0.333 mole H_2 (c) 0.167 mole CO (d) 66.7% (e) 0.317 mole CO_2 (f) 0.47 mole CO
9. (c) 0.78 mole (d) 0.302 mole
10. (a) 2.0 (b) 0.58 mole
13. (a) increase (c) decrease (e) increase (g) decrease
14. (a) for CCl_4: $\Delta H^0 = -14.1$ kcal, $\Delta G^0 = -57.1$ kcal, $\Delta S^0 = 144$ cal/deg; for $SiCl_4$: $\Delta H^0 = -4.2$ kcal, $\Delta G^0 = -33.3$ kcal, $\Delta S^0 = 97.7$ cal/deg (b) very low reaction rate
16. $\Delta H^0 = -669.5$ kcal/mole; $\Delta G^0 = -688.4$ kcal/mole; $\Delta S^0 = 62.1$ cal/deg-mole; $K = 5 \times 10^{504}$
17. for reaction (1): (a) -139 kcal; -138.4 kcal (b) ~25 cal/deg; 30 cal/deg (c) from ΔG_f^0 values -147.2 kcal
    for reaction (2): (a) -29.5 kcal (b) ~ -25 cal/deg
19. ~55 kcal/mole; unstable; ΔG^0 is positive
20. (a) $CH_3\overset{\displaystyle O}{\overset{\|}{C}}CH_3$ (c) $CH_3\overset{\displaystyle O}{\overset{\|}{C}}-NH_2$
22. (a) (1) -18 kcal (2) -29.5 kcal (3) -29.4 kcal (b) -37.1 cal/deg (c) -18.4 kcal; $K_p = 2.6 \times 10^{13}$

CHAPTER 14

1. Reactions occur in b, c, d, f, g, h, i, j, k
3. 2.74×10^{-8}
4. (a) 8.8×10^{-7} mole/liter (b) 1.65×10^{-4} mole/liter
5. 3.9×10^{-3} mole/liter Mg^{2+}; 2.6×10^{-3} mole/liter PO_4^{3-}
6. 8.65×10^{-3} g/liter
8. (a) 1.65×10^{-4} mole/liter (b) 1.5×10^{-5} mole/liter (c) 4.5×10^{-8} mole/liter
10. $7 \times 10^{-8} M$
12. $c < b < a < d < e$
16. (a) $1.0 \times 10^{-7} M$
17. (a) yes (b) no (c) yes
18. 2.77 g
19. (a) $0.50N$ $CdSO_4$ (b) $0.50N$ Na_2SO_4 (c) $1.5N$ $Al_2(SO_4)_3$
22. $0.1153M$

CHAPTER 15

1. (a) PH_3 (d) H_2O (h) $ClCH_2CH_2COOH$ (j) (n) ClOH
2. (a) $Al(OH)_3$ (d) NH_2^- (g) PO_4^{3-} (j) CH_3NH_2 (n) PH_3
6. (a) less (c) less (e) greater (g) less
8. (a) 3.82
9. (b) 8.83
10. (b) $2.4 \times 10^{-4} M$
11. 1.0×10^{-5}

13. 5.0×10^{-10}
14. (a) $[HCOO^-] = 4.2 \times 10^{-3}M$; $[HCOOH] = 0.10M$; $[H_3O^+] = 4.2 \times 10^{-3}M$; $[OH^-] = 2.4 \times 10^{-12} M$; (c) $[(C_2H_5)_3NH^+] = 1.8 \times 10^{-3}M$; $[(C_2H_5)_3N] = 8 \times 10^{-3}M$; $[OH^-] = 1.8 \times 10^{-3}M$; $[H_3O^+] = 5.6 \times 10^{-12} M$
15. (a) 2.17
19. $[S^{2-}] = 5.2 \times 10^{-19} M$; 1.9×10^{-27} mole CuS
20. $2.5M$; $[Cd^{2+}] = 5 \times 10^{-7}M$
21. (a) $[NH_4^+] = 1.0M$; $[Cl^-] = 1.0M$; $[NH_3] = 2.4 \times 10^{-5}M$; $[H_3O^+] = 2.4 \times 10^{-5}M$; $[OH^-] = 4.2 \times 10^{-10}M$
22. (a) acidic (c) basic (e) neutral (g) acidic (i) acidic (k) acidic (m) neutral
24. 15 g
25. (a) 3.15; (b) 3.09; 3.22
27. (a) 7.0 (b) 5.28 (c) 11.15
28. 1.8×10^{-7} m/liter
29. (a) 45.00 g/equiv (b) 90.0; $C_2H_2O_4$
31. (a) 1.0×10^{-4} (d) $[HX] = 1.0M$; $[X^-] = 0.10M$; $[Na^+] = 0.10M$; $[H_3O^+] = 1.0 \times 10^{-3}M$; $[OH^-] = 1.0 \times 10^{-11} M$ (f) 8.34 (h) 18 ml; 64% error

CHAPTER 16

6. (a) less (d) less (f) greater
7. less; $\Delta H^0 = 35.1$ kcal
10. (a) oxide (c) oxide (f) sulfide
11. (a) $CH_3OCH_3 + I^-$ (b) $CH_3CH_2NH_2 + Br^-$ (c) $CH_3C(=O)O$—(phenyl)
    $+ Cl^-$ (d) $CH_3C(=O)OH + CH_3O^-$ (e) (phenyl)$-C(=O)OH + Cl^-$

CHAPTER 17

1. (a) $\overset{+1}{Ag}$, $\overset{+5}{N}$, $\overset{-2}{O}$ (d) $\overset{-3}{C}$, $\overset{+1}{H}$ (p) $\overset{+7}{S}$, $\overset{-2}{O}$
2. (a) HNO_3 is oxidant, CuS is reductant (e) none (g) $KClO_3$ is both oxidant and reductant
7. (a) A, +; B, − (b) A, −; B, + (c) A, +; B, − (d) A, +; B, − (e) A, −; B, +
9. (a) 1.2×10^{24} (c) 3.8×10^{14} (e) 1.0×10^{20}
11. (a) 0.75 V (c) 0.53 V (e) 2.05 V
13. yes
14. The precipitate is $BaSO_4$.
15. $N_2H_4 + 2O_2 \rightarrow N_2 + 2H_2O_2$
18. (a) cathode: $2H^+ + 2e^- \rightarrow H_2$
        anode: $2H_2O \rightarrow O_2 + 4H^+ + 4e^-$
    (e) cathode: $Cu^{2+} + 2e^- \rightarrow Cu$
        anode: $2Br^- \rightarrow Br_2 + 2e^-$

    (g) cathode: $2H_2O + 2e^- \rightarrow H_2 + 2OH^-$
          anode: $2I^- \rightarrow I_2 + 2e^-$

19. 7.8 g

20. 1.9 hr

21. 107 amp

23. (a) $K_2S_2O_8$
    (b) cathode: $2H^+ + 2e^- \rightarrow H_2$
          anode: $2SO_4^{2-} \rightarrow S_2O_8^{2-} + 2e^-$

CHAPTER 18

1. (a) $_{11}^{24}Na$ (c) $_{15}^{32}P$ (e) $_{6}^{12}C$ (g) $_{30}^{63}Zn$

2. $_{19}^{40}K \xrightarrow{K\text{-capture}} \,_{18}^{40}Ar$

    $_{19}^{40}K \longrightarrow \,_{20}^{40}Ca + _{-1}^{0}e$

3. 5.33 MeV

5. 1.72×10^7 kcal; 7.18×10^7 kJ

6. (b) $_{11}^{24}Na \rightarrow _{-1}^{0}e + _{12}^{24}Mg$
    (d) stable
    (f) $_{4}^{7}Be \rightarrow _{+1}^{0}e + _{3}^{7}Li$

7. $_{26}^{53}Fe \rightarrow _{+1}^{0}e + _{25}^{53}Mn$

    $_{26}^{59}Fe \rightarrow _{-1}^{0}e + _{27}^{59}Co$

11. 2.33 tons

12. 0.13 g

20. for $H_2O(l)$, $\Delta G^0 = -1.7$ kcal; for $H_2O(g)$, $\Delta G^0 = -3.8$ kcal

CHAPTER 19

8. B_2H_6

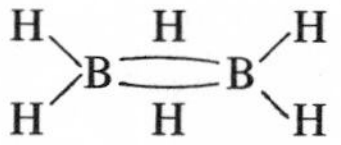

9. 174 g

10. 9.5×10^2 g

13. $[Ca^{2+}] = 9.3 \times 10^{-5}$; 10.9 g NaCl

CHAPTER 20

6. CuCl(s) $\Delta H_f^0 = -32$ kcal/mole; $CuCl_2$(s) $\Delta H_f^0 = -49$ kcal/mole;
    MgCl(s) $\Delta H_f^0 = -26$ kcal/mole; $MgCl_2$(s) $\Delta H_f^0 = -150$ kcal/mole
    $2CuCl \rightarrow Cu + CuCl_2$ $\Delta H^0 = +15$ kcal
    $2MgCl \rightarrow Mg + MgCl_2$ $\Delta H^0 = -98$ kcal

7. MgS, $\Delta H_f^0 = -70$ kcal/mole; CaS, $\Delta H_f^0 = -106$ kcal/mole

11. -13 kcal/mole

12. The precipitate is $Mg(OH)_2$.

13. $\Delta G^0 = +2.4$, therefore unfavorable

CHAPTER 21

4. (a) sp^3d^2 (b) sp^3 (c) sp^3d^2 (d) sp^3d (e) sp^3 (f) sp^3d^2 (g) sp^3d
   (h) sp^3
6. 30.2 kcal
9. $\Delta H^0 = -10.7$ kcal/mole; $\Delta G^0 = -28.3$ kcal/mole;
   $\Delta S^0 = 59.1$ cal/mole-deg; $\Delta G(400°C) = -50.5$ kcal/mole
15. (b) GaF_3 (c) 260 kcal (d) yes; standard emf is +0.52 V
17. $\Delta H^0 = -12.8$ kcal; $\Delta G^0 = -69.6$ kcal
20. N_2H_4

CHAPTER 22

1. (a) potassium hexanitrocobaltate(III)
   (b) sulfatopentaamminechromium(III) chloride
   (c) sodium tetrachloroaurate(III)
   (d) chlorotetraammineaquochromium(III) chloride
   (e) potassium hexacyanoferrate(II)
2. (a) $[Fe(H_2O)_5OH]^{2+}$ (c) $[Co(SCN)_4]^{2-}$ (e) $K_3[Al(C_2O_4)_3]$
4. 1d, 1f, 1j, 2d, 2f
5. CFSE: (a) 0 (c) -2.4Δ (e) -2.0Δ
   magnetic susceptibility: (a) 5.91 (c) 0 (e) 1.73
14. (d^1) .4Δ (d^2) .8Δ (d^3) 1.2Δ (d^4) .6Δ and -1.6 Δ
    (d^5) 0 and -2Δ (d^6) .4Δ and -2.4Δ (d^7) .8Δ and -1.8Δ (d^8) 1.2Δ
    (d^9) .8Δ (d^{10}) 0
16. A = $[Cr(H_2O)_6]Cl_3$ B = $[Cr(H_2O)_5Cl]Cl_2 \cdot H_2O$
    C = $[Cr(H_2O)_4Cl_2]Cl \cdot 2H_2O$

Index

Index